Cerebral Sinus Thrombosis

Experimental and Clinical Aspects

Cerebral Sinus Thrombosis

Experimental and Clinical Aspects

Edited by

Karl Einhäupl
Ludwig-Maximilians-University
Klinikum Grosshadern
Munich, Germany

Oliver Kempski
Johannes Gutenberg-University
Institute for Neurosurgical Pathophysiology
Mainz, Germany

and

Alexander Baethmann
Ludwig-Maximilians-University
Klinikum Grosshadern
Munich, Germany

Plenum Press • New York and London

Library of Congress Cataloging-in-Publication Data

Cerebral sinus thrombosis : experimental and clinical aspects / edited by Karl Einhäupl, Oliver Kempski, and Alexander Baethmann.
p. cm.
Proceedings based on an international symposium held Sept. 20-22, 1987, in Nymphenburg, Germany.
Includes bibliographical references and index.
ISBN 978-1-4684-8201-0 ISBN 978-1-4684-8199-0 (eBook)
DOI 10.1007/978-1-4684-8199-0
1. Sinus thrombosis--Congresses. I. Einhäupl, Karl.
II. Kempski, Oliver. III. Baethmann, A.
[DNLM: 1. Sinus Thrombosis--diagnosis--congresses. 2. Sinus Thrombosis--physiopathology--congresses. WL 355 C414 1987]
RC388.5.C3976 1991
616.2'12--dc20
DNLM/DLC
for Library of Congress 91-2456
CIP

Proceedings based on an international symposium on Experimental and Clinical Aspects of Cerebral Sinus Thrombosis, held September 20-22, 1987, in Nymphenburg, Germany

ISBN 978-1-4684-8201-0

Softcover reprint of the hardcover 1st edition 1990

A Division of Plenum Publishing Corporation
233 Spring Street, New York, N.Y. 10013

PREFACE

Contrary to the neurological manifestations of arterial cerebral blood flow disturbances, respective conditions resulting from obstruction of the cerebro-venous system are far less well understood. Hence, cerebral sinus vein thrombosis (CSVT) ranks prominently among the group of neglected diseases of the brain. This might be attributable (a) to the diagnostic difficulties of the disorder and (b) to the fact that CSVT is associated with a host of heterogeneous neurological symptoms which often are not specific for the underlying venous flow disorder. Another complicating aspect is that CSVT is a consequence of other diseases as disparate as focal infection, trauma, neoplasia, or a thrombosis disposition caused by oral contraceptive use.

Although progress has been made in establishing the correct diagnosis of the syndrome, many problems remain as the discussions contained within this volume vividly demonstrate. The same is true for the present understanding of the pathophysiological basis of the disease, eg, concerning the cerebro-venous circulation, the hemodynamic and neuropathological consequences in particular. Part of these deficits may be attributed to a scarcity of solid experimental data due to the limited availability of animal models. However, relevant experimental models are required for an in-depth analysis of the pathophysiological mechanisms, eg, causing brain tissue damage in relationship with the topographical distribution of the venous flow obstruction and, most importantly, for the testing of specific methods of treatment.

This volume is based on an international symposium which covered all the important aspects of SVT. These are, among others, the structural and functional basis of the cerebro-venous system, and the neuropathology of SVT, particularly emphasizing specificities of the resulting brain tissue damage. Further topics concern novel experimental models for induction of an SVT in animals, together with a comprehensive morphological and functional analysis. The physiology and pathophysiology of the venous circulation of the brain is reviewed as well as recent findings of cerebral blood flow studies in patients. General discussions on the mechanisms of brain damage are included, eg, concerning smooth muscle reactivity under abnormal conditions, the nature of ischemic brain injury, fluid translocation between the extra- and intracellular space as the cause of cytotoxic cell swelling in ischemia and, finally, the role of vasoactive mediator compounds.

The basic science issues are balanced by detailed analyses of the clinical aspects reaching from historical perspectives to the most recent experiences with modern methods of diagnosis. Yet, recent technical progress notwithstanding, the conclusion appears unavoidable that a prompt diagnosis of SVT is still difficult to achieve. Therefore, in spite of modern technologies such as contrast enhanced computertomography, subtraction angiography, magnetic resonance imaging or ultrasound sonography, assessment of the disorder is still based on a careful clinical analysis of the patient.

An update of the subject would be incomplete without discussions of treatment. Both neurosurgical interventions as well as conservative methods using heparin are competently covered. A variety of neurological disorders require neurosurgery of the cerebral sinus and of the cerebro-venous system. On the other hand, thrombotic occlusion of the sinus or of the cerebral veins is a domain of medical treatment. High-dose heparin appears currently to be the method of choice. Since heparin is not expected to dissolve the thrombotic obstruction but rather to prevent further spread of thrombosis, the effectivity of heparin provides additional evidence for SVT as a gradual process. This is well reflected by the clinical course.

This volume will fulfil its purpose if it raises the level of attention of clinicians to consider SVT in patients with a subacute, often insidious onset of neurological symptoms and progressing deterioration as opposed to the acute symptomatology of an arterial occlusion. Since effective therapeutical methods are available, prompt establishment of the diagnosis is important.

The editors would like to express their gratitude to the sponsors who have provided the basis for publication of this volume. We gratefully acknowledge support by Drs I. Bauer and P. Hansen, UCB-Chemie GmbH, Kerpen/Sindorf, FRG, and by Mr Bartusch, Carl-Friedrich von Siemens-Stiftung. We would also like to acknowledge the valuable secretarial support provided by Isolde Juna and Helga Kleylein and the editorial assistance offered by Ms M. Carter and J. Curtis of Plenum Publishing Co., London, United Kingdom.

K. Einhäupl
O. Kempski
A. Baethmann

Munich, 1990

CONTENTS

CLINICAL ASPECTS OF SINUS VEIN THROMBOSIS

STRUCTURAL AND FUNCTIONAL BASIS

HISTOMORPHOLOGY AND ULTRASTRUCTURE OF THE CEREBRO-VENOUS SYSTEM

W. Roggendorf

Institute for Brain Research, University of Tübingen
Calwerstraβe 3, 7400 Tübingen, FRG

INTRODUCTION

During the last decade our knowledge and understanding of the structure and organization of the cerebral venous system has dramatically increased. Detailed descriptions of the ultrastructure of the venous system in various species including man are available. Different types of innervation have been reported and refined methods using corrosion casts and Indian ink injections have allowed investigation of the angioarchitecture of the intracerebral and meningeal venous vessels. This chapter focuses on the angioarchitecture, ultrastructure, and specialized structures of the venous vessel wall as a basis for the understanding of thrombosis of the dural sinuses.

Angioarchitecture of Cranial Venous Vessels and of the Dura Mater

The cranial venous blood flow has been characterized according to several criteria including developmental aspects, vascular hemodynamics, and angiographic patterns. This led to three subdivisions: (1) the superficial segment which drains the scalp, muscles, and tendons; (2) an intermediate segment draining the skull, diploe and dura; and (3) the cerebral segment draining the brain.

The detailed anatomy of the venous system was recently reviewed by Capra and Anderson [1], while this chapter focuses on the *cerebral parenchymal* segments and the *dura mater.* The venous vessels of cerebral cortex attract attention for various reasons. The different size of veins and arteries and their branching patterns were studied by Saunders et al. [2]. Intracortical anastomoses were described by Hasegawa et al. [3] and Ravens [4]. The remarkable work of Pfeiffer [5] gives detailed descriptions of the angioarchitecture of various brain areas. His work, however, has a fundamental error, namely the author permanently confused arteries and veins. Duvernoy et al. [6] reinvestigated in detail cortical blood vessels of human brain, and the following description is based mainly on this study.

Cerebral Sinus Thrombosis, Edited by K. Einhäupl *et al.*
Plenum Press, New York, 1990

In general, the *meningeal venous network* is very dense and composed of vessels with a larger diameter than the corresponding arteries. In contrast to arteries, the collecting venous vessels remain at the cortical surface in their entire course until emptying into the sinus. These venous vessels form many superficial anastomoses constituting a large mesh. However, in certain areas, (eg, gyrus angularis or gyrus supramarginalis) only a few venous anastomoses are found. No arterio-venous anastomoses and virtually no capillary network is present at the cortical surface. Therefore, blood flow must pass through the cortex before reaching the sinus. This is confirmed also by the studies of Forbes [7], Scharrer [8] and Wolff [9]. Only Rowbotham and Little [10] report arterio-venous shunts at the cortical surface.

There have been many investigations of the angioarchitecture of *venous vessels in the cortex* of various species. Bär [11,12] investigated the cerebral vasculature during development in rats. In collaboration with Wolff [13] he developed a model system of the angioarchitecture of the neocortex. Hodde and Nowell [14] studied by scanning electron microscopy branching of venous vessels using micro-corrosion casts. The terminal vascular bed of the superficial cortex in rats was investigated on corrosion casts by Motti et al. [15]. From these studies and the investigations of Duvernoy et al. [6] the following picture can be drawn.

Five types of intracortical veins can be distinguished: types 1, 2 and 3 are short venous vessels draining lamina 4, 3, 2, and 1, while types 4 and 5 are larger veins draining the middle layers of the cortex and the subcortical white matter. As in the case of arteries, veins of the intermediate length types 2 and 3 are the most frequent (Fig. 1). In cortex there are virtually no anastomoses between venous trunks while arterio-venous anastomoses occur [4,6,16]. On the precapillary level, anastomoses are predominantly seen in the superficial and deep layers [3,6]. For the distribution pattern of arteries and veins three categories were suggested by Duvernoy [6]: (1) large venous vessels (principal veins) surrounded by two arteriolar rings, (2) two principal veins surrounded by one arterial ring, and (3) small venous vessels containing only one arteriolar ring (Fig. 1).

Dura Mater

The dural venous sinuses are unique vascular structures in the fibrous tissue of the cranial dura mater. Most of the cerebral venous blood flows either in a posterior direction and empties through the sinuses into the internal jugular veins, or through the cavernous sinus at the base of the brain into the internal jugular veins, the internal vertebral venous plexus, or the veins of the orbit. These venous systems are described in detail by Capra and Anderson [1].

The angioarchitecture of the human dura was investigated by Hammersen [17,18] and by Kerber and Newton [19]. Two veins were found to accompany a primary artery on the external or periosteal dural surface. Anastomoses are found between the arterial and venous system in the midparietal part of the dura. Hammersen [18] observed so called "Bügelkapillaren" (arch-like capillaries) between the smaller arterioles and so called "Entenschnabelgefäße" (ie, vessels opening up like a trumpet) at the venous side. These anatomical details may be of interest

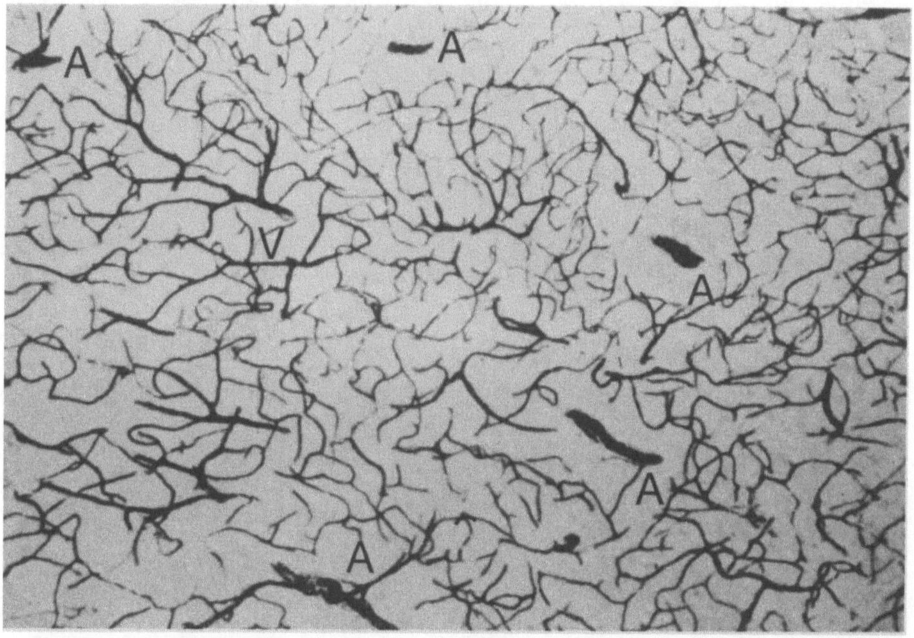

Fig. 1. Indian ink injection of normal human brain. Arterioles are characterized by a capillary-free perivascular space. The arterioles (A) are arranged around a central vein (V) 100 μ Paraffin section (x 100).

with regard to arterio-venous malformations or fistulas as a cause of sinus thrombosis.

The innervation and vasculature including lymphatic drainage of the dura were recently investigated by Andres et al. [20]. The authors studied the dura mater of rats on an ultrastructural basis and observed a rich supply with myelinated and unmyelinated nerve fibers. Venules were found to contain fenestrated endothelium and nerve bundles close to the periendothelial cells. Nerve terminals localized in the connective tissue in close contact to the sinus contain various peptides and transmitters such as VIP, substance P and neuropeptide Y and are, therefore, regarded as stretch- or pain receptors [21].

Ultrastructure of Intracranial Venous Vessels

The general ultrastructure of venules was studied by Rhodin [22] in detail concerning dermal veins of rabbits. The classification of the venous system established by this author is applicable for most of the venous vessels of the brain. Along the direction of flow the first segments are the *postcapillary venules*, which are microvessels with a diameter of 8-30 μm contiguous with the venous capillaries. With an increase in luminal diameter the number of pericytes and Veil cells is also gradually increased. *Collecting venules* (Fig. 2) follow, which have a diameter of 30-50 μm. Collecting venules have a complete layer of pericytes and Veil cells. Occasionally, primitive forms of smooth muscle cells are found. The next venous segments are *muscular venules* with a diameter of 50-100 μm. Here the periendothelial cells are represented by overlapping smooth muscle cells. Sometimes the smooth muscle cells form two layers, but myoendothelial junctions occur rarely.

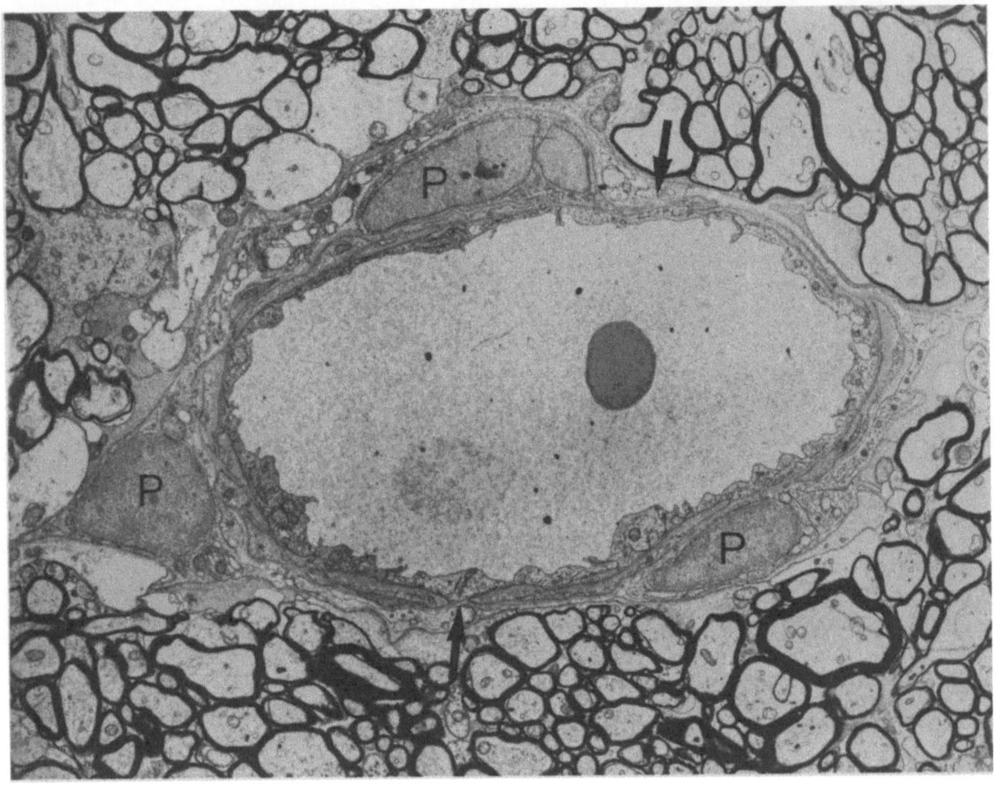

Fig. 2. Collecting venule from pons of cat with incomplete ensheathment (⟶) and several periendothelial cells (P) (x 2400).

Intracerebral Venules

The ultrastructure of the cerebral venous system has been investigated in various species including man [23-27]. The *endothelium* of cerebral venules forms a continuous layer without pores. Zonulae adherentes and zonulae occludentes are found between adjacent cells with overlapping of protoplasmic processes and protrusions into the lumen. Differences in frequency and distribution of tight junctions are observed between the different parts of the microvasculature as reported by Simionescu et al. [28]. In peripheral organs tight junctions are numerous in the endothelium of arterioles, less frequent in venules, and absent in capillaries. The ultrastructure of the endothelium does not differ between the various types of venules and is similar to that of capillaries.

There are slight differences between some animal species and humans. Endothelial cells of human venules have lipid-like vacuoles near the nuclear region (Fig. 5b). Only in human cortex could we find the third type of venules (cf above) containing true smooth muscle cells as observed in other organs [22,29,30].

In contrast to endothelium, the *perithelium* is different in various subtypes of venules and differs from that of other organs. In correspondence to "perithelial cell", which we think is more appropriate as a term than "pericyte", we use "perithelium" for the whole ensheathment of the endothelium. Postcapillary venules are difficult to distinguish from capillaries. In postcapillary venules, however, perithelial cells are more branched resulting in a more complete periendothelial ensheathment. Cerebral perithelial cells of postcapillary venules have a dense grouping of organelles in the perinuclear region and a loose formation of filaments in the cellular processes (Fig. 3). The perithelium of collecting venules (Fig. 2) appears to be more compact. The cells contain more filaments and semidesmosome-like structures. Contradictory views concerning morphology, origin, and function of these cells are discussed in detail in a previous communication [27]. There are two types of contacts between endothelium and perithelium. First: zonulae adherentes, where cell membranes of both cells are not separated by basal lamina, and second: protrusions of endothelium into smooth muscle cells and vice versa (Fig. 4). The second type is regularly found in arterioles but is rare in the venous part of the microcirculation.

Meningeal Venules and Veins

Meningeal venules and veins differ from the cerebral veins with respect to the perithelium. Meningeal venules have no particular space filled with an extracellular matrix. Perithelial cells of meningeal venules contain filaments adjusted in parallel. These filaments are more numerous than in collecting venules but do not correspond to those of typical smooth muscle cells of muscular venules of other organs. Thus, in different *animal species*, both the intraparenchymal and meningeal vessels do not have typical smooth muscle cells as opposed to venules of other organs (Fig. 5a). Elastic material found in small subdural veins (200 μ in diameter) or in large venules of omentum and mesenterium was not present in the meningeal vessels of corresponding size. Similar results were reported on 200 μ-diameter veins of the retina of man or monkeys [31].

A most important criterion to distinguish *human* meningeal venules from animal venules is presence of smooth muscle cell processes with characteristic structural elements such as many filaments, pinocytotic vesicles, and attachment devices (Figs 5b and 6).

Specialized Structures and Functional Aspects of the Venous Vessel Wall

A histomorphological review of cerebral veins would be incomplete without data of immunohistologic investigations, especially on the *extracellular matrix* and *cytoskeleton*. Since only limited information on cerebral vessels was available, we have recently studied desmin and vimentin and extracellular matrix components, as, eg, collagen I, III, IV, V and VI, laminin, fibronectin, and undulin of brain vessels [32]. Desmin is an intermediate filament characteristic of smooth and striated muscles. In the venous system it could be detected only in muscular vessels of the meninges. Endothelium or periendothelial cells of venules do not contain desmin on a light-microscopical level. Immunoelectron-microscopical observations of Fujimoto and Singer [33] have shown that desmin filaments occur in various endothelial cells of organs other than the brain, eg, in capillaries of renal cortex and pancreas. Vimentin is present in endothelial and smooth muscle cells of arterioles, whereas it is less prominent in endothelium of capillaries and veins. Thus desmin and vimentin may be used to distinguish arterial vessels from the rest of the microcirculation.

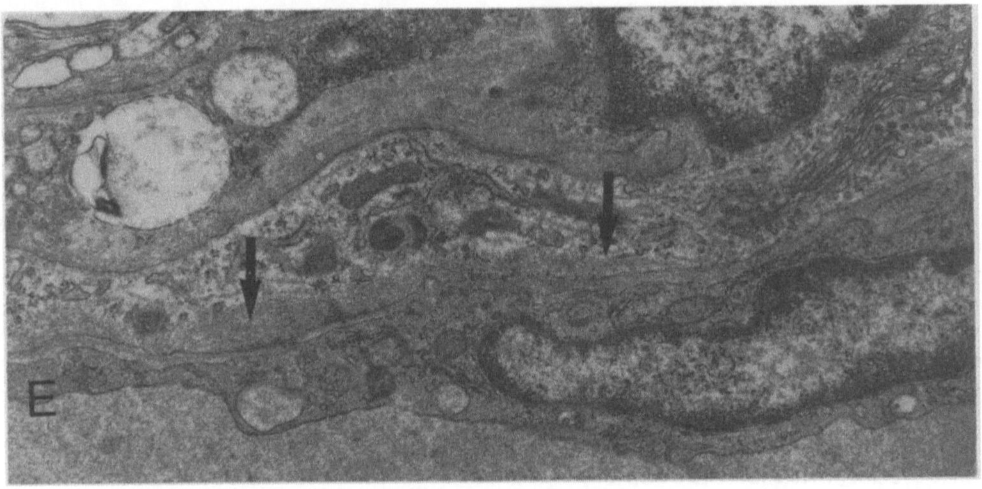

Fig. 3. Collecting venule from cat with a thin endothelial cell layer (E) and scanty filaments predominantly in periendothelial processes (⟶). Note collagen bundles in the perivascular space (x 8400).

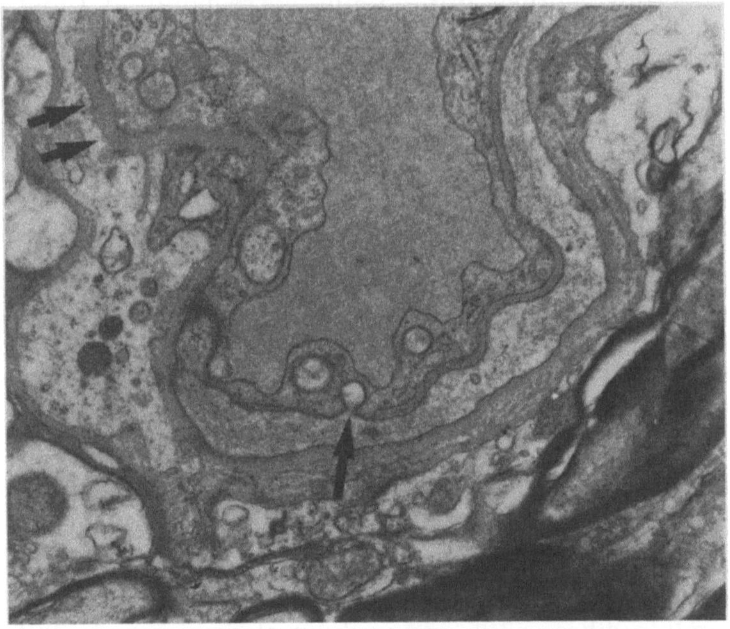

Fig. 4. Postcapillary venule from human subcortical white matter with incomplete ensheathment of the endothelium (⇉) and protrusion of a periendothelial cell into the endothelium (⟶) (x 8100).

The distribution of basal lamina components in venous vessels has been studied by various authors [34-36]. Accordingly, laminin, collagen type IV, and fibronectin are found in the basal lamina, while fibronectin was also seen in the interstitial space. As to subtypes of collagen, their distribution varies only in the arterial part of the microcirculation. In our material, type V collagen was mostly found on the luminal side of arterial

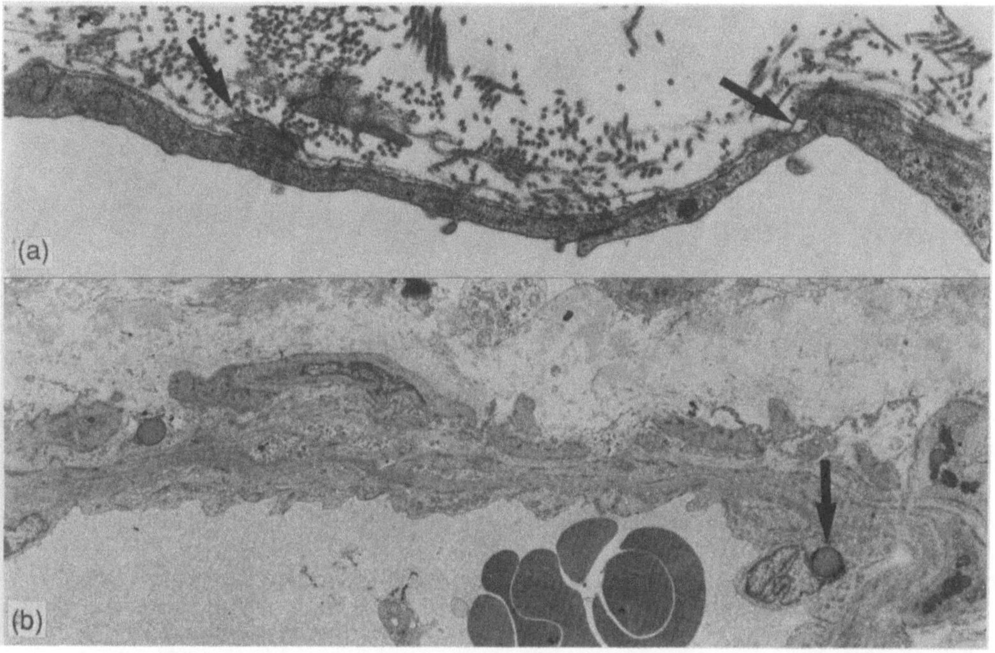

Fig. 5. (a) Collecting meningeal venule from cat. Only small periendothelial processes surrounded by a basal lamina (⟶) are detectable. (x 8400). (b) Human muscular meningeal vein. Note endothelium (⟶) with lipid vacuole surrounded by small smooth muscle cells. Collagen bundles are seen in the perivascular space (x 4600).

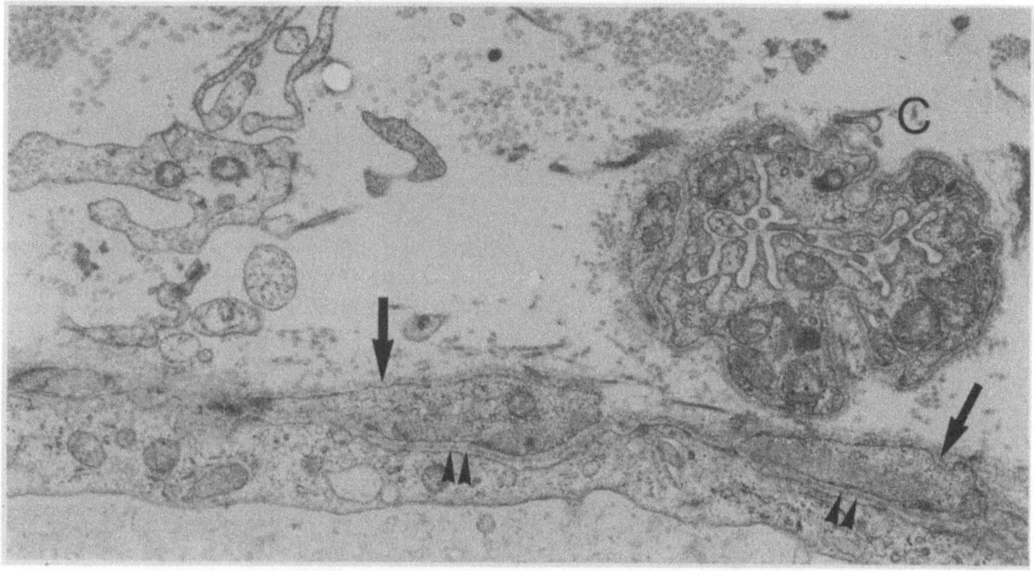

Fig. 6. Meningeal collecting venule of human with typical smooth muscle processes containing pinocytic vesicles (⟶), dense filaments (⇉), and a collapsed capillary (C) in the surrounding meninges (x 16,500).

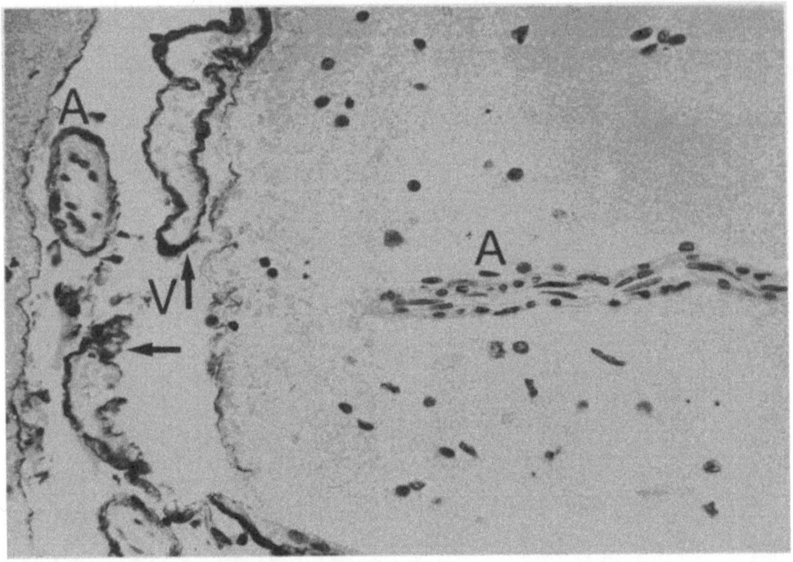

Fig. 7. Meninges and upper cortical layers with arterioles (A) and venules (V). Note intensive staining for collagen type VI in the meninges. The intracerebral arteriole does not contain collagen type VI. Peroxidase-antiperoxidase technique (x 200).

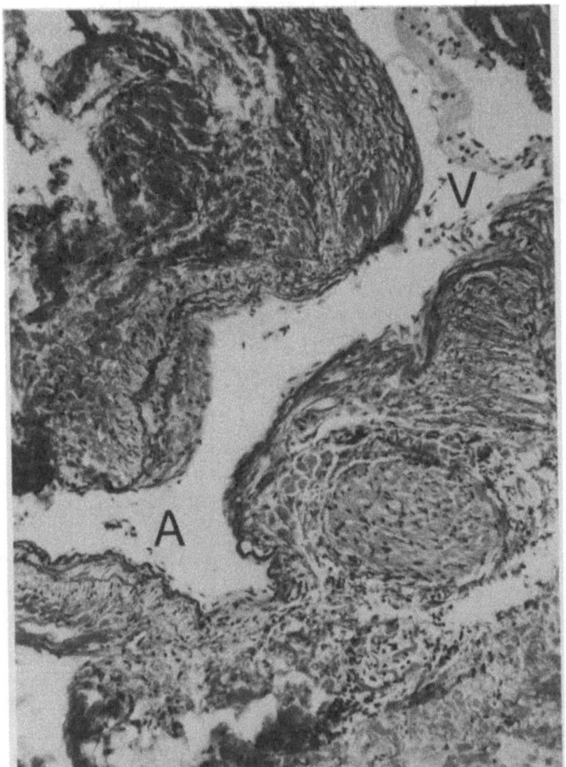

Fig. 8. Arterio-venous dural fistula showing the arterial segment (A) and the draining vein (V). Paraffin section (x 125).

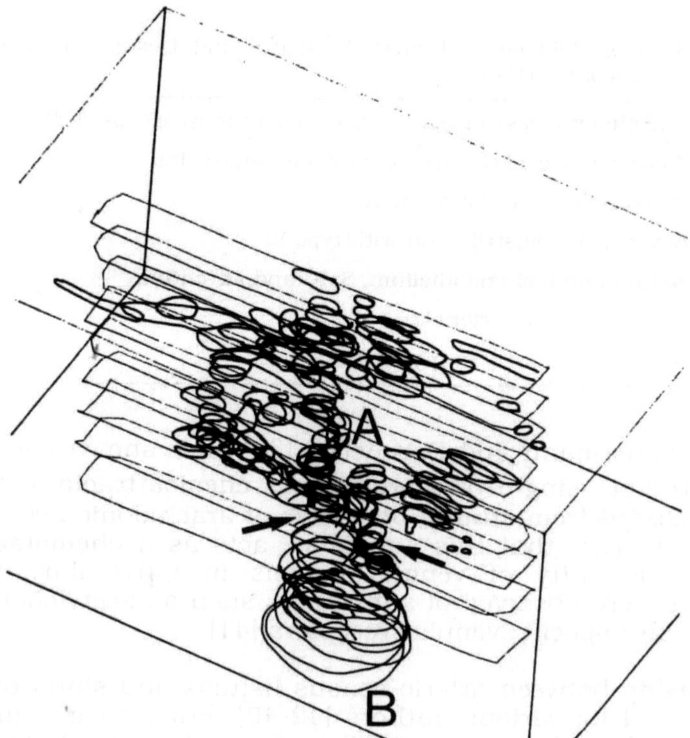

Fig. 9. Superimposed serial sections through an arterio-venous dural fistula showing the transition (⟶) from the intradural arterial system (A) to the draining vein (B). (Computer assisted drawing).

vessels, while undulin and collagen types VI and III were found only in the adventitia. At the abluminal side of the venous endothelium all collagen types were co-distributed (Fig. 7). Based on these findings it is suggested that upon injury, endothelium matrix components assembled in the subendothelial space could comprise attractive components for adhesion of circulating cells. This might be an initial step of thrombosis induction, as supported by the studies of Legrand et al. [37]. The authors report that collagen type VI microfibrils elicit a much more rapid response of platelets than do other collagen types, eg, type III. The venous system is attractive not only for platelets but also for monocytes and phagocytes. Voss et al. [38] using umbilical veins showed that phagocytes attach to extracellular matrix in vitro. Yet further investigations are necessary to support such an "injury theory".

An important question is whether specific endothelium markers or receptors are present in the venous vessel wall. A specific role of the venous segments in the *immune response* or in *cell penetration* during inflammation is well accepted also for the brain. Venous endothelium is a target of lymphocytes in experimental allergic encephalomyelitis (EAE). But it is not fully understood how lymphocytes recognize endothelium of cerebral venules. Leukocytes penetrate venous endothelium during

Table 1. Distribution of Collagen Subtypes, Vimentin and Desmin in Arterioles and Venous Vessels of the Brain

Vimentin:		Endothelium, less in venules and smooth muscle cells (SMC)
Desmin:		Smooth muscle cells of arterioles and large veins
Collagen	**I:**	Intima, media and adventitia
	III:	Adventitia, codistribution with type VI
	IV:	Basal lamina of endothelium, SMC and adventitia
	V:	Luminal part of arterioles
	VI:	Adventitia, mostly in meningeal vessels

inflammation. Faustmann and Dermietzel [39] have shown two modes of leukocyte migration using α-bungarotoxin as chemoattractant. Unterberg et al. [40] concluded from studies on effects of arachidonic acid on blood-brain barrier function that this substance acts as a chemotactic agent increasing permeability of venous vessels in particular. Moreover, localized effects were observed of substances such as acetylcholine which was shown to have specific venular receptors [41].

A relationship between arterio-venous fistulas and sinus thrombosis has been reported by various authors [42-45], but its consequences for etiology and the clinical course are still unclear. We prefer "arterio-venous fistula" (AV-angioma and AV-shunts are common synonyms) as a term, because connections between the arterial and venous system have been confirmed by histology (Fig. 8). Since these malformations are subjected to neurosurgery we had an opportunity to study four cases with cranial arterio-venous dural fistulas. Serial paraffin sections were made of the specimen to visualize transitions between the arterial and venous part. The serial sections were then subjected to computerized reconstruction (Fig. 9). Thereby the entire course could be followed from the arterial to venous vasculature. In these studies it was shown that there was only one draining vein to the sinus or to the meningeal surface of the brain. Since in most cases clinical symptoms have a late onset, we assume that these lesions develop under pathological conditions rather than being congenital in origin.

ACKNOWLEDGEMENTS

The author thanks Dr J. Iglesias (Berlin) for providing Figure 1, and Drs J. Becker and R. Hengstenberg (Tübingen) for their assistance in doing the reconstructions (Fig. 9). The excellent technical assistance of Mrs G. Mundle and U. Lang is gratefully acknowledged.

REFERENCES

1. N. F. Capra and K. V. Anderson, *in*: "The Cerebral Venous System and Its Disorders", J. P. Kapp and H. H. Schmidek, eds., Grune & Stratton Inc., New York (1984).
2. R. L. de C. H. Saunders, W. H. Feindel and V. R. Carvalho, X-ray microscopy of the blood vessels of the human brain, Part I, *Med. Biol. Illus.*, 15:108-122 (1965).

3. T. Hasegawa, J. R. Ravens and J. F. Toole, Precapillary arterio-venous anastomoses. "Thoroughfare channels" in the brain, *Arch. Neurol.*, 16:217-224 (1967).
4. J. R. Ravens, Anastomoses in the vascular bed of the human cerebrum, *in*: "Pathology of Cerebral Microcirculation", J. Cervos-Navarro, ed., De Gruyter, Berlin and New York (1974).
5. R. A. Pfeiffer, "Grundlegende Untersuchungen für die Angioarchitektonik des menschlichen Gehirns", Springer Verlag, Berlin (1930).
6. H. M. Duvernoy, S. Delon and J. L. Vannson, Cortical blood vessels of the human brain, *Brain Res. Bull.*, 7:519-579 (1981).
7. H. S. Forbes, The cerebral circulation. 1: Observation and measurement of pial vessels, *Arch. Neurol. Psychiat.*, 19:751-761 (1928).
8. E. Scharrer, Arteries and veins in the mammalian brain, *Anat. Rec.*, 78:173-196 (1940).
9. H. G. Wolff, The cerebral blood vessels. Anatomical principles, *in*: "The Circulation of the Brain and Spinal Cord", Proc. Ass. Res. Nerv. Ment. Dis. 1937, Vol. XVIII, Williams and Wilkins, Baltimore (1938).
10. G. F. Rowbotham and E. Little, A new concept of the circulation and the circulation of the brain. The discovery of surface arterio-venous shunts, *Br. J. Surg.*, 52:539-542 (1965).
11. Th. Bär, "The Vascular System of the Cerebral Cortex. Advances in Anatomy, Embryology and Cell Biology", Vol. 59, Springer Verlag, Berlin, Heidelberg and New York (1980).
12. Th. Bär, Distribution of radially penetrating arteries and veins in the neocortex of rat, *in*: "Cerebral Microcirculation and Metabolism", J. Cervos-Navarro and E. Fritschka, eds., Raven Press, New York (1981).
13. J. R. Wolff, An ontogenetically defined angioarchitecture of the neocortex, *Arzneimittel-Forsch.*, 26:1239 (1976).
14. K. C. Hodde and J. A. Nowell, SEM of micro-corrosion casts, *Scan Electron. Microsc.*, 2:89-106 (1980).
15. E. D. F. Motti, H. G. Imhof and M. G. Yasargil, The terminal vascular bed in the superficial cortex of the rat. An SEM study of corrosion casts, *J. Neurosurg.*, 65:834-846 (1986).
16. A. C. P. Campbell, Variation in vascularity and oxidase content in different regions of the brain of the cat, *Arch. Neurol. Psychiat.*, 41:223-242 (1939).
17. F. Hammersen, Die Angioarchitektonik der dura mater encephali. I. Zur Problematik der sog. Mäander-Gefässe und Knäuelarterien, *Zeitsch. für Zellforschung*, 59:153-164 (1963).
18. F. Hammersen, Über das feinere Gefässmuster in der dura mater encephali des Menschen, *Anatomischer Anzeiger*, 113:375-377 (1964).
19. C. W. Kerber and T. H. Newton, The macro- and microvasculature of the dura mater, *Neuroradiology*, 6:175-179 (1973).
20. K. H. Andres, M. von During, K. Muszynski and R. F. Schmidt, Nerve fibers and their terminals of the dura mater encephali of the rat, *Anat. Embryol.*, 175:289-301 (1987).
21. T. Matsuyama, S. Shiosaka, A. Wanaka, S. Yoneda, K. Kimura, T. Hayakawa, P. C. Emson and M. Tohyama, Fine structure of peptidergic and catecholaminergic nerve fibers in the anterior cerebral artery and their inter-relationship: an immunoelectron microscopic study, *J. Comp. Neurol.*, 223:46-56 (1985).
22. I. A. G. Rhodin, Ultrastructure of mammalian venous capillaries, venules and small collecting veins, *J. Ultrastruct. Res.*, 452-500 (1968).
23. W. Drommer and L. C. Schulz, Feinstruktur der normalen Kapillaren und venolen im Rüchenmark des Schweins, *Anat. Anz.*, 128:232-247 (1971).
24. W. Roggendorf, J. Cervos-Navarro and M. D. Lazaro-Lacalle, Ultrastructure of venules in the cat brain, *Cell Tissue Res.*, 192:461-474 (1978).
25. W. Lange and Z. Halata, Comparative studies on the pre- and post-terminal blood vessels in the cerebellar cortex of rhesus monkey, cat and rat, *Anat. Embryol.*, 158:51-62 (1979).

26. K. Nakai, H. Imai, I. Kamai et al., Microangioarchitecture of rat parietal cortex with special reference to vascular "sphincters". Scanning microscopic and dark field microscopic study, *Stroke*, 12:653-659 (1981).
27. W. Roggendorf and J. Cervos-Navarro, Normal and pathologic ultrastructure of human cerebral venules, *in*: "Cerebral Venous System and Its Disorders", J. P. Kapp and H. H. Schmidek, Grune & Stratton Inc. (1984).
28. M. Simionescu, N. Simionescu and G. Palade, Segmental differentation of cell junctions in the vascular endothelium, *J. Cell Biol.*, 67:863-885 (1975).
29. H. Z. Movat and V. P. Fernando, The fine structure of the terminal vascular bed. 4. The venules and their perivascular cells, *Exp. Molec. Pathol.*, 3:98-114 (1964).
30. F. Hammersen, "Anatomie der Terminalen Strombahn", Urban und Schwarzenberg Verlag, Berlin (1971).
31. J. M. Hogan and L. Feeney, The ultrastructure of the retinal vessels. The small vessels, *J. Ultrastruct. Res.*, 9:29-46 (1963).
32. W. Roggendorf, H. Opitz and D. Schuppan, Altered expression of collagen type VI in brain vessels of patients with chronic hypertension. A comparison with the distribution of collagen IV and procollagen III. *Acta Neuropathol. (Berl.)*, 77:55-60 (1988).
33. T. Fujimoto and S. J. Singer, Immunocytochemical studies of endothelial cells in vivo. I. The presence of desmin only, or of desmin plus vimentin, or vimentin only, in the endothelial cells of different capillaries of the adult chicken, *J. Cell Biol.*, 103:2775-2786 (1986).
34. M. T. Giordana, I. Germano, G. Giaccone, A. Mauro, A. Migheli and D. Schiffer, The distribution of laminin in human brain tumors: an immunohistochemical study, *Acta Neuropathol. (Berl.)*, 67:51-57 (1985).
35. R. D. McComb, J. M. Moul and D. D. Bigner, Distribution of type VI collagen in human gliomas; comparison with fibronectin and glioma-mesenchymal matrix glycoprotein. *J. Neuropathol. Exp. Neurol.*, 46:623-633 (1987).
36. R. Mayne, Collagenous proteins of blood vessels, *Arteriosclerosis*, 6:585-593 (1986).
37. Y. J. Legrand, F. Fauvel, B. Arbeille, D. Leger, H. Mouhli, N. Gutman and J. P. Muh, Activation of platelets by microfibrils and collagen. A comparative study, *Lab. Invest.*, 54:566-573 (1986).
38. B. Voss, T. Kerenyi, B. Herrenpoth, H. Jellinek and U. Gerlach, The influence of monokines on endothelial cells from human umbilical veins, *Path. Res. Pract.*, 182:576 (1987).
39. P. M. Faustmann and R. Dermietzel, Extravasation of polynuclear leukocytes from the cerebral microvasculature. Inflammatory response induced by alpha-bungarotoxin, *Cell Tissue Res.*, 242:399-407 (1985).
40. A. Unterberg, M. Wahl, F. Hammersen and A. Baethmann, Permeability and vasomotor response of cerebral vessels during exposure to arachidonic acid, *Acta Neuropathol. (Berl.)*, 73:209-219 (1987).
41. W. Felix, Unterschiedliche Reaktionen prä- und post-kapillärer gefässe auf Pharmaka, *in*: "Gefäßwandelemente in vivo und in vitro", H. Fischer and E. Betz, eds., Herausgeber, Tübingen (1983).
42. J. Dichgans, M. Gottschaldt and K. Voigt, Arteriovenöse Dura-angiome am sinus transversus. Klinische Symptome, charakteristische arterielle Versorgung und häufige venose äbflubstörungen, *Zbl. Neurochir.*, 33:1:1-18 (1972).
43. O. W. Houser, J. K. Campbell, R. J. Campbell and T. M. Sundt, Arterio-venous malformation affecting the transverse dural venous sinus - an acquired lesion, *Mayo Clin. Proc.*, 54:651-661 (1979).
44. A. Thron, K. Wessel, D. Linden, G. Schroth and J. Dichgans, Superior sagittal sinus thrombosis: neuroradiological evaluation and clinical findings, *J. Neurol.*, 233:283-288 (1986).
45. A. Thron and K. Voigt, Vaskuläre kraniozerebrale Erkrankungen arteriovenose Durafisteln, *in*: "Radiologische Diagnostik in Klinik und Praxis", George Thieme Verlag, Stuttgart and New York (1986).

NEUROPATHOLOGICAL FINDINGS IN THE THROMBOSIS OF CEREBRAL VEINS AND SINUSES: VASCULAR ASPECTS

J. Cervos-Navarro and S. Kannuki

Institute of Neuropathology
Free University of Berlin
Hindenburgdamm 30, 1000 Berlin 45, FRG

INTRODUCTION

Cerebral sinus vein thrombosis (CSVT) is a well recognized clinical and neuropathological entity. Recent progress of neuroradiology made intra vitam diagnosis and visualization of gross parenchymal damage possible [3,19,20,22]. Previously the disease was thought to have a poor prognosis and high mortality [14]. In recent neuroradiological and/or clinical studies on CSVT, however, many patients exhibited a good prognosis [3,10,13]. Neuroradiology, however, provides only limited information on the nature of the thrombus, even when using high-resolution CT or NMR scans [19,22]. Up to now examinations of vascular aspects of CSVT in autopsy series are rare. In this article we review retrospectively 41 cases of CSVT which were neuropathologically examined in our institute between 1969 and 1987.

INCIDENCE, AGE AND SEX

The 41 cases with CSVT form 0.44% of all autopsies in our department. The incidence of CSVT reported in various autopsy series has a wide range. Ehlers [7] found only 0.1% in his series, and Barnett [1] reported 39 autopsy cases in 20 years. In contrast to these older reports, CSVT had an incidence of 9% in 182 consecutively studied autopsies by Towbin [23]. It is confirmed in recent clinical studies that the true incidence is higher than formerly presumed on the basis of autopsies [3,8,20,23]. However, angiography or other neuroradiological methods may occasionally fail to demonstrate occlusion of cerebral veins and sinuses, particularly in cerebral vein occlusion without involvement of the sinus. Kawase et al. [13] reported four cases of cortical vein thrombosis suggestive of subcortical haemorrhage. In three patients diagnosis could be made only by intraoperative fluorescein angiography.

Pregnancy, postpartum, and oral contraceptives enhance the probability of thromboembolic disorders [8,17]. Anatomical continuities in man have been demonstrated between the uterovaginal and vertebral veins and

Cerebral Sinus Thrombosis, Edited by K. Einhäupl *et al.*
Plenum Press, New York, 1990

the intracranial veins and the dural venous sinuses [2]. Thus, it is possible that during a period of high intra-abdominal pressure a small embolus passes from the pelvic plexus along this route into intraspinal and intracranial veins and the dural venous sinuses forming a nidus there for a large thrombus [17]. Such an embolus could arise from thrombosed veins normally found in the pelvis after parturition or abortion. In general, no significant differences in the frequency of CSVT between men and women were found in our series. Yet four women aged between 20 and 40 years were affected, as compared with no male patients of this age. The age distribution of our patients was from 1 month to 86 years and involved four infants (1 year old), four children (15 years old), and 14 elderly patients (60 years old). Age distribution has a close relationship with the underlying disease causing thrombosis.

ETIOLOGY

CSVT is a "syndrome" accompanying various diseases. There are also idiopathic cases in which no etiology can be detected.

Generally, three factors may predispose to the development of thrombosis:

1. Changes of the endothelial lining.
2. Changes of blood flow, eg, slowing, stasis, and eddying of flow currents.
3. Changes of blood constituents.

Although one or the other may predominate, often more than one factor contributes to thrombus formation. We grouped our cases according to the three factors named above. Besides these categories, six cases of our series were idiopathic.

(a) In the 14 cases resulting from changes of the vessel wall, the underlying diseases were meningitis, leukemia (invasion), or carcinoma (invasion). Other diseases reported in the literature are chronic otitis media, mastoiditis, etc. In one case a fresh thrombus was attached to an old mural thrombosis. Normal endothelium of the vascular system forms a smooth surface which inhibits coagulation. However, if the integrity of the endothelium is disturbed, platelets adhere to the vessel wall initiating thrombosis.

(b) In 11 cases with changes of the blood composition, the underlying conditions were leukaemia (without direct invasion), cachexia, carcinoma of another organ, polycythemia, and a postpartum state. Other diseases belonging to this category, such as thrombocythemia, antithrombin-III deficiency, or disseminated intravascular coagulation, etc, have also been reported.

(c) Eight cases had disturbances of the cerebral blood flow. The underlying diseases were intracranial mass hemorrhage, cerebral arterial occlusion (main trunk), and circulatory shock. Other conditions, such as congenital heart disease, or chronic obstruction of the respiratory tract, etc, are also mentioned in this context.

In older series, infections used to be the most common cause of CSVT with primary thrombosis (non-infectious) and thrombosis secondary

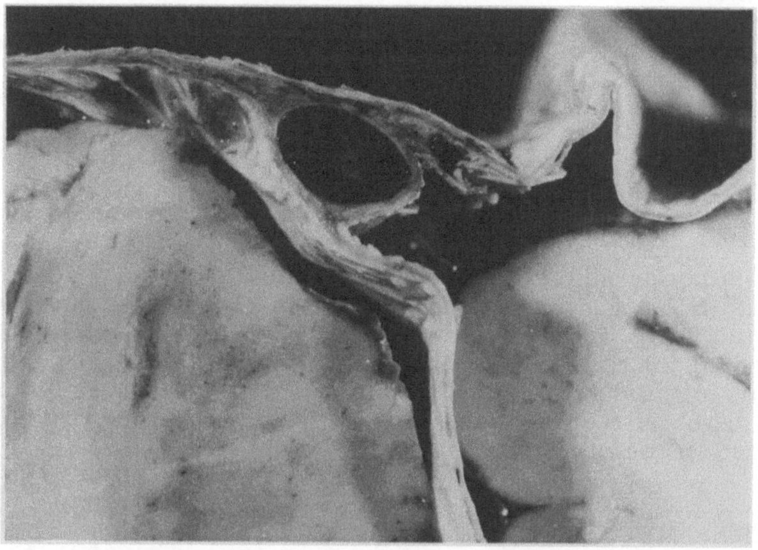

Fig. 1. Thirteen-year-old boy/man with leukemia, pulmonary arterial thromboembolism and thrombophlebitis of the upper limb. No leukemic cell infiltration was found in the central nervous system. Thrombosis of the SSS and cerebral cortical veins is shown.

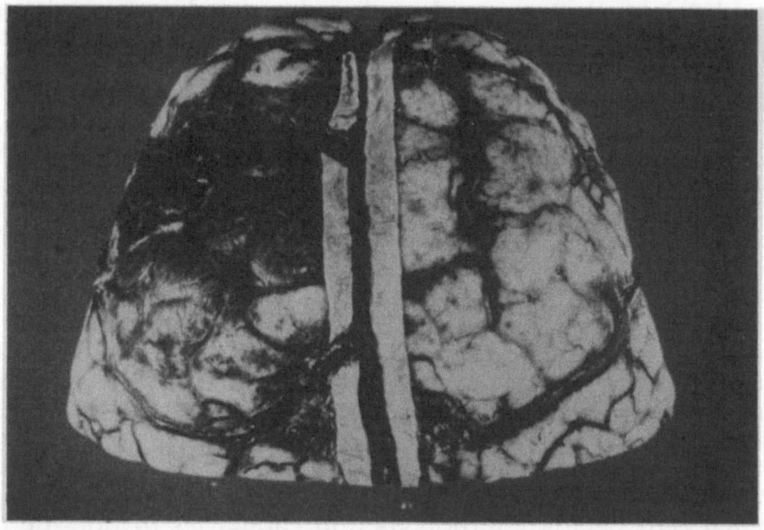

Fig. 2. Seventy-four-year-old man with occlusion of the internal carotid and medial cerebral arteries leading to infarction 12 days before death. Total occlusion of SSS by a fresh thrombus was found.

to pyogenic infection as principal terminologies. In Krayenbühl's large series, the majority of cases were due to infections [14]. The incidence of infections as a cause of CSVT has decreased, while that of idiopathic thrombosis has relatively increased constituting approximately 15-25% in recent series [3,10]. Our findings confirm this tendency: six of 41 cases (14.6%) were idiopathic while six of 41 cases (14.6%) resulted from meningitis. No otitis media was found in our series.

LOCALIZATION OF THROMBOSIS

Twenty-six cases in our series had a single occluded sinus. In 19 cases the superior sagittal sinus (SSS), and in seven cases the sinus transversus (ST) was involved. Eleven cases had two or more occluded sinuses, ten with involvement of the SSS and ST. The SSS was thrombosed in 29 cases (70.7%), and the ST in 17 (41.5%). Four cases had venous occlusion only without sinus thrombosis; in two of this subgroup, cerebral cortical veins were affected, while deep cerebral veins were affected in the other two cases.

TYPE OF THROMBOSIS RELATED TO TIME COURSE

We could distinguish three groups according to age and course. In many cases one patient could have thrombi at different stages. The most common pattern were fresh thrombi in cortical veins together with older thrombi of the sinus. The cases were categorized according to the oldest thrombus.

(a) At the beginning of thrombosis, platelets adhere to the endothelial lining of a vessel or heart. By electron microscopy it can be seen that platelets at first are loosely aggregated but retain a normal morphologic structure. With the deposition of more platelets, the aggregates become closely packed and some platelets lose their granules. Leukocytes tend to adhere to the granulated platelets. In new thrombosis, platelets form a system of lamellae on the vessel wall, one behind the other in more or less parallel rows running obliquely or transversely to the longitudinal axis of the vessel. The fibrin strands produced by the primary platelet thrombi trap large numbers of erythrocytes and leukocytes. Eighteen cases of our series had this type of thrombus.

(b) The formed elements of a recent thrombus soon disintegrate and fibrin loses its fibrillar appearance. Thereby, the thrombus mass becomes more pale and homogeneous. Hyalinized thrombi can be seen within 2 or 3 days. Cellular debris is removed by macrophages. The fibrin strands become less fibrillar and more amorphous. Ten cases of our series were in this stage.

(c) When a venous thrombus persists for a longer time, ingrowth of mesenchymal cells begins. Fibroblasts and capillaries invade the thrombus mass at the site of attachment. Spaces forming in the thrombus by shrinkage or dissolution of thrombus components subsequently become lined with endothelial cells. Endothelium also tends to cover the exposed surface of the thrombus. Gradually, the thrombus is entirely replaced by vascularized connective tissue and converted into a dense fibrous mass. During organization some degree of circulation may be reestablished, in

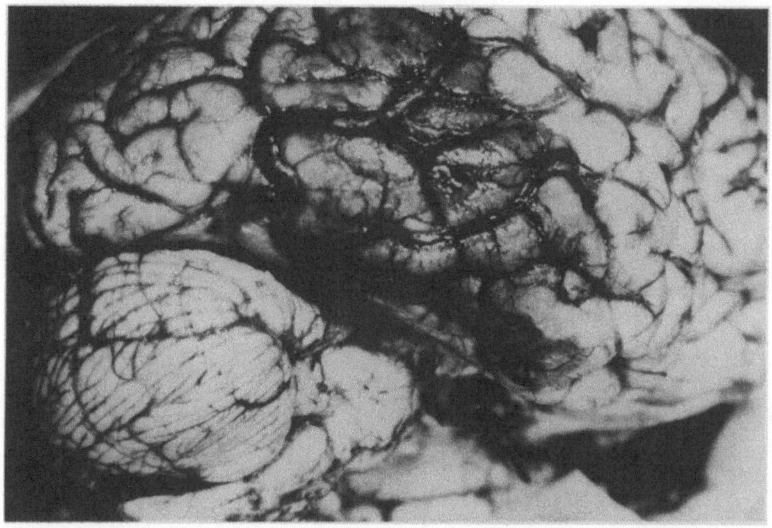

Fig. 3. Seventy-year-old woman with peritonitis after surgery for colon cancer. No brain metastasis. Thrombosis was located in the transverse sinus and inferior cerebral vein.

the case of an occlusive thrombus by formation of new channels (recanalization). Retraction of the thrombus from the vessel wall is a way of accomplishing this. Thirteen cases of our series were in different stages of organization.

Although some of the cases were asymptomatic, or their onset was clinically obscured due to the severe symptomatology of the underlying disease, we tried to correlate duration of occlusion from the beginning of symptoms with the type of thrombus.

(a) Four cases of recent thrombosis had an obvious clinical onset. In two cases, 2 days passed between onset and death while 3 days passed in the other two. In other organs thrombus organization with homogenization of fibrin strands and removal of cell debris by activated macrophages is known to begin after 2 or 3 days. In our cases no signs of organization were found.

(b) Five cases with hyalinized thrombi had an obvious onset. Duration was 5 days in two cases, 10 days in another two cases, and 1 week in one patient. Proliferation of endothelial cells and of connective tissue may begin 1 or 2 weeks after thrombus formation. Thrombosis with these characteristics belongs to the next category.

(c) Thrombus organization could be identified in five cases with clear onset. It had a course of 8 days (two cases), 3 weeks (1 case), or 6 weeks (two cases), respectively. The relationship between the clinical symptoms and stage of thrombosis was more complex. Three cases were found with clinical symptoms of apparently short duration with various stages of thrombus formation, not only at different sites but also in the same location. Two cases with acute leukemia had both partly hyalinized and partly organized thrombi in different cerebral veins. In one case the internal cerebral vein as well as the paraventricular veins were

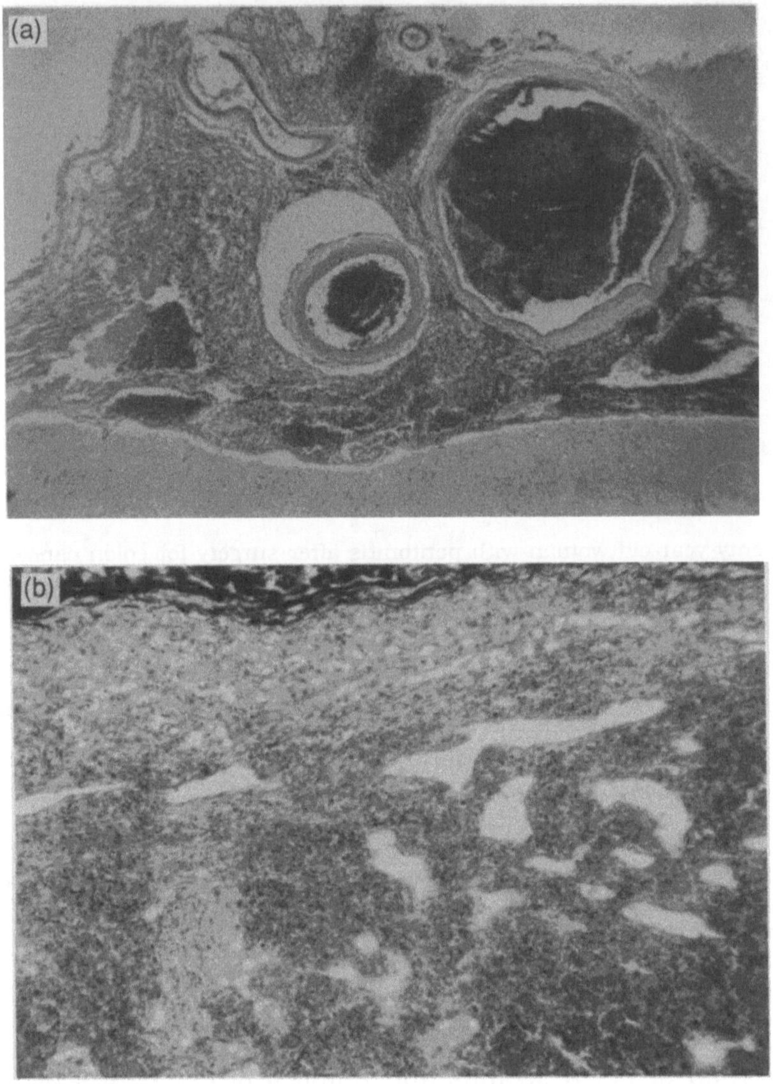

Fig. 4. Specimen of a 53-year-old man with meningitis. Eight days before death the patient became somnolent. (a) Fresh thrombosis of cortical veins without involvement of neighboring arteries. Hematoxylin and Eosin (HE) stain (x80). (b) Diapedesis in neighboring parenchyma.

thrombosed. The duration of symptoms of 8 days corresponded to the evolution time of partly hyalinized thrombi. Older "silent" thrombi probably resulted from blood hyperviscosity, while leukemic cell infiltration of the CNS was not found. According to the literature, thrombosis of the deep cerebral vein (vein of Galen) has a poor prognosis [6]. However, existence of "silent" thrombi in the paraventricular veins suggests that thrombosis of the deep cerebral vein is not always symptomatic. Some patients may have a good recovery or remain asymptomatic.

When comparing clinical onset with the stage of thrombosis it seems that organization in CSVT is delayed as compared to other organs.

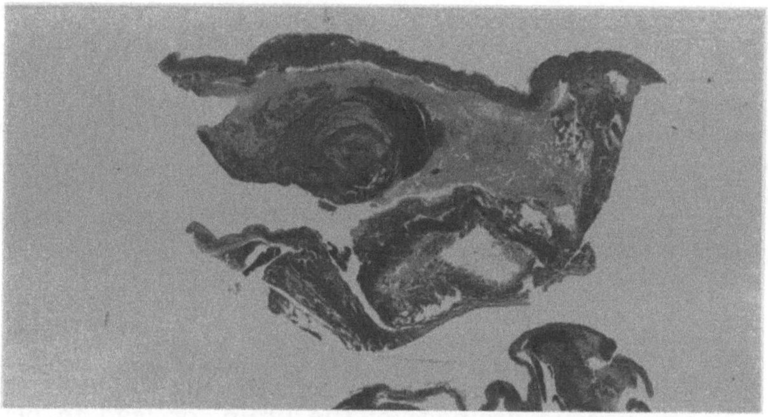

Fig. 5. Specimen of a 41-year-old woman with pulmonary thromboembolism and left hemiparesis 4 weeks before death. An old organized thrombus of the SSS with fibrin lamellae and fresh bleeding is shown. van Gieson stain (x 33).

However, one has to be aware (a) that appearance of clinical symptoms does not always coincide with the beginning of thrombus formation, and (b) that old, organized thrombi are rare in CSVT [10].

Other cases showed completely different stages of organization in the same thrombus. One component was old and organized, another relatively fresh, partly laminated, or hyalinized. The old thrombus component was associated with recent hemorrhage indicative of recanalization. In the same area hemosiderin-laden macrophages were found. This case had a clinical course of 6 weeks. Therefore, it must be assumed that a recent thrombus formed above an old mural one, and that clinical symptoms

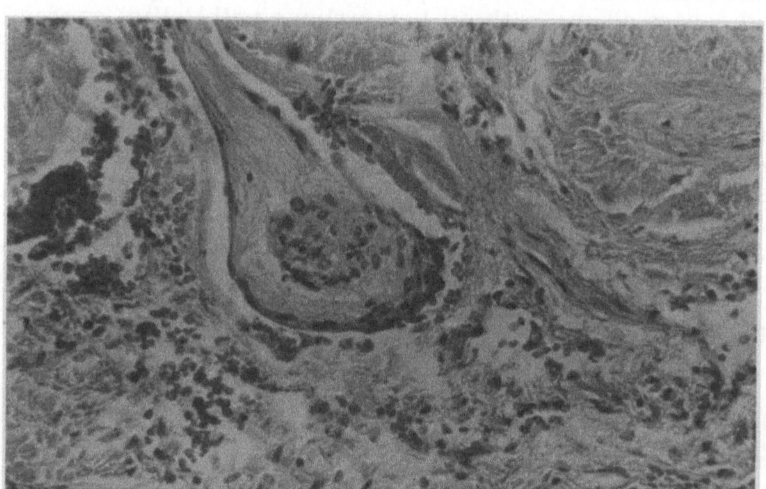

Fig. 6. Asymptomatic old thrombus in a 64-year-old man. Pacchionian granulations are seen in an old organized thrombus. HE stain (x 330).

appeared only after the sinus was completely occluded by the new thrombus.

The existence of different stages of thrombosis in one patient suggests that an initial thrombus may secondarily progress into the cerebral veins, or other sinuses, forming new thrombi there. Propagation of thrombosis in CVST may proceed symptomatically or asymptomatically, as observed in a patient of our series with an asymptomatic mural thrombus.

PARENCHYMAL CHANGES

Localized thrombus formation in the cerebro-venous system is unlikely to impair venous drainage of the brain, or lead to permanent damage because of an abundant collateralization. On the other hand, if one sinus and a portion of its tributaries are obstructed venous infarction may ensue. Hemorrhage can be found mainly affecting the white matter ranging from petechiae to large confluent bleedings sometimes involving the subarachnoid space. Vasospasm caused by SAH may contribute to the ischemic damage of cerebral cortex. Recently thrombosed venous channels appear distended and bluish on a macroscopical basis. Necrotic cortex and white matter disintegrate rapidly resulting in a shrunken, cystic and yellowish-brown material. When a thrombosis of sinuses and veins has been organized and recanalized, it is difficult to demonstrate remnants, even by microscopic examination.

Eighteen cases in our series had parenchymal damage. All patients had hemorrhagic infarction, and nine had intracerebral hematomas. In 13 cases there were multiple and diffuse changes with hemorrhagic infarction and/or intracerebral hematomas. In one case with total occlusion of the SSS and right ST, acute hydrocephalus and diffuse cerebral edema was found together with small hemorrhagic infarctions in the parasagittal cortex. CSVT thus seems to cause severe brain edema and acute hydrocephalus. A similar pattern of parenchymal changes with diffuse and severe edema without hemorrhage with lethal course has been reported [10]. As in our case, thrombosis obstructed both the SSS and ST including the torcular sinus. The pathogenetic mechanisms leading to hydrocephalus are controversially discussed. It is obvious that CSVT can block major CSF absorption pathways. However, only a few cases with CSVT and ventricular enlargement were reported even in acute or in chronic stages. Symonds [21] concluded that venous obstruction may cause swelling of the brain and intracranial hypertension without ventricular dilatation. Bousser [3] reported 25 cases with all stages of CVST. However, only three cases had a ventricular enlargement in the CT scan. By all means, disturbances of the CSF circulation should be mild due to secondary pathways for reabsorption of CSF.

RELATIONSHIP BETWEEN THE SITE OF THROMBOSIS AND PARENCHYMAL DAMAGE

Seventeen cases with parenchymal damage from disturbances of venous drainage (Group 1) had cerebral vein thrombosis with or without involvement of the sinus. Thirteen cases had total or partial thrombosis of SSS and of the cerebral cortical veins with or without involvement of the ST. Four other cases of this group had vein thrombosis only. One previously mentioned case had complete thrombosis of the SSS and right ST without occlusion of cerebral veins. In contrast, 20 of 23 cases with sinus thrombosis only had no parenchymal damage (Group 2).

The difference between these two groups, the findings on venous obstruction without sinus thrombosis, confirms the important role of venous occlusion for induction of parenchymal damage by disturbances of the venous drainage. Hemorrhagic infarction or intracerebral hematoma usually appear in the territory of occluded cortical and/or deep cerebral veins. However, the topographical relationship between venous occlusion and tissue damage is more variable than in arterial infarction because of the complexity of the venous collateral circulation.

TYPE OF THROMBUS AND PARENCHYMAL DAMAGE

As already pointed out, clinical symptoms as well as parenchymal damage in CSVT has a large variability. Among six cases with recent thrombus and parenchymal damage, four had multiple diffuse, while two had localized hemorrhagic infarction. One case had a small hematoma in an hemorrhagic infarct. In five cases with hyalinized thrombus, three had multiple lesions and two had intracerebral hematoma. On the other side, all six cases with an organized thrombus had intracerebral hematomas and multiple lesions. These findings may indicate that propagation of thrombosis into cortical veins occurred as a cause of gradual clinical worsening after the beginning of symptoms. Hemorrhagic infarction with SAH appeared first whilst intracerebral bleeding appeared later.

CAUSE OF DEATH

Among the 18 cases with CSVT in association with brain tissue damage, 13 cases died from an increased intracranial pressure. Two cases suffered from cardiac failure, two cases had peritonitis with ileus, and in another pulmonary embolism was found. In 23 further cases without brain tissue damage death was attributable to the disease underlying formation of CSVT eventually resulting in cardiac failure.

SUMMARY

Forty-one cases of CSVT diagnosed at autopsy are reported representing 0.44% of all autopsies in our department. Six cases (14.6%) were idiopathic, 15 cases resulted from changes of the vessel walls by meningitis, leukemia with invasion, etc, 12 cases from abnormal changes of blood composition in leukemia, cachexia etc, and eight cases from blood flow abnormalities due to an intracranial mass or circulatory shock. In 29 cases SSS was affected with or without other lesions, and in 17 cases thrombosis was occluding the ST. Four cases had venous occlusions without sinus involvement.

Hemorrhagic infarcts and/or intracerebral hematoma were present in 18 cases. Seventeen of these had venous occlusion, underscoring the importance of the involvement of veins in the formation of brain damage.

Thrombosis was divided into three groups according to the histological stage:

Recent Thrombus: 18 cases. six cases were symptomatic with clinical onset 2 or 3 days before death. These patients had hemorrhagic infarction, with multiple or diffuse parenchymal lesions in four cases. One case had a small intracerebral hematoma.

Hyalinized Thrombus: 10 cases. In six cases symptoms appeared 5 to 10 days before death. All cases had hemorrhagic infarction, three cases with multiple localizations. Two cases had an intracerebral hematoma.

Organized Thrombus: 13 cases. Many of the six symptomatic cases had a discrepancy between the histological maturity of the thrombus and the clinical course indicating existence of cases with "silent" CSVT. Comparing thrombus organization in CSVT with that in other organs, the former seems to progress more slowly, although the opposite may occur as well.

REFERENCES

1. H. J. M. Barnett and H. H. Hyland, Non-infective intracranial venous thrombosis, *Brain*, 76:36 (1953).
2. O. V. Bason, Function of vertebral veins and their role in spread of metastases, *Ann. Surg.*, 112:138 (1940).
3. M. B. Bousser, J. Chiras, J. Bories and P. Castaigno, Cerebral venous thrombosis - a review of 38 cases, *Stroke*, 16:199 (1985).
4. F. Debruyne, Lateral sinus thrombosis in the eighties, *J. Laryngol. Otol.*, 99:91 (1985).
5. J. Cervós-Navarro, Verschlüsse der Hirnvenen, *in*: "Pathologie des Nervensystems", J. Cervós-Navarro and H. Schneider, Springer Verlag, Berlin, Heidelberg and New York (1980).
6. J. J. Eick, K. D. Miller, K. A. Bell and R. H. Tutton, Computed tomography of deep cerebral venous thrombosis in children, *Radiology*, 140:399 (1981).
7. H. Ehlers, C. B. Courville and L. A. Calif, Thrombosis of internal cerebral veins in infancy and childhood - a review of literature and report of five cases, *J. Pediat.*, 8:600 (1936).
8. B. Estanol, A. Rodrigues, G. Conte, J. M. Anderson, M. Loyo and J. Pizzuto, Intracranial venous thrombosis in young women, *Stroke*, 10:680 (1979).
9. T. O. Gabrielsen, J. F. Seeger, J. E. Knake and E. W. Stilwill, Radiology of cerebral vein occlusion without dural sinus occlusion, *Radiology*, 140:403 (1981).
10. P. C. Gates and H. J. M. Barnett, Venous disease: cortical veins and diseases, *in*: "Stroke", vol 2, H. J. M. Barnett, B. M. Stein, J. P. Mohr and F. M. Yatsu, eds., Churchill Livingstone Inc., New York (1986).
11. R. A. Goldenberg, Lateral sinus thrombosis. Medical or surgical treatment?, *Arch. Otolaryngol.*, 111:56 (1985).
12. V. A. Huhn, Die Hirnvenen-und Sinusthrombose, *Fortschr. Neurol. Psychiat.*, 25:440 (1957).
13. T. Kawase, T. Tazawa and K. Mizukami, Cerebral venous thrombosis - computed tomography and operative findings in 4 cases of simulated subcortical hemorrhage, *Prog. Comput. Tomog.*, 5:19 (1983).
14. H. A. Krayenbühl, Cerebral venous and sinus thrombosis, *Clin. Neurosurg.*, 14:1 (1966).
15. P. L. Lavin, I. Bone, J. T. Lamb and M. Swinburne, Intracranial thrombosis in the first trimester of pregnancy, *J. Neurol. Neurosurg. Psychiat.*, 41:726 (1978).

16. P. J.Macchi, R. I. Grossman, J. M. Gomori, H. I. Goldberg, R. A. Zimmermann and L. T. Bilaniuk, High field MR imaging of cerebral venous thrombosis, *J. Comp. Assist. Tomog.*, 10:10 (1986).
17. J. P. Martin, Thrombosis in superior longitudinal sinus following childbirth, *Br. Med. J.*, 2:537 (1941).
18. H. Noetzel and F. Jerusalem, "Die Hirnvenen- und Sinusthrombose", Springer Verlag, Berlin, Heidelberg and New York (1965).
19. K. C. V. G. Ral, H. C. Knipp and E. J. Wagner, Computed tomography findings in cerebral sinus and venous thrombosis, *Radiology*, 140:391 (1981).
20. P. Rousseaux, M. H. Bernard, B. Scherpereel and J. F. Guyot, Intracranial venous sinus thrombosis - 22 cases, *Neurochirurg.*, 24:197 (1978).
21. C. Symonds, Otitic hydrocephalus, *Neurology*, 6:681 (1956).
22. A. Thron, K. Wessels, D. Linden, G. Schroth and J. Dichgans, Superior sagittal sinus thrombosis. Neuroradiological evaluation and clinical findings, *J. Neurol.*, 233:283 (1986).
23. A. Towbin, The syndrome of latent cerebral venous thrombosis: its frequency and relation to age and congestive heart failure, *Stroke*, 4:419 (1973).

THROMBOSIS OF CRANIAL VEINS AND SINUSES:

BRAIN PARENCHYMAL EFFECTS

J. H. Garcia

Department of Pathology (Neuropathology)
Henry Ford Hospital
2799 West Grand Boulevard
Detroit, MI 48236-2689, USA

INTRODUCTION

The blood contained within the intracranial sinuses and veins comprises 70% of the total cerebral blood volume [6]; therefore, interrupting the flow of blood through one of the large intracranial sinuses could have disastrous consequences for the entire brain circulation. Nevertheless, because of abundant interconnections among sinuses and some of the larger veins, some portions of the sinuses may be surgically occluded without undesirable consequences. In contrast, thrombosis of a cerebral cortical vein almost always is followed by serious effects [21,22]; this is because these cortical cerebral veins lack collateral anastomoses [6,23]. The region of the superior sagittal sinus (SSS) extending caudal to the Rolandic veins is critical; sudden interruption of flow in this region, including the vein of Galen and major tributaries, may have disastrous results. Gradual dural sinus occlusion of the type that is induced by slow growing tumors, such as a meningioma, is much better tolerated than acute occlusion secondary to either trauma or surgical intervention [24].

Spontaneous thrombosis of intracranial sinuses and veins (ISV) was, before the antibiotic era, primarily associated with infectious processes involving the face, paranasal sinuses, and skull [3,5]. Non-infectious conditions, associated with ISV thrombosis, are very heterogeneous and include: tumor invasion of a sinus, cardiac abnormalities especially congenital heart defects [29], cerebral arteriovenous malformations [17], dehydration, hypercoagulable conditions and numerous other situations that do not appear to have a common unifying feature [5].

The traditional assumption that ISV thrombosis is nearly always a fatal condition has been tempered by recent publications in which a favorable outcome was demonstrated after a long-term follow-up of a large group of patients. In all of these 38 patients the diagnosis of ISV thrombosis had been established by appropriate angiographic studies [4]. Two additional facts about ISV thrombosis were confirmed in the study published by Bousser and associates [4], namely: (1) most patients' initial

Cerebral Sinus Thrombosis, Edited by K. Einhäupl *et al.*
Plenum Press, New York, 1990

complaints include: either headache (of the subarachnoid-hemorrhage type) or blurred vision, secondary to papilledema; and (2) risk factors for ISV thrombosis remain largely unrecognized in most patients; Behçet's disease, systemic neoplasia (such as lymphoma) and renal disorders are among the unexpected putative causes of ISV thrombosis [4]. Among the invasive vascular procedures associated with ISV thrombosis, implantation of a transvenous pacemaker through the right jugular vein has resulted in at least one angiographically confirmed instance of thrombosis of the right transverse and lateral dural sinuses and the posterior third of the SSS [12].

The objective of the present study is to analyze the features of the brain parenchymal alterations among cases in which death occurred a few days after the ISV occlusion. The cases selected are representative of four separate clinical situations, involving: (1) the SSS; (2) the Galenic-venous system; (3) either the cortical or intraparenchymal cerebral veins; and (4) the cavernous sinus. An additional instance of confirmed arterial embolism is included for the purpose of contrasting the features of arterial and venous infarcts.

PRESENTATION OF CASES

Case 1: Intracerebral Venous Thrombosis (UAB 16,970)

A 21-year-old man, known to be homozygous for the sickle-cell gene (HbSS), was neurologically normal at the time he developed a sickle-cell crisis about 3 days before death. At autopsy there were no subarachnoid hemorrhages, the dural sinuses were patent, the brain was diffusely swollen and coronal sectioning showed multiple sites of softening in the cerebral hemispheres, none of which was larger than 1.5 cm in maximum diameter. Most of these sites of softening were situated parasagittally in the cerebral hemispheres.

The histologic evaluation revealed multiple, thrombosed intraparenchymal veins in association with areas of infarction where three features were prominent: neutrophil infiltrates were abundant, vacuolation of neuropil was prominent and there was "incomplete" brain tissue necrosis, a term referring to the fact that only neurons are destroyed while glial and vascular elements survive.

Case 2: Superior Sagittal Sinus Thrombosis (UAB 15,746)

A 7-year-old child, known to be homozygous for the sickle-cell gene (HbSS), developed an acute crisis characterized by nausea, vomiting, hemiparesis and, beginning approximately 5 days before death, enlarging lateral ventricles; subarachnoid hemorrhage and small subcortical hemorrhages were demonstrated by CT scan. A cranial ventriculostomy was completed a few hours before death.

At autopsy there was extensive thrombosis of the SSS and other intracranial sinuses; there was extensive subarachnoid hemorrhage, especially over the convexity of the right frontal lobe. The brain was diffusely swollen, heavier than normal (by 25% over the expected weight for this age group) and the lateral ventricles were narrowed. Parenchymal hemorrhages were confined to the subcortical white matter of the cerebral hemispheres; there were no softenings in the diencephalon or in the infratentorial structures. The white matter of the corona radiata was swollen, prominent and friable, while the cerebral cortex appeared slightly dusky (Fig. 1).

The main histologic features, observed primarily in the sections of the cerebral hemispheres, included: focal coagulative necrosis involving both cortex and white matter; extensive vacuolation of neuropil, abundant neutrophil infiltrates, periadventitial hemorrhages, thrombosed intra-parenchymal veins, ring hemorrhages, subarachnoid hemorrhage both acute and subacute and parenchymal extravasation of plasma as well as red blood cells. Many of the extravasated red blood cells frequently showed characteristic sickle deformity.

Case 3: Thrombosis of the Vein of Galen, Superior Sagittal Sinus and Cerebral Cortical Vein (UAB 15,518)

A 26-year-old diabetic woman developed severe headache about 5-7 days before death; terminally she lapsed into coma. Two CT scans of the head suggested first occlusion of a parieto-temporal cortical vein and next thrombosis of the vein of Galen and its tributaries. Precipitating factors could not be identified from a review of the clinical records, and because the postmortem examination was limited to the intracranial contents, systemic factors could not be completely ruled out.

The brain was massively and symmetrically swollen, it was heavier (by 20%) than the average for persons in her age group. Subarachnoid hemorrhage was localized to an area of softening over the surface of the right superior temporal gyrus. There was thrombosis of the vein of Galen and major tributaries as well as thrombosis of the SSS.

The outstanding microscopic features included: confluent pattern of parenchymal hemorrhage sometimes in areas free of ischemic necrosis but with massive sponginess of the neuropil; ring hemorrhages, relative absence of glial nuclear swelling, marked hypertrophy of endothelial cells especially in capillaries and venules, and abundant neutrophil infiltrates.

Case 4: Bilateral Cavernous Sinus Thrombosis (UAB 16,474)

A 49-year-old man with end-stage renal disease was treated over a period of 9 years with hemodialysis and more recently with deferoxamine (because of secondary hemosiderosis). About 9 days before death he developed rhinorrhea, fever and retro-orbital headaches; blindness and coma were the final events. There was no facial cellulitis or proptosis.

At autopsy, the brain was massively swollen and considerably softer than normal. Slight subarachnoid exudate made the leptomeninges opaque and cloudy, especially around the temporal tips. Minimal, spotty subarachnoid hemorrhage was noted and on cut surface there was extensive, non-hemorrhagic brain infarction which was especially noticeable in the temporal and frontal lobes, where there were abundant microabscesses, many containing fungal hyphae. The intracavernous portion of both internal carotid arteries was thrombosed. The lumen of the carotid arteries was filled with abundant hyphae of a microorganism having the features of mucoracea, including the size (up to 100 μm in length) and the nonseptate hyphae branching at 90° angles.

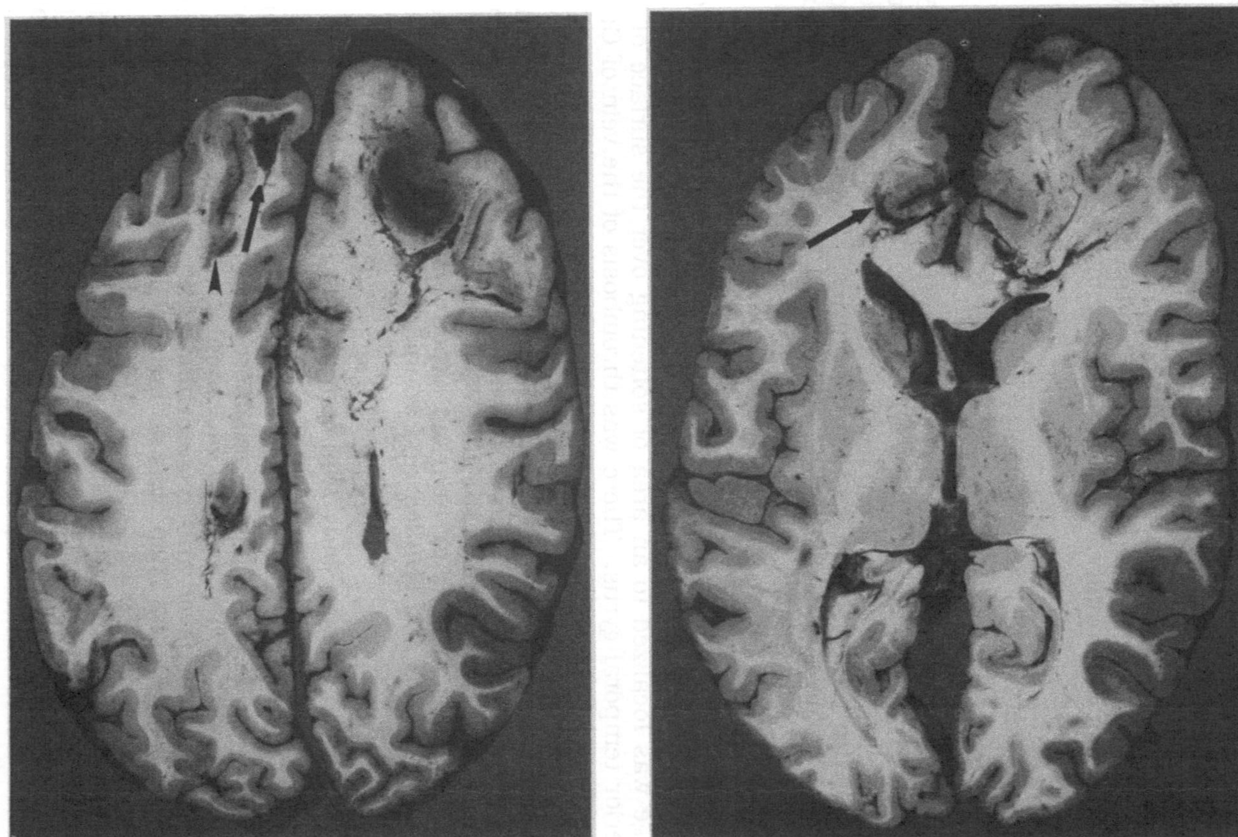

Fig. 1. (UAB 15,746) SSS thrombosis after a sickle-cell crisis in a 7-year-old patient (Case 2 in the text). Several features suggest occlusion of a major, midline sinus: bilateral hemispheric swelling, dusky appearance of the cortex, subcortical hemorrhages (arrows) and a large subarachnoid hemorrhage especially abundant over the convexity of the right frontal lobe.

Case 5: Sagittal Sinus Associated with a Neoplasm (UAB 16,911)

The patient was a 70-year-old man with a 3-months history of left leg weakness; this symptom was followed by paraparesis, papilledema and waxing and waning symptoms including psychiatric problems, various localized neurologic deficits and Parkinsonian complaints appearing during the last 3 weeks of his life.

At autopsy, multiple sites of hemorrhagic infarctions, several weeks old, were found in the left occipital lobe, left thalamic nuclei and right superior parietal lobule. A small (2.7 cm in diameter) meningioma was found on the mesial surface of the right superior frontal gyrus, next to an organizing thrombus that occluded approximately a 5.0 cm segment of the SSS. Many of the morphological features in this case were those of a venous infarct, including the thrombosis of the intraparenchymal veins, and the prominent vacuolation of neuropil in areas of softening. There were no intracranial arteries occluded and the arterial walls in all branches examined were unremarkable. Extensive pulmonary thromboembolism, old and recent, was demonstrated at autopsy. Castaigne and associates [7] have reported an instance of SSS thrombosis associated with a parasagittal meningioma in a patient who recovered following the administration of heparin.

Case 6: Multiple Arterial Embolism (UAB 15,310)

An 88-year-old man suffered skin burns about 36 days before death. He developed hypercoagulability, as a part of a disseminated intravascular syndrome, and, at autopsy, multiple nonbacterial vegetations (also called "marantic endocarditis") were found in all four cardiac valves.

At postmortem evaluation of the brain, numerous, small, hemorrhagic infarctions were visible in the territory of both middle cerebral arteries. Microscopically these lesions were circumscribed to one or two adjacent gyri, and involved the bottom more consistently than the crest of the gyrus. Brain hemorrhages were invariably associated with sites of ischemic coagulative necrosis, neutrophilic infiltrates were minimal, there was no large subarachnoid hemorrhage and there were no signs of hypertrophy of the endothelial cells in the venules or capillaries.

COMMENT

The clinical manifestations of ISV thrombosis are protean but in most instances headaches, signs of raised intracranial pressure (such as papilledema), localized neurologic deficits and seizures are the predominant features of this condition [4]. Headache is the most constant and usually the earliest symptom; the frequent pleocytosis in the cerebrospinal fluid in patients complaining of severe headache may mimic a leptomeningeal infectious process [21].

Likewise, the morphologic expressions of ISV thrombosis are extremely varied and range from multiple, small lesions of almost microscopic size (Case 1) to massive brain swelling of the type demonstrated in Case 2. Some patients with confirmed SSS thrombosis, such as Case 22 of Kalbag, had minimal histologic brain abnormalities in spite of the demonstrated papilledema [21]. There is no consensus among neurosurgeons about the consequences of tying off deep cerebral veins but it is well established that sinus thrombosis does not always cause hemorrhagic brain infarction [29], as illustrated by our Case 2 and some of the cases reported by others [4]. All seven cases of SSS thrombosis diagnosed by angiography at a single medical center occurred in women younger than 36 years of age. Five of these patients survived and were followed for several months during which time they remained symptom free [18].

Causes

Causes of ISV thrombosis are so numerous and varied [2,4] that it is difficult to identify a unifying factor that could be responsible for the formation of a venous/sinus thrombus. Hemoglobinopathy S (sickle cell disease) is a condition in which the mechanism of thrombosis in the venous part of the circulation is understandable; when erythrocytes containing hemoglobin S are exposed to low oxygen tension, polymerized fibers of abnormal hemoglobin are formed, resulting in sickle-shaped cells. This structural change is accompanied by a marked increase in blood viscosity that promotes stasis and intravascular thrombosis [11]. Sickle-cell crises are frequently accompanied by neurologic symptoms attributable to disorders that are broadly classified into: subarachnoid hemorrhages, said to be more common in adults, and vascular thrombosis, either arterial or venous [27]. Intracranial arterial thromboses in two young adults with SS hemoglobinopathy have been illustrated by McCormick and Schochet [25].

In the mid-nineteenth century the most common associated factor, among patients with ISV thrombosis, was osteomyelitis of the petrous portion of the temporal bone [20]. Cavernous sinus thrombosis is most often associated with infections secondary to streptococcus, staphylococcus, or pneumococci; if untreated, the process promptly becomes bilateral and complicated by meningitis and multiple brain abscesses [31]. On the other hand, thrombosis of the intracavernous portion of the carotid artery, as illustrated in our Case 4, is almost always a complication of facial mucormycosis [1].

Several authors have drawn attention to an association between ISV thrombosis and oral contraceptives [26]; however, in a series of cases of ISV thrombosis, among women of reproductive age, Estanol and associates [10] did not find statistical evidence to suggest that the condition of ISV thrombosis may be more frequent among women who ingest oral contraceptives over a prolonged period of time. Towbin [30] maintains that ISV thrombosis is frequently overlooked as a potentially significant condition, especially among elderly patients afflicted with congestive heart failure, pulmonary thromboembolism and related circulatory disorders.

In patients with either systemic or localized neoplasia, ISV thrombosis may reflect invasion of the dura mater as it occurs in some leukemias and carcinomas that tend to metastasize to the skull. Nonmetastatic sagittal sinus thrombosis has been documented by arteriography in seven cancer patients: five with hematologic malignancies and two with solid tumors. The authors attributed the ISV thrombosis in these patients to the development of a "hypercoagulable state" associated with cancer [28].

Among patients with thrombosis of the transverse dural sinus, arteriovenous malformations (primarily localized to the dura) may represent an acquired lesion that results from neovascularization after thrombosis of the sinus [19].

Sinus/Venus Thrombosis: Parenchymal Lesions

The following is a summary of the structural changes thought to be characteristic of venous infarctions. This compilation reflects a blend of features described in the medical literature, especially by Kalbag and Woolfe [22] and McCormick and Schochet [25], as well as those seen in cases of our own laboratory.

The brain is usually swollen, heavier than normal, and the sulci and ventricles are generally narrowed. Subarachnoid hemorrhages, especially those primarily located over the convexity of the cerebral hemispheres, suggest venous or sinus occlusion.

The distribution of the lesions vary according to the occluded sinus or vein. Cortical and subcortical areas of softening either hemorrhagic or pale may be visible in the parasagittal areas, and these may be accompanied by isolated lesions in the centrum semiovale in association with partial or total occlusion of the SSS. The characteristic lesions following intracranial phlebothrombosis in children show considerable variation in severity and extent ranging from absence of parenchymal changes to disseminated cortical scars, cavitary lesions and diffuse gliosis (incomplete necrosis). Parasagittal distribution is most common, but cerebral and cerebellar lesions frequently coexist with subarachnoid hemorrhage [13]. Softenings, usually hemorrhagic, on both sides of the midline, involving the thalamic nuclei and the upper portions of the brain stem are to be expected in cases of occlusion of the vein of Galen or its major tributaries [25].

Following occlusion of a cerebral cortical vein, one commonly sees large hemorrhagic brain softenings that usually do not correspond to an arterial branch distribution and have marked hemorrhagic components extending widely into the subcortical white matter. In some cases the hemorrhage in the white matter may be sufficiently large as to suggest a lobar bleed of arterial origin [25]. The hemorrhagic infarction of arterial origin by comparison characteristically has petechiae confined to the cortex or gray matter structures [15,16] (Fig. 2).

Microscopically, parenchymal lesions secondary to venous or sinus occlusion are characterized by large lakes of plasma extravasation, pronounced sponginess of the neuropil (Fig. 3) even in the absence of neuronal necrosis, periadventitial and parenchymal hemorrhages (also visible in the absence of tissue necrosis), ring hemorrhages, profuse infiltration by neutrophils (Fig. 4), usually associated with occluded veins, markedly distended veins, collapsed arteries, thrombosed veins, large subarachnoid hemorrhages, which are generally absent in infarctions of arterial origin, and pronounced hypertrophy of the venous endothelial cells. Poltera [26] reported two cases of ISV thrombosis in which abundant neutrophils were observed in the walls of the veins and the surrounding parenchyma; hemorrhages were prominent, especially in perivascular location. The hypertrophy of the endothelial cells in the capillaries and venules was attributed by the author to the effects of oral

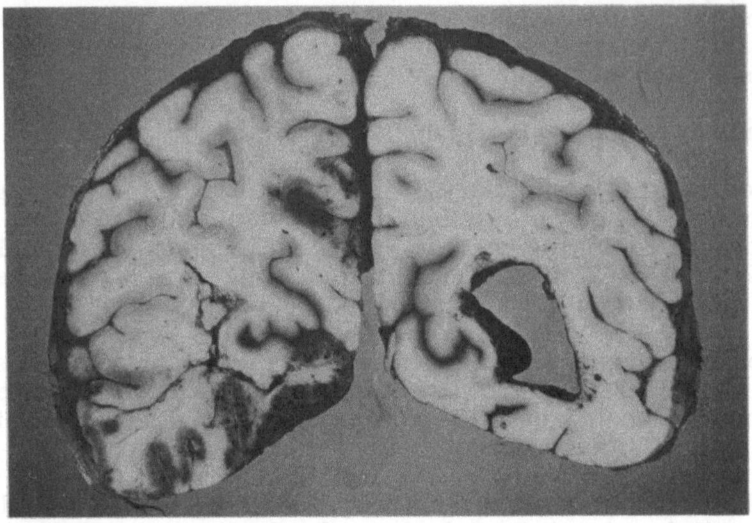

Fig. 2. (UAB 15,585) Hemorrhagic arterial infarction in the territory of the left posterior cerebral artery. In this instance, the hemorrhages are pinpointed and limited to the cortical ribbon. A large hemorrhage had accumulated in the left frontal lobe (3 days before) and had led to massive left uncinate herniation.

contraceptives, because administration of estrogens produce endothelial hyperplasia in the endometrium. Incomplete (or selective) necrosis has been used to designate the features of ischemic lesions in which *only* the neurons are destroyed [8]. In a series of subhuman primates with occlusion of a MCA, selective neuronal necrosis (with preservation of glia and vasculature) was observed in those animals in which the local CBF values remained closer to the normal values [8]. Selective necrosis in cases of ISV thrombosis may indicate that the degree of ischemia is less severe than after an arterial occlusion.

A case of SSS thrombosis with long-term survival (about 25 years) has been used to illustrate the fact that the resulting parenchymal lesions can be markedly asymmetric and that, except for the unusual distribution (which does not correspond to that of an arterial vessel), the healing process progresses in a similar manner in both arterial and venous infarcts [9].

In an autopsy series of 50 cases of ISV thrombosis, among children younger than 30 months at the time of death, there were 26 cases associated with infections and 24 cases attributed to malnutrition and dehydration. Microscopically the parenchymal brain lesions (even in the cases without infection) featured abundant neutrophilic infiltrates, especially in and around the thrombosed vessels [5].

Among survivors of neonatal meningitis, infarctions of the brain were attributed to the formation of fibrin thrombi in leptomeningeal, cortical and subependymal veins in nine cases. SSS thrombosis was not common in these patients [14].

In summary, ISV thrombosis may be more common than previously suspected. The diagnosis requires prompt angiographic evaluation. The parenchymal lesions may be microscopic or may consist of large,

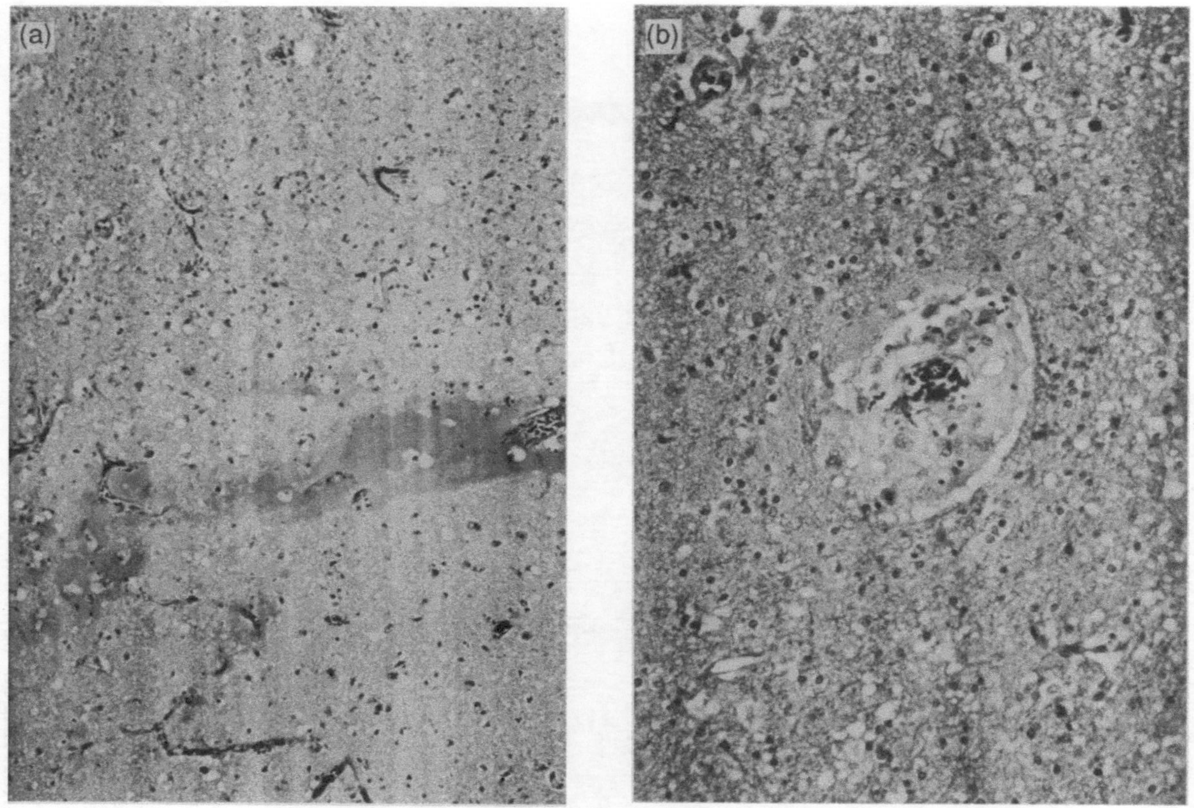

Fig. 3. (UAB 15,746) Venous infarctions. (a) Large lake of protein-rich fluid reflects extravasation of plasma from venules. Most of these vessels (arrowheads) are dilated and filled with red blood cells that impart the dusky appearance to the involved tissues. Hematoxylin and Eosin (HE) stain (x 165). (b) Another field of the same venous infarction showing marked tissue vacuolation in the absence of coagulative necrosis. In the center a blood vessel is surrounded by transudate. Hematoxylin and Eosin (HE) stain (x 330).

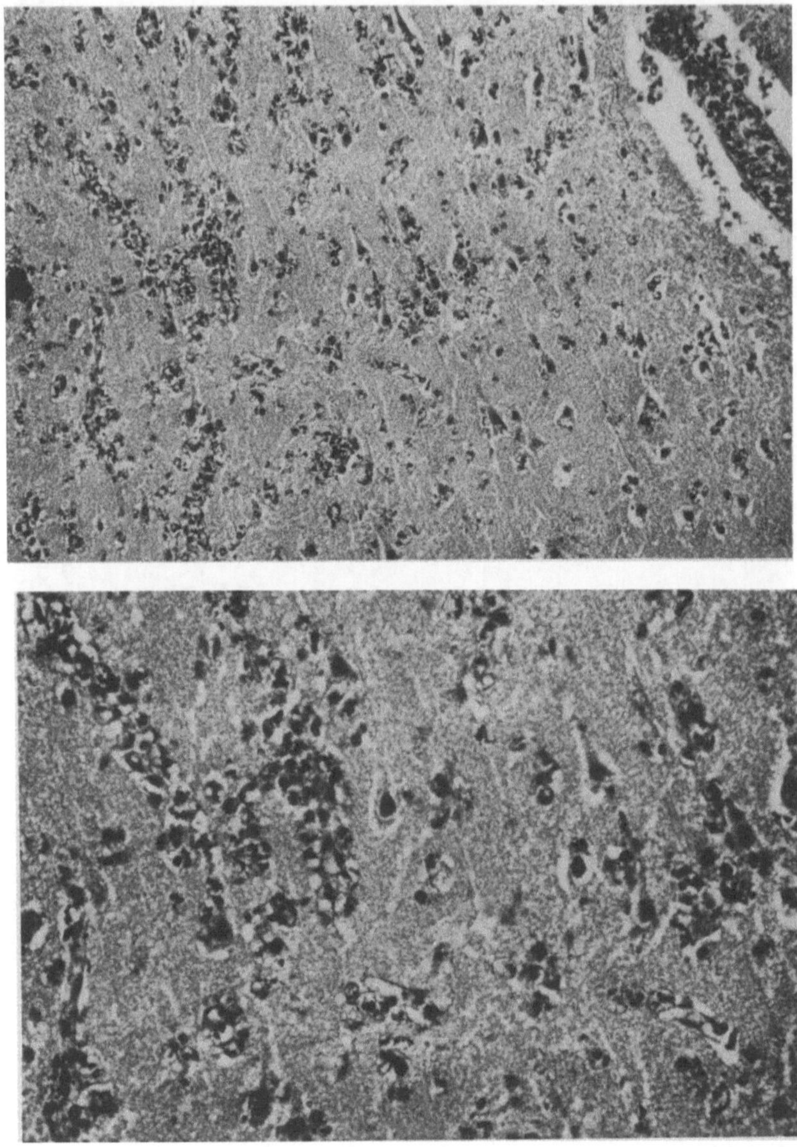

Fig. 4. (UAB 15,746) Venous infarction. Two views of an area where extravasation of polymorphonuclear leukocytes (mostly necrotic) is abundant. This pattern of inflammation is not visible in arterial infarctions. Necrotic neuronal perikarya are still recognizable because of their size and pyramidal shape (arrowheads). Hematoxylin and Eosin (HE) stain (x 165 and x 330, respectively).

confluent hemorrhages especially common in the white matter of the cerebral hemispheres (Fig. 1). Brain swelling and localized subarachnoid hemorrhages (especially those located over the convexity of the cerebral hemispheres) are common features of ISV thrombosis (Fig. 1). Microscopically, the absence of neuronal necrosis, the marked sponginess of the neuropil, the profuse neutrophil infiltration (Fig. 4), the large plasma lakes (Fig. 3) and the endothelial hypertrophy of venules and capillaries are all features of the parenchymal lesions commonly observed in cases of ISV thrombosis.

ACKNOWLEDGEMENT

I am very grateful to Mrs Victoria Jones for her excellent secretarial support.

REFERENCES

1. E. J. Anaissie and A. H. Shikhari, Rhinocerebral mucormycosis with internal carotid occlusion: report of two cases and review of the literature, *Laryngoscope*, 85:1107-1113 (1985).
2. P. Averback, Primary cerebral venous thrombosis in young adults. The diverse manifestations of an underrecognized disease, *Ann. Neurol.*, 3:81-86 (1978).
3. H. J. M. Barnett and H. H. Hyland, Non-infective intracranial venous thrombosis, *Brain*, 76:36-49 (1953).
4. M. G. Bousser, J. Chiras and J. Bories, Cerebral venous thrombosis. A review of 38 cases, *Stroke*, 16:199-213 (1985).
5. R. K. Byers and G. M. Hass, Thrombosis of dural sinuses in infancy and childhood, *Am. J. Dis. Child*, 45:1161-1183 (1933).
6. N. F. Capra and J. P. Kapp, Anatomic and physiologic aspects of venous system, *in*: "Cerebral Blood Flow", J. H. Wood, ed., McGraw Hill Inc., New York, pp 37-58 (1987).
7. P. Castaigne, D. Laplane and M. G. Bousser, Superior sagittal sinus thrombosis, *Arch. Neurol.*, 34:788-789 (1977).
8. U. De Girolami, R. M. Crowell and F. W. Marcoux, Selective necrosis and total necrosis in focal cerebral artery occlusion in the Macaque monkey, *J. Neuropathol. Exper. Neurol.*, 43(1):57-71 (1984).
9. A. S. Dekaban and R. M. Norman, Hemiplegia in early life associated with thrombosis of sagittal sinus and its tributary veins in the hemisphere, *J. Neuropathol. Exper. Neurol.*, 17:461-470 (1958).
10. B. Estanol, A. Rodriguez and G. Conte, Intracranial venous thrombosis in young women, *Stroke*, 10:680-684 (1979).
11. J. A. Feldenzer, M. J. Bueche, J. L. Venes and S. S. Begarski, Superior sagittal sinus thrombosis with infarction in sickle-cell trait, *Stroke*, 18:656-660 (1987).
12. W. L. Floyd and M. S. Mahaley, Cerebral dural venous sinus thrombosis following cardiac pacemaker implantation, *Arch. Intern. Med.*, 124:368-372 (1979).
13. R. L. Friede, Cerebral lesions complicating neonatal leptomeningitis, acute and residual lesions, *Acta Neuropathol. (Berl.)*, 23:245-253 (1973).
14. R. L. Friede, Residual lesions of infantile cerebral phlebothrombosis, *Acta Neuropathol. (Berl.)*, 22:319-332 (1972).
15. J. H. Garcia, Circulatory disorders and their effects on the brain, *in*: "Textbook of Neuropathology", R. L. Davis and D. M. Robertson, eds., Williams & Wilkins, Baltimore, pp 548-631 (1985).
16. J. H. Garcia, A. S. Lossinsky and F. C. Kauffman, Neuronal ischemic injury: light microscopy, ultrastructure and biochemistry, *Acta Neuorpathol. (Berl.)*, 43:85-95 (1978).
17. J. H. Garcia, J. P. Williams and J. Tanaka, Spontaneous thrombosis of deep cerebral veins: a complication of arteriovenous malformation, *Stroke*, 6:164-171 (1975).
18. D. M. Gettelfinger and E. Kokmen, Superior sagittal sinus thrombosis, *Arch. Neurol.*, 34:2-6 (1977).

19. O. W. Houser, J. K. Campbell and R. J. Campbell, Arteriovenous malformations affecting the transverse dural venous sinus. An acquired lesion, *Mayo Clin. Proc.*, 54:651-661 (1979).
20. S. Jarcho, Von Dusch on thrombosis of the cerebral sinuses (1861), *Am. J. Cardiol.*, 20:555-560 (1967).
21. R. M. Kalbag, Cerebral venous thrombosis, *in*: "The Cerebral Venous System and its Disorders", Grune & Stratton Inc., New York, pp 505-536 (1984).
22. R. M. Kalbag and A. L. Woolf, eds., "Cerebral Venous Thrombosis with Special Reference to Primary Aseptic Thrombosis", Oxford University Press, New York (1967).
23. B. Kaplan and J. Parker Mickle, Cavernous sinus thrombosis, *in*: "The Cerebral Venous System and its Disorders", Grune & Stratton Inc., New York, pp 537-546 (1984).
24. H. A. Krayenbühl, Cerebral venous and sinus thrombosis, *Clin. Neurosurg.*, 14:1-24 (1967).
25. W. F. McCormick and S. S. Schochet, Cerebral infarction in sickle-cell disease, *in*: "Atlas of Cerebrovascular Disease", W. B. Saunders Co., Philadelphia, pp 226-230 (1976).
26. A. A. Poltera, The pathology of intracranial venous thrombosis in oral contraception, *J. Pathol.*, 106:209-219 (1972).
27. D. Powars, B. Wilson and C. Imbus, The natural history of stroke in sickle-cell disease, *Am. J. Med.*, 65:461-471 (1978).
28. B. Sigsbee, M. D. F. Deck and J. B. Posner, Nonmetastatic superior sagittal sinus thrombosis complicating systemic cancer, *Neurology*, 29:139-146 (1979).
29. R. R. Smith and R. A. Sanford, Disorders of the deep cerebral veins, *in*: "The Cerebral Venous System and its Disorders", Grune & Stratton Inc., New York, pp 547-555 (1984).
30. A. Towbin, The syndrome of latent cerebral venous thrombosis. Its frequency in relation to age and congestive heart failure, *Stroke*, 4:419-430 (1973).
31. C. J. Yarington, Cavernous sinus thrombosis revisited, *Proc. R. Soc. Med.*, 70:456-459 (1977).

A NEW EXPERIMENTAL MODEL OF SINUS VEIN THROMBOSIS

M. Deckert, K. Frerichs*, P. Mehraein†, O. Kempski*, A. Baethmann* and K. Einhäupl

Department of Neurology, *Institute of Surgical Research and †Department of Neuropathology
Ludwig-Maximilians-University, Klinikum Großhadern
8000 Munich 70, FRG

INTRODUCTION

Several experimental attempts have failed to induce thrombosis of the cerebral sinus using methods such as ligation, coagulation, or injection of thrombogenic material [1-3]. Ligation and injection of lard-oil or cyanoacrylate into the superior sagittal sinus (SSS) produced obstruction of the SSS, or of the superior cerebral veins (SCV) yet without evidence of thrombosis [4-7]. Materials such as scandium chloride, sodium morrhuate, or thrombin injected into the SSS may cause sinus or vein thrombosis. However, histological changes of the brain were either absent or not systematically investigated [8-10].

The current report describes a novel experimental method to induce a sinus vein thrombosis (SVT) in rats. Frerichs and coworkers [11] using the same method for thrombosis induction reports on corresponding functional and behavioral changes. This paper communicates both acute and chronic morphological changes of the brain tissue resulting from induction of sinus vein thrombosis.

MATERIALS AND METHODS

Sprague Dawley rats of both sexes of 250-350 g body weight were used. Anesthesia was induced and maintained by ip administration of chloralhydrate. The SSS was exposed by removing the calvarium with a dental drill and was rostrally ligated, as well as close to the torcular, using fine silk. After ligation, 100 μl of a kaolin-cephalin suspension (Boehringer Mannheim, Inc.) were injected into the SSS over a period of 10 minutes. Formation and spread of thrombosis was assessed by fluorescence angiography. The animals were sacrificed at the end of the experiment by bleeding and were subjected to perfusion fixation with 5% formalin after survival for 30 minutes, 6 hours, 24 hours, 4 days or

Cerebral Sinus Thrombosis, Edited by K. Einhäupl *et al.*
Plenum Press, New York, 1990

5 weeks, respectively post-thrombosis induction. Frontal sections of the brain were prepared and stained by conventional histological methods.

RESULTS

Induction of thrombosis of the SSS was successful in all the animals (n = 14) examined. Thrombosis was limited to the ligated portion of the SSS in six animals, and to the SSS plus the adjacent superior cortical veins (SCV) in the remaining eight. Five of eight animals with additional thrombosis of the SCV had petechial subarachnoidal hemorrhage surrounding the thrombosed veins. Bleeding appeared immediately after induction of thrombosis and increased over a period of 20 minutes.

Eleven animals had extensive brain tissue damage including edematous (n = 3) and ischemic changes of the parenchyma (n = 8), or intracerebral petechial bleedings (n = 3). Three remaining animals had no apparent structural damage of the brain. Histological examination of cerebral cortex revealed perivascular vacuolization already 30 minutes after induction of thrombosis. At 6 hours, ischemic nerve cell damage characterized by shrinkage and triangular deformation of the cell nuclei together with appearance of a status spongiosus in the center of the lesion was seen. Enlarged vacuoles were found in the corpus callosum. Twenty-four hours post-thrombosis, pallor of nerve cells or shrunken triangular-shaped cells became visible with condensed nuclei and status spongiosus. Petechial hemorrhage was restricted to the damaged areas proper. Advanced necrosis with proliferation of capillaries and of perivascular astroglial cells was found at 4 days. At that time fat-containing granule cells and monocytes became abundantly present. By 4 weeks, an extensive loss of nerve cells could be observed with only few glial nuclei remaining. An increased number of blood vessels and few macrophages were found. The area of venous infarction appeared completely necrotic, eventually transforming into a cystic scar.

DISCUSSION

A model of SSS thrombosis was established in rats. Histological brain tissue changes were apparent in 11 of 14 animals with sinus occlusion. The pathological changes consisted of, among others, subarachnoidal and intracerebral hemorrhage together with marked edema in cerebral cortex. The most prominent finding, however, was sharp demarcation of a parasagittal ischemic necrosis involving all layers of cerebral cortex.

The results correspond to respective findings in humans. Several authors have reported brain edema, subarachnoidal and intracerebral hemorrhage [12-14]. Ischemic nerve cell damage in association with SVT has been also observed in man [12,13,15].

Nevertheless, ischemic nerve cell damage appeared more extensive and severe in the current experiments than that found in humans, while brain edema was less intensive. It has been discussed for awhile now, whether SVT produces ischemic nerve cell damage. Meesen [14] believes that ischemic nerve cell changes occur only if additional factors complicate SVT. Our results indicate that SVT alone is sufficient to cause an ischemic necrosis. Systemic factors, such as arterial hypoxia, can be ruled out as a mechanism of necrosis because no ischemic damage was seen in selectively vulnerable areas, eg, hippocampus. Spread of necrosis was restricted to areas with topographical relation to thrombosed vessels.

Absence of brain tissue lesions in some animals with proven SSS thrombosis may be explained by venous collateralization providing for drainage of blood to inferior cerebral veins immediately after obstruction of the SSS. This is concluded from observations that reversal of blood flow in SCV occurred immediately after ligation of the SSS. Absence of brain tissue lesions would agree with respective observations in man. Not all human cases with SVT are associated with severe neurological symptoms. Probably these are the patients where no evidence of structural brain tissue damage is present [12].

REFERENCES

1. D. J. Beck and D. S. Russel, Experiments on thrombosis of the superior longitudinal sinus, *Neurosurgery*, 3:337 (1946).
2. G. Owens, G. Stahlman, J. Capps and A. M. Meirowsky, Experimental occlusion of dural sinuses, *J. Neurosurg.*, 14:640 (1957).
3. S. Sato, Y. Miyahara, Y. Dohmoto, T. Kawase and A. Toya, Cerebral microcirculation in experimental sagittal sinus occlusion in dogs, *in*: "Cerebral Vein", L. M. Auer and F. Loew, eds., Springer, Wien, pp 111-117 (1983).
4. T. J. Putnam, Studies in multiple sclerosis. V "Encephalitis" and sclerotic plaques produced by venular obstruction, *Arch. Neurol. Psych.*, 39:929 (1935).
5. A. L. Woolf, Experimentally produced cerebral venous obstruction, *J. Pathol. Bact.*, 67:1 (1954).
6. K. Fujita, S. Kose, N. Tamaki and S. Matsumoto, A new experimental method for producing sagittal sinus occlusion and the change of CSF dynamics in sinus occlusion, *in*: "Intracranial Pressure V", S. Ishii, H. Nagai and M. Brock, eds., Springer, Berlin, pp 693-996 (1983).
7. K. Fujita, N. Kojima, N. Tamaki and S. Matsumoto, Brain edema in intracranial venous hypertension, *in*: "Brain Edema", Y. Inaba, I. Klatzo and M. Spatz, eds., Springer, Berlin, pp 228-234 (1985).
8. H. Selye, S. Varga and G. Winandy, Production and prevention of intracranial venous thrombosis in the rat, *Exp. Neurol.*, 15:329 (1966).
9. E. R. Heinz, D. Geeter and T. P. Gabrielsen, Cortical vein thrombosis in the dog with a review of aseptic intracranial venous thrombosis in man, *Acta Radiol. Diagn.*, 13:105 (1972).
10. M. Sarwar, C. Virapongse and P. Carbo, Experimental production of superior sagittal sinus thrombosis in the dog, *Am. J. Neuroradiol.*, 6:19 (1985).
11. K. Frerichs, M. Deckert, O. Kempski, K. M. Einhäupl and A. Baethmann, Brain function, pathophysiology and heparin therapy in experimental cerebral sinus vein thrombosis, *in*: "Cerebral Sinus Thrombosis: Experimental and Clinical Aspects", K. Einhäupl, O. Kempski and A. Baethmann, eds, Plenum Press, New York and London, pp 53-58 (1990).
12. J. Escola, Die Gewebsveränderungen bei Thrombosen der Sinus und zerebralen Venen, *Arch. Psychiat. Ztsch. Neurol.*, 203:342 (1982).
13. A. Huhn, ed., "Thrombosen der intrakraniellen Venen und Sinus; Klinische und Pathologisch-Anatomische Untersuchungen", Schattauer, Stuttgart (1965).
14. H. Meesen and O. Stochdorph, Die Thromboembolie, die arterielle und die venöse Thrombose, *in*: "Handbuch der speziellen pathologischen Anatomie und Histologie, Bd. 13: Nervensystem", O. Lubarsch, F. Henke and R. Rössle, eds., Springer, Berlin (1957).

15. H. Noetzel and F. Jerusalem, eds., "Die Hirnvenen- und Sinusthrombosen unter besonderer Berücksichtigung der Topographie des hämorrhagischen Infarkts", Springer, Berlin (1965).
16. E. C. Greene, ed., "Anatomy of the Rat", Hafner Publishing Co., New York (1959).

EXPERIMENTAL MODEL IN THE CAT FOR CEREBRAL SINO-VENOUS OCCLUSION

S. Kannuki, J. Cervos-Navarro, K. Matsumoto*
and Y. Nakagawa*

Institute of Neuropathology, Klinikum Steglitz
Free University of Berlin, FRG and
*Department of Neurological Surgery
Tokushima University, Japan

INTRODUCTION

Cerebral sinus vein thrombosis (CSVT) leads to a wide range of changes in brain tissue and to various clinical syndromes. This variability is due to the many different collateral connections of the venous circulation. It was formerly believed that CSVT has consistently a poor prognosis and high mortality [8]. Although recent publications report CSVT cases with a favorable prognosis, severe hemorrhagic venous infarction still has a poor outcome. Further, treatment of CSVT is still controversial. In the present study, the superior sagittal sinus (SSS) of 22 cats was occluded by embolization using a polymer. Extravasation of tracers indicating changes of blood-brain barrier (BBB) permeability was studied and correlated with histological changes induced by SSS occlusion. In addition, regional cerebral blood flow (rCBF) as well as brain edema measured as increase in water content of brain tissue using a freeze-drying method were studied. Brain edema was also correlated with changes of the BBB permeability.

MATERIALS AND METHODS

Adult cats of both sexes weighing about 3-4 kg were used. In anesthesia induced by intramuscular injection of ketamine (50 mg/kg), the cats were paralyzed by pancuronium and artificially ventilated via an endotracheal tube using a Harvard animal respirator. After catheterization of both the femoral artery and vein, the cats were fastened onto a sterotaxic frame in a sphinx-like position. Arterial blood gases and body temperature were monitored periodically to maintain physiological conditions. For measurement of the regional cerebral blood flow (rCBF) with labelled microspheres, an additional catheter was introduced into the left cardiac ventricle through the brachial artery. Localization of the catheter tip in the left cardiac ventricle was roentgenographically confirmed. Thereafter, a craniectomy of about 0.5 x 1.5 cm was made in

Cerebral Sinus Thrombosis, Edited by K. Einhäupl *et al.*
Plenum Press, New York, 1990

the parietal region of the skull above the SSS. The SSS was exposed and punctured with a 27 G cannula. A 1:1 mixture of 0.2-1.0 ml isobutyl-2-cyanoacrylate and iophendilate was then injected into the SSS. The mixture polymerized 3 seconds after injection. As sham operation craniotomy was made above the SSS without injection of polymer.

rCBF and Water Content Determination

For determination of rCBF and of the cerebral water content, six cats with sinus occlusion and three sham-operated cats were employed. Microspheres labelled with ^{51}Cr, ^{57}Co or ^{95}Nb suspended in 10% dextran were used. Five milliliters of a suspension containing 1×10^6 microspheres were injected over 15 seconds through the cardiac catheter into the left ventricle. Reference samples of arterial blood were obtained at a rate of 2.06 ml/minute for 4 minutes from the thoracic aorta via the catheter placed into the femoral artery. The regional CBF was measured prior to, and 2 or 6 hours after, SSS occlusion. Six hours after SSS occlusion, the cats were killed by intracardiac injection of KCl. The brain was immediately removed, coronally cut, and dissected into 0.5-2.0 g specimens. Gray and white matter was studied in 16 regions, basal ganglia in six regions, while brain stem and cerebellum where studied in three regions. The brain specimens were weighed and counted in a well-type gamma counter for 1 minute. rCBF was calculated using the equation of Marcus et al. [9]:

$$rCBF = C_t . Q_{ar} / A_{ar} \text{ (ml/100 g wet weight x minutes)},$$

where C_t is radioactivity of the tissue (per 100 g), Q_{ar} the withdrawal rate (ml/minute) and A_{ar} the total radioactivity of the arterial reference sample.

Subsequently, the brain tissue specimens were frozen and dried in a freeze vacuum device for 72 hours. The percentage brain water content was calculated according to Shigeno et al. [15]:

$$\text{\% tissue weight} = \text{(wet weight-dry weight/wet weight) x 100}.$$

BBB Permeability

In order to study the topographical distribution of permeability changes of the BBB, two tracers were intravenously administered. Evans-blue (EB) (2%) was used as a protein tracer, and Na^+-fluorescein (F) (10%) as a small molecular-weight marker. One-fifth of the tracer dose was administered before occlusion, and similar doses were injected at equivalent time intervals until death of the animal. The total dose was 20 ml in the case of Evans-blue and 5 ml in the case of Na^+-fluorescein. Localization of both tracers was macroscopically studied using UV-illumination. Distribution of Evans-blue was also microscopically investigated after perfusion fixation of the brain.

Histology

Sixteen cats with sinus occlusion and two sham-operated cats were used for histology. One, 3 or 6 hours after sinus occlusion, the cats were killed by perfusion fixation through the cardiac route with 10% formalin or Karnovsky's solution following previous rinsing of the vasculature with 500 ml physiological saline. Dehydration with alcohol, or acetone and embedding in paraffin or epon was made by routine methods. HE, Nissl, PTAH, and elastica van Gieson stains were performed for light microscopy. Ultrathin sections stained with uranyl- and lead acetate were used for transmission EM examination.

X-Ray Studies

After perfusion fixation the brain was subjected to roentgenography. Thrombosis formation could be visualized because of the radiopaque nature of the polymer material given for sinus occlusion.

RESULTS

Extent of Sinus and Venous Occlusion

According to severity, the cats were divided into groups with partial and total occlusion. In the partial occlusion group, the sagittal sinus only was involved. If cortical veins were affected in addition, cats were considered to have total occlusion (Fig. 1). As a rule, partial occlusion seemed to occur when smaller amounts (0.2-0.3 ml) of polymer were injected, larger amounts causing total occlusion.

A total of 22 sinus occluded cats and five control cats were studied. In 14 animals, sinus occlusion was partial, while in eight cats it was total. Three cats with total and three with partial occlusion were used for measurements of the cerebral water content. Determinations of rCBF were made in the same number of cats of each group. Five cats with total and 11 with partial occlusion were used for BBB and histological studies. Three control cats were used for measurements of rCBF, and two controls for distribution studies of BBB markers and histology.

Water Content and rCBF

In the partial occlusion group, the cerebral water content was not significantly increased as compared to the control group. In the total occlusion group, one animal had severe subarachnoidal hemorrhage and diffuse hemorrhagic infarction. The cerebral water content was decreased due to an increased amount of blood. In two other cats the water content was significantly increased in cortex and white matter. The water content was particularly high in cortical and white matter near the SSS (Table 1). In both groups CBF was diffusely decreased after 2 hours in all regions except cerebellum and brain stem in the animals with partial occlusion, where rCBF was not changed. After 6 hours, rCBF continued to decrease in the white matter in the total occlusion group. In cerebral cortex and basal ganglia of both groups, rCBF was found at the low level, where it already was 2 hours after occlusion. In the partial occlusion group, rCBF of cerebellum and brain stem, although unchanged after 2 hours, was decreased after 6 hours (Table 2).

Distribution of BBB Markers

In the two control cats, leakage of tracer did not occur during the entire observation period. In 11 cats of the partial occlusion group virtually no leakage of EB was found within 3 or 6 hours after sinus

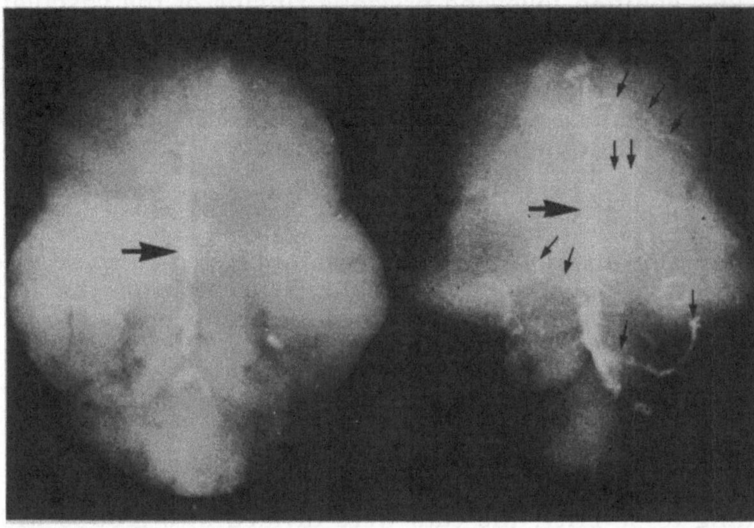

Fig. 1. Roentgenogram of partial (left) and total (right) sinus vein occlusion. In partial occlusion, distribution of embolization material is mainly limited to the SSS, while in the total occlusion group many cortical veins are also involved.

occlusion. On the other hand, moderate leakage of Na^+-fluorescein developed which was confined to the parasagittal gray matter. In animals with total occlusion, EB leakage was conspicuous in different cerebral areas, however, to a variable degree depending on the survival time after sinus occlusion. One hour after occlusion, EB was mainly found in the parasagittal gray matter (Fig. 2). Using fluorescence microscopy EB was also seen in the perivascular spaces of the parasagittal subcortical white matter. After 3 hours, large amounts of EB extravasated in association with small hemorrhages (see below) into the parenchyma of the parasagittal cortex and spread into the subcortical white matter (Fig. 3a). Small amounts of EB were detected by fluorescence microscopy in the

Table 1. Water Content Change

		Control	Partial Occlusion	Total Occlusion	
				Without H	With H
Prox. GM	(8)	79.35 ± 1.04	79.13 ± 0.79	*82.52 ± 0.91	80.18 ± 0.85
Dist. GM	(8)	79.35 ± 1.04	79.57 ± 0.77	*81.48 ± 1.40	79.98 ± 0.80
Prox. WM	(8)	69.56 ± 2.24	69.64 ± 1.73	*75.11 ± 2.36	70.28 ± 2.72
Dist. WM	(8)	69.56 ± 2.24	69.55 ± 1.63	*73.01 ± 1.90	68.33 ± 2.09
Caudate	(2)	80.97 ± 0.34	80.28 ± 1.27	81.85 ± 1.30	79.15
Thalamus	(2)	76.56 ± 2.20	75.31 ± 1.78	*79.03 ± 0.35	74.50
Hippocampus	(2)	77.75 ± 0.74	77.12 ± 2.05	77.83 ± 3.86	77.41
Cerebellum	(2)	77.83 ± 0.45	76.53 ± 0.29	78.43 ± 1.33	76.82
Pons	(3)	70.60 ± 0.40	70.25 ± 0.75	70.65 ± 1.95	69.30

Values are means ± standard deviation, * $p < 0.01$.
( ) Are sampling number of each lesion.
H: Hematoma, GM: Gray matter of cerebral cortex, WM: white matter of subcortex, Prox.: parasagittal lesion.

Table 2. Regional CBF (ml/100g/minute)

		Control	Partial Occlusion (n=3)		Total Occlusion (n=3)	
		(n=3)	2 Hours	6 Hours	2 Hours	6 Hours
Prox. GM	(8)	52.8 ± 10.5	*36.0 ± 4.3	*28.9 ± 5.8	*26.9 ± 8.2	*25.4 ± 11.1
Dist. GM	(8)	48.9 ± 9.8	*33.3 ± 8.0	*26.3 ± 9.8	*30.1 ± 6.4	*26.4 ± 7.4
Prox. WM	(8)	39.5 ± 10.8	*24.7 ± 5.7	*18.0 ± 3.2	*22.6 ± 5.1	*14.8 ± 4.7
Dist. WM	(8)	37.8 ± 9.8	*21.2 ± 6.6	*16.3 ± 4.3	*19.8 ± 4.3	*14.2 ± 2.6
Caudate	(2)	65.4 ± 14.7	45.2 ± 10.1	*39.4 ± 10.6	*42.7 ± 4.8	*34.6 ± 5.6
Thalamus	(2)	48.4 ± 11.3	36.9 ± 6.9	*35.5 ± 6.9	*25.3 ± 6.1	*21.0 ± 5.5
Hippocampus	(2)	32.9 ± 8.0	*23.6 ± 4.3	*19.3 ± 3.3	*14.4 ± 2.5	*14.2 ± 1.7
Cerebellum	(3)	37.0 ± 7.0	31.0 ± 1.5	*22.4 ± 2.3	*21.9 ± 4.8	*18.8 ± 3.8
Brain stem	(3)	30.0 ± 3.7	25.6 ± 2.2	*22.1 ± 2.5	*22.3 ± 3.5	*15.0 ± 2.9

Values are means ± standard deviation, * $p < 0.01$.
( ) Are sampling numbers of each region in one cat.
n Indicates the number of cats in each group.
GM: Gray matter of cerebral cortex, WM: white matter of subcortex.

perivascular white matter distant from SSS occlusion, while the small molecular-weight tracer, Na^+-fluorescein, spread through the parenchyma of the whole gray matter. Six hours after sinus occlusion, intracerebral hematoma were observed in parasagittal cortex in three animals spreading into the subcortical white matter together with a surrounding massive extravasation of Evans-blue (Fig. 4).

Histological Findings

Cats of the partial occlusion group had no histological changes at 3 and 6 hours after SSS occlusion except for mild focal spongiosis which, however, could not be attributed to occlusion, since it was also present in the sham-operated cats. No ischemic neuronal changes were found in cortex. In cats with total sinus and venous occlusion, focal spongiosis was found in the parasagittal white matter after 1 hour, however, without ischemic changes of neurons. After 3 hours, perivascular hemorrhage was mainly found in the parasagittal cortex (Fig. 3b). The surrounding subcortical white matter had a conspicuous spongiosis. Distribution of hemorrhage in cortex as well as of spongiosis in white matter corresponded to the extravasation of Evans-blue. Occasionally there were focal ischemic neuronal changes in cortex, but most of the neurons appeared normal. In animals with total occlusion, large intracerebral hematomas were observed in the parasagittal gray and white matter at 6 hours.

EM Findings

In the partial occlusion group, edematous changes could not be observed in the regions which were electron microscopically studied at 3 or 6 hours. On the other hand, prominent edematous changes of the parasagittal region were seen after 3 hours in the total occlusion group (Fig. 5). The perivascular space of venules in the parasagittal gray and white matter was markedly widened. In subcortical white matter, the extracellular space had considerably expanded and contained blood cell components, while no edematous changes were found in distal regions or in thalamus.

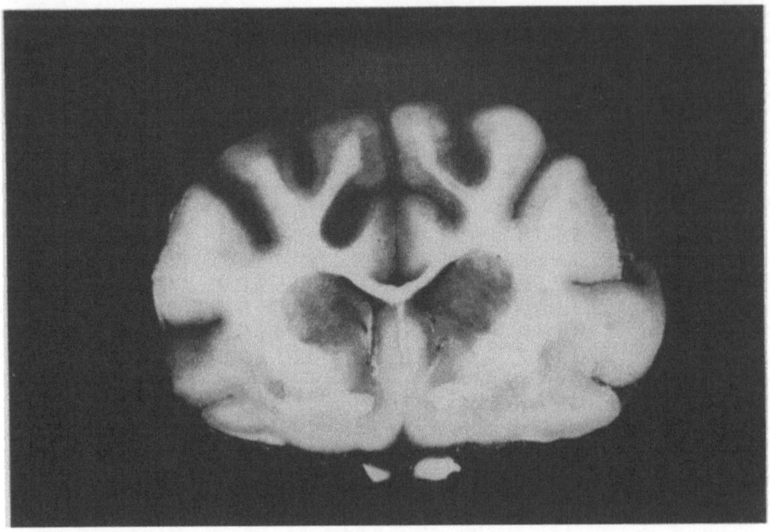

Fig. 2. Macroscopical preparation obtained 1 hour after occlusion. Extravasation of Evans-blue mainly affects parasagittal cortex.

DISCUSSION

Various models of SSS occlusion have been developed [2-6,10-14, 16]. Several weeks or a few months after SSS ligation in dogs with injection of oil [12,16], white matter had plaques of diffuse nerve fiber degeneration and gliosis but no degeneration of neurons in the gray matter. One dog in a series of Woolf had typically a small hemorrhagic infarct in the parasagittal cortex. Beck and Russell [2] tried to induce sinus thrombosis by ligation or injection of different materials in rabbits, dogs, and kittens. However, only one rabbit injected with ethamoline had sinus occlusion after 2 days, mild subarachnoidal hemorrhage, and occasional perivascular hemorrhages. All other methods used in this series failed to produce pathological changes. Miyagami et al. [10,11] reported acute and chronic changes of the CSF circulation after ligation of the SSS in dogs. They found occasional perivascular hemorrhages in the parasagittal cortex and periventricular white matter in the acute (2 or 3 days) and chronic (3 weeks) stage. Sato et al. [14] found mild extravasation of Na^+-fluorescein from parasagittal pial veins a few minutes after SSS clipping in dogs. Sarwar et al. [13] inducing total SSS occlusion with a balloon catheter and injection of thrombin did not observe retrograde spread of thrombosis into neighboring cortical veins nor hemorrhagic changes in the cerebral parenchyma after 4 hours.

Fujita et al. [5,6] were among the first to use a polymer in dogs as we are utilizing for SSS occlusion. The authors emphasized the significance of an involvement of cortical veins for induction of hemorrhagic infarction. All dogs with additional cortical vein occlusion had massive hemorrhagic infarction and subarachnoidal hemorrhage. The material employed for occlusion in the present study is commonly used for embolization in vascular neurosurgery because the time of polymerization can be controlled and because the substance is radiopaque. In vitro polymerization requires 3 seconds if a 50% mixture of isobutyl-2-cyanoacrylate

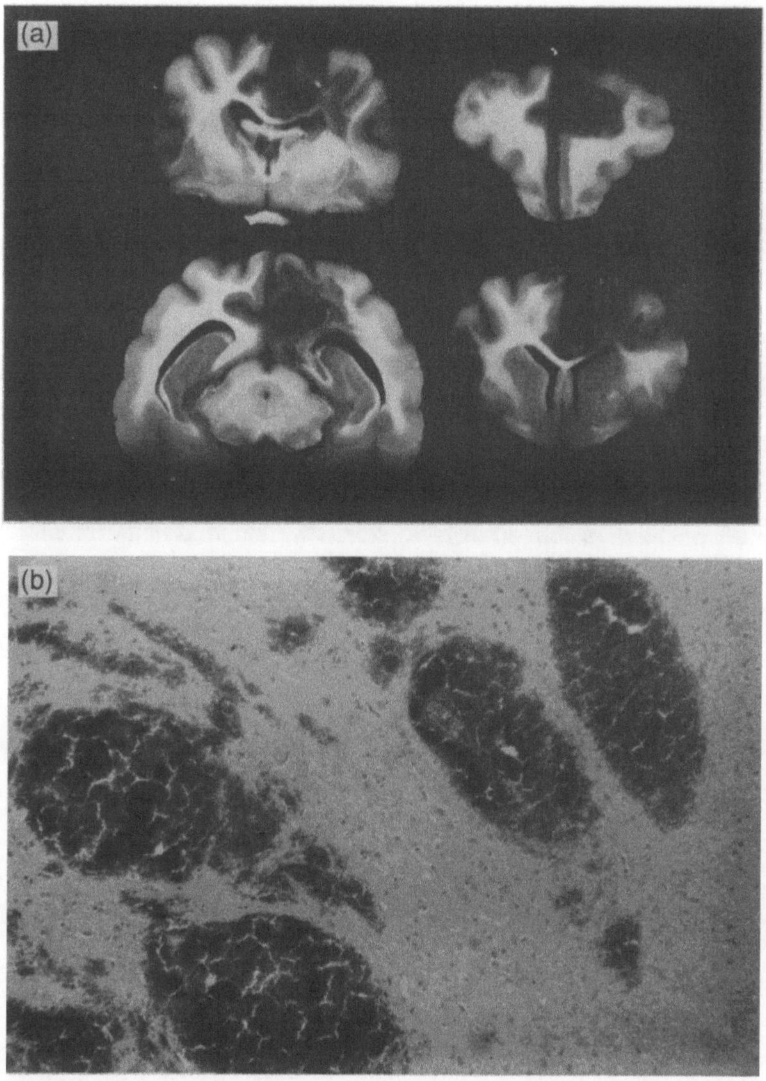

Fig. 3. Three hours after occlusion. (a) Evans-blue extravasation in both gray and subcortical white matter. (b) EB extravasation is also found around numerous small hemorrhages in gray matter. Hematoxylin and Eosin (HE) stain (x 80).

and iophendylate is added to citrated human plasma. The extent of occlusion seems to depend on the amount of polymer injected into the venous system.

From autopsy studies and clinical observations [7,8] it becomes obvious that occlusion of the cortical veins plays an important role for the development of parenchymal hemorrhage in cerebral sinus thrombosis. The difference in frequency of hemorrhage between animals with partial and total occlusion confirms this notion. Experimental occlusion of the SSS and veins led to parenchymal changes which were similar to those seen in human CSVT. In CT images of human CSVT, "gyral enhancement" can be found which is attributable to an increased BBB permeability in the

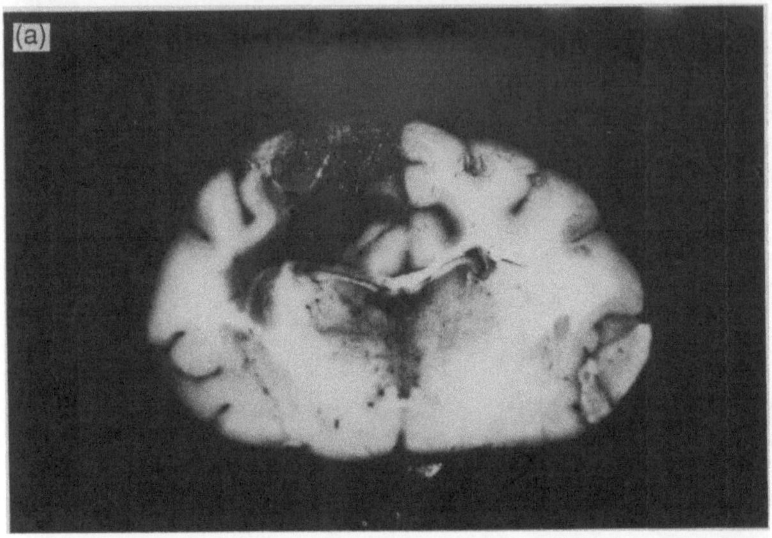

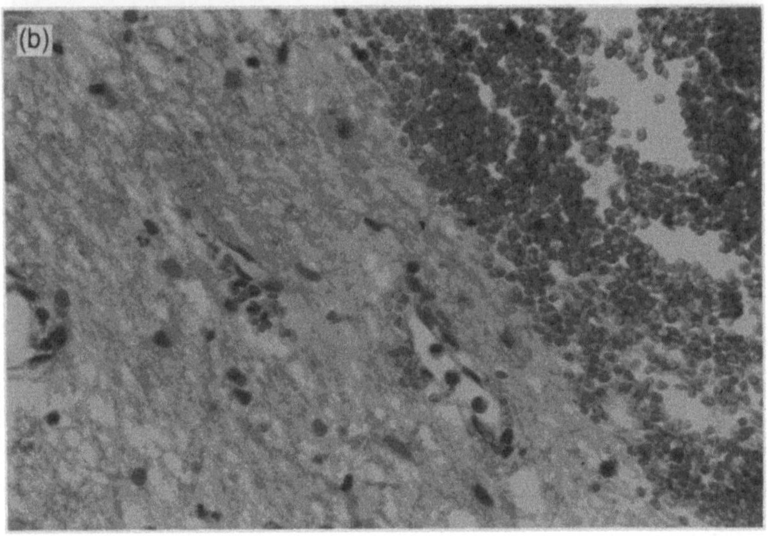

Fig. 4. Six hours after occlusion. (a) Intracerebral hematoma in the parasagittal region. (b) Microscopical findings in association with intracerebral hematoma. Hematoxylin and Eosin (HE) stain (x 330).

acute stage. Hemorrhagic infarction and/or intracerebral hematoma with marked EB leakage observed in the present study may correspond to diffuse low-density areas surrounding hematomas in human CT scans. The interval between occlusion and the beginning of parenchymal changes may constitute a major discrepancy between the current model and human CSVT. In the present model parenchymal changes were rapidly evolving, probably due to the speed of occlusion, as compared to human CSVT, where thrombosis develops along a more protracted time scale.

Sato et al. [13] studied changes in rCBF after clipping of the SSS and found a gradual flow decrease in gray matter, with hyperemia in the subcortical white matter. Using a different occlusion technique, Fujita et

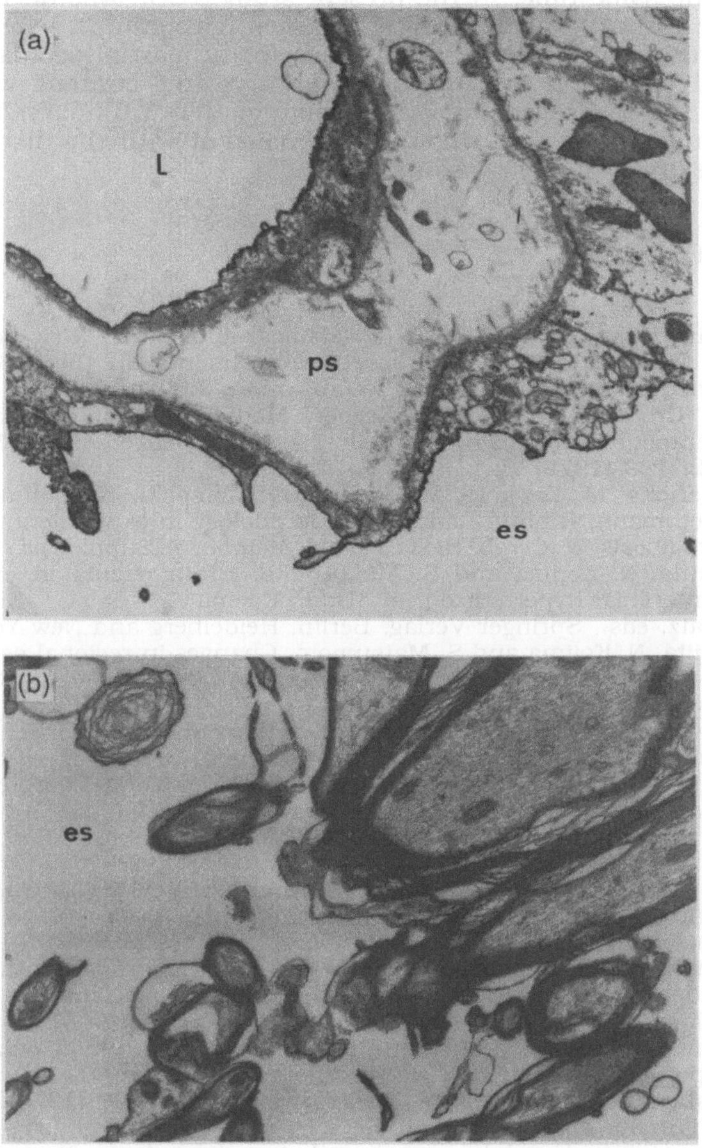

Fig. 5. EM findings in parasagittal subcortical white matter 3 hours after occlusion. Top left the lumen of a venule (L) is seen. (a) Note the extremely dilated perivascular (ps) and extracellular (es) spaces [(a) and (b)].

al. [6] reported absence of CBF changes after SSS occlusion if the cortical veins were not involved, while rCBF was gradually found to decrease to 30% of the normal value at 6 hours after occlusion if the cortical veins were affected. In both studies rCBF was measured by the H_2-clearance method. A major problem with this technique, however, is that the presence of small hemorrhages around the H_2-electrode, particularly in congested brain tissue, may influence the result. The microsphere method is not affected by these changes. It allows measurement of the topographical distribution of CBF as well as the changes of brain water content at the same time. In the present study, rCBF was decreased at 2 hours in the cortical gray and white matter in both groups, whereas it was relatively preserved in the posterior fossa in the partial occlusion group. In the total occlusion group, the cerebral water content was mostly increased in the parasagittal gray and white matter at 6 hours. This agrees with the pattern of edema, which was consistent with the distribution of Evans-blue extravasation at 6 hours.

REFERENCES

1. J. H. Adams, J. A. N. Corsellis and L. W. Duchen, eds, "Greenfield's Neuropathology", 4th ed, Wiley-Medical Publication, New York (1984).
2. D. J. K. Beck and D. S. Russell, Experiments on thrombosis of the superior longitudinal sinus, *Neurosurgery* 3:693 (1946).
3. L. D. Cromwell and C. W. Kerber, Modification of cyanoacrylate for therapeutic embolization: preliminary experience, *Am. J. Roentgenol.*, 132: 799 (1979).
4. K. Frerichs, M. Deckert, M. Jansen, O. Kempski, K. Einhäupl and A. Baethmann, Brain function and morphology in experimental sinus-vein thrombosis, *J. Cereb. Blood Flow & Metabl.*, 7 (Suppl. 1):57 (1987).
5. K. Fujita, N. Kojima and S. Matsumoto, Brain edema in experimental intracranial hypertension, *in*: "Brain Edema", Y. Inaba, I. Klatzo and M. Spatz, eds., Springer Verlag, Berlin, Heidelberg and New York (1986).
6. K. Fujita, N. Kojima and S. Matsumoto, Changes in regional cerebral blood flow in intracranial venous hypertension, *Neurol. Med. Chir., Tokyo*, 26:111 (1986).
7. P. G. Gates and H. J. M. Barnett, Venous disease: cortical veins and sinuses, *in*: "Stroke", vol. 2, H. J. M. Barnett, B. M. Stein, J. P. Mohr and F. M. Yatsu, eds., Churchill Livingstone, New York, Edinburgh, London and Melbourne (1986).
8. H. A. Krayenbühl, Cerebral venous and sinus thrombosis, *Clin. Neurosurg.*, 14:1 (1966).
9. M. L. Marcus, D. D. Heistad, J. C. Ehrhardt and F. M. Abboud, Total and regional cerebral blood flow measurement with 7, 10, 15, 25 and 50 μm microspheres, *J. Appl. Physiol.*, 40:501 (1976).
10. M. Miyagami, S. Nakamura and N. Moriyasu, Hydrodynamic of the CSF under experimental occlusion of the superior sagittal sinus, *Neurosurg. Tokyo*, 3:739 (1975).
11. M. Miyagami, S. Nakamura and N. Moriyasu, Ventricular enlargement in experimental occlusion of superior sagittal sinus - in reference to histopathological findings, *Neurosurg. Tokyo*, 3:947 (1975).
12. T. J. Putnam, Studies in multiple sclerosis. "Encephalitis" and sclerotic plaques produced by venular obstruction, *Arch. Neurol. Psychiat.*, 33:929 (1935).
13. M. Sarwar, C. Virapongse and P. Carbo, Experimental production of superior sagittal sinus thrombosis in the dog, *Am. J. Neuroradiol.*, 6:19 (1985).
14. S. Sato, M. Ohtani and T. Kawase, The effect of sagittal sinus occlusion on blood-brain barrier permeability and cerebral blood flow in the dog, *in*: "Brain Edema", Y. Inaba, I. Klatzo and M. Spatz, eds., Springer Verlag, Berlin, Heidelberg, New York and Tokyo (1986).
15. T. Shigeno, M. Brock, S. Shigeno, E. Fritschka and J. Cervos-Navarro, The determination of brain water content: microgravimetry versus drying-weighing method, *J. Neurosurg.*, 57:99 (1982).
16. A. L. Woolf, Experimentally produced cerebral venous obstruction, *J. Path. Bact.*, 67:1 (1954).

BRAIN FUNCTION, PATHOPHYSIOLOGY AND HEPARIN THERAPY IN EXPERIMENTAL CEREBRAL SINUS VEIN THROMBOSIS

K. Frerichs, M. Deckert, O. Kempski, K. Einhäupl
and A. Baethmann

Institute for Surgical Research and
Department of Neurology
Ludwig-Maximilians-University
Klinikum Groβhadern, 8000 Munich 70, FRG

INTRODUCTION

Diagnosis of cerebral sinus vein thrombosis (SVT) is difficult to establish and, thus, rarely made by clinicians. For this reason it is frequently observed only at autopsy raising the suspicion that SVT often is a lethal disorder. Recent experience indicates, however, that SVT has a wide clinical spectrum reaching from discrete neurological deficits to most severe courses with coma [1]. Contrary to findings for arterial stroke, even severe neurological deficits arising from SVT have a likelihood of disappearing again. Due to the elusive nature and diagnostical problems of the disease, understanding of underlying mechanisms as well as therapeutical concepts are fragmentary and controversial. This may be attributable in part to the as yet limited experimental approaches and availability of animal models. To our knowledge, no experimental studies using small laboratory animals have been conducted so far for the investigation of pertinent functional and structural aspects of SVT. The current report analyzes functional parameters obtained in a new experimental model of SVT established by our group.

MATERIALS AND METHODS

Experiments were carried out in male Wistar rats. Intravenous chloralhydrate (1 ml/100 g bw of a 2.6% solution) was given for anesthesia. Catheters were inserted into the femoral artery and vein for monitoring of arterial blood pressure, for blood gas analysis, and administration of drugs. Two cranial windows were made for exposure of the superior sinus sagittalis (SSS), in front of the bregma and near the torcular. Sinus thrombosis was induced by ligation of the rostral and posterior parts of the sinus with 9.0 prolene together with injection of 100 µl of a kaolin-cephalin solution (PTT reagent, Boehringer Mannheim, Inc). Interruption of the circulation from occlusion was confirmed by fluorescence angiography using iv injection of Na+-fluorescein.

Cerebral Sinus Thrombosis, Edited by K. Einhäupl *et al.*
Plenum Press, New York, 1990

Effects of SVT were studied on: (1) spontaneous locomotor activity employing running-wheel cages [3,4,5], (2) the electroencephalogram [6,7], and (3) electrical tissue impedance as well as intracranial pressure acutely during occlusion of the sinus. Wheel-running was quantified in special cages [3,4] consisting of a living compartment attached with a revolving drum. The animals had free access to the running-wheel and were supplied with food and drinking water ad libitum. Revolutions of the running-wheel were recorded as well as drinking and uptake of food pellets. The running-wheel cages were placed in a room which was isolated from external stimulation to avoid interference with the spontaneous activity of the experimental animals. The room was climatized and kept at constant humidity as well as in a regular light-dark cycle of 12 hours each. SVT was induced in six animals, four were subjected to sham operation as controls. At termination of the experiment the animals were subjected to perfusion fixation for histological investigation of the brain.

In order to investigate acute effects of thrombosis on the spontaneous electrical activity of the brain (EEG), sinus occlusions were made in additional experiments after the animals had recovered from anesthesia. In this group an indwelling 25-gauge catheter was implanted into the ligated part of the sinus and attached to the occipital skull with dental cement. The EEG was recorded from two bifrontal and biparietal electrodes. Sinus thrombosis was induced 24 hours later by ligature of the exposed sinus together with injection of PTT reagent within 5 minutes. EEG activity was recorded prior to, during, and immediately after thrombosis induction. The EEG was subjected to fast-Fourier transformation for power spectrography of the different frequency bands [7].

Measurements of the specific electrical resistance (impedance) of the brain were made to gather information on acute alterations of the extra-intracellular fluid distribution indicative of ischemia from interruption of the cerebro-venous drainage by thrombosis. The electrical impedance was measured between two needle electrodes vertically inserted into the cerebral parenchyma with a stereotactic device. Both electrodes were separated by approximately 0.5 mm. Impedance measurement was made by a compensated bridge using an alternating current of 1000 Hz and a voltage of maximally 4.5 mV. Under these conditions, electrical resistance of brain tissue is largely determined by the conductivity of the extracellular compartment. Thus, changes in impedance can be interpreted to reflect changes of the extracellular fluid volume, eg, from a fluid shift into the intracellular space [2,8,9,11].

At the end of the impedance experiments, the rats were bled from the abdominal aorta to investigate the remaining ischemic response of cerebral impedance due to global interruption of the cerebral blood flow. This was made to assess the ischemic component of sinus thrombosis in the breakdown of the extra-intracellular fluid homeostasis. Finally, intracranial pressure was studied in two animals by stereotactic placement of a catheter into the left ventricle [21] which was connected to manometric recording (Statham). Furthermore, in a limited number of animals heparin was administered in a dose of 1 IE/g bw intravenously within 30 minutes starting at 10 or 30 minutes after induction of thrombosis. The therapeutical efficiency of heparin was assessed by analyzing electrical tissue impedance.

RESULTS

Locomotion and Spontaneous Behavior

SVT in man has many clinical symptoms [1]. Observations of the motor drive by studying locomotion was used in the present study as an approach to analyze the effects of blocking cerebro-venous outflow and its sequelae on a higher and complex central nervous function. Immediately after thrombosis the animals became somnolent, while unconsciousness or coma, did not evolve. Gross neurological deficits, such as paresis or disturbances of the motor coordination did not appear after recovery from anesthesia. Marked decrease of spontaneous wheel-running was a particularly prominent finding. Since rats are night-active, running occurred largely throughout the night. Therefore recording was made only during that period. Wheel-running under control conditions was studied during a period of 3 weeks prior to induction of SVT. Taken together, the data demonstrate that SVT markedly affects spontaneous locomotor drive without a tendency towards recovery during the subsequent 4 weeks. On the other hand, disturbances of neurological function, or of eating and drinking, were not found after occlusion of the sinus.

Electroencephalogram

In order to study the electrical activity of the brain under physiological conditions, EEG recordings were made in awake rats prior to and immediately after occlusion of the sinus sagittalis superior. Sinus occlusion was immediately followed by severe depression of the EEG almost reaching isoelectricity. Yet the animals were not unconscious and had at best a reduced vigilance. The EEG started to recover 20 minutes later with appearance of delta-wave-dominated activity. Depression of the electroencephalogram was quantified by power spectrography. Initially after sinus occlusion, the EEG power was down in all frequency bands, which later was followed by an increase of delta-power during the next 4 hours. In a single experiment the EEG was monitored over 6 months after thrombosis. Even then delta-wave complexes remained the predominant EEG feature of the entire period. In some animals EEG changes resembling seizure activity, such as spikes, were found. This, however, was not associated with respective clinical symptoms.

Electrical Brain Tissue Impedance

Since acute interruption of the cerebral venous drainage might have caused cerebral ischemia, changes in the electrical tissue impedance were utilized to assess potential effects on maintenance of the intra-extracellular electrolyte and fluid homeostasis as a measure of blood flow impairment. It is known that acute interruption of cerebral blood flow results in a characteristic increase of the electrical brain tissue impedance, this being attributable to movements of fluid and electrolytes from the extra- to intracellular space [8,9,11,12].

With a latency of a few minutes, induction of thrombosis indeed led to an increase of tissue impedance to approximately 170% of the normal value, which is close to what is found after arrest of the cerebral circulation. No tendency towards recovery could be recognized within the ensuing 4 hours. In order to assess the remaining cerebral impedance change which could be elicited by complete cerebral ischemia, the animals were sacrificed by bleeding from the aorta. This was followed by a further exponential increase of brain tissue impedance, suggesting that parts of the brain still were able to maintain a normal electrolyte and fluid homeostasis, which collapsed only after global interruption of cerebral blood flow. The manometrical measurements of intracranial pressure conducted in a few experiments were indicative of an acute increase after sinus occlusion.

For treatment heparin was given either 10 or 30 minutes after thrombosis over a period of 30 minutes using tissue impedance to measure its efficacy. Contrary to the untreated control animals, treatment was found to reverse the ischemic impedance response towards the normal, pre-occlusion level. Infusion with heparin let the electrical tissue impedance of the brain fall below the 95% confidence range of untreated experimental animals. The findings may indicate that heparin, at least in part, facilitated reestablishment of venous drainage after sinus occlusion.

DISCUSSION AND CONCLUSIONS

Only a few attempts have so far succeeded in producing a reproducible model of SVT in animals [13-17]. Experimental problems were related, eg, with unphysiological manifestations of thrombosis, failure to report histological or functional changes, or the use of large animals such as dogs or monkeys thus making it difficult to attain a statistically sufficient number of experiments. Here a reproducible model of SVT is presented using small laboratory animals which allows repeated investigations of functional as well as structural aspects [18]. As has been demonstrated, SVT acutely causes a variety of functional disturbances involving spontaneous locomotion, the EEG, as well as the extra-intracellular electrolyte and fluid homeostasis as in other forms of brain edema associated with a functional impairment [10]. Depression of spontaneous wheel-running may reflect loss of the spontaneous drive due to damage of cerebral areas by venous infarction [18]. Failure to recover from the locomotor depression within the ensuing observation period may indicate either irreversibility of the lesion, or that restoration of the motor drive required longer recuperation periods for the training of compensatory centers. Irreversibility of the defect might be concluded from histology, where extensive structural damage was found in both hemispheres within 4 weeks leading to bilateral hydrocephalus [19].

The marked depression of the EEG acutely after thrombosis is similar to what is observed after complete interruption of the cerebral circulation. It may suggest either that interruption of venous drainage had led to a flow-dependent failure of energy metabolism inhibiting normal nerve cell function, or the acute induction of spreading depression by thrombosis. Although development of spreading depression could in part be supported by the characteristic impedance changes, it may not suffice to explain the EEG findings. In normal animals without cerebral lesion, induction of spreading depression is usually followed by complete recovery, eg, of the electrical tissue impedance, which was not observed in our experiments. Appearance of spikes in the EEG recordings was found in approximately one-third of experimental animals which would confirm observations in man with SVT.

The experimental model may allow analysis of the underlying hemodynamic pathophysiology causing brain tissue damage in greater detail than was previously possible. The question of how far occlusion of the venous system must spread to induce venous infarction could perhaps be answered by utilizing quantitative histology and regional blood flow measurements. Finally, the model lends itself to investigations of novel forms of treatment. This is a particularly interesting aspect, since neurology may be more amenable to therapy in SVT than after arterial stroke [1,20].

ACKNOWLEDGEMENTS

The secretarial assistance of Isolde Juna and Helga Kleylein is gratefully acknowledged. We are grateful to Professor Dr J. Aschoff and Christiane von Götz of the Max-Planck-Institute of Behavioural Physiology, Erling who made available the activity cages.

REFERENCES

1. K. Einhäupl, Klinik und Pathophysiologie von blanden Hirnvenen- und Sinusthrombosen, Thesis, Ludwig-Maximilians-University, Munich (1986).
2. J. Bures, O. Buresova and J. Huston, eds., "Techniques and Basic Experiments for the Study of Brain and Behavior", Elsevier Scientific Publishing, Amsterdam, pp 37-84 (1975).
3. C. P. Richter and G. H. Wang, New apparatus for measuring the spontaneous motility of animals, *J. Lab. Clin. Med.*, 12:289-292 (1926).
4. C. P. Richter, Animal behavior and internal drives, *Quarterly Rev. Biol.*, Vol. II, No. 3 (1927).
5. B. F. Skinner, "The Behavior of Organisms - An Experimental Analysis", Appelton-Century-Crofts Inc., New York (1936).
6. H. Künkel, Pharmaco-electroencephalography-methods and problems, *in*: "Pain Measurement in Man. Neurophysiological Correlates of Pain", B. Bromm, ed., Elsevier Science Publishers BV, Amsterdam (1984).
7. I. Sulg, Quantitative EEG as a measure of brain dysfunction, *in*: "Brain Ischemia: Quantitative EEG and Imaging Techniques. Progress In Brain Research", Vol. 62, G. Pfurtscheller, E.J. Jonkman and G. H. Lopes da Silva, Elsevier Science Publishers BV, Amsterdam (1984).
8. A. Baethmann, Das metabolische Hirnödem, Thesis, Ludwig-Maximilians-University, Munich (1974).
9. A. Baethmann, Pathophysiological and pathochemical aspects of cerebral edema, *Neurosurg. Rev.*, 1:85-100 (1978).
10. A. Baethmann, K. Maier-Hauff, O. Kempski, A. Unterberg, M. Lange and U. Gross, Hirnödem und Bewußtsein, *Zeitschr. Allg. Med.*, 57:2121-2126 (1981).
11. O. Kempski, Die Lokalisation des Glutamat-induzierten Hirnödems, Inauguraldissertation, Ludwig-Maximilians-University, Munich (1982).
12. I. Klatzo, Pathophysiological aspects of brain edema, *Acta Neuropathol.*, 72:236-239 (1987).
13. D. J. K. Beck and D. Russell, Experiments on thrombosis of the superior longitudinal sinus, *J. Neurosurg.*, 3:337-347 (1946).
14. G. Owens, G. Stahlman, J. Capps and A. Meirowsky, Experimental occlusion of dural sinuses, *J. Neurosurg.*, 14:640-647 (1957).
15. K. Fujita, N. Kojima, N. Tamaki and S. Matsumoto, Brain edema in intracranial venous hypertension, *in*: "Brain Edema", Y. Inaba, I. Klatzo and M. Spatz, eds., Springer, Berlin, pp 228-234 (1985).

16. E. R.Heinz, D. Geeter and T.O. Gabrielsen, Cortical vein thrombosis in the dog with a review of aseptic intracranial venous thrombosis in man, *Acta Radiol. Diagn.*, 13:105-114 (1972).
17. S. Sato, Y. Miyahara, Y. Dohmoto, T. Kamase and S. Toya, Cerebral microcirculation in experimental sagittal sinus occlusion in dogs, *in:* "The Cerebral Veins", L. M. Auer and F. Loew, eds., Springer, New York (1983).
18. M. Deckert, K. Frerichs, P. Mehraein, O. Kempski, A. Baethmann and K. Einhäupl, A new experimental model of sinus vein thrombosis, in: "Cerebral Sinus Thrombosis: Experimental and Clinical Aspects", K. Einhäupl, O. Kempski and A. Baethmann, eds, Plenum Press, New York and London, pp 39-42 (1990).
19. D. D'Avella, R. P. Greenberg, S. Mingrino, M. Scanarini and K. Pardatscher, Alterations in ventricular size and intracranial pressure caused by sagittal sinus pathology in man, *J. Neurosurg.*, 53:656-661 (1980).
20. A. Thron, K. Wessel, D. Linden, G. Schroth and J. Dichgans, Superior sagittal sinus thrombosis: neuroradiological evaluation and clinical findings, *J. Neurol.*, 233:283-288 (1986).
21. J. F. R. König and R. A. Klippel, "The Rat Brain. A Stereotaxic Atlas of the Fore-Brain and Lower Parts of the Brain Stem", The Williams and Wilkins Company, Baltimore (1963).

MECHANISMS OF BRAIN TISSUE DAMAGE

PHYSIOLOGY AND PATHOPHYSIOLOGY OF THE CEREBRO-VENOUS CIRCULATION

Ludwig M. Auer

Department of Neurosurgery
University of Homburg/Saar
6650 Homburg/Saar, FRG

The different cortical layers of the brain are drained by deep, intermediate and superficial branches of postcapillary venules (nomenclature of H. Duvernoy) [1]. Most of the cortical venous blood reaches the superficial pial veins, which have a rich network of short and long veno-venous connections [2]. The latter are draining into bridging veins, which enter the dural sinus. A number of large pial collateral veins, the Rolandic, Sylvian, Trolard's vein, the superficial middle cerebral veins, and Labbé's vein usually join in the temporo-parietal region. Another system of large draining veins collects blood from the deep structures and empties into the great vein of Galen. It is known from neurosurgical operations that the redundancy of the collateral venous system prevents circulatory disturbances if major collecting veins are occluded. Exceptions, however, are the Rolandic vein and, thus, the posterior part of the superior sagittal sinus (SSS) including the confluens sinuum (Torcular herophili).

The cerebral veins have received little attention until recently. Due to the low intraluminal pressure and contrary to veins of other organs, cerebral veins are believed to be pressure-passive collapsible tubes without contractile elements. This is concluded from an extremely thin vessel wall and the comparatively little changes in caliber under physiological or pathological circumstances. Recent histochemical studies have revealed, however, that endothelial and periendothelial cells of cerebral veins have contractile elements, such as actin and myosin fibers, as a basis of an active adjustment of the vessel caliber [3]. Moreover, immunofluorescence staining of pial veins has demonstrated the presence of adrenergic nerve fibers [4]. Although less numerous than in arteries, the adrenergic fibers may be sufficient to induce contraction of the venous vessels in view of their comparatively low intraluminal pressure.

Experiments in cats have shown that pial veins constrict vividly upon electrical stimulation of the cervical sympathetic nerve [5]. The response was maintained under α-adrenergic blockade, while it could be inhibited by α-adrenergic receptor blockers (Fig. 1) [5]. Moreover, subarachnoid microapplication of norepinephrine into the perivascular space of cat pial

Cerebral Sinus Thrombosis, Edited by K. Einhäupl *et al.*
Plenum Press, New York, 1990

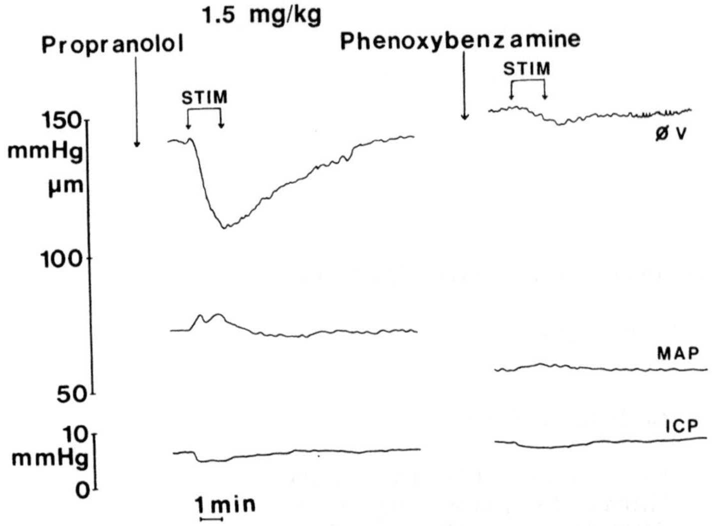

Fig. 1. Electrical stimulation (STIM) of the cervical sympathetic nerve in cat causing marked constriction of a 140 and 70 μm pial vein during β-adrenergic blockade with propranolol. The response is abolished after α-adrenergic blockade with phenoxybenzamine. Venous constriction is paralleled by slight reduction of ICP [5]. MAP, ICP: Mean arterial pressure or intracranial pressure in mmHg, ØV: caliber of pial veins in micrometers.

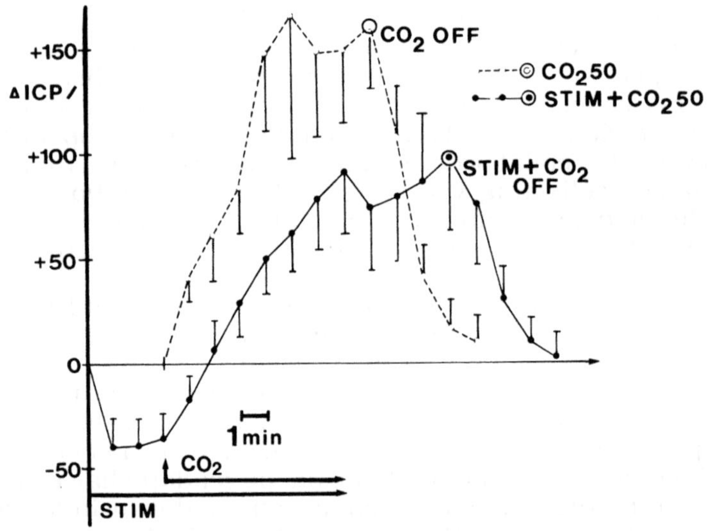

Fig. 2. Changes of intracranial pressure (ICP) in two series of cats. Dotted line: increase of ICP (cisterna magna) during a brief episode of hypercapnia to $PaCO_2$ of 50 mmHg by administration of CO_2. Solid line: decrease of ICP during electrical stimulation of the cervical sympathetic nerve. As stimulation continued hypercapnia was induced as in the first series. The rise of ICP from hypercapnia is more slow and less intense during sympathetic stimulation (with permission: cf Ref. 35).

veins induced marked constriction [6]. The pial veins were found to simultaneously constrict, as arteries of similar size, which can not be explained by mere passive reduction of the venous lumen following

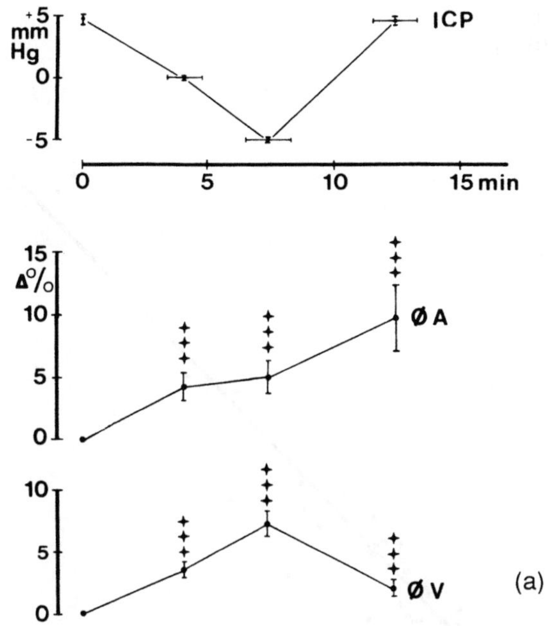

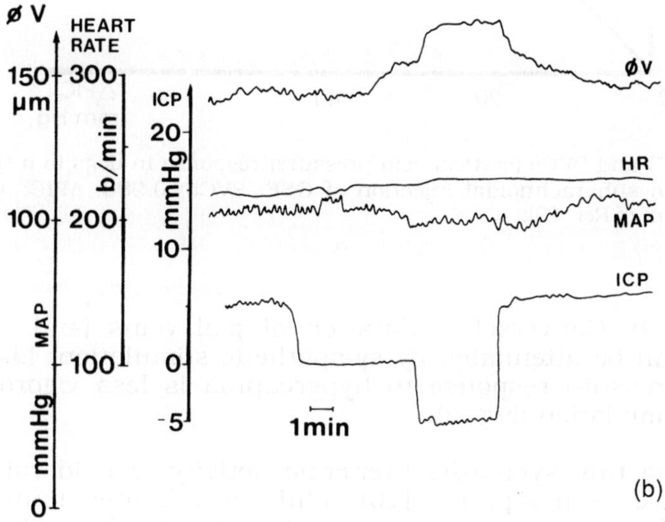

Fig. 3. (a) Response (mean ± SEM) of pial arteries (ØA) and veins (ØV) in percent of resting level to a stepwise decrease of ICP from +5 mmHg to 0 mmHg, or -5 mmHg, and back to the resting level of +5 mmHg. +++: p< 0.0005 (with permission: cf Ref. 25). (b)Individual experiment with stepwise lowering of ICP by withdrawal of CSF from the cisterna magna. Recordings of mean arterial pressure, heart rate (HR) and caliber of a pial vein (ØV) are included. The vein dilated with decreasing ICP, but returned to the previous caliber when the initial ICP level was restored.

arterial constriction. The sympathoadrenergic response of the cerebral veins seems to play a role in the regulation of the cerebral blood volume [7], and thereby of intracranial pressure [5,8]. Sympathoadrenergic discharges are increased in hypercapnia or during intracranial

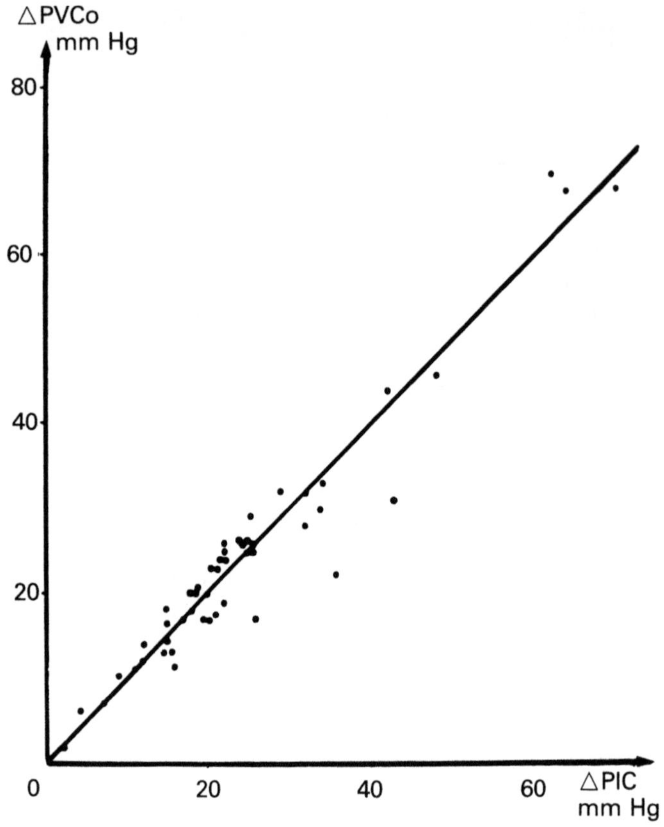

Fig. 4. ICP- (ΔPIC) and PVCo (cortical vein pressure) response in dogs to a Queckenstedt test, or to subarachnoidal injection of CSF. PVCo: 0.986, ΔPIC: +0.112 (with permission: cf Ref. 19).

hypertension [9]. Conversely, dilatation of pial veins (and arteries) to hypercapnia can be attenuated by sympathetic stimulation. Likewise, the intracranial pressure response to hypercapnia is less vigorous during sympathetic stimulation (Fig. 2).

Concerning this sympathoadrenergic activity, considerable species differences have been reported [10], while only limited information on man is available. In rats the network of aminergic fibers is denser in intracranial cerebral veins than in extracranial veins [11]. Transmitters, such as substance P, vasoactive intestinal peptide (VIP), and serotonin have little, if any, effect on cerebral veins [12]. The CO_2-induced increase in cerebral blood flow in rats, however, was found to depend on the presence of 5-HT fibers in the vessel wall. While effective in arteries, changes of the perivascular concentration of K^+, or pH have virtually no effect on cerebral veins [14]. In vitro studies suggest human pial veins [15] to have more α-adrenergic receptors than arteries, and arteries to have more 5-HT receptors than veins. Acetylcholine causes dilatation of arteries and constriction of veins. Bradykinin, neuropeptide Y, PGE_1, $PGF_{2\alpha}$, and calcitonin-gene-related peptides induce similar responses in arteries and veins. However, the venous response seems to be independent of the presence of calcium ions. Thus, evidence is available on active

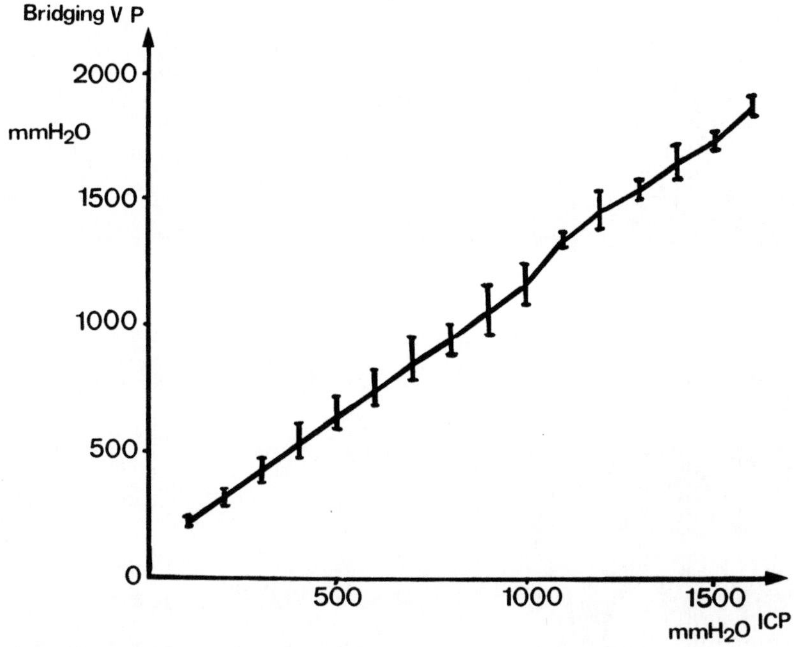

Fig. 5. Average bridging vein pressure obtained in 10 measurements and plotted against intracranial pressure in six dogs. Vertical bars indicate upper and lower values of each ICP level. The pressure in the bridging veins increased in an almost linear relationship with ICP during intracranial hypertension (with permission: cf Ref. 26).

vasomotor control of cerebral veins largely by sympathoadrenergic constriction in order to control cerebral blood volume to some extent.

Brain tissue, the CSF spaces, and the cerebral blood vessels are localized in a closed cavity. Consequently, the pressures in the CSF space (P_{CSF}) and cerebral veins may interfere with each other. The intracranial compartment is a closed cavity filled with incompressible fluid, although limited escape of fluid through the foramen magnum to the spinal canal, or through other routes, is possible. Pressure changes in one compartment, therefore, may induce respective changes in another. The venous transmural pressure, ie, the difference between intraluminal and perivascular pressure (CSF-pressure) is close to zero and remains unchanged over a wide range of the intracranial pressure [16-19].

According to the Starling resistor theorem, the complex interaction of pressures in collapsible tubes, eg, the pial veins, and their environment, the CSF spaces localized in a cavity, prevents the cerebral veins from collapsing, even if the venous pressure falls below zero [16,20,21, 23,24]. Since the cerebral venous pressure is more dependent on P_{CSF} than on the sagittal sinus or jugular vein pressure, venous outflow from the brain remains constant, irrespective of the position of the head above or at the cardiac level. This may fail only under highly pathological circumstances. Withdrawal of CSF is likely to induce similar changes as increasing the cerebral venous pressure, namely dilation of cerebral veins from an increased venous transmural pressure. Dilation of cerebral

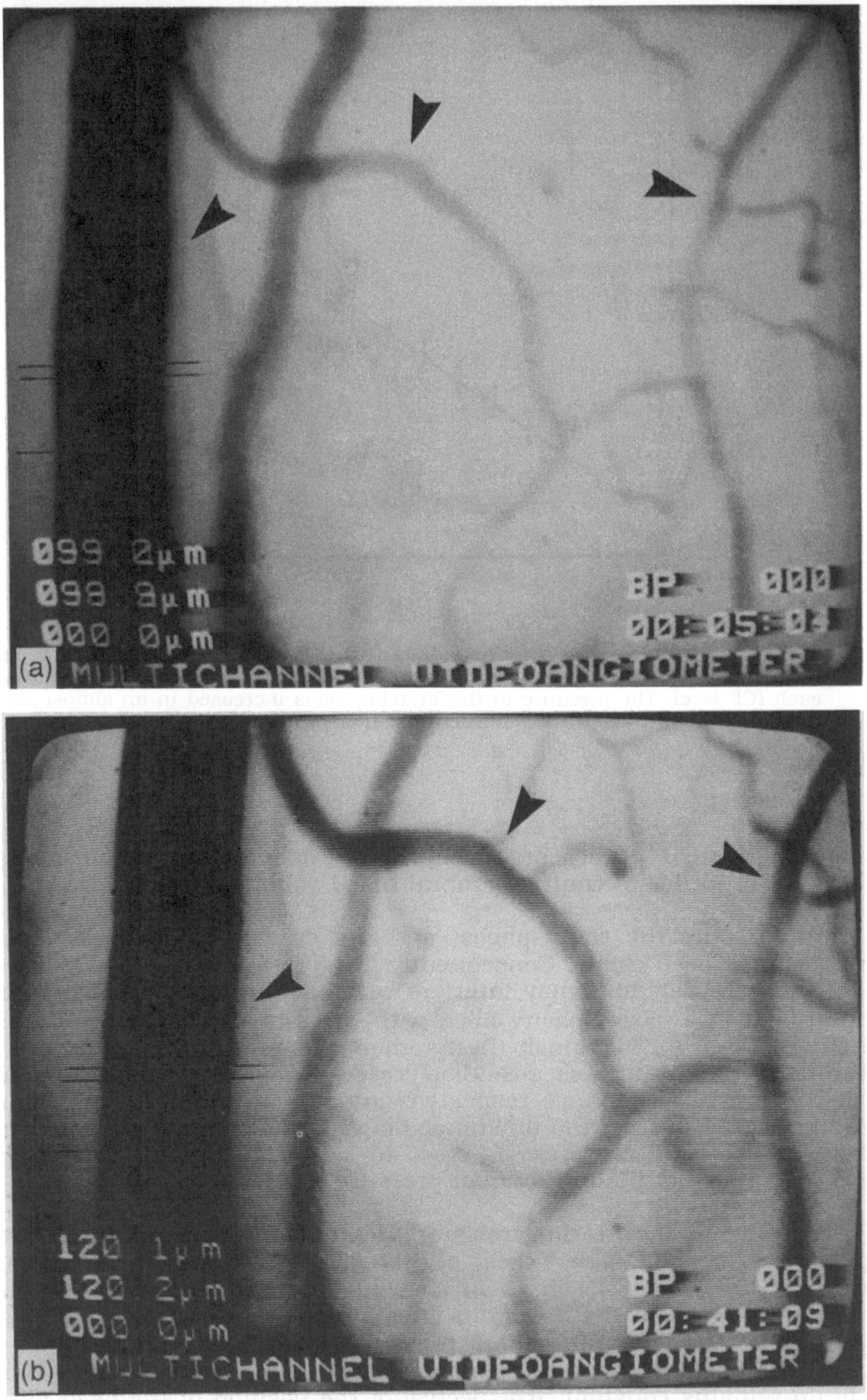

Fig. 6. Cat pial vessels studied by multichannel videoangiometry. (a) Normal vessels; arrows indicate pial artery with branches. A pial vein is seen to the right of a big artery. (b) Same picture during cisternal infusion of CSF raising intracranial pressure to 50 mmHg, (c) at ICP of 114 mmHg (MAP: 106 mmHg), or (d) at ICP of 145 mmHg (with permission: cf Ref. 36).

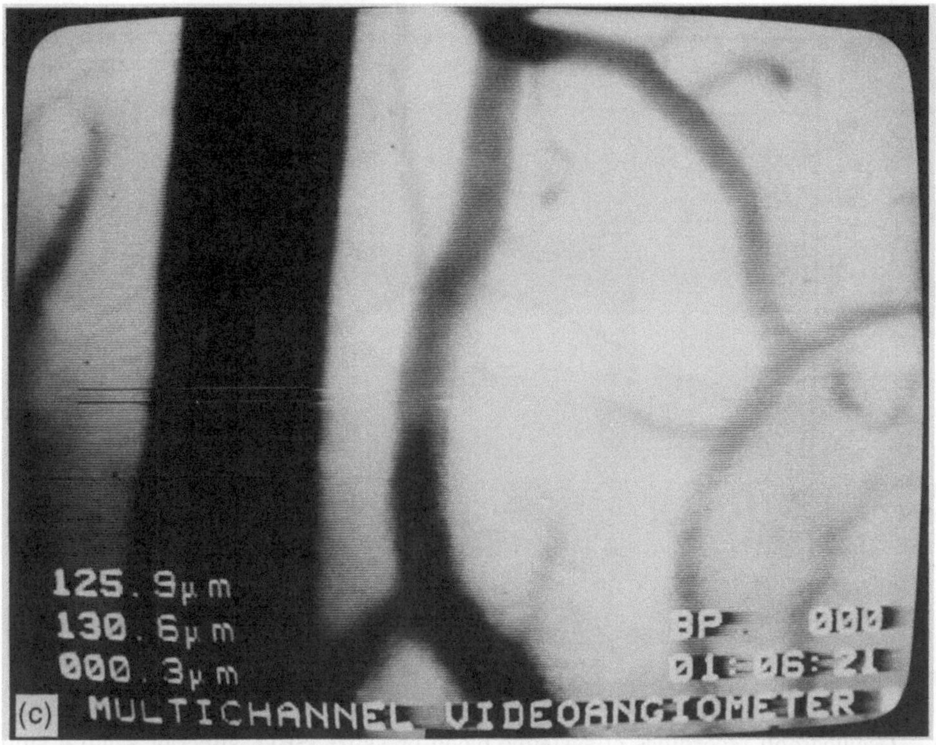
125.9μm
130.6μm
000.3μm
BP 000
01:06:21
(c)
MULTICHANNEL VIDEOANGIOMETER

113.9μm
115.1μm
000.1μm
BP 000
01:06:02
(d)
MULTICHANNEL VIDEOANGIOMETER

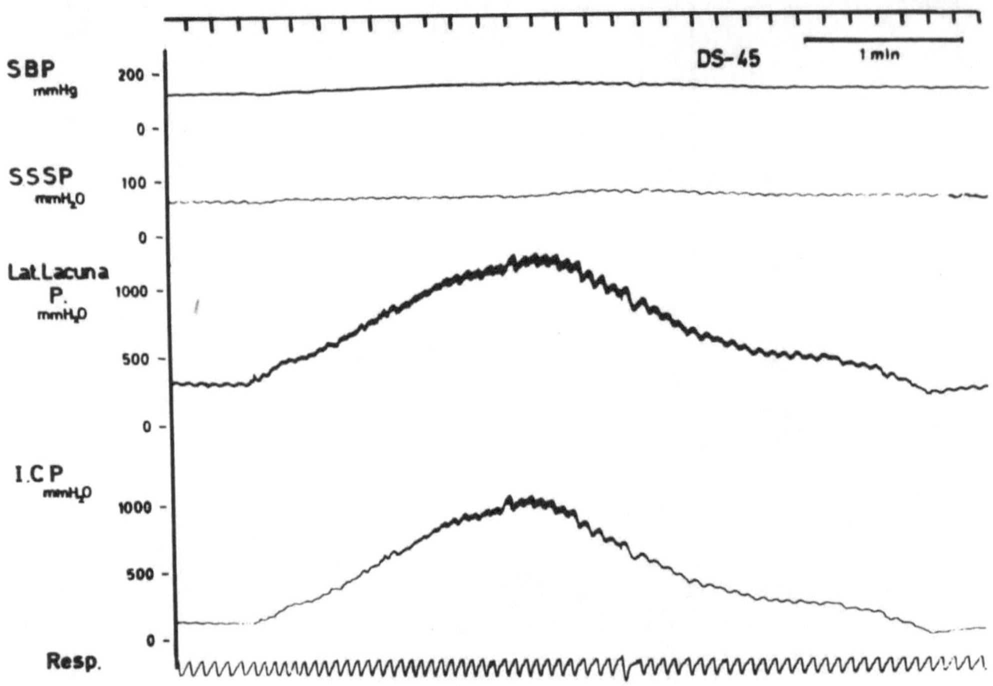

Fig. 7. Oscillograph recordings of systemic blood pressure (SBP), superior sagittal sinus pressure (SSSP), lateral lacunae pressure (Lat Lacuna P), intracranial pressure (ICP), and respiration (Resp) during intracranial hypertension of 1000 mmH_2O. Lat Lacuna P was about 180 mmH_2O above ICP at an ICP resting level of 140 mmH_2O. Note the synchronous rise of Lat Lacuna P with elevations of ICP, while SSSP remains unchanged (with permission: cf Ref. 26).

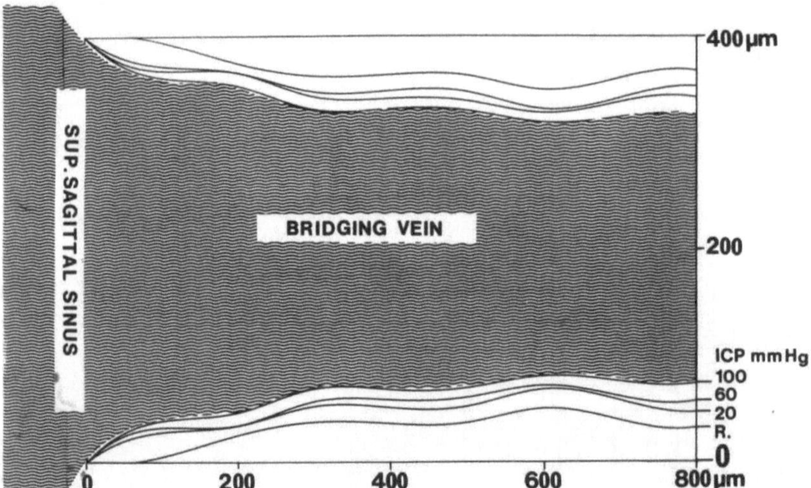

Fig. 8. Schematic drawing of superior sagittal sinus (SSS) and bridging vein showing mean caliber and shape at various levels of intracranial pressure (ICP). The abscissa indicates distance from entry of the bridging vein into the SSS, the ordinate the horizontal diameter in micrometers. The hatched area shows the mean bridging vein caliber at an ICP of 100 mmHg. The additional lines indicate mean vessel diameters at ICPs of 60, 20 mmHg, or at the resting level (R) at 4.15 mmHg (with permission: cf Ref. 28).

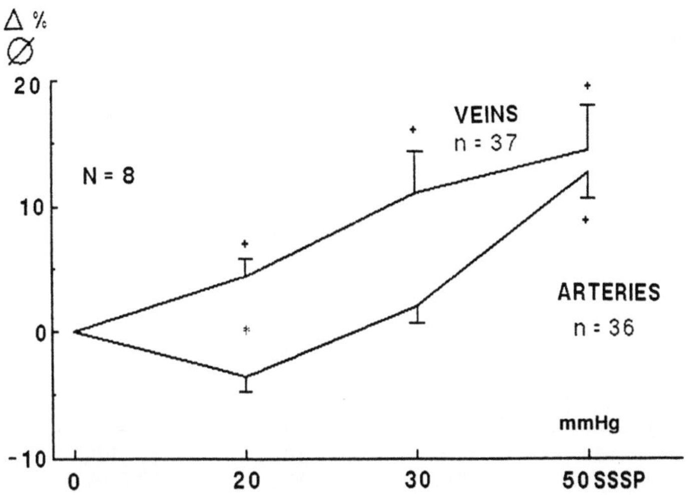

Fig. 9. Dilatation of pial veins during pressure elevation in the superior sagittal sinus (SSSP) by infusion of Ringer's solution (cf Fig. 10) (with permission: cf Ref. 29).

arteries under these conditions is indicative of cerebrovascular dysregulation as a consequence of sudden decompression. Cerebral arteries may remain dilated after CSF pressure has been normalized, whereas cerebral veins immediately readjust to their previous caliber (Fig. 3) [25].

On the other hand, a CSF pressure rise induces parallel increases of pressure in the pial as well as bridging veins (Figs 4 and 5) [19,26]. An increased CSF pressure does not compress pial veins as is widely believed but leaves the venous diameter largely unaffected (Fig. 6) [27] until venous and CSF pressure reach arterial pressure. Cerebral blood flow stops at this point because the cerebral perfusion pressure, the difference between arterial and venous pressure, approaches zero. While the pressures in pial and bridging veins rise in parallel with an increase in CSF pressure, measurements of the sagittal sinus pressure demonstrate the latter to remain unchanged (Fig. 7) [26]. This led to the conclusion that the pressure in the final part of the bridging veins and lacunae is below the ambient CSF pressure. Cuffing at this point would cause venous congestion and, thus, reduction of blood flow. Already under normal conditions a sudden drop in pressure is found between the lacunae and sagittal sinus [26]. Recent measurements of calibers of the bridging veins at the entrance into the superior sagittal sinus in rats did not reveal major narrowing of the lumen during elevation of the CSF pressure (Fig. 8) [28]. Upstream pial veins were not congested nor distended [27]. An explanation of the pressure difference between the lateral lacunae and sagittal sinus may be given from morphological findings of the entry zone. A jet-like narrowing is found at the transition between the thin wall of the bridging vein and thick wall of the sinus. Thus, the pressure decrease between lacunae and sinus may have a morphological rather than physiological basis. The pressure decrease may occur probably closer to the sinus wall than previously assumed.

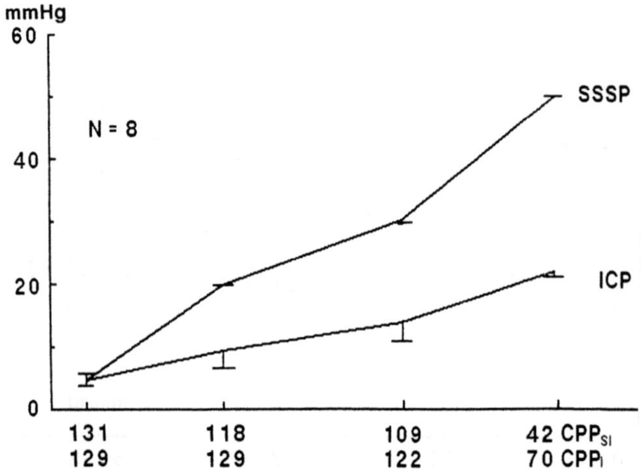

Fig. 10. Elevation of superior sagittal sinus pressure (SSSP) by infusion of Ringer's solution, followed but not paralleled by a rise of CSF pressure (ICP). Abscissa: Cerebral perfusion pressures calculated as difference between arterial pressure and SSSP (CPP_{SI}), or ICP (CPP_I).

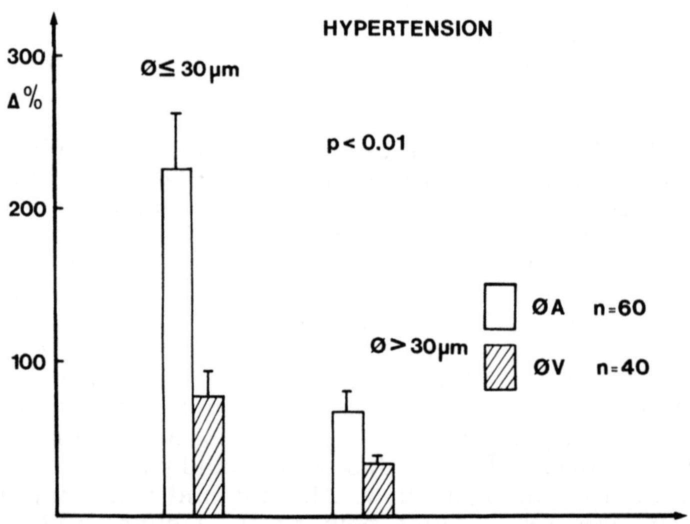

Fig. 11. Response of cat pial vein (ØV) or artery (ØA) diameters in percent (Δ%) of resting level during acute, drug-induced arterial hypertension. Mean values ±SEM are given for vessels up to 30 μm resting caliber (Ø < 30 μm), or larger vessels (Ø > 30 μm). Dilatations were significant at $p < 0.01$ (with permission: cf Ref. 30).

Raising the superior sagittal sinus pressure (SSSP) causes distension of pial vessels due to the rise in intraluminal pressure (Fig. 9) [29] until the structural elements of the vessel wall inhibit further expansion. This may be reached at a SSSP of 50 mmHg or slightly above. Both a rise in venous pressure together with an increase in cerebral blood volume

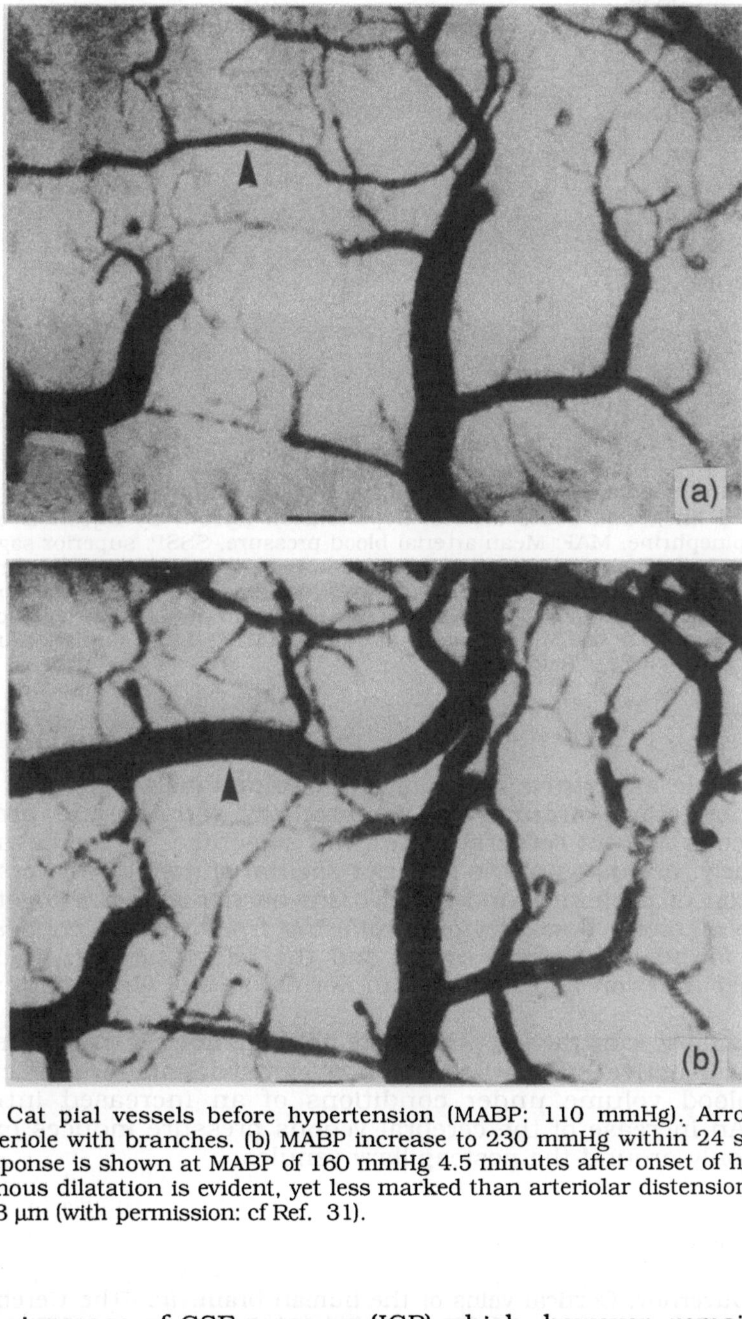

Fig. 12. (a) Cat pial vessels before hypertension (MABP: 110 mmHg). Arrow indicates arteriole with branches. (b) MABP increase to 230 mmHg within 24 seconds. The response is shown at MABP of 160 mmHg 4.5 minutes after onset of hypertension. Venous dilatation is evident, yet less marked than arteriolar distension. Black bar: 183 μm (with permission: cf Ref. 31).

causes an increase of CSF pressure (ICP) which, however, remains significantly below SSSP (Fig. 10). Consequently, the transmural venous pressure increases. Similarly, SSSP and venous transmural pressure are elevated during acute arterial hypertension. The cerebral arteries are distended by the sudden increase of intraluminal pressure. A decrease in arterial resistance, ie, a "breakthrough" of the autoregulation (Figs 11 and 12) [30,31] causes the venous pressure to increase and distension of veins. The cerebral venous pressure may exceed the CSF pressure in the supra- and infratentorial compartment, probably due to the limited dilatory capacity of the veins (Fig. 13) [32]. Less than 20 seconds after onset of acute arterial hypertension, disruption of the blood-brain barrier

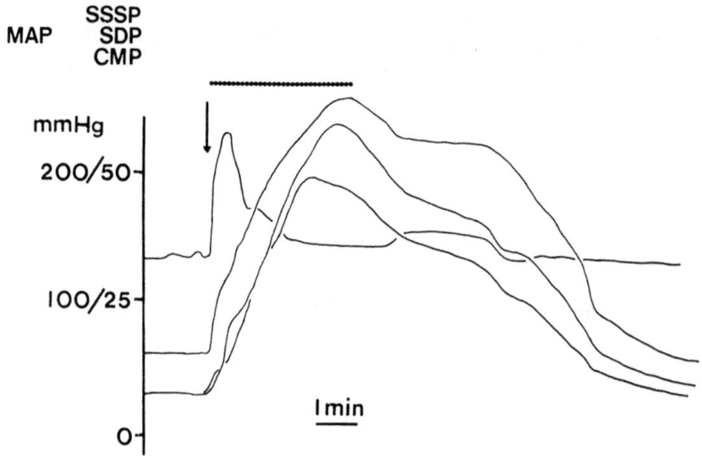

Fig. 13. Pressure response curves of cat during hypercapnia by inhalation of 20% CO_2 (dotted line), and acute arterial hypertension (arrow) by injection of 5 μg/kg norepinephrine. MAP: Mean arterial blood pressure, SSSP: superior sagittal sinus pressure (high peak), SDP: subdural pressure in the supratentorial compartment (middle peak), CMP: cisterna magna pressure (low peak). SSSP is shown to rise more extensively than CSF pressure and to remain at a higher level for a longer period than SDP or CMP. Left value at ordinate: MAP; right value at ordinate: SSSP (with permission: cf Ref. 32).

becomes visible as extravasation of Evans-blue-labeled plasma proteins. Extravasation starts around the postcapillary venules and produces a patchy discoloration of cerebral cortex and sometimes of the basal ganglia [33]. Similarly, occlusion of the superior sagittal sinus also results in rapid extravasation of protein-bound blood-brain barrier markers together with a decrease of blood flow. Regions suffering from venous congestion are frequently found in cerebral cortex and the basal ganglia, while blood-brain barrier function and flow remain normal in the white matter [34].

In summary, contrary to previously held opinions, cerebral veins have a capacity of active sympathoadrenergic constriction for the control of cerebral blood volume under conditions of an increased intracranial pressure. An increase of the cerebral venous pressure induces extravasation commonly around the postcapillary venules.

REFERENCES

1. H. Duvernoy, Cortical veins of the human brain, *in*: "The Cerebral Veins: An Experimental and Clinical Update", L. M. Auer and F. Loew, eds., Springer, New York, pp 3-38 (1983).
2. L. M. Auer, Intravitalmikrokopische Beobachtung der pialen Gefäße im Tierexperiment, *Zentralbl. Nchrg.*, 38:175-184 (1977).
3. R. Kleinert and L. M. Auer, Actin and myosin positive immunostaining of contractile elements in cerebral veins and venules, in preparation.
4. L. Edvinsson, L. M. Auer and R. Uddman, Autonomic nerves and morphological organization of cerebral veins, *in*: "The Cerebral Veins: An Experimental and Clinical Update", L. M. Auer and F. Loew, eds., Springer, New York, pp 73-79 (1983).

5. L. M. Auer, L. Edvinsson and B. B. Johansson, Effect of sympathetic nerve stimulation and adrenoreceptor blockade on pial arterial and venous calibre and intracranial pressure in the cat, *Acta Physiol. Scand.*, 119:213-217 (1983).
6. K. Ulrich, L. M. Auer and W. Kuschinsky, Cat pial venoconstriction by topical microapplication of norepinephrine, *J. Cereb. Blood Flow Metab.*, 2:109-111 (1982).
7. L. Edvinsson, K. C. Nielsen, C. H. Owman, H. and K. A. West, Sympathetic adrenergic influence on brain vessels as studied by changes in cerebral blood volume of mice, *Eur. Neurol.*, 6:193-202 (1971/72).
8. L. M. Auer and B. B. Johansson, Pial venous constriction during cervical sympathetic stimulation in the cat, *Acta Physiol. Scand.*, 110:203-205 (1980).
9. S. Sadoshima, M. Thames and D. Heistad, Cerebral blood flow during elevation of intracranial pressure: role of sympathetic nerves, *Am. J. Physiol.*, 10:H78-H84 (1981).
10. D. Heistad and M. Marcus, Response to "Do vasomotor nerves significantly regulate cerebral blood flow?", *Circ. Res.*, 43:494-495 (1978).
11. T. Itakura, K. Nakakita, I. Kamel et al., Aminergic innervation of cerebral veins. Histochemical comparison with extracranial veins, *J. Neurosurg.*, 60:140-144 (1984).
12. L. Edvinsson, J. McCulloch and R. Uddman, Feline cerebral veins and arteries: comparison of autonomic innervation and vasomotor responses, *J. Physiol.*, 325:161-173 (1982).
13. T. Itakura, H. Yokote, H. Kimura et al., 5-Hydroxytryptamine innervation of vessels in the rat cerebral cortex, *J. Neurosurg.*, 62:42-47 (1985).
14. J. McCulloch, L. Edvinsson and P. Watt, Comparison of the effects of potassium and pH on the calibre of cerebral veins and arteries, *Pflügers Arch.*, 393:95-98 (1982).
15. J. E. Hardebo, J. Kahrström, C. H. Owman and L. G. Salford, Vasomotor effects of neurotransmitters and modulators on isolated human pial veins, *J. Cereb. Blood Flow Metabl.*, 7:612-618 (1987).
16. J. M. Luce, J. S. Huseby, W. Kirk and J. Butler, A Starling resistor regulates cerebral venous outflow in dogs, *J. Appl. Physiol. Respirat. Environ. Exercise Physiol.*, 53:1496-1503 (1982).
17. K. Shulman and G. R. Verdier, Cerebral vascular resistance changes in response to cerebrospinal fluid pressure, *Am. J. Physiol.*, 213:1084-1088 (1967).
18. J. M. Luce, J. S. Huseby, W. Kirk and J. Butler, Mechanism by which positive end-expiratory pressure increases cerebrospinal fluid pressure in dogs, *J. Appl. Physiol. Respirat. Environ. Exercise Physiol.*, 52:231-235 (1982).
19. A. L. Benabid, J. C. Persat, J. de Rougemont et al., Intracranial pressure IV - relationship with post-capillary vascular pressures, *J. Physiol.*, 74:369-378 (1978).
20. J. P. Holt, The collapse factor in measurement of venous pressure, *Am. J. Physiol.*, 134:292 (1941).
21. E. H. Starling, "The Law of the Heart", The Linacre lecture delivered at St John's College, Cambridge 1915, Longmans, Green, London (1918).
22. A. Noordergraaf, "Circulatory System Dynamics", Academic Press, New York (1978).
23. S. Perlmutt and R. L. Riley, Hemodynamics of collapsible vessels with tone: the vascular waterfall, *J. Appl. Physiol.*, 18:924-932 (1963).
24. K. Yada, Y. Nakagawa and M. Tsuru, Circulatory disturbance of the venous system during experimental intracranial hypertension, *J. Neurosurg.*, 39:723-729 (1973).
25. L. M. Auer and N. Ishiyama, Cerebrovascular response to elevated intracranial pressure, *in*: "Intracranial Pressure VI", J. D. Miller, G. M. Teasdale, J. O. Rowan, S. L. Galbraith and A. D. Mendelow, eds., Springer, Berlin and Heidelberg, pp 399-403 (1986).

26. Y. Nakagawa, M. Tsuru and K. Yada, Site and mechanism for compression of the venous system during experimental intracranial hypertension, *J. Neurosurg.*, 41:427-434 (1974).
27. L. M. Auer, N. Ishiyama and R. Pucher, Cerebrovascular response to intracranial hypertension, *Acta Neurochir.*, 84:124-128 (1987).
28. L. M. Auer, N. Ishiyama, K. C. Hodde, R. Kleinert and R. Pucher, Effect of intracranial pressure on bridging veins in rats, *J. Neurosurg.*, 67:263-268 (1987).
29. Y. Kato, M. Mokry and L. M. Auer, Cerebrovascular response to elevation of venous pressure, in preparation.
30. L. M. Auer and E. T. MacKenzie, Phsiology of the cerebral venous system, *in*: "The Cerebral Venous System and Its Disorders", J. P. Kapp and H. M. Schmidek, eds., Grune and Stratton Inc., New York (1984).
31. L. M. Auer, The pathogenesis of hypertensive encephalopathy, *Acta Neurochir.*, (Suppl.) 23:1-111 (1978).
32. L. M. Auer, B. B. Johansson and E. T. MacKenzie, Cerebral venous pressure during actively induced hypertension and hypercapnia in cats, *Stroke*, 11:180-183 (1980).
33. B. B. Johansson, Blood-brain barrier dysfunction in acute arterial hypertension, Dissertation, Göteborg (1974).
34. S. Sato, S. Toya, M. Ohtani and T. Kawase, The effect of sagittal sinus occlusion on BBB permeability and cerebral blood flow in the dog, *in*: "Brain Edema", Y. Inaba, I. Klatzo, M. Spatz, eds, Springer, Berlin, pp 235-239 (1986).
35. L. M. Auer, Sympathetic control of pial vessels under in vivo conditions, *in*: "Neural Regulation of Brain Circulation", C.H. Owman and J. E. Hardebo, eds., Elsevier Science Publishers BV., Amsterdam, pp 497-513 (1986).
36. L. M. Auer, I. Sayama, B. B. Johansson et al., Pial venous reaction to sympathetic stimulation during elevated intracranial pressure, *in*: "The Cerebral Veins: An Experimental and Clinical Update", L. M. Auer and F. Loew, eds., Springer, New York, pp 143-153 (1983).

CEREBRAL BLOOD FLOW IN PATIENTS WITH SINUS VEIN THROMBOSIS

P. Schmiedek, K. Einhäupl, E. Moser,
U. Büll, G. Leinsinger and T. Kreisig

Departments of Neurosurgery, Neurology
and Radiology/Nuclear Medicine
Ludwig-Maximilians-University
Klinikum Großhadern, 8000 Munich 70, FRG

INTRODUCTION

Sinus vein thrombosis (SVT) of the brain is a rare disease. Recent evidence, however, suggests that it may be diagnosed more often once it is included into differential diagnostic considerations which then will lead to appropriate studies, eventually allowing the correct diagnosis to be made [4]. Apart from diagnostic difficulties, there is still a considerable lack of evidence to explain the underlying pathophysiologic mechanisms of cerebral SVT. For example, there is little known on the effect of SVT on cerebral blood flow (CBF). In fact, there is only one clinical report on regional CBF (rCBF) in SVT [9]. In this study, Shinohara et al. investigated three patients with SVT using the intracarotid ^{133}Xe injection technique.

Therefore, it was the purpose of the present study to investigate in a larger group of patients rCBF under resting and under stimulated conditions both in the acute and chronic stage of SVT. In addition, a concept is proposed on differential aspects of cerebral ischemia either due to arterial or cerebral venous obstruction.

PATIENTS

The study includes 10 patients in whom serial rCBF measurements were performed during the acute and chronic stage of SVT. There were four male and six female patients with a mean age of 26.9 years and an age range between 17 and 53 years. All patients presented with a characteristic variety of clinical symptoms, including depressed level of consciousness, paresis of various degree, seizures, headaches and papilledema. Computerized tomography revealed hemorrhagic lesions in four of these patients. The diagnosis of SVT was based on cerebral angiographic findings showing clear-cut evidence of cerebral venous or sinus obstruction during the late venous phase of the angiogram. Once the diagnosis had been established, treatment with heparin was started.

Cerebral Sinus Thrombosis, Edited by K. Einhäupl *et al.*
Plenum Press, New York, 1990

Measurements of rCBF were carried out as soon as the patients were neurologically stable, usually within 1 week after admission to the hospital. Follow-up rCBF studies were done 6 to 28 weeks (mean 14 weeks) after the initial study. At this time all patients had recovered from the acute stage of SVT except for minor deficits in two cases and all were on a regimen of anticoagulation treatment.

MEASUREMENT OF rCBF

For rCBF measurements the ^{133}Xe inhalation technique with a rapidly rotating detector system of 64 separate Na^+-iodide crystals was used (Tomomatic 64-Medimatic Inc., Copenhagen, Denmark). Details of this technique have been described in previous reports and also those modifications for the evaluation of data which were developed by our own laboratory [2,3,5,8]. Therefore, only the essentials of this method are reported here.

The technique allows the simultaneous quantitative measurement of rCBF in three transaxial tomographic slices of the brain which are centered 1, 5 and 9 cm above the orbitomeatal line. The slices thus obtained were each 2 cm thick with a resolution of each slice of 1.7 cm. Following a 1 minute period of ^{133}Xe inhalation, reaching a maximum concentration of 10 mCi/liter, the washout of the isotope is measured for the subsequent 3 minutes and an arterial input curve is estimated from the output of a single collimated detector which is positioned over the right lung. For calculation of rCBF a conventional filtered back projection algorithm is used and flow values are then displayed on a color-coded screen. With the use of a semi-automatic computer program, it is then possible to obtain information on regional CBF in 2 x 6 axially symmetric areas of each slice. In the present study only flow values of the two uppermost slices were taken into consideration because the lowest slice provides mainly information on blood flow of infratentorial structures. At the end of each study the arterial blood pressure and the arterial pCO_2 were measured.

In our earlier study for standardization of the technique, a group of 32 normal volunteers was investigated. The normal mean value of rCBF was found to be 58 ± 8 ml/100 g/minute which is in good agreement with results reported by others [10].

Following an initial study under resting conditions, the measurement was then repeated 20 minutes following the injection of 1 g of acetazolamide (Diamox) which is a carboanhydrase inhibitor and is known to result in a marked increase of rCBF [12]. In order to allow a comparative analysis of Diamox stimulated rCBF in patients, the Diamox response was again tested in a group of 32 normal volunteers without evidence of brain disease. The normal response following an iv injection of 1 g of Diamox consisted of an increase from 58 ± 8 ml/100 g/minute at rest to 73 ± 10 ml/100 g/minute. It was also found that the Diamox induced increase of rCBF is dependent on the respective resting rCBF value. Thus, in cases with a relatively low resting rCBF, Diamox induced a more pronounced increase of rCBF as compared with cases of a relatively high resting rCBF. When plotting the rCBF at rest against the respective increase of rCBF following Diamox, a regression line was found with a good correlation (correlation coefficient r = -0.77).

The potential of rCBF to increase following Diamox is termed herein as cerebrovascular reserve capacity (CVRC). It is defined as the difference between the Diamox-induced rCBF and the rCBF at rest:

rCBF (DMX) - rCBF (rest) = CVRC ml/100 g/minute [11].

Finally, it should be emphasized that the application of 1 g of Diamox was well tolerated in all patients and usually resulted only in transient headaches or in circumoral tingling and numbness. However, no serious reactions were noted. Arterial blood pressure and the arterial pCO_2 remained stable during the Diamox test in all patients.

RESULTS

Resting rCBF in Acute SVT

Measurement of rCBF in 10 patients during the acute phase of SVT in the resting state revealed subnormal flow values in each case for both tomographic brain slices included for data analysis (Fig. 1). Although the degree of cerebral ischemia was less pronounced when compared with findings which are usually seen in patients with arterial cerebrovascular disease, there were five patients where the flow decrease reached a level of two standard deviations below the normal value (58 ± 8 ml/100 g/minute). In the other five patients mean resting flow values were in the range between one and two standard deviations below the level of rCBF obtained in 32 normal volunteers. The regional distribution of ischemic regions showed a pattern of bilateral involvement except of one case in whom only one parietal area revealed definite signs of brain ischemia. There was no correlation between the clinical condition of the patient and the result of the rCBF study.

Diamox-activated rCBF in Acute SVT

In five of these 10 patients the rCBF measurement was repeated following injection of 1 g of Diamox. An increase of rCBF was found in all patients (Fig. 2). When compared with the normal data for Diamox-activated rCBF in volunteers, the increase of flow in these five patients during the acute stage SVT was found to reach the expected normal range in 10 out of 20 regions that were analyzed. In the remaining 10 regions including hemispheric flow of slice 2 and 3 of both sides, Diamox activation resulted in an increase of rCBF slightly below the normal range. Therefore, despite evidence that in acute SVT there is ischemia of the brain as demonstrated during flow studies under resting conditions, the vasoregulatory capacity is less affected than might have been expected.

Resting rCBF in Chronic SVT

All 10 patients who were studied during the acute phase of SVT underwent follow-up rCBF studies when all clinical symptoms had subsided and in fact they were regarded to have completely recovered. Separate analysis of 10 hemispheric regions of both slice 2 and 3 revealed an increase of rCBF in almost all counting regions with a mean value approaching the normal reference value (Fig. 3). There were also no disturbances of the regional flow pattern, which in fact was in good agreement with what was found in the control group.

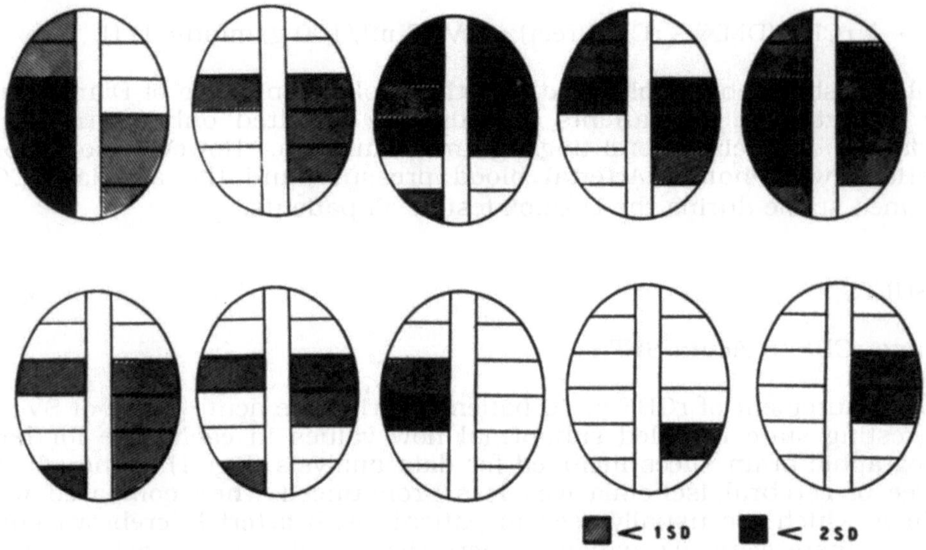

Fig. 1. Regional rCBF under resting conditions in 10 patients with acute SVT. Tomographic slice 2. Black areas indicate flow values lower than two standard deviations below normal; grey areas of one standard deviation below normal.

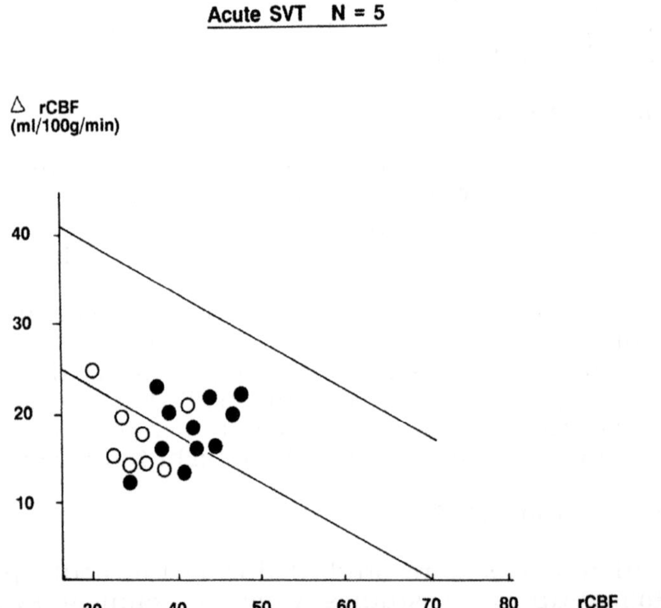

Fig. 2. Diamox-induced increase of rCBF in five patients with acute SVT plotted against the corresponding flow under resting conditions. Black circles (one circle for each hemisphere) represent mean flow of slice 2, open circles of slice 3. The two lines represent the range of the cerebrovascular reserve capacity (CVRC) as determined in 32 controls.

FOLLOW-UP rCBF IN 10 PATIENTS WITH SVT - SLICE 2

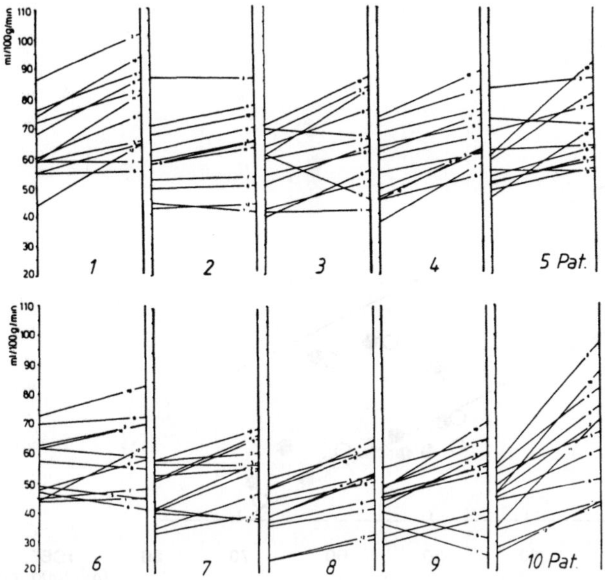

Fig. 3. Follow-up of rCBF findings in 10 patients with SVT. Of each patient 12 areas of slice 2 were compared. Left: acute SVT; right: chronic stage.

Interestingly, in two patients who were not included in this study, a comparison of acute and late follow-up rCBF studies failed to show a normalization of rCBF at the chronic stage measurement. In both patients, anticoagulation therapy was stopped prematurely and they developed a subsequent episode of recurrent SVT. In none of the 10 patients reported herein was another deterioration due to SVT observed over a long-term follow-up period.

Diamox-activated rCBF in Chronic SVT

In accordance with the normalization of rCBF under resting conditions in patients who had recovered from acute SVT, Diamox-activated studies which were performed in five patients revealed a normal cerebrovascular reserve capacity in 13 out of 20 regions under investigation (Fig. 4). In the other seven regions the Diamox-induced flow increase was close to the lower limit of the normal range of the values which were obtained in controls. When comparing the cerebrovascular reactivity in acute and chronic SVT, a shift on the normogram from the left to the right side can be noted which, however, is easily explained by the different resting flow values in the two situations.

DISCUSSION

In the only previous clinical study on the influence of cerebral venous obstruction on rCBF, Shinohara et al. [9] found in three patients a slight decrease of rCBF at resting conditions and a normal response during hyperventilation, whereas rCBF did not increase secondary to an

Chronic SVT N = 5

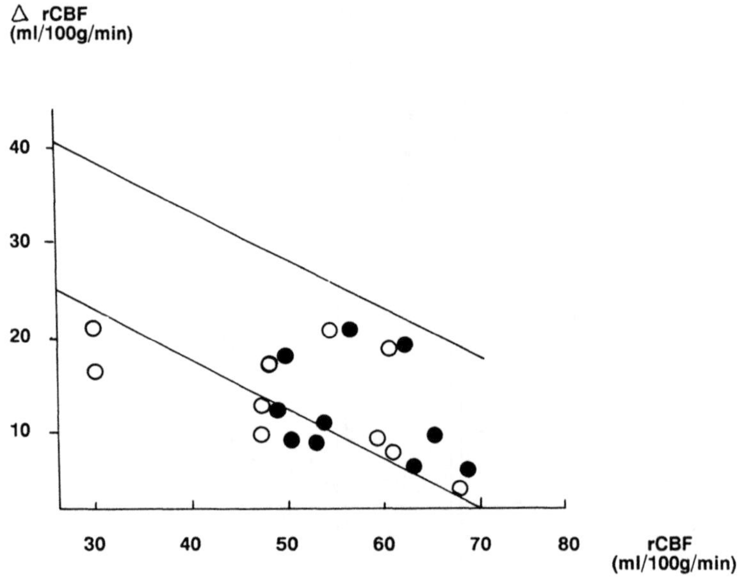

Fig. 4. Diamox-activated increase of rCBF in five patients during chronic stage SVT plotted against blood flow obtained under resting conditions.

increased arterial carbon dioxide tension. However, these authors used the intracarotid ^{133}Xe injection technique with only one large stationary external detector, thus obtaining only information on hemispheric blood flow of one side.

The present study not only includes a considerably larger clinical sample, namely 10 patients, but also reports on follow-up rCBF measurements during the acute stage of the disease and later when the patients had recovered from their clinical symptoms. Furthermore, the method for measurement of rCBF, which was used in this study, namely the tomographic D-SPECT technique with ^{133}Xe inhalation, provides more accurate information, particularly because it allows bilateral quantitative, regional determinations of rCBF. This is important, because in contrast to cerebral ischemia due to arterial cerebrovascular disease which usually affects only one hemisphere, SVT is likely to result in an impairment of bilateral rCBF. In fact, with one exception all patients included in this study did show a bilateral impairment of rCBF. Nevertheless, our results are in agreement with the earlier observations by Shinohara et al. [9] in that we also found a slight but definite decrease of rCBF in acute SVT during resting conditions.

Interestingly, there was no correlation of the clinical condition of the patient and the degree of brain ischemia. In other words, patients with marked neurologic deficits during acute stage SVT did not show more pronounced abnormalities on their rCBF findings when compared with those showing only minor clinical symptoms.

With regard to the cerebrovascular reactivity during SVT, we did not investigate the influence of blood pressure changes on rCBF and, therefore, have no information on the autoregulation. However, these studies that were performed following stimulation with Diamox did show that the cerebrovascular reserve capacity of rCBF was essentially preserved during acute SVT. This is in contrast to the observations of Shinohara et al. [9] who found a decrease of rCBF during hyperventilation but no increase during inhalation of CO_2. Although the exact mechanism by which Diamox causes an increase of rCBF is still unknown, experimental evidence suggests that this effect is mediated by a retention of CO_2 in the brain tissue, which in turn results in a decrease of the brain tissue pH [6]. This condition then is similar to the effect of CO_2 inhalation which is usually used for the study of rCBF responsiveness. The advantage of the Diamox test, however, is that it seems to affect exclusively brain blood flow and does not exhibit any systemic effects which are known to acompany inhalation of a CO_2/air mixture.

Although the rCBF measurements during acute stage SVT do not provide conclusive evidence as yet on the pathophysiologic mechanisms leading to the impairment of brain blood flow, the findings can be used to explain the fact that these patients despite their severe clinical condition have a good prognosis provided appropriate treatment is used. The characteristic finding of rCBF in SVT, therefore, is a marginal and transient degree of brain ischemia with an overall preservation of the capacity to increase blood flow in response to functional stimulation. The reversibility of brain ischemia in SVT is in contrast to the changes of rCBF which are known to occur in patients with arterial cerebral ischemia (Fig. 5). If the collateral circulation is insufficient, occlusion of one of the main cerebral arteries usually will result in a much more pronounced ischemia of brain tissue with a concomitant derangement of the vasoregulatory capacity of the cerebral arteries.

Experiments have shown that brain function disintegrates reversibly at a rCBF level of about 25 ml/100 g/minute, but irreversibly at approximately 15 ml/100 g/minute, which then results in structural changes of the brain tissue and ischemic neuronal loss [1]. The level of ischemia that we have observed in our patients with acute SVT never reached the critical borderline beyond which permanent neuronal damage will occur. It has to be admitted, however, that this conclusion does not concern all patients with acute SVT. For methodological reasons rCBF studies could only be performed in patients being awake and fairly cooperative. Therefore, the situation might be quite different in those patients with more severe neurologic symptoms due to acute SVT who have not been investigated as yet. When considering the still poor prognosis of patients with massive SVT, it can be anticipated that impairment of rCBF in these patients is much more pronounced, eventually resembling the changes due to acute arterial cerebral ischemia.

Another limitation of the present study is that we have not investigated the autoregulatory capacity of the cerebral circulation in acute SVT following induced changes of the arterial blood pressure. This needs to be done in order to develop a more reliable concept on the impairment of rCBF in SVT.

Notwithstanding these shortcomings, our results indicate that the distrubances of the cerebral circulation in SVT are different from those

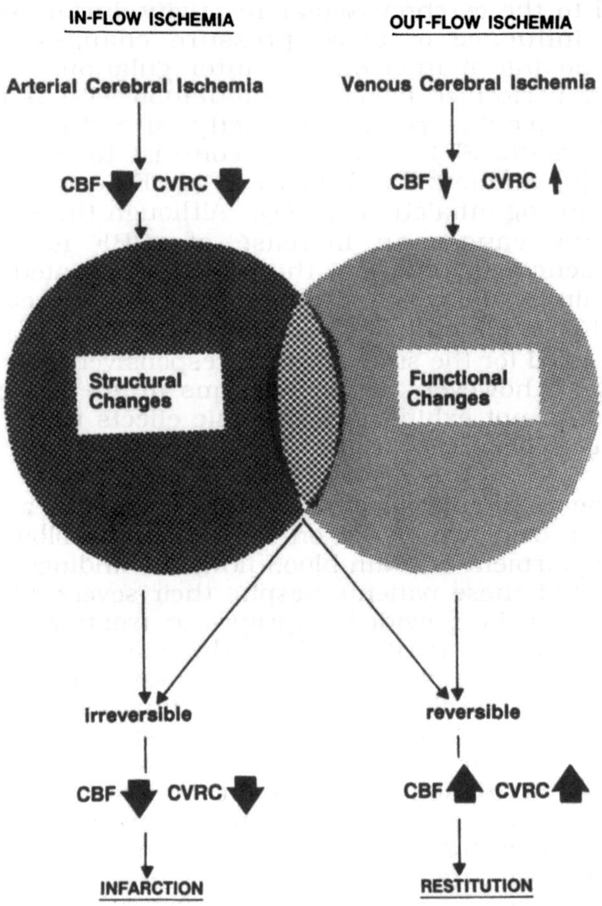

Fig. 5. Pathophysiological concept to explain differences in function and outcome between arterial inflow and venous outflow cerebral ischemia. For details see text.

seen in arterial brain ischemia. Therefore, although being of less clinical significance when compared with arterial cerebral ischemia, the reversibility of brain ischemia in SVT, which is the essential finding of this study, may add an important facet to our understanding of cerebral ischemia in general.

REFERENCES

1. J. Astrup, B. K. Siesjö and L. Symon, Thresholds in cerebral ischemia. The ischemic penumbra, *Stroke*, 12:723-725 (1981).
2. U. Büll, E. Moser, C. M. Kirsch and P. Schmiedek, 133-Xe-DSPECT. Ergebnisse einer neuen nicht invasiven Methode zur Bestimmung der regionalen Hirndurchblutung (rCBF), *Fortschr. Röntgenstr.*, 139:351-358 (1983).
3. U. Büll, E. Moser, P. Schmiedek, G. Leinsinger and T. Kreisig, Dynamic SPECT with 133-Xenon: regional cerebral blood flow in patients with unilateral cerebrovascular disease, *J. Nucl. Med.*, 25:441-446 (1984).
4. K. M. Einhäupl, A. Villringer, R. L. Haberl, W. Pfister, M. Deekert, H. Steinhoff and P. Schmiedek, Clinical spectrum of sinus venus

thrombosis, *in*: "Cerebral Sinus Thrombosis: Experimental and Clinical Aspects", K. Einhäupl, O. Kempski and A. Baethmann, eds, Plenum Press, New York and London, pp 149-155 (1990).
5. L. Henriksen, N. A. Lassen, O. B. Paulson and P. Rommer, Evaluation of a new multi-crystal ECT-system, *Nucl. Med.*, 22-24 (1980).
6. D. Heuser, J. Astrup, N. A. Lassen and E. Betz, Brain carbonic acid acidosis after acetazolamide, *Acta Physiol. Scand.*, 93:385-390 (1975).
7. T. Kreisig, P. Schmiedek, G. Leinsinger, K. Einhäupl and E. Moser, 133-Xe-DSPECT: Normalwerte von zerebraler Ruhedurchblutung und Reservekapazität, *Nucl. Med.*, 261:192-197 (1987).
8. N. A. Lassen, L. Henriksen and O. Paulson, Regional cerebral blood flow in stroke by 133-Xenon inhalation and emission tomography, *Stroke*, 12:284-288 (1981).
9. Y. Shinohara, S. Takagi, K. Kobatake and F. Gotoh, Influence of cerebral venous obstruction on cerebral circulation in humans, *Arch. Neurol.*, 39:479-481 (1982).
10. N. Shirahta, L. Henriksen, S. Vorstrup, S. Holm, M. Lauritzen, O.B. Paulson and N. A. Lassen, Regional cerebral blood flow assessed by 133-Xe inhalation and emission tomography: normal values, *J. Comp. Ass. Tomogr.*, 5:861-866 (1985).
11. P. Schmiedek, T. Kreisig, K. Einhäupl, V. Olteanu-Nerbe, E. Moser and F. Marguth, Improved reserve capacity of the cerebral circulation following bypass surgery, *in*: "Controversies in EIAB for Cerebral Ischemia", R. Gagliardi and L. Benvenuti, eds., Montuzzi Ed., pp 109-116 (1987).
12. S. Vorstrup, L. Henriksen and O. B. Paulson, Effect of acetazolamide on cerebral blood flow and cerebral metabolic rate of oxygen, *J. Clin. Invest.*, 74:1634-1639 (1984).

VASCULAR SMOOTH MUSCLE REACTIVITY IN HYPOXIA

J. Grote and G. Siegel*

Institute of Physiology I, University of Bonn
Nußallee 11, 5300 Bonn 1, FRG and
*Institute of Physiology, Free University of Berlin
Arnimallee 22, 1000 Berlin 33, FRG

INTRODUCTION

Arterial as well as tissue hypoxia induces vasodilation and a subsequent increase of blood flow in several organs [2,5,7,11,13]. In brain tissue regional blood flow increases significantly if the arterial oxygen falls below 60 mmHg. In the presence of hypoxia severe enough to cause anoxia in single cells, characteristic metabolic changes develop in cerebral cortex. Therefore, adjustment of the vascular system under these conditions is usually attributed to metabolic factors [9]. Among these K^+, H^+ ions, and adenosine seem to play an important role [2,7,13].

However, in vitro measurements of mechanical tension or of vascular diameter using strips of different arteries and veins or perfused vessel segments have demonstrated that hypoxia directly causes relaxation of vascular smooth muscle cells [3,4,21,22,24,26]. Different vasoactive substances released at low oxygen tensions from endothelial or smooth muscle cells, such as prostacyclin or endothelium-derived relaxing factor (EDRF), may mediate these reactions [1,3,6,14-16,18,24-26]. Concurrent investigations of vascular smooth muscle tension and intracellularly recorded membrane potential during oxygen deficiency have not been conducted yet. Hellstrand and coworkers [10] merely observed in rat portal veins a reduction of spontaneous spike discharges during hypoxia.

MATERIALS AND METHODS

To study electromechanical properties of vascular smooth muscle cells during hypoxia, measurements of membrane potential and of mechanical tension were performed simultaneously using isolated canine carotid arteries. The carotid arteries were surgically excised from dogs within 3-5 minutes after sacrifice. Adventitial connective tissue was removed and strips of the vessels were prepared, which were mounted in an organ bath and perfused with a modified Krebs solution at 37°C

Cerebral Sinus Thrombosis, Edited by K. Einhäupl *et al.*
Plenum Press, New York, 1990

[19,20,22]. Under control conditions, the perfusate was equilibrated with a gas mixture of 95% O_2 and 5% CO_2. For the experiments, the oxygen tension was adjusted to lower levels, while CO_2 was maintained at a normal value of about 40 mmHg. The PO_2 of the perfusate was stepwise decreased from approximately 550 mmHg by changing perfusion solutions. Application time of each was 15 minutes. The oxygen tension of the perfusate was continuously recorded close to the vessel preparation. In some experiments, additional measurements of the PO_2 of endothelium and superficial smooth muscle cell layers were performed using platinum microelectrodes [8]. For determination of smooth muscle tension, the artery segments were attached to an isometric force transducer. The membrane potential of the smooth muscle cells was measured intracellularly with glass microelectrodes [20,22]. At each oxygen tension level at least 10 impalements were performed, and the values obtained were averaged. In two additional groups of experiments, similar measurements were performed after administration of indomethacin (10^{-5}M), or following complete removal of the endothelium.

RESULTS AND DISCUSSION

In the first series of control experiments with oxygen tensions of about 550 mmHg, a mean membrane potential of -65 mV was found. The corresponding smooth muscle tension was 2 g (Fig. 1). Since the initial mechanical tensions differed slightly in various preparations, normalization of the data was performed as previously described [23]. Figure 1 demonstrates measurements of membrane potential and mechanical tension carried out at varying ambient oxygen pressures. The stepwise lowering of the PO_2 in the Krebs solution induced hyperpolarization of the cell membranes and a decrease of smooth muscle tone, as soon as values below 120 mmHg were reached. Maximal hyperpolarization and relaxation of -74 mV and 1.3 g, respectively, were found at oxygen tensions between 30 and 40 mmHg. Between 150 and 35 mmHg, the decrease in oxygen tension caused a dose-dependent hyperpolarization of the cell membranes as well as smooth muscle relaxation. Hyperpolarization and reduction of the muscle tone as a function of oxygen tension run parallel through all experiments. The PO_2 measurements in the upper cell layers of the preparations performed under comparable conditions yielded oxygen tensions of 10-15 mmHg below the corresponding ambient oxygen pressure. Up to PO_2 values of about 30 mmHg, anoxic endothelial or smooth muscle cells were not found.

The further decrease in oxygen tension induced attenuation of hyperpolarization, or even depolarization of the cell membranes with a subsequent increase in smooth muscle tone up to or above the initial value. However, membrane potential and force development remained strictly coupled, also at oxygen tensions below 35 mmHg. Comparable smooth muscle contractions at very low oxygen tensions have been observed in canine coronary arteries by Rubanyi and Vanhoutte [17]. When increasing again the oxygen tension of the Krebs solution, the changes in membrane potential and smooth muscle tone found during hypoxia were reversible. A first reaction of the vessel preparations to the stepwise decrease or increase of PO_2 occurred after 1-2 minutes, the total effect became visible after approximately 10 minutes.

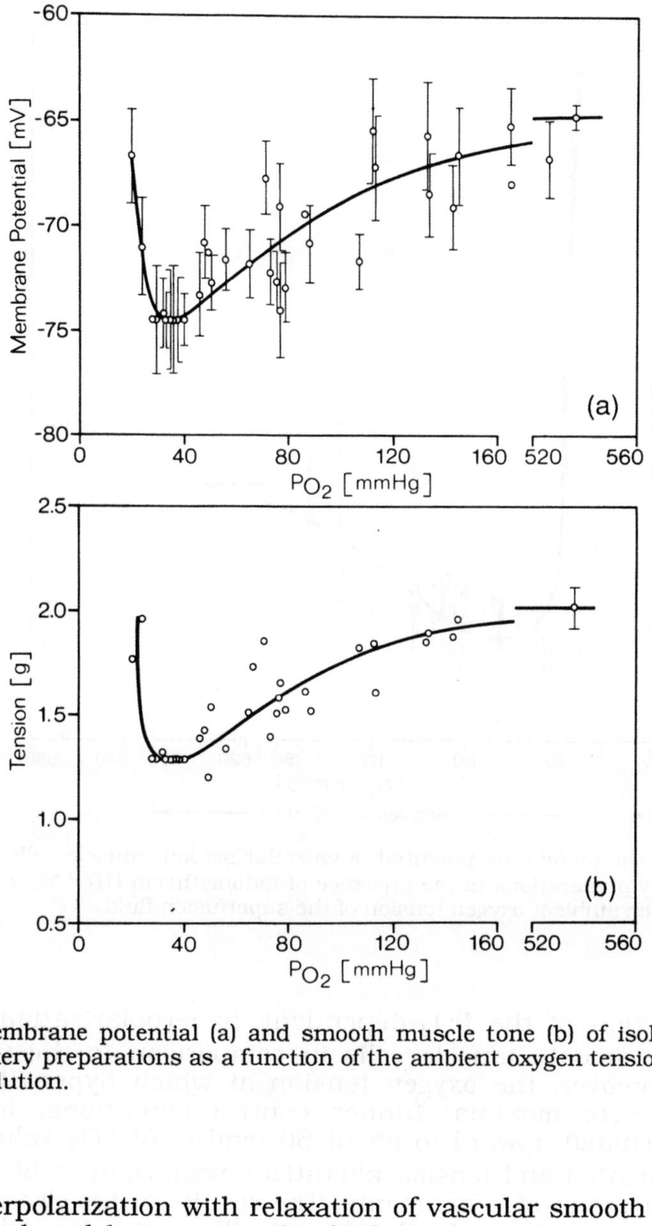

Fig. 1. **Membrane potential (a) and smooth muscle tone (b) of isolated canine carotid artery preparations as a function of the ambient oxygen tension in the superfusion solution.**

Hyperpolarization with relaxation of vascular smooth muscle cells can also be induced by prostacyclin [23]. Since production of prostanoids is enhanced at low oxygen tensions [3,12], the hypoxia experiments were repeated in the presence of the cyclooxygenase inhibitor indomethacin. The aim of the experiments was to test whether the changes in membrane potential and smooth muscle tension with decreasing ambient oxygen tension were mediated by prostacyclin, or by other vasodilatory prostanoids. Under control conditions administration of indomethacin (10^{-5} M) resulted in a slight depolarization of the cell membrane by about 2 mV and an augmentation of the basal tension by about 0.3 g (Figs 2 and 3). A similar increase in vascular smooth muscle tone by indomethacin has been reported by Rubanyi and Vanhoutte [17]. When lowering the oxygen tension of the superfusion solution, indomethacin led to a small but

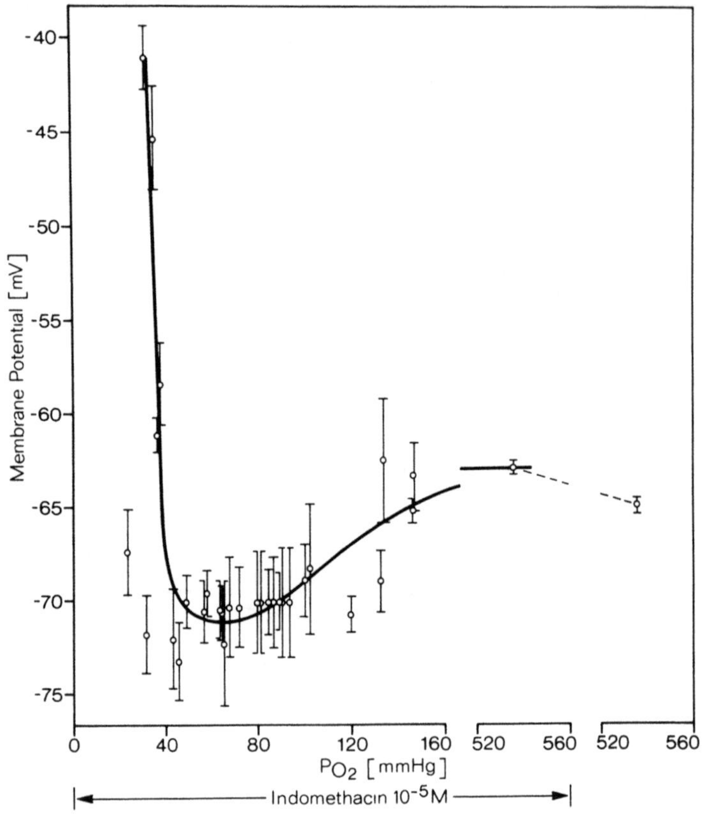

Fig. 2. Response of the membrane potential of vascular smooth muscle cells of isolated carotid artery preparations in the presence of indomethacin (10^{-5} M) to a stepwise decrease of the ambient oxygen tension of the superfusion fluid.

significant reduction of the PO_2-dependent hyperpolarization and to relaxation of the smooth muscle cells in comparison to the untreated preparations. Moreover, the oxygen tension at which hyperpolarization and relaxation were maximal (under control conditions, ie, 30 to 40 mmHg), was shifted upward to about 60 mmHg. At PO_2 values below this level, the potential and tension alterations were comparable to those found in the first group of experiments. The results of the two groups of experiments are summarized in Figure 4 to illustrate the influence of indomethacin on the hypoxia-induced changes of both parameters. The upper curves give the smooth muscle potentials and tensions found in the indomethacin-treated preparations, while the response of the untreated controls is symbolized by the lower curves. The shaded areas represent the effect of prostaglandins.

Complete removal of the endothelium was carried out to test whether the changes in membrane potential and tension of vascular smooth muscle cells were mediated by EDRF. Removal of endothelium caused, first of all, instantaneous depolarization of cell membranes and contraction of smooth muscle cells. Under control conditions, depolarization of about 4 mV and an increase in tension of 1.3 g was found

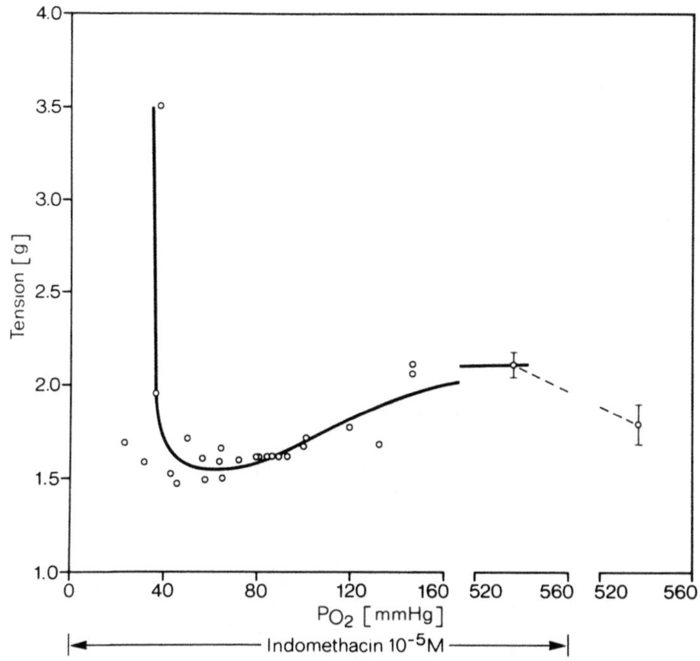

Fig. 3. Smooth muscle tension of isolated carotid artery preparations administered with indomethacin (10^{-5} M) at decreasing oxygen tensions.

(Fig. 5). In the first three experiments, the decrease of ambient oxygen tension to about 65 mmHg led to a slight hyperpolarization of 1.7 mV with simultaneous relaxation of 0.2 g. At PO_2 values below 50 mmHg, a decrease in hyperpolarization and vasoconstriction occurred.

When comparing the results of the three groups of investigations, one may conclude that the hypoxia-induced hyperpolarization of the cell membrane and relaxation of smooth muscle cells is mainly effected by EDRF, while prostaglandins are of minor importance. The present results provide no explanation of the causal mechanisms underlying the changes in membrane potential and tone of the smooth muscle preparations with varying PO_2. According to our results and to reports in the literature, a release of either prostacyclin, EDRF, or of oxygen-free radicals may all be related to membrane hyperpolarization and smooth muscle relaxation during oxygen deficiency. As a mechanism of hyperpolarization these substances may induce changes of passive membrane properties of smooth muscle cells such as an increase in K^+-permeability or decrease in Na^+-permeability, or of active transport processes, eg, stimulation of the electrogenic Na^+-outward pump [4,19,20]. The effector mechanism of prostacyclin has been elucidated. A strong increase in K^+-permeability and stimulation of the electrogenic Na^+-outward transport are responsible for the hyperpolarization induced by prostacyclin [23]. The simultaneous relaxation of smooth muscle cells can be attributed to a decrease in intracellular Ca^{2+}-activity resulting from membrane hyperpolarization [19,20]. In addition, a decrease in smooth muscle tone may be facilitated by elevated levels of intracellular cyclic GMP and activation of cyclic GMP-dependent protein kinase. Since EDRF stimulates the soluble guanylate

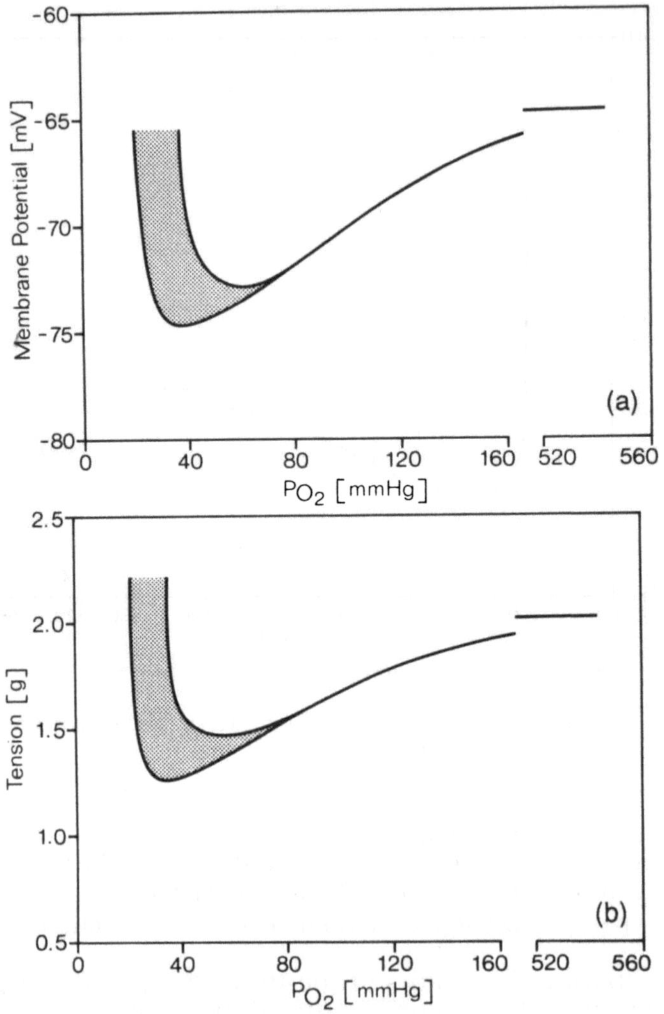

Fig. 4. Membrane potential (a) and mechanical tension (b) of isolated carotid artery preparations with (upper curves) or without (lower curves) indomethacin (10^{-5} M) at varying oxygen tensions. The shaded areas illustrate the effect of prostaglandins.

cyclase, an increased production of cyclic GMP is held responsible for the relaxation caused by endothelium-dependent relaxants [25,26]. Whether membrane depolarization and contraction of the preparations at very low oxygen tensions is caused by release of an endothelium-derived contracting factor [17] or by norepinephrine cannot yet be decided on the basis of the presently existing investigations.

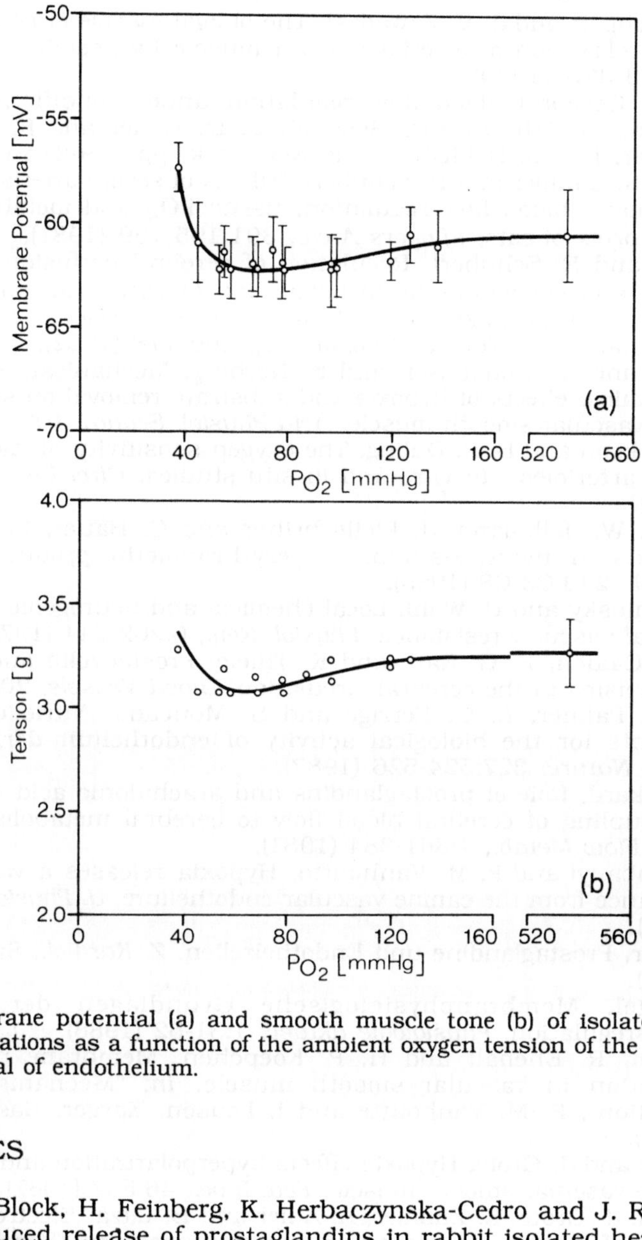

Fig. 5. Membrane potential (a) and smooth muscle tone (b) of isolated carotid artery preparations as a function of the ambient oxygen tension of the superfusate after removal of endothelium.

REFERENCES

1. A. J. Block, H. Feinberg, K. Herbaczynska-Cedro and J. R. Vane, Anoxia-induced release of prostaglandins in rabbit isolated hearts, *Circ. Res.*, 36:34-42 (1975).
2. D. W. Busija and D. D. Heistad, Factors involved in the physiological regulation of the cerebral circulation, *Rev. Physiol. Biochem. Pharmacol.*, 101:161-211 (1984).
3. R. Busse, U. Förstermann, H. Matsuda and U. Pohl, The role of prostaglandins in the endothelium-mediated vasodilatory response to hypoxia, *Pflugers Arch.*, 401:77-83 (1984).
4. R. Detar, Mechanism of physiological hypoxia-induced depression of vascular smooth muscle contraction, *Am. J. Physiol.*, 238:H761-H769 (1980).
5. B. R. Duling, Oxygen sensitivity of vascular smooth muscle. II. In vivo studies, *Am. J. Physiol.*, 227:42-49 (1974).

6. R. F. Furchgott and J. V. Zawadzki, The obligatory role of endothelial cells in the relaxation of arterial smooth muscle by acetylcholine, *Nature*, 288:373-376 (1980).
7. J. Grote, Cerebral blood flow regulation under conditions of arterial hypoxia, *in*: "The Arterial System", R. D. Bauer and R. Busse, eds., Springer, Berlin, Heidelberg and New York, pp 209-215 (1978).
8. J. Grote, K. Zimmer and R. Schubert, Effects of severe arterial hypocapnia on regional blood flow regulation, tissue PO_2 and metabolism in the brain cortex of cats, *Pflugers Arch.*, 391:195-199 (1981).
9. J. Grote and R. Schubert, Regulation of cerebral perfusion and PO_2 in normal and edematous brain tissue, *in*: "Oxygen Transport to Human Tissue", J. A. Loeppky and M. L. Riedesel, eds., Elsevier North Holland, Amsterdam, New York and Oxford, pp 169-178 (1982).
10. P. Hellstrand, B. Johansson and K. Norberg, Mechanical, electrical and biochemical effects of hypoxia and substrate removal on spontaneously active vascular smooth muscle, *Acta Physiol. Scand.*, 100:69-83 (1977).
11. W. F. Jackson and B. R. Duling, The oxygen sensitivity of hamster cheek pouch arterioles. In vitro and in situ studies, *Circ. Res.*, 53:515-525 (1983).
12. A. Kurtz, W. Jelkmann, J. Pfeilschrifter and C. Bauer, Role of prostaglandins in hypoxia-stimulated erythropoietin production, *Am. J. Physiol.*, 249:C3-C8 (1985).
13. W. Kuschinsky and U. Wahl, Local chemical and neurogenic regulation of cerebral vascular resistance, *Physiol. Rev.*, 6:202-211 (1978).
14. T. A. McCalden, R. G. Nath and K. Thiele, Prostacyclin and vasodilator mechanisms in the cerebral circulation, *Blood Vessels*, 20:202 (1983).
15. R. M. J. Palmer, A. G. Ferrige and S. Moncada, Nitric oxide release accounts for the biological activity of endothelium-derived relaxing factor, *Nature*, 327:524-526 (1987).
16. J. D. Pickard, Role of prostaglandins and arachidonic acid derivatives in the coupling of cerebral blood flow to cerebral metabolism, *J. Cereb. Blood Flow Metab.*, 1:361-384 (1981).
17. G. M. Rubanyi and P. M. Vanhoutte, Hypoxia releases a vasoconstrictor substance from the canine vascular endothelium, *J. Physiol.*, 364:45-56 (1985).
18. K. Schrör, Prostaglandine und Endothelzellen, *Z. Kardiol.*, Suppl. 7:93-97 (1984).
19. G. Siegel, Membranphysiologische Grundlagen der peripheren Gefässregulation, *Physiologie aktuell*, 1:31-52 (1986).
20. G. Siegel, R. Ehehalt and H. P. Koepchen, Membrane potential and relaxation in vascular smooth muscle, *in*: "Mechanisms of Vasodilatation", P. M. Vanhoutte and I. Leusen, Karger, Basel, pp 56-72 (1978).
21. G. Siegel and J. Grote, Hypoxia effects hyperpolarization and relaxation in canine vascular smooth muscle, *Fed. Proc.*, 46:507 (1987).
22. G. Siegel, J. Grote, K. Zimmer, A. Adler and B. Litza, Electrophysiological effects of hypoxia on vasocular smooth muscle, *in*: "Vasodilatation", P. M. Vanhoutte, ed., Raven Press, New York, pp 371-376 (1988).
23. G. Siegel, G. Stock, F. Schnalke and B. Litza, Electrical and mechanical effects of prostacyclin in the canine carotid artery, *in*: "Prostacyclin and its Stable Analogue Iloprost", R. J. Gryglewski and G. Stock, Springer, Berlin, pp 143-149 (1987).
24. P. M. Vanhoutte, Effects of anoxia and glucose depletion on isolated veins of the dog, *Am. J. Physiol.*, 230:1261-1268 (1976).
25. P. M. Vanhoutte, The end of the quest?, *Nature*, 327:459-460 (1987).
26. P. M. Vanhoutte, G. M. Rubanyi, V. M. Miller and D. S. Houston, Modulation of vascular smooth muscle contraction by the endothelium, *Ann. Rev. Physiol.*, 48:307-320 (1986).

MECHANISMS OF ISCHEMIC BRAIN DAMAGE

O. Kempski, F. Staub, G.-H. Schneider
M. Zimmer and A. Baethmann

Institute for Surgical Research
Ludwig-Maximilians-University
Klinikum Großhadern
8000 Munich 70, FRG

INTRODUCTION

One of the pathophysiological consequences of cerebral sinus vein thrombosis (SVT) is hypoperfusion of circumscript brain areas drained by the occluded veins. In some of these territories the supply of blood may fall below a critical threshold, resulting in an ischemic or hypoxemic lesion. Experimental SVT is indeed characterized by a significant rise of tissue electrical impedance [1] as an indication of ischemic cell swelling. Therefore, treatment of patients with SVT may require anti-ischemic measures. Unfortunately, after decades of research on cerebral ischemia, there has been little progress in the development of specific treatment protocols. New information is now available on the pathophysiological mechanisms activated during and after an ischemic insult. Thus, based on recent findings, novel therapeutic approaches may become available in the near future. A thorough understanding of pathophysiological events should help to design specific therapeutic schemes. The subject of ischemic brain damage, however, is too complex to cover it completely in a short review. Therefore, this chapter intends to summarize recent concepts of pathophysiological interactions implicated in ischemic brain damage.

Stroke therapy in most instances reaches the patient too late. Five minutes of normothermic ischemia is considered a maximum which can be tolerated without damage. In the gerbil, 5 minutes of ischemia are sufficient to selectively destroy the CA1 sector of the hippocampus [2]. Cell volume control is lost already after 1-3 minutes of complete cerebral ischemia. This is accompanied by an inverse change of extracellular (ec) Na^+ and K^+ concentrations [3-5], and a drop of intra- and extracellular pH [6]. Obviously, many parameters are affected simultaneously.

The development of cytotoxic edema, ie, swelling of glial cells, dendrites and postsynaptic endings, has been considered an early sign of damage due to the failure of active transport after breakdown of the cerebral energy supply. This interpretation, however, may be too

Cerebral Sinus Thrombosis, Edited by K. Einhäupl *et al.*
Plenum Press, New York, 1990

simplistic: brain tissue can survive prolonged periods of ischemia with good recovery, providing the experimental conditions during recirculation are excellent. Current research, therefore, must discriminate damaging factors from tissue specific homeostatic mechanisms.

In order to avoid misunderstandings, the term "brain damage" should be used very carefully. Pathophysiological events like cytotoxic edema have to be understood in every detail before a conclusion is permitted as to whether these are attributes of brain damage, or instead reflect an activation of homeostatic mechanisms by viable cells.

IN VITRO STUDIES

Studies on mechanisms of brain damage are hampered by the fact that the determinants of cell death in the brain are not really known. One definition assumes that breakdown of the normal ion gradients signals beginning cell death. Therefore, the rapid onset of potassium release and cell swelling after cerebral ischemia has often been considered an early sign of brain damage. During spreading depression, however, ion gradients are also reduced just as in ischemia but without the detrimental consequences. Currently, influx of Ca^{2+} ions is thought to constitute a final common pathway of toxic cell death [7].

These considerations suggested that we should study related problems under controlled in vitro conditions which allowed us

(1) to separately examine individual cell types, ie, glial, nerve, and endothelial cells, and

(2) to manipulate the extracellular environment for the study of individual pathophysiological conditions with respect to cell swelling and cytotoxicity.

This laboratory has been engaged in respective investigations for several years [8,9,10,13]. An in vitro model has been established using suspended C6 glioma cells as model cells with glial characteristics [14]. The system permits a close control of pertinent extracellular parameters (Fig. 1). A large suspension medium serves to dilute released intracellular material down to comparatively ineffective concentrations. Cell volume is measured very accurately by flow cytometry.

Since glial swelling is among the very early and prominent events after cerebral ischemia, one of the first objectives was to study the kinetics of the swelling process using this model. Surprisingly, however, glial swelling was never observed during anoxia, even in combination with inhibition of anaerobic metabolism by iodoacetate over periods of up to 2 hours [11; (Fig. 2)]. The cells gained Na^+ and lost K^+ ions in equal amounts which explains the absence of any volume increase. Similar results were obtained by Ames and coworkers [15,16] who used the isolated retina to study the effect of anoxia and substrate deprivation on nervous tissue. The authors tested (1) tissue swelling, and (2) the incorporation of radioactive leucine as a measure of protein synthesis, ie, viability. Anoxia plus substrate deprivation was tolerated for 40 minutes without swelling or irreversible damage, provided the extracellular compartment was large. Retinal water content did increase if the extra- to intracellular volume

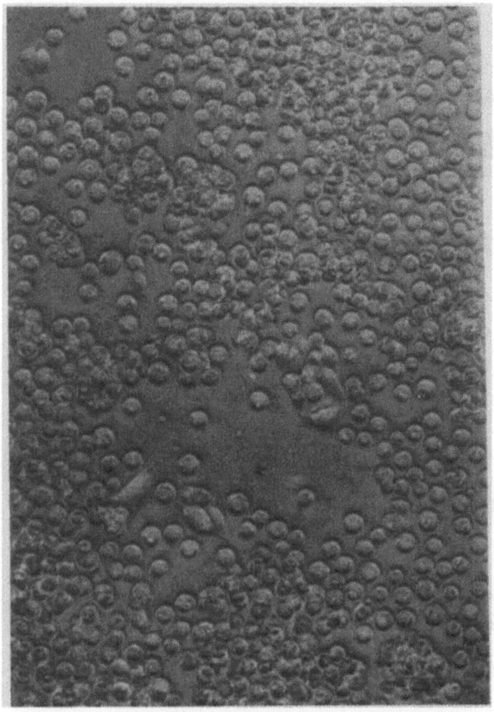

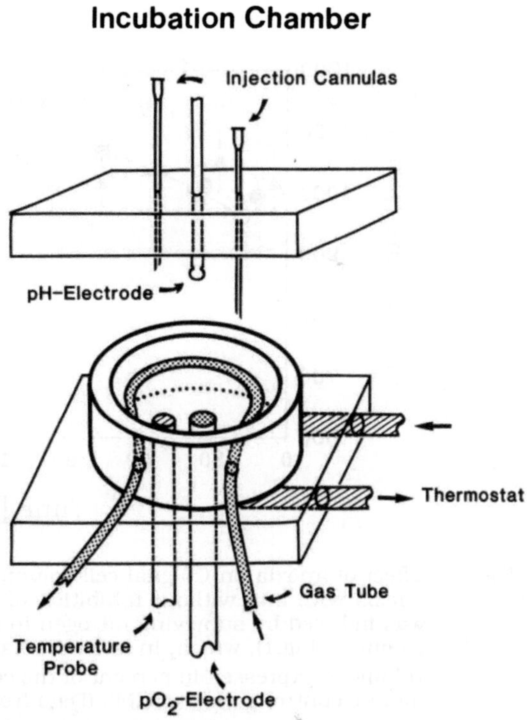

Fig. 1. In vitro model for the study of glial swelling. C6 glioma cells are cultivated as monolayers in Petri dishes using Dulbecco's modified minimal essential medium + 100 IU/ml penicillin G, 50 μg/ml streptomycin and 10% fetal calf serum (FCS) at 37°C in a humified atmosphere of 5% CO_2 and 95% room air. For the experiments, confluent cultures obtained 2 days after subcultivation are used. The cells are suspended in FCS-free medium as described previously [9]. The cell suspension is transferred to an incubation chamber equipped with electrodes for continuous registration of pH, pO_2 and temperature. A permeable tubing inside the chamber serves as a membrane oxygenator allowing variation of the oxygen and carbon dioxide tension of the incubation medium. A magnetic stirrer prevents cell sedimentation. Cell volume is determined by flow cytometry using Metricell®. This system is based on the Coulter principle and employs hydrodynamic focusing of the cell suspension during electrical sizing [56]. The system is sensitive enough to detect volume changes of <1%.

ratio was reduced to 4:1 or 1:1. Under these conditions, a decreased protein synthesis rate accompanied tissue swelling. Moreover, the abnormal changes of the retina in anoxia were considerably enhanced if the tissue was exposed to conditioned medium from retinas undergoing a period of anoxia.

MEDIATORS OF SECONDARY BRAIN DAMAGE

Both experiments described above indicate that neural tissue can well survive prolonged episodes of oxygen or substrate deprivation, provided the extracellular space is large enough to dilute substances released from the cells. Therefore, the current interest focuses on the

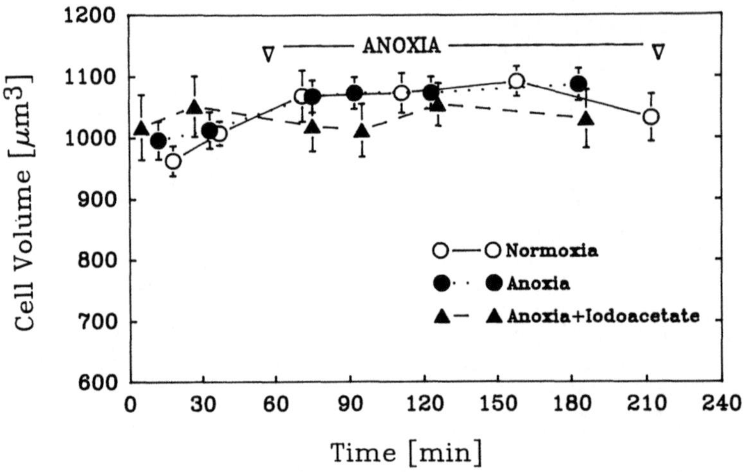

Fig. 2. Effect of anoxia on C6 glial cell volume. C6 glioma cells were exposed to anoxia or anoxia with and without inhibition of anaerobic glycolysis by iodoacetate. Anoxia was induced by supplying nitrogen to the membrane oxygenator of the incubation chamber (Fig. 1), which, in addition, had been placed into a tent filled with N_2. Cell volume is expressed in percent of the control volume obtained during the initial 45-minute control period (±SEM). (Data from Kempski et al. [10].

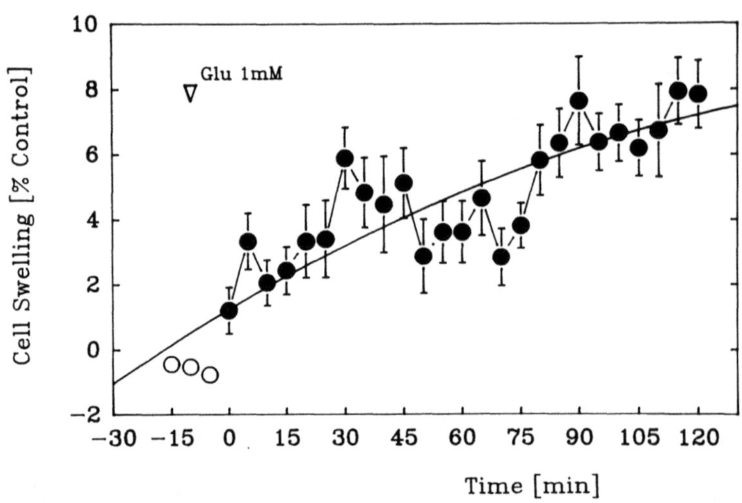

Fig. 3. Effect of glutamate (1 mM) on glial cell volume. To evaluate the effect of the excitatory neurotransmitter glutamate on C6 glial volume, the experiments were carried out in amino acid free media (Dulbecco's salts, glucose and DMEM vitamins) and subsequent addition of glutamate. In control experiments using this medium, a slight but significant decrease of cell volume within 120 minutes was observed. Therefore, data from experiments using glutamate (1 mM) were corrected for this drift (n=6). Cell volume is given in percent of the control volume established during the initial 45-minute control period (±SEM).

role of mediator compounds after ischemic injury [17,18]. Candidates for such a mediator function are excitotoxic amino acids, tissue acidosis, free fatty acids and free radicals.

Glutamate

During the past 5 years several studies have been published which stress the detrimental effect of an abundant release of the excitatory neurotransmitter, glutamate [19-23]. An "excitotoxic" [17] effect of glutamate was first suggested by Van Harreveld [29,30] after Ames and others had reported tissue swelling following exposure to the amino acid [24-27]. Olney examined the effect of oral or subcutaneous glutamate on hypothalamic neurons in newborn mammals, and found a substantial loss of neurons in animals with an incomplete blood-brain barrier [28]. Van Harreveld injected glutamate electrophoretically into the cortex of rats, and observed a halo of swollen cells, predominantly glia and dendrites, together with dying neurons around the injection site [29,30].

Meanwhile, far more is known about the role of glutamate in ischemia. A release of glutamate into the extracellular space during ischemia has been shown by perfusion-dialysis techniques [31,32]. Rothman [33,34] has demonstrated in hippocampal cultures that excitatory transmitters mediate neuronal death, and that damage from anoxia can be prevented if synaptic activity is inhibited.

Glutamate and related substances act on specific receptors located on nerve cells throughout the brain. At least three types of glutamate receptors have been found. The NMDA receptor is currently under investigation regarding its involvement in ischemic and epileptic neuronal damage. Two different mechanisms are considered for the early and delayed functional and morphological alterations: a short-term increase of the Na^+-permeability causing osmotic dendritic swelling, and a prolonged influx of Ca^{2+} ions initiating a final pathway of cell death [23]. First experiments suggest indeed that inhibition of the NMDA receptor can prevent neuronal death after an episode of ischemia [35].

The pathophysiological significance of glial cells in this process is often underestimated. Normally, glial cells accumulate glutamate by active transport. Glial glutamine synthetase catalyzes the formation of glutamine which diffuses back to the neuron as a glutamate precursor [36,37].

The uptake of glutamate into glial cells goes along with a net influx of Na^+ ions [38,39], and, consequently, cell swelling [8,40]. Experimental evidence for this concept was obtained using the in vitro model described previously. The first results are given in Figure 3. The data show that by increasing the ec concentration of glutamate, glial cell volume homeostasis is disturbed (Fig. 3). Concomitantly, the glutamate concentration in the medium was gradually reduced (data not shown), suggesting cellular uptake of the amino acid. Experiments on the kinetics of glutamate removal at different glutamate concentrations are currently in progress. Additional experiments indicate that removal of the extra-intracellular sodium gradient by preincubation of the cells with ouabain can prevent glutamate-induced swelling (manuscript in preparation).

Glutamate-induced glial swelling, therefore, appears to be a side-effect of the homeostatic function of these cells. The removal of excess glutamate from the interstitial space is potentially beneficial and should not be interfered with by therapeutic efforts, which should rather focus on antagonizing the effects of glutamate on neuronal receptors or on the prevention of glutamate release. Nevertheless, glial swelling could become a problem as soon as the increase of tissue water content exhausts the intracranial compliance, thereby raising intracranial pressure to a critical level.

Acidosis

Extra- as well as intracellular acidosis is regularly observed after an interruption of the cerebral energy supply [6]. In an ischemic focus and in the surrounding penumbra zone, lactic acid is produced due to anaerobic metabolism. Actual measurements have revealed pH drops down to 6.0 and below in ischemic brain tissue [6]. Tissue acidosis and lactate accumulation are increasingly considered important causes of brain damage [41,42]. In a previous study [13], the effects of reduced pH on glial cell volume at levels between 7.4 and 6.2 have been investigated following a similar protocol as used in Fig. 4. It was found that glial cell volume remained normal between pH 7.4-7.0. At pH 6.8 and below, cell volume increased by about 10%. There was no dose-dependency, ie, more pronounced swelling at lower pH levels, and no effect of time if an "open" system was used where CO_2 could gradually evaporate through the membrane oxygenator (Fig. 4). In the "closed" system, ie, with pCO_2 maintained elevated at 180 mmHg, glial swelling progressed to 118% of normal after 1 hour incubation at pH 6.2. Experiments using amiloride or sodium-free media suggest an involvement of the Na^+/H^+-antiporter in glial swelling induced by extracellular acidosis [13].

From these results the following concept has emerged. In short, extracellular buffering of H^+ ions by bicarbonate leads to the formation of carbonic acid with subsequent dissociation into CO_2 and water. CO_2 readily enters the intracellular compartment, where carbonic anhydrase catalyzes the generation of carbonic acid which dissociates into bicarbonate and protons. The latter (catalyzed) reaction is faster than the respective extracellular (noncatalyzed) reaction. Therefore, an extra-/intracellular CO_2 gradient is formed, since bicarbonate as well as protons may be eliminated from the cell via specific antiporters. The antiporters participate in the regulation of intracellular pH by transporting Na^+ against H^+, or HCO_3^- against Cl^-, respectively. Therefore, maintenance of a normal ic pH is achieved by influx of osmotically active cations and anions as long as the CO_2 gradient persists. These considerations may explain why swelling of glial cells was more pronounced in the closed system (Fig. 4) where the pCO_2 remained elevated over the whole experimental period.

Taken together, pH-induced glial swelling can be viewed as a result of homeostatic mechanisms regulating ic pH under ischemic conditions. This function may finally become damaging since it accelerates the breakdown of the transcellular Na^+ gradient. The drop of ic pH becomes a threat to glial survival if values below 5.3 are reached [42]. The detrimental effect of extracellular acidosis on intracellular pH can be attenuated if the rise of the pCO_2 from buffering by bicarbonate is inhibited in time. Therefore, treatment of tissue acidosis should be effective in patients with ischemic stroke if buffers like tromethamine [43] are used. Bicarbonate should be avoided, since this would further increase the formation of CO_2.

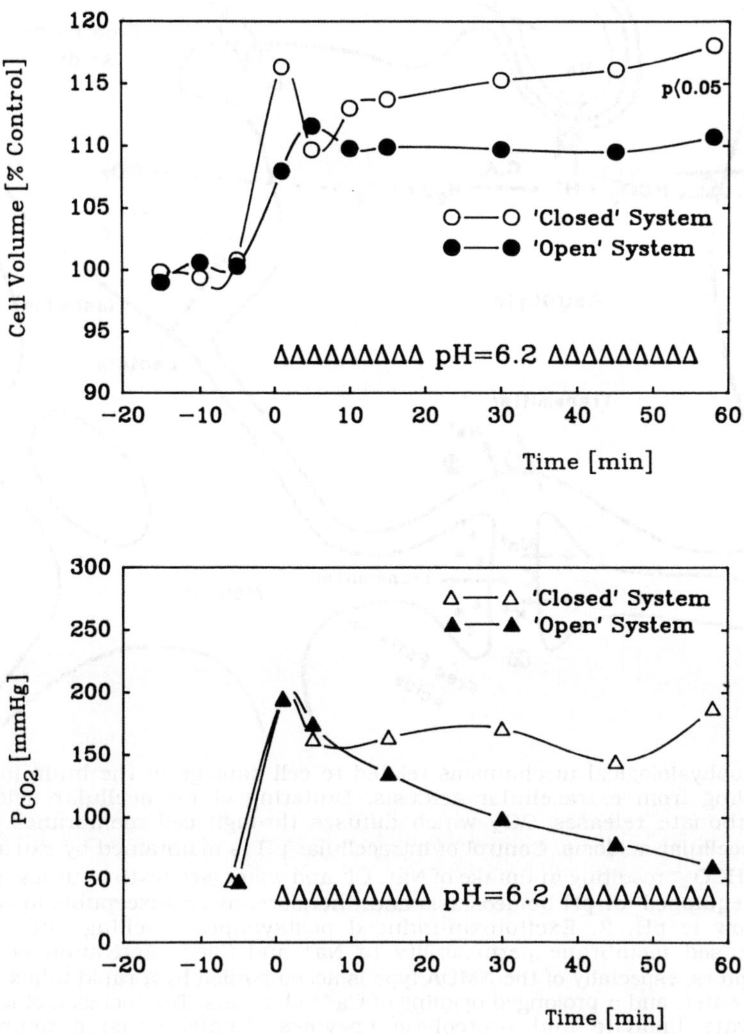

Fig. 4. Glial cell volume and pCO_2 during acidosis. Acidosis (pH=6.2) was induced by titration of the medium with sulphuric acid made isotonic by addition of NaCl. All experiments were carried out using bicarbonate buffered media. In some of the experiments, CO_2 generated from buffering with bicarbonate was eliminated through the membrane oxygenator ("open" system). This made it necessary to readjust pH by repeated injections of sulphuric acid. In additional experiments a closed system was used under maintenance of the pCO_2 at levels found immediately after addition of sulphuric acid. pCO_2 (bottom) was measured by a Radiometer Copenhagen Acid Base Laboratory ABL 300. All experiments were performed after an initial 45-minute control period to obtain normal cell sizes, viability and the osmolality of the medium. Cell volume is expressed in percent of the normal value. When using the "open" system, glial swelling reached a maximum of 10% 5 minutes after induction of acidosis and remained stable for the remaining 60 minutes of the experiment. In the "closed" system cell swelling progressed to a maximum of 18% after 60 minutes.

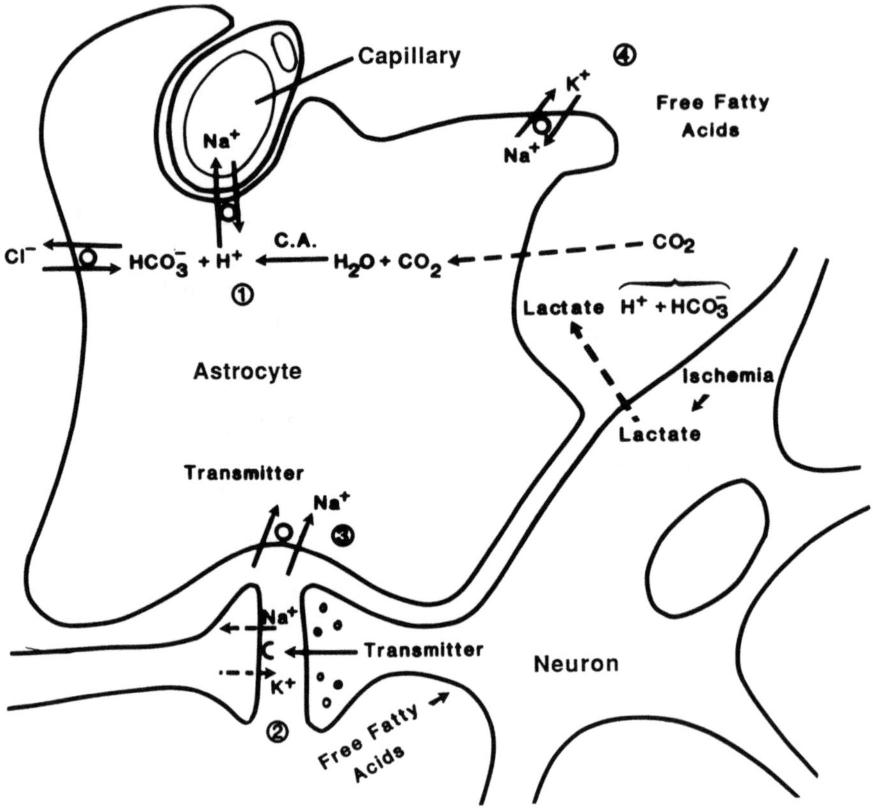

Fig. 5. Pathophysiological mechanisms related to cell damage in the brain [57] 1. Glial swelling from extracellular acidosis. Buffering of extracellular acidosis with bicarbonate releases CO_2 which diffuses through cell membranes leading to intracellular acidosis. Control of intracellular pH is maintained by extrusion of H^+ and HCO_3^- resulting in uptake of Na^+, Cl^- and water (see text). Neurons may be less well equipped for pH homeostasis and, therefore, more susceptible to cell damage by low ic pH. 2. Excitotoxin-induced postsynaptic swelling and damage by increased membrane permeability to Na^+ and Ca^{2+}. Activation of glutamate receptors, especially of the NMDA type is accompanied by a rapid influx of Na^+, Cl^- and water, and a prolonged opening of Ca^{2+} channels. The increase of ic Ca^{2+} may activate lipolytic and proteolytic enzymes, finally causing neuronal self-destruction. 3. Glial swelling accompanying transmitter uptake. Active transport of glutamate into the glial cell must be considered primarily a protective mechanism, preventing neuronal death (see 2). Inactivation of glutamate, however, is energy-dependent, and, in ischemia, may exhaust the limited energy stores. 4. Interaction of free fatty acids with structural or functional components of the cell membrane, eg, inhibition of Na^+/K^+-ATPase or direct membrane lesions.

Free Fatty Acids and Free Radicals

Free fatty acids are another class of putative mediators of secondary brain damage. They are formed early in the course of ischemia [12,44,45]. Fatty acids, especially arachidonic acid, have been described as enhancing cell swelling and eventually causing cell death in brain tissue [46,47]. The toxic mechanisms of arachidonic acid (AA) in brain tissue are not completely understood, although several adverse effects of AA have been described. So, AA uncouples oxydative phosphorylation, inhibits Na^+/K^+-

ATPase, and interferes with the extracellular clearance of glutamate and GABA [46,47]. In addition, AA may directly increase BBB permeability [12,48]. Moreover, AA is substrate for a variety of reactions, causing formation of prostaglandins, leukotrienes and free radicals, all of which are produced in high amounts after recirculation, when (1) AA is mobilized from membrane lipids, and (2) oxygen is again available for the necessary oxygenation reactions. Free radicals are formed by lipid peroxidation promoted in the postischemic brain by iron ion delocalization [49,50]. Xanthine oxidase is another source of reactive oxygen in ischemia [51,52]. Production of eicosanoids as well as free radicals occurs during the recirculation period [12,53,54]. Damaging mechanisms of these substances include vascular injury as well as membrane perturbation [55].

CONCLUSIONS

Evidence is accumulating that ischemic damage in the brain is caused by more than one single mechanism. Reversible phenomena, like some aspects of glial swelling, should be discriminated from neuronal extinction. In Figure 5 some of the known and discussed aspects related to cell damage in cerebral ischemia are summarized: glial swelling in acidosis as the consequence of controlling ic pH, the excitotoxic action of glutamate, and the effect of free fatty acids on the structure and function of cell membranes. A variety of other mechanisms may still as yet be unknown. Of particular interest are the final minutes or seconds of a dying nerve cell, ie, the nature of mechanisms activated upon cell death. Although influx of Ca^{2+} ions is discussed in this context, its molecular interactions are not fully understood. Active measures to interfere with these final pathways of neuronal demise ultimately promise to open a late therapeutic window for the treatment of stroke patients.

REFERENCES

1. K. Frerichs, M. Deckert, O. Kempski, K. Einhäupl and A. Baethmann, Brain function, pathophysiology and heparin therapy in experimental cerebral sinus vein thrombosis, *in*: "Cerebral Sinus Thrombosis: Experimental and Clinical Aspects", K. Einhäupl, O. Kempski and A. Baethmann, eds., Plenum Press, New York and London, pp 53-58 (1990).
2. T. Kirino, Delayed neuronal death in the gerbil hippocampus following ischemia, *Brain Res.*, 239:57-69 (1982).
3. K.-A. Hossmann, Development and resolution of ischemic brain swelling, *in*: "Dynamics of Brain Edema", H. Pappius and W. Feindel, eds., Springer, Berlin, Heidelberg and New York, pp 219-227 (1976).
4. K.-A. Hossmann, S. Sakaki and V. Zimmermann, Cation activities in reversible ischemia of the cat brain, *Stroke*, 8:77-81 (1977).
5. A. J. Hansen, Effect of anoxia on ion distribution in the brain, *Physiol. Res.*, 65:101-148 (1985).
6. R. von Hanwehr, M.-L. Smith, B. K. Siesjö, Extra- and intracellular pH during near-complete forebrain ischemia in the rat, *J. Neurochem.*, 46:331-339 (1986).
7. F. A. X. Schanne, A. B. Kane, E. E. Young and J. L. Farber, Calcium dependence of toxic cell death: a final common pathway, *Science*, 206:700-702 (1979).
8. O. Kempski, U. Gross and A. Baethmann, An in vitro model of cytotoxic brain edema: cell volume and metabolism of cultivated glial and nerve cells, *in*: "Computerized Tomography. Brain Metabolism. Spinal

Injuries", W. Driesen, M. Brock and M. Klinger, eds., Springer, Berlin, Heidelberg and New York, pp 254-258 (1982).
9. O. Kempski, L. Chaussy, U. Gross, M. Zimmer and A. Baethmann, Volume regulation and metabolism of suspended C6 glioma cells: an in vitro model to study cytotoxic brain edema, *Brain Res.*, 279:217-228 (1983).
10. O. Kempski, M. Zimmer, A. Neu, F. von Rosen, M. Jansen and A. Baethmann, Control of glial cell volume in anoxia - in vitro studies on ischemic cell swelling, *Stroke*, 18:623-628 (1987).
11. O. Kempski, D. Shohami, D. von Lubitz, J. M. Hallenbeck and G. Feuerstein, Postischemic production of eicosanoids in gerbil brain, *Stroke*, 18:111-119 (1987).
12. O. Kempski, A. Villacara, M. Spatz, R. F. Dodson, C. Corn, N. Merkel and J. Bembry, Cerebromicrovascular endothelial permeability, *Acta Neuropath.*, 74:329-334 (1987).
13. O. Kempski, F. Staub, M. Jansen, F. Schödel and A. Baethmann, Glial swelling during extracellular acidosis in vitro, *Stroke*, 19:385-392 (1988).
14. P. Benda, J. Lightbody, G. Sato, L. Levine and W. Sweet, Differentiated rat glial cell strain in tissue culture, *Science*, 161:370-371 (1968).
15. A. Ames and F. B. Nesbett, Pathophysiology of ischemic cell death: I. Time of onset of irreversible damage; importance of different components of the ischemic insult, *Stroke*, 14:219-226 (1983).
16. A. Ames and F. B. Nesbett, Pathophysiology of ischemic cell death. III. Role of extracellular factors, *Stroke*, 14:233-240 (1983).
17. A. Baethmann, Pathophysiological and pathochemical aspects of cerebral edema, *Neurosurg. Rev.*, 1:85-100 (1978).
18. A. Baethmann, W. Oettinger, W. Rothenfusser, O. Kempski, A. Unterberg and R. Geiger, Brain edema factors. Current state with particular reference to plasma constituents and glutamate, *in*: "Advances in Neurology, Vol. 28, Brain Edema", J. Cervos-Navarro and R. Ferszt, eds., Raven Press, New York, pp 171-195 (1980).
19. S. B. Prusiner, Disorders of glutamate metabolism and neurological dysfunction, *Ann. Rev. Med.*, 32:521-542 (1981).
20. J. W. Olney, Excitotoxins: an overview, *in*: "Excitotoxins", K. Fuxe, P. Roberts and R. Schwarcz, eds., Plenum Press, New York, pp 82-96 (1983).
21. S. M. Rothman and J. W. Olney, Glutamate and the pathophysiology of hypoxic-ischemic brain pathway, *Ann. Neurol.*, 19:105-111 (1986).
22. R. C. Collins, Selective vulnerability of brain: new insights from the excitatory synapse, *Metab. Brain Dis.*, 1:231-240 (1986).
23. M. L. Mayer and G. L. Westbrook, Cellular mechanisms underlying excitotoxicity, *Trends Neurosci.*, 10:59-61 (1987).
24. A. Ames, Studies on water and electrolytes in nervous tissue. II. Effect of glutamate and glutamine, *J. Neurophysiol.*, 19:213-223 (1956).
25. A. Ames, Y. Tsukada and B. Nesbett, Intracellular Cl^-, Na^+, K^+, Ca^{2+}, Mg^{2+} and P in nervous tissue, response to glutamate and to changes in extracellular Ca^{2+}, *J. Neurochem.*, 14:145-159 (1967).
26. J. A. Harvey and H. McIlwain, Excitatory amino acids and cation content and Na^+-ion flux of isolated tissues from the brain, *Biochem. J.*, 108-269-274 (1968).
27. H. Lund-Andersen and L. Hertz, Effect of potassium and of glutamate on swelling and on sodium and potassium content in brain cortex slices from adult rats, *Exp. Brain Res.*, 11:199-212 (1970).
28. J. W.Olney, Brain lesions, obesity and other disturbances in mice treated with monosodium glutamate, *Science*, 164:719-721 (1969).
29. A. Van Harreveld and E. Fifkova, Light- and electron microscopic changes in central nervous tissue after electrophoretic injection of glutamate, *Exp. Mol. Pathol.*, 15:61-81 (1971).
30. A. Van Harreveld and E. Fifkova, Effects of glutamate and other amino acids on the retina, *J. Neurochem.*, 18:2145-2154 (1971).

31. H. Benveniste, J. Drejer, A. Schousboe and N. H. Diemer, Elevation of the extracellular concentrations of glutamate and aspartate in rat hippocampus during transient cerebral ischemia monitored by intracerebral microdialysis, *J. Neurochem.*, 43:1369-1374 (1984).
32. J. Drejer, H. Benveniste, N. H. Diemer and A. Schousboe, Cellular origin of ischemia-induced glutamate release of brain tissue in vivo and in vitro, *J. Neurochem.*, 45:145-151 (1985).
33. S. M. Rothman, Synaptic activity mediates death of hypoxic neurons, *Science*, 220:536-537 (1983).
34. S. Rothman, Synaptic release of excitatory amino acid neurotransmitter mediates anoxic neuronal death, *J. Neurosci.*, 4:1884-1891 (1984).
35. R. P. Simon, J. H. Swan, T. Griffiths and B. S. Meldrum, Blockade of N-methyl-D-aspartate receptors may protect against ischemic damage in the brain, *Science*, 226:850-852 (1984).
36. S. Berl, W. J. Nicklas and D. D. Clarke, Compartmentation of glutamic acid metabolism in brain slices, *J. Neurochem.*, 15:131-140 (1968).
37. A. M. Benjamin and J. H. Quastel, Fate of glutamate in the brain, *J. Neurochem.*, 23:457-464 (1974).
38. F. A. Henn, M. N. Goldstein and A. Hamburger, Uptake of the neurotransmitter candidate glutamate by glia, *Nature*, 249:663-664 (1974).
39. A. Schousboe, G. Svenneby and L. Hertz, Uptake and metabolism of glutamate in astrocytes cultured from dissociated mouse brain hemispheres, *J. Neurochem.*, 29:999-1005 (1977).
40. P. H. Chan, R. A. Fishman, J. L. Lee and L. Candelise, Effects of excitatory neurotransmitter amino acids on swelling of rat brain cortical slices, *J. Neurochem.*, 33:1309-1315 (1979).
41. B. K. Siesjö, Cell damage in the brain: a speculative synthesis, *J. Cereb. Blood. Flow Metab.*, 1:155-185 (1981).
42. R. Kraig, C. K. Petito, F. Plum and W. A. Pulsinelli, Hydrogen ions kill brain at concentrations reached in ischemia, *J. Cereb. Blood Flow Metab.*, 7:379-386 (1987).
43. M. J. Rosner and D. P. Becker, Experimental brain injury: successful therapy with the weak base, tromethamine. With an overview of CNS acidosis, *J. Neurosurg.*, 60:961-971 (1984).
44. N. G. Bazan, Effects of ischemia and electroconvulsive shock on free fatty acid pool in the brain, *Biochem. Biophys. Acta*, 218:1-10 (1970).
45. S. Yoshida, S. Inoh, T. Asano, K. Sano, M. Kubata, H. Shimazaki and N. Ueta, Effect of transient ischemia on free fatty acids and phospholipids in the gerbil brain, *J. Neurosurg.*, 53:323-331 (1980).
46. P. H. Chan, R. A. Fishman, J. L. Lee and S. C. Quan, Arachidonic acid-induced swelling in incubated rat brain cortical slices, *Neurochem. Res.*, 5:629-640 (1980).
47. P. H. Chan and R. A. Fishman, Alterations of membrane integrity and cellular constituents by arachidonic acid in neuroblastoma and glioma cells, *Brain Res.*, 248:151-157 (1982).
48. A. Unterberg, M. Wahl, F. Hammersen and A. Baethmann, Permeability and vasomotor response of cerebral vessels during exposure to arachidonic acid, *Acta Neuropath.*, 73:209-219 (1987).
49. N. R. Nayini, B. C. White, S. D. Aust, R. R. Huang, R. J. Indrieri, A. T. Evans, H. Bialek, W. A. Jacobs and J. Komara, Post resuscitation iron delocalization and malondialdehyde production in the brain following prolonged cardiac arrest, *J. Free Radicals Biol. Med.*, 1:111-116 (1985).
50. B. C. White, G. S. Krause, S. D. Aust and G. E. Eyster, Post-ischemic tissue injury by iron-mediated free radical lipid peroxidation, *Ann. Emerg. Med.*, 14:804-809 (1985).
51. D. A. Parks and D. N. Granger, Ischemia-induced vascular changes: role of xanthine oxidase and hydroxyl radicals, *Am. J. Physiol.*, 245:G285-G289 (1983).

52. D. E. Chambers, D. E. Parks, G. Patterson, R. Roy, J. M. McCord, M. Yoshida, L. F. Parmley and J. M. Downey, Xanthine oxidase as a source of free radical damage in myocardial ischemia, *J. Mol. Cell. Cardiol.*, 17:145-152 (1985).
53. R. J. Gaudet and L. Levine, Transient cerebral ischemia and brain prostaglandins, *Biochem. Biophys. Res. Commun.*, 86:893-901 (1979).
54. H. S. Mickel, Y. N. Vaishnav, O. Kempski, D. von Lubitz, J. Weiss and G. Feuerstein, Breathing 100% oxygen after global brain ischemia in Mongolian gerbils results in increased lipid peroxidation and increased mortality, *Stroke*, 18:426-430 (1987).
55. B. K. Siesjö, Acid-base homeostasis in the brain: physiology, chemistry and neurochemical pathology, *in*: "Molecular Mechanisms of Ischemic Brain Damage", K. Kogure, K. A. Hossmann, B. K. Siesjö and F. A. Welsh, eds., Elsevier, Amsterdam, pp 121-154 (1981).
56. V. Kachel, Basic principles of electrical sizing of cells and particles and their realization in the new instrument "Metrizell", *J. Histochem. Cytochem.*, 24:211-230 (1976).
57. O. Kempski, Cell swelling mechanism in brain, *in*: "Mechanisms of Secondary Brain Damage", A. Baethmann, G. Go and A. Unterberg, eds., Plenum Press, New York and London, pp 203-220 (1986).

CELLULAR MECHANISMS INVOLVED IN GENERATION OF BRAIN EDEMA

U. Heinemann and J. Sarvey

Institute for Normal and Pathological
Physiology, University of Köln
Robert-Koch-Straße, 5000 Köln 41, FRG

INTRODUCTION

Brain edema is a general complication of many central nervous system disorders of infectious, vascular, tumor, hypoxic or epileptic origin. Two major causes determine the pathophysiology of brain edema. These can be distinguished by changes in the size of the extracellular space (ES). The first results from breakdown of the blood-brain barrier with transfer of proteins from blood vessels into the ES [5] with a consequent increase in the size of the ES, in brain volume and in intracranial pressure. Interestingly, such edema is often more marked in white than in grey matter, which may point to active volume regulatory properties in astrocyte and nerve cell rich grey matter.

The other major cause of edema is considered to be cytotoxic in origin, with a primary increase in intracellular osmolarity due to anaerobic accumulation of lactate, phosphates, etc, and a consequent watershift from the ES to the IS [35,36].

The end result of both processes is the same, ie, an increase in intracranial pressure with consequent disorders in neuronal functioning and eventually coma. This scheme appears to permit a simple differentiation of a principal manifestation in brain edema, that is the size of the ES. However, the discrimination is in many respects probably too simplistic. This may explain why popular therapeutic approaches often fail. One reason for a sometimes inadequate therapy could be that each form of edema can to some extent develop into the other. For example, local vasogenic edema ultimately impairs local blood flow with the consequence of local hypoxia and subsequent cytotoxic edema. On the other hand, maintenance of the integrity of the blood-brain barrier is an energy-requiring process, and local hypoxia or tissue pH as well as ionic changes may disrupt the blood-brain barrier. Consequently, a vasogenic component may appear in cytotoxic edema. In this review an attempt is made to summarize some general principles of central nervous system electrolyte homeostasis and some of the mechanisms involved in cytotoxic brain edema.

Cerebral Sinus Thrombosis, Edited by K. Einhäupl *et al.*
Plenum Press, New York, 1990

BASELINE LEVELS OF ELECTROLYTES IN THE CEREBRAL SPINAL FLUID AND THE EXTRACELLULAR SPACE AND PHYSIOLOGICAL VARIATIONS IN CSF ELECTROLYTE CONSTITUTION

The extracellular fluid (ECF) and the cerebral spinal fluid (CSF) appear to be rather similar in constitution but differ from plasma. This applies particularly to K^+, Ca^{2+}, and Mg^{2+}. For example, the K^+ concentration $[K^+]_o$ is about 4.2 mM in plasma but not more than 3 mM in CSF.

There is some dispute over the actual size of the ES. Electron microscopic investigations with freeze substitution agree today that the width of the ES ranges from 20 to 50 nm [67]. Altogether the volume fraction of the ES in the CNS is, due to a large tortuosity, near 20%. This has been determined by impedance measurements with low frequencies, where capacity loading of cell membranes can be neglected [58,68], as well as by various clearance methods, provided a correction for cellular uptake is made. Measurements with local probes of the ES have shown more recently that the extracellular volume fraction is near 16%, and that the tortuosity factor (a measure of the prolongation of the diffusion path for cell impermeable substances) is about 1.5 [53]. This implies that almost all molecules, including divalent cations, can freely diffuse within the ES. From this calculated size of the ES it was estimated that about 5% of extracellularly applied currents pass through cells, with a preference for low resistance glial cells. In general, most parts of the CNS show little systematic variation in baseline size of the ES. However, in the pyramidal cell layer of the hippocampus the ES is particularly small with a volume fraction of 5-7% [32]. This small volume fraction, together with the small reserves in blood supply for the hippocampus, may be a factor which makes this area particularly vulnerable in hypoxic situations.

Neither the ES nor the electrolyte constitution is constant over time. Small variations in $[K^+]_o$ occur, for example, during spindle activity in the EEG, which is associated with increase in $[K^+]_o$ by 0.2 mM or more [30]. Slightly larger variations in $[K^+]_o$ are noted in sensory systems during physiological stimuli [27,65,63]. In primary relay areas, such as the dorsal horn of the spinal cord, the cuneate nucleus, the retina, etc, stimuli can increase $[K^+]_o$ considerably, sometimes even by more than 1 mM. These changes are associated with no or only minor changes in the ES. The situation becomes different, however, if the thalamic relay nucleus or the cortical surface itself is stimulated (Fig. 1). Then cortical and thalamic $[K^+]_o$ can increase to 10 or even 12 mM [46,63,64]. Similar alterations are also observed during seizure activity. Comparative analysis in many cortical areas and different species has shown that this ceiling level is not easily surpassed. Following the end of stimulation or ictal activity $[K^+]_o$ decreases below baseline activity and recovers only slowly to resting values. The limitation of rises in $[K^+]_o$ as well as the recovery is largely determined by the Na-K-ATPase activated both by intracellular Na^+ and extracellular K^+ accumulation. The undershoots in $[K^+]_o$ result from the fact that some K is not available for the exchange pump due to uptake of K^+ into glial cells and spatial K^+ redistribution by diffusion and glial buffering (see p 109). Consequently the amplitude and duration of undershoots in $[K^+]_o$ correlates with the amplitude of the preceding rises.

Stimulus and seizure activity is generally associated with other ionic shifts. Typical observations are illustrated in Figures 1 and 3. Parallel to rises in $[K^+]_o$, changes in $[Na^+]_o$, $[Ca^{2+}]_o$ and $[Cl^-]_o$ are observed. In this particular example from cat neocortex, which shows recordings from

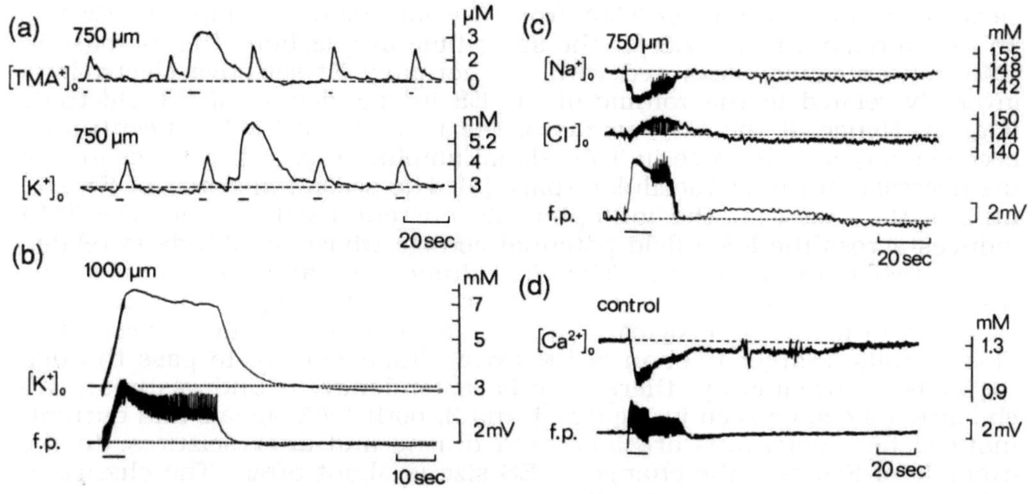

Fig. 1. Stimulus-induced changes in ionic concentration and extracellular space in cat cortex. (a) and (b), recordings with potassium-sensitive electrodes. FP is field potential. (a) Upper record shows pulses of iontophoretic application of tetramethylammonium. Calibration on the right. The iontophoretic increase in TMA concentration is enhanced following a stimulus-induced rise in extracellular potassium concentration. The increase in the TMA signal indicates a shrinkage of the extracellular space. Lower record: iontophoretically-induced rises in potassium concentration are decreased following a rise in extracellular potassium concentration, indicating enhanced cellular uptake of potassium. Note that rises in extracellular potassium concentration were induced by repetitive cortical surface stimulation (20 per second, 0.05 millisecond pulses, duration of stimulus train as indicated by bar). (b) Shows again a stimulus induced rise in extracellular potassium concentration followed by self-sustained after discharges. Cortical surface stimulation, 20 per second, 0.05 millisecond pulses, number refers to recording depth below cortical surface. (c) Simultaneous recording of sodium and chloride concentration as well as of field potential at a recording depth of 750 µm below the cortical surface. Stimulus parameters similar as in (a). (d) Stimulus-induced decrease in extracellular calcium concentration induced by a similar stimulus. Recording depth approximately 300 µM below cortical surface.

medial cortical layers, rises in $[K^+]_o$ are associated with about 10 mM decreases in $[Na^+]_o$ which turn into rises above normal towards the end of the stimulus train. Similar kinetics are seen for changes in $[Ca^{2+}]_o$ (see also Refs 21,24,27-29,31,50,64). Both during stimulation and during seizures $[Cl^-]_o$ concentration rises as well. The increase in the sum of $[Na^+]_o$, $[K^+]_o$ and $[Cl^-]_o$ at the end of seizures points to an increase in overall tissue osmolarity, which probably reflects an increase in intracellular osmolarity as well. Hence, this is a paradigm of cytotoxic edema which results in a decrease of the ES.

Indeed, changes in ES size appear to accompany both stimulus and seizure activity. In the examples of Figures 1 and 3, this was determined in two ways (see Fig. 2). An iontophoresis electrode was glued in close approximation to a normally K^+-sensitive microelectrode. The iontrophoresis electrode was filled with 100 mM tetramethylammonium (TMA)-Cl which could be ejected by passing positive current through the iontophoresis barrel. The resulting increase in $[TMA]_o$ could be measured with the nominally K^+-selective electrode filled with Corning 477317 resin. This is actually more sensitive to TMA and related substances than it is to K^+. Then constant current pulses at regular intervals were used to eject constant amounts of TMA from the iontophoresis pipette. TMA is known to remain in the ES for the short time in question. The increase in TMA concentration measured with the normally K^+-selective electrode is inversely related to the volume of the ES in the vicinity of the electrode (Fig. 2). Hence, if the ES decreases, the increase in TMA concentration becomes larger. This rise in TMA signal amplitude is a direct measure of the decrease in the extracellular space [9]. The second measure of ES size utilizes the fact that the iontophoresis current used to eject the TMA induces across the ES a field potential change whose amplitude is related to the resistivity of the ES. The ES volume ultimately determines this resistance, as long as the cell membrane resistance is not too low. However, in many pathological states the membrane resistance decreases. This permits a larger fraction of the extracellular current to pass through the cells. Consequently, there may be a tendency to underestimate ES changes. As can be seen in Figures 1 and 3, both TMA signals and current-induced field potentials are increased during and after seizures. In the example of Figure 1 the change in ES size is about 50%. The change in field potentials indicates a smaller change.

The most interesting point of this example is that the ES size change is much larger than that expected from the increase in extracellular osmolarity, which indicates an ES decrease of about 10%. Hence, the increase in tissue osmolarity alone cannot account for the seizure-related changes in the ES size. Thus the question arises as to which other mechanisms are involved in the generation of this form of cytotoxic edema. Morphological analysis of tissue during hypoxia or seizures shows that glial cells, dendritic spines and often presynaptic endings are swollen [55]. Nerve cell bodies can show decreases and/or increases in size. The swelling of dendritic spines is particularly interesting, since they are often the end point of excitatory synaptic connections. The fact that glial cells are swollen points to an important contribution of glial cells in the decrease of the ES.

MECHANISMS INVOLVED IN GLIAL CELL SWELLING

At the moment a number of mechanisms are discussed. A very important mechanism is probably KCl uptake into these cells [4]. The precise mechanism of this uptake is presently not known. In addition there may be uptake of glutamate, aspartate, GABA, glucose and metabolites into glial cells. Usually this uptake is Na^+-dependent and often associated with a sodium cotransport [38]. In such cases the uptake process will lead to an increase in intracellular osmolarity. The known affinity constants and kinetic properties suggest that such a mechanism of glial cell swelling is slow. Moreover, the transported quantities may be small. Consequently, it may be assumed that uptake processes contribute little to the initial glial cell swelling. Some attention has been paid to a Na^+-H^+/bicarbonate-Cl counter transport which together with the Na^+-K^+-ATPase activity [19,71] could well account for the net KCl uptake into glial cells [60]. Again, this process appears to be rather slow and may not account in total for the glial cell swelling. A further mechanism of glial cell swelling is related to glial cell function in K^+ regulation.

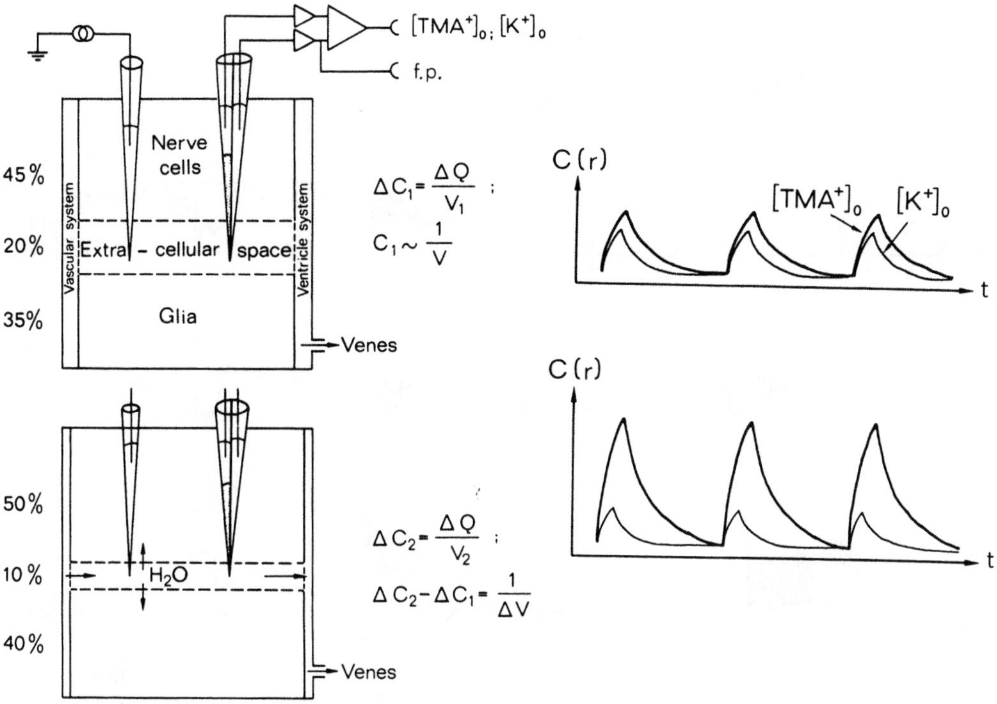

Fig. 2. Experimental design showing the methods used to evaluate changes in extracellular space size. TMA-containing micropipettes were glued in close proximity to a nominally potassium-selective electrode prepared with the Corning 477313 ion exchange resin. A constant amount of tetramethylammonium is injected into the extracellular space. A nearby ion-sensitive electrode senses the rise in extracellular TMA concentration. When water moves into nerve and glial cells and the extracellular space shrinks, the same current induces a larger change in extracellular TMA concentration.

SPATIAL K^+ BUFFERING AS A CAUSE OF CHANGES IN GLIAL SIZE

One of the most important and best understood functions of glial cells within the nervous tissue is their role in spatial redistribution of K^+ by which they limit local rises in $[K^+]_o$ as long as elevations in $[K^+]_o$ are spatially inhomogeneous [16-18,69]. This is usually the case during seizures [9,11,12,29,30] and initially during hypoxia [22]. Spatial buffering utilizes the fact that the transmembrane potential of glial cells is almost exclusively determined by the transmembrane K^+ concentration gradient [56] and that glial cells possess long processes and are electrically coupled amongst each other [20,62,63]. Consequently, if K^+ ions accumulate locally in the ES, glial cells respond with a depolarization which spreads along the processes of astro- and oligodendrocytes and also into neighboring cells. This charge transfer from one cell to the other results in a somewhat smaller depolarization than that expected from the local

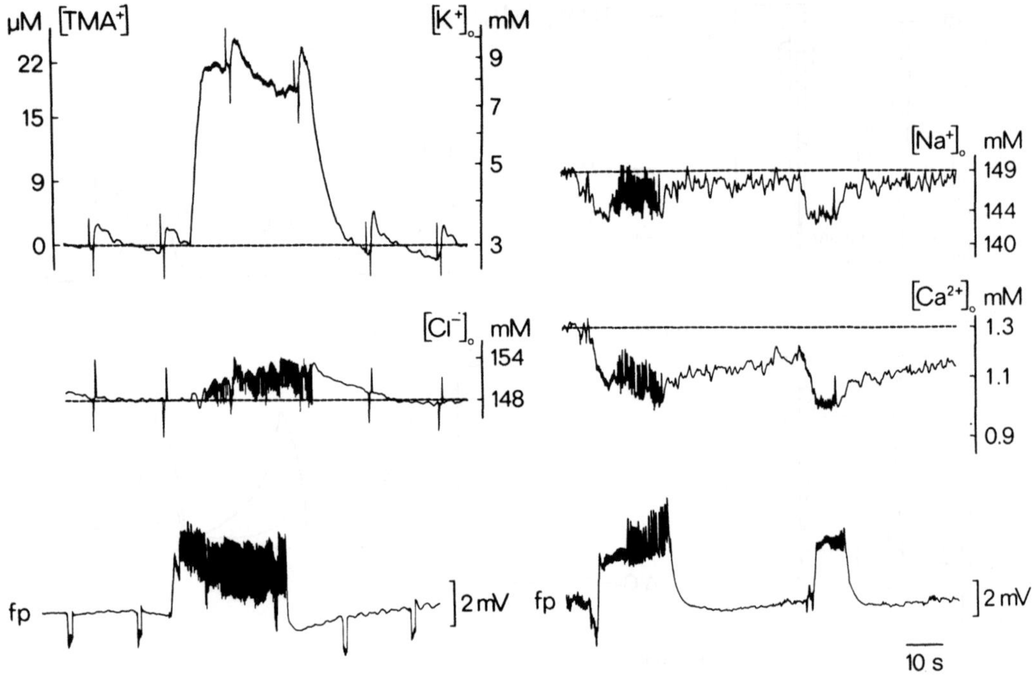

Fig. 3. Ionic changes and extracellular space size changes during pentylenetetrazole-induced seizures. Simultaneous record of potassium and chloride concentration as well as iontophoretically induced TMA signals. Note that the iontophoretically induced TMA signals are enhanced during and following the seizure. On the field potential the iontophoretic current applications induce positive going field potential shifts. Due to the increase in extracellular resistivity these field potential transients increase. On the right a simultaneous record of sodium and calcium concentration changes together with the field potential.

extracellular K^+ accumulation. This implies that the Nernst equilibrium is disturbed. Therefore, K^+ flows from the ES into glial cells and the extracellular potential becomes negative at the site of maximal K^+ accumulation. In remote areas, where the increase in $[K^+]_o$ is smaller, glial cells are more depolarized than expected from the local K^+ accumulation [15]. This again leads to a disturbance in K^+ equilibrium with the result that K^+ flows out of glial cells and positive field potentials are generated. As a net effect potassium is rapidly redistributed within the tissue.

It has been recognized only recently [9-12] that spatial buffering leads to changes in the size of the ES. The flux of K^+ into and out of glia represents an ionic current which is extracellularly compensated by a current to the site of maximal K^+ accumulation (eg, Fig. 4). This extracellular current is carried by Na^+ and Cl^- ions which predominate in the ES. Consequently, Na^+ is brought to the place where extracellular K^+ accumulation and Na^+ uptake into nerve cells is greatest. However, about half of the current is carried by Cl^- which moves away from the site of K^+ accumulation. The depletion of Cl^- and K^+ is not equilibrated by the Na^+ flux to this site and the decreased extracellular osmolarity causes water movement from the ES into glial cells. Consequently, glial cells

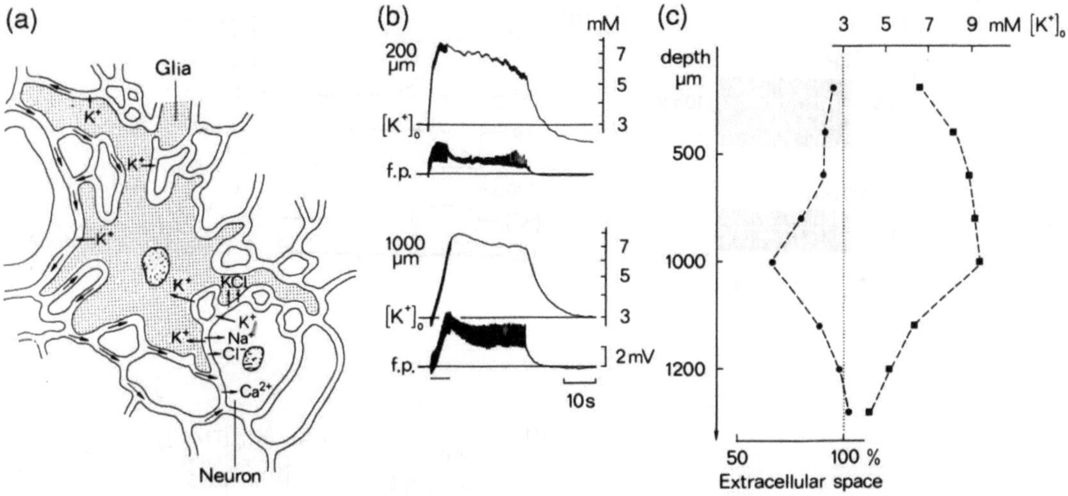

Fig. 4. (a) Diagram of ion movements between nerve cells, interstitium and glial cells. (b) Stimulus-induced changes in extracellular field potential and simultaneous measured rises in extracellular potassium concentration. Note that the rise at a recording depth of 1000 µm below cortical surface is larger than the one at 200 µm below cortical surface. The rise in potassium was induced by repetitive cortical surface stimulation (20 per second, 0.05 millisecond pulses, stimulus train duration approximately 10 seconds). (c) Laminar profile of average changes in extracellular potassium concentration during cortical surface stimulation and laminar profile of extracellular space size changes. Note that at depths below 1200 µm the extracellular space can become wider, although potassium is still elevated.

swell at sites of maximal K^+ accumulation while the size of the ES decreases. In remote areas, the opposite happens. Glial cells decrease in size and the ES increases. Such mechanisms, of course, will have effects also on nearby nerve cells. At the site where the water content increases, water moves not only into glial cells but also into nerve cells causing cell swelling. In remote areas where the extracellular space becomes wider, water moves out of nerve cells and, hence, nerve cells shrink there. This could well explain different observations on nerve cell size during various pathophysiological states.

EVIDENCE FOR SPATIAL K^+ BUFFERING DURING ELECTRICAL STIMULATION AND DURING SEIZURE GENERATION IN VARIOUS AREAS OF THE CENTRAL NERVOUS SYSTEM

When stimulus-induced or seizure-related ionic shifts are studied in relation to the changes in ES size, it is seen that the ES decreases only at the site where K^+ accumulation is maximal. In remote areas the ES size changes are smaller, or the ES can even increase. This has been noted during seizure activity in cat sensorimotor cortex as well as during thalamus and cortical surface repetitive stimulation [10,12]. An example of this is shown in Figure 4. Similar observations have also been made in hippocampal slices. There both ortho- and antidromic stimulation leads

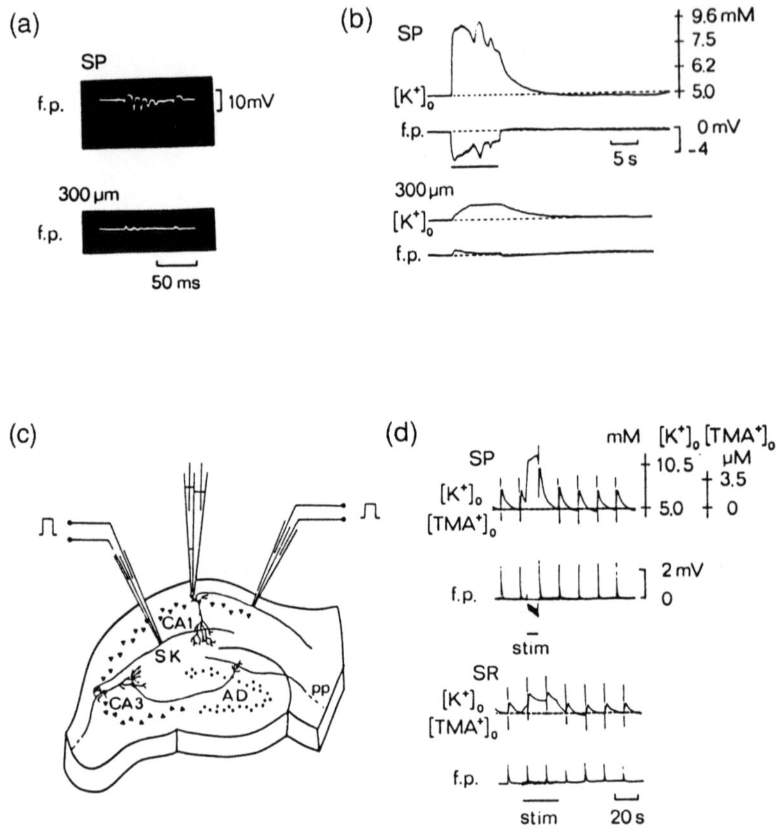

Fig. 5. Changes of field potentials and extracellular potassium concentration during antidromic stimulation of the CA1 hippocampal pyramidal cells in the presence of low calcium medium in order to suppress synaptic transmission. (a) Shows field potential recordings in stratum pyramidale and 300 μm away in stratum radiatum/moleculare. Note, the repetitive discharge evoked by the first stimulus of train. (b) Changes in extracellular potassium concentration and field potentials in stratum pyramidale and in stratum radiatum/moleculare evoked by antidromic stimulation. Train duration 10 seconds, 20 per second, 0.05 millisecond pulses. (c) Shows the experimental paradigm. (d) Iontophoretically induced TMA signals and field potential changes as well as rises in potassium concentration, recorded in stratum pyramidale and in stratum radiatum, under conditions of blocked synaptic transmission. Note the decrease in extracellular space size in stratum pyramidale and a widening of extracellular space size in stratum radiatum evoked by antidromic stimulation.

to a maximal K^+ accumulation in the pyramidal cell layer of hippocampal area CA1 [3]. In this layer the K^+ accumulations are associated with decreases in ES size [32]. When ES size changes are monitored in the synaptic input layer the ES size change is much smaller, and with antidromic stimulation often a widening of the ES is noted. This persists when synaptic transmission is blocked by lowering extracellular Ca^{2+} concentration (Albrecht et al., manuscript in preparation). Under this condition antidromic stimulation still leads to maximal K^+ concentration increases in the pyramidal cell layer. Some K^+ accumulation is also seen in the stratum radiatum and stratum moleculare. When one proceeds with the ion selective electrodes into the neighboring part of the dentate gyrus, even there some K^+ accumulation is noted. While the K^+ accumulation is associated with negative field potential changes in the hippocampal pyramidal cell layer, it is associated with positive field potential changes in stratum radiatum, moleculare and in the dentate gyrus. In these areas the positive field potential changes are also associated with a widening of the ES (Fig. 5).

A similar effect has been noted when recordings were made at the border of an alumina-cream-induced epileptogenic focus in cat sensorimotor cortex. The area consists of gliotic tissue [26]. In the border between gliotic and normal tissue, degenerative and regenerative alterations of nerve cells are noted. In this border area, spontaneous epileptogenesis may occur. If one stimulates such potentially epileptogenic areas, one observes rather marked increases in K^+ concentration. These are associated with large decreases in ES. When the electrode is moved to a site in the gliotic area where no nerve cells exist, increases of the ES size are associated with the local K^+ accumulation there. Since neuronal activity cannot account for the K^+ accumulation inside the gliotic tissue, it is rather likely that the glial cells mediate the K^+ accumulation. This also explains the widening of the ES in these areas.

At first sight, of course, these spatial redistribution mechanisms are not producing an edema-like situation because this is only a redistribution process, and the amount of cell swelling occurring at one site is compensated by a widening of the ES and a decrease in cell size at others. However, there are some reasons why this must not necessarily hold true. For example, when the blood-brain barrier breaks down, at every capillary interface there is an elevated K^+ level, which rapidly leads to K^+ redistribution into the tissue. The KCl uptake mechanism will then additionally lead to cell swelling. Similarly, when there is a local area of decrease in ES size with an increase in tissue osmolarity, then there should be water redistribution from the blood vessels into the ES and into nerve and glial cells as well. This local swelling may well impair local blood supply and cause local hypoxia, with the result that the situation is aggravated. Such processes could be active during status epilepticus.

EXCITATORY AMINO ACIDS AND CHANGES IN THE EXTRACELLULAR IONIC MICROENVIRONMENT

Nerve cell swelling during seizures applies particularly to dendrites and there particularly to dendritic spines [55]. Spines are frequently the contact zone for excitatory synapses. This finding points to an involvement of excitatory neurotransmitters in cytotoxic edema. This led us to study the changes in ionic composition and in ES size induced by excitatory amino acids [25,33,43,57]. The most common excitatory transmitters in the central nervous system are probably glutamate and aspartate. These neurotransmitters bind theoretically to three different types of amino acid transmitter receptors [72]. Based on potent agonists, these can be divided into receptors activated by kainate (kainate receptors), quisqualate (quisqualate receptors), or N-methyl-D-aspartate (NMDA receptors). In addition, there are binding sites for cellular uptake. Specific antagonists are available for NMDA receptors. 2-Aminophosphonovalerate (2APV), a competitive antagonist, and ketamine and related non-competitive antagonists are examples [8,72,66].

The NMDA receptor has a few peculiar properties which are summarized briefly (see Fig. 6). The ionophore operated by the NMDA receptor has a relatively wide diameter and permits passage of Na^+ and K^+ ions as well as Ca^{2+} and Mg^{2+}. However, the diameter of the pore is believed to be regulated by the transmembrane voltage. This means that the diameter is small at least at some point of the pore when the nerve cell membrane is near its resting potential. Consequently Mg^{2+} and Ca^{2+} ions tend to block the channel [1]. However, it becomes wider when the membrane is depolarized. This gives NMDA receptors an amplifying capability similar to that of Na^+ channels involved in the generation of action potentials [47,54].

Hence NMDA receptors permit not only the passage of large ions, such as Ca^{2+} and Mg^{2+}, but also are capable of amplifying EPSPs into giant excitatory synaptic potentials, so-called paroxysmal depolarization shifts which are the landmark of epileptogenesis [2]. Indeed, NMDA receptor antagonists are good anticonvulsants [6,13,18,40,49,51,70]. The amplifying potential of NMDA receptors plays also an important role in the generation of other pathological states (see below). The Ca^{2+} permeability of the NMDA-operated ionophores [43,48] may explain why NMDA receptor antagonists seem to have a protective action against nerve cell degeneration which is in many cases believed to be Ca^{2+}-dependent [14,59,60]. Quisqualate and kainate receptor ionophores, in common with NMDA receptor ionophores, permit the passage of Na^+ and K^+. However, they do not have voltage-dependent properties. Specific antagonists for quisqualate and kainate are presently not available. When glutamate, aspartate, quisqualate, kainate, or NMDA is iontophoretically or directly applied to slice preparations or iontophoretically applied in intact preparations, large changes in the electrolyte composition are observed (see Fig. 7). Na^+ can decrease by more than 50 mM, Cl^- by more than 35 mM, K^+ can easily increase by 15 mM or more. These large ionic shifts are associated with decreases in the ES size by 50% and greater. Such a large decrease in ES size should increase the Na^+ and Cl^- concentration by a factor of two. The observed decreases in Na^+ and Cl^- concentration therefore point to a considerable intracellular uptake of Na^+ and Cl^- which certainly is involved in the generation of dendritic and dendritic spine swelling.

An interesting observation is that the ES decreases transiently by 80% and more while there appears to be a large and unbalanced drop in extracellular solute concentration. We explain this finding with a possible restriction of shrinkage due to mechanical properties of the ES. This is also suggested by the fact that ES decreases both during stimulation and excitatory amino acid applications are smaller in the hippocampal pyramidal cell layer where the ES is physiologically small. An alternative possibility, however, is that other extracellular solutes increase in concentration. This would most noticeably be bicarbonate. However, in view of the almost balanced relationship between cations and anions this is not too likely (eg, also Ref. 22). The large decrease in ES size is usually not reflected by measurements of the extracellular resistance. This is explained by the fact that the membrane resistance of nerve cells decreases to a fraction of its original size and becomes unmeasureably small [43]. As a result a larger fraction (normally about 5%) of extracellularly applied current passes through nerve cells and leads to an underestimate of ES size change. During perfusion of excitatory amino

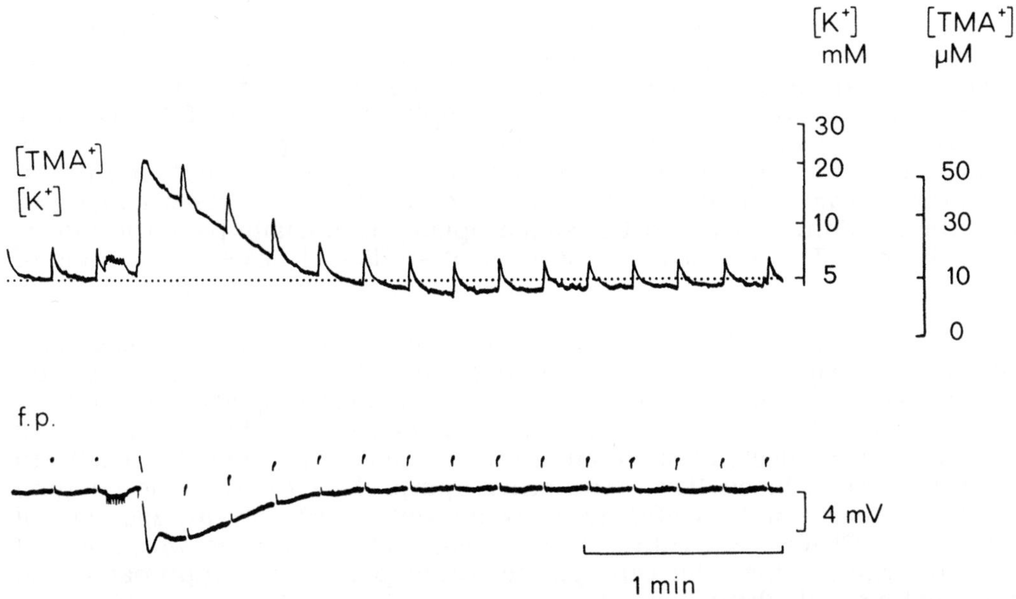

Fig. 8. Measurement of changes in extracellular potassium concentration and associated field potential as well as change in the size of the ES as indicated by the TMA method. Note that the sensitivity of the electrode towards TMA decreases largely as the extracellular potassium concentration increases. The SD developed after a spontaneous seizure induced by bicuculline (50 μM) applied to a neocortical slice from a young rat (age about 20 days).

a large increase in extracellular K^+. This produces an increase in nerve cell firing with greatly enhanced transmitter release. Consequently, extracellular Na^+, Cl^- and Ca^{2+} concentrations decrease. The ES decreases usually by more than 50% (Fig. 8). Such decreases in ES size are a consequence of glial buffer mechanisms and of NaCl uptake into nerve cells as well as of KCl uptake into glial cells. Again there is a strong net decrease in extracellular Na^+ and Cl^-, and it is not clear whether it is balanced by a bicarbonate increase [22,37,42,52].

The process is reversible within some minutes provided pump mechanisms are still intact. However, during hypoxia, hypoglycemia and other disorders of oxygen supply this is not the case [22,23]. Consequently, the ionic shifts do not return to baseline. Moreover, due to the large accumulation of Ca^{2+} inside cells, it must be assumed that mitochrondria become overburdened with Ca^{2+}. This may compromise other functions. Moreover, mitochondrial accumulation of Ca^{2+} is a rather energy consuming process [7] and the lost energy is not available for the production of ATP. Therefore, this process may rapidly lead to a decrease in intracellular energy supply, below the level required for maintenance of the structural integrity of nerve cells and glia. As a result of large Ca^{2+} accumulation [34], proteolytic enzymes, such as the capping proteins for actin filaments, which are essential for the cell structure, may be activated. Other proteolytic and lipolytic enzymes may be activated as well. As a result, cells may be irreversibly damaged. This raises the problem how to protect against this tissue damage. One strategy involves, of course, protection against [Ca^{2+}] overload of cells [34]. This can be accomplished in two ways. The NMDA-mediated Ca^{2+} uptake can be prevented to some extent by the application of NMDA receptor antagonists. Unfortunately, currently available specific competitive antagonists do not pass the blood-brain barrier. A way out of this is, however, is that non-competitive NMDA receptor antagonists belonging to the phencyclidine class and the sigma opiate class easily pass the blood-brain barrier. This has led to a number of studies of protective actions of NMDA receptor antagonists with some promising results, but also some disappointment. A second way is to use organic Ca^{2+}-entry blockers. There again, the problem exists that the substances do not easily pass through the blood-brain barrier. The best blood-brain barrier permeability is ascribed to flunarizine, whose calcium antagonist properties are not yet well studied. Nevertheless, Ca^{2+}-entry blockers have been reported to have beneficial effects in such situations, some of which may be related to improvement of blood flow. The other approach is to treat the ensuing edema as early and as effectively as possible. Certainly antagonists of glutamate, Cl^--flux-preventing substances, and substances which would help to regulate the glial cell volume would provide new approaches to the treatment of edema.

REFERENCES

1. B. Ault, R. H. Evans, A. A. Francis, D. J. Oakes and J. C. Watkins, Selective depression of excitatory amino acid induced depolarization by magnesium ions in isolated spinal cord, *J. Physiol. London*, 307:413-428 (1980).
2. G. F. Ayala, M. Dichter, R. J. Gumnit, M. Matsumoto and W. A. Spencer, Genesis of epileptic interictal spikes. New knowledge of cortical feedback systems suggest a neurophysiological explanation of brief paroxsyms, *Brain Res.*, 52:1-17 (1973).
3. C. Benninger, J. Kadis and D. A. Prince, Extracellular calcium and potassium changes in hippocampal slices, *Brain Res.*, 187:165-182 (1980).
4. R. S. Bourke and K. M. Nelson, Further studies on the K^+-dependent swelling of primate cerebral cortex in vivo, *J. Neurochem.*, 19:663-685 (1972).
5. J. Cervos-Navarro and R. Ferszt, eds, "Advances in Neurology 28, Brain Edema", Raven Press, New York (1980).
6. E. J. Coan and G. L. Collingridge, Magnesium ions block an N-methyl-D-aspartate receptor mediated component of synaptic transmission in rat hippocampus, *Neurosci. Lett.*, 53:21-26 (1985).
7. R. C. Collins, C. Kennedy, L. Sokoloff and F. Plum, Focal cortical seizures induce distant thalamic lesions, *Arch. Neurol.*, 33:356-542 (1976).
8. M. Croucher, J. F. Collins and B. S. Meldrum, Anticonvulsant action of excitatory amino acid antagonists, *Science*, 216:899-901 (1982).
9. I. Dietzel, U. Heinemann, G. Hofmeier and H. D. Lux, Transient changes in the size of extracellular space in the sensorimotor cortex of cats in relation to stimulus-induced changes in potassium concentration, *Exp. Brain Res.*, 40:432-439 (1980).
10. I. Dietzel, U. Heinemann, G. Hofmeier and H. D. Lux, Stimulus induced changes in extracellular Na^+ and Cl^--concentration in relation to changes in the size of the extracellular space, *Exp. Brain Res.*, 46:73-84 (1982).
11. I. Dietzel and U. Heinemann, Extracellular electrolyte changes during enhanced neuronal activity can be explained by spatial glial K^+ buffering in addition to small decreases in osmotic-pressure possibly induced by

increases in metabolic activity, *in*: "Cerebral Blood Flow, Metabolism and Epilepsy", M. Baldy-Moulinier, D. H. Ingvar and B. S. Meldrum, eds., John Libbey, London, pp 195-201 (1983).

12. I. Dietzel and U. Heinemann, Dynamic variations of the brain cell microenvironment in relation to neuronal hyperactivity, *Ann. NY Acad. Sci.*, 481:72-86 (1986).
13. R. Dingledine, Involvements of N-methyl-D-aspartate receptors in epileptiform bursting in rat hippocampal slice, *Trends Neurosci.*, 9:47-49 (1986).
14. J. L. Farber, The role of calcium in cell death, *Life Science*, 29:1289-1295 (1981).
15. K. J. Futamachi and T. A. Pedley, Glial cells and extracellular potassium. Their relationship in mammalian-cortex, *Brain Res.*, 109:311-322 (1976).
16. A. R. Gardner-Medwin, Possible roles of vertebrate neuroglia in potassium dynamics, spreading depression and migraine, *Exp. Biol.*, 95:111-127 (1981).
17. A. R. Gardner-Medwin, Analysis of potassium dynamics in mammalian brain tissue, *J. Physiol. London*, 335:393-426 (1983).
18. A. R. Gardner-Medwin, A study of the mechanisms by which potassium moves through brain tissue in the rat, *J. Physiol.*, 335:353-374 (1983).
19. T. M. Grisar, Neuron glial relationship in human and experimental epilepsy: a biochemical point of view, *Adv. Neurol.*, 44:1045-1073 (1986).
20. M. J. Gutnick, B. W. Connors and B. R. Ransom, Dye-coupling between glial cells in the guinea pig neocortical slice, *Brain Res.*, 213:486-492 (1981).
21. B. Hamon and U. Heinemann, Effects of GABA and bicuculine on N-methyl-D-aspartate and quiqualate induced reduction in extracellular free calcium in area CAl of the hippocampal slice, *Exp. Brain Res.*, 64:27-36 (1986).
22. A. J. Hansen, Effect of anoxia on ion distribution in the brain, *Physiol. Rev.*, 65:101-148 (1985).
23. R. J. Harris and L. Symon, Extracellular pH, potassium and calcium activities in progressive ischemia of rat cortex, *J. Cereb. Blood Flow Metab.*, 4:178-186 (1984).
24. U. Heinemann, Basic mechanisms of the epilepsies, *in*: "Textbook of Clinical Neurophysiology", A. M. Halliday, S. R. Butler and R. Paul, eds., John Wiley, pp 497-534 (1987).
25. U. Heinemann, *in*: "The Epileptic Focus", Wieser et al., eds., John Libbey, London, Paris, pp 27-44 (1987).
26. U. Heinemann and I. Dietzel, Extracellular potassium concentration in chronic aluminia cream foci of cats, *J. Neurophysiol.*, 52:421-434 (1984).
27. U. Heinemann and M. J. Gutnick, Relation between extracellular potassium concentration and neuronal activities in cat thalamus (VPL) during projection of cortical epileptiform discharges, *Electroenceph. Clin. Neurophysiol.*, 47:345-357 (1979).
28. U. Heinemann and B. Hamon, Calcium and epileptogenesis, *Exp. Brain Res.*, 65:1-10 (1986).
29. U. Heinemann, A. Konnerth, R. Pumain and W. Wadman, Extracellular calcium and potassium concentration changes in chronic epileptic tissue, *Adv. Neurol.*, 44:641-661 (1986).
30. U. Heinemann, H. D. Lux, M. G. Marciani and G. Hofmeier, Slow potentials in relation to changes in extracellular potassium activity in the cortex of cats, *in*: "Origin of Cerebral Field Potentials", E. J. Speckmann and H. Caspers, eds., Georg Thieme, Stuttgart, pp 33-48 (1979).
31. U. Heinemann, H. D. Lux and M. J. Gutnick, Extracellular free calcium and potassium during paroxysmal discharges in the cerebral cortex of the cat, *Exp. Brain Res.*, 7:237-243 (1977).

32. U. Heinemann, S. Neuhaus and I. Dietzel, I. Aspects of K^+ regulation in normal and gliotic brain tissue, *in*: "Cerebral Blood Flow, Metablism and Epilepsy", M. Baldy-Moulinier, D. H. Ingvar and B. S. Meldrum, eds., John Libbey, London, pp 271-278 (1983).
33. U. Heinemann and R. Pumain, Extracellular calcium activity in cat sensorimotor cortex induced by iontophoretic application of amino acids, *Exp. Brain Res.*, 40:247-250 (1980).
34. N. Hori, J. M. H. French-Mullen and D. O. Carpenter, Kainic acid responses and toxicity show pronounced Ca^{++} dependence, *Brain Res.*, 358:380-384 (1985).
35. K. A. Hossmann, Cortical steady potentials, impedance and excitability changes during and after total ischemia of the cat brain, *Exp. Neurol.*, 32:163-175 (1971).
36. K. A. Hossmann, S. Sakaki and V. Zimmermann, Cation activities in reversible ischemia of the cat brain, *Stroke*, 8:77-81 (1977).
37. J. Janus, E. J. Speckmann and A. Lehmenkühler, Relations between extracellular K^+ and Ca^{2+} activities and local field potentials in the spinal cord of the rat during focal and generalized seizure discharges, *in*: "Ion-Selective Microelectrodes and their Use in Excitable Tissues", E. Sykova, P. Hnik and L. Viklicky, eds., Plenum Press, New York and London, pp 181-185 (1981).
38. H. Kettenmann and M. Schachner, Pharmacological properties of gamma aminobuytric acid, glutamate and aspartate induced depolarizations in cultured astrocytes, *J. Neurosci.*, 5:3295-3301 (1985).
39. H. Kettenmann, U. Sonnhof and M. Schachner, Exclusive potassium dependence of the membrane potential in cultured mouse oligodendrocytes, *J. Neurosci.*, 3:500-505 (1983).
40. G. Köhr and U. Heinemann, Anticonvulsive properties of ketamin and 2-aminophosphonovalerate in the low magnesium epilepsy in rat hippocampal slices, *Neurosci. Res. Commun.*, 1:17-21 (1987).
41. A. Konnerth, Y. Yaari and U. Heinemann, Nonsynaptic epileptogenesis in the mammalian hippocampus in vitro, *J.Neurophysiol.*, 56:409-423 (1986).
42. R. P. Kraig and C. Nicholson, Extracellular ionic variations during spreading depressions, *Neurosci.*, 3:1045-1059 (1978).
43. J. D. C. Lambert and U. Heinemann, Extracellular calcium changes accompanying the action of excitatory amino acids in area CA1 of the hippocampus. Possible implications for the initiation and spread of epileptic discharges, *in*: "Epilepsy and Calcium", E. J. Speckmann et al., eds., Urban und Schwarzenberg, München, pp 35-62 (1986).
44. A. A. P. Leao, Spreading depression of activity in the cerebral cortex, *J. Neurophysiol.*, 7:359-390 (1944).
45. A. Lehmenkühler, W. Zidek and H. Caspers, Changes of extracellular Na^+ and Cl^--activities in the brain cortex during seizure discharges, *in*: "Physiology and Pharmacology of Epileptogenic Phenomena", M. R. Klee, H. D. Lux and E. J. Speckmann, eds., Raven Press, New York, pp 37-45 (1982).
46. H. D. Lux, U. Heinemann and I. Dietzel, Ionic changes and alterations in the size of the extracellular space during epileptic activity, *Adv. Neurol.*, 44:619-639 (1986).
47. M. L. Mayer, G. L. Westbrook and P. B. Guthrie, Voltage-dependent block of Mg^{2+} of NMDA-responses in spinal cord neurones, *Nature*, 309:250-263 (1984).
48. A. B. MacDermott, M. L. Mayer, G. L. Westbrook, S. J. Smith and J. L. Barker, NMDA-receptor activation increases cytoplasmic calcium concentration in cultured spinal cord neurones, *Nature*, 321:519-522 (1986).
49. B. S. Meldrum, J. J. Croucher, S. J. Czucwar, J. F. Collins, K. Curry, M. Joseph and T. W. Stone, A comparison of the anticonvulsant potency of (+-)2-amino-phosphonovaleric acid and (+-)2-amino-7-phosphonoheptanoic acid, *Neuroscience*, 9:925-930 (1983).

50. I. Mody and U. Heinemann, Laminar profile of the changes in extracellular calcium concentration induced by repetitive stimulation and excitatory amino acids in the rate dentate gyrus, *Neurosci. Lett.*, 69:137-142 (1986).
51. I. Mody, J. D. C. Lambert and U. Heinemann, Low extracellular magnesium induces epileptiform activity and spreading depression in rat hippocampal slices, *J. Neurophysiol.*, 57:869-888 (1987).
52. C. Nicholson, Dynamics of the brain cell microenvironment, *NRP Bulletin*, 18:172-322 (1980).
53. C. Nicholson and J. M. Phillips, Ion diffusion modified by tortuosity and volume fraction in the extracellular microenvironment of the rat cerebellum, *J. Physiol. London*, 321:225-258 (1981).
54. L. Nowak, O. P. Bregestowski, P. Ascher, A. L. Herbet and A. Prochiantz, Magnesium gates glutamate-activated channels in mouse central neurones, *Nature*, 307:462-464 (1984).
55. J. W. Olney, R. C. Collins and R. S. Sloviter, Excitotoxic mechanisms of epileptic brain damage, *Adv. Neurol.*, 44:857-878 (1986).
56. R. F. Orkand, Functional consequences of ionic changes resulting from electrical activity, *Fed. Proc.*, 39:1514-1543 (1980).
57. R. Pumain and U. Heinemann, Stimulus and amino-acid induced calcium and potassium changes in the rat neocortex, *J. Neurophysiol.*, 53:1-16 (1985).
58. J. B. Ranck, Specific impedance of cerebral cortex during spreading depression and an analysis of neuronal, neuroglial and interstitial contributions, *Exp. Neurol.*, 9:1-16 (1964).
59. B. K. Siesjö, Cell damage in the brain: a speculative synthesis, *J. Cerebral Blood Flow Metab.*, 1:155-185 (1981).
60. B. K. Siesjö and T. Wieloch, Epileptic brain damage: pathophysiology and neurochemical pathology, *Adv. Neurol.*, 44:813-848 (1986).
61. S. M. Rothman, The neurotoxicity of excitatory amino acid is produced by passive chloride influx, *Neuroscience*, 5:1483-1489 (1985).
62. G. G. Somjen, Electrogenesis of sustained potentials, *Prog. Neurobiol.*, 1:201-232 (1973).
63. G. G. Somjen, Neuroglia and spinal fluids, *J. Exp. Biol.*, 95:129-133 (1981).†
64. G. G. Somjen, Stimulus evoked and seizure related responses of extracellular calcium activity in spinal cord compared to those in cerebral cortex, *J. Neurophysiol.*, 44:617-632 (1980).
65. E. Sykova, Extracellular K^+ accumulation in the central nervous system, *Prog. Biophys. Mol. Biol.*, 42:135-189 (1983).
66. A. M. Thomson, D. C. West and D. Lodge, A N-methyl-D-aspartate receptor mediated synapse in rat cerebral cortex: a site of action of ketamine, *Nature*, 313:479-481 (1985).
67. A. Van Harreveld, J. Crowell and S. K. Malhotra, A study of extracellular space in central nervous tissue by freeze substitution, *J. Cell. Biol.*, 25:117-137 (1985).
68. A. Van Harreveld, T. Murphy and K. W. Nobel, Specific impedance of rabbit cortical tissue, *Am. J. Physiol*, 205:203-207 (1963).
69. S. S. Varon and G. G. Somjen, Neuron-glia interaction, *Neurosci. Res. Prog. Bull.*, 17:1-238 (1979).
70. H. Walther, J. D. C. Lambert, R. S. G. Jones, U. Heinemann and B. Hamon, Epileptiform activity in combined slices of the hippocampus, subiculum and entorhinal cortex during perfusion with low magnesium medium, *Neurosci. Lett.*, 69:156-161 (1986).
71. W. Walz and L. Hertz, Ouabain-sensitive and ouabain-resistant net uptake of potassium into astrocytes and neurones in primary cultures. II. Role of extracellular potassium, *J. Neurochem.*, 39:70-77 (1982).
72. J. C. Watkins and R. H. Evans, Excitatory amino acid transmitters, *Ann. Rev. Pharmacol. Toxicol.*, 21:165-204 (1981).

73. Y. Yaari, A. Konnerth and U. Heinemann, Nonsynaptic epileptogenesis of the mammalian hippocampus in vitro, *J. Neurophysiol.*, 56:424-438 (1986).

EICOSANOIDS, PEPTIDES AND AMINES AND THE CEREBRAL BLOOD VESSELS

G. Feuerstein

Department of Neurology
Uniformed Services University of the Health Sciences
4301 Jones Bridge Road
Bethesda, MD 20814-4799, USA

INTRODUCTION

The brain, like every other organ, is dependent on continuous blood flow to maintain its functions. However, the brain is unique among the organs in its tight relationships between blood flow, oxygen and glucose metabolism and neuronal activity. Of primary importance in this regard are ionic species and metabolic products such as potassium, calcium, carbon dioxide and lactate which in spite of careful regulation might also be subjected to rapid and profound fluctuations.

The role of metabolic and ionic factors in regulation of the cerebral blood flow has been extensively reviewed previously [1]. In brief, acidity has been shown to relax while alkalinity to contract cerebral arterioles [2-8]. In this regard, a decrease in the pH is considered to be a primary factor yet CO_2 might also act in a similar way [9]. However, the relationships of pH and CO_2 are more complex and CO_2 levels may produce contraction of cerebral vessels in certain experimental conditions [1]. Potassium has also been shown to affect the blood vessel tone in a complex mode of actions. While high potassium concentrations traditionally are viewed to contract isolated cerebral arteries and pial arterioles, small increases in extracellular potassium were also shown to relax the cerebrovascular smooth muscle due to electrogenic stimulation of the Na^+-K^+ pump [1,4,6,10,11].

Calcium is part of the final common pathway in any smooth muscle contraction. Increase in intracellular calcium activates the actomyosin contractile mechanism which leads to muscle contraction. Increase in calcium concentration in the extracellular fluid produces vasoconstriction in the cerebral circulation [12] while reducing extracellular calcium produces relaxation [3,13]. These data taken together clearly show that the cerebral circulation is subjected to influences from factors directly associated with the ongoing metabolic and neural activity of the brain tissue.

Cerebral Sinus Thrombosis, Edited by K. Einhäupl *et al.*
Plenum Press, New York, 1990

Inasmuch as the ionic and metabolic status of the brain is closely associated to the cerebral blood flow, it has equally been recognized that several neurogenic and humoral factors can also regulate cerebral vessel tone and blood flow. The cerebral vessels are exposed to many metabolic variables besides CO_2, lactate or pH; they are densely innervated by sympathetic and parasympathetic nerves which contain numerous vasoactive substances. It is the primary goal of this brief review to highlight the vast potential of the more recent discoveries related to vasoactive substances which are intimately associated with the cerebral circulation.

EICOSANOIDS AND THE CEREBRAL CIRCULATION

Arachidonic acid is a major fatty acid present in the phospholipid bilayer of every biological membrane. Arachidonic acid is constantly cleaved and reacylated in the membranal milieu by enzymatic reactions which involve calcium and ATP. Ninety-nine percent (99%) of the arachidonate released is effectively reacylated to the membranal phospholipids. A small fraction of free arachidonate is continuously converted to a host of metabolites, the eicosanoids (Fig. 1), many of which are extremely potent vasoactive substances. This process takes place in neuronal cells, glial cells and blood vessels [14]. The possible involvement of the eicosanoids in maintenance of physiological blood vessel tone and cerebral blood flow can be appreciated from studies showing reduction in cerebral blood flow by drugs (eg, indomethacin) which effectively block brain prostaglandin (PG) production [14]. However, other studies failed to demonstrate a change in basal cerebral blood flow or in hypercapnia induced increase in cerebral blood flow (for review see Refs 1,126). Both small and large cerebral vessels are capable of producing a variety of PGs including PGE_2, $PGF_{2\alpha}$, PGI_2 and PGD [15]. Furthermore, all known PGs and thromboxane A (TXA) affect cerebral vascular tone. Generally, $PGF_{2\alpha}$ and TXA_2 produce vasoconstriction [16-18], PGI_2 vasodilation [19-21]; PGE_2 produces vasoconstriction in some species and dilation in others [21]. Only a few studies examined the effect of PGD_2 [21] and responses to PGD_2 seem to vary [1].

More recently, a new series of 5-lipoxygenase metabolites of arachidonate have been identified: the leukotrienes (LTs) [22]. This family of peptido-lipids possess outstanding vasomotor activities on peripheral blood vessels of many species (for review see Refs 23-25). Release of leukotrienes from cerebral arteries has been demonstrated in vitro [26,27]. Furthermore, leukotrienes have been shown to constrict isolated human cerebral vessels [28], but this issue is still controversial [23,24]. Animal studies also produced conflicting data. In the rabbit, LTC_4, LTD_4 and 12-HETE failed to produce any appreciable change in pial arterial diameter [29]; in cats, pial vessels superfused through a cranial window responded with vasoconstriction (venous and arterial) to LTC_4, LTD_4 and LTE_4 [30]. The latter study also failed to demonstrate any significant effect of the LTs on the blood-brain barrier [30]. However, other studies have demonstrated potent vasoconstriction by all LTs (LTC_4, LTD_4 and LTE_4) in the newborn pig pial circulation [31]; furthermore intra-parenchymal application of LTD into the rat brain resulted in increased microvascular permeability and brain edema [32]. These findings are summarized in Table 1.

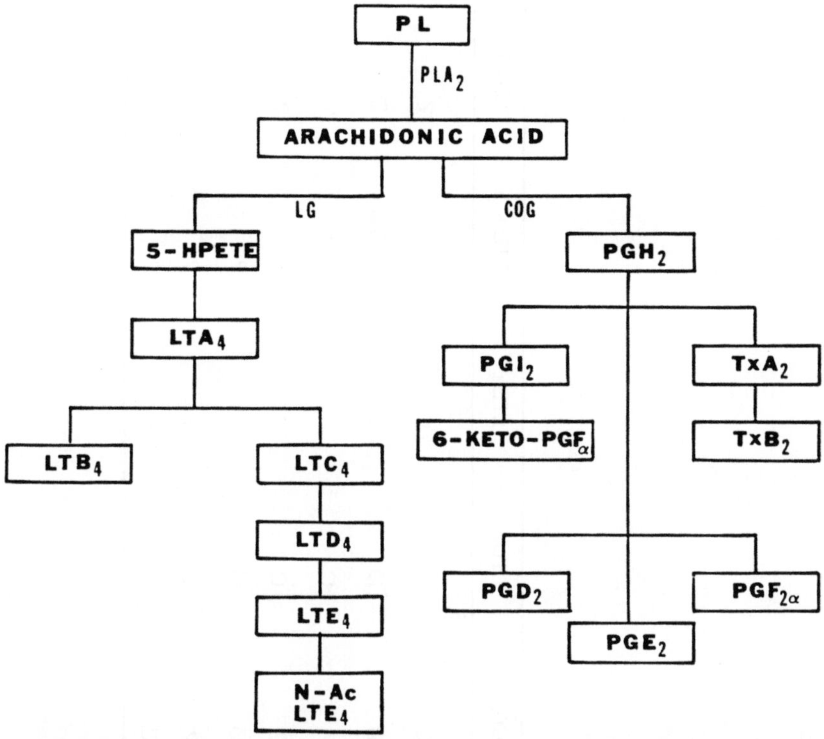

Fig. 1. Metabolic pathways of formation of prostanoids and leukotrienes secondary to activation of phospholipase A_V, cyclooxygenase and lipoxygenase.

In summary, while it is evident that numerous potent vasomotor arachidonate metabolites are produced by brain tissue and vessels, their physiological and pathophysiological role is far from clear.

PEPTIDES AND THE CEREBRAL CIRCULATION

Bradykinin

Bradykinin is one of the most active peptide products of the kallikrein-kinin system. Bradykinin is known for its proinflammatory actions in the peripheral circulation [33,34]. In the brain, all the components of an intracerebral kallikrein-kinin system have been demonstrated [35,36]. Bradykinin has also been implicated in pathological events of the central nervous system such as trauma, where uptake of kininogen from the intravascular space into the brain and formation of bradykinin has been suggested to contribute to brain edema [37,38].

The inducing effect of bradykinin in the brain is the result of its actions on the microcirculation. Thus, contracted human and feline pial and middle cerebral arteries relax via the B_2-kinin receptor [35,39–41]. However, opposite responses, ie, constriction of cerebral vessels, were also reported (for review see Ref. 42). More consistent data are derived from in situ experiments. Bradykinin has been shown to be a potent vasodilator of pial arteries of cats [35] and rabbits [29] through a

Table 1. Effect of Leukotrienes on Cerebral Vessels

Species	Preparation	Method	LTC_4	LTD_4	LTE_4	LTB_4	12-HETE
Cat	Parietal pial vessels	Open window	C	C	C	C	?
Rabbit	Parietal pial vessels	Closed window	Ø	Ø	Ø	C (5%)	Ø
Rat	Carotid infusion	In vivo	?	C	?	?	?
Human	(1) Intracranial arteries	In vitro	?	C	?	?	?
	(2) Cerebral artery	In vitro	Ø	Ø	?	?	?
Pig	(1) Carotid infusion	In vivo	?	C	?	?	?
	(2) Pial arteries	In situ	C	C	C	?	?

C = Constriction; ? = not known; Ø = no response.

perivascular site of action [41]. However, prolonged superfusion of bradykinin on pial veins produced venoconstriction [43]. This action of bradykinin is of special significance, since bradykinin was shown to produce leakage of intravascular sodium fluorescin into the brain parenchyma [44,45]. The bradykinin-induced extravasation takes place at the post-capillary venules, is endothelial dependent and quite selective for a lower molecular weight range [44].

While these data provide circumstantial evidence for a putative role of bradykinin in modulation of cerebral vessel tone, further studies will be necessary to clarify the role of bradykinin in physiological and pathophysiological states of the cerebral circulation. Of special importance in this regard would be studies utilizing selective bradykinin receptor antagonists in an attempt to modify pathological situations of cerebral blood flow, blood-brain barrier permeability and inflammatory processes. Also, specific assays of bradykinin and findings on its precise localization and function are largely missing.

Neuropeptide Y

Neuropeptide Y (NPY) is a 36-amino-acid peptide originally identified by Tatemoto et al. [46]. Immunocytochemical studies have revealed NPY-containing nerve fibers in cerebral vessels [47,48]. These fibers are found in the muscular part of major arteries, the pial arterioles of the cortex and even in penetrating arterioles. Since NPY immunoreactivity virtually disappears upon removal of the sympathetic nerves (eg, superior ganglion resection) and since NPY immunoreactivity coexists with that of dopamine-β-hydroxylase (a marker of adrenergic neurons), it is commonly believed that NPY is functionally related with norepinephrine in the perivascular nerves of the cerebral circulation [49].

NPY is one of the most potent constrictors of cerebral vessels. Substantial contraction of human cerebral arteries in vitro can be elicited by nanomolar concentrations of NPY [47], and the constriction of the vessels is calcium dependent. In feline pial arteries, NPY produces strong and sustained constriction [48] while venular constriction is somewhat less pronounced. Future studies should be focused on the potential role of NPY in regulation of cerebral vessel tone by direct assays in situations where the vascular tone is modified by physiological stimuli (eg, CO_2) or pathophysiological (eg, hypertension, hemorrhage) conditions. Furthermore, the development of potent and specific NPY antagonists would be a crucial step in further elucidation of the role of NPY in cerebrovascular regulation. In addition, the relationships of NPY to the adrenergic neurotransmitters (primary norepinephrine) in modulation of the cerebral circulation is still obscure.

Vasoactive Intestinal Polypeptide

Vasoactive intestinal polypeptide (VIP) is a 33-amino-acid peptide which is widely distributed in organs of humans and animals. VIP immunoreactivity (VIP-ir) is also found in the cerebral vasculature [50-53], the choroid plexus [54] and penetrating vessels. Some VIP-ir-containing fibers which are found in cortical pial arterioles originate from VIP neurons [55,56], while most of the VIP fibers in the cerebral vasculature originate from the sphenopalatine ganglia [57]. A unique situation has been described in the cat where VIP fibers in pial vessels are believed to originate from local microganglia [58].

VIP is one of the most potent dilatory substances known to date [50,53,59]. VIP-induced vasodilation is obtained in some vascular preparations at subnanomolar concentrations; the vasodilation is mediated by a specific VIP receptor, and is independent of the endothelium [59,60]. VIP-induced vascular relaxation is probably associated with formation of c-AMP [60,61]. Superfusion of VIP in situ of pial arterioles and veins elicits dilation; this effect is particularly obvious in small arterioles (60 μm) rather than larger (150 μm) ones [62]. Furthermore, VIP-induced relaxation is more pronounced after application to the adventitial rather than the luminal side of the vessel [62,63].

The effect of VIP on the microcirculation has also been shown to be associated with profound metabolic activation. Thus, administration of VIP into the cortex increases metabolic activity and glucose consumption [64,65]. Since VIP neurons can communicate with both cortical neurons and cortical blood vessels it is tempting to speculate that the VIP-ergic system may play a pivotal role in regulation of both local neuronal activity and blood flow. However, further studies are needed to confirm this hypothesis.

Tachykinins

The tachykinins are a group of peptides with a common sequence at the C-terminus of the peptide. Substance P (SP) was the first of the tachykinins to be found in the mammalian central nervous system but two other tachykinins, neurokinin A (NKA) and neurokinin B (NKB), have recently been discovered. The tachykinins, especially SP and SK, are widely distributed in the central nervous system [66,67]. SP immunoreactive nerve fibers have been observed in pial arteries and veins from humans and various mammals [68-71]. These nerve fibers originate from sensory ganglia such as the trigeminus. Lesions of this sensory nucleus depletes much of the SP-ir from the pial vessels [72,73], and retrograde tracing techniques have indicated that nerve fibers in cerebral vessels originate from this nucleus [71]. A functional role of SP in regulation of the pial-arachnoid vessels has been suggested by studies demonstrating the release of SP from these vessels in vitro using high potassium [69] or electrical [74] stimuli.

Substance P is one of the most potent vasoactive substances in the peripheral circulation [75]. In the cerebral circulation, SP does not elicit significant vasomotor responses from isolated resting cerebral vessels; however, pre-contracted (by $PGF_{2\alpha}$ or 5-HT) vessels respond with relaxation to both SP and NKA [76,77]. The relaxation produced by SP is likely to be mediated by specific SP receptors. None of the classical antagonists of the β-adrenergic (propranolol), cholinergic (atropine) or histaminergic (metiamide) receptors affect SP induced vasodilation [76]. However, SP-induced dilation seems to be mediated by an endothelium-derived relaxing factor [60].

It is also noteworthy that SP is co-localized with another tachykinin, NKA [78] in sensory nerve terminals and perikaria. NKA is derived from the same precursor: beta-preprotachykinin [79]. Both tachykinins are present in perivascular sensory fibers [78] and produce similar effects, although NKA is much less potent probably due to a lower affinity to the SP-P receptors in blood vessels [75].

Calcitonin Gene-related Peptide

Calcitonin gene-related peptide (CGRP) is a 37-amino-acid peptide that has been identified in humans [80,81] and in the nervous system of other mammals [82,83]. CGRP is widely distributed in the brain, but especially high levels are found in hindbrain and sensory spinal cord areas suggesting a primary involvement in processing of peripheral nociceptive or other sensory information. In addition to its distribution in brain and spinal cord nuclei, CGRP-ir has been detected also in major cerebral arteries and pial arterioles and veins [73]. It has been suggested that the trigeminal ganglion is a major source of CGRP-containing fibers where it occurs together with other putative peptide neurotransmitters such as SP or NKA [73,84]. Furthermore, nerve fibers surrounding cerebral vesels were also found to contain both SP and CGRP [73]. The major contribution of the trigeminal ganglion to CGRP- and SP-containing nerves is suggested by findings on almost complete disappearance of these substances from blood vessels of animals where the trigeminal ganglion has been surgically lesioned [73].

CGRP is an extremely potent vasodilator of peripheral (eg, coronary, [85]) and cerebral vessels [86]. The vascular relaxation produced by CGRP, unlike that of SP, is independent of the vascular endothelium [86] and is probably mediated by c-AMP [60]. CGRP is even more potent in dilation of pial arterioles in situ than SP, yet its vasoactivity seems to be confined to arteriolar vessels [87,88]. The physiological significance of CGRP in cerebral vessel tone regulation is further suggested by studies showing enhanced responses of cerebral vessels to a variety of constrictor stimuli ($PGF_{2\alpha}$, norepinephrine) in the absence of the sensory innervation of these vessels [88]. Taken together, these data suggest that the putative peptide neurotransmitters, tachykinins and CGRP, might serve as an anticonstrictive defense mechanism.

Miscellaneous Peptides

Gastrin-releasing Peptide and Bombesin. Gastrin-releasing Peptide (GRP) is a 27-amino-acid peptide which belongs to the bombesin-like family of peptides. These "skin-gut-brain" peptides, originally isolated from the skin of amphibians, are also found in the mammalian nervous system [89]. GRP and bombesin not only share a significant homology in the biologically active C-terminus but also similar biological activities. GRP-ir is found around cerebral and spinal cord vessels of both large and small caliber [90]. The origin of the bombesin/GRP fibers are still unknown as is their functional role in cerebral blood flow regulation. Attempts to study the direct effect of bombesin or GRP on cerebral vesels have failed to show significant vasoactivity thus far.

Neurotensin. Neurotensin is a tridecapeptide of known vascular activity upon central [91] or peripheral [92] administration. Neurotensin is widely distributed in the central nervous system of many species, including humans [93]. Some neurotensin-containing fibers have been described in close proximity to brainstem capillaries [94] but no appreciable vasomotor effect of neurotensin has been described in the cerbral vasculature.

Cholecystokinin. Cholecystokinin (CCK) is a putative neuropeptide transmitter containing eight amino acids. CCK is widely distributed in the brain [95,96]. CCK neurons have been shown to be intimately associated with the cortical arteries and capillaries [97]; furthermore, large cerebral vessels possess CCK-containing nerve fibers [98]. In spite of this strong antomical evidence in favor of a possible role of CCK in regulation of the brain macro- and microcirculation, there is substantial difficulty in demonstrating functional significance since CCK has no appreciable vasomotor activity either on resting, dilated or contracted vessels [99].

Opioid Peptides. Several families of opioid peptides are found in the central nervous system of all mammals, including humans [100,101]. These opioid peptides belong to one of three main groups of peptides: the opiomelanocortins, preproenkephalins and preprodynorphins. The opioid peptides have been implicated in multiple behavioral and autonomic functions of which analgesia behavior and respiratory functions are widely cited. However, some information suggests that opioid peptides might be involved in cerebral vessel regulation in spite of the difficulties of demonstrating direct innervation of cerebral arteries and arterioles by opioid peptide-containing fibers [102]. Opioid receptors have been demonstrated in bovine cerebral microvessels [103].

The difficulties in demonstrating the presence of opioid peptides or receptors in the cerebral vessels extend also to their vascular activity. While some in vitro studies show the activity of opioid peptides on peripheral vascular smooth muscles [104-107], other studies fail to show any effect of a wide range of opioid peptides or morphine [108,109]. In the cerebral circulation, comprehensive testing of several natural and synthetic enkephalins as well as morphine did not reveal any vasomotor activity on a wide range of diameters of feline pial arterioles up to pharmacological concentrations of 10^{-5} M. Only concentrations of 10^{-4}-10^{-3} M caused mild vasodilation but at 10^{-3} M this effect was also produced by the opiate antagonist naloxone [110]. Since the enkephalins used in the above mentioned study are preferential agonists of the μ and δ opiate receptors, the author suggested a role for these receptors in opioid-mediated microvessel dilatation at very high doses. The possibility of a more physiological role of the opioid peptides in cerebral blood vessel regulation has not been suggested as yet.

AMINES AND THE CEREBRAL CIRCULATION

Norepinephrine

Cerebral vessels are richly innervated by sympathetic nerves [111,112] which contain norepinephrine as the primary neurotransmitter [113], which is known to contract the cerebral vessels of many species [114-116]. An exception to this rule are pig cerebral vessels in which sympathetic activity results in vasodilation [113]. Substantial evidence has been generated by in situ perivascular application of substances which show that norepinephrine constricts microvessels, arteries and veins [117-119]. The in situ studies, as well as many in vitro studies [113], clearly indicate that α-adrenoceptors are the mediators of virtually all of the adrenergic effects on cerebral vessels [113,120,121]. The α-receptor-mediated vasoconstriction produced by norepinephrine is also subjected to modulation by metabolic or ionic factors; eg, acidity attenuates and increased potassium augments the norepinephrine responses [117,122].

Of primary importance in elucidating the physiological role of the sympathetic nerves on cerebral vessel tone are in vivo studies where sympathetic stimulation results in endogenous transmitter release. Such studies reveal lesser responses of small (< 120 μm) versus larger vessels [121,123]. However, in many cases, in spite of an increase in cerebral vessel resistance little changes are observed in cerebral blood flow; in a variety of studies sympathetic stimulation resulted in approximately 20% reduction in the cerebral blood flow (for review see Ref. 42). Large variation between studies is probably the result of species differences, systemic hemodynamic and metabolic conditions as well as the anesthetic agents used in the various experiments [124-126]. However, in spite of the anatomic and pharmacological evidence provided above, it is still unclear whether noradrenergic tone influences normal cerebral vascular tone and blood flow at resting conditions [18-20,117,122]. Yet, noradrenergic influences could be more pronounced and even detrimental to the integrity of the microcirculation in pathophysiological conditions such as hypertension [128-131].

Serotonin

Serotonin (5-hydroxytryptamine, 5-HT) is a well-established neurotransmitter in the nervous system which has been characterized in view of its cellular localization, brain pathways and receptor distribution. The origin of serotonin found in nerves within the cerebral vasculature is believed to originate in the brain stem [132,133].

The pharmacological effects of serotonin on the cerebral circulation are extremely complex and variable. Species difference, vessel size, type of preparation (in vitro vs in situ), and site of injection (perivascular vs systemic) are well recognized factors for the variable responses. The common effect of 5-HT on isolated cerebral arteries of various species is constriction [134-136]. However, a β-adrenergic receptor mediated relaxation (evidenced by blockade with propranolol) has been reported [136]. In in situ preparation, relaxation of small feline pial vessels has been reported [137-139], which is vessel-tone-dependent [7] and in part β-adrenoceptor-mediated [139]. In contrast to the small pial vessels, the larger ones (>200 μm) exhibit more variable responses including constriction [137-139]. In summary, serotonin localization in nerve endings, in the adventitial-medial side of the cerebral vessels, and its effects upon perivascular application, support the possibility of it being a potential modulator of cerebral vessel tone. Yet, the precise effects are probably highly specialized to species, type of vessels and interaction with other transmitter systems.

Histamine

Histamine is a well-established neurotransmitter in the central nervous system and its perikarial localization, pathways and receptors are well described [140,141]. The histaminergic system is unique among the potential neurotransmitters implicated in cerebral vascular tone regulation by its localization not only in nerve endings, but also in non-neuronal elements like mast cells which are abundant at perivascular sites [142-144].

The primary effect of histamine when applied perivascularly in situ on pial arteries is vasodilation [145-147]; furthermore, histamine relaxes pre-contracted cerebral vessels in vitro [148]. In this regard, it is the arterial rather than the venular side of the capillaries which is responsive to histamine-induced vasodilation [146].

Although both H_1 and H_2 histamine receptors were found in the cerebral vessels [145,147], it is currently believed that H receptors are primarily involved in cerebral vessel dilation. An important aspect of histamine actions on the cerebral circulation includes the blood-brain barrier. Histamine was shown to increase the permeability of the blood-brain barrier and to promote edema formation [149]. Thus, histamine could be involved in both normal and pathophysiological processes of cerebral blood flow and the integrity of the microcirculation.

SUMMARY

This brief review has presented over 25 substances which are potentially involved in cerebral vessel regulation. These substances are either metabolic products (eg, lactate, CO_2), ionic constituents of the intracellular and extracellular environment (eg, K^+, Ca^{2+}), preformed and prestored substances such as neurotransmitters (eg, catecholamines, SP) or secondary mediators formed in response to a primary stimulus (eg, prostaglandins, leukotrienes). The evidence in support for any substance to be a modulator of cerebral vessel tone consists of: (1) anatomical - presence of the substance in the blood vessel or in close proximity to influence its smooth muscle or endothelium; (2) pharmacological - demonstration of vasoactivity of the substance upon application to the cerebral vessels in vivo or in vitro; (3) physiological - evidence of release of the substance from cerebral vessels in response to proper stimuli and preferably in the in vivo milieu; and (4) pathological - blocking of the substance synthesis or action to reverse or prevent pathophysiological processes associated with the cerebral vessels.

While using these guidelines in the process of establishing factors in cerebral vessel function seems to be self evident, none of the substances discussed in this review (nor others, eg, acetylcholine) fulfil these prerequisites. For most of the substances the evidence is limited to anatomical or pharmacological findings; physiological data are sparse and pathophysiological proofs almost never attempted. The latter is in fact difficult to achieve since no specific antagonists or synthesis inhibitors are available and, therefore, a critical tool with which to investigate the potential role of the putative modulator is missing. This is particularly true for most of the neuropeptides, for which no adequate antagonists or synthesis inhibitors exist. Therefore, the evidence presented in this review is far from being complete and at best is only suggestive.

Of particular importance are pathophysiological states where numerous changes occur simultaneously and follow a complex sequence. Interactions of a multitude of factors rather than a particular substance might be of critical importance. Only limited attention is given to this aspect by investigators studying selected factors. With these considerations in mind, the reader is encouraged to exercise caution and a critical attitude to the above presented data and views.

ACKNOWLEDGEMENTS

This work was supported by USUHS protocol No. GM9229. The opinions or assertions contained herein are the private ones of the author(s) and are not to be construed as official or as necessarily reflecting the views of the Department of Defense or the Uniformed Services University of the Health Sciences. There is no objection to its presentation and/or publication. The experiments reported herein were conducted according to the principles set forth in the Guide for Care and Use of Laboratory Animals, Institute of Laboratory Animal Medicine, National Research Council, DHEW Publication No. (NIH) 80-23, 1980. I wish to thank Mrs Laura L. Garza for the preparation of this paper.

REFERENCES

1. J. D. Pickard, Ionic and eicosanoid regulation of cerebrovascular smooth muscle contraction, *in*: "Cerebral Blood Flow", J. H. Wood, ed. , McGraw Hill, New York, pp 131-144 (1987).
2. E. Betz, Cerebral blood flow - its measurement and regulation, *Physiol. Rev.*, 52:595-630 (1972).
3. E. Betz, H. G. Enzenrob and V. Vlahov, Interaction of H^+ and Ca^{++} in the regulation of local pial vascular resistance, *Pflügers Arch.*, 34:79-88 (1973).
4. I. R. Cameron and J. Caronna, The effect of local changes in potassium and bicarbonate concentration on hypothalamic blood flow in the rabbit, *J. Physiol.*, 262:415-430 (1976).
5. R. G. Dacey and B. R. Duling, A study of rat intracerebral aterioles: methods, morphology and reactivity, *Am. J. Physiol.*, 243:H598-H606 (1982).
6. W. Kuschinsky, M. Wahl, O. Bosse and K. Thurau, Perivascular potassium and pH as determinants of local pial arterial diameter in cats: a microapplication study, *Circ. Res.*, 31:240-247 (1972).
7. J. L. Pannier, J. Weyne, G. Demeester and I. Leusen, Influence of changes in the acid-base composition of the ventricular system on cerebral blood flow in cats, *Pflügers Arch.*, 333:337-351 (1972).
8. M. Wahl, P. Deetjen, K. Thurau, D. H. Ingvar and N. A. Lassen, Micropuncture evaluation of the importance of perivascular pH for the arteriolar diameter on the brain surface, *Pflügers Arch.*, 316:152-163 (1970).
9. H. A. Kontos, A. J. Raper and J. L. Patterson, Analysis of vasoactivity of local pH, pCO_2 and bicarbonate on pial vessels, *Stroke*, 8:358-360 (1977).
10. N. Toda, Potassium induced relaxation in isolated cerebral arteries contracted with prostaglandin $F_{2\alpha}$, *Pflügers Arch.*, 364:235-242 (1976).
11. N. Toda, Responsiveness to potassium and calcium ions of isolated cerebral arteries, *Am. J. Physiol.*, 227:1206-1211 (1974).
12. C. Carivin, R. Leutzenhiser and C. Van Breemen, Mechanisms of calcium antagonist induced vasodilation, *Ann. Rev. Pharmacol. Toxicol.*, 23:373-396 (1983).
13. M. C. Evans, R. A. F. Linton and I. R. Cameron, The effect of local changes in calcium concentration on hypothalamic blood flow in the anesthetized rabbit, *Acta Neurol. Scand.*, 56 (Suppl. 64), 21:8-9 (1977).
14. J. D. Pickard, Role of prostaglandins and arachidonic acid derivatives in the coupling of cerebral blood flow to cerebral metabolism, *J. Cereb. Blood Flow Metab.*, 1:361-384 (1981).

15. L. S. Wolfe, Eicosanoids - prostaglandins thromboxanes leukotrienes and other derivatives of carbon 20 unsaturated fatty acids, *J. Neurochem.*, 38:1-4 (1982).
16. D. J. Boullin, S. Bunting, W. P. Blaso, T. M. Hunt and S. Moncada, Responses of human and baboon arteries to prostaglandin endoperoxides and biologically generated and synthetic prostacyclin, *Br. J. Clin. Pharmacol.*, 7:138-147 (1979).
17. N. Toda, Responses to prostaglandin H_2 and I_2 of isolated dog cerebral and peripheral arteries, *Am. J. Physiol.*, 238:H111-H117 (1980).
18. N. Toda and M. Miyorzaki, Responses of isolated dog cerebral and peripheral arteries to prostaglandins after application of aspirin and polyphloretin phosphate, *Stroke*, 9:490-498 (1978).
19. K. S. Paul, E. T. Whalley, C. H. Forster, R. Lye and J. Dutton, Prostacyclin and cerebral vessel relaxation, *J. Neurosurg.*, 57:334-340 (1982).
20. T. Uski, K. E. Andersson, L. Brandt, L. Edinsson and B. Ljunggren, Responses of isolated feline and human cerebral arteries to prostacyclin and some of its metabolites, *J. Cereb. Blood Flow Metab.*, 3:238-245 (1983).
21. E. F. Ellis, E. P. Wei and H. A. Kontos, Vasodilation of cat cerebral arterioles by prostaglandin D_2, E_2, G_2 and I_2, *Am. J. Physiol.*, 237:H381-H385 (1979).
22. B. Samuelsson, Leukotrienes: mediators of immediate hypersensitivity reactions and inflammation, *Science*, 220:568-575 (1983).
23. G. Feuerstein, Autonomic pharmacology of leukotriences, *J. Autonom. Pharmacol.*, 5:149-168 (1985).
24. G. Feuerstein and J. M. Hallenbeck, Leukotrienes in health and disease, *FASEB*, 1:186-192 (1987).
25. P. J. Piper, "The Leukotrienes", Raven Press, New York (1986).
26. A. Dembinska Kiec, T. Simmet and B. A. Peskar, Formation of leukotriene C_4 like material by rat brain tissue, *Eur. J. Pharmacol.*,99:57-62 (1984).
27. P. J. Piper and A. W. B. Stanton, Actions of leukotrienes in special circulations of the pig, *Adv. Prostag. Thromb. Leukot. Res.*, 16:255-260 (1980).
28. D. Tagari, G. H. DuBoulay, V. Aiken and D_4 J. Boullin, Leukotriene D and the cerebral vasculature in vivo and in vitro, *Prostag. Leukot. Med.*, 11:281-297 (1983).
29. T. Kawitami, M. A. Little and E. F. Ellis, Effect of leukotrienes, 12-HETE, histamine, bradykinin and 5-hydroxytryptamine on in vivo rabbit cerebral arteriolar diameter, *J. Cereb. Blood Flow Metab.*, 5:554-559 (1985).
30. A. Unterberg, W. Schmidt, T. Polk, M. Wahl, E. Ellis, A. Marmarou and A. Baethmann, Evidence against leukotrienes as mediators of brain edema, *J. Cereb. Blood Flow Metab.*, 7 (Suppl. 1):5625 (1987).
31. D. W. Busija, C. W. Leffler and D. G. Beasley, Effects of leukotriene C_4, D_4 and E_4 on cerebral arteries of newborn pigs, *Pediat. Res.*, 20:973-976 (1986).
32. K. L. Balck, Leukotriene C_4 induced vasogenic edema in rats, *Prostag. Leukot. Med.*, 14:339-340 (1984).
33. D. Regoli and J. Barabe, Pharmacology of bradykinin and related kinins, *Pharmacol. Rev.*, 32:1-46 (1980).
34. D. H. Gawlowski, A. B. Ritter and W. Duran, Reproducibility of microvascular permeability responses to successive topical applications of bradykinin in the hamster cheek pouch, *Microvascular Res.*, 24:354-363 (1982).
35. M. Wah, A. R. Young, L. Edvinsson and F. Wagner, Effects of kininase II inhibitors on the vasomotor response to bradykinin of feline intracranial and extracranial arteries in vitro and in situ, *J. Cereb. Blood Flow Metab.*, 3:39-45 (1983).
36. E. G. Erdös, "Bradykinin, Kallidin and Kallikrein", Handbook Exp. Pharmacol., Vol. 25 (Suppl), Springer-Verlag, Berlin, Heidelberg and New York (1979).

37. A. Baethmann, W. Oettinger, W. Rothenfusser, O. Kempski, A. Unterberg and R. Geiger, Brain edema factors: current state with particular reference to plasma constituents and glutamate, *in*: "Advances in Neurology 28, Brain Edema", J. Cervos-Navarro and R. Ferszt, eds., Raven Press, New York, pp 171-195 (1980).
38. K. Maier-Hauff, A. J. Baethmann, M. Lange, L. Schürer and A. Unterberg, The Kallikrein-kinin system as mediator in vasogenic brain edema, Part II, *J. Neurosurg.*, 61:97-106 (1984).
39. E. T. Whalley and M. Wahl, The effect of kininase II inhibitors on the response of feline cerebral arteries to bradykinin and angiotensin, *Pflügers Arch.*, 398:175-177 (1983).
40. N. Toda, Actions of bradykinin on isolated cerebral and peripheral arteries, *Am. J. Physiol.*, 232:H267-H274 (1977).
41. E. T. Whalley and M. Wahl, Analysis of bradykinin receptor mediating relaxation of cat cerebral arteries in vivo and in vitro, *Naunyn Schmiedebergs Arch. Pharmacol.*, 323:66-71 (1983).
42. M. Wahl, Local chemical neural and humoral regulation of cerebrovascular resistance vessels, *J. Cardiovascular Pharmacol.*, 7 (Suppl. 3):536-546 (1985).
43. A. Unterberg, M. Wahl and A. Baethmann, Effects of bradykinin on cerebrovascular permeability and resistance, *J. Cereb. Blood Flow Metab.*, 3 (Suppl. 1):5234-5235 (1983).
44. A. Unterberg, M. Wahl and A. Baethmann, Effects of bradykinin on permeability and diameter of pial vessels in vivo, *J. Cereb. Blood Flow Metab.*, 4:574-585 (1984).
45. M. Wahl, A. Unterberg and A. Baethmann, Intravital fluorescence microscopy for the study of blood brain barrier function, *Int. J. Microcirc. Clin. Exp.*, 4:3-18 (1985).
46. K. Tatemoto, M. Carlquist and V. Mutt, Neuropeptide Y - a novel brain peptide with structural similarities to peptide YY and pancreatic polypeptide, *Nature*, 296:659-660 (1982).
47. L. Edvinsson, P. Emsson, J. McCulloch, K. Tatemoto and R. Uddman, Neuropeptide Y: cerebrovascular innervation and vasomotor effects in the cat, *Neurosci. Lett.*, 43:79-84 (1983).
48. L. Edvinsson, P. Emsson, J. McCulloch, K. Tatemoto and R. Uddman, Neuropeptide Y: immunocytochemical localization to and effect upon pial arteries and veins in vitro and in situ, *Acta Physiol. Scand.*, 122:155-163 (1984).
49. E. Edblad, L. Edvinsson, C. Wahlestedt, R. Udalman, R. Hakanson and F. Sundler, Neuropeptide Y co-exists and co-operates with noradrenaline in perivascular nerve fibers, *Regulatory Peptides*, 8:225-235 (1984).
50. L. Edvinsson, and R. Ekman, Distribution and dilatory effect of vasoactive intestinal polypeptide (VIP) in human cerebral arteries, *Peptides*, 5:329-331 (1984).
51. L. Edvinsson, J. Fahrenkrug, J. Hanko, C. Owman, F. Sundler and R. Uddman, VIP (Vasoactive Intestinal Polypeptide)-containing nerves of intracranial arteries in mammals, *Cell Tissue Res.*, 208:135-142 (1980).
52. S. Kobayashi, K. Koshima, J. A. Olschowska and D. M. Jacobowitz, Vasoactive intestinal polypeptide immunoreactive and cholinergic nerves in the whole mount preparation of the major cerebral arteries of the rat, *Histochemistry*, 79:377-381 (1983).
53. L.-I. Larson, L. Edvinsson, J. Fahrenkrug, R. Hakanson, C. Owman, O. B. Schaffolitzky de Muckadell and F. Sundler, Immunohistochemical localization of a vasodilator polypeptide (VIP) in cerebrovascular nerves, *Brain Res.*, 113:400-404 (1976).
54. M. Lindvall, J. Alumets, L. Edvinsson, J. Fahrenkrug, R. Hakanson, J. Hanko, C. Owman, O. B. Schaffolitzky de Muckadell and F. Sundler, Peptidergic (VIP) nerves in the mammalian choroid plexus, *Neurosci. Lett.*, 9:77-82 (1978).

55. F. Eckenstein and R. W. Baughman, Two types of cholinergic innervation in cortex, one co-localized with vasoactive intestinal polypeptide, *Nature*, 309:153-155 (1984).
56. I. Loren, P. C. Emson, J. Fahrenkrug, A. Bjorklund, J. Alumets, R. Hakanson and F. Sundler, Distribution of vasoactive intestinal polypeptide in the rat and mouse brain, *Neuroscience*, 4:1953-1976 (1979).
57. H. Hara, G. S. Hamil and D. M. Jacobowitz, Origin of cholinergic nerves to the rat major cerebral arteries: co-existence with vasoactive intestinal polypeptide, *Brain Res. Bull.*, 14:179-188 (1985).
58. I. L. Gibbins, J. E. Brayden and J. A. Bevan, Distribution and origins of VIP immunoreactive nerves in the cephalic circulation of the cat, *Peptide*, 5:209-212 (1984).
59. J. J.-F. Lee, A. Saito and I. Berezin, Vasoactive intestinal polypeptidic-like substance: the potential transmitter for cerebral vasodilation, *Science*, 224:898-901 (1984).
60. L. Edvinsson, B. B. Fredholm, E. Hamel, I. Janson and C. Verecchia, Perivascular peptides relax cerebral arteries concomitant with stimulation of cyclic adenosine monophosphate accumulation or release of an endothelium-derived relaxing factor in the cat, *Neurosci. Lett.*, 58:213-217 (1985).
61. Y. Suzuki, D. McMaster, K. Lderis and O. P. Rorstad, Characterization of the relaxant effect of vasoactive intestinal polypeptide (VIP) and PHI on isolated brain arteries, *Brain Res.*, 322:9-16 (1984).
62. J. McCulloch and L. Edvinsson, Cerebral circulatory and metabolic effects of vasoactive intestinal polypeptide, *Am. J. Physiol.*, 238:H449-H456 (1980).
63. D. A. Wilson, J. T. O'Neill, S. I. Said and R. J. Traystman, Vasoactive intestinal polypeptide and the canine cerebral circulation, *Circ. Res.*, 48:138-148 (1981).
64. J. McCulloch, P. A. T. Kelly, R. Uddman and L. Edvinsson, Functional role for vasoactive intestinal polypeptide in the caudate nucleus, *Proc. Nat. Acad. Sci.*, 80:1472-1476 (1983).
65. J. McCulloch and P. A. T. Kelly, A functional role for vasoactive intestinal polypeptide in anterior cingulate gyrus, *Nature*, 304:438-440 (1983).
66. B. Pernow, Substance P, *Pharmacol. Rev.*, 35:85-141 (1983).
67. H. Nawa, M. Dotcuchi, K. Igano, K. Inouye and S. Nakanishi, Substance K: a novel mammlian tachykinin that differs from substance P in its pharmacological profile, *Life Sci.*, 34:1153-1160 (1984).
68. L. Edvinsson, J. McCulloch and R. Uddman, Substance P: immunohistochemical localization and effect upon cat pial arteries in vitro and in situ, *J. Physiol.*, 318:251-258 (1981).
69. L. Edvinsson, S. Rosenthal-Helgesen and R. Uddman, Substance P: localization, concentration and release in cerebral arteries, choroid plexus and dura mater, *Cell Tissue Res.*, 234:1-7 (1983).
70. R. Uddman, L. Edvinsson, C. Owman and F. Sundler, Perivascular substance P occurrence and distribution in mammalian pial vessels, *J. Cereb. Blood Flow Metab.*, 1:227-231 (1981).
71. K. Yamamoto, T. Matsuyama, S. Shioska, S. Inagaki, E. Senba, Y. Shimizu, I. Yshimoto, T. Hayakawa, M. Matsumoto and M. Tohyama, Overall distribution of substance P containing nerves in the wall of the cerebral arteries of the guinea pig and its origin, *J. Comp. Neurol.*, 215:421-426 (1983).
72. L.-Y. Liu-Chen, D. H. Han and M. A. Moskowitz, Pia arachnoid contains substance P originating from trigeminal neurons, *Neuroscience*, 9:803-808 (1983).
73. R. Uddman, L. Edvinsson, R. Ekman, T. A. Kingman and J. McCulloch, Innervation of the feline cerebral vasculature by nerve fibers containing calcitonin gene-related peptide: trigeminal origin and co-existence with substance P, *Neurosci. Lett.*, 62:131-136 (1985).

74. M. A. Moskowitz, M. Brody and L.-Y. Liu-Chen, In vitro release of immunoreactive substance P from putative afferent nerve endings in bovine pia arachnoid, *Neuroscience*, 9:804-814 (1983).
75. D. Ezra, F. R. M. Laurindo, J. Eimerl, C. Peck, G. Goldstein and G. Feuerstein, Tachykinin modulation of coronary blood flow, *Eur. J. Pharmacol.*, 122:135-138 (1986).
76. L. Edvinsson, J. McCulloch, S. Rosell and R. Uddman, Antagonism by D-Pro^2, D-Trip^{7,9})-substance P of the cerebrovascular dilation induced by substance P, *Acta Physiol. Scand.*, 116:411-416 (1982).
77. L. Edvinsson and R. Uddman, Immunocyochemical localization and dilatory effect of substance P on human cerebral vessels, *Brain Res.*, 232:466-471 (1982).
78. F. Sundler, F. Brodin, E. Ekblad, R. Hakanson and R. Uddman, Sensory nerve fibers distribution of substance P, neurokinin A and calcitonin gene-related peptide, *in*: "Tachykinin Antagonists", R. Hakanson and F. Sundler, eds., Elsevier, Amsterdam, pp 3-14 (1985).
79. H. Nawa, T. Hirose, H. Takashima, S. Inayama and S. Nakanisti, Nucleotide sequences of cloned cDNAs for two types of bovine brain susbtance P precursor, *Nature*, 306:32-36 (1983).
80. H. R. Morris, M. Panico, T. Etienne, J. Tippins, S. I. Girgis and I. MacIntyre, Isolation and characterization of human calcitonin gene-related peptide, *Nature*, 308:746-748 (1984).
81. S. G. Amara, V. Tonas, M. G. Rosenfeld, E. S. Ong and R. M. Evans, Alternative RNA processing in calcitonin gene expression generates mRNAs encoding different products, *Nature*, 298:240-244 (1982).
82. M. G. Rosenfeld, J. J. Mermod, S. G. Amara, L. W. Swanson, P. E. Sawchenko, J. Rivier, W. W. Vale and R. M. Evans, Productions of a novel neuropeptide encoded by the calcitonin gene via tissue specific RNA processing, *Nature*, 304:129-135 (1983).
83. D. M. Jacobowitz and G. Skofitsch, Calcitonin gene-related peptide in the central nervous system, *in*: "Neural and Endocrine Peptides and Receptors", T. W. Moody, ed., Plenum Press, New York, pp 247-288 (1986).
84. T. Lee, Y. Kawai, S. Shiosaka, K. Takami, H. Kiyama, C. J. Hillyar, S. Girgis, I. MacIntyre, P. C. Emson and M. Tohyama, Co-existence of calcitonin gene-related peptide and substance P-like peptide in single cells of the trigeminal ganglion of the rate: immunohistochemical analysis, *Brain Res.*, 330:194-196 (1985).
85. D. Ezra, F. R. M. Laurindo, D. S. Goldstein, R. E. Goldstein and G. Feuerstein, Calcitonin gene-related peptide: a potent modulator of coronary flow, *Eur. J. Pharmacol.*, 137:101-105 (1987).
86. J. Hanko, J. E. Hardebo, J. Kahrstrom, C. Owman and F. Sundler, Calcitonin gene-related peptide is present in mammalian cerebrovascular nerve fibers and dilates pial and peripheral arteries, *Neurosci. Lett.*, 47:91-95 (1985).
87. L. Edvinsson, J. McCulloch and R. Uddman, Feline cerebral veins and arteries comparison of autonomic innervation and vasomotor responses, *J. Physiol.*, 325:161-173 (1982).
88. J. McCulloch, R. Uddman, T. A. Kingman and L. Edvinsson, Calcitonin gene-related peptide: functional role in cerebrovascular regulation, *Proc. Nat. Acad. Sci.*, 83:5731-5735 (1986).
89. N. Minamino, K. Kangaweka and M. Matsuo, Neuromedin C: a bombesin like peptide identified in porcine spinal cord, *Biochem. Biophys. Res. Comm.*, 119:14-20 (1984).
90. R. Uddman, L. Edvinsson, C. Owman and F. Sundler, Nerve fibers containing gastrin releasing peptide around pial vessels, *J. Cereb. Blood Flow Metab.*, 3:386-390 (1983).
91. C. Sumners, M. I. Phillips and E. M. Richards, Central pressor action of neurotensin in the conscious rat, *Hypertension*, 4:888-893 (1982).

92. R. Quirion, F. Rioux, D. Regoli and S. St Pierre, Selective blockade of neurotensin induced coronary vessel constriction in perfused rat heart by a neurotensin analog, *Eur. J. Pharmacol.*, 61:309-312 (1980).
93. L. Jennes, W. E. Stumpf and P. W. Kalivas, Neurotensin: topographical distribution in rat brain by immunohistochemistry, *J. Comp. Neurol.*, 210:211-224 (1982).
94. V. Chan-Palay, Innervation of cerebral blood vessels by norepinephrine, indoleamine, substance P, and neurotensin fibers and the leptomeningeal indoleamine axons: their role in vasomotor activity and local alterations of brain blood composition, *in*: "Neurogenic Control of Brain Circulation", C. Owman and L. Edvinsson, eds., Pergamon Press, Oxford, pp 39-54 (1977).
95. L.-I. Larsson and J. F. Rehfeld, Localization and molecular heterogeneity of cholecystokinin in the cerebral and peripheral nervous system, *Brain Res.*, 165:201-218 (1979).
96. I. Loren, J. Alumets, R. Hakanson and F. Sundler, Distribution of gastrin and CCK-like peptides in rat brain, *Histochemistry*, 59:249-257 (1979).
97. S. H. C. Hendry, E. G. Jones and M. C. Beinfield, Cholecystokinin immunoreactive neurons in rat and monkey cerebral cortex make symmetric syrapses and have intimate associations with blood vessels, *Proc. Nat. Acad. Sci.*, 80:2400-2404 (1983).
98. J. McCulloch, Perivascular nerve fibers and the cerebral circulation, *Trends Neurosci.*, 7:135-138 (1984).
99. J. McCulloch and P. A. T. Kelly, Effects of cholecystokinin octapeptide on pial arteriolar diameter, *J. Cereb. Blood Flow Metab.*, 4:625-628 (1984).
100. G. W. Pasternak, R. Goodman and S. H. Snyder, Endogenous morphine like factor in mammalian brain, *Life Sci.*, 16:1765-1769 (1975).
101. J. Huges, T. W. Smith, H. W. Kosterlitz, L. A. Fothergill, B. A. Morgan and H. R. Morris, Identification of two related pentapeptides from the brain with potent opiate agonist activity, *Nature*, 258:577-579 (1975).
102. S. E. Kapadia and N. C. de Lanerolle, Immunohistochemical and electron microscopic demonstration of vascular innervation in the mammalian brain stem, *Brain Res.*, 292:33-40 (1984).
103. S. J. Peruoutka, M. A. Moskowitz and S. H. Snyder, Neurotransmitter receptor binding in bovine cerebral microvessels, *Science*, 208:610-613 (1980).
104. J. Knoll, Neuronal peptide (enkephalin) receptors in the ear artery of the rabbit, *Eur. J. Pharmacol.*, 39:403-407 (1976).
105. J. H. Hanko and J. E. Hardebo, Enkephalin induced dilatation of pial arteries in vitro probably mediated by opiate receptors, *Eur. J. Pharmacol.*, 51:295-297 (1978).
106. T. M. Wong, A. Koo and C. H. Li, β-endorphin vasodilation effect on the microcirculatory system of hamster cheek pouch, *Int. J. Pept. Protein Res.*, 18:420-422 (1981).
107. Y. Yamamoto, K. Hotta and T. Matsuda, Effect of methionine-enkephalin on the spontaneous electrical and mechanical activity of the smooth muscle of the rat portal vein, *Life Sci.*, 34:993-999 (1984).
108. G. Feuerstein and Z. Zukowska-Grojec, Effect of dermorphin and morphine on the symapthetic and the cardiovascular system of the pithed rat, *Neuropeptides*, 9:139-150 (1987).
109. J. Eimerl and G. Feuerstein, The effect of μ, δ, κ and ε opioid receptor agonists on heart rate and blood pressure of the pithed rat, *Neuropeptides*, 8:351-358 (1986).
110. M. Wahl, Effects of enkephalins morphine and naloxone on pial arteries during perivascular microapplication, *J. Cereb. Blood Flow Metab.*, 5:451-457 (1985).
111. L. Edvinsson and E. T. MacKenzie, Amine mechanisms in the cerebral circulation, *Pharmacol. Rev.*, 28:275-348 (1977).
112. W. Kuschinsky and M. Wahl, Local chemical and neurogenic regulation of cerebral vascular resistance, *Physiol. Rev.*, 58:656-689 (1978).

113. J. A. Bevan, J. Duckworth, I. Laher, M. A. Oriowo, G. A. McPherson and R. D. Bevan, Sympathetic control of cerebral arteries: specialization in receptor type, reserve, affinity and distribution, *FASEB*, 1:193-198 (1987).
114. N. Toda and Y. Fujita, Responsiveness of isolated cerebral and peripheral arteries to serotonin, norepinephrine and transmural electrical stimulation, *Circ. Res.*, 33:98-104 (1973).
115. L. Edvinsson and C. H. Owman, Pharmacological characterization of adrenergic alpha and beta receptors mediating the vasomotor responses of cerebral arteries in vitro, *Circ. Res.*, 35:835-849 (1974).
116. I. C. Medgett and S. Z. Langer, Characterization of smooth muscle α-adrenoceptors and of responses to electrical stimulation in the cat isolated perused middle cerebral artery, *Naunyn Schmiedeberg's Arch. Pharmacol.*, 323:24-32 (1983).
117. M. Wahl, W. Kuschinsky, O. Bosse, J. Olesen, N. A. Lassen, D. H. Ingvar, J. Michaelis and K. Thurau, Effect of 1-norepinephrine on the diameter of pial arterioles and arteries in the cat, *Circ. Res.*, 31:248-256 (1972).
118. K. Ulrich, L. M. Auer and W. Kuschinsky, Cat pial venoconstriction by topical microapplication of norepinephrine, *J. Cereb. Blood Flow Metabl.*, 2:109-111 (1987).
119. P. M. Gross, A. M. Harper and G. M. Teasdale, Interactions of histamine with noradrenergic constrictory mechanisms in cat cerebral arteries and vein, *Can. J. Physiol. Pharmacol.*, 61:756-763 (1983).
120. M. Wahl, W. Kuschinsky, O. Bosse and A. Neiss, Micropuncture evaluation of β-receptors in pial arteries of cats, *Pflügers Arch.*, 348:293-303 (1974).
121. W. Kuschinsky and M. Wahl, Alpha receptor stimulation by endogenous and exogenous norepinephrine and its blockade by phentolamine in pial arteries of the cat, *Circ. Res.*, 37:168-174 (1975).
122. W. Kuschinsky and M. Wahl, Interaction between perivascular norepinephrine and potassium or osmolarity on pial arteries of cats, *Microvasc. Res.*, 14:173-1980 (1977).
123. L. M. Auer, B. B. Johansson and S. Lund, Reaction of pial arteries and veins to sympathetic stimulation in the cat, *Stroke*, 12:528-531 (1981).
124. D. D. Heistad, M. L. Marcus, S. Sandberg and F. M. Abboud, Effect of sympathetic nerve stimulation on cerebral blood flow and on large cerebral arteries of dogs, *Circ. Res.*, 41:342-350 (1977).
125. M. L. Marcus and D. D. Heistad, Effects of sympathetic nerves on cerebral blood flow in awake dogs, *Am. J. Physiol.*, 236:H549-H553 (1979).
126. J. F. Vatner, L. L. Priano, J. D. Rutherford and W. T. Manders, Sympathetic regulation of the cerebral circulation by the carotid chemoreceptors reflex, *Am. J. Physiol.*, 238:H597-H598 (1980).
127. M. Beausang Linder and A. Bill, Cerebral circulation in acute arterial hypertension - protective effects of sympathetic nervous activity, *Acta Physiol. Scand.*, 111:193-199 (1981).
128. A. Bill and J. Linder, Sympathetic control of cerebral blood flow in acute arterial hypertension, *Acta Physiol. Scand.*, 96:114-121 (1976).
129. L. Edvinsson, C. H. Owman and B. K. Siesjö, Physiological role of cerebrovascular sympathetic nerves in the autoregulation of cerebral blood flow, *Brain Res.*, 117:519-523 (1976).
130. D. D. Heistad and M. L. Marcus, Effect of sympathetic stimulation on the permeability of the blood brain barrier to albumin during acute hypertension in cats, *Circ. Res.*, 45:331-338 (1979).
131. E. T. MacKenzie, A. P. McGeorge, D. I. Graham, W. Fitch, L. Edvinsson and A. M. Harper, Effect of increasing arterial pressure on cerebral blood flow in the baboon: influence of the sympathetic nervous system, *Pflügers Arch.*, 379:157-163 (1979).
132. V. Chan-Palay, Serotonin axons in the supra and sub-ependymal plexuses and in the leptomeninges; their role in local alterations of cerbrospinal fluid and vasomotor activity, *Brain Res.*, 132:103-130 (1976).

133. B. Scatton, L. Edvinsson, A. Degeueurce, D. Duverger, R. L'Heurex and E. T. MacKenzie, Histochemical and biochemical evidence for perivascular 5-hydroxytryptamine nerves in cerebral blood vessels, *J. Cereb. Blood Flow Metab.*, 3 (Supp 1):5230-5231 (1983).
134. C. Forster and E. T. Whalley, Analysis of the 5-hydroxytryptamine induced contraction of the human basilar arterial strip compared with the rat aortic strip in vitro, *Naunyn Schmiedebergs Arch. Pharmacol.*, 319:12-17 (1982).
135. J. E. Hardebo, L. Edvinsson, C. H. Owman and N. A. Svendgaard, Potentiation and antagonism of serotonin effect on intracranial and extracranial vessels, *Neurology*, 28:64-70 (1978).
136. L. Edvinsson, J. E. Hardebo and C. H. Owman, Pharmacological analysis of 5-hydroxytryptamine receptors in isolated intracranial and extracranial vessels of cat and man, *Circ. Res.*, 42:143-151 (1978).
137. L. M. Auer, I. Sayama, K. Haselsberger and K. Leber, Reaction of pial vessels and cerebral blood volume to intravenous methohexital and to topical serotonin, *J. Cereb. Blood Flow Metab.*, 3 (Suppl 1):S568-S569 (1983).
138. A. M. Harper and E. T. MacKenzie, Effects of 5-hydroxytryptamine on pial arteriolar calibre in anesthetized cats, *J. Physiol.*, 271:735-746 (1977).
139. L. Edvinsson, J. E. Hardebo, E. T. MacKenzie and M. Stewart, Dual action of serotonin on pial arterioles in situ and the effect of propranolol on the response, *Blood Vessels*, 14:366-371 (1977).
140. C. R. Calcutt, The role of histamine in the brain, *Gen. Pharmacol.*, 7:15-26 (1976).
141. J. C. Schwartz, Histaminergic mechanisms in the brain, *Ann. Rev. Pharmacol. Toxicol.*, 17:325-340 (1977).
142. P. M. Gross, Cerebral histamine: indications for neural and vascular regulation, *J. Cereb. Blood Flow Metab.*, 2:3-23 (1982).
143. L. Edvinsson, C. H. Owman and N. O. Sjoberg, Autonomic nerves, mast cels and amine receptors in human brain vessels; a histochemical and pharmacological study, *Brain Res.*, 115:377-393 (1976).
144. A.-L. Ronnberg, L. Edvinsson, L. I. Larsson, K. C. Nielsen and C. H. Owman, Regional variation in the presence of mast cells in the mammalian brain, *Agents and Actions*, 3:191-192 (1973).
145. L. Edvinsson, P. M. Gross and A. Mohamed, Characterization of histamine receptors in cat cerebral arteries in vitro and in situ, *J. Pharmacol. Exp. Ther.*, 225:168-175 (1983).
146. P. M. Gross, A. M. Harper and G. M. Teasdale, Cerebral circulation and histamine: 2. Responses of pial veins and arterioles to receptor agonists, *J. Cereb. Blood Flow Metab.*, 1:219-225 (1981).
147. M. Wahl and W. Kuschinsky, The dilating effect of histamine on pial arteries of cats and its mediation by H_2-receptors, *Circ. Res.*, 44:161-165 (1979).
148. L. Edvinsson and C. H. Owman, A pharmacologic comparison of histamine receptors in isolated extracranial and intracranial arteries in vitro, *Neurology*, 25:271-276 (1975).
149. P. M. Gross, G. M. Teasdale, D. I. Graham, W. J. Angerson and A. M. Harper, Intra-arterial histamine increases blood-brain barrier transport in rats, *Am. J. Physiol.*, 243:H307-H317 (1982).

CLINICAL ASPECTS OF SINUS VEIN THROMBOSIS

HISTORICAL PERSPECTIVES ON THE PATHOGENESIS OF THROMBOSIS IN THE CEREBRAL VEINS AND SINUSES

H. J. M. Barnett

The John P. Robarts Research Institute
POB 5015, London
Ontario N6A 5K8, Canada

INTRODUCTION

My invitation to contribute a chapter to this volume is based on an early communication on this fascinating topic which dates back to the days prior to the introduction of computed tomography (CT), magnetic resonance imaging (MRI) and even routine cerebral angiography [1]. As a neurological resident my interest was roused by the admission of a 32-year-old patient comatose with neck stiffness, a third nerve palsy and grossly bloody cerebrospinal fluid. Assuming that the patient had suffered a ruptured posterior communication aneurysm, Canada's pioneer neurosurgeon, Dr K. G. McKenzie, had performed an exploratory and negative craniotomy. At the time the patient was 10 days postpartum, and failed to survive the operation. At postmortem examination multiple sinus and vein thromboses were encountered with widespread areas of hemorrhagic infarction and gross blood within the subarachnoid space and the ventricles.

My mentor, Dr H. H. Hyland, at that time recently had published an account of another postpartum patient who developed seizures, aphasia and mild hemiparesis [2]. She recovered sufficiently well to be preparing herself for discharge when she was afflicted suddenly with a fatal pulmonary embolus, the result of an extension of thrombus down the superior vena cava via the jugular vein from the lateral sinus.

Historically, French, English and German writers had recognized this condition beginning in 1825 when Ribes described a 45-year-old male patient with superior sagittal sinus thrombosis accompanying disseminated malignancy [3]. Abercrombie, in 1828, recognized the association of sinus thrombosis with the puerperium [4]. Gowers described it in association with marasmic and cachectic conditions and with a failing circulation [5]. Quincke and Nonne identified the clinical syndrome of pseudotumor cerebri (a term coined by the latter) as a clinical counterpart to sinus thrombosis [6,7].

Cerebral Sinus Thrombosis, Edited by K. Einhäupl *et al.*
Plenum Press, New York, 1990

In the 1940s Martin [8], and then Martin with Sheehan [9], focusing on postpartum venous sinus and cerebral venous thrombosis, evoked a number of explanations including air embolism, embolism by way of the paravertebral venous plexuses, and damage to vein walls, secondary to the venous engorgement of pregnancy.

Provoked by the uncertainties surrounding the pathogenesis of aseptic venous and sinus thrombosis, a review of the files of the Banting Institute in Toronto was undertaken. Here Professor Linell had kept meticulous records for a 20-year period from a major general hospital (Toronto General Hospital) and the material which was scrutinized indicated that aseptic cerebral vein and sinus thrombosis occurred under the same circumstances as crural phlebo-thrombosis [1]. During the conduct of approximately 8000 postmortem examinations (over the 20-year period, 1932-1951), 39 patients died with, but not necessarily because of, venous and sinus thrombosis. Cervos-Navarro encountered very similar numbers (41 cases in 19 years) [10]. It is interesting to note that using modern methods Einhäupl [11] has made the clinical diagnosis 71 times at a rate of eight cases per year and Bousser et al. [12] encountered 38, and the rate of five cases per year contrasted with two per year in our own and in Cervos-Navarro's postmortem series. Despite the possible disparity in the reference pattern and size of the centers given, it is evident that our diagnostic ability has been sharpened by newer techniques of imaging and that a heightening sense of awareness probably has resulted in an increased search for this entity.

The disorders with which the 39 patients in our previously published review were associated are detailed in Table 1.

It is to be noted that in this previous era heart disease and cardiac failure were the most common associations with non-septic venous and sinus thrombosis. It was at a time when massive anasarca was more common because it was more difficult to treat than by modern methods. The postoperative patients had been submitted to a variety of procedures including hysterectomy, cholecystectomy and thyroidectomy. The post-traumatic cases were usually accompanied by serious head injuries as described by Sargeant during World War I but others were associated with non-cerebral trauma. This series included four patients with intracerebral arterial lesions, two with posterior cerebral artery territory infarction, one with a thrombosed internal carotid artery; one of the patients with an

Table 1. Non-infective Intracranial Venous Thrombosis - Associated Disorders in 39 Fatal Cases

Heart disease	10
Cachexia and marasmus	7
Postoperative	5
Posttraumatic	5
Major cerebral arterial occlusion or hemorrhage	4
Postpartum	3
Blood dyscrasias	2
Fever therapy and hyperpyrexia	2
No associated disorder	1

Table 2. Recognized Coagulation Defects

Antithrombin III
Cryofibrinogenemia
Disseminated intravascular coagulopathy
Idiopathic thrombocytosis
Polycythemia
Paroxysmal nocturnal hemoglobinemia

extensive longitudinal sinus occlusion had suffered a hypertensive intra-occipital hemorrhage.

Contemporary studies indicate that the most important additional factor in delineating etiology has been the extending ability to detect coagulation abnormalities. Table 2 indicates the known defects in coagulation which have been associated with aseptic cerebral vein and sinus thrombosis. References related to these disorders are available in a recent review [13] and in a paper by Bousser [12].

There are other conditions in which no measurable defect can be detected; this may be a reflection of the imperfections of our ability to measure all coagulation abnormalities. More often we are able to identify the clinical circumstances under which coagulation will be abnormal than to detect the specific hematologic abnormality. Table 3 lists a number of the conditions which are presumed to be associated with coagulation abnormality and have been recognized as coexisting with cerebral vein and sinus occlusions.

The literature including reviews presented in this volume indicate that the commonest situations in which aseptic vein and sinus thrombosis will be recognized are with the use of the oral contraceptives, during pregnancy, in the postpartum state, and in association with systemic malignancy. In all of these situations there is a common association with crural vein thrombosis as well.

CLINICAL FEATURES

Clinical features of cerebral vein and sinus thrombosis are dependent upon a number of factors. This includes the site of thrombosis within the cerebral venous system, the rate at which it has progressed, the extent to

Table 3. Coagulation Abnormalities Presumed in Cerebral Vein and Sinus Thrombosis

Oral contraceptives
Pregnancy, postpartum
Ulcerative colitis, Crohn's disease
Carcinoma, leukemia
Nephrotic syndrome
Androgen therapy
Trauma, fractures, postoperative

which it has progressed and the presence or absence of recanalization. Recanalization is a feature of cerebral vein and sinus thrombosis. By comparison this process is much less commonly recognized in arterial thrombosis. It occurs with greatly reduced frequency when the arterial wall is diseased. It will occur in the presence of thrombo-embolism to an artery which is not the site of disease. In this way the comparison is to venous and sinus thrombosis: in-site thrombosis within a vessel which has intact endothelium.

CLINICAL SYNDROMES AND THEIR PATHOLOGICAL CORRELATES

Pseudotumor Cerebri

With modern diagnostic methods, including venous angiography and MRI, patients with this syndrome related to sinus thrombosis are much more readily detected. The clinical condition consisting of headache, papilledema, and the false localizing phenomena of any oculomotor nerve palsies (III, IV, VI, with the VIth the most common) are sufficient to justify consideration of this syndrome. It will occur with both lateral and superior sagittal sinus thrombosis and the accompanying cerebral edema may be without infarction or hemorrhage.

Cavernous Sinus Syndrome

This is associated occasionally with an aseptic, as opposed to a septic, condition as a sequel to head or facial trauma. The characteristic painful proptosis, chemosis, ocular palsy visual loss and progression from one side to the other will be seen. Blindness of one or both eyes may develop, as may an associated hemiparesis related to an accompanying internal carotid artery occlusion.

Cortical Vein Thrombosis

The onset of the clinical features related to this pathological process are often abrupt and may be stepwise. It is commonly associated with headache, and focal or generalized seizures. Deep lesions produce early coma and the most important prerequisite to the making of the diagnosis is the suspicion that it may occur. This requires an alerting of the clinician triggered by the presence of the clinical settings in which venous and sinus thrombosis develop. As a working rule, cerebral vein thrombosis should be taken into consideration in any patient who presents with apoplexy associated with either severe headache, convulsion or both. The suspicion should be even higher if the patient is affected with any condition that may predispose to venus thrombosis.

Hemorrhagic Syndromes

Subarachnoid hemorrhage and intracerebral hematoma(ta) may be major features of venous and sinus thrombosis. The diagnosis will be made provided the observer attends to the circumstances under which the neurological disability has developed. The clinical syndrome with subarachnoid hemorrhage may well be no different from that encountered in rupture of an aneurysm in some of the patients. Progressive and bilateral signs will be encountered with intra-arterial hematomata.

PROGNOSIS

The prognosis will be calculated variously, depending on the accuracy and frequency with which the diagnosis is made. If it is diagnosed only when a most florid clinical condition exists, it is readily apparent that only those with the most severe forms of the disease will be known and the outlook will be considered more grave. Nowadays when the clinical diagnosis can be established with accuracy more frequently, and thereby recognized in less severe clinical circumstances, patients will more often be encountered who survive without serious sequelae. The early occurrence of grand mal seizures is more likely to carry a poor prognosis; if the average interval from onset of clinical symptoms to seizure activity is less than 4 days, fatalities occur more frequently compared with patients who experience the later development of seizures. Often with late seizures, a majority recover. Hemorrhagic infarction and intracerebral hematomata carry a poor prognosis. However, the recent and exciting work of Bousser and of Einhäupl suggest that an extension of a hemorrhagic infarction of the brain and the risk of pulmonary embolism are both decreased if heparin therapy is administered. New prognostic standards must now be developed which account for better ability to diagnose milder cases and for this substantive advance in therapy.

REFERENCES

1. H. J. M. Barnett and H. H. Hyland, Non-infective intracranial venous thrombosis, *Brain*, 76(1):36 (1953).
2. H. H. Hyland, Intracranial venous thrombosis in the puerperium, *J. Am. Med. Assoc.*, 142:707-710 (1950).
3. F. Ribes, Des recherches faites sur la phlebite, *Rev. Med.*, 3:5 (1825).
4. J. Abercrombie, "Pathological and Practical Researches of the Brain and Spinal Cord", John Carfrae and Son, Edinburgh, p 83 (1828).
5. W. R. Gowers, "A Manual of Diseases of the Nervous System", 1st Ed., Vol. 2, Churchill, London (1888).
6. H. Quincke, Über Meningitis serosa, Volkmann's Sammlung klinischer Vorträge, *Inn. Med. Neue Reihe*, 23:655-694 (1893).
7. M. Nonne, Über Fälle vom Symptomenkomplex "tumor cerebri" mit Ausgang in Heilung; über letal verlaufende Fälle von "Pseudotumor cerebri" mit Sektionsbefund, *Dtsch. Z. Nervenheilk.*, 27:169 (1904).
8. J. P. Martin, Thrombosis in the superior longitudinal sinus following childbirth, *Br. Med. J.*, 2:537 (1941).
9. J. P. Martin and H. L. Sheehan, Primary thrombosis of cerebral veins following childbirth, *Br. Med. J.*, 1:349 (1941).
10. J. Cervos Navarro and S. Kannuki, Neuropathological findings in the thrombosis of cerebral veins and sinuses: vascular aspects, *in:* "Cerebral Sinus Thrombosis: Experimental and Clinical Aspects", K. Einhäupl, O. Kempski and A. Baethmann, eds, Plenum Press, New York and London, pp 15-25 (1990).
11. K. M. Einhäupl, A. Villringer, R. L. Haberl, W. Pfister, M. Deckert, H. Steinhoff and P. Schmiedek, The clinical spectrum of sinus venus thrombosis, *in*: "Cerebral Sinus Thrombosis: Experimental and Clinical Aspects", K. Einhäupl, O. Kempski and A. Baethmann, eds, Plenum Press, New York and London, pp 149-155 (1990).
12. M. G. Bousser, Cerebral venous thrombosis - a review of 38 cases, *Stroke*, 16:199 (1985).
13. P. C. Gates and H. J. M. Barnett, Venous disease: cortical veins and sinuses, *in*: "Stroke Pathophysiology, Diagnosis and Management", Churchill Livingstone, New York, pp 731-743 (1986).

CLINICAL SPECTRUM OF SINUS VENOUS THROMBOSIS

K. M. Einhäupl, A. Villringer, R. L. Haberl
W. Pfister, M. Deckert, H. Steinhoff* and P. Schmiedek†

Departments of Neurology, *Radiology and
†Neurosurgery, Ludwig-Maximilians-University
Klinikum Großhadern, 8000 Munich 70, FRG

INTRODUCTION

The true incidence of sinus venous thrombosis (SVT) is unknown because many patients remain undiagnosed or are misdiagnosed. Unfortunately, the symptoms and clinical signs of SVT are unspecific and resemble those of other more common or better known neurological disorders such as ischemic stroke, meningitis, encephalitis, or subarachnoid hemorrhage (SAH). Furthermore, the gold standard in the diagnosis of SVT is cerebral angiography, an invasive method with considerable risk of severe side effects, which many physicians try to avoid in the diagnostic process.

In order to establish a diagnostic protocol for this disease the present study was undertaken. It is a prospective clinical study of aseptic sinus venous thrombosis in man started in 1977.

PATIENTS AND METHODS

The inclusion criterion was an angiographically confirmed diagnosis of SVT.

The angiographic diagnosis required direct signs, such as non-visualization of a sinus or part of a sinus, plus indirect signs, such as broken or corkscrew-like ascending veins, dilated collateral venous channels, thrombosed cortical veins and regionally or globally delayed venous emptying (Figs 1 and 2).

Excluded were patients with mechanical obstruction of the sinus due to skull impression or tumors, patients with cavernous thrombosis, septic SVT, tumors affecting the sinus, children (below 15 years of age), as well as patients with angiographically dubious findings.

Cerebral Sinus Thrombosis, Edited by K. Einhäupl *et al.*
Plenum Press, New York, 1990

The clinical status of each patient was assessed once daily. In addition, all but two of the patients had one or more CT, EEG and CSF examinations.

A detailed protocol containing more than 250 items was used for each patient. The mode of onset, symptoms, acuity, severity, temporal profile and the clinical recovery were analyzed.

RESULTS AND DISCUSSION

Age and Sex

Of the 71 patients, 42 were female (59%) and 29 were male (41%). The age range was between 15 and 72 years, the mean age was 34.3 years (women) and 36.4 years (men). The age distribution showed a peak in the third decade for both women and men.

We did not notice a second peak of incidence in the higher age range as reported by Noetzel and Jerusalem [14]. Also the widely held opinion that SVT is a disease mainly of neonates, puerperal women and the elderly is not supported by our data, rather SVT can occur in all age groups and in both sexes with a peak incidence in the third decade and a 3:2 female:male sex ratio.

Risk Factors and Etiology

Possible causal factors for the thrombotic process in our SVT patients are listed in Table 1.

Two main risk factors emerged: in 32 of 42 (76%) women, hormonal influences like the use of hormonal contraceptives or pregnancy were found [1-8]. The second important risk factor was infectious or inflammatory diseases which occurred in 15 (21%) of our patients [9].

Besides these a variety of other less frequent causal factors have been found (Table 1). Sixteen patients had an extracranial thrombosis in the previous year (four patients), during (five patients) or after SVT (seven patients), indicating that in many instances SVT is not a local disease but rather part of a more generalized process.

Mode of Onset

The mode of onset may be sudden, subacute or lingering.

Twenty-five of our patients had a sudden onset of the disease. In 14 of these patients the initial symptom was a sudden headache resembling the symptoms of subarachnoid hemorrhage. Only two patients, however, had blood in the CSF. Such a beginning of the disease has been noted already by Huhn [10]. In nine patients the initial symptom was an epileptic seizure, in two patients a TIA-like transient aphasia or hemiparesis occurred.

In the 17 patients with subacute onset (< 24 hours), headache, confusional states and increasing hemiparesis suggestive of a progressive stroke were described.

Table 1. Possible Causal Factors in Sinus Venous Thrombosis

Causal Factor	Single Cause	Combined Cause
Pregnancy	2	3
Hormonal contraceptives	10	13
Estrogens after menopause	3	1
Local infection	3	3
Viral infection	3	4
Sepsis in recent anamnesis	1	1
Polyglobulia	1	-
AT3-deficiency	3	-
Lues cerebrospinalis	1	-
Vasculitis	1	-
Trauma	1	1
Surgery in small pelvis	1	-
Shunted hydrocephalus	1	2
Oral corticosteroids	-	2
Intrathecal corticosteroids	-	1
Cytostatic therapy	-	2
Myelography	-	1
Leukemia	-	1
Teratocarcinoma	-	2
Multiple sclerosis	-	2
No causal factor found	7	

The remaining 31 patients had a lingering onset (> 24 hours), in most of these patients headache was also the initial symptom.

Six patients had a history of headache more than 3 months before admission to the hospital. This is comparable to the finding of Bousser [11] who observed seven of 38 patients with a history of symptoms attributable to SVT of more than 3 months duration (up to 8 months), and Huhn [10] noted two cases with a duration of the disease of more than 6 years. Similar reports about extremely long histories have been given by Gutschera-Wang [12], Buchanan and Brazinski [13] and Noetzel and Jerusalem [14].

In conclusion, all three types of onset do occur frequently from sudden onset comparable to the onset of subarachnoid hemorrhage to prodromal like symptoms lasting months or even years. The most common initial symptom is headache (54/71 cases = 76%).

Acuity

Acuity was defined as the time between the onset of the first symptom and admission to the hospital. As with onset, the acuity of the disease is also very variable. Only 20% of the patients were admitted to the hospital within the first 24 hours after the first symptom appeared. After 6 days, 50% of the patients had been admitted. By 3 weeks post-onset, 15% of the patients remained ambulatory. This suggests that unlike arterial disease, SVT in many cases is a slowly progressive disease. We have not found a similar analysis in the literature.

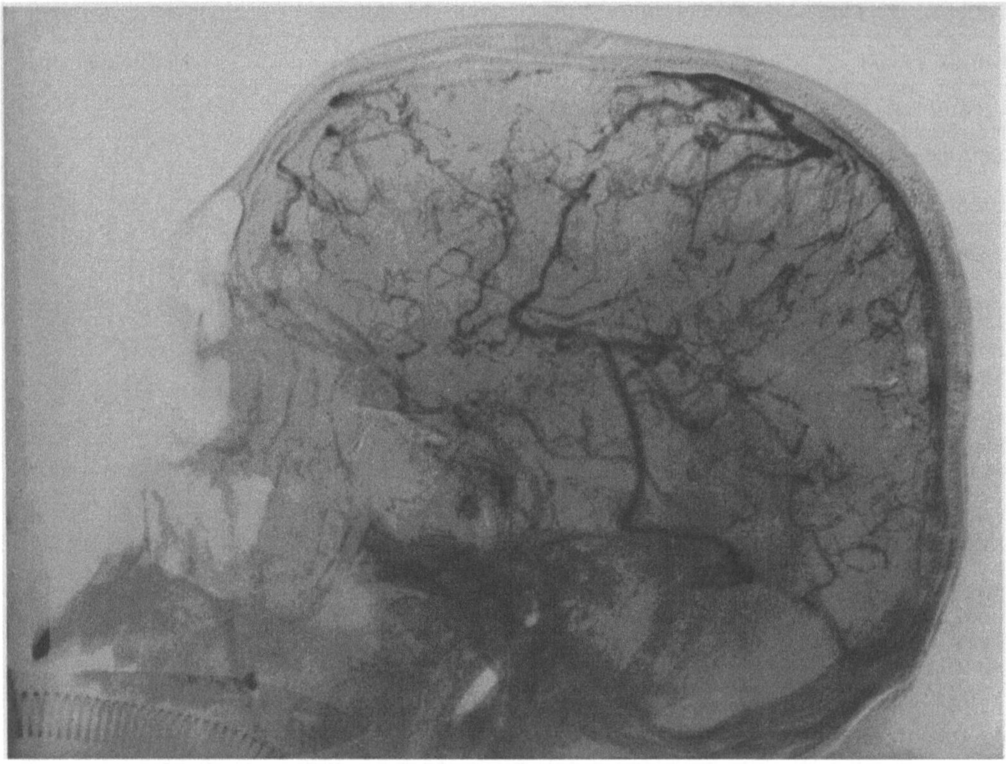

Fig. 1. Right carotid phlebogram, lateral projection. Non-opacification of the frontal and parietal superior sagittal sinus and adjacent cortical veins. Multiple tortuous veins, small corkscrew-like veins, and dilated collateral channels draining into the basal sinus.

Frequency of Symptoms

The most frequent symptom in our series was headache with an occurrence in 91% of cases. This figure is higher than those reported in the literature which range from 38 to 86% [10,11,15]. We believe that the lower percentage of headache in these studies is mainly due to their retrospective nature which makes it likely to overlook a non-specific and non-vital symptom like headache especially when the headache is not severe (as in 11, ie, 16%, of our patients).

A variety of other symptoms were observed, however, there was no specific symptom or constellation of symptoms (see Table 2). The conclusion from the clinical findings is that the coincidence of headache with any other neurological symptom should always lead to consideration of sinus venous thrombosis in the differential diagnosis, but lack of headache does not exclude SVT.

The frequent occurrence of intracranial hemorrhages in SVT is well known [14,16,17]. The incidence in our study was 28%. The finding of an ICH in an untypical location for hypertensive bleeding should always raise suspicion of SVT.

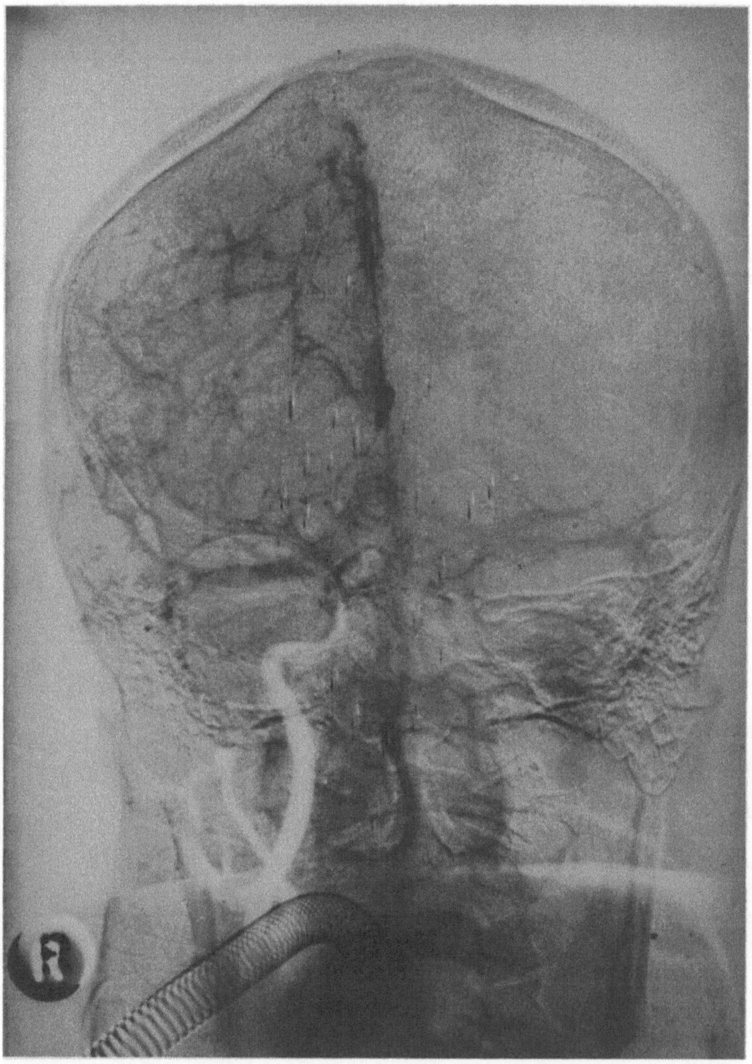

Fig. 2. Right carotid phlebogram, frontal projection. Non-filling of the superior sagittal sinus, visualization of the right lateral sinus break off of ascending veins.

CSF and EEG often are pathologic in SVT; however, a normal finding in one of these never excludes SVT. The same is true for findings in cranial computed tomography which are reviewed separately (Villringer et al., unpublished results). Conversely, 68 patients (94%) had a pathological finding with at least one of these techniques. However, the final diagnosis will always require cerebral angiography as no specific findings in CT, CSF or EEG can establish the diagnosis.

Temporal Profile

The temporal pattern was evaluated using a newly defined sinus venous thrombosis severity score (Einhäupl et al., unpublished results). No typical profile was seen. There were both acute cases with rapid relief of symptoms as well as with long lasting fluctuating symptoms.

Table 2. Frequency of Symptoms and Other Clinical Findings in the Course of Sinus Venous Thrombosis in 71 Patients

Clinical Symptoms	Patients	Percentage
Headache	63 (of 69)*	91
Seizure	34	48
Todd's paresis after seizure	14 (of 34)	41
Focal sign	47	66
Impairment of consciousness	40	56
Papilledema	19	27
Meningismus	9	13
Other clinical findings:		
Intracranial hemorrhage	27	38
Pathologic CSF	34 (of 68)	50
Pathologic EEG	52 (of 64)	81

* In the remaining two patients no case history was obtained.

Table 3. Outcome in 71 Patients with Sinus Venous Thrombosis

Outcome	Patients	Percentage
Complete clinical recovery	42	59
Minimal neurological deficit	13	19
Significant neurological deficit	3	4
Severe neurological deficit	3	4
Dead	10	14

Prognosis

The outcome of our patients is summarized in Table 3. The prognosis is generally good considering that 78% of all patients recover completely or with a slight neurological deficit. Mortality in our study was 14%.

The literature gives very different numbers for the prognosis of SVT. Reported mortality ranges from 0 to 100%; particularly older reports mention mortalities between 70 and 100% (eg, see Refs 10,14,18,19,24) whereas newer reports indicate that many cases might be more benign, mortality in these reports ranges from 5 to 27% [11,20-22].

We think that these differences mainly reflect the different selection criteria favoring more severe or less severe cases, respectively.

However, there is no doubt that there are many benign cases of SVT with good prognosis. On the other land, the fact that there is still a significant mortality in this disease which affects many young people emphasizes the need of immediate diagnosis and appropriate treatment [23].

REFERENCES

1. F. R. Stansfield, Puerperal cerebral thrombophlebitis treated by heparin, *Br. Med. J.*, 4:436-439 (1942).
2. J. D. Carroll, D. Leak and H. A. Lee, Cerebral thrombophlebitis in pregnancy and the puerperium, *Q. J. Med.*, 11:347-368 (1966).
3. J. Dubois, Les thrombophlebites cerebrales du post-partum, *Gynecologie et Obstetrique*, 5:472-493 (1956).
4. E. A. Atkinson, B. Fairburn and K. W. G. Heathfield, Intracranial venous thrombosis as complication of oral contraception, *Lancet*, 1:914-918 (1970).
5. K. J. Endtz and O. G. Mulder, Hirnvenenthrombose und Antikonzeptiva, *Radiologe*, 11:416-419 (1971).
6. A. A. Poltera, The pathology of intracranial venous thrombosis in oral contraception, *J. Pathol.*, 1:209-219 (1972).
7. M. C. Shende and H. Lourie, Sagittal sinus thrombosis related to oral contraception, *Neurosurgery*, 33:714-717 (1970).
8. C. Sisson, Intracranial venous thrombosis complicating oral contraception, *Lancet*, 2:419 (1970).
9. A. Hromadka and M. Hohenegger, Primäre Hirnvenen- und Sinusthrombose bei Erwachsenen, *Arch. Psychiat. Nervenkr.*, 209:79-86 (1967).
10. A. Huhn, "Die Thrombosen der intrakraniellen Venen und Sinus", F. K. Schattauer Verlag, Stuttgart and New York (1965).
11. M. G. Bousser, J. Chiras, J. B. Bories and P. Castaigne, Cerebral venous thrombosis - a review of 38 cases, *Stroke*, 16:199-213 (1985).
12. L. Gutschera-Wang, Zur Klinik von letalen Hirnvenen- und Sinusthrombosen anhand von 102 Fällen Erwachsener aus der Literatur, Inaugural Dissertation, Ludwig-Maximilians University of München (1982).
13. D. S. Buchanan and A. Brazinsky, A dural sinus and cerebral venous thrombosis, *Arch. Neurol.*, 28:174-178 (1978).
14. H. Noetzel and F. Jerusalem, "Die Hirnvenen- und Sinusthrombosen", Springer Verlag, Berlin and New York (1965).
15. R. D. Nagpal, Dural sinus and cerebral venous thrombosis, *Neurosurg. Rev.*, 6:155-160 (1983).
16. R. M. Kalbag and A. L. Woolf, Thrombosis and thrombophlebitis of cerebral veins and dural sinuses, *in*: "Vascular Diseases of the Nervous System, Part 2", P. J. Vinken and G. W. Bruyn, eds., North Holland, Amsterdam (Handbook of Clinical Nuerology, vol. 12) (1972).
17. W. Krücke, Pathologie der cerebralen Venen- und Sinusthrombosen, *Radiologe*, 11:370-377 (1971).
18. T. Hahn, Die Elektroencephalographie bei zerebralen Thrombophlebitiden und Thrombosen, *Schweiz Arch. Neurol. Psychiat.*, 73:57-99 (1954).
19. G. Weber, Zerebrale Venen- und Sinusthrombosen, *Schweiz Arch. Neurol. Psychiat.*, 82:320-344 (1958).
20. P. Rousseaux, M. H. Bernard, N. Scherpereel and J. F. Guyot, Thrombose des sinus veineux intracraniens, *Neurochirurgie*, 24:197-201 (1978).
21. F. S. Buonanno, D. M. Moody, M. R. Ball and D. W. Later, Computed cranial tomographic findings in cerebral sinovenous occlusion, *Comp. Assist. Tomography*, 2:282-291 (1978).
22. B. Estanol, A. Rodriguez, G. Conte, J. M. Aleman, M. Loyo and J. Pizzuto, Intracranial venous thrombosis in young women, *Stroke*, 10:680-684 (1979).
23. W. Meister, K. M. Einhäupl, A. Villringer, P. Schmiedek, R. Haberl, H. W. Pfister, M. Pellkofer, H. Steinhoff, M. Deckert and F. Anneser, Treatment of patients with cerebral sinus and vein thrombosis with heparin, in: "Cerebral Sinus Thrombosis: Experimental and Clinical Aspects", K. Einhäupl, O. Kempski and A. Baethmann, eds, Plenum Press, New York and London, pp 225-230 (1990).
24. R. M. Kalbag and A. L. Woolf, "Cerebral Venous Thrombosis", University Press, London and Oxford (1967).

ASEPTIC CEREBRAL VENOUS THROMBOSIS

M. Jacewicz and F. Plum

Cerebrovascular Disease Research Center
Department of Neurology, Cornell University
Medical College, 1300 York Avenue
New York, NY 10021, USA

INTRODUCTION

Although aseptic thrombosis of the cerebral veins and dural sinuses has been clinically recognized for more than a century, physicians often overlook the diagnosis [1-3]. Several reasons may explain this deficiency: cerebral venous thrombosis (CVT) is relatively rare; it produces subtle and often ambiguous symptoms, and it may accompany severe and diverse systemic illnesses [1,3-6]. The advent of angiography and increasingly accurate CT and MRI imaging, however, have increased the capability for successful antemortem diagnosis [5,7-10]. The growing evidence that the use of anticoagulants enhances the likelihood of clinical recovery [6] adds urgency to the need for prompt and accurate diagnosis before brain hemorrhage and infarction develop.

From the clinical and autopsy records of New York Hospital, we identified 14 adults with CVT diagnosed since 1974, when CT scanning became routinely available. Analysis ρf one patient who died from CVT associated with chronic myelogenous leukemia was omitted because her chart was unavailable. The diagnosis was confirmed by CT (5), angiography (10), MRI (1) or autopsy (5). Only a minority of cases were diagnosed before serious neurologic complications had developed. This review attempts to identify the features that would have allowed earlier recog-nition and more effective therapy.

RESULTS AND DISCUSSION

The Clinical Spectrum of CVT

As others have found [2,3,6], the clinical spectrum of CVT proved diverse. The eight men and six women (ages 24-69 years) in the current series featured a multiplicity of risk factors, venous pathology, treatments and clinical outcomes (Table 1). A hypercoagulable predisposition in nine patients constituted the most frequent risk factor for CVT and was

Cerebral Sinus Thrombosis, Edited by K. Einhäupl *et al.*
Plenum Press, New York, 1990

Table 1. Cerebral Venous Thrombosis in 14 Patients

Case	Age	Sex	Risk Factors	Thrombosis	Treatment	Outcome
BN	35	F	Oral contraceptives, hypercoagulability	SSS, CV R&L LS SS	Steroids	Massive brain infarction, death
TM	45	F	Dehydration, hyperglycemia, hypercoagulability, breast cancer	SSS CV	Hydration, insulin, antibiotics, steroids	Brain infarction and autolysis, death
HG	69	F	Dehydration, hyperglycemia	CV	Hydration, insulin, antibiotics	Massive brain infarction, death
BR	52	F	Dehydration, hyperglycemia, hypercoagulability, DVT, pheochromocytoma	SSS	Hydration, insulin, antibiotics	A-MCA watershed, hemorrhagic infarcts, brain autolysis, death
VG	62	F	Dehydration, hyperglycemia, CVA	SSS	Hydration, insulin	R hemispheric infarction, occluded R IC, death
RS	24	M	DVT	SSS Bilat LS	Heparin, steroids	Full recovery
IH	49	M	Hypercoagulability	SSS	Heparin	Full recovery
JM	27	M	? Recurrent R otitis	SSS Bilat LS	Heparin	Full recovery
NDV	53	M	Hypercoagulability, occult glioma	SSS SS Bilat LS	Heparin	Partial recovery, dural AVM, dementia
CP	43	M	Thrombocytosis, leukocytosis, cerebral arteritis	SSS CV	Heparin, steroids	Vegetative

AF	57	M	Hypercoagulability	SSS	Steroids	Full recovery
JB	29	M	None	SSS R LS	Frequent LP, furosemide	Pseudotumor, full recovery
NC	32	F	Puerperium, occult glioma, lupus anticoagulant	L LS	Steroids, acetazolamide	Full recovery
LC	38	M	? Occult glioma	SSS	Subtemporal craniectomy	Dementia, pseudotumor

SSS, Superior sagittal sinus; LS, lateral sinus; CV, cerebral vein; SS, straight sinus; DVT, deep venous thrombosis; IC, internal carotid; A-MCA, anterior and middle cerebral A; R, right and L, left.

associated with an occult glioma in two and systemic tumor in three patients. All five deaths arose from cerebral infarction, edema and herniation in association with dehydration (4/5) and hyperglycemia (5/5). Four of five heparinized patients improved, while only three of the nine non-heparinized patients recovered neurologically.

Headache initiated the illness (10/14) and ultimately affected all patients who were not too obtunded to talk sensibly (11/14). Additional symptoms at onset also included lethargy (7/14), confusion (6/14), fever (6/14) and focal seizures (2/14). Following these early changes, several days to months often elapsed without major neurologic events (Tables 2 and 3). However nonspecific, these "prodromal" symptoms distressed most of the patients sufficiently to seek medical attention before serious focal deficits (2/14) or brain hemorrhage (1/14) had occurred. Secondary cerebral ischemia or infarction, whenever it appeared, was heralded by focal (2/14) and generalized seizures (4/14), hemiparesis (6/14) and a further worsening in mental status (12/14). Even these changes, however, were potentially reversible, and four patients given heparin after such a deterioration improved.

Role of Radiologic Investigations in Early Diagnosis of CVT

The admission CT, obtained on average 4 (range 1-12) days after first symptoms, directly visualized thrombosis in only three patients. This required contrast administration to outline the clot in the superior sagittal sinus (empty delta sign). The early images otherwise provided little direct diagnostic help, appearing entirely normal in four patients, disclosing edema or nonspecific hemispheric hypodensities in six and hemorrhagic infarction in one (IH). In four non-heparinized patients, a follow-up CT detected evidence of new hemorrhage that coincided with further neurologic deterioration. Three of these patients died. In two patients (IH and NDV), hemorrhage resolved within 2 weeks, and a course of heparin was then given without adverse consequences. Follow-up CT revealed an empty delta sign in two additional patients and documented its resolution over 4 weeks in one patient (JM). The presence or absence of an empty delta sign, however, correlated poorly with the severity of the neurologic deficits. Bousser et al. [6] report similar findings: CT scans were normal in five of 25 patients with CVT, and specific signs existed in only three instances. Kalbag [2] observes that the clinical diagnosis of CVT has actually declined since CT has displaced frequent angiography.

Angiography established CVT in seven cases and confirmed the diagnosis in three patients who had an empty delta sign on the admission CT. The site and degree of venous occlusion correlated poorly with outcome, except in one patient (BN) who died with extensive thrombosis of her intracranial venous structures (Table 1). Angiography had to be repeated in three instances when the venous phase was not sufficiently delayed to demonstrate CVT unequivocally. Subarachnoid hemorrhage from a ruptured aneurysm had been strongly suspected in two cases (RS and NDV) and herpes encephalitis in the third (BN). Recanalization of a superior sagittal thrombosis occurred by angiographic criteria within 2 months in one patient (JB).

In one case (AF), MRI clearly imaged the thrombus in the superior sagittal sinus 26 days after the onset of symptoms (Fig. 1). In two others (NC, LC) MRI confirmed dural sinus patency established earlier by

Table 2. Temporal Profile of Cerebral Venous Thrombosis (Subacute Encephalopathy)

Case	Presenting Symptoms	"Prodromal" Interval	Signs of Progression	Initial Diagnosis	Time before CVT Diagnosed
BN	Headache, lethargy, fever, confusion	8 days	Status epilepticus, RCD (17)*	Herpes encephalitis	16 days (Angio)
TM	Polyuria, polydipsia, fever, confusion	2 days	RCD (partial), seizures (5), RCD (8)	NKH coma, sepsis, CVT	2 days (CT delta sign)
HG	Lethargy, fever, confusion	3 days	RCD	NKH coma, sepsis	8 days (Autopsy)
BR	Mute, fever, lethargy	6 days	Hemiparesis, seizures (7), RCD (8)	Bacterial endocarditis	14 days (Autopsy)
VG	Headache, confusion, L hemiplegia	3 days	Fever, mute, RCD (6)	Stroke, metabolic encephalopathy	7 days (Autopsy)
RS	Headache, fever, neck stiffness, lethargy	11 days	Hemiparesis, aphasia, focal seizure, heparin (11)	Subarachnoid hemorrhage	11 days (Angio)
IH	Headache	3 days	Lethargy, focal seizure	CVT	5 days (Angio)
			Seizures (5), mild R hemiparesis (5), fever, confusion (16 d), worse hemiparesis (16), heparin (16)		
JM	Headache, retro-orbital pain, neck stiffness	11 days	Fever, confusion, agitation, heparin (14)	Migraine, viral encephalitis	12 days (CT delta sign)
AF	Headache, neck stiffness	10 days	Episodic limb incoordination	Subarachnoid hemorrhage	7 days (CT delta sign)

*( ) Days after symptom onset; NKH, nonketotic hyperglycemic coma; RCD, rostral-caudal deterioration [14].

Table 3. Temporal Profile of Cerebral Venous Thrombosis (Chronic Course)

Case	Presenting Symptoms	Time Before Progression	Signs of Progression	Initial Diagnosis	Time Before CVT Diagnosed
JB	Focal seizure	3 months	Headache, retro-orbital pain, CSF pressure waves, lethargy	CVT	3 months (Angio)
CP	Headache	2 months	Recurrent TIA (dysarthria), mild hemiparesis	CVT	2 months (Angio)
			Transient aphasia (5 days later), heparin		
			Profound hemiparesis (9 days later), hypotension, coma, decerebration		
NC	Headache, visual blurring, lethargy, confusion	4 months	Symptoms worse with steroid taper	CVT	<1 week* (Angio)
NDV	Headache, lethargy	1 week	Hemiparesis, heparin (14 days later)	Subarachnoid hemorrhage	1 week (Angio)
			Worse headache (6 months later), worse memory		
LC	Focal seizure, headache, lethargy, fever	3 months	Increased ICP, stupor, hemiparesis	Herpes encephalitis	4 months (Angio)

* Details of CVT diagnosis from another hospital are not available.

angiography. Currently MRI appears to be the study of choice for CVT because: (1) it is non-invasive; (2) it unambiguously visualizes intravascular thrombosis [11,12] (Moots et al., unpublished results); (3) it can discriminate between the progression and resolution of CVT [10]; and (4) it offers reliable early diagnosis for herpes encephalitis [13], one of the major misdiagnoses in CVT.

Role of CSF Analysis in CVT

The subacute phase of CVT produced a CSF formula that was often distinct from the chronic phase. Seven patients lacked papilledema early in their illness, and the concurrent CSF opening pressure ranged from 46 to 230 mm CSF. Papilledema in one patient (RS) coincided with an opening pressure of 550 mm CSF. The early CSF profile was characterized by red blood cells (5/9), predominantly mononuclear pleocytosis (6/9) and an elevated protein (6/9). Papilledema with CSF pressure of 215-550 mm CSF evolved in at least four of the nine patients during their subacute illness, and in all patients in their chronic phase. CSF contents in the latter often were normal (4/5). The exception involved an elevated protein in CP that predated his CVT and was attributed to a severe neuropathy of "paraneoplastic" origin.

Classification of CVT

Previous authors [1,4-6] have stressed the diverse signs and symptoms that accompany CVT and have proposed a number of classifications. Bousser et al. [6] recently have suggested six main patterns, each resembling another neurologic disease: (1) pseudotumor; (2) arterial stroke; (3) abscess; (4) tumor; (5) subarachnoid hemorrhage; and (6) TIA. We have taken a broader syndromic approach and divided the pattern of illness in our patients into two broad categories: (1) a subacute illness marked by death or continuous improvement by 3-4 weeks (Table 2); and (2) a chronic illness expressed by months-long symptoms of increased intracranial pressure (Table 3). The two groups presented strikingly different diagnostic and management problems.

Subacute Phase. Lack of papilledema (7/9) or of an elevated CSF opening pressure (4/7) suggested that increased intracranial pressure was absent; as a result, a hallmark of CVT was often overlooked. Instead, the presence of red and white blood cells in the CSF suggested subarachnoid hemorrhage, encephalitis or another inflammatory process. The diagnostic perplexity was highlighted by the early clinical impression in the current study: viral encephalitis (three patients), berry aneurysm rupture (three patients), sepsis and hyperglycemic coma (two patients), bacterial endocarditis (one patient), migraine (two patients) and thromboembolic stroke (one patient) (Tables 2 and 3). A diagnosis of CVT was not made until autopsy in three cases. As the above suggests, the differential diagnosis for acute and subacute (febrile) encephalopathy is varied [14] and should also include disseminated intravascular coagulation, thrombotic thrombocytopenic purpura, non-bacterial thrombotic endocarditis, cerebral vasculitis and brain abscess. With recovery, some of the CVT patients complained of headache and malaise months later, but the symptoms were of little medical consequence and resolved.

Chronic Phase. After an acute onset of symptoms, five patients showed incomplete abatement, a recrudescence or a progression of symptoms that began or lasted over many months (Table 3). Papilledema

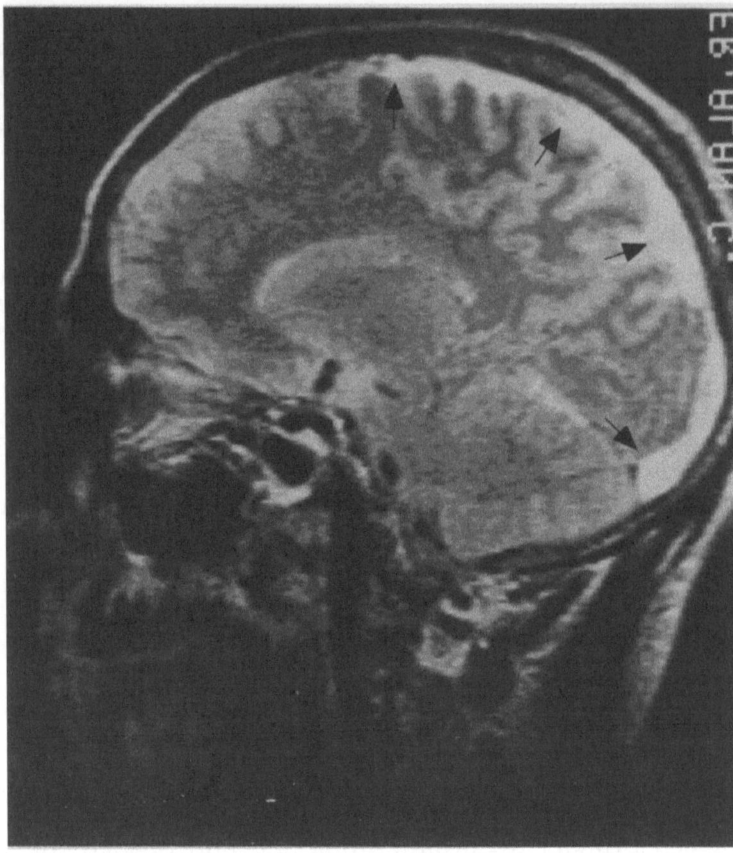

Fig. 1. MRI (T2 weighted) demonstrates an increased signal in the superior sagittal sinus (arrows).

and an elevated CSF opening pressure with a benign profile were characteristic. Radiologic studies showed that CVT was resolving (1/5) or had completely resolved (3/5). A brain tumor was eventually discovered in two of these patients and strongly suspected in a third. A dural AVM was found in one patient (NDV) who also had a glioma. Arteriography in one patient (CP) also showed a probable cerebral arteritis, which may have aggravated ischemic injury to produce a vegetative state. Relief of intracranial pressure was associated with the use of steroids (3/5), multiple lumbar punctures (1/5), and craniectomy (1/5). The persistence of signs lasting beyond 1 or 2 months should stimulate a search for other intracranial pathology.

Role of Coagulation Studies in CVT

Six patients demonstrated evidence of a hypercoagulable state in coagulation studies (presence of fibrin split products, increased reptilase activity), and three had (one previous, two concurrent) extracranial venous thrombosis. One patient's blood transiently contained a lupus anticoagulant, a recently described risk factor for CVT [15]. One patient (CP) with an unproven, but strongly suspected, lymphoproliferative disease had thrombocytosis and leukocytosis as the putative risk factor for CVT. CVT has been previously associated with idiopathic thrombocytosis [16,17], blood dyscrasias [4] and remote cancer [18,19]. Bousser et al. [6] found coagulation studies of no help for CVT diagnosis, but advocated their need for detecting "possible causative abnormalities" in idiopathic cases. The presence of malignancy, known and occult, in five of our nine hypercoagulable patients indicates that coagulation studies are appropriate in all CVT patients.

Anticoagulation in CVT

Four of the five heparinized patients made good recoveries from CVT (within 24-48 hours in two instances) despite evidence of parenchymal involvement (ie, seizures, hemiparesis) in four patients. One heparinized patient (CP) was left in a vegetative state. The results compare favorably with the nine non-heparinized patients, of whom only three recovered, but the prominence of deleterious factors (eg, dehydration, hyperglycemia) in non-heparinized patients may not make the two groups strictly comparable.

The use of anticoagulants in CVT has been controversial. They were first reported useful in CVT by Stansfield [20], but his case of puerperial CVT was not documented by angiography. Barnett and Hyland [4] cautioned against the use of anticoagulants because of the hemorrhagic nature of CVT infarctions, a view shared by others [21,22]. Gettelfinger and Kokmen [23] found that the use of heparin in three patients (plus urokinase in one) led to massive hemorrhage and death in two instances; in contrast, four patients not receiving anticoagulants made satisfactory recoveries. Hickey et al. [24] describe one anticoagulated patient, who at autopsy showed hemorrhagic infarction. CT scans, however, were obtained in none of these patients, who may have had brain hemorrhage before anticoagulation was initiated.

Table 4 summarizes the results of anticoagulant usage in several series. Kalbag and Woolf [1,25] favored early anticoagulation, before seizures began, which they believed indicated extension of sinus thrombosis to the cerebral veins and made hemorrhagic infarction more likely. Krayenbühl [5] observed that patients treated with anticoagulation and antibiotics suffered a 7% mortality, whereas deaths occurred in 37% of those treated with antibiotics alone and in 70% of untreated patients. From 1959 to 1966, he never saw "intracranial hemorrhage with well controlled anticoagulant treatment in cases of venous and sinus thrombosis". Among 16 young women with CVT treated with aspirin and dipyridamole, Estanol et al. [26] reported that none bled and all survived with only two suffering neurologic sequelae. Di Rocco et al. [27] treated five patients with heparin and urokinase early after symptom onset (mean 2 days, range 1-4 days) with excellent recovery in all, despite evidence of parenchymal involvement (ie, hemiparesis, seizures) in four patients. Steinherz et al. [28] reported that their two anticoagulated children with CVT and acute lymphoblastic leukemia recovered completely, whereas the literature showed that only two of six similar patients survived without heparin, one of whom was left with neurologic residua. Bousser et al. [6] stressed that heparinization at the very least was not harmful, since among their 23 heparinized patients, none died and 19 completely recovered. In addition, they witnessed dramatic improvement in three patients soon after heparin initiation, including in one patient whose CT already suggested hemorrhagic infarction. Our associates at Memorial

Table 4. Anticoagulation in Cerebral Venous Thrombosis

	Anticoagulated		Not Anticoagulated	
	Survived(*) Improved	Died	Survived(*) Improved	Died
Krayenbühl (1966)*	16	1	32	24
Bousser (1985)	23	0	11	4
Case reports (1942-1987)†	25	3 bled	25	44
Walker (unpublished results)	6	0	7	1
Current study	4	1 veg	4	5
Total	74 (94%)	5 (6%)	79 (50%)	78 (50%)

* Krayenbühl described mortality.
† Selected from the English literature [1,9,15,16,18,20-24,27,28,32,42-56].

Sloan-Kettering Hospital report that none of their six recently anticoagulated children with CVT suffered bleeding complications (R. Walker, personal communication). It should be emphasized that among 79 anticoagulated patients, only three bled and 94% improved (Table 4).

These results recommend the early use of heparin, or at very least, antiplatelet agents, to treat CVT and indicate that the advent of seizures, hemiparesis and other signs of cerebral injury need not contraindicate heparin provided that the CT shows no blood.

Long-term Complications of CVT

CVT can produce sufficient ischemic damage to leave the survivor hemiparetic or with impaired cognitive capacity [1] but our experience suggests that the persistence or progression of signs after the subacute phase (3-4 weeks) should raise the suspicion of additional intracranial pathology (glioma in three, cerebral arteritis in one, new dural AVM in one). Our survivors who had no systemic disease (6/9) did well, and CVT in isolation carries a good prognosis [6]. Seizures, a relatively frequent complication (7/14) of CVT, were easily controlled in the four survivors and did not necessarily indicate a lifelong disorder. One case (JB) had no further seizures after phenytoin was stopped 6 months after onset of epilepsy. Only one of eight patients had residual epilepsy in Bousser's series [6]. Pseudo-tumor is a well-known sequel of dural sinus thrombosis [6,29-32]. Frequent lumbar punctures, as in case JB, or lumboperitoneal shunting [6,31] can alleviate the papilledema and threatened visual loss. A subtemporal craniotomy successfully decompressed patient LC who, at the time, exhibited impending cerebral herniation. There was no recurrence of CVT in our series and was uncommon (2/26) in Bousser's (1985) series [6].

The development of a dural arterio-venous malformation (AVM) as a complication of CVT has not been adequately appreciated. In patient NDV, an occipital dural AVM (with patent dural sinuses) was detected by angiography 2 years after the same procedure had documented thrombosis of the superior sagittal and both transverse sinuses. Houser et al. [31] found that eight of 14 patients with a dural AVM affecting the transverse sinus had occlusion of one or both transverse or sigmoid sinuses. Thrombosis preceded the development of the AVM by 1 and 4 years in two patients. They hypothesized the thrombosis to be the primary event followed by the development of a direct artery-to-sinus communication via indigenous dysplastic dural vessels as the intravenous thrombus became organized. The subsequent entry of high pressure arterial blood into the sinuses could impair venous drainage, increase cerebral venous congestion, produce edema and account for the persistence of both focal and diffuse cerebral dysfunction [31].

Influence of Hyperglycemia

All five patients who died suffered from hyperglycemia at the time that cerebral infarction occurred (range 151-1466, mean 514 mg/dL), and two were sufficiently hyperosmolar (increased by 75 and 30 mOsm/kg above baseline) for that alone to have produced symptoms [14]. In contrast, only one surviving patient had a prominent blood sugar elevation (LC, 200 mg/dL). In hyperosmolar, nonketotic hyperglycemic coma, uncomplicated by other illness, focal stroke-like symptoms are common, but few patients develop permanent residual neurologic deficits [33-35].

Dehydration accompanying hyperglycemia could have increased the degree and extent of venous occlusion and caused a greater ischemic insult than that suffered by the survivors. The pathologic venous anatomy, however, did not distinguish four of the five who died from the nine who survived (Table 1). The single exception (BN) developed extensive thrombosis of cerebral venous structures but also had the lowest blood sugar (151 mg/dL) in the hyperglycemic group and was minimally dehydrated at the time of infarction (BUN 23). In the present instance of CVT-related cerebral infarction and death, we suspect that in the presence of venous occlusion, increased glucose levels in the brain predisposed to the severe degree of tissue injury.

In experimental animals, the presence of high carbohydrate stores in brain at the onset of ischemia greatly accentuates the ensuing degree of neuropathological damage [36]. Furthermore, hospital studies suggest that acute strokes produce greater neurological devastation in hyperglycemic patients (blood sugar >150 mg/dL) than in normoglycemics [37]. The intensified degree of injury among hyperglycemics is thought to result from the high levels of lactic acid that accumulate in the ischemic, high carbohydrate brain. In the absence of available oxygen, glucose in brain converts in a 1:2 ratio to non-metabolizable lactate via glycolysis. Brain lactate concentrations that exceed 15-17 μmoles/g during ischemia appear to be critically dangerous [38] and require pre-ischemic tissue glucose levels of no more than about 8 μmoles/g. Sustained hyperglycemia of greater than 150-170 mg/dL (8-9.5 mM) appears to be sufficient to create such concentrations in brain. In brain tissue, the toxic effect of lactic acid appears to be directed primarily at glial cells, producing irreversible damage at intraglial pH levels at or below approximately 5.3 [39-41]. This degree of intraglial acidosis is predictably reached during complete or near-complete cerebral ischemia in hyperglycemic experimental animals.

SUMMARY

Fourteen patients with cerebral venous and dural sinus thrombosis (CVT) formed a diverse group in etiology, treatment and outcome. A prodrome that featured headache and lethargy (2-11 days) was sufficiently distressful for most patients to obtain early medical attention, usually before the appearance of brain infarction (2/14) or hemorrhage (1/14). The admission CT, however, was diagnostic of CVT in only 3/13 cases.

Two major clinical categories stood out: (1) a subacute, often febrile, encephalopathy that led to death or initiation of recovery within 3-4 weeks (9/14); and (2) a more chronic course characterized by symptoms of increased intracranial pressure (5/14). Of the five patients who died, massive cerebral infarction and herniation was the principal cause of death. All five had elevated blood sugars (range 151-1466, mean 514 mg/dL) at the time of infarction, and none had received heparin. Of the five heparinized patients, four improved with three showing full recovery. In the chronic group, five patients had an acute onset of symptoms, but suffered a prolonged (> 2 months) course of increased intracranial pressure despite resolution of CVT (4/5). A glioma was later found in two cases and strongly suspected in a third. One patient developed "pseudotumor" with threatened visual loss, but recovered completely after 9 months of weekly spinal taps; the last patient suffered from several months of headache before undergoing a subacute deterioration in association with cerebral arteritis that left him in a vegetative state.

This study suggests that: (1) hyperglycemia may aggravate ischemic damage in CVT; (2) anticoagulant therapy is beneficial; and (3) a "prodromal" interval frequently precedes brain infarction and hemorrhage that should prompt MRI or angiography for accurate diagnosis.

REFERENCES

1. R. M. Kalbag and A. L. Woolf, "Cerebral Venous Thrombosis", University Press, London and Oxford (1967).
2. R. M. Kalbag, Cerebral venous thrombosis, *in*: "The Cerebral Venous System and Its Disorders", J. P. Kapp and H. H. Schmideck, eds., Grune and Stratton, Orlando, Fla., pp 505-536 (1984).
3. P. C. Gates and H. J. M. Barnett, Venous disease: cortical veins and sinuses, *in*: "Stroke: Pathophysiology, Diagnosis and Management", H. J. M. Barnett, J. P. Mohr, B. M. Stein and F. M. Yatsu, eds., Churchill Livingstone, New York, pp 731-743 (1986).
4. H. J. M. Barnett and H. H. Hyland, Noninfective intracranial venous thrombosis, *Brain*, 76:36-49 (1953).
5. H. A. Krayenbühl, Cerebral venous and sinus thrombosis, *Clin. Neurosurg.*, 14:1-24 (1966).
6. M.-G. Bousser, J. Chiras, J. Bories and P. Castaigne, Cerebral venous thrombosis - a review of 38 cases, *Stroke*, 16:199-213 (1985).
7. F. S. Buonanno, D. M. Moody, M. R. Ball and D. W. Laster, Computed cranial tomographic findings in cerebral sinovenous occlusion, *J. Comput. Assist. Tomog.*, 2:281-290 (1978).
8. T. O. Gabrielsen, J. F. Seeger, J. E. Knake and E. W. Stilwill, Radiology of cerebral vein occlusion without dural sinus occlusion, *Radiology*, 140:403-408 (1981).
9. A. L. Goldberg, A. E. Rosenbaum, H. Wang, W. S. Kim, V. L. Lewis and D. F. Hanley, Computed tomography of dural sinus thrombosis, *J. Comput. Assist. Tomog.*, 10:16-20 (1986).
10. P. J. Macchi, R. I. Grossman, J. M. Gomori, H. I. Goldberg, R. A. Zimmerman and L. T. Bilaniuk, High field MR imaging of cerebral venous thrombosis, *J. Comput. Assist. Tomog.*, 10:10-15 (1986).

11. S. K. McMurdo, M. Brant-Zawadzki, W. G. Bradley, G. Y. Chang and B. O. Berg, Dural sinus thrombosis: study using intermediate field strength MR imaging, *Radiology*, 161:83-86 (1986).
12. W. M. Bauer, K. Einhäupl, S. H. Heywang, T. Vogl, M. Seiderer and D. Clados, MR of venous sinus thrombosis, *AJNR*, 8:713-715 (1987).
13. G. Schroth, J. Gawehn, A. Thron, A. Vallbracht and K. Voigt, Early diagnosis of herpes encephalitis by MRI, *Neurology*, 37:184-189 (1987).
14. F. Plum and J. Posner, "The Diagnosis of Stupor and Coma", F. A. Davis Co., Philadelphia, pp 253-255 (1980).
15. S. R. Levine, S. Kieran, K. Puzio, H. Feit, S. C. Patel and K. M. A. Welch, Cerebral venous thrombosis with Lupus anticoagulant: report of two cases, *Stroke*, 18:801-804 (1987).
16. I. Iob, M. Scanarini, G. C. Andrioli and K. Pardatscher, Thrombosis of the superior sagittal sinus associated with idiopathic thrombocytosis, *Surg. Neurol.*, 11:439-441 (1979).
17. D. D'Avella, R. P. Greenberg, S. Mingrino, M. Scanarini and K. Pardatscher, Alterations in ventricular size and intracranial pressure caused by sagittal sinus pathology in man, *J. Neurosurg.*, 53:656-661 (1980).
18. B. Sigsbee, M. D. F. Deck and J. B. Posner, Non-metastatic superior sagittal sinus thrombosis complicating systemic cancer, *Neurology*, 29:139-146 (1979).
19. W. D. F. Smith, J. Sinar and M. Carey, Sagittal sinus thrombosis and occult malignancy, *J. Neurol. Neurosurg. Psychiat.*, 46:187-188 (1983).
20. F. R. Stansfield, Puerperal cerebral thrombophlebitis treated by heparin, *Br. Med. J.*, 1:436-438 (1942).
21. D. S. Buchanan and J. H. Brazinsky, Dural sinus and cerebral venous thrombosis. Incidence in young women receiving oral contraceptives, *Arch. Neurol.*, 22:440-444 (1970).
22. F. Dindar and M. E. Platts, Intracranial venous thrombosis complicating oral contraception, *Can. Med. Assoc. J.*, 111:545-548 (1974).
23. D. M. Gettelfinger and E. Kokmen, Superior sagittal sinus thrombosis, *Arch. Neurol.*, 34:2-6 (1977).
24. W. F. Hickey, M. B. Garnick, I. C. Hendersen and D. M. Dawson, Primary cerebral venous thrombosis in patients with cancer - a rarely diagnosed paraneoplastic syndrome, *Am. J. Med.*, 73:740-750 (1982).
25. R. M. Kalbag and A. L. Woolf, Thrombosis and thrombophlebitis of cerebral veins and dural sinuses, *in*: "The Handbook of Clinical Neurology", P. J. Vinken and G. W. Bruyn, eds., North-Holland, Amsterdam, pp 422-446 (1972).
26. B. Estanol, A. Rodriguez, G. Conte, J. M. Aleman, M. Loyo and J. Pizzuto, Intracranial venous thrombosis in young women, *Stroke*, 10:680-684 (1979).
27. C. L. Di Rocco, A. Iannelli, G. Leone, M. Moschini and V. M. Valori, Heparin-urokinase treatment in aseptic dural sinus thrombosis, *Arch. Neurol.*, 38:431-435 (1981).
28. P. G. Steinherz, L. P. Miller, F. Ghavimi, J. C. Allen and D. R. Miller, Dural sinus thrombosis in children with acute lymphoblastic leukemia, *JAMA*, 246:2837-2839 (1981).
29. J. Foley, Benign forms of intracranial hypertension - "toxic" and "otitic" hydrocephalus, *Brain.*, 78:1-41 (1955).
30. M. Greer, Benign intracranial hypertension: mastoiditis and lateral sinus obstruction, *Neurology*, 12:472-476 (1962).
31. O. W. Houser, J. K. Campbell, R. J. Campbell and T. M. Sundt, Arterio-venous malformation affecting the transverse dural venous sinus. An acquired lesion, *Mayo Clin. Proc.*, 54:651-661 (1979).
32. R. E. Kaplan, J. E. Springate, L. G. Feld and M. E. Cohen, Pseudotumor cerebri associated with cerebral venous sinus thrombosis, internal jugular vein thrombosis and systemic lupus erythematosus, *J. Pediat.*, 107:266-267 (1985).
33. M. Maccario, Neurological dysfunction associated with nonketotic hyperglycemia, *Arch. Neurol.*,19:525-534 (1968).

34. A. I. Arieff and H. J. Carroll, Nonketotic hyperosmolar coma with hyperglycemia: clinical features, pathophysiology, renal function, acid-base balance, plasma-cerebrospinal fluid equilibria and the effects of therapy in 37 cases, *Medicine*, 51:73-94 (1972).
35. B. M. Singh, D. R. Gupta and R. J. Strobos, Nonketotic hyperglycemia and epilepsia partialis continua, *Arch. Neurol.*, 29:187-190 (1973).
36. W. A. Pulsinelli, S. Waldman, D. Rawlinson and F. Plum, Moderate hyperglycemia augments ischemic brain damage: a neuropathological study in the rat, *Neurology*, 32:1239-1246 (1982).
37. W. A. Pulsinelli, D. E. Levy, B. Sigsbee, P. Scherer and F. Plum, Increased damage after ischemic stroke in patients with hyperglycemia with or without established diabetes mellitus, *Am. J. Med.*, 74:540-544 (1983).
38. F. Plum, What causes infarction in ischemic brain? The Robert Wartenberg Lecture, *Neurology*, 33:222-233 (1983).
39. R. Kraig, W. A. Pulsinelli and F. Plum, Carbonic acid buffer changes during complete ischemia, *Am. J. Physiol.*, 250:R348-R357 (1986).
40. R. Kraig, C. K. Petito, F. Plum and W. A. Pulsinelli, Hydrogen ions kill brain at concentrations reached in ischemia, *J. Cereb. Blood Flow Metab.*, 7:379-386 (1987).
41. R. Kraig and C. Nicholsen, Profound acidosis in presumed glia during ischemia, *in*: "Cerebral Vascular Disease" (15th Princeton-Williamsburg Conference), E. Raichle and J. W. Powers, eds., Raven Press, New York, pp 97-102 (1987).
42. P. Averback, Primary cerebral venous thrombosis in young adults: the diverse manifestations of an under-recognized disease, *Ann. Neurol.*, 3:81-86 (1978).
43. I. Bank and C. Weart, Dural sinus thrombosis in Behcet's disease, *Arth. Rheum.*, 27:816-818 (1984).
44. B. D. Barnes, M. Brant-Zawadzki and W. Mentzer, Digital subtraction angiography in the diagnosis of superior sagittal sinus thrombosis, *Neurology*, 33:508-510 (1983).
45. M. F. Beal, L. R. Wechsler and K. R. Davis, Cerebral vein thrombosis and multiple intracranial hemorrhages by computed tomography, *Arch. Neurol.*, 39:437-438 (1982).
46. M. Brant-Zawadzki, G. Y. Chang and G. E. McCarty, Computed tomography in dural sinus thrombosis, *Arch. Neurol.*, 39:446-447 (1982).
47. S. L. Bridgers, E. Strauss, E. O. Smith, D. Reed and M. D. Ezekowitz, Demonstration of superior sagittal sinus thrombosis by Indium-111 platelet scintigraphy, *Arch. Neurol.*, 43:1079-1081 (1986).
48. B. Fairburn, Intracranial venous thrombosis complicating oral contraception: treatment by anticoagulant drugs, *Br. Med. J.*, 2:647 (1973).
49. J. A. Feldenzer, M. J. Bueche, J. L. Venes and S. S. Gebarski, Superior sagittal sinus thrombosis with infarction in sickle cell trait, *Stroke*, 18:656-660 (1987).
50. R. A. Fishman, D. Cowen and M. Silbermann, Intracranial venous thrombosis during the first trimester of pregnancy, *Neurology*, 7:217-220 (1957).
51. W. K. Imai, F. R. Everhart and J. M. Sanders, Cerebral venous sinus thrombosis: report of a case and review of the literature, *Pediatrics*, 70:965 (1982).
52. H. M. Ingstrup and P. S. Jorgensen, Tentorial changes in sigmoid sinus thrombosis, *J. Comput. Assist. Tomog.*, 5:760-762 (1981).
53. S. O. Lau, G. H. Bock, J. R. Edson and A. F. Michael, Sagittal sinus thrombosis in the nephrotic syndrome, *J. Pediatr.*, 97:948-950 (1980).
54. R. N. Nishimura, D. Stepanek, J. Howieson and J. Hammerstad, Internal cerebral vein thrombosis, *Arch. Neurol.*, 39:439-440 (1982).
55. B. Sigsbee and D. A. Rottenberg, Sagittal sinus thrombosis as a complication of regional enteritis, *Ann. Neurol.*, 3:450-452 (1978).
56. W. L. Stringer and S. J. Peerless, Superior sagittal sinus thrombosis after closed head injury, *Neurosurgery*, 12:95-97 (1983).

CLINICAL FINDINGS AND DIFFERENTIAL DIAGNOSIS OF CEREBRAL VEIN THROMBOSIS

A. Hartmann, J. Wappenschmidt*, L. Solymosi* and F. Brassel*

Neurology Clinic and *Neurological Department
of the Neurosurgical University Clinic
5300 Bonn, FRG

Compared to cerebral arterial insults, diseases of the brain due to partial or complete occlusion of cerebral veins are rare. In one-fifth of all patients with cerebral vein thrombosis (CVT) no cause may be detected. Computerized tomography of the brain (CCT) can be performed in most neurologic centers to establish a diagnosis.

With CT and MRI, diagnosis may be proven easily. In most cases with CVT and a severe clinical picture one sinus is completely occluded, without or with concomitant occlusion of cortical veins. Isolated cortical vein thrombosis with clinical signs and symptoms is a rarity [15]. The first detailed description of CVT was made by Ribes [28] who in 1825 referred to a case with thrombosis of the superior sagittal sinus (SSS) and lateral sinus and a metastatic carcinoma in the brain.

Causes of CVT which are mentioned frequently in the literature are skull injury, dehydration, congestive heart failure, blood dyscrasias, anemia, otitis and mastoiditis, steroids, pregnancy and puerperium [2,12,30] and others (Tables 1 and 2). Nowadays reports on CVT in adults focus mostly on pregnancy and puerperium and paraneoplastic diseases [1,13,29]. Overalimentation and oral contraceptives have brought new causes of CVT as well as intensive care medicine with pacemaker implantation [14] and long-remaining intravenous catheters [31]. In rare diseases like Wegener's granulomatosis [26] or Behçet's syndrome [7], it must be kept in mind that central nervous symptoms may be due to development of CVT.

The particular situation of the cerebral veins which are easily compressible by mechanical stress from the outside, the attachment of the sinus to the neighborhood and the trabeculae which thwart the sinus and may cause turbulence of the blood stream contribute to the development of CVT. In addition the inability of the sinus to adapt the lumen to the cardiac output might favor stasis of blood flow under pathologic conditions. In many cases the anastomotic network prevents development of neurologic symptoms (Fig. 1). Occlusion of the major

Cerebral Sinus Thrombosis, Edited by K. Einhäupl *et al.*
Plenum Press, New York, 1990

Table 1. Causes of Thrombosis of Cerebral Veins and Sinus

Thrombophlebitis migrans
Thrombophlebitis obliterans
Pregnancy and puerperium
Oral contraceptives
Postsurgical intervention
Postneurosurgical intervention
Blood dyscrasias/sickle cell anemia
Malignant processes
Meningitis, meningoencephalitis
Sinusitis, septicemia
Intracranial perivenous space occupying lesions
Dehydration
Cerebral arterial occlusion
Hemolytic anemia

Table 2. Diseases Rarely Associated with Cerebral Vein/Sinus Thrombosis

Ulcerative colitis	Borda et al. (1973)
Behçet's disease	Bousser (1982)
Periarteritis nodosa	Kulawik and Hoppe (1971)
Polycythemia rubra vera	Barnett and Hyland (1953)
Enteritis (regional)	Sigsbee and Rottenberg (1978)
Wegener's granulomatosis	Mickle et al. (1977)
Sickle cell trait	Schenk (1964), Feldenzer et al. (1987)
Paroxysmal nocturnal hemoglobinuria	Johnson et al. (1970)
Budd-Chiari syndrome	Kalbag and Woolf (1967)
Cryofibrinogenemia	Dunsker et al. (1970)
Trichinosis	Evand and Patten (1982)
Diabetes mellitus	Askenasy et al. (1962)
Nephrotic syndrome	Lau et al. (1980)
Intrauterine instillation of hypertonic saline solution	Goldman and Eckerling (1972)
Propagation from jugular vein after pacemaker transplantation or neck dissection	Floyd and Mahaley (1969) Fielding et al. (1973)
Leukemic vein infiltration	David et al. (1975)

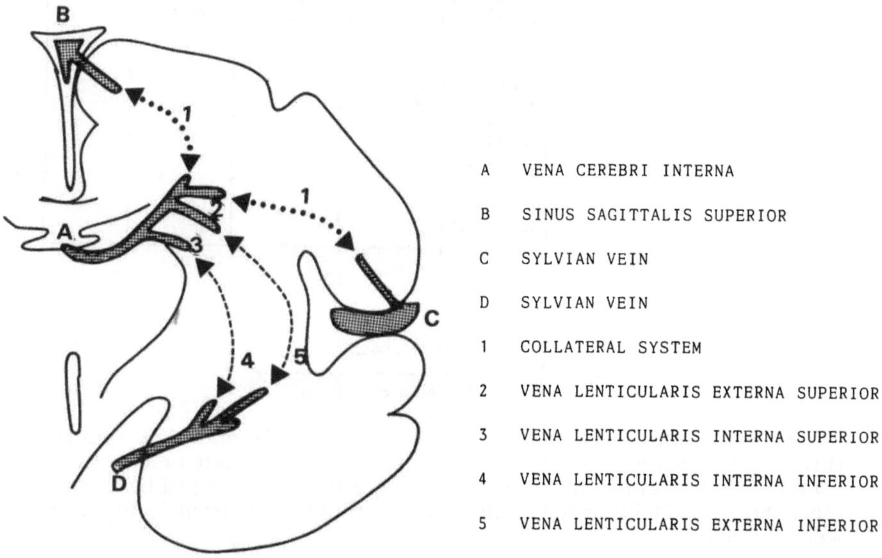

Fig. 1. Anastomotic network between inner and outer cerebral veins. These anastomotic vessels and other anastomoses permit bypassing of blood in the case of cerebral vein thrombosis and make neurologic symptoms less common than in arterial occlusion.

sinus leads to moderate impairment of blood flow with consecutive congestion and increase of intracranial pressure (ICP). The typical signs of increased ICP such as papilledema, headache, neck stiffness, nausea, vomiting and alteration of the level of consciousness [33] may be seen in many neurologic disorders and the role of the physician is to think of a possible CVT.

Additional signs may facilitate the diagnosis. However, very often they are not typical. Some signs are characteristic: dilatation of the scalp veins in infants with tenderness of the fontanelle points to occlusion of the SSS. Tenderness over the mastoid (posterior portion) due to edema should be attributed to occlusion of the transverse/lateral sinus with involvement of the jugular vein. Forehead edema with eventual swelling of the upper eyelid is a consequence of major obstruction of the underlying sinus. If in young infants focal seizures follow or are preceded by development of hemiparesis, occlusion of either a cortical vein or an artery must be considered [4,27]. Isolated SSS thrombosis without occlusion of a cortical vein in children is only rarely indicated by epileptic attacks and focal signs.

In adults one of the earliest signs is subacute headache (Fig. 2). It is not reported by all patients with clinical significant occlusion of a sinus [24]. Pain is persistent, slowly progressing, does not respond to medication, and may be localized or generalized. Convulsions may accompany pain or follow after days or even weeks. Often seizures develop at the height of the headache. The slow progressive course of headache in CVT distinguishes it clearly from most cases with sub-arachnoid hemorrhage, where headache is at maximum at the beginning of the

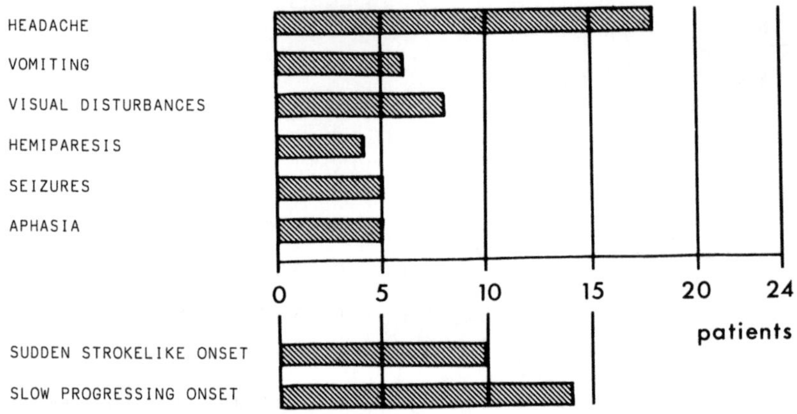

Fig. 2. Clinical symptoms in 24 patients treated between 1976 and 1986 with cerebral vein thrombosis. Some patients were diagnosed and treated before introduction of computer tomography. In these cases diagnosis was confirmed by angiography. On admission headache and hemiparesis were the most prominent symptoms and signs observed.

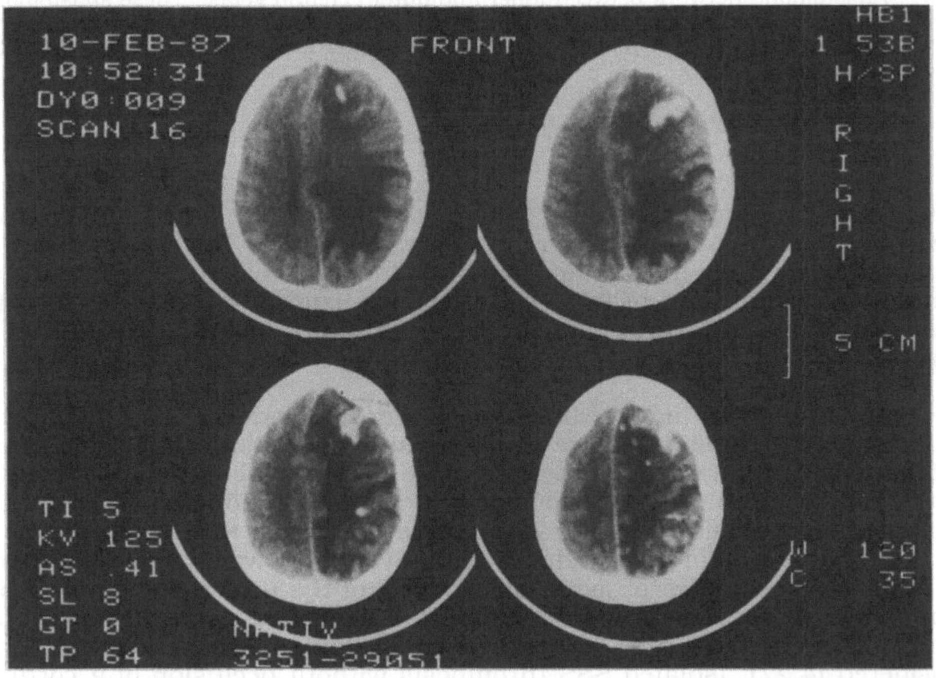

Fig. 3. Thrombosis of the superior sagittal sinus with retrograde thrombosis of the ascending veins of the right hemisphere. Multiple hemorrhages and development of brain edema led to significant midline shift.

disease. However, severe headache, seizures and focal neurologic signs or symptoms may also be seen in cerebral neoplastic processes and therefore are not exclusively indicative of CVT. Postictal paresis for less than 24 hours (Todd's paresis) may be seen in both disorders. Alternating paresis between both sides of the body, however, are more frequently seen in occlusion of a major intracerebral vein. Furthermore, fluctuating paresis of one extremity is more often seen in CVT than in cerebral tumors.

Besides signs of increased ICP (due to development of edema, hemorrhage and CSF-malabsorption, Fig. 3) development of seizures and focal signs, mental changes like slight irritability, lack of concentration and nervousness are early signs of cerebral venous obstruction. Insomnia is reported by many patients. Later on lethargy and development of delirium and even coma can be observed. Since in many diseases which favor development of CVT mental alterations may be present without CVT (like puerperal psychosis, steroid psychosis or post-traumatic psychosis), again one has to think of CVT.

Headache is reported by more than half of the patients [18]. Epileptic attacks range from focal to generalized forms. The Jacksonian type is common [5]. The seizures may change sides and may be accompanied by ascending paresis which is typical for SSS thrombosis, and status epilepticus is characteristic for an advanced state of SSS thrombosis or other CVT [16]. In most cases the paresis does not precede the epileptic attack [22]. It develops slowly, is first of flaccid, later of spastic character. Sometimes hemiparesis may be lateral or alternating [23,26] or of repeated and transient nature [10].

Cranial nerves (CN) are involved only if sinuses other than the SSS are affected. Thrombosis of the cavernous sinus (CS) affects CN III, IV, V and VI. In CS thrombosis inflammatory diseases of the paranasal sinus or a facial furuncle must be considered. Head trauma is less common as a cause of CS thrombosis. Diagnosis is easily made by the typical clinical presentation with headache, unilateral proptosis (eventually bilateral), chemosis and restriction of eye movements. If amaurosis develops, an ophthalmic artery is affected. A contra-lateral hemiparesis is due to involvement of the internal carotid artery which passes the CS. In only a few cases, the CS thrombosis spreads to the connecting transverse sinus or to the jugular veins. If thrombosis is affecting the jugular bulb, possible vagus compression results in aphonia, hoarseness, respiratory paralysis and bradycardia. Vernet's syndrome is the combined affectation of the IXth, Xth and XIth CN by the same mechanism.

Paresis of the abductor nerve may be caused by thrombosis of the inferior petrosal sinus [32]. Gradenigo's syndrome (combined paresis of the Vth and VIth CN, apex of the petrous bone) may be due to thrombosis of the basal veins [35]. The inferior petrosal sinus thrombosis causes a paresis of the VIth CN since both are located close together in Dorell's canal at the petrous bone, which leads the VIth CN from the posterior to the middle fossa.

Thrombosis of the sigmoid sinus may be diagnosed by tenderness and swelling of the skin along the posterior border of the mastoid bone if the mastoid emissaries are involved (Griesinger's sign [9]). Papilledema may be present in only half of the patients with SSS thrombosis [3,11,19]. Martin explained complete visual disorientation by SSS thrombosis [25]. Occlusion of the anastomotic vein of Labbé leads to a hemianopic defect. Combined symptoms of the supratentorial cortex and the cerebellum are due to obstruction of the transverse sinus.

Thrombosis of the deeper cerebral veins (Fig. 4) are rarer than SSS thrombosis (Fig. 5) [21]. Decerebrate rigidity and other brainstem signs are indicative of the progressed state. They might be confused with obstruction of the basilar artery. In addition to a paraparesis or side-

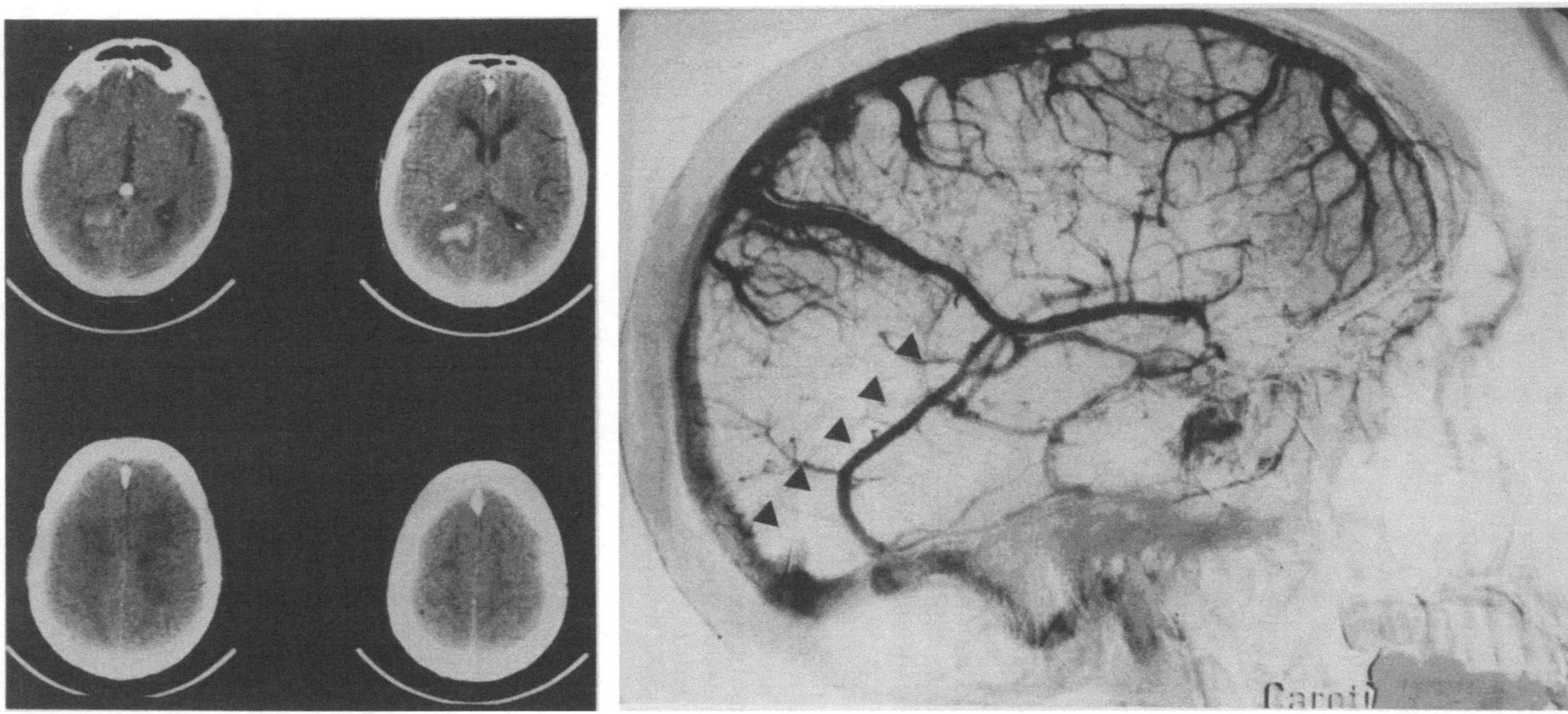

Fig. 4. Computerized tomogram of a patient with intracerebral bleeding in the left posterior lobe. This hemorrhage was due to complete occlusion of the rectus sinus, as indicated in the angiogram (arrows). The basal vein of Rosenthal and the internal cerebral vein drain into the lateral mesencephalic vein.

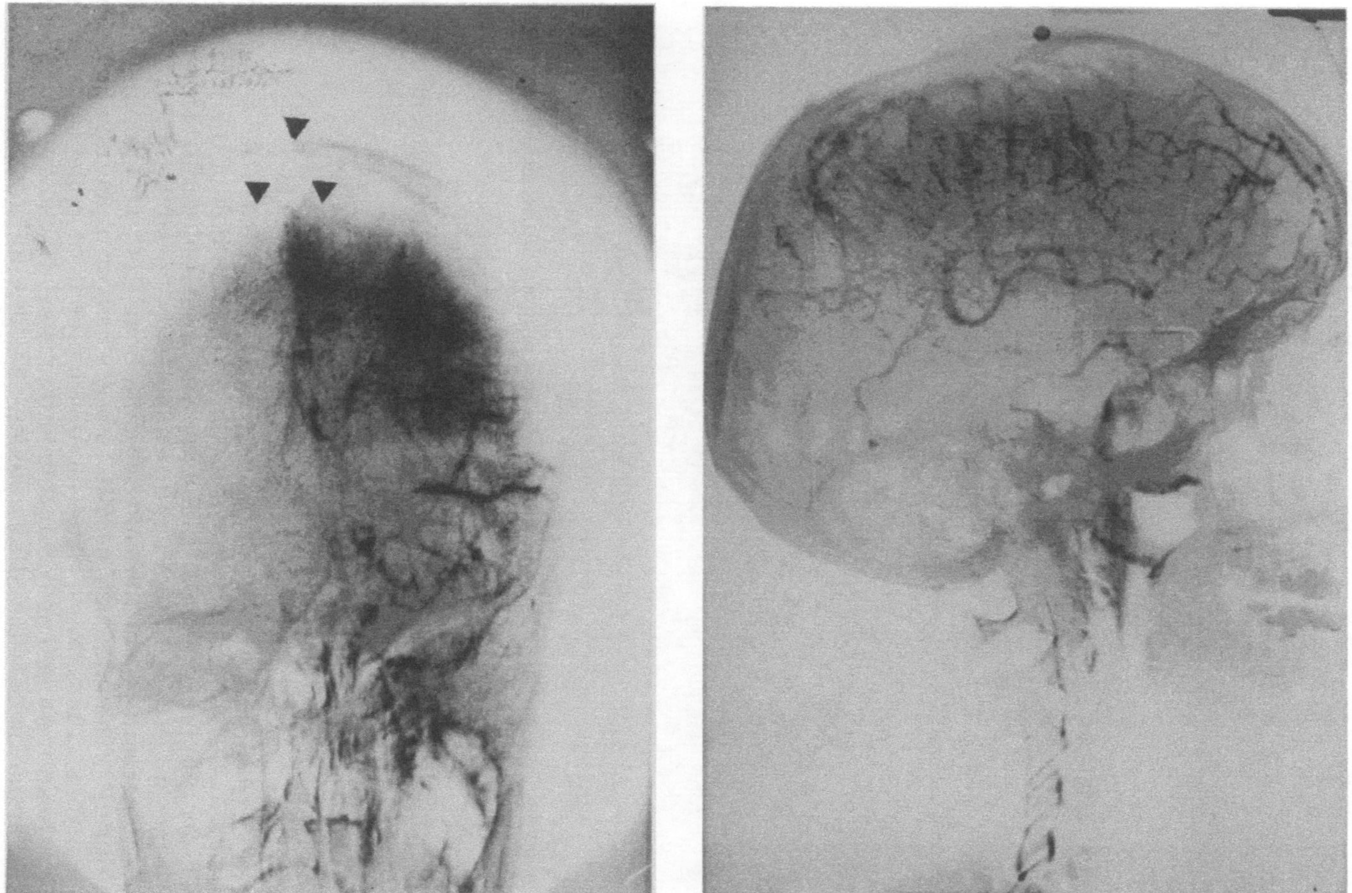

Fig. 5. Complete occlusion of the superior sagittal sinus (subtraction angiography) with typical empty triangle sign in a-p projection (arrows). Typical corkscrew appearance of the ascending veins in the lateral projection.

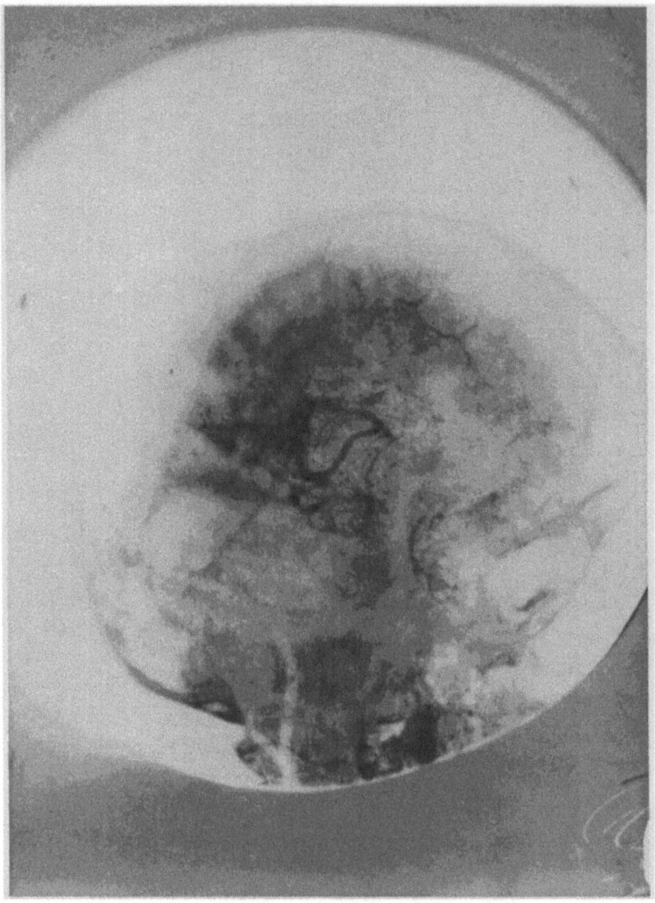

Fig. 6. Complete occlusion of the superior sagittal sinus without anastomotic network as shown in the oblique projection.

alternating, focal symptoms may be caused by both a thrombosis of the SSS or the vertebrobasilar arterial system.

DIFFERENTIAL DIAGNOSIS

The increase of the erythrocyte sedimentation rate is a rather frequent sign in CVT as well as leucocytosis. More than half of the patients with CVT do not present with blood in cerebrospinal fluid. The most important fact in recognizing CVT is to consider its possibility in diseases which may present with neurologic complications. Development of seizures during pregnancy may be due to a gestosis or a genuine epilepsy but also to CVT. The same is true for development of epileptic attacks after severe head trauma. Postcontusional attacks or attacks due to venous obstruction must be considered as causative factors and require different diagnostic methods for recognition of the true etiology.

Headache, focal or generalized seizures without or with focal deficits accompanying sinusitis, mastoiditis or otitis media should remind one of the complication of thrombophlebitis, an intracranial abscess or

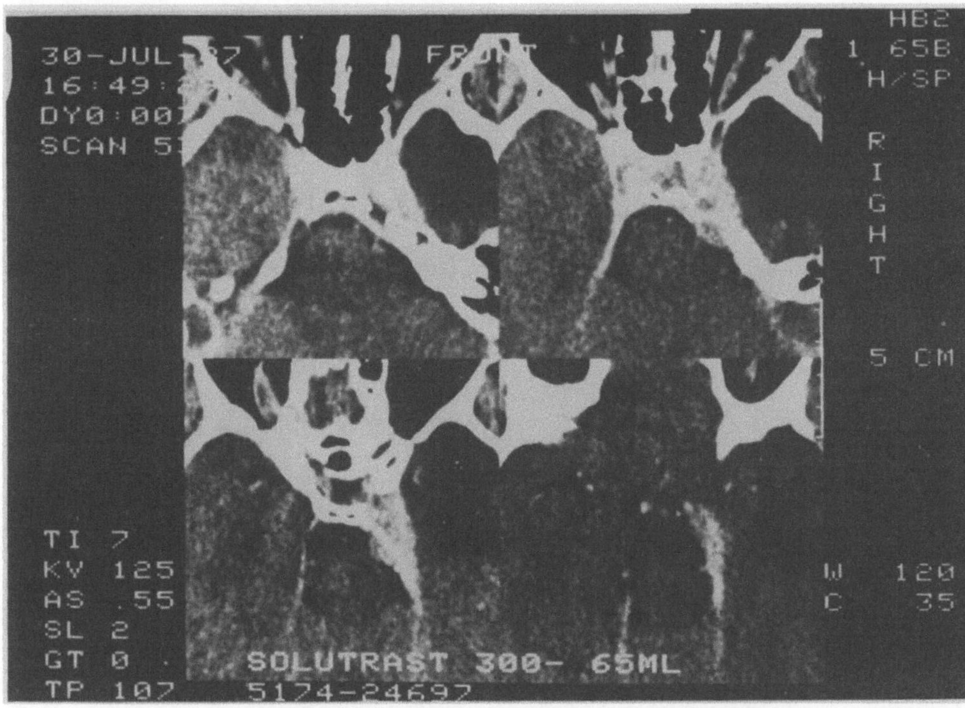

Fig. 7. Cavernous sinus meningeoma with the typical neurological picture of the tumor in the middle cerebral fossa involving the cavernous sinus. The tumor led to attachment of the tentorium at the clinoid processes.

meningitis. Focal signs are less common in meningitis and more common in abscess. In CVT headache belongs to the initial signs, whereas in abscesses focal deficits are more typical. With respect to the differentiation between arterial and venous occlusion in the brain, one may speculate that a sudden beginning is more typical for arterial ischemia whereas a progressive paresis occurs more often in venous thrombosis. Variations of the level of paresis is seen more frequently in venous thrombosis. Epileptic activity is observed more often in CVT than in arterial ischemia.

In thrombotic occlusion of a major cerebral sinus and even of the vein of Galen neurologic symptoms and pathologic alteration of cerebral tissue may not be detected if the anastomotic network bypasses the blood from the occluded vessel (Fig. 1). If this mechanism fails (Fig. 6), sinus or cerebral vein thrombosis is accompanied by clinical pictures which may resemble other neurologic diseases. The classic combination of symptoms for CVT are headache, signs of increased intracranial pressure, fever, epileptic seizures and focal or general neurologic deficits. Part of this combination or the total picture may go along with other cerebral vascular accidents, inflammatory diseases of the brain and space-occupying lesions. However if the clinical presentation of a particular patient is characterized by all these symptoms, the experienced neurologist will consider CVT as one of the possible causes.

In arterial ischemic diseases of the brain, signs of increased ICP and particularly papilledema will be observed far less frequently and not as early as in CVT. In embolic arterial diseases the fully developed clinical picture will be reached within seconds or minutes. Alternating symptoms

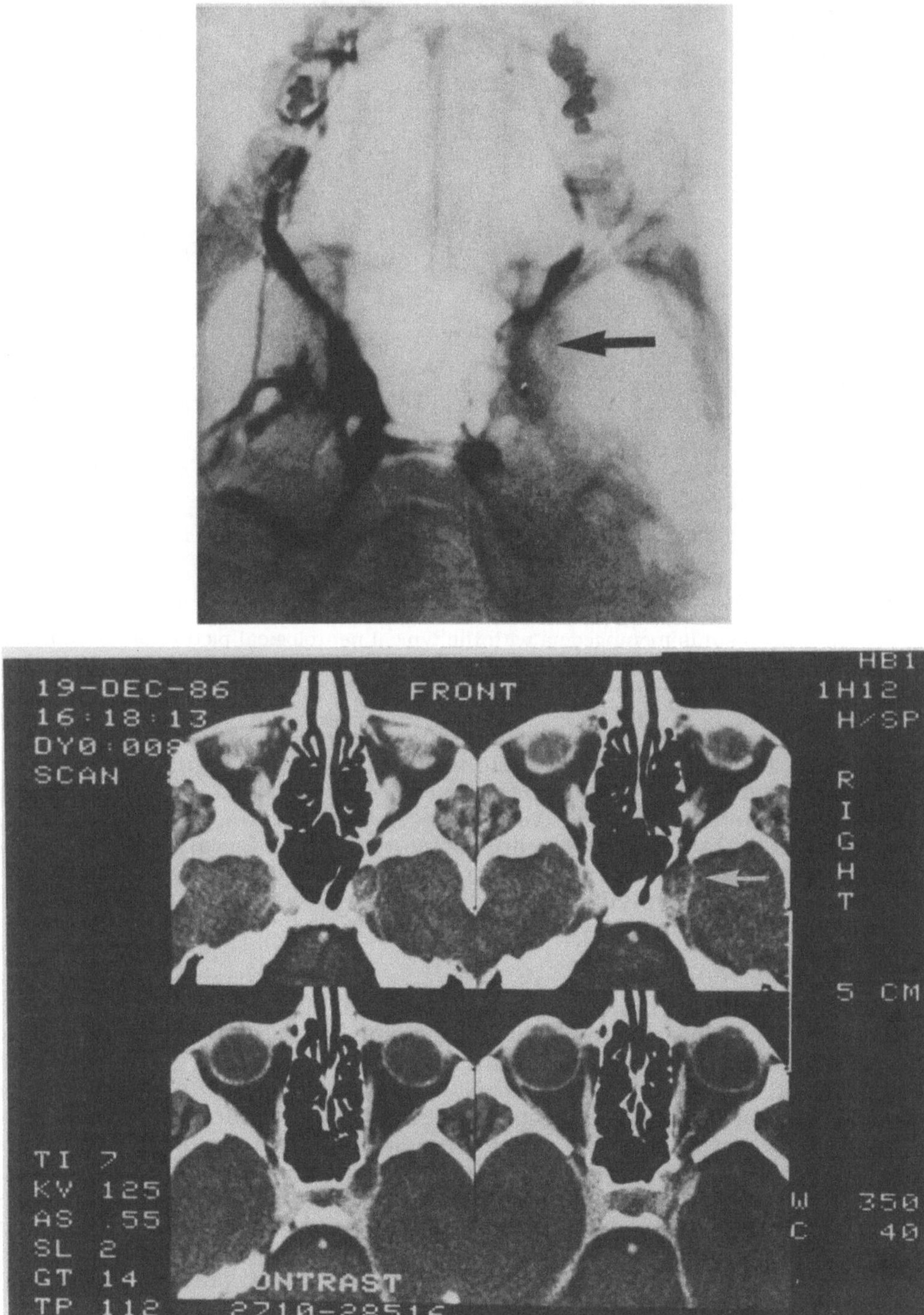

Fig. 8. Cranial computer tomogram with contrast medium enhancement in a patient with Tolosa-Hunt syndrome who responded well to therapy. Neurologic symptoms on admission: severe right-sided temporal pain, slight paresis of the VIth cranial nerve and reduced sensation over the Vth cranial nerve. The orbital phlebography and the CCT revealed impaired filling of the cavernous sinus (arrow).

or a changing severity of paresis are highly uncommon. Fever is typical in embolic endocarditis, seizures are not uncommon for embolic strokes. In cerebral hemorrhage long-lasting headache preceding the acute attack, epileptic activity, signs of increased ICP and focal symptoms are part of the symptomatology. However, clouding of consciousness occurs much earlier in the clinical course.

Subarachnoid hemorrhage (SAH) in most cases is characterized by sudden onset of headache. However, many patients report the occurrence of preceding symptoms like light headache or tinnitus. As in sinus thrombosis of the brain, neurologic symptoms may develop after a few days (due to vasospasms) and fever is not uncommon. Signs of intracranial hypertension will be present if a communicating hydrocephalus develops due to blockade of cerebral spinal fluid passage in the extraventricular space. Epileptic attacks are not as frequent in CVT. Despite similar symptoms of CVT and SAH with complicated hydrocephalus, in many patients the different course of the headache may separate both diseases clearly.

The differentiation between CVT and subdural hematomas is not easy. Both disorders share similar symptoms; however, fever is rare in hematomas. The course may be identical in both diseases. Radiological studies are necessary to differentiate between CVT and subdural hematoma in many cases. Certainly the underlying diseases are quite different for CVT (Table 1) and subdural hematoma.

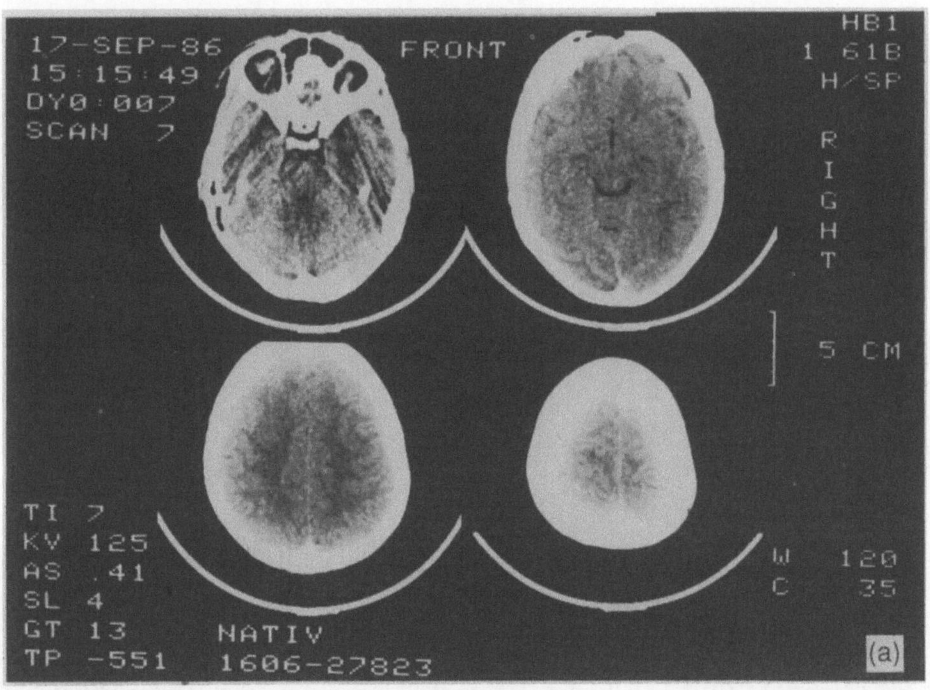

Fig. 9. (a) CCT (without contrast) on Day 1 of admission indicating slight hyperdense appearance of the confluens sinuum (upper right scan). (b) Contrast scan on Day 1 showing the typical delta sign. (c) The delta sign is significant (upper right scan), direct visualization of the thrombus in the right transverse (upper left scan) and the superior sagittal sinus (lower right scan). CCT was done 9 days after first examination. (d) Third CCT (15 days after first examination) indicated partial recanalization in the SSS, but no change of the delta sign. (e) Last scan (6 months after admission) indicated complete recanalization of the SSS, without delta sign.

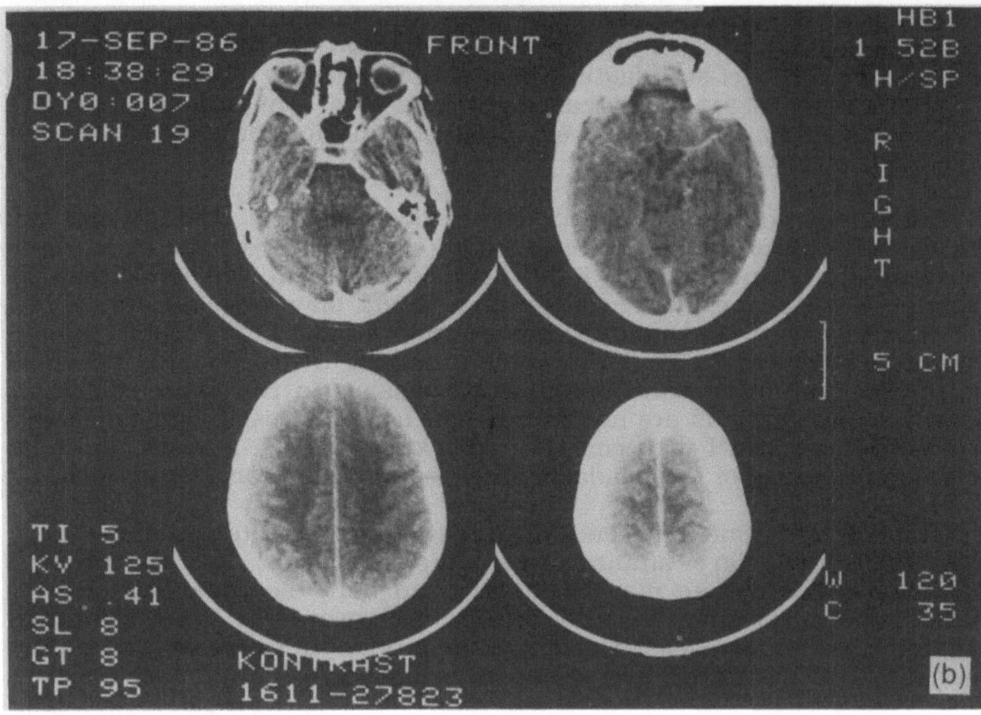

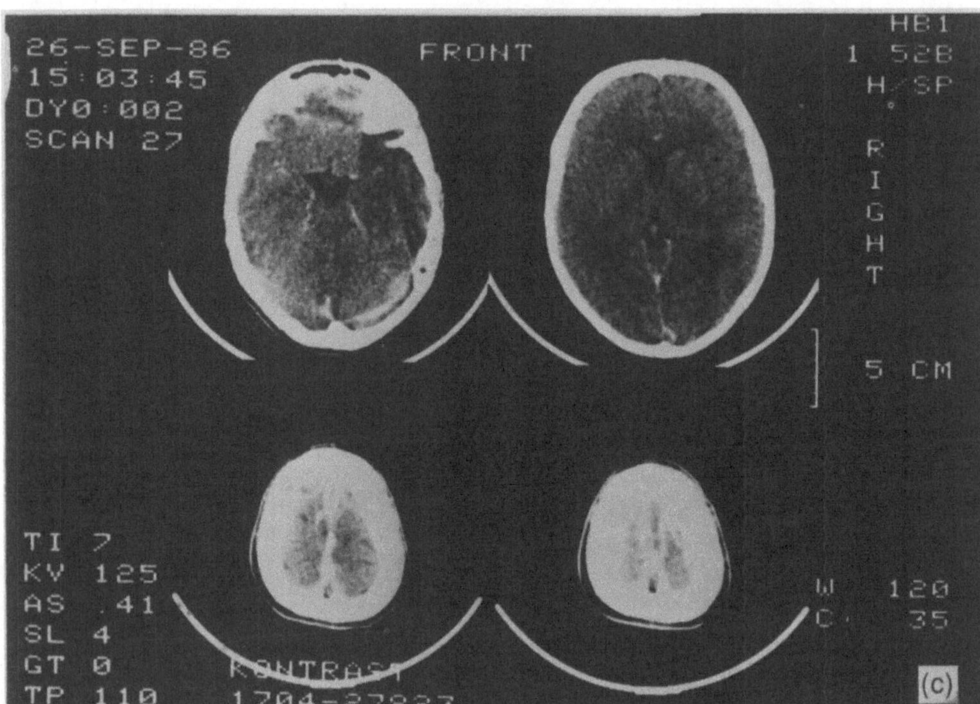

Fig. 9. (continued).

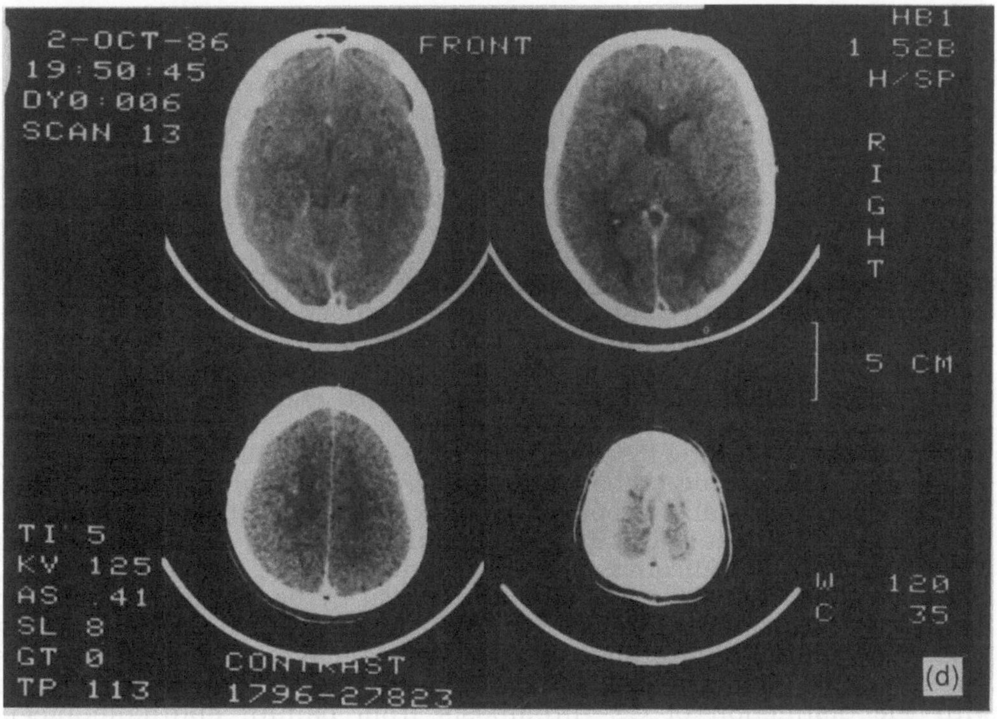

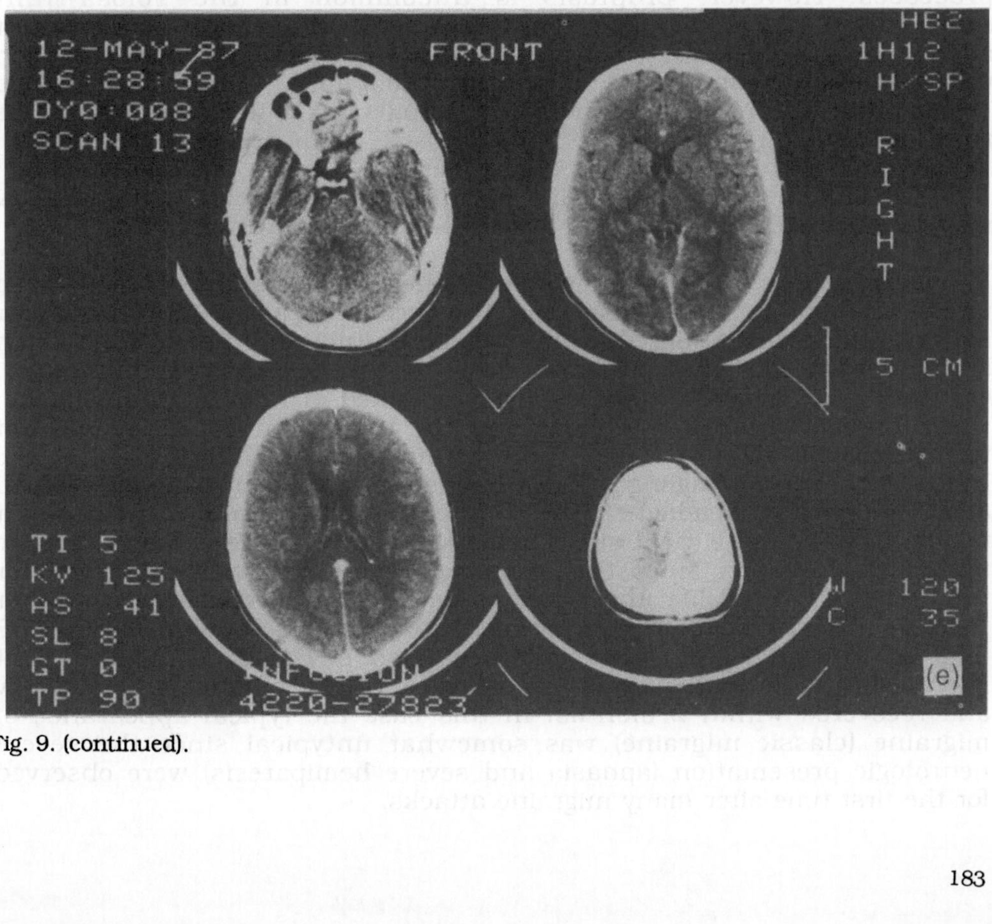

Fig. 9. (continued).

The separation of CVT from inflammatory diseases of the brain requires examination of the cerebral spinal fluid (CSF) and neuroradiologic studies. The clinical course of an abscess may be similar, but a hemorrhagic CSF points to CVT while pleocytosis points to an abscess. Abscesses do not provoke alternating paresis but this may occur with disseminated inflammatory diseases of the brain tissue. However, the latter are easily separated from CVT by other clinical signs, the patient's history, the lack of headache, and so on. Fever, headache, epileptic activity and signs of intracranial hypertension may be seen in both abscesses and CVT. If, however, high fever develops with signs of septicemia, this clearly indicates an inflammatory process. It should be kept in mind that some causes of CVT like infectious processes of the paranasal sinus or the petrous bone may cause abscesses also. The term otitic hydrocephalus [32] refers to this combined etiology.

In diseases affecting the cavernous sinus, mostly thrombosis has to be considered. A pulsating eyeball is characteristic more of an arteriovenous fistula, whereas a stable proptosis points more to the thrombosis. Sometimes space-occupying lesions like meningiomas may cause symptoms which seem to be indicative of a CS thrombosis (Fig. 7). Paratrigeminal neuralgia (Raeder's syndrome) is characterized by tic-like pain attacks in the first or second branch of the Vth CN. The sympathetic nerves causing ptosis and miosis but without a sweat deficit may be involved too. This should not be confused with CS thrombosis. More difficulties may arise with the distinction of CS thrombosis from painful ophthalmoplegia (Tolosa-Hunt syndrome) (Fig. 8). The pain may be long term in both disorders, the cranial nerves passing the CS may be affected by both processes. However, proptosis is uncommon in the Tolosa-Hunt syndrome. Fever, nausea and vomiting may occur in both disorders but additional involvement of the optic nerve (20% of patients with paratrigeminal neuralgia) is a rare symptom in CS thrombosis. Since Tolosa-Hunt syndrome is treated with steroids, a CS thrombosis must be ruled out with certainty.

The difficulties in recognizing a CVT from the clinical point of view may be elucidated by a patient's history as in the following example.

Case 1. This 23-year-old female, a chemical assistant, worked in an industrial laboratory and was exposed to certain chemical substances. For several years she suffered from repeated unilateral headache, which started mostly with photopsias over both eyes, some nausea and ill feeling. On the day of admission she again complained of right-sided headache, which spread within a few hours over the total cranium. She could not tolerate bright light, felt sick and within 2 hours after onset of the headache observed tingling in both hands. On admission she complained about increasing headache. She was obtunded, presented with slight motor aphasia and a right-sided hemiparesis. The diagnosis was "classic migraine". However, since the neurologic symptoms were observed for the first time in her life, the young woman was subjected to a computer tomogram of the head, which revealed the typical delta sign in the SSS (Fig. 9). The angiogram confirmed occlusion of the SSS and the right lateral sinus. The patient was treated with 1000 IU/hours of heparin iv and recovered within 2 months. In this case the typical appearance of migraine (classic migraine) was somewhat untypical since the severe neurologic presentation (aphasia and severe hemiparesis) were observed for the first time after many migraine attacks.

REFERENCES

1. E. A. Atkinson, B. Fairburn and K. W. Healthfield, Intracranial venous thrombosis as complication of oral contraception, *Lancet*, 1:914 (1970).
2. J. A. Bargen and N. W. Barker, Extensive arterial and venous thrombosis complicating chronic ulcerative colitis, *Arch. Intern. Med.*, 58:17 (1936).
3. D. B. Barnes and D. P. Winestock, Dynamic radionuclide scanning in the diagnosis of thrombosis of the superior sagittal sinus, *Neurology*, 27:656 (1977).
4. E. R. Bickerstaff, Aetiology of acute hemiplegia in childhood, *Br. Med. J. (Clin. Res.)*, 2:82 (1964).
5. A. Biemond, Clinical symptoms of so-called spontaneous thrombosis of superior longitudinal sinus, *Ned. T. Geneesj.*, 79:5422 (1935).
6. G. Boots, Thrombosis of the galenic systems veins in the adult, *Acta Neuropathol.*, 17:227 (1971).
7. M. G. Bousser, Cerebral vein thrombosis in Behçet's syndrome, *Arch. Neurol.*, 39:322 (1982).
8. P. Bradshaw, Benign intracranial hypertension, *J. Neurol. Neurosurg. Psychiat.*, 19:28 (1956).
9. A. Braun, "Sinus Thrombophlebitis. Inflammatory Diseases of the Venous Sinuses of the Dura Mater", New York (1928).
10. J. Brismar, Computer tomography in superior, sagittal sinus thrombosis, *Acta Radiol. Diagn.*, 21:321 (1980).
11. F. S. Buonanno, D. M. Mody and M. R. Ball, Computed cranial tomographic findings in cerebral sinovenous occlusion, *J. Comput. Assist. Tomog.*, 2:281 (1978).
12. F. Dindar and M. E. Platts, Intracranial venous thrombosis complicating oral contraception, *CMA*, 3:545 (1974).
13. B. Estanol, A. Rodrigues, G. Conte, J. M. Aleman, M. Loyo and J. Fezzuto, Intracranial venous thrombosis in young women, *Stroke*, 10:680 (1979).
14. W. L. Floyd and M. S. Mahaley, Cerebral dural venous sinus thrombosis following cardiac pacemaker implantation, *Arch. Intern. M.*, 124:368 (1979).
15. T. O. Gabrielsen, J. F. Seeger, J. E. Knake and E. W. Stilwill, Radiology of cerebral vein occlusion without dural sinus occlusion, *Radiology*, 140:403 (1981).
16. D. M. Gettelfinger and E. Kokmen, Superior sagittal sinus thrombosis, *Arch. Neurol.*, 34:2 (1977).
17. S. Heilbronn, Über Hirnherderkrankungen bei otitischer Sinus Thrombose, *Arch. Ohrenheilk*, 89:17 (1912).
18. E. Hervieux, "Traité Clinique et Practique des Maladies Puerpérales", Paris, p 771 (1870).
19. A. Huhn, Die Hirnvenen- und Sinus Thrombose, *Fortschr. Neurol. Psychiat.*, 25:440 (1957b).
20. A. Hunter and J. Lipscomb, Congenital pulmonary atresia with cerebral thrombosis and hemiplegia, *Br. Heart J.*, 4:124 (1942).
21. S. Johnson, R. Greenwood and M. A. Fishman, Internal cerebral vein thrombosis, *Arch. Neurol.*, 28:205 (1973).
22. A. B. King, Neurologic conditions occurring as complications of pregnancy, *Arch. Neurol. Psychiat. (Chic).*, 63:471 (1950).
23. D. P. E. Kingsley, B. E. Kendall and I. F. Moseley, Superior sagittal sinus thrombosis: an evaluation of the changes demonstrated on computed tomography, *J. Neurol. Neurosurg. Psychiat.*, 31:1065 (1978).
24. H. Krayenbühl, Cerebral venous thrombosis, diagnostic value of cerebral angiography, *Schweiz Arch. Neurol. Neurochr. Psychiat.*, 74:261 (1954).
25. J. P. Martin, Venous thrombosis in the central nervous system, *Proc. Roy. Soc. Med.*, 37:383 (1944).
26. J. P. Mickle, J. E. McLennan and C. W. Lidden, Cortical vein thrombosis in Wegeners granulomatosis, *J. Neurosurg.*, 46:248 (1977).
27. R. G. Mitchell, Venous thrombosis in acute infantile hemiplegia, *Arch. Dis. Child*, 27:95 (1952).

28. F. Ribes, Des recherches faites sur la phlebité, *Rev. Méd.*, 3:5 (1825).
29. B. Sigsbee, M. D. F. Deck and J. B. Posner, Non-metastatic superior sagittal sinus thrombosis complicating systemic cancer, *Neurology*, 29:139 (1979).
30. B. Sigsbee and D. A. Rottenberg, Sagittal sinus thrombosis as a complication of regional enteritis, *Ann. Neurol.*, 3:450 (1978).
31. R. G. Souter and A. Mitchell, Spreading cortical venous thrombosis due to infusion of hyperosmolar solution into the internal jugular vein, *Br. Med. J. (Clin. Res.)*, 285:936 (1982).
32. C. P. Symonds, Intracranial thrombophlebitis. Otolaryngology Lecture, *Ann. Roy. Coll. Surg. Eng.*, 10:347 (1952).
33. A. Towbin, The syndrome of latent cerebral venous thrombosis. Its frequency and relation to age and congestive heart failure, *Stroke*, 4:419 (1973).
34. G. Wilson, Crural monoplegia and paraplegia of cortical origin with a discussion of the cortical centers for the return, bladder and sexual functions, *Arch. Neurol. Psychiat. (Chic.)*, 10:669 (1923).
35. K. J. Zülch, Neurologische Diagnostik bei endokraniellen Komplikationen von otorhinologischen Erkrankungen, *Arch. Oto-Rhino-Laryngol.*, 183:1 (1964).

DIAGNOSTIC STRATEGIES IN CEREBRAL SINUS VEIN THROMBOSIS

Marie-Germaine Bousser, Colette Goujon*,
Valeria Ribeiro and Jacques Chiras†

Service de Neurologie, Hôpital Saint Antoine
184 rue de Faubourg Saint-Antoine
75571 Paris, France
*Service de Neurochirurgie, Hôpital Henri Mondor
Avenue du Marechal, Delattre de Tassigny
94000 Creteil, France and
†Service de Neuroradiologie, Hôpital de la Salpetriere
47 Bd de L'hôpital, 75013 Paris, France

INTRODUCTION

For about 130 years after the initial description by Ribes in 1825 [1], cerebral venous thrombosis (CVT) was mostly diagnosed at autopsy. This has led to the classical description [2-7] of a rare and severe disease characterized clinically by headache, papilledema, seizures, bilateral or alternating neurological deficits, progressive coma and death, and anatomically by an hemorrhagic and often bilateral cerebral infarction prohibiting the use of anticoagulants. In the last 35 years, angiography has made possible the diagnosis in patients [8-10]. Evidence has accumulated that the clinical spectrum of CVT is much wider than previously thought, as a major cause of the diagnostic problems [11-17].

The following discussion of diagnostic strategies in CVT is based mainly on a series of 62 adult patients with angiographically proven CVT of the dural sinuses. Reports on 38 patients of this series have already been published [16]. The diagnostic strategies are divided in four steps:

1. Clinical diagnosis
2. Role of various investigations
3. Topographic diagnosis
4. Etiologic diagnosis.

CLINICAL DIAGNOSIS

The main neurological signs found in the present series are indicated in Table 1. The classical symptoms are dominated by headache (79%),

Cerebral Sinus Thrombosis, Edited by K. Einhäupl *et al.*
Plenum Press, New York, 1990

which in all series is the most frequent and often the earliest sign [4,7,12,13,16,17]. Papilledema which is present in 50% of our cases is more frequent than in other series, where a range of 15-34% is reported [12]. By contrast, motor or sensory deficits (31%), seizures (29%), confusion or coma (26%), and dysphasia (8%) are less common than in other reports. In the literature, focal deficits, seizures and disorders of consciousness are present in about 50% of the cases. Other neurological manifestations were rare. There was not a single case with bilateral or alternating cortical signs as classical diagnostic indication of superior sagittal sinus (SSS) thrombosis.

Grouping of these symptoms was extremely variable. Twenty-eight patients (45%) had a well-defined isolated syndrome of intracranial hypertension (ICH), with headache in 26 (93%), papilledema in 26 (89%), and VIth nerve palsy in three cases. Thirty-five patients (55%) mainly had focal symptoms with or without signs of intracranial hypertension, 16 had focal deficits, five had only focal seizures, 22 had seizures and deficits, one had typical transient ischemic attacks (TIA), and another had short lasting (1 or 2 minutes) left arm paresthesia, which could not be differentiated as TIA or seizure.

The mode of onset varied considerably from patient to patient. An attempt was made to take into account the time elapsing between appearance of the first symptom and entry into the hospital. Three main modes of onset were arbitrarily defined:

Acute: <48 hours.
Subacute: longer than 48 hours, but less than 1 month.
Chronic: longer than 1 month.

Onset was acute in 17 patients (27%), most often in those with focal signs. Onset was chronic in 29 patients (47%), more frequently without [19] than with [10] focal signs. The clinical presentations were extremely variable with a wide spectrum of neurological signs and modes of onset. Six major patterns could be distinguished, each simulating another pathologic condition. The most frequent (42%) and homogeneous was a subacute or progressive onset of isolated signs of *intracranial hypertension,* corresponding to "benign intracranial hypertension", or "pseudotumor cerebri" [18-21]. This indicates that sinus thrombosis is one of the leading causes of this syndrome [16,17,23,24]. We suggest that the definition of this syndrome should include not only the classical tetrade [24,25] of:

1. Isolated signs of intracranial hypertension
2. Normal CT scan, or small ventricles
3. Increased CSF pressure (> 20 mmHg)
4. Normal CSF composition, except for low protein,

but also *a normal four-vessel angiogram or MRI* as a fifth criterion. The second pattern (24%) was a *sudden onset of focal deficits* simulating arterial stroke. However, the presence of seizures and the absence of well-defined arterial syndromes should alert towards the possibility of CVT, particularly if the patient steadily deteriorates instead of improving. The third presentation (13%) was *subacute onset of deficits* (over a few days) and/or seizures with or without signs of intracranial hypertension simulating a brain abscess or encephalitis, particularly in patients with fever, increased ESR (red blood cell sedimentation), and CSF pleocytosis.

Table 1. Symptoms Observed in 62 Patients with Cerebral Venous Thrombosis

Main Neurological Symptoms Observed	Number
Headache	49 (79%)
Papilledema	31 (50%)
Motor or sensory deficits	19 (31%)
Seizures	18 (29%)
Drowsiness, confusion, or coma	16 (26%)
Dysphasia	5 (8%)
Multiple cranial nerve palsies	3 (5%)
Cerebellar incoordination	2 (3%)
Nystagmus	2 (3%)
Hearing loss	1
Bilateral or alternating cortical signs	0

Only neuroimaging may differentiate between these conditions. The fourth pattern (14%) simulated a brain tumor with corresponding clinical signs but an onset over weeks or months. Other less common patterns were headache of sudden onset with red blood cells in CSF mimicking subarachnoid hemorrhage (two patients) and TIA, which was typical in one patient and atypical, ie, probably presenting with focal seizures in another. From the above description it is obvious that diagnosis of CVT, early diagnosis in particular, cannot be made on clinical grounds only. The correct diagnosis can only be suspected from the variety of clinical presentations. This is why it is necessary to be aware of CVT in such patients, requiring appropriate investigations in order not to overlook this condition.

INVESTIGATIONS

General Investigations and Hemostasis Studies

Temperature was higher than 37.5°C in 25 patients (40%). The white blood cell (PMN) count was increased (>7250 mm^3) in 23 patients (37%). ESR was >20 mm at the first hour in 32/56 (59%) cases, eight of which had an ESR >100. Of these, five had an infection, two had Behçet's disease, and one had systemic lupus.

Thus, about half the patients had fever and/or increased ESR and/or an elevated PMN count, usually in association with infective, inflammatory, or malignant causes. These abnormalities were occasionally found also in idiopathic cases. Their absence, therefore, does in no way preclude the diagnosis of CVT, while their presence might point to CVT in patients presenting with pseudotumor cerebri.

Detailed hemostasis studies have only rarely been performed in CVT and have yielded conflicting results. Some have reported a hypercoagulability state [26-27], an increase in platelet adhesiveness and aggregability [12,26], and a decrease in fibrinolytic activity [12]. However, this was associated with pregnancy, postpartum state, or found in women taking oral contraceptives. Others [28] could not confirm these findings in women of childbearing age. In the present series, two studies were performed. The first on 17 patients was examining mainly the role of platelets including: platelet count, platelet retention on glass beads, spontaneous platelet aggregation, ADP-, epinephrine- and collagen-induced aggregation. The only anomalies found were an increase in platelet count in four patients with an inflammatory syndrome and an increased platelet aggregation by the lowest dose of epinephrine which, however, is non-specific. It was therefore concluded that detailed studies on platelets are not helpful in CVT, thus allowing us to abandon these tests. A second study on the coagulatory properties was performed more recently in 18 patients involving: aPTT, fibrinogen, ATIII antigen and -activity, plasminogen, protein C antigen and -activity, protein S, and factor VIII. The main finding was an increase in the levels of fibrinogen, plasminogen, and factor VIII in half the number of patients. These changes could again be associated with inflammatory processes. Another anomaly was a decrease in ATIII-activity in one patient suggestive of an increased ATIII consumption. It should, however, be noted that most of the other tests are difficult to interpret in patients treated with heparin or inhibitors of vitamin K synthesis. On the basis of these findings it appears that studies on hemostasis are not useful for diagnosis of CVT, and that this parameter is often modified by anticoagulation. Nevertheless, hemostasis investigations are necessary in all apparently "idiopathic" cases for the discovery of underlying, possibly causal, mechanisms such as ATIII-deficiency [30], protein C deficiency [29], or presence of circulating antibodies, such as lupus anticoagulants [31].

Standard Electroencephalogram (EEG)

Several reports have emphasized the importance of EEG changes in CVT. The most common pattern is a generalized slowing more marked at one side with frequent epileptic activity [32]. Indeed, in the present series of 57 cases with EEG studies, diffuse slow activity was the most common finding, occurring in 27 patients (47%) including cases with focal signs. Focal slow activity was observed in 14 patients (24%) and epileptic activity in 11 (19%). The EEG was normal in one-quarter of the cases, mainly in patients with intracranial hypertension. The most interesting observation was a generalized slowing of the EEG in patients with focal lesions indicative of a more diffuse nature of the neurological impairment as apparent from the clinical symptoms.

Isotope Studies

Isotope brain scanning with ^{99m}Tc is considered a useful diagnostic tool in CVT [33,34]. However, in two of 13 of our patients the isotope scan was normal, even in the posterior oblique projection with one SSS and one lateral sinus occlusion. Angiography disclosed an intensive collateral circulation in the sinus walls, which may have accounted for the false negative results. Isotope studies as a non-invasive tool have now been replaced by DIVA and MRI. However, a recent report [35] emphasizes the value of ^{111}In platelet scintigraphy in the diagnosis of CVT, particularly for the distinction between recent and old thrombosis, which might be of therapeutic relevance.

CT Scan

A CT scan is the first neuroradiological examination which should be performed if CVT is suspected. It serves (1) to rule out disorders other than CVT, and (2) to visualize characteristic abnormalities indicative of, or even diagnostic of, CVT [17, 36-46]. CT scans were performed in 49 patients in the present series. The results are given in Table 2. The findings differ slightly from the data reported in the literature as, eg, summarized by Virapongse et al. [40] on the basis of 24 publications. The CT scans in our series were much more often normal (22%) than those reported in the literature, mainly because our patients often presented with isolated intracranial hypertension. In contrast the CT scan was rarely normal (11%) in patients with focal signs.

Direct CT signs indicating CVT were uncommon, eg, "cord sign" [36-38] or "dense triangle" [39,43] were not found. The "delta sign" was found only in 18% of our cases, while it is reported at a rate of 10-70% in the literature [36, 38-42]. The discrepancy might have several explanations. Some CT scans were probably performed too early or too late. The delta sign is seen only 1-4 weeks after onset [41]. The posterior part of the sinus was not involved in one-third of our patients. Finally, technical refinements recommended to improve sensitivity and specificity of CT diagnosis, such as orthogonal sectioning, different window and level settings, and multiplanar reformation [40], were rarely utilized.

Indirect, but non-specific, CT abnormalities were more frequently found. The most common finding was presence of narrow ventricles (51%). It was the only finding in eight patients, while enlarged ventricles were found in three cases. Thus, large ventricles do not exclude CVT. Parenchymal lesions were present in 51% of our cases, but the classical spontaneous hyperdensity suggestive of hemorrhagic infarction [39,44] was observed in six cases only. Focal hypodensity with or without contrast enhancement was more frequent [16] which, however, is less indicative of CVT.

The priority of CT in the diagnostic strategy of CVT is high because it can rule out other conditions, such as arterial stroke, brain abscess, brain tumor, or SAH. CT should initially be performed without contrast then with contrast in the absence of hemorrhagic infarction. If changes pathognomonic of CVT are seen, the patient might be spared the potential side-effects of angiography [42], although this has been questioned [17]. If, however, the changes are non-specific, or if the CT scan is normal, subsequent procedures should include CSF examination, MRI and angiography.

Cerebrospinal Fluid (CSF)

Examination of CSF is still an important diagnostic tool, since its composition and pressure are rarely normal (7%) in CVT. Evaluation of CSF pressure should be performed because of its diagnostic and prognostic value, particularly in patients with benign intracranial hypertension. In the present series of 55 CVT patients CSF pressure was increased (> 20 mmHg) in 63%, while normal in 29%, or at a borderline level, respectively, in the remaining cases. The CSF composition was abnormal in 60% of cases, with an increased protein content in 45%, increased white blood cell count in 24%, and more than 20 red blood cells in 45%. Such findings are in accordance with the classical literature

Table 2. CT Scan in 49 Patients with Cerebral Venous Thrombosis

CT Scan	Number
Normal	11 (22%)
Ventricles	
Small	25 (51%)
Large	3 (7%)
Focal Signs	25 (51%)
Hypodensity	16 (33%)
Hyperdensity	6 (12%)
Isodense localized swelling	3
Postcontrast Enhancement	
Delta sign	9 (18%)
Gyral	17 (35%)
Tentorial	10 (20%)
Combination of various signs	20 (41%)

[4,7,13]. Pleocytosis and presence of red blood cells are mainly seen in patients with focal signs [47]. However, such signs were also found in patients with benign intracranial hypertension suggesting that abnormalities of the cellular CSF composition should alert towards cerebral sinus thrombosis. On the other hand, a normal CSF composition does not exclude CVT.

Angiography

Angiography is still the best method of a positive diagnosis of CVT [7-9,16,45,48]. It should be performed in all cases without pathognomonic changes in the CT scan. Although controversial, we perform angiography even in cases with characteristic CT changes because of the precise information on the extension of thrombosis and on an involvement of cortical veins. This attitude, however, may change in the future, if more experience is available with MRI. Although not pathognomonic [46,48], the partial or complete lack of filling of the sinuses is the best symptom of CVT. It requires, however, excellent angiography, ie, a three- or four-vessel angiogram by conventional or intra-arterial digitalized techniques together with visualization of the entire venous phase and at least two projections. Two anatomical variations may be confused with sinus thrombosis. The first is absence of the anterior one-third of the superior sagittal sinus. Under these conditions, a large draining cortical vein will be found [8]. For establishment of occlusion of the anterior part of the SSS, involvement of another sinus should be seen or non-equivocal, indirect signs of CVT, such as delayed emptying and dilation of collateral veins, particularly of cortical veins. The latter are more often involved than the dural or scalp veins. Another misleading anatomical variation is congenital hypoplasia of a lateral sinus (LS). This is found in up to 25% of cases, more frequently on the left side [49]. A main argument indicative of sinus thrombosis is absence of filling of the total lateral sinus, whereas hypoplasia usually affects only the transverse portion [49]. This is in contrast with presence of the sinus groove and a normal jugular foramen on a plain X-ray of the skull. In cases where these symptoms are lacking, diagnosis by MRI is necessary to distinguish between thrombosis and hypoplasia of the lateral sinus.

Magnetic Resonance Imaging (MRI)

MRI offers major advantages for evaluating patients with suspected CVT because of its sensitivity to blood flow, its high intrinsic contrast, and its non-invasiveness. For many authors MRI is the imaging modality of choice for diagnosis and follow-up of CVT because in contrast to angiography, MRI visualizes the thrombosis itself. A variety of MRI findings have been described and related to the evolution of thrombosis [50]. In high field MRI, early thrombosis appears isointense when using T1WI, while hypointense with T2WI. The normally occurring flow void is not seen in any plane, or imaging sequence. In the intermediate phase, the thrombus converts to a high intensity signal in T1WI, thereafter in T2WI. In the case of large vessels, high intensity starts in the periphery and proceeds centrally. In the late phase, recanalization may occur with reappearance of a flow void, but a hyperintensive signal may persist for a year or more [60]. The diagnostic criteria of CVT in MRI, therefore, vary with time. Late diagnosis may be almost impossible if recanalization has occurred.

MRI is not only useful in visualizing thrombosis, it also allows detection of parenchymal lesions, such as solid hematomas or haemorrhagic infarcts, which sometimes cannot be seen on the CT scan (Fig. 1).

TOPOGRAPHIC DIAGNOSIS

The present series confirms that the SSS (69%) and LS (72%) are most frequently involved, and that occlusion is rarely (21%) limited to one sinus. In most cases occlusion involved other sinuses (40%) and/or cerebral veins (39%). Cortical veins were more commonly affected than the internal cerebral veins (three cases) or cerebellar veins (three cases). The frequent associations may, at least partly, explain why no good clinico-radiological correlations can be established. Other reasons are the well-known difficulty of diagnosing cerebral vein occlusion [54], and the inability of CT scanning to detect all the parenchymal lesions. Thus, whereas most (26/28) patients presenting with isolated intracranial hypertension had thrombosis restricted to the sinuses only, an involvement of cerebral veins could not be firmly established in 22/34 of those presenting with focal signs.

MRI studies may help to improve a clinico-radiological correlation to define clinical patterns of different involvements of the sinuses. Although, a topographic diagnosis does not seem to be crucial, thrombosis limited to the sinuses seems to bear a better prognosis than involvement of the cerebral veins [16]. Diagnosis of internal cerebral vein thrombosis remains difficult. One patient presented with a 1-month history of headache, hemiparesis, and slight confusion. She recovered spontaneously within 6 weeks illustrating a benign course, which was also emphasized in recent reports [55-56]. Two other patients had a more classical severe course [2,4,57,58] with headache, hemiplegia, aphasia, and coma in one case, and confusion, hemiparesis, vertical gaze palsy and bladder disturbances in the other. Observations of cerebellar vein thrombosis were almost

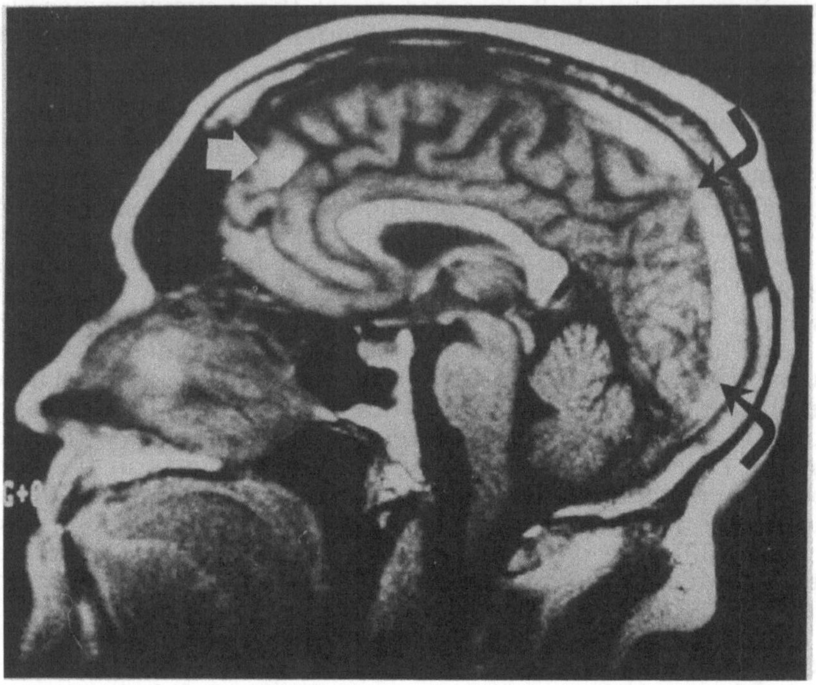

Fig. 1. T1-Weighted sagittal MR image. Black arrows indicate signal resulting from thrombus in the posterior half of the superior sagittal sinus. White arrow shows a small high-intensive signal area in the frontal region suggestive of a small hemorrhagic infarct not visible on the CT scan.

exclusively made in autopsy studies [6], its clinical and angiographical diagnosis is extremely difficult. One patient presented with a 3-month history of cranial nerve palsies, cerebellar coordination problems, and papilledema simulating a brain tumor. Another had severe headache of sudden onset simulating SAH. The third patient had a 6-month history of coordination disturbances of the lower limbs and bilateral hearing loss. This patient had also meningitis from sarcoidosis, which could partly explain the clinical findings.

ETIOLOGY

Search for a cause remains one of the most vexing problems in the diagnostic strategy of CVT. Etiologies of patients in the present series are summarized in Table 3. Like in other series it presents a selected group, which is not representative of the numerous causes that have been described (Table 4). An obvious example for a referral bias is the great number of patients with Behçet's disease [9/62] reflecting our interest in this condition. However, we think that - at least in our own country - Behçet's disease is an underrecognized cause of CVT, probably because CVT can be the only initial manifestation of the disease [59-61].

The diagnostic strategy in the search for a cause differs from patient to patient. In some cases the cause is obvious, eg, in patients with septic complications from head injury, local infections, end stage malignancy,

Table 3. Etiology in 62 Patients with Cerebral Venous Thrombosis

Etiology	Number	
Infective	7	(11%)
ENT	4	
Abdominal	2	
Generalized	1	
Local	6	(10%)
Minor head injury	5	
Arachnoid cyst	1	
Obstetrical	4	(6%)
Postpartum	3	
Pregnancy	1	
General Diseases	25	(40%)
Behçet's disease	9	
Malignancies	5	
Carcinomatous meningitis	2	
Malignant lymphoma	1	
Breast carcinoma	1	
Myeloid leukemia	1	
Venous thromboembolic disease	3	
Uveomeningitis	2	
Sarcoidosis	1	
Chronic anemia (bleeding fibroma)	1	
Primary thrombocytaemia	1	
Decreased antithrombin III activity	1	
Systemic lupus	1	
Nephrotic syndrome	1	
Oral Contraceptives	5	(8%)
"Idiopathic"	15	(24%)

and so on. In these conditions CVT is usually not the main problem. In other cases there is no obvious cause, and one has to look systematically for all the known causes of deep vein thrombosis, ie, surgical, obstetrical, or medical, including the number of local or regional infective or non-infective causes (Table 4). According to our experience it can be noted that the frequency of septic CVT has markedly declined (7/62) since the advent of antibiotics. This, however, is not the case in all series [62]. Use of oral contraceptives is a well-recognized cause of CVT [15, 26, 63, 64]. It was the only etiologic factor in five patients (8%) in the present series. However, in another patient, where oral contraceptives were the only initially assessed factor, typical symptoms of Behçet's disease appeared 8 months later. One should, therefore, be cautious and always consider other causes as well. The complication from taking contraceptives, nevertheless, has led us - as others [15, 63, 64] - to stop presenting contraceptives and look for CVT in women presenting with any of the above mentioned neurological manifestations.

In 15/62 cases no cause could be found, despite a follow-up of more than a year. These "idiopathic" or "primary" cases need extensive examinations with detailed studies of coagulation, search for malignancy, connective tissue disease, and so on, and close follow-up. In six cases which were initially interpreted as idiopathic, general underlying diseases

Table 4. Recognized Causes of Cerebral Venous Thrombosis

Infective Causes

- *Local*
  - Trauma with septic complications
  - Intracranial infections: abscess, empyema, meningitis ...
  - Regional infections: otitis, tonsillitis, sinusitis, stomatitis ...
- *General*
  - Bacterial: septicemia, endocarditis, typhoid, tuberculosis ...
  - Viral: measles, hepatitis, encephalitis (herpes, HIV) ...
  - Parasitic: malaria, trichinosis ...
  - Fungal: torulosis ...

Non-infective Causes

- *Local*
  - Head injury (open or closed, with or without fracture)
  - Neurosurgical operation
  - Cerebral infarcts and hemorrhages
  - Tumors (meningioma, metastasis, glomus tumor)
  - Porencephaly, arachnoid cysts
- *General*
  - Surgical: any surgery with or without deep vein thrombosis
  - Gyneco-obstetrical:
    - Pregnancy and puerperium
    - Oral contraceptives (estrogens, progestogens)
  - Medical:
    - Cardiac: congenital heart disease, cardiac insufficiency, pacemaker ...
    - Malignancies: any visceral carcinoma, lymphoma, leukemia, carcinoid ...
    - RBC disorders: polycythaemia, post-hemorrhagic anaemia, sickle cell disease, paroxysmal nocturnal haemoglobinuria ...
    - Thrombocythaemia (primary or secondary)
    - Coagulation disorders: AT_{III}, protein C deficits, circulating anticoagulants ...
    - Severe dehydration of any cause
    - Digestive: cirrhosis, Crohn, ulcerative colitis ...
    - Connective tissue: systemic lupus, temporal arteritis, Wegener's granulomatosis ...
    - Venous thrombo-embolic disease, Hughes Stovin syndrome
    - Others: Behçet's disease, sarcoidosis, uveomeningitis, nephrotic syndrome ...

Idiopathic

were discovered 6-12 months later, ie, Behçet's disease in five, and malignancy in one.

CONCLUSIONS

Neuroimaging techniques now allow early diagnosis of CVT which appears to be more common, less severe, and more diverse in its current clinical presentations than previously thought. In patients with pseudotumor cerebri in the absence of other obvious causes, CVT should be considered and confirmed by angiography or MRI. In patients with focal signs, CVT may mimic arterial stroke, brain abscess, brain tumor, or even SAH. The CT scan is of major significance in ruling out these conditions, but angiography or MRI are needed, nevertheless, to confirm CVT and to determine its extension. Etiologic diagnoses encompass a multitude of causes indicating the need for detailed examinations, particularly in the search for general underlying diseases. However, about one-quarter of CVT cases remain idiopathic, which then require close and prolonged follow-up.

REFERENCES

1. M. F. Ribes, Des recherches faites sur la phlébite. Revue médicale et étrangère et Journal de clinique de l'Hôtel-Dieu et de la Charité de Paris, 3:5-41 (1825).
2. H. Ehlers and C. B. Courville, Thrombosis of internal cerebral veins in infancy and childhood. Review of literature and report of five cases, *J. Pediat.*, 8:600-623 (1936).
3. C. P. Symonds, Cerebral thrombophlebitis, *Br. Med. J.*,2:348-352 (1940).
4. R. Garcin and M. Pestel, "Thrombophlébites Cérébrales", Masson et Cie, Paris (1949).
5. H. J. M. Barnett and H. H. Hyland, Non-infective intracranial venous thrombosis, *Brain*, 76:36-49 (1953).
6. H. Noetzel and F. Jerusalem, Die Hirnvenen- und Sinusthrombosen. Monographien aus dem Gesamtgebiet der Neurologie und Psychiatrie, 106:1-63 (1965).
7. R. M. Kalbag and A. L. Woolf, "Cerebral Venous Thrombosis", University Press, London and Oxford (1967).
8. H. Krayenbühl, Cerebral venous thrombosis. The diagnostic value of cerebral angiography, *Schweiz Arch. Neurol. Neurochir. Psychiat.*, 74:261-287 (1954).
9. B. Montrieul and P. Janny, Contribution à l'étude angiographique des thromboses veineuses cérébrales, *Neurochirurgie*, 8:175-188 (1962).
10. M. G. Yasargil and M. Damur, Thrombosis of the cerebral veins and dural sinuses, *in*: "Radiology of the Skull and Brain. Angiography", T. H. Newton and D. G. Potts, eds., The CV Mosby Company, Saint-Louis (1974).
11. P. Averback, Primary cerebral venous thrombosis in young adults. The diverse manifestations of an under-recognized disease, *Ann. Neurol.*, 3:81-86 (1978).
12. B. C. Bansal, R. R. Gupta and C. Prakash, Stroke during pregnancy and puerperium in young females below the age of 40 years as a result of cerebral venous/venous sinus thrombosis, *Jpn. Heart J.*, 21:171-183 (1980).
13. G. Coquillat and J. M. Warter, Thromboses veineuses cérébrales. Rapport de neurologie présenté au Congrès de psychiatrie et de neurologie de langue Francaise, 1 vol., Masson, Paris (1976).
14. D. M. Gettelfinger and E. Kokmen, Superior sagittal sinus thrombosis, *Arch. Neurol.*, 34:2-6 (1977).
15. P.Rousseaux, M. J. Bernard, B. Scherpereel and J. F. Guyot, Thrombose des sinus veineux intra-crâniens (à propos de 22 cas), *Neurochirurgie*, 24:197-203 (1978).
16. M. G. Bousser, J. Chiras, B. Sauron, J. Bories and P. Castaigne, Cerebral venous thrombosis. A review of 38 cases, *Stroke*, 16:199-213 (1985).
17. A. Thron, K. Wessel, D. Linden, G. Schroth and J. Dichgans, Superior sagittal sinus thrombosis: neuroradiological evaluation and clinical findings, *J. Neurol.*, 233:283-288 (1986).
18. J. Foley, Benign forms of intracranial hypertension: "toxic" and "otitic" hydrocephalus, *Brain*, 78:1-41 (1955).
19. B. Guidetti, B. Giuffre and D. Gambacorta, Follow-up study of 100 cases of pseudo-tumor cerebri, *Acta Neurochir.*, 18:259-267 (1968).
20. H. G. Boddie, M. Banna and W. G. Bradley, "Benign" intracranial hypertension. A survey of the clinical and radiological features and long term prognosis, *Brain*, 97:313-326 (1974).
21. C. Bulens, J. de Vries and H. Vancrevel, "Benign" intracranial hypertension, a retrospective and follow up study, *J. Neurol. Sci.*, 40:145-157 (1979).
22. C. Symonds, Otitic hydrocephalus, *Neurology*, 6:681-685 (1956).
23. D. D'Avella, R. P. Greenberg, S. Mingrino, M. Scanarini and K. Pardatscher, Alterations in ventricular size and intracranial pressure caused by sagittal sinus pathology in man, *J. Neurosurg.*, 53:656-661 (1980).

24. M. Greer, Management of benign intracranial hypertension (pseudo-tumor cerebri), *Clin. Neurosurg.*, 15:161-174 (1967).
25. J. J. Corbett, Problems in the diagnosis and treatment of pseudo-tumor cerebri, *Can. J. Neurol. Sci.*, 10:221-229 (1983).
26. F. Dindar and M. E. Platts, Intracranial venous thrombosis complicating oral contraception, *Can. Med. Assoc. J.*, III:545-548 (1974).
27. A. A. Poltera, The pathology of intracranial venous thrombosis in oral contraception, *J. Pathol.*, 106:209-219 (1972).
28. B. Estanol, A. Rodriguez, G. Conte, J. M. Aleman, M. Loyo and J. Pizzuto, Intracranial venous thrombosis in young women, *Stroke*, 10:680-684 (1979).
29. I. Iob, M. Scanarini, G. C. Andrioli and K. Pardatscher, Thrombosis of the superior sagittal sinus associated with idiopathic thrombocytosis, *Surg. Neurol.*, 11:439-441 (1979).
30. A. R. Wintzen, A. W. Broekmans, R. M. Bertina, E. Briet, P. E. Briet, A. Zecha, G. J. Vielvoye, G. Th. A. M. Bots, Cerebral hemorrhagic infarction in young patients with hereditary protein C deficiency: evidence for "spontaneous" cerebral venous thrombosis, *Br. Med. J.*, 290:350-352 (1985).
31. S. R. Levine, S. Kieran, K. Puzio, H. Feit, S. C. Patel and K. M. A. Welch, Cerebral venous thrombosis with lupus anti-coagulants. Report of two cases, *Stroke*, 18:801-804 (1987).
32. S. Burkhardt and F. Regli, EEG-Veränderungen bei 27 Fällen von zerebraler Thrombophlebitis, *Schweiz Arch. Neurol. Psychiat.*, 94:1-14 (1964).
33. R. T. Go, C. L. Chiu and L. A. Neuman, Diagnosis of superior sagittal sinus thrombosis by dynamic and sequential brain scanning. Report of one case, *Neurology*, 23:1199-1204 (1973).
34. B. D. Barnes and D. P. Winestock, Dynamic radionuclide scanning in the diagnosis of thrombosis of the superior sagittal sinus thrombosis, *Neurology*, 27:656-661 (1977).
35. S. L. Bridgers, E. Strauss, E. O. Smith, D. Reed and M. D. Ezekowitz, Demonstration of superior sagittal sinus thrombosis by Indium-111 platelet scintigraphy, *Arch. Neurol.*, 43:1079-1081 (1986).
36. F. Buonanno, D. M. Moody, M. R. Ball and D. W. Laster, Computed cranial tomographic findings in cerebral sino-venous occlusion, *J. Comput. Assist. Tomog.*, 2:281-290 (1978).
37. K. Ford and M. Sarwar, Computed tomography of dural sinus thrombosis, *Am. J. Neuroradiol.*, 2:539-543 (1981).
38. K. C. V. G. Rao, H. C. Knipp and E. J. Wagner, CT findings in cerebral sinus and venous thrombosis, *Radiology*, 140:391-398 (1981).
39. F. S. Buonanno, D. M. Moody and R. M. Ball, CT scan findings in cerebral sinovenous occlusion, *Neurology*, 12:288-292 (1962).
40. C. Virapongse, C. Cazenave, R. Quisling, M. Sarwar and S. Hunter, The empty delta sign: frequency and significance in 76 cases of dural sinus thrombosis, *Radiology*, 162:779-785 (1987).
41. Y. Shinohara, M. Yosmitoshi and F. Yoshii, Appearance and disappearance of empty delta sign in superior sagittal sinus thrombosis, *Stroke*, 17:1282-1284 (1986).
42. A. L. Goldberg, A. E. Rosenbaum, H. Wang, S. W. Kim, V. L. Lewis and D. F. Hanley, Computed tomography of dural sinus thrombosis, *J. Comput. Assist. Tomog.*, 10:16-20 (1986).
43. N. J. Patronas, E. E. Duda, M. Mirfakhraee and R. L. Wollmann, Superior sagittal sinus thrombosis diagnosed by computed tomography, *Surg. Neurol.*, 15:11-14 (1981).
44. J. Brismar, Computed tomography in superior sagittal sinus thrombosis, *Acta Radiol. (Stockh.)*, 21:321-326 (1980).
45. F. S. Vines, D. O. Davis, Clinical radiological correlation in cerebral venous occlusive disease, *Radiology*, 98:9-22 (1971).
46. J. Chiras, M. G. Bousser, J. F. Meder, A. Koussa and J. Bories, CT in cerebral thrombophlebitis, *Neuroradiology*, 27:145-154 (1985).

47. C. P. Symonds, Hydrocephalic and focal cerebral symptoms in relation to thrombophlebitis of the dural sinuses and cerebral veins, *Brain*, 60:531-550 (1937).
48. H. M. Askenasy, I. Z. Kosary and J. Braham, Thrombosis of the longitudinal sinus. Diagnosis by carotid angiography, *Neurology*, 12:288-292 (1962).
49. H. Hacker, Normal supratentorial veins and dural sinuses, *in*: "Radiology of the Skull and Brain. Angiography", T. H. Newton and D. G. Potts, eds., CV Mosby Company, St. Louis (1974).
50. P. J. Macchi, R. I. Grossman, J. M. Gomori, H. I. Goldberg, R. A. Zimmerman and L. T. Bilaniuk, High field MR imaging of cerebral venous thrombosis, *J. Comput. Assist. Tomog.*, 10:10-15 (1986).
51. W. C. Hanigan, L. J. Rossi, J. M. McLean and R. M. Wright, MRI of cerebral vein thrombosis in infancy. A case report, *Neurology*, 36:1354-1356 (1986).
52. P. J. Savino, R. I. Grossman, N. J. Schatz, R. C. Sergott and T.M. Bosley, High field magnetic resonance imaging in the diagnosis of cavernous sinus thrombosis, *Arch. Neurol.*, 43:1081-1082 (1986).
53. S. K. McMurdo, M. Brant-Zawadzki, W. G. Bradley, G. Y. Chang and B. O. Berg, Dural sinus thrombosis. Study using intermediate field strength MR imaging, *Radiology*, 161:83-86 (1986).
54. T. O. Gabrielsen, J. F. Seeger, J. E. Knake and E. W. Stilwill, Radiology of cerebral vein occlusion without dural sinus occlusion, *Radiology*, 140:403-408 (1981).
55. S. Johnsen, R. Greenwood and M. A. Rischman, Internal cerebral vein thrombosis, *Arch. Neurol.*, 28:205-207 (1973).
56. R. N. Nishimura, D. Stepanek, J. Howieson and J. Hammerstad, Internal cerebral vein thrombosis. A case report, *Arch. Neurol.*, 39:439-440 (1982).
57. G. A. M. Bots, Thrombosis of the galenic system veins in the adult, *Acta Neuropathol.*, 17:227-233 (1971).
58. J. J. Eick, K. D. Miller, K. A. Bell and R. H. Tutton, Computed tomography of deep cerebral venous thrombosis in children, *Radiology*, 140:399-402 (1981).
59. M. G. Bousser, O. Bletry, M. Launay, E. Portier, A. Guillard and P. Castaigne, Thrombose veineuse cérébrale au cours de la maladie de Behçet, *Rev. Neurol.*, 11:753-762 (1980).
60. D. Rougemont, M. G. Bousser, B. Wechsler, O. Bletry, P.Castaigne and P. Godeau, Manifestations neurologiques au cours de la maladie de Behçet. 24 observations, *Rev. Neurol. (Paris)*, 138:493-505 (1982).
61. M. N. Pamir, T. Kansu, A. Erbengi and T. Zileli, Papilledema in Behçet's syndrome, *Arch. Neurol.*, 38:643-645 (1981).
62. P. C. Gates, Cerebral venous thrombosis. A retrospective review, *Aust. NZ J. Med.*, 16:766-770 (1986).
63. D. S. Buchanan and J. H. Brazinsky, Dural sinus and cerebral venous thrombosis. Incidence in young women receiving oral contraceptives, *Arch. Neurol.*, 22:440-444 (1970).
64. B. Fairburn, Intracranial venous thrombosis complicating oral contraception: treatment by anticoagulant drugs, *Br. Med. J.*, 2:647 (1973).

ULTRASOUND DIAGNOSIS OF CEREBRO-VENOUS FLOW DISTURBANCES

M. Hennerici

Department of Neurology
University of Heidelberg
Theodor-Kutzer-Ufer
6800 Mannheim 1, FRG

Various ultrasound methods have been developed during the last decade for the diagnosis of cerebrovascular diseases affecting the extracranial carotid and vertebral arteries, the major intracranial cerebral arteries at the base of the skull and intrathoracic segments of the innominate, carotid and vertebral arteries near the aorta. These methods have also been used for the detection of heart diseases like valve disorders and abnormal movements of the cardiac walls, and for the diagnosis of peripheral vascular diseases affecting the aorta, its major branches (eg, the renal arteries) and the limb vessels. Finally laser Doppler techniques have been used to study capillary microcirculation (Table 1). The major advantage of these methods is that they can be used for non-invasive investigations of vascular processes during treatment and in long-term prospective trials.

The large variety of ultrasound methods available reflects the different demands associated with such heterogeneous diagnostic tools: continuous wave (CW) Doppler (Fig. 1a) probably represents the best method for the study of hemodynamically significant stenosis and occlusion in large arteries of the body and the extracranial carotid system in particular. Documentation of the Doppler signals recorded can either be established by means of an audioanalysis, a zero-crossing detector or by a more sophisticated computer-assisted Fourier-transformed spectrum analysis. Whereas CW Doppler techniques have only occasionally been combined with ultrasound imaging systems, such as echotomography (B-mode), pulsed wave (PW) Doppler (Fig. 1b) devices have long since been incorporated in so-called duplex systems. Their major advantage is the ability to focus on spatial volume components within the artery as defined by the time interval between the emitted and received Doppler signals. Sequential increase of the time gates adjusts the transition of the sample volume across the vessel lumen and enables selective measure-ment of flow velocity from the anterior to the posterior vessel wall.

This principle also forms the basis for transcranial PW Doppler analysis of the basal cerebral arteries. Intracranial studies have long been

Cerebral Sinus Thrombosis, Edited by K. Einhäupl *et al.*
Plenum Press, New York, 1990

Table 1. Common Cardiovascular Applications of Various Ultrasound Methods

Ultrasound Methods	Cerebral Arteries		Thoracic/Abdominal System			Peripheral Arteries	
	Extracranial	Intracranial	Heart	Aorta	Major Arteries	Limb Arteries	Cutaneous Arteries
Continuous wave Doppler	++	Ø	Ø	Ø	Ø	++	Ø
Pulsed wave Doppler							
Single-gate	+	++	+	+	+	+	Ø
Multi-gate	+	Ø	Ø	+	Ø	Ø	Ø
Echotomography (B-mode)	++	Ø	++	+	+	+	Ø
Duplex-system Analysis							
CW Doppler	+	Ø	+	Ø	Ø	+	Ø
PW Doppler	+	Ø	++	+	++	++	Ø
DCFI	++	(+)	++	+	+	++	Ø
Laser Doppler	Ø	Ø	Ø	Ø	Ø	Ø	+

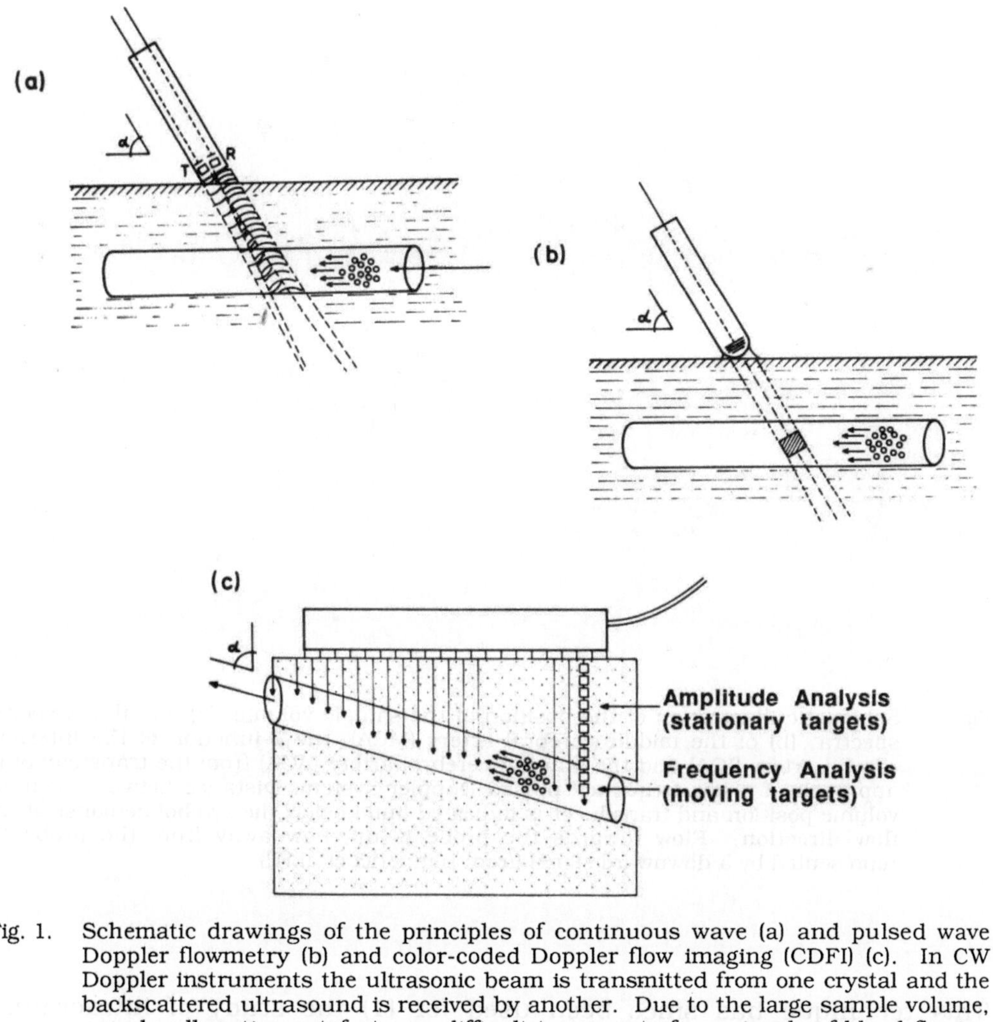

Fig. 1. Schematic drawings of the principles of continuous wave (a) and pulsed wave Doppler flowmetry (b) and color-coded Doppler flow imaging (CDFI) (c). In CW Doppler instruments the ultrasonic beam is transmitted from one crystal and the backscattered ultrasound is received by another. Due to the large sample volume, vessel wall motion artefacts are difficult to separate from signals of blood flowing at low velocities, while high frequency signals are favorably recorded. In pulsed Doppler flowmeters usually one single crystal operates alternately as transmitter and receiver; single-channel and multichannel instruments can be distinguished. The latter provide instantaneous flow velocity profiles across the artery and enable localization of flow alterations. CDFI combines simultaneous displays of hemo-dynamic and morphologic information and facilitates diagnosis of complex vascular diseases. However, for quantitative analysis its major disadvantage consists in the inability to select an optimal angle of insonation: maximal echotomograms are due to perpendicular insonation whereas for optimal Doppler recordings 0° or 180° angles are optimal.

restricted by the marked attenuation of the ultrasound energy at the skull bone. Since the introduction of PW Doppler methods operating at low frequencies (1-2 MHz) with a considerably increased energy of the emitting ultrasound beam (10-120 mW/cm2), this method has been used for examination of flow velocity within the large basal cerebral arteries. Using the sample volume time-gate controls the stationary probe placed at the temporal bone can trace the course of the middle and anterior cerebral arteries by making selective recordings at various depths (Fig. 2).

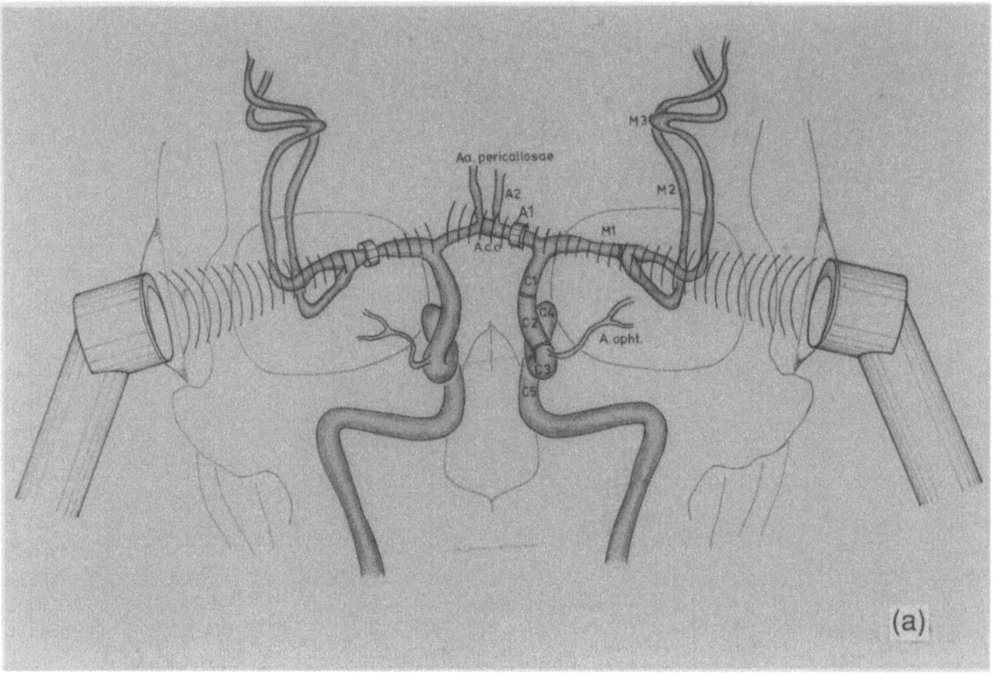

Fig. 2. Schematic illustration of the position of the sample volume (a) and flow velocity spectra (b) of the middle cerebral artery (MCA), the T-junction of the internal carotid artery (ICA) and the anterior cerebral artery (ACA) from the transtemporal approach of a low frequency pulsed Doppler system. Distance between sample volume position and transducer is indicated (mm) below the symbol demonstrating flow direction. Flow towards the probe is up, flow away from the probe is represented by a downward signal (from Hennerici et al.[1])

This technique has since been used for (1) the study of intracranial stenosis and occlusion, (2) the evaluation of intracranial collateral circulation in the presence of extracranial arterial disease, (3) the analysis of arteriovenous malformations and dolichoectasis, (4) the monitoring of arterial spasms after subarachnoid hemorrhage as well as (5) the measurement of vascular flow in patients treated medically and surgically after cerebrovascular events. In the evaluation of anatomic variations, major difficulties of the technique, particularly in the absence of angiograms which can be used for topographic comparison, include problems in the correct identification of the various arterial segments and in the separation of crossing or parallel vessels. These difficulties result in part from the large size of the sample volume which in turn results from the high ultrasound energy.

Although duplex-systems have long been used for the diagnosis of significant obstructions within the extracranial carotid system, they have not been able to reliably detect small plaques because the spatial resolution of former echotomographs was rather limited. The introduction of a multichannel rather than a single-gate PW Doppler system considerably improved the quality of these instruments and offers the chance for a detailed analysis of flow velocity profiles across a particular vessel. In combination with high-resolution echotomograms, the clinical

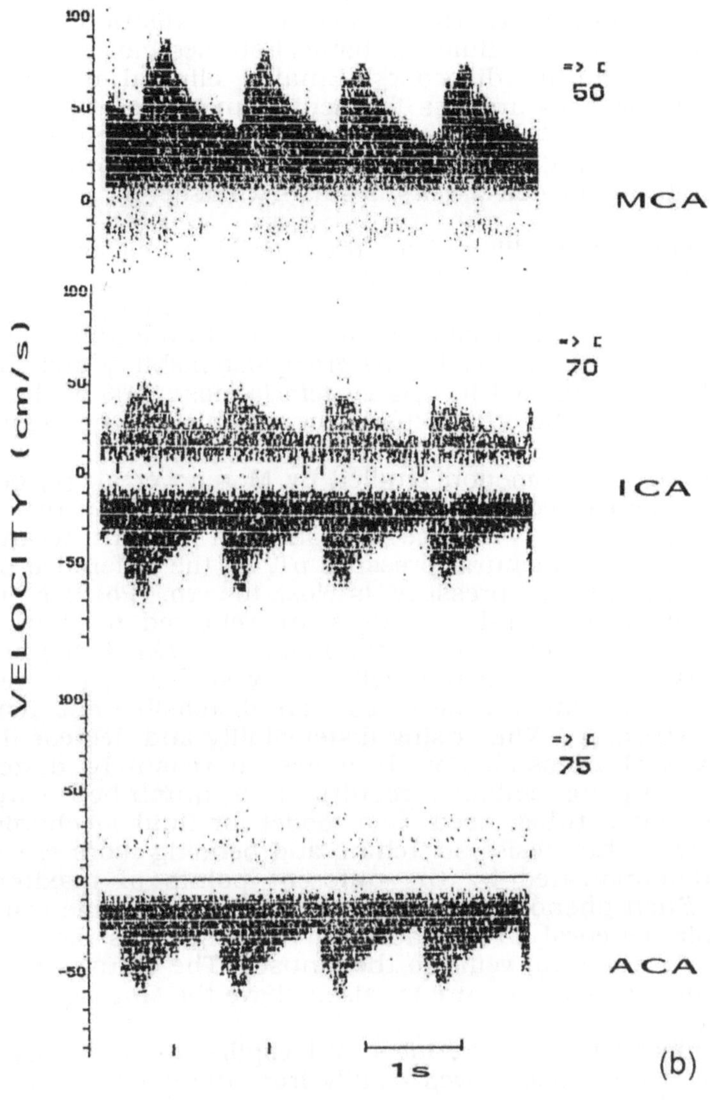

(b)

application of these new instruments includes the diagnosis of the initial stages of atherosclerosis. A recently designed new method (Fig. 1c) combines amplitude and frequency analysis of a single echoreflection within the ultrasound beam and displays stationary and moving targets simultaneously (color-coded Doppler flow imaging).

In contrast to the considerable progress which has been made for non-invasive ultrasound evaluation of arterial diseases, the application of these methods is limited for the detection and analysis of venous and cerebrovenous disorders in particular. This is not only due to technical problems, which are similar to those for the arterial circulation but which also reflect the greater variability in the anatomy of the cerebral veins and their collateral networks, and which make identification and display of the venous system difficult. In addition, separation of normal from abnormal conditions is difficult due to still insufficient rheologic knowledge. In contrast to the excessive investigation of the arterial hemodynamics, only a few fluid mechanical studies have been done on the venous system and its disorders. Finally, clinical manifestations of cerebrovenous diseases are less frequent than those of arterial diseases and are often obscured by the preservation of compensatory collateral blood drainage, leaving some patients with venous disorders without any history or clinical evidence of venous stasis whatsoever.

The venous system differs from the arterial system in many ways and venous hemodynamics are much more complex and less well understood. For example, transmural pressures which are normally negative in veins may vary with subsequent changes in shape. In addition, the particular architecture of the venous walls, the great distensibility and the changing amount of blood contained in this system (almost 80% of the total blood volume) are a few of the other differences which could be mentioned.

A particular phenomenon studied by Moreno et al. [2] occurs when flow causes a vein to collapse. This happens normally in the veins above the level of the heart with the exception of the intracranial venous sinuses, when the transmural pressure pT, ie, the difference between the internal and the external pressure, is close to zero. Positive and negative changes of the transmural pressures are reflected by marked contour alterations (Fig. 3): when pT exceeds about 1.5 k/Nm^2, the cross-section of the vein is circular, and the wall is very stiff. As pT is reduced, the vessel remains circular, but becomes more distensible and tends to form an elliptical structure. With rising distensibility and decreasing pT below zero, the vessel cross-section becomes increasingly distorted until an almost complete collapse results in a dumb-bell configuration. Interestingly, latex tubes used as a model for fluid mechanical studies behave similarly, but resist stretching and bending more effectively than veins, as demonstrated by the different points of maximum in the curvatures. Such phenomena are important hemodynamic issues relating to extracranial cerebral and intracranial veins, such as the bridging veins connecting the subdural veins to the sinuses. The sinuses never collapse even if the intracranial pressure is raised above the sinus pressure.

As compared to the arterioles and capillaries, the pressure in the veins is low. Thus blood moves readily from the small veins to the right atrium through medium-sized and large veins which offer little resistance to flow. Alterations of intrathoracic pressure with respiration and phasic changes due to cardiac activity may be transmitted to the venous system and represent important parameters for non-invasive studies (Table 2). Spontaneous venous flow changes with respiratory effects and superimposed pulsatile alterations during each cardiac cycle are characteristics of normal venous function (Figs 4 and 5). In addition, alterations in flow in response to changes in posture may affect the venous systems especially of the lower extremities with the peculiar venous valves.

The changes in large veins during the cardiac cycle are characterized by three events: first, increased flow resulting from a drop in the atrial pressure after opening of the atrioventricular valves, when blood rushes into the ventricles. Second, the atrial pressure rises during initial ventricular systole. Third, a shortening of the ventricular muscles follows, which pull the atrium towards the apex of the heart. As the atrial volume increases, pressure decrease and flow in large veins increases. The temporal sequence of these periods depends on several factors, especially

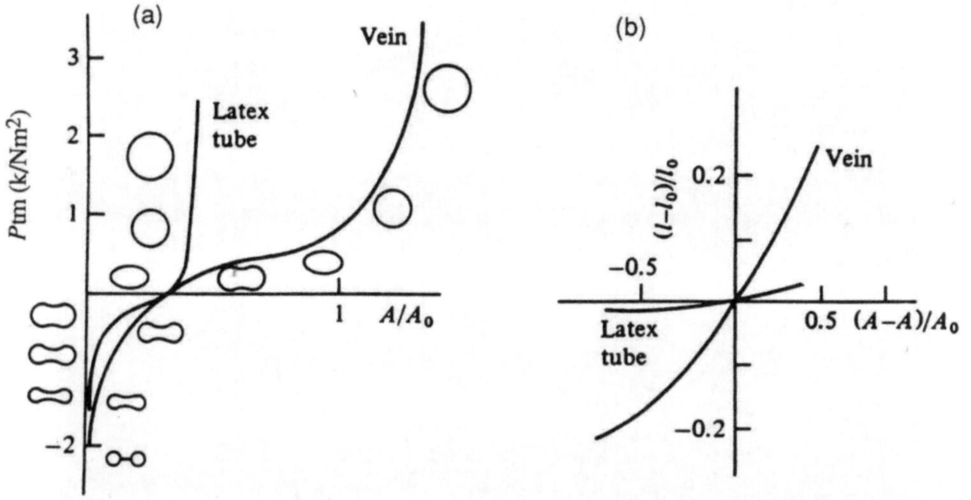

Fig. 3. Comparison of the elastic properties of a latex tube with those of an excised segment of a canine vena cava. (a) Transmural pressure as a function of area, A, related to the area at zero transmural pressure, Ao. Transverse cross-sections are shown for various areas. (b) Perimeter as a function of a cross-sectional area (from Moreno et al. [2]).

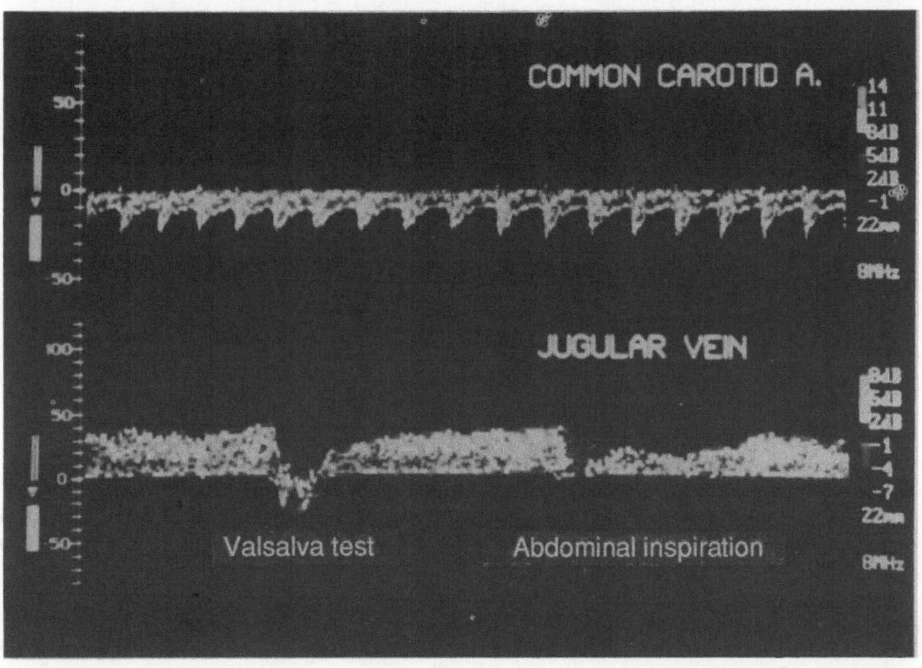

Fig. 4. Fourier spectrum of the Doppler signals recorded from the common carotid artery and the jugular vein in the neck. Note the superimposed respiratory alterations particularly during the Valsalva maneuver.

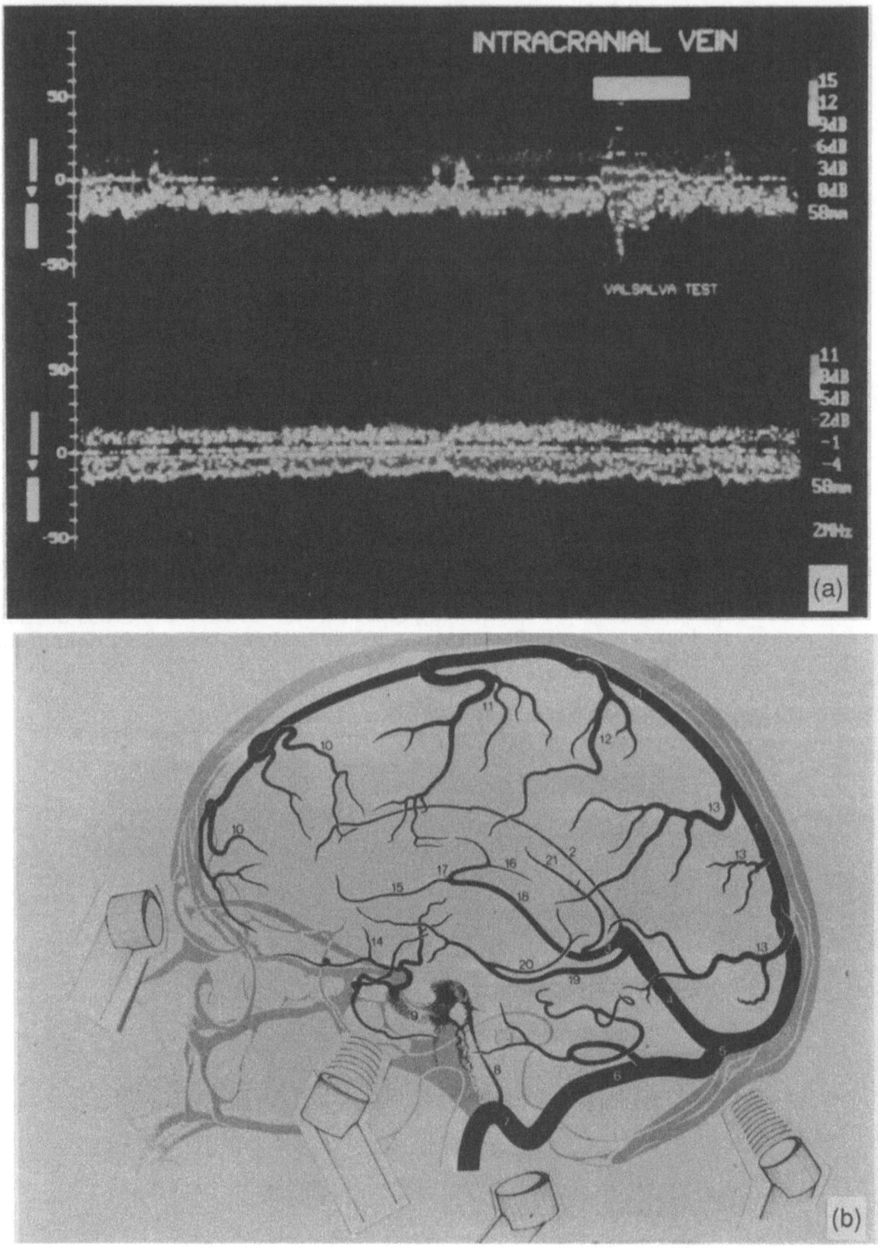

Fig. 5. Fourier spectrum of the Doppler signals recorded from intracranial veins in the cavernous sinus (a) through a transtemporal approach (b); note the smaller respiratory-dependent changes of venous blood flow including a weaker response to the Valsalva maneuver.

on the actual heart rate. Thus venous blood may either continuously flow, temporarily decrease, or even stop.

Flow in medium-sized and small veins is less affected by the cardiac cycle but is, as in large veins, considerably affected by respiration (Fig. 5). During inspiration the venous volume increases but flow decreases. The opposite effect may be observed during expiration when venous volume

Table 2. Venous Flow Determinants

	Extracranial	Intracranial
Respiratory	++	+
Cardiac cycle	+	+
Position	+	+
Peripheral resistance	+	+
Adjacent arteries	+	++
AVM	++	++

decreases and flow increases. While these effects on the Doppler signal of large veins like the jugular vein are small during normal thoracic respiration, they become more pronounced during abdominal respiration. A phasic response may be exaggerated by respiratory maneuvers such as the Valsalva test, which increases intrathoracic and abdominal pressure and stops or even reverses venous flow.

In addition, effects of posture on peripheral vascular resistance (vasconstriction or vasodilatation) influence venous flow because they alter capillary loading, and therefore venous filling and flow rates. Retro-grade filling due to incompetent valves is another mechanism by which changes in posture affect venous circulation.

In contrast to the peripheral and extracranial veins, blood flow and flow velocity in intracranial veins are less influenced by these parameters. However, intrathoracic venous obstructions and severe chronic pulmonary disease may affect the characteristic changes of blood flow in the cerebral veins and sometimes cause reversed flow in the presence of severe proximal stasis. Furthermore, pulsatile movements from adjacent arteries can be transmitted to intracranial veins and change flow more than in peripheral veins.

REFERENCES

1. M. Hennerici, W. Rautenberg, G. Sitzer and A. Schwartz, Transcranial Doppler ultrasound for the assessment of intracranial arterial flow velocity - Part 1. Examination technique and normal values, *Surg. Neurol.*, 27:439-448 (1987).
2. A. H. Moreno, A. I. Katz, L. D. Gold and R. V. Reddy, Mechanics of distension of dog veins and other very thin-walled tubular structures, *Circ. Res.*, 27:1069-1080 (1970).

DELTA (EMPTY TRIANGLE) SIGN IN PATIENTS WITHOUT THROMBOSIS OF THE SUPERIOR SAGITTAL SINUS

C. Hamburger, A. Villringer*, M. Bauer†, and T. Lorz

Departments of Neurosurgery, *Neurology
and †Radiology, Ludwig-Maximilians-University
Klinikum Großhadern, 8000 Munich 70, FRG

INTRODUCTION

Since the first published case reports with CT findings in patients with sinus venous thrombosis (SVT) [1-3] and the first systematic work based on a series of 11 patients by Buonanno [4], many different CT signs of SVT have been described. Most are unspecific, such as intracerebal bleeding, small ventricles, gyral enhancement, and so on (for review see Refs 4-7).

Two signs, however, have been regarded pathognomonic for SVT allowing diagnosis by CT alone: the direct visualization of a hyperdense thrombosed vein (the chord sign) or sinus in unenhanced CT scan and the Delta sign after contrast agent application. In contrast-enhanced CT the isodense thrombosis of the dorsal superior sinus sagittalis (SSS) is surrounded by enhancing collateral veins in the walls of the sinus [7]. The chord sign is relatively rare, and therefore of minor diagnostic importance [6,7 and Villringer et al., unpublished results].

The Delta sign, however, is a frequent finding. A recent literature survey of 71 cases with SSS thrombosis reports a frequency of 28.6% [7], in Chiras' series (1985) of 39 patients (not included in the literature survey of Virapongse) with SSS it is 23.1%, and in our series of 71 cases it is 20% [8]. No patients with false-positive Delta signs have so far been reported. Careful examination of the literature, however, reveals one case (case number 6 in Ref. 5) where an empty triangle was reported without thrombosis of SSS yet thrombosis of the transverse and straight sinus. We think this is a false-positive symptom, since a true Delta sign is not compatible without thrombosis in the dorsal part of SSS which causes the central hyperdensity of the sign.

In this paper we describe three additional cases with a false-positive Delta sign. In all patients, despite the presence of an unambiguous Delta sign, subsequent angiography showed no thrombosis of the SSS.

Cerebral Sinus Thrombosis, Edited by K. Einhäupl *et al.*
Plenum Press, New York, 1990

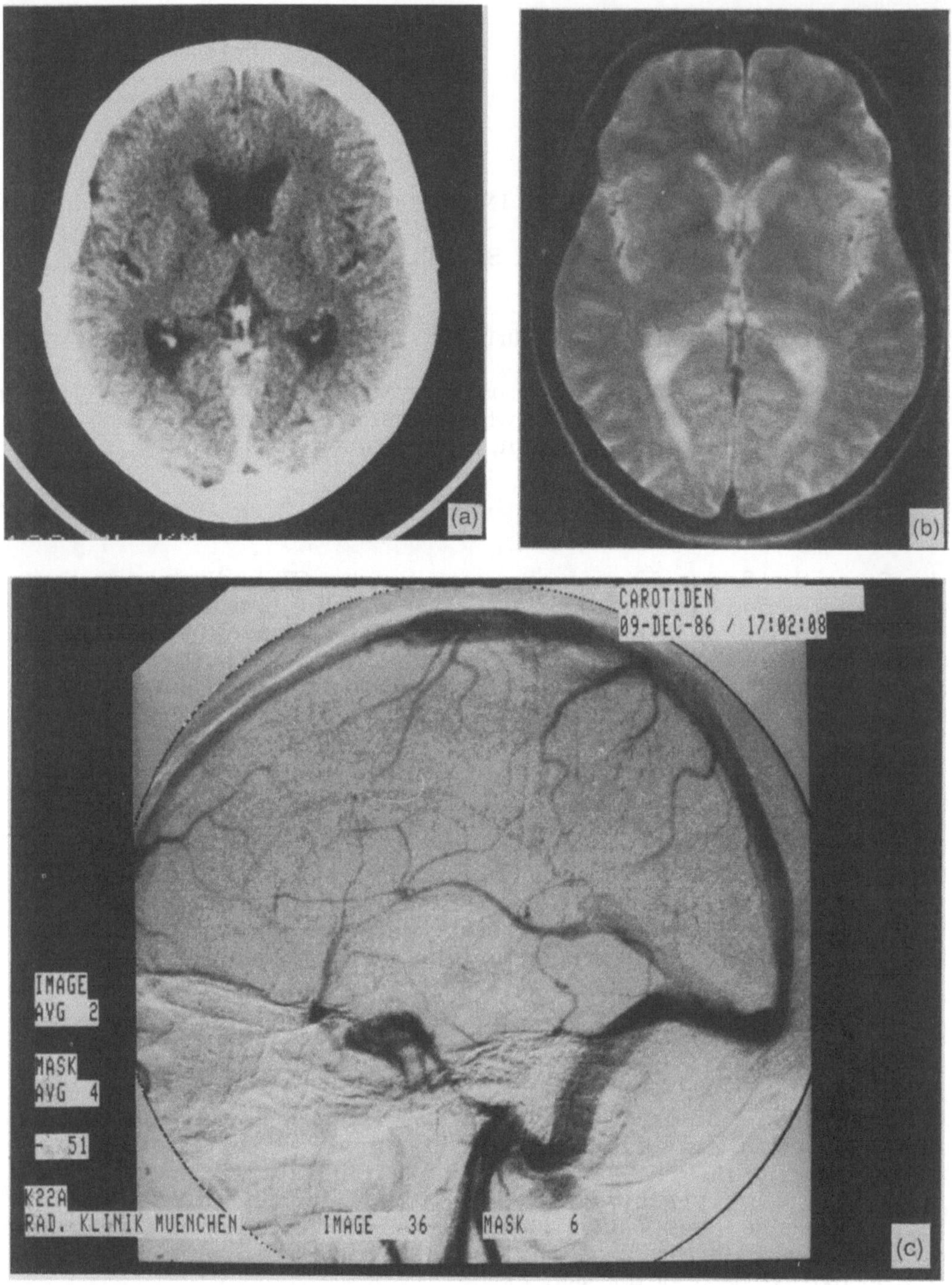

Fig. 1. Patient 1. (a) Contrast CCT axial section showing Delta sign; (b) NMR (spin echo x/y) axial section corresponding to CCT section; (c) DSA lateral view.

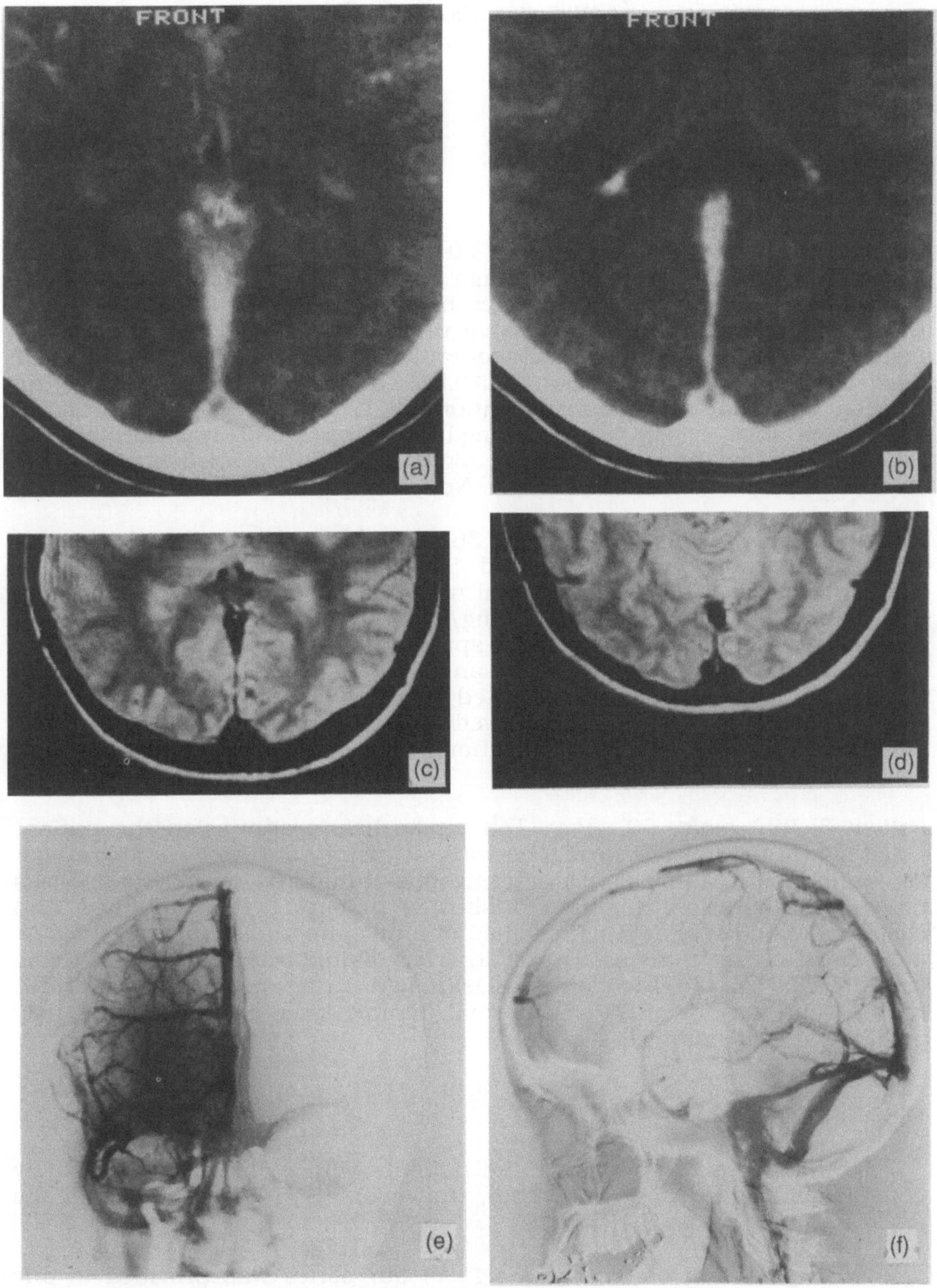

Fig. 2. Patient 2. (a) and (b) Contrast CCT showing Delta sign; (c) and (d) NMR scan (spin echo x/y) showing a split sinus; (e) and (f) angiography showing free SSS.

METHODS

All CT scans were performed using a Siemens Somatom DRH CT scanner. As contrast agent Ultravist 300 (Schering AG, Berlin) was infused in a dose of 1 ml/kg body weight.

RESULTS

Case Reports

Patient 1: CR, female, 45 years. A 45-year-old woman with a history of pulmonary tuberculosis (more than 20 years ago) presented with ablatio mammae on the right side 2 years ago and underwent radiotherapy. Four months before a CT study, a right cerebellar metastasis was removed, and a local recurrent tumor was operated upon 1 month before the CT scan demonstrated a Delta sign. Subsequently, the patient underwent total head radiotherapy. A NMR scan (23.10.86) showed a normal SSS, and the CT scan with contrast agent application (04.12.86) a Delta sign. A digital subtraction angiography was performed (09.12.86) which was normal. The patient never did have any clinical signs of SVT. She died on 01.01.87 because of peritonitis carcinomatosa. No autopsy was performed.

Patient 2: WS, female, 36 years. A 36-year-old woman with a history of common migraine suffered in mid September 1987 from progressive diffuse headache. There was no focal neurological deficit. Examinations of CSF revealed 44/3 cells and 72 mg/dl protein. In the CT scan small ventricles and small external CSF spaces were found. The contrast CT (16.10.87) showed a Delta sign. By angiography (23.11.87) a normal SSS was found, while NMR demonstrated very high splitting of the SSS. A second CT (10.11.87) again showed a Delta sign. A second lumbar puncture revealed liquor concentrations of 300 mg/dl protein. The final diagnosis was encephalitis.

Patient 3: WH, male, 50 years. Ten weeks before admission the patient noted slowly progressive headache without history of a precedent trauma. six weeks before admission progressive exophthalmus on both sides was noted. At admission the patient showed pulsating exophthalmus and chemosis of both eyes. He complained of a pulsating noise in both ears. As a diagnosis a cavernous sinus-carotid artery fistula was established. Using contrast CT (14.10.87) we found a Delta sign. In an angiogram performed on the same day the SSS was normal. Symptoms ceased spontaneously and surgery was not necessary.

DISCUSSION

In this paper we present three patients with a Delta sign found by contrast-enhanced CT in whom, however, SVT was excluded by angiography.

An explanation of the Delta sign in our series was provided in one case (WS), where T_2-weighted NMR imaging as well as angiography showed abnormal high splitting of the sinus. High splitting has been reported in the literature as a possible pitfall of the diagnosis of SVT. It

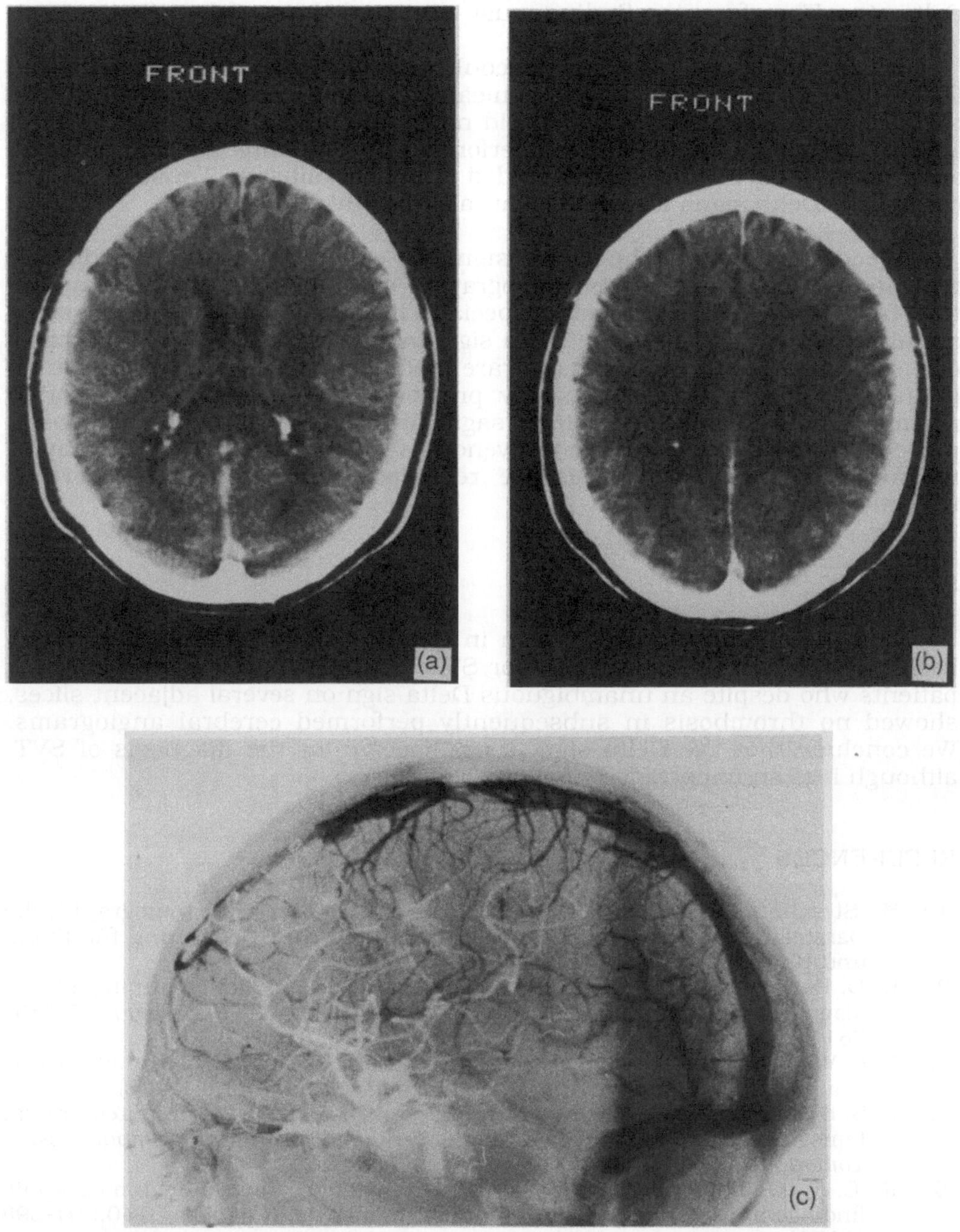

Fig. 3. Patient 3. (a) and (b) Contrast CCT with Delta sign; (c) angiography lateral view showing normal SSS.

has been stated that presence of a Delta sign on several adjacent slices should allow the distinction against high splitting. It must be emphasized, however, that the Delta sign was visible in this patient in several adjacent slices, yet the CT was not able to differentiate high splitting and a true Delta sign. Therefore, the finding must be regarded as a false-positive.

In the other two patients we could not find an explanation for the Delta sign. In both patients the clinical symptoms could be attributed to other neurological diseases which did not have clinical features of SVT. In addition, angiography which was performed on the same day in one case, and 5 days later in the other excluded sinus thrombosis. Thus, these two cases with Delta signs also represent false-positive findings.

We conclude that the Delta sign, as can other direct or indirect indicators of SVT in computed tomography, can be an important symptom to establish a correct diagnosis, especially if other direct or indirect signs are also present. However, the Delta sign cannot be regarded as final proof of a sagittal sinus thrombosis, the rare instances mentioned by Buonanno [4] notwithstanding. In that study previous contrast CT demonstrated normal enhancement of the sinus sagittalis superior. Further diagnostic procedures to directly visualize the venous system, such as angiography or magnetic resonance imaging, are required in all cases (for further references see Refs 9-14).

SUMMARY

The Delta (empty triangle) sign in cranial computed tomography has been regarded as pathognomonic for SVT. In this report we present three patients who despite an unambiguous Delta sign on several adjacent slices, showed no thrombosis in subsequently performed cerebral angiograms. We conclude that the Delta sign is not specific for the diagnosis of SVT, although it is an important symptom.

REFERENCES

1. H. Steinhoff, W. Lanksch and E. Kazner, Die computer-tomographische Darstellung encephalomalacischer Herde und ihre Bedeutung für Praxis und Klinik, *Internist.*, 17:32-37 (1976).
2. B. D. Barnes and D. P. Winestock, Dynamic radionuclide scanning in the diagnosis of thrombosis of the superior sagittal sinus, *Neurology*, 27:656-661 (1977).
3. L. R. Wendling, Intracranial venous sinus thrombosis: diagnosis suggested by computed tomography, *Am. J. Roentgenol.*, 130:978-980 (1978).
4. F. S. Buonanno, D. M. Moody, M. R. Ball and D. W. Laster, Computed cranial tomographic findings in cerebral sino-venous occlusion, *J. Comput. Assist. Tomog.*, 2:281-290 (1978).
5. K. C. V. G. Rao, H. C. Knipp and E. J. Wagner, Computed tomographic findings in cerebral sinus and venous thrombosis, *Radiology*, 140:391-398 (1981).
6. J. Chiras, M. G. Bousser, J. F. Meder, A. Koussa and J. Bories, CT in cerebral thrombophlebitis, *Neuroradiology*, 27:145-154 (1985).
7. C. Virapongse, C. Cazenave, R. Quisling, M. Sarwar and S. Hunter, The empty Delta sign: frequency and significance in 76 cases of dural sinus thrombosis, *Radiology*, 162:779-785 (1987).

8. K. M. Einhäupl, A. Villringer, R. L. Haberl, W. Pfister, M. Dekert, H. Steinhoff and P. Schmiedek, Clinical spectrum of sinus venous thrombosis, in: "Cerebral Sinus Thrombosis: Experimental and Clinical Aspects", K. Einhäupl, O. Kempski and A. Baethmann, eds, Plenum Press, New York and London, pp 149-155 (1990).
9. D. D'Avella, A. Russo, G. Santoro, N. Pandolfo, F. Cafarelli, F. De Blasi, R. Eugenio Del Vivo, Diagnosis of superior sagittal sinus thrombosis by computerized tomography, *J. Neurosurg.*, 61:1129-1131 (1984).
10. A. L. Goldberg, A. E. Rosenbaum, H. Wang, W. S. Kim, V. L. Lewis and D. F. Hanley, Computed tomography of dural sinus thrombosis, *J. Comput. Assist. Tomog.*, 10:16-20 (1986).
11. D. P. E. Kingsley, B. E. Kendall and I. F. Moseley, Superior sagittal sinus thrombosis: an evaluation of the changes demonstrated on computed tomography, *J. Neurol. Neurosurg. Psychiat.*, 41:1065-1068 (1978).
12. W. Matthes, D. Doerstelmann, Hirnvenen- und Sinusthrombose, *Dtsch. Med. Wschr.*, 106:744-747 (1981).
13. Y. Shinohara, S. Takagi, K. Kobatake and F. Gotoh, Influence of cerebral venous obstruction on cerebral circulation in humans, *Arch. Neurol.*, 39:479-481 (1982).
14. R. Wodarz, M. Ratzka and M. Nadjmi, Zur Sicherheit der CT-Diagnose bei zerebralen Sinusthrombosen, *Radiologe*, 22:383-388 (1982).

NEUROSURGERY OF THE CEREBRAL VENOUS AND SINUS SYSTEM

O. Gratzl

Department of Neurosurgery
University of Basel, Kantonsspital
Spitalstraße 21, 4031 Basel, Switzerland

Neurosurgeons are confronted with the cerebral venous and sinus system (CVSS) in three circumstances: following injuries (laceration, post-traumatic fistulae), invasion of tumors into the CVSS, and cerebral sinus thrombosis and sequelae (ie, post-thrombotic fistulae).

PRINCIPLES

Based on anatomical-physiological grounds the following principles of CVSS surgery are considered:

1. Avoidance of air embolism by an adequate non-sitting position (20-30°), Doppler monitoring and cardiac catheter.
2. Quick management, digital compression and temporary clipping to cope with blood loss.
3. Intraoperative Doppler sonography: this method seems to be helpful for the determination of the blood flow direction and the decision as to whether or not a sinus is patent [1].
4. Points of resection or ligation: it would seem that ligation or resection of the superior longitudinal sinus can be performed at any point that does not require ligation of the unobstructed rolandic inflow of cortical veins into the sinus [2].
5. Temporo-occipital veins draining into the transverse sinus (venae cerebri inferiores and vena anastomotica parva/vena Labbé) cannot be interrupted without causing an ensuing neurological deficit. Surgical ligation and interruption of the transverse sinus has to be preceded by preoperative studies to confirm the function of the contralateral transverse sinus and the patency of the Torcular herophili for flow in the transverse direction [3].
6. Temporary shunt in case of postrolandic longitudinal or lateral sinus reconstruction without natural collaterals.
7. Reconstruction of the two-fold layers of the sinus structure: internal venous and external non-compressible dura or implant.

Cerebral Sinus Thrombosis, Edited by K. Einhäupl *et al.*
Plenum Press, New York, 1990

METHODS

The methods of surgery in the cerebral venous and sinus system are nearly the same as in arterial vessel procedures. We use suture, microsuture, patches (dura, muscle, veins), clips on a lateral angle, reconstruction with implants and nowadays also replacement of segments including a bridging vein by a suture or non-suture technique. Heparin is administered locally. As an illustration we provide some descriptions of the methods mentioned.

Yasargil reported in 1969 the repair of venous sinuses in two patients [4]. He used microsurgical suturing in one patient and a fascia patch in the other one. In the early 1970s there was some experimental work published concerning the replacement and repair of the superior longitudinal sinus using veins [5]. Donaghy was successful in some clinical cases when using this method. His procedure involved immediate implantation of a vascular T-tube which was later replaced by an intima line (saphenous vein graft). In 1976 Sindou [6] began bypass surgery for sagittal sinus repair by performing experimental studies in dogs. The patency rate was 54% in the venous autograft group. Reconstruction of cerebral cortical veins using silicone tubing during an interhemispheric surgical approach was first described by Sakaki [7].

CLINICAL EXPERIENCE

The mortality of injuries to the sinus system is related to the localization. In the Vietnam War Kapp [8] observed 78 cases of injuries to the sagittal sinus. Anterior sagittal sinus injury had a mortality of 5%, and posterior of 43%, in comparison to 24% in lacerations of the transverse sinus. In 1979 Hassler [9] presented a summary of sinus surgery in traumatology. His survey included the use of a Fogarty catheter, the prerolandic ligation of the sagittal sinus, direct suture and microsuture, Dandy suture, patch, clipping of the angle of the sinus and venous implants reinforced with lyophilized dura or silicone tubes.

In the treatment of spontaneous or post-traumatic carotid cavernous fistula we have the possibility of embolization with acrylates, we have the detachable balloon technique, embolization with particles (Ivalon) and some surgical methods such as trapping or the direct approach. In 1986 Isamat [10] reported direct intracavernous obliteration of high-flow carotid cavernous fistulae by the introduction of muscle fragments and fibrin sealant. In three cases he succeeded in closing the fistula while maintaining carotid flow. In 1983 Dolenc published four cases of traumatic carotid cavernous fistulae which were obliterated by direct microsurgical repair in the sinus. However, these surgical techniques are rarely just direct occlusion techniques of the fistulae in the wall of the internal carotid artery as attempted by Parkinson [11]. Rather, these are venous trapping techniques which were also tried indirectly via the ophthalmic vein by introducing steel coils through a catheter.

Some pioneers approach basal cerebral aneurysms directly in or near the cavernous sinus. Perneczky [12] clipped infraclinoidal aneurysms approaching the anterior part of the cavernous sinus roof. Four internal carotid artery aneurysms with an intracavernous localization were treated by Matsuoka [13]. He used a semi-sitting position, Doppler monitoring and temporary clipping of the carotid artery. In a similar procedure Dolenc [3] occluded three aneurysms of the internal carotid artery after opening the cavernous sinus and intracavernous d :tion.

Tumors may invade or obliterate dural venous sinuses. This may affect the chances of curative surgery of the patient. A frequent opinion is that the incidence of recurrence of parasagittal meningeomas is reduced up to three-fold when it is possible to resect the invaded portions of the sinus. This is the reason why some authors tried to radically resect tumors with intracavernous or intrasagittal sinus invasion. In 1979 Hakuba [14] published a case report on the total removal of a parasagittal meningeoma of the posterior third of the sagittal sinus and its repair by an autologous vein graft. Sekhar [15] resected seven tumors involving the cavernous sinus. Whereas invasion of the cavernous sinus by pituitary adenomas clearly requires combination of treatment with radiotherapy, in meningeomas the question remains to be answered whether radical surgery with or without radiotherapy is better than radiotherapy alone. The study of Sekhar [15] demonstrates that total or subtotal resection of intra-cavernous tumors can be achieved with low morbidity.

Nowadays neurosurgeons try to achieve a radical resection in cases of parasagittal meningeomas. Fear of recurrency and the necessity of preserving venous blood flow lead us to remove meticulously all tumor tissue from the sinus and, if necessary, use an autologous vein graft to reconstruct the sagittal sinus. If a rolandic vein is invaded by the meningeoma it is sometimes possible to choose an autologous vein with a collateral branch that can be sutured to the rolandic vein. This was published by Bonnal [16] in 1982 who used the procedure in 21 parasagittal meningeomas. As far as tumors in the cavernous sinus are concerned, a direct microsurgical approach is possible via an extradural basal access. Using this method Lesoin [17] operated on 21 cases, but he concluded that it is not possible to completely remove the intracavernous tumor without aggravating oculomotor symptoms. Thus in the case of an invasion of tumor into the sinus system, there are some limitations on the surgical procedure. We all try to employ radical tumor removal by microsurgical means, but a critical evaluation of the results is necessary. In a control study a direct surgical procedure to resect the invading tumor was found to be of no benefit. Benoist and Pasztor [18] reported that either total removal of the tumor matrix or only its coagulation did not make any noticeable difference with respect to recurrencies.

SURGERY OF SINUS VEIN THROMBOSIS

Special methods in surgery of sinus vein thrombosis (Table 1) are:

1. Disobliteration of the sinus or cortical veins by thrombectomy or phlebectomy and repair of the vessel by suture.
2. Extra-anatomical bypass surgery in cases of tinnitus or chronic elevation of intracranial pressure.
3. Decompressive surgery of hematomas and ventriculo-peritoneal shunt after the acute period of sinus thrombosis.
4. Surgical methods dealing with the later follow up in cases of arterio-venous fistulae following sinus thrombosis.

In the literature we found only one case of successful disobliteration and repair of the superior longitudinal sinus after sinus thrombosis with coagulation of a bridging vein and postoperative angiographic control [16].

Table 1. Surgery of SVT

Reference	Technique	Site	Number of Cases
Sindou (1980)	Venous bypass	Transverse sinus	1
Hitchcock (1981)	Venous bypass	Transverse sinus	1
Bonnal (1982)	Disobliteration	Sagittal sinus	1
Kawase (1982)	Exploration	Cortical veins	4
George (1984)	Venous bypass	Transverse sinus	10

Surgical intervention in thrombosis of cortical veins was reported by Kawase [19]. In four cases with subcortical hemorrhages a diagnostic craniotomy was performed. During surgical exploration a white thrombus was found in the cortical vein corresponding to the lesion in the angiogram. No therapeutic intervention was made. Conventional treatment of persisting intracranial hypertension irrespective of medical treatment is CSF shunting, usually by a ventricular or lumbar peritoneal shunt. In respective cases so-called otitic hydrocephalus, ie, ventricular enlargement following cerebral venous outflow obstruction, has not been frequently observed in clinical practice [20]. However, all cases had an elevated CSF pressure.

Unconventional methods were reported by Sindou [21]. Bilateral thrombosis of the transverse sinus was treated by microsurgical revascularization with an extra-anatomical venous bypass. In one case with intracranial hypertension secondary to thrombophlebitis of the two transverse sinuses and internal jugular veins microsurgical revascularization was attempted. This was carried out by means of a bypass graft inserted between the right transverse sinus and a superficial jugular vein. The technique had previously been developed in an experimental animal study. A second case using the same method has been published by Hitchcock [22]. In a patient with pseudotumor cerebri a venous bypass graft was inserted between the left lateral sinus and a jugular vein. This was followed by rapid resolution of the raised intracranial pressure. Surgical intervention should be considered if medical treatment fails to relieve intracranial hypertension. In a follow-up study concerning extra-anatomical bypass, George et al. [23] reported 10 cases of venous bypass for revascularization of the lateral sinus. This technique proved to be efficient in normalizing hypertension of venous origin. The etiology was lateral sinus thrombosis following otitis (three cases), congenital anomaly (five cases) and surgical ligation during tumor surgery in two cases. A saphenous vein graft was used in eight cases and a synthetic graft (Goretex) in two cases. Bypass patency was confirmed by angiography in eight cases, graft thrombosis occurred in two cases, and no morbidity was observed.

Arterio-venous malformations (AVMs) of lateral and sigmoid sinuses are acquired lesions evolving from a previously thrombosed dural sinus [24]. There are preformed arterio-venous channels.Sequential angiograms have documented the progression of a thrombosed sinus to a dural AVM, and the concurrence of an occluded sinus and dural AVM in our experience is common. As a mechanism, it is proposed that an inflammatory or thrombotic process within a major sinus leads to partial or total occlusion and subsequent opening of pre-existing arterio-venous fistulae in the sinus walls. The arterial supply consists of:

1. The occipital artery.
2. Vascularized petrous bone.
3. Meningeal arteries in the dura overlying the cerebellum.
4. Branches of the middle meningeal artery.
5. Branches of the meningeo-hypophyseal artery to the tentorium.
6. In some cases a recurrent meningeal branch of the ophthalmic artery.

Sundt [25] preferred complete excision coupled with packing of the sigmoid sinus in 27 patients. During the surgical intervention isolation of the entire length of the sinuses including most of the mastoid may be necessary. The best results in our experience are obtained by combination of embolization and surgery. Fistulae of the superior sagittal sinus are less frequent and often occur with cortical venous drainage. The best treatment in these cases seems to be embolization with isobutylacrylate.

In order to evaluate the surgical methods in sinus vein thrombosis we favor in the future an extra-anatomical bypass procedure in cases of pseudotumor following sinus thrombosis. Success with this method may be limited by the extent of an accompanying thrombosis of the cortical veins. Hitherto there is no promising concept on the surgical therapy of cortical phlebothrombosis.

Post-thrombotic AV-fistulae give rise to interesting considerations from a surgical point of view. Is the fistula some sort of a self-healing reaction to accelerate flow in the venous low-flow system to prevent further thrombosis and start recanalization? If so, an adequate therapy would be to surgically create AV-fistulae, for example, between the superficial temporal artery and transverse sinus. This would be contradictory to the concept of eliminating the fistulae by surgical intervention. Or is it more efficient to subject a post-thrombotic dural AV-fistulae to thrombectomy of the sinuses or extra-anatomical bypass than to blocking and packing of the sinuses? Obviously surgery of sinus vein thrombosis is technically as well as conceptionally still in its infancy.

REFERENCES

1. H. Friedrich, G. Hänsel-Friedrich and W. Seeger, Intraoperative Dopplersonographie an Hirngefässen, *Neurochirurgia*, 23:89-98 (1980).
2. R. Jaeger, Observations on resection of the superior longitudinal sinus at and posterior to the rolandic venous inflow, *J. Neurosurg.*, 8:103-109 (1951).
3. V. Dolenc, Direct microsurgical repair of intracavernous vascular lesions, *J. Neurosurg.*, 58:824-831 (1983).
4. M. G. Yasargil, "Microsurgery", Thieme, Stuttgart (1969).
5. P. Donaghy, L. J. Wallman, M. J. Flanagan and M. Numoto, Sagittal sinus repair, *J. Neurosurg.*, 38:244-248 (1973).
6. M. Sindou, J.-F. Mazoyer, G. Fischer, J. Pialat and C. Fourcade, Experimental bypass for sagittal sinus repair, preliminary report, *J. Neurosurg.*, 44:325-330 (1976).
7. T. Sakaki, T. Morimoto, K. Takemura, S. Miyamoto, K. Kyoi and S. Utsumi, Reconstruction of cerebral cortical veins using silicone tubing, technical note, *J. Neurosurg.*, 66:471-473 (1987).

8. J. P. Kapp, I. Gielchinsky and St. L. Deardourff, Operative techniques for management of lesions involving the dural venous sinuses, *Surg. Neurol.*, 7:339-342 (1977).
9. W. Hassler, Traumatische Sinusverletzung und ihre operative Therapie unter Berücksichtigung der topographischen Anatomie und der Hirnsinus-varianten, *Neurochirurgia*, 22:165-180 (1979).
10. F. Isamat, E. Ferrer and J. Twose, Direct intracavernous obliteration of high-flow carotid-cavernous fistulas, *J. Neurosurg.*, 65:770-775 (1986).
11. D. Parkinson, Carotid cavernous fistula: direct repair with preservation of the carotid artery, technical note, *J. Neurosurg.*, 38:99-106 (1973).
12. A. Perneczky, E. Knops, P. Vorkapic and Th. Czech., Direct surgical approach to infraclinoidal aneurysms, *Acta Neurochirurgica*, 76:36-44 (1985).
13. Y. Matsuoka, A. Hakuba, H. Kishi and S. Nishimura, Direct surgical treatment of intracavernous internal carotid artery aneurysm: report of four cases, *Surg. Neurol.*, 26:360-364 (1986).
14. A. Hakuba, C. W. Huh, S. Tsujikawa and S. Nishimura, Total removal of a parasagittal meningeoma of the posterior third of the sagittal sinus and its repair by autogenous vein graft, *J. Neurosurg.*, 51:379-382 (1979).
15. L. N. Sekhar and A. R. Moller, Operative management of tumors involving the cavernous sinus, *J. Neurosurg.*, 64:879-889 (1986).
16. J. Bonnal, La chirurgie conservatrice et réparatrice du sinus longitudinal supérieur (SLS), *Neurochirurgie*, 28:147-172 (1982).
17. F. Lesoin and M. Jomin, Direct microsurgical approach to intracavernous tumors, *Surg. Neurol.*, 28:17-22 (1987).
18. G. Benoist and E. Pasztor, Analysis of 177 operated cases of parasagittal meningeoma from the point of view of venous disturbances, *in*: "Cerebral Veins", L. M. Auer and F. Loew, eds., Springer, Vienna and New York, pp 357-364 (1982).
19. T. Kawase, T. Tazawa and M. Mizukami, Cerebral venous thrombosis: findings from computer tomography and fluorescein angiography, *in*: "Cerebral Veins", L. M. Auer and F. Loew, eds., Springer, Vienna and New York, pp 327-336 (1982).
20. D. D'Avella, R. P. Greenberg, S. Mingrino, M. Scanarini and K. Pardatscher, Alterations in ventricular size and intracranial pressure caused by sagittal sinus pathology in man, *J. Neurosurg.*, 53:656-661 (1980).
21. M. Sindou, Ph. Mercier, J. Bokor and J. Brunom, Bilateral thrombosis of the transverse sinuses: microsurgical revascularization with venous bypass, *Surg. Neurol.*, 13:215-220 (1980).
22. E. R. Hitchcock and R. A. Cowle, Sino-jugular venous graft in otitic hydrocephalus, *Acta Neurochirurgica*, 59:187-193 (1981).
23. B. George, C. Sainte-Roset, M. Sindou and E. R. Hitchcock, Lateral sinus reconstructive surgery: treatment of intracranial hypertension by venous anastomosis, *Neurosurg. Rev.*, 6:203-206 (1984).
24. O. W. Houser, H. L. Baker and A. L. Rhoton, Intracranial dural arterio-venous malformations, *Radiology*, 105:55-64 (1972).
25. T. Sundt and D. G. Piepgras, The surgical approach to arterio-venous malformations of the lateral and sigmoid dural sinuses, *J. Neurosurg.*, 59:32-39 (1983).

TREATMENT OF PATIENTS WITH CEREBRAL SINUS AND VEIN THROMBOSIS WITH HEPARIN

W. Meister§, K.M. Einhäupl, A. Villringer, P. Schmiedek*, R. Haberl, H.W. Pfister, M. Pellkofer, H. Steinhoff†, M. Deckert, and F. Anneser

Departments of Neurology, *Neurosurgery, †Radiology and §Internal Medicine, Ludwig-Maximilians-University Klinikum Großhadern, 8000 Munich 70, FRG

INTRODUCTION

Cerebral sinus vein thrombosis (CSVT) is a possibly fatal disease and demands effective treatment. There are, however, no controlled clinical trials. Treatment recommendations are based on pathophysiological delibera-tions, case reports, or case report series [1-9].

The use of heparin in this disease was first described in 1942 by Stansfield [10]. Ever since, this topic has been discussed controversially in the medical literature, with the advocates stressing the clinical improvement seen after heparin, the opponents calling attention to a supposed increased risk of intracranial hemorrhage.

This paper is a preliminary report of two studies addressing the role of high-dose heparin in the acute phase of CSVT: a placebo-controlled, prematurely terminated trial and a retrospective analysis of all patients treated in our institution between 1977 and 1985 with special reference to the effect of heparin treatment on intra-cerebral hemorrhage.

PLACEBO-CONTROLLED TRIAL

The prospective, randomized trial compared the effect of heparin with that of placebo in addition to standard therapy on the prognosis of CSVT. It considered all consecutive patients with an angiographically confirmed diagnosis who were referred to the Department of Neurology of the University of Munich. The exclusion criteria were: pretreatment with anticoagulants or platelet-inhibiting drugs: extracranial indications for, or contraindications against, heparin; or a serious concomitant disease. The partial thromboplastin time (PTT) was used as a guideline for the heparin doses; a doubling of the baseline PTT value was the therapeutical goal. Because of this policy, the physicians conducting treatment were aware of the therapy, whereas the observer of the clinical course remained

Cerebral Sinus Thrombosis, Edited by K. Einhäupl *et al.*
Plenum Press, New York, 1990

blinded. The primary endpoint was the clinical outcome according to a severity score from 1 to 9. A secondary endpoint was the incidence of intracranial hemorrhage.

The design of the study required enrolment of 60 patients. After the first 20 patients, however, an intermediate analysis prompted investigators to terminate the study prematurely.

During the time of recruiting patients for the study, eight patients diagnosed at our department were not entered into the study: two because of pretreatment with heparin, one because of uncontrollable hypertension, one because of an extracranial indication for heparin, and three because of treatment with platelet-inhibiting analgetics; one patient refused his consent.

RESULTS

The baseline data were very similar in the two groups. Notably, the day of starting treatment after the initial symptom and the initial, clinical severity score were the same.

Impressive differences, however, were seen 3 months after starting treatment (Table 1). All patients under heparin improved, eight of them even to score 0, ie, they were symptom-free. Under placebo, the course varied, and three patients died in this group. The effect of heparin was significant ($p<0.05$, sign-test), whereas there was obviously no significant change under placebo.

Three patients in the heparin group already had a cerebral hemorrhage, but all recovered completely under treatment. No hemorrhage occurred in the other eight patients of the group. Two patients in the placebo group presenting with a hemorrhage died; in addition, two new hemorrhages occurred in the placebo group. This evidence, albeit limited, suggests an increase of intracranial hemorrhage under placebo and a favorable and protective action of heparin.

Extracranial complications were one fatal case of pulmonary embolism in the placebo group, and two minor bleeding complications in the heparin group, not requiring withdrawal of treatment, however.

DISCUSSION

The current study shows a superiority of heparin over placebo in the acute treatment of CSVT as assessed by the severity score at 3 months. In addition, heparin may not only prevent intracerebral hemorrhage, but also improve prognosis of this complication.

The two groups were very similar despite the small number of randomized patients. There are two important aspects of this study. First, because of the requirement of titration of the heparin dose, blinding of the treating physicians was not attempted. The observer of the clinical outcome, however, remained blinded throughout the study. Second, premature termination of the study was necessary because of the favorable results in the heparin group. This was not a decision by an independent committee but by the investigators.

Table 1. Clinical Sinus Venous Thrombosis Severity Score

Heparin Group		Placebo Group	
Start of Treatment	After 3 months	Start of Treatment	After 3 months
7	0	7	9
7	0	5	0
6	1	4	2
5	0	3	9
3	1	3	9
3	0	3	3
3	0	3	2
2	0	3	2
2	0	2	2
2	0	2	2
2	0	2	2

0 = No symptoms, 9 = dead.

Nevertheless, this pioneering controlled clinical trial in sinus vein thrombosis demonstrates a clear benefit of heparin.

RETROSPECTIVE ANALYSES

Starting in 1977, a registry with a standardized prospective documentation in all patients with angiographically confirmed diagnoses was started in our institution. The following analyses of therapeutic issues are based upon the first 71 cases. Patients were excluded with occlusion by extravascular causes (five patients), accompanying intracerebral tumors or angioma (seven patients), septic sinus vein thrombosis (seven patients), cavernous sinus thrombosis (12 patients), children under 10 years (two patients), angiographically doubtful cases (38 patients) and postmortem diagnosis (three patients).

Twenty-five patients received no heparin, 21 a low (15,000 IE/d) or intermediate (>15,000 IE/d ≤21,000 ID/d) dose therapy and 25 a high or full-dose heparin therapy, comparable to that used in the controlled trial.

The groups were comparable as to age, onset of therapy after the first symptoms, and severity of symptoms.

A major problem of the current retrospective analysis is a potential time- and experience-bias. The median year of patients not receiving heparin is 1978, that of the patients treated with full-dose heparin 1984. This may reflect the current practice of our department to heparinize all patients with CSVT including those with intracerebral hemorrhage.

As in the prospective study, the results favor full-dose heparin therapy. The majority of patients in the group without heparin

Table 2. Influence of Anticoagulation on the Course of Sinus Venous Thrombosis with Intracranial Hemorrhage. Retrospective Data from 28 Patients

Patients Treated Without Heparin After ICH						
		ICH		Heparin		
Age	Sex	Type	Day	Before ICH	After ICH	Clinical Course in the First Week After ICH
28	F	ICERH	20	?	No	Worse
56	M	ICERH	5	No	No	Better
69	F	ICERH	1	No	No	Worse
57	F	ICERH	6	No	No	Unchanged
20	F	ICERH	5	No	No	Worse
29	M	SDH	32	No	No	Worse
24	F	ICERH/SDH	19	Low	No	Worse
48	M	ICERH	3	No	No	Worse
56	F	ICERH	7	No	No	Better
51	M	ICERH	37	?	No	Worse
72	M	ICERH	1	No	No	Worse
67	M	ICERH	4	No	No	Better
30	F	SDH	61	Low	No	Better

ICH, Intracranial hemorrhage; ICERH, intracerebral hemorrhage; SAH, subarachnoid hemorrhage; SDH, subdural hemorrhage

experienced intracranial hemorrhage, whereas this complication was documented in only one case of the patients treated with heparin.

INTRACRANIAL HEMORRHAGE AND HEPARIN

Twenty-seven of the 71 patients had intracerebral hemorrhage. Thirteen did not receive anticoagulants thereafter (Table 2). Most of them had no heparin therapy before the event either. The majority of this group died. This is in contrast with the prognosis of patients who received anticoagulation despite intracranial hemorrhage (Table 3). Most of these patients recovered completely.

Both the randomized, controlled trial, and retrospective studies of a prospectively conducted registry endorse the use of full-dose heparin therapy in CSVT.

CONCLUSIONS

Heparin favourably affects prognosis in CSVT. Many neurologists still hesitate to administer heparin because of a suspected risk of intracerebral hemorrhage. Our results refute this assumption. Heparin not only tended to decrease the risk of intracerebral hemorrhage, but also improved the prognosis in patients with an already established diagnosis of this complication before starting heparin.

Table 3. Influence of Anticoagulation on the Course of Sinus Venous Thrombosis with Intracranial Hemorrhage. Retrospective Data from 28 Patients

Patients Treated With Full-dose Heparin After ICH						
		ICH		Heparin		
Age	Sex	Type	Day	Before ICH	After ICH	Clinical Course in the First Week After ICH
22	M	ICERH	12	Low	Low	Unchanged
24	F	SAH	6	No	Low	Better
36	M	ICERH	19	No	High	Better
17	M	ICERH	102	Low	High	Better
47	F	ICERH	33	No	High	Better
28	M	ICERH	10	Low	High	Better
17	M	ICERH/SDH	5	Low	High	Better
50	M	ICERH	13	No	Low	Better
44	F	ICERH	9	No	High	Better
46	F	ICERH	3	No	High	Better
43	F	SDH	35	No	High	Better
53	F	ICERH	13	No	High	Better
47	M	ICERH	33	High	high	Worse
21	M	ICERH	1	No	High	Better

ICH, Intracranial hemorrhage; ICERH, intracerebral hemorrhage; SAH, subarachnoid hemorrhage; SDH, subdural hemorrhage

Pathophysiological considerations may explain this effect. Although heparin cannot dissolve a clot, it may prevent a further growth of the thrombus, thus supporting endogenous lytic processes. Relief of the venous occlusion may normalize outflow out of the cerebral capillary bed and thus decrease the capillary pressure and risk of hemorrhage.

REFERENCES

1. U. Bogdahn, D. Dommasch and R. Wodarz, Thrombolytische Therapie der Sinusthrombose, *in*: "Pathologische Erregbarkeit des Nervensystems und ihre Behandlung", H. G. Mertens and H. Przuntek, Springer, Berlin, Heidelberg and New York, pp 675-678 (1980).
2. M. G. Bousser, J. Chiras, J. B. Bories and P. Castaigne, Cerebral venous thrombosis - a review of 38 cases, *Stroke*, 16:199-210 (1985).
3. C. Di Rocco, A. Ianelli, G. Leone, M. Moschini and V. M. Valori, Heparin-urokinase treatment in aseptic dural sinus thrombosis, *Arch. Neurol.*, 38:431-435 (1981).
4. D. Doerstelmann, H. Dobisch, W. Mattes and R. Reuther, Hirnvenen- und Sinusthrombose, *Nervenarzt.*, 52:243-246 (1981).
5. D. M. Gettelfinger and E. Kokmen, Superior sagittal sinus thrombosis, *Arch. Neurol.*, 34:1-6 (1977).
6. H. Hacker and B. May, Zur Behandlung der Hirnvenenthrombose, *Nervenarzt.*, 40:440-442 (1969).
7. J. E. Harvey, Streptokinase therapy and cavernous sinus thrombosis, *Br. Med. J.*, 5:84-91 (1974).

8. J. A. Scott, R. M. Pascuzzi, P. V. Hall and G. J. Becker, Treatment of dural sinus thrombosis with local urokinase infusion.Case report, *J. Neurosurg.*, 68:284-287 (1988).
9. M. Sindou, P. H. Mercier, J. Bokor and J. Brunnon, Bilateral thrombosis of the transverse sinuses: microsurgical revascularization with venous bypass, *Surg. Neurol.*, 13:215-220 (1980).
10. F. R. Stansfield, Puerperal cerebral thrombophlebitis treated by heparin, *Br. Med. J.*, 4:436-439 (1942).

SUBJECT INDEX

MIX
Papier aus verantwortungsvollen Quellen
Paper from responsible sources
FSC® C105338

If you have any concerns about our products,
you can contact us on
ProductSafety@springernature.com

In case Publisher is established outside the EU,
the EU authorized representative is:
Springer Nature Customer Service Center GmbH
Europaplatz 3, 69115 Heidelberg, Germany

Printed by Libri Plureos GmbH
in Hamburg, Germany

Volumes already published

Volume 1: Trees I (1986)
Volume 2: Crops I (1986)
Volume 3: Potato (1987)
Volume 4: Medicinal and Aromatic Plants I (1988)
Volume 5: Trees II (1989)
Volume 6: Crops II (1988)
Volume 7: Medicinal and Aromatic Plants II (1989)
Volume 8: Plant Protoplasts and Genetic Engineering I (1989)
Volume 9: Plant Protoplasts and Genetic Engineering II (1989)
Volume 10: Legumes and Oilseed Crops I (1990)
Volume 11: Somaclonal Variation in Crop Improvement I (1990)
Volume 12: Haploids in Crop Improvement I (1990)
Volume 13: Wheat (1990)
Volume 14: Rice (1991)
Volume 15: Medicinal and Aromatic Plants III (1991)
Volume 16: Trees III (1991)
Volume 17: High-Tech and Micropropagation I (1991)
Volume 18: High-Tech and Micropropagation II (1992)
Volume 19: High-Tech and Micropropagation III (1992)
Volume 20: High-Tech and Micropropagation IV (1992)

Volumes in preparation

Volume 21: Medicinal and Aromatic Plants IV (1993)
Volume 22: Plant Protoplasts and Genetic Engineering III (1993)
Volume 23: Plant Protoplasts and Genetic Engineering IV
Volume 24: Medicinal and Aromatic Plants V
Volume 25: Maize
Volume 26: Medicinal and Aromatic Plants VI
Volume 27: Somatic Hybridization in Crop Improvement I

Biotechnology in Agriculture and Forestry 20

High-Tech and Micropropagation IV

Edited by Y. P. S. Bajaj

With 212 Figures

Springer-Verlag
Berlin Heidelberg New York
London Paris Tokyo
Hong Kong Barcelona Budapest

Professor Dr. Y.P.S. Bajaj
A-137
New Friends Colony
New Delhi 110065, India

ISBN 3-540-53661-2 Springer-Verlag Berlin Heidelberg New York
ISBN 0-387-53661-2 Springer-Verlag New York Berlin Heidelberg

Library of Congress Cataloging-in-Publication Data. (Revised for volume 4). High-tech and micropropagation. (Biotechnology in agriculture and forestry ; 17–) Includes bibliographical references and indexes. 1. Plant propagation—In vitro. I. Bajaj, Y.P.S., 1936– . II. Series: Biotechnology in agriculture and forestry ; 17, etc. SB123.6.H54 1991 631.5′3 91-22304

Printed in Germany

Production Editors: Ingrid Samide and Jacob Sabu Mathew, Heidelberg

Typesetting: Macmillan India Limited, Bangalore
31/3145-5 4 3 2 1 0 – Printed on acid-free paper

Dedicated to Professor R. L. M. Pierik
with whom I had the privilege of working
at the Agriculture University of Wageningen,
The Netherlands, from 1973–1974

Preface

Micropropagation of plants is a multibillion dollar industry being practiced in hundreds of small and large nurseries and commercial laboratories throughout the world. At the present, it is the only component of plant biotechnology which has been commercially exploited on such a large scale, especially for the production of ornamentals. Now micropropagation of trees and medicinal plants is assuming great importance. With recent progress made in the propagation of fruit and forest trees, and the immediate need for afforestation and planting of orchards, propagules and plantlets are required quickly and in large numbers. Although the basic methodology of in vitro culture of plant tissues has been fairly mechanized over the last 25 years, conventional micropropagation methods are relatively slow, time-consuming, expensive, and labor-intensive; up to 50–80% of the cost of plant production is attributed to labor. In order to reduce the cost and to increase the efficiency and scale up of plant production, automation is being incorporated. Robots, microcomputers, and other automated systems of artificial intelligence have been employed to surface-sterilize tissues, dispense media, and transfer plants to pots and field. The automatic control of environments in greenhouses to protect cultivation is another area of tremendous commercial importance.

The use of artificial intelligence, image analysis, video cameras, bioreactors, mistifiers, automatic production lines, synthetic seed, storage and cryopreservation of elite genetic stocks, and the computerization of information management will facilitate micropropagation and protected cultivation. Judging from the enthusiasm and interest of biotechnology enterprises, it is envisaged that it will not be long before we see the routine jobs being done by robots for industrial micropropagation.

Taking the above-mentioned points into consideration, it appeared necessary to review the literature and to present the state of the art on micropropagation of plants. Thus, 118 chapters contributed by experts from 30 countries are compiled in a series composed of the following four parts:

High-Tech and Micropropagation I comprises 29 chapters on various aspects of micropropagation and automation, i.e., setting up a commercial laboratory, meristem culture, somatic embryogenesis, graft incompatibility, vitrification, acclimatization, nature and origin of explants, problems with explant exudates, disposable culture vessels, induction of rooting, artificial substrates, rockwool, autotrophic micropropagation, storage and cryopreservation of cultures, quiscent somatic embryos, synthetic seed, automated systems, artificial intelligence, use of robots in transplantation, information management, bioreactors, and computerized greenhouses.

High-Tech and Micropropagation II comprises 26 chapters on American sweetgum, black locust, birches, cedar, cocoa, larches, Persian walnut, poplars, *Quercus*, sandalwood, *Sorbus*, apple, pear, peach, plum, cherry, *Citrus*, raspberry, strawberry, grapevine, kiwi, Japanese persimmon, papaya, pineapple, date palm, and olive tree.

High-Tech and Micropropagation III comprises 34 chapters on vegetable and tuber crops, medicinal and aromatic plants, fiber and oilseed crops, and grasses. The specific plants dealt with include: asparagus, lettuce, horseradish, cucumber, potato, sweet potato, artichoke, white yam, cardamom, fennel, celery, thyme, leek, mentha, turmeric, lavender, agave, yucca, cotton, jute, sunflower, ryegrass, zoysiagrass, and various species of *Aconitum, Artemisia, Camelia, Centaurium, Digitalis, Dioscorea, Glehnia, Parthenium,* and *Pinella.*

High-Tech and Micropropagation IV comprises 29 chapters on ornamental species of *Alstroemeria, Antirrhinum, Begonia, Chrysanthmum, Cornus, Euphorbia, Gardenia, Gladiolus, Hyacinthus, Impatiens, Iris, Lycoris, Nematanthus, Phalaenopsis, Paeonia, Pelargonium, Rhododendron, Ruscus, Saintpaulia, Senecio, Syringa,* orchids, cacti, roses, and Boston ferns.

These books will be of use to advanced students, research workers, and teachers in horticulture, forestry, tissue culture, and plant biotechnology in general, and also to individuals interested in commercial micropropagation.

New Delhi, Summer 1992

Professor Y. P. S. Bajaj
Series Editor

Contents

X Micropropagation of *Gladiolus*
Y. P. S. Bajaj, M. M. S. Sidhu, and A. P. S. Gill
(With 4 Figures)

XI Micropropagation of Hyacinths
(*Hyacinthus orientalis* L.)
A. Bach (With 8 Figures)

XII Micropropagation of *Impatiens* spp.
L. C. Stephens, J. L. Weigle, S. L. Krell, and K. Han
(With 2 Figures)

XIII Micropropagation of Iris with Special Reference
to *Iris × hollandica* Tub.
P. C. G. van der Linde and J. A. Schipper (With 9 Figures)

XIV Micropropagation of Surprise Lily (*Lycoris squamigera*)
F. H. Huang, G. L. Klingaman, and H. H. Chen
(With 8 Figures)

XV Micropropagation of *Nematanthus*
C. L. Lê (With 7 Figures)

XVI Micropropagation of Orchids – *Spathoglottis plicata* and *Epidendrum radicans*
F. Singh (With 9 Figures)

XVII Micropropagation of *Phalaenopsis* spp.
M. Tanaka (With 22 Figures)

XVIII Micropropagation of Peony (*Paeonia* spp.)
J. A. T. Buchheim and M. M. Meyer, Jr. (With 7 Figures)

XXVII Micropropagation of Some Parasitic Plants
S. N. C. Okonkwo (With 6 Figures)

XXVIII In Vitro Culture of *Arabidopsis thaliana*
K. A. Feldmann (With 4 Figures)

XXIX Micropropagation of Boston Ferns (*Nephrolepis* spp.)
W. Amaki and H. Higuchi (With 4 Figures)

List of Contributors

AITKEN, E. A. B., School of Biological Sciences, University of Birmingham, P.O. Box 363, Birmingham B15 2TT, United Kingdom

AMAKI, W., Laboratory of Floriculture, Department of Agriculture, Tokyo University of Agriculture, 1-1-1 Sakuragaoka, Setagaya-ku, Tokyo 156, Japan

ATKINSON, N. J., School of Biological Sciences, University of Birmingham, P.O. Box 363, Birmingham B15 2TT, United Kingdom

BACH, A., Department of Horticulture, University of Agriculture of Kraków, al. 29 Listopada 54, Kraków, Poland

BACKHAUS, R. A., Department of Botany, Arizona State University, Tempe, AZ 85287-1601, USA

BAJAJ, Y. P. S., Former Professor of Tissue Culture, Punjab Agricultural University, Ludhiana, India. (Present address: A-137 New Friends Colony, New Delhi 110065, India)

BEATY, R. M., Department of Ornamental Horticulture and Landscape Design, Institute of Agriculture, University of Tennessee, Knoxville, TN 37901-1071, USA

BRIGGS, B. A., Briggs Nurseries Inc., 4407 Henderson Blvd., Olympia, WA 98501, USA

BUCHHEIM, J. A. T., Department of Horticulture, University of Illinois, Urbana, IL 61801, USA. (Present address: Department of Horticulture, Louisiana State University, Baton Rouge, LA 70803, USA)

CASSELLS, A. C., Department of Plant Science, University College, Cork, Ireland

CHEN, H. H., Department of Horticulture and Forestry, University of Arkansas, Fayetteville, AR 72701, USA

CLAYTON, P. W., Department of Agronomy and Horticulture, Plant Genetic Engineering Laboratory, New Mexico State University, Las Cruces, NM 88003-0003, USA

CURIR, P., Istituto Sperimentale per la Floricoltura, 18038, Sanremo, Italy

ECONOMOU, A. S., Department of Horticulture, Aristotle University, P.O. Box 281, 54006 Thessaloniki, Greece

EDICK-CATON, L., Briggs Nurseries Inc., 4407 Henderson Blvd., Olympia, WA 98501, USA

FELDMANN, K. A., Central Research and Development, Plant Sciences, E. I. du Pont de Nemours & Company, Wilmington, DE 19898, USA (Present address: University of Arizona, Plant Sciences, Forbes 303, Tucson, AZ 85721, USA)

FORD-LLOYD, B. V., School of Biological Sciences, University of Birmingham, P.O. Box 363, Birmingham B15 2TT, United Kingdom

GERTSSON, U. E., The Swedish University of Agricultural Sciences, Department of Horticulture, Box 55, 23053 Alnarp, Sweden

GILL, A. P. S., Punjab Agricultural University, Ludhiana, India

HAN, K., Department of Horticulture, Iowa State University, Ames, IA 50011, USA

HIGUCHI, H., Laboratory of Floriculture, Department of Agriculture, Tokyo University of Agriculture, 1-1-1 Sakuragaoka, Setagaya-ku, Tokyo 156, Japan

HO, J., Shihezi Agricultural College, Shihezi, Xinjiang, P.R. China

HORN, W. A. H., Lehrstuhl für Zierpflanzenbau, Technische Universität München-Weihenstephan, W-8050 Freising 12, FRG

HUANG, F. H., Department of Horticulture and Forestry, University of Arkansas, Fayetteville, AR 72701, USA

HUBSTENBERGER, J. F., Department of Agronomy and Horticulture, Plant Genetic Engineering Laboratory, New Mexico State University, Las Cruces, NM 88003-0003, USA

JUNGNICKEL, F., Friedrich-Schiller-University Jena, Institute of General Botany, von-Hase-Weg 3, O-6900 Jena, FRG

KLINGAMAN, G. L., Department of Horticulture and Forestry, University of Arkansas, Fayetteville, AR 72701, USA

KRELL, S. L., Department of Horticulture, Iowa State University, Ames, IA 50011, USA

LÊ, C. L., Station fédérale de Recherches Agronomiques de Changins (RAC), 1260 Nyon, Switzerland

LOWE, K. W., Department of Ornamental Horticulture and Landscape Design, Institute of Agriculture, University of Tennessee, Knoxville, TN 37901-1071, USA

MCCULLOCH, S. M., Briggs Nurseries Inc., 4407 Henderson Blvd., Olympia, WA 98501, USA

MEYER, M. M., JR., Department of Horticulture, University of Illinois, Urbana, IL 61801, USA

MONETTE, P. L., Agriculture Canada, Plant Quarantine Station, 8801 East Saanich Road, Sidney, British Columbia, V8L 1H3, Canada

NEWBURY, H. J., School of Biological Sciences, University of Birmingham, P.O. Box 363, Birmingham B15 2TT, United Kingdom

OKONKWO, S. N. C., Department of Botany, University of Nigeria, Nsukka, Nigeria

PHILLIPS, G. C., Department of Agronomy and Horticulture, Plant Genetic Engineering Laboratory, New Mexico State University, Las Cruces, NM 88003-0003, USA

PIERIK, R. L. M., Department of Horticulture, Agricultural University, Haagsteeg 3, 6708 PM Wageningen, The Netherlands

PREECE, J. E., Department of Plant and Soil Science, Southern Illinois University, Carbondale, IL 62901-4415, USA

RIPLEY, K. P., Department of Plant and Soil Science, Southern Illinois University, Carbondale, IL 62901-4415, USA

RYBALKO, A. E., The Enterprise of the Meristematic Cultures, Ministry of Housing and Municipal Economy, RSFSR Sochi, A-354354, USSR

SCHIPPER, J. A., Bulb Research Centre, Central Research Laboratory for Tissue Culture of Horticultural Crops, P.O. Box 85, 2160 AB Lisse, The Netherlands

SCHIVA, T., Istituto Sperimentale per la Floricoltura, 18038, Sanremo, Italy

SIDHU, M. M. S., Carewel Nursery, Parkhowal Road, Ludhiana (Punjab), India

SIMMONDS, J., Agriculture Canada Research Branch, Plant Research Centre, Ottawa, Ontario, K1A 0C6, Canada

SINGH, F., Orchid Laboratory, Indian Institute of Horticultural Research, Post Box No. 8025, Bangalore 560080, India

SPRENKELS, P. A., Department of Horticulture, Agricultural University, Haagsteeg 3, 6708 PM Wageningen, The Netherlands

STEEGMANS, H. H. M., Department of Horticulture, Agricultural University, Haagsteeg 3, 6708 PM Wageningen, The Netherlands

STEPHENS, L. C., Department of Horticulture, Iowa State University, Ames, IA 50011, USA

TANAKA, M., Laboratory of Floriculture, Faculty of Agriculture, Kagawa University, Miki-cho, Kagawa 761-07, Japan

TRIGIANO, R. A., Department of Ornamental Horticulture and Landscape Design, Institute of Agriculture, University of Tennessee, Knoxville, TN 37901-1071, USA

VAN DER LINDE, P. C. G., Bulb Research Centre, Central Research Laboratory for Tissue Culture of Horticultural Crops, P.O. Box 85, 2160 AB Lisse, The Netherlands. (Present address: RZ Biofleur C.V., P.O. Box 2, 9123 ZR Metslawier, The Netherlands)

WEIGLE, J. L., Department of Horticulture, Iowa State University, Ames, IA 50011, USA

ZAID, S., Damascus University, Faculty of Sciences, Department of Botany, Damascus, Syria

I Micropropagation of Inca Lily (*Alstroemeria* spp.)

P.L. MONETTE[1]

1 Introduction

Alstroemeria (Peruvian or Inca lily), a genus of herbaceous rhizomatous perennials with colorful flowers (Fig. 1), is generally considered to belong to the monocot family Amaryllidaceae. About 60 species have been described, all of which are from South America (Baker 1888; Uphof 1952; Robinson 1963). They grow in very diverse habitats, ranging from dry coastal sand dunes (*A. pelegrina* L.), desert sites (*A. violacea* Phil.), river valleys (*A. haemantha* Ruiz), and semi-woodlands (*A. aurantiaca* D. Don) to the snowline of the Andes (*A. nivalis* Phil. in Lin.) (Heins and Wilkins 1979).

Fig. 1. An inflorescence of *Alstroemeria* Alsaan showing cymes and flowers. (Lin and Molnar 1983)

[1] Agriculture Canada, Plant Quarantine Station, 8801 East Saanich Road, Sidney, British Columbia, Canada V8L 1H3

Biotechnology in Agriculture and Forestry, Vol. 20
High-Tech and Micropropagation IV (ed. by Y.P.S. Bajaj)

The plant consists of a white subterranean rhizome from which arise aerial shoots and a moderately fibrous root system (Fig. 2). The shoots can be either vegetative or reproductive depending on soil temperature, daylength, and light intensity (Vonk Noordegraaf 1975; Heins and Wilkins 1979; Healy et al. 1982; Lin 1985; Healy and Wilkins 1986; Powell and Bunt 1986). The inflorescence is a whorl

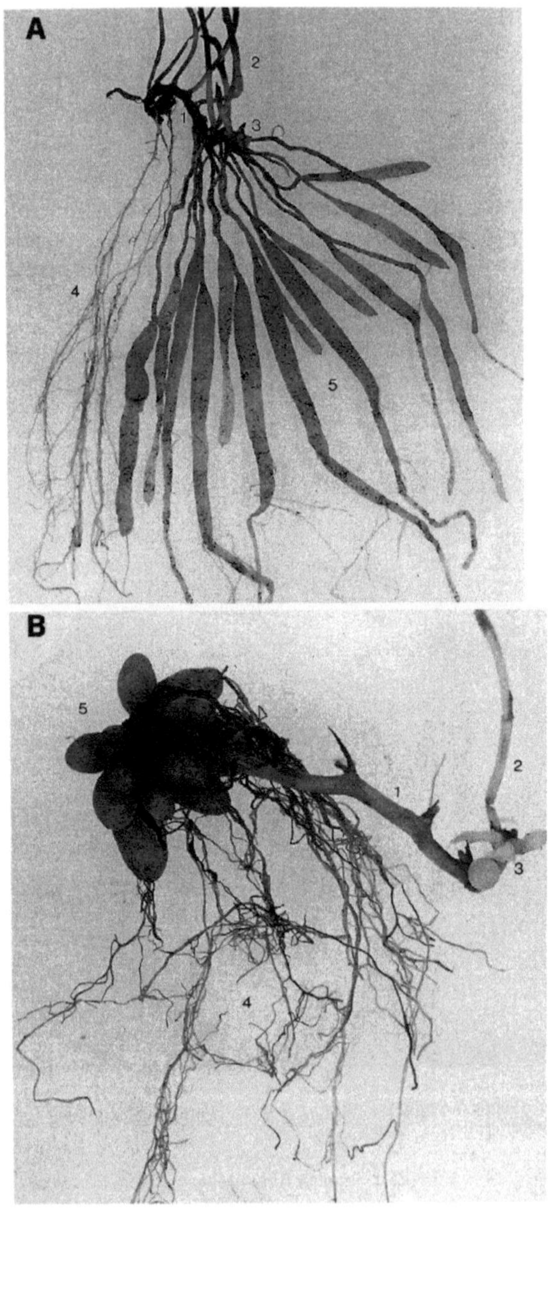

Fig. 2A, B. *Alstroemeria* cv. Regina (**A**) and *A. pulchella* (**B**) rhizomes (*1*), shoots (*2*), rhizome tips (*3*), roots (*4*), and storage roots (*5*)

of cymes forming an umbel at the top of the stem. Each cyme has two to four buds depending on the season, the light intensity, and the size and vigor of the plant (Hughes et al. 1986).

Alstroemeria species, which come in many different shades of reds and yellows, are diploid with a chromosome number of 2n = 16 (Darlington and Wylie 1955). Among the cultivars there are also numerous triploids and tetraploids (Tsuchiya and Hang 1987).

1.1 Importance of the Plant

Alstroemeria species are generally not suitable for commercial cut flower production. They tend to have a short flowering period and their leaves are often of poor quality. The commercial cultivars and hybrids have been obtained through interspecific hybridization, and radiation treatments (Goemans 1962; Broertjes and Verboom 1974). The modern *Alstroemeria* are valued for the large size of their flowers, the wide range of flower colors, an interesting flower form and a vase life of up to 14 days. The crop grows well at low greenhouse temperatures (Healy and Wilkins 1981a), and some cultivars flower almost all year round.

1.2 Distribution and Some Production Figures

The commercial development of *Alstroemeria* was initiated in England. Further breeding and development followed in The Netherlands, which is today the major producer of *Alstroemeria* plants for cut-flower production. Some idea of the growth of this crop can be gained from Table 1, which shows a 139-fold increase in the number of stems sold at auctions in The Netherlands in an 18-year period; 275 ha were in production, world-wide, in 1986 (Pierik et al. 1988). Most of the commercial cultivars have been patented by Dutch propagators and are sold under licence to foreign growers. A list of some major cultivars is presented in Table 2. Others are listed in Verboom (1980) and Healy and Wilkins (1981b). North Americans have only recently started growing *Alstroemeria* commercially (Wilkins 1985). Columbia and Kenya also produce *Alstroemeria*.

Table 1. *Alstroemeria* cut flower production in The Netherlands, 1968–1986

Year	Stems (in millions) sold at auctions	Reference
1968	0.9	Verboom (1980)
1972	4.4	Verboom (1980)
1976	10.5	Verboom (1980)
1977	15	Verboom (1980)
1978	22	Verboom (1980)
1983	65	Hakkaart and Versluijs (1985)
1986	125	Pierik et al. (1988)

Table 2. Overview of *Alstroemeria* cultivars by color, breeding line, and plant breeder. (Hughes et al. 1986)

White/yellow	Orange	Red	Salmon	Pink	Purple	Plant breeder
Minas (O)					Paramo (O)	Konst
Rio (O)					Sonja	Konst
Orchid (O)						Konst
Paloma (O)						Konst
Canaria (O)	Mandarin (Y)	King Cardinal (H)	Apple Blossom (H)	Rita (B)	Jubilee (Y)	Van Staaveren
Mona Lisa (H)		Red Sunset (C)	Atlas (C)	Rosello (B)	Luciana (?)	Van Staaveren
Orchid (O)		Rosita (C)	Regina (?)	Pink Triumph (H)	Tango (?)	Van Staaveren
Yellow King (H)				Annabel (B, Y)		Van Staaveren
Zebra (O)				Isabella (?)		Van Staaveren
				Manon (B)		Van Staaveren
				Monika (B)		Van Staaveren
				Somora (Y)		Van Staaveren
				Ramona (Y)		Van Staaveren
				Libelle (Y)		Van Staaveren
Sun Valley (C)	Sunrise (C)	Red Valley (C)	Vera (O)	Pink Panther (B)	Westland (H)	Van Duyn
Cyprus (O)	Summerland (C)	Red Bird (C)		Jacqueline (B)	Ontario (B)	Van Duyn
Moonlight (H)	Orange Glow (?)	Purple Joy (B)		Lilac Glory (B)	Evening Glow (H)	Van Duyn
Yellow Queen (O)	Victorine (O)	Valeria (O)		Rose Wings (B)	Orlando (B)	Van Duyn
				Pink Tiger (B)		Van Duyn
				Ohio (B)		Van Duyn
				Sapporo (B)		Van Duyn
				Oscar (B)		Van Duyn
Friendship	Valiant	Carmen	Capitol			Wulfinghoff/Parigo
Eleanor		Marina				Wulfinghoff/Parigo

O: Orchid flowering B: Butterfly type C: Carmen type Y: Yeararound H: Hybrid ?: unknown

1.3 Diseases and Pests

A few viruses have been isolated from *Alstroemeria* (Table 3). *Alstroemeria* mosaic virus (AMV), a potyvirus, is the most widespread. Depending on the cultivar or species, symptoms of AMV infection vary from slight chlorotic spotting to severe

Table 3. Pathogens and pests of *Alstroemeria*

A) Viruses

Virus	Observation	Reference
Alstroemeria mosaic virus (a potyvirus)	First report of a potyvirus	Maat (1977)
	Virus named	Brunt and Phillips (1981)
	No breaking of flower color in the Netherlands	Hakkaart and Versluijs (1985)
	Slight leaf chlorosis and "flower break" in *A. caryophylla*. Severe leaf chlorosis in cvs. Canaria and Rosario.	Phillips and Brunt (1986)
Alstroemeria carlavirus	First report	Brunt and Phillips (1981)
	Antiserum prepared and virus named	Maat (1983)
	Close serological relationship to lily symptomless virus	Phillips and Brunt (1983)
	Detected in several cvs.	Hakkaart and Versluijs (1985)
	Detected in virtually symptomless *Alstroemeria*	Phillips and Brunt (1986)
Tobacco rattle virus	First report	Van Hoof (1971)
	Severe leaf necrosis in only one severely affected imported crop	Brunt and Phillips (1981); Phillips and Brunt (1986)
	Isolated once from Canaria with malformed leaves, yellow flecks	Hakkaart and Versluijs (1985)
Tomato spotted wilt virus	Rare occurrence	Hakkaart and Versluijs (1985)
Ilarvirus	Rare occurrence	Phillips and Brunt (1986)

B) Fungi

Disease	Causal agent
Botrytis	*Botrytis cinerea* Pers.
Pythium root rot	*Pythium* sp.
Rhizoctonia root rot	*Rhizoctonia* sp.

C) Pests

Aphids
Whiteflies
Caterpillars
Red spider mites
Slugs
Snails

leaf chlorosis and stunting; flower color breaking may also occur. A carlavirus has been detected in several *Alstroemeria* cultivars. It has little effect on the growth and flower yield of infected plants and appears to be an *Alstroemeria*-adapted strain of lily symptomless virus. Tobacco rattle virus, tomato spotted wilt virus, and an ilarvirus have also been detected in *Alstroemeria*, but they occur only rarely.

Alstroemeria are subject to a few fungal diseases. Botrytis can be a problem when damp conditions are present during poor light periods. This problem can be remedied by the application of fungicides, but the utmost care must be taken as *Alstroemeria* foliage is quite susceptible to pesticide injury. The root diseases pythium and rhizoctonia can occur in young plants. These diseases can be prevented by proper soil pasteurization or controlled by soil drenching.

Pests do not pose a serious problem on *Alstroemeria*. The most common are aphids, whiteflies, and slugs (Hughes et al. 1986). A few others have occasionally been reported (Verboom 1980; Anonymous 1983). These pests can be controlled by standard chemicals and biological control agents. Certain cultivars are susceptible to iron deficiency (Anonymous 1983).

1.4 Conventional Practices for Propagation and Improvement

1.4.1 Propagation

Alstroemeria plants are vegetatively propagated by division of the rhizome (Healy and Wilkins 1981a). The plants are severely pruned 1 or 2 weeks before division. Plants are dug from the ground, usually in September or October, care being taken to include the growing points of the rhizomes and to lift the brittle storage roots intact. The plant is divided so that each new division consists of a single rhizome with an undamaged blunt growing point, some new aerial shoots, and also some large fleshy storage roots. Only the youngest 1 to 3 inches of rhizome is kept. Pots, soil, or ground beds should be prepared before plants are divided so that new divisions can be planted immediately. A fungicidal drench with Lesan and Benlate (8 oz. each per 100 gal. water) is recommended at the time of planting (Healy and Wilkins 1981a). The plants are then watered only as they become dry, as excess watering will rot the rhizomes. They are grown at 18 °C until they become well established, generally in 4 to 8 weeks. The temperature is then reduced to 4.5 °C for 4 to 6 weeks to vernalize the rhizomes and induce flowering.

1.4.2 Improvement

Hybridization. Hybridization programs rely heavily on *Alstroemeria* species for improvement. The most important of these, historically, have been *A. ligtu*, *A. aurantiaca*, and *A. pelegrina* (Oosthoek 1967), as well as *A. haemantha* and *A. violacea*. The once very popular Ligtu hybrids resulted from crosses between *A. ligtu* var. *augustifolia* and *A. haemantha* (Coats 1969; Perry 1972). One important interspecific hybrid is Orchid, which may be a cross of *A. violacea* × *A. aurantiaca* (Heins and Wilkins 1979). Another important cultivar is Regina, which is a cross between *A. pelegrina* and *A. aurantiaca* (Heins and Wilkins 1979).

Interspecific sterility often presents a problem in *Alstroemeria* breeding programs. Tsuchiya and Hang (1987), who conducted chromosome studies on nine species of *Alstroemeria*, found that all were diploid with a chromosome number of 2n = 16, but that there was a considerable difference in their karyotypes. Some species were heterozygous, with one or more nonhomologous pairs of chromosomes. It thus appears likely that gametes from some of these species could carry deleterious or lethal recessive genes into interspecific crosses.

Most of the modern cultivars have unpublished or uncertain parentage. Almost all of them are sterile. The diploid cultivars Orchid, Canaria, Zebra, and Eureka and the 12 triploid cultivars studied by Tsuchiya and Hang (1987) all had very low pollen fertility. A hypertriploid (2n = 25) cultivar, Orange Beauty, had about 40% viable pollen grains. This cultivar has been crossed with *A. aurantiaca* D. Don to produce the Canadian cultivar Alsaan (Lin and Molnar 1983). Four of the 25 cultivars studied by Tsuchiya and Hang (1987) were tetraploid.

Irradiation. A conventional method of introducing genetic variation in sterile cultivars or species of vegetatively propagated crops is to induce mutations by exposure of plant material to X-rays. Such "mutation breeding" has been an important part of *Alstroemeria* improvement programs. Broertjes and Verboom (1974) subjected the actively growing rhizomes of young plants to irradiation. Mutated characters included flower color, plant height, flower productivity, and flower quality. One disadvantage of irradiation which is a serious problem with some crops is the production of chimeras. This apparently was not a problem with *Alstroemeria.*

2 In Vitro Approaches

2.1 Need to Incorporate Unconventional Methods

The most serious disadvantage of propagating *Alstroemeria* by division of rhizomes is that it takes a long time to produce a large number of plants, once a commercially valuable hybrid or mutant has been obtained. A large amount of space is also required. The development of media and procedures for the micropropagation of *Alstroemeria* would alleviate these problems. Plant tissue culture techniques also offer the opportunity of producing virus-free plants.

2.2 Summary of Tissue Culture Work

There is little published information on the in vitro culture of *Alstroemeria*; a summary of the tissue culture work is presented in Table 4.

2.2.1 Explant Selection and Disinfestation

Ziv et al. (1973), working with Parigo and Ligtu hybrids, obtained callus from ovaries larger than 5 mm in diameter and root growth from flower pedicels at the

Table 4. Summary of in vitro studies with *Alstroemeria*

Species or cultivar	Explant	Medium	Cytokinin(s)	Auxin	Response	Reference
Ligtu hybrids	Segments of subapical inflorescence stem	White's (1963) or MS	Kinetin	NAA or IAA	Plantlets regenerated	Ziv et al. (1973)
Orchid	Rhizome vegetation point	Linsmaier and Skoog (1965)			Plantlets regenerated (120)	Quak (1974)
Unspecified	Rhizome tips	Unspecified	BA	IBA	Rooted plantlets	Hussey et al. (1979)
Zebra	Rhizome segments	MS	BA	NAA	Growth regulator requirements defined for organogenesis	Gabryszewska and Hempel (1985)
Rosario, Toledo, Jubilee	Rhizome axillary meristems	MS	2-iP	IAA or NAA	Plantlets regenerated; *Alstsroemeria* carlavirus and AMV eliminated	Hakkaart and Versluijs (1985)
Three hybrid cvs. and a cv. of *A. pulchella*	Immature inflorescences and shoot tips; ovules	Unspecified	Unspecified	Unspecified	Callus, shoot and root production; haploid somatic embryos from ovules of hybrids.	Bridgen (1986)

Unspecified	Peduncle sections proximal to inflorescence; pedicels of inflorescence; rhizome tips	MS	BA and/or kinetin	Unspecified	Unspecified	King and Bridgen (1987)
Alsaan	Rhizome tips	MS	Zeatin and kinetin	IBA	True-to-type plantlets regenerated	Lin and Monette (1987)
Alsaan	Rhizome tips	MS	Zeatin and kinetin for initiation: then no growth regulator	NAA	75% rooting	Monette and lin (1991)
Toledo and others	Rhizome tips	MS	BA	NAA	Plantlets regenerated	Pierik et al. (1988)

point where the inflorescence stalk branches into several pedicels. No further differentiation was observed on these explants and they eventually died. These authors were successful in regenerating plantlets of Ligtu hybrids when the explants were taken from undifferentiated inflorescence stems 1–2 mm below the apical meristem. Hussey et al. (1979), working with unspecified cultivars, reported that rhizomes could be initiated adventitiously at the base of cultured vegetative shoots or on parts of the inflorescence but that they were more easily cultured directly from rhizome tips. Hakkaart and Versluijs (1985) regenerated plantlets of the cvs. Rosario, Toledo, and Jubilee from axillary meristems excised from rhizomes and Gabryszewska and Hempel (1985) studied the influence of cytokinins and auxins in vitro on rhizome cultures of Zebra. Lin and Monette (1987) reported that rhizome tips (Fig. 3) of field-grown seedlings of *A. aurantiaca* were more suitable as explants than either flower pedicels or subapical segments of inflorescence or vegetative stems. They also reported that, with the cv. Alsaan, the quality of the cultures was superior when intact rhizome tips rather than longitudinal sections were used for culture initiation (Table 5). Pierik et al. (1988) used terminal and lateral rhizome tips for the micropropagation of various *Alstroemeria* hybrids. They reported that terminal tips had a higher rhizome multiplication rate than lateral tips.

Ziv et al. (1973) surface sterilized their explants by stirring them for 15 min in a 9% filtered solution of calcium hypochlorite. Hakkaart and Versluijs (1985) rinsed their rhizomes with tap water, dipped them for a few seconds in 70% alcohol and then submerged them for 20 min in 5% calcium hypochlorite. Lin and Monette (1987) stirred rhizome tips vigorously for 20 min in a 0.6% solution of sodium hypochlorite containing 0.1% Tween 20. Pierik et al. (1988) dipped their rhizome segments in 70% alcohol for 2–3 s and then placed them for 20 min in a 1.5%

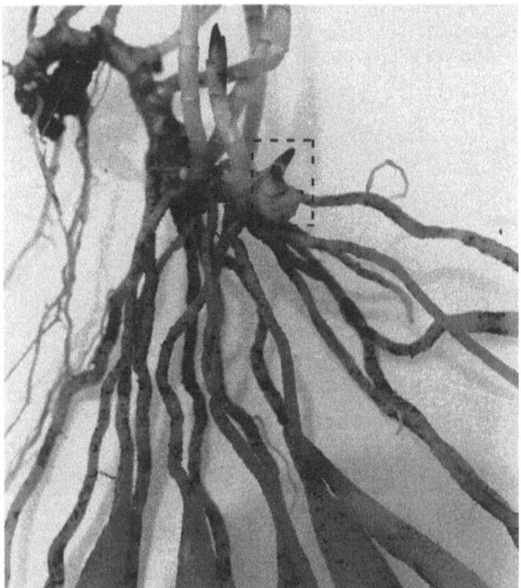

Fig. 3. Rhizome tip of *Alstroemeria* cv. Regina selected as explant for micropropagation. *Dashed line* shows portion to be excised

Table 5. Influence of type of rhizome tip and temperature on the initiation of rhizome tip cultures of *Alstroemeria* Alsaan. [a](Lin and Monette 1987)

Temperature (°C)	Type of rhizome tip		
	1/4 longitudinal section	1/2 longitudinal section	Intact
22	1.0	1.9	2.7
8	2.3	2.8	3.2

SE of each mean = 0.39

Source of variation	Level of significance[b]
Temperature (T)	**
Type of rhizome tip (TRT)	**
T × TRT	NS

[a] Ten replicates (tubes) per treatment combination.
[b] NS, **: NS = not significant, ** = significant at 1% level.

sodium hypochlorite solution containing Tween 20. All groups rinsed their explants with three changes of sterile water. Pierik et al. (1988) then excised apical and axillary tips from their rhizome segments, resterilized them for 10 min in 0.05% sodium hypochlorite, and rinsed them with water prior to culture initiation.

2.2.2 Medium Composition and Culture Conditions

Medium for Culture Initiation and Rhizome Multiplication. The macro- and micronutrients of Murashige and Skoog (MS; 1962) at various strengths have generally been used for in vitro studies with *Alstroemeria.* Ziv et al. (1973) obtained similar results with either MS (1962) or White's (1963) medium. These authors reported that subapical inflorescence stem segments of Ligtu hybrids produced several bud primordia after 6–8 weeks on White's (1963) medium supplemented with $FeSO_4$ and EDTA according to Murashige and Skoog (1962), and containing 0.4 mg/l thiamine, 0.5 mg/l kinetin (6-furfurylaminopurine), 0.1 mg/l napthaleneacetic acid (NAA), and 2% sucrose. These cultures were then transferred to White's (1963) medium lacking growth regulators and several more buds developed. The influence of the cytokinins 6-gamma-gamma-dimethylallyl aminopurine (2-iP) and benzyladenine (BA) on rhizome branching and shoot formation was studied by Gabryszewska and Hempel (1985). Their medium consisted of MS macro- and micronutrients, 40.3 mg/l NaFeEDTA, 100 mg/l meso-inositol, 0.4 mg/l thiamine-HCl and 3% sucrose. The medium, at a pH of 5.6, was solidified with 0.8% agar. The growth of axillary rhizome initials (rhizome branching) was stimulated by either 2-iP or BA concentrations of 0.5, 1, 2, 4, and 8 mg/l. BA had a more pronounced effect than 2-iP (on a mg/l basis) on the development of secondary rhizomes. Shoot formation was also stimulated by either BA or 2-iP. More shoots were produced with BA than with 2-iP, but they were shorter. Increasing the concentration of either cytokinin led to a decrease in shoot length. Optimum results in terms of both rhizome branching and the maximum number of shoots produced were obtained with 2 mg/l BA. A concentration of 8 mg/l BA resulted in rhizomes with very short internodes and fasciated shoots. Hussey et al. (1979) also reported that the production of branched structures was promoted by 1–4µg/ml BA.

Hakkaart and Versluijs (1985) used a medium consisting of MS macronutrients at half-strength, MS micronutrients at full strength, MS organic substances, 0.4 mg/l thiamine and 3% sucrose. Their medium, at a pH of 5.8, was solidified with 0.6% agar. These authors used various combinations of the growth regulators 2-iP (0.01, 1, 2, 5, and 10 mg/l) and indole-3-acetic acid (IAA; 0, 0.1, and 1 mg/l), but found no clear correlation between medium composition and explant performance. The medium used by Lin and Monette (1987) for initiating rhizome tip cultures of Alsaan (Fig. 4A) consisted of MS salts at three-quarter-strength supplemented with 0.22 M D-glucose, 1.48 µM vitamin B1 (thiamine-HCl), 1.43 µM IAA, 9.12 µM zeatin and 2.32 µM kinetin. This medium, at a pH of 5.7, was solidified with 0.7% agar. Their cultures were then transferred (Fig. 4B) to a liquid medium, at a pH of 5, of the same composition. The growth regulators had no clear effect on the cultures and they were omitted from the liquid medium in later experiments, with no adverse effects (Monette and Lin 1991). Pierik et al. (1988) used a basal medium consisting of MS macronutrients at full strength, MS micronutrients (except iron) at full strength, 25 mg/l NaFeEDTA, 3% sucrose, 0.4 mg/l vitamin B_1, and 2–4 mg/l BA. The pH was 6 and agar was added to 0.7%. These authors reported that MS macronutrients at full strength led to a higher rhizome multiplication rate than at half-, three-quarter-, or double strength. Sucrose was preferable to glucose, and the optimal concentration was 3%. No benefit was derived from peptone, tryptone, casein hydrolysate, banana homogenate, or coconut milk at the concentrations tested. NAA or indole-3-butyric acid (IBA) had no stimulatory effect on rhizome multiplication, but BA had a strong influence. BA was more effective than 2-iP, kinetin, or zeatin riboside.

Medium for Rooting. Ziv et al. (1973) reported that the regeneration of root primordia occurred on inflorescence stem sections grown in White's (1963) medium containing 0.05–2 ppm NAA and that the development of these roots took place in medium containing no growth regulator. These authors successfully regenerated plantlets of Ligtu hybrids by subculturing explants onto a medium containing 1.5% sucrose and 0.5 ppm filter-sterilized IAA. Hussey et al. (1979) reported that rooting of rhizome cultures readily took place on media containing 0.03 µg/ml IBA. Hakkaart and Versluijs (1985) obtained rooted plantlets of Rosario by culturing rhizome axillary meristems excized in August in medium containing 0.01 mg/l 2-iP and 1 mg/l IAA. No rooting was observed in this medium with axillary meristems excized in December. These authors had very little success in rooting plantlets of Toledo and Jubilee using medium containing IAA, NAA, or no growth regulator. Gabryszewska and Hempel (1985) reported that roots were produced on rhizomes of Zebra grown in medium containing 1, 2, 4, 8, and 16 mg/l of either NAA or IBA. More roots were obtained with NAA than with IBA, but the roots obtained with 4–16 mg/l NAA were fasciated. The growth of the roots was suppressed by NAA at all concentrations tested and by IBA at a concentration of 16 mg/l. Lin and Monette (1987) attempted rooting of Alsaan rhizome tip cultures

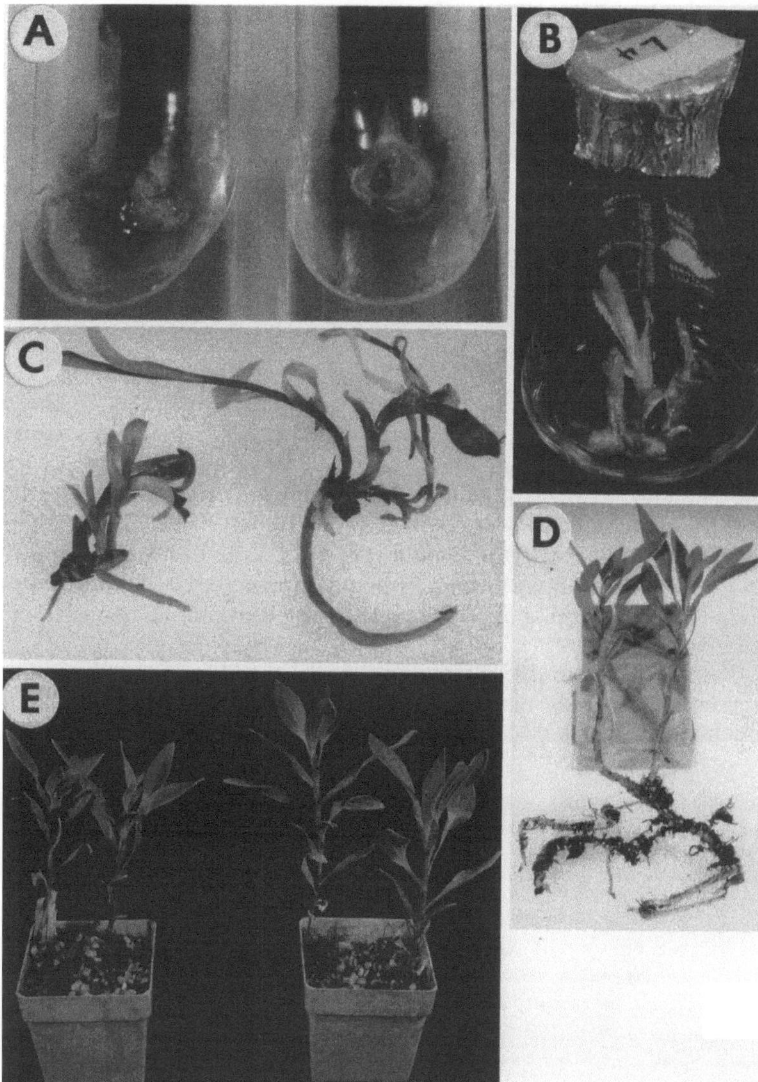

Fig. 4A–E. In vitro propagation of *Alstroemeria* rhizome tips. **A** Rhizome tip cultures showing growth of rhizome and shoot primordia, 4 weeks after initiation into a medium consisting of three-quarter-strength MS salts, 0.22 M D-glucose, 1.48 μM vitamin B1, 1.43 μM IAA, 9.12 μM zeatin, and 2.32 μM kinetin. The medium, pH 5.7, was solidified with 0.7% Difco Bacto agar. **B** Rhizome multiplication and shoot development 2 weeks after transfer of a culture from the initiation medium (8 weeks) to a liquid medium, pH 5, of the same composition. **C** Root development of rhizome tip cultures 4 weeks after transfer into a liquid medium, pH 5, consisting of quarter-strength MS salts, 0.06 M D-glucose, 1.48 μM vitamin B1, and 1.48 μM IBA. **D** Continued root development of plantlets 4 weeks after transfer to a potting mix. **E** Regenerated Alsaan plantlets 8 weeks after being removed from intermittent mist. Survival was 90%. All plants flowered true to type. (Lin and Monette 1987)

(Fig. 4C) in an MS-based medium containing 0, 0.15, 1.48, or 14.75 μM IBA, and found no clear effect of IBA on rooting. They demonstrated that rhizome segments containing a rhizome terminal meristem rooted better than rhizome segments lacking one (Table 6). These authors later determined that some 75% of Alsaan rhizome tip cultures rooted if they were exposed for 1 week to 21.5 μM NAA (Monette and Lin 1991). For rooting Toledo rhizome tip cultures, Pierik et al. (1988) omitted BA from their basal medium, added 0.5 mg/l NAA and increased the sucrose concentration to 5%. They reported that MS salts at full strength were preferable to one quarter- or half-strength, that a pH of 6 was preferable to 5 or 7, that sucrose was preferable to glucose and that NAA was preferable to IBA. The cytokinins kinetin, BA, and 2-iP inhibit rooting and should be omitted from rooting medium (Ziv et al. 1973; Hussey et al. 1979; Gabryszewska and Hempel 1985; Pierik et al. 1988).

Temperature and Light. The micropropagation of *Alstroemeria* has been conducted at temperatures of 17 °C (Hakkaart and Versluijs 1985), 23 °C (Gabryszewska and Hempel 1985; Lin and Monette 1987), 25 °C, and 27 °C (Ziv et al. 1973). Lin and Monette (1987) showed that the quality of the rhizome tip cultures was superior when initiation was carried out at 8 rather than 22 °C (Table 5). Pierik et al. (1988) reported that temperatures of 15–21 °C had no significant influence on the rhizome multiplication rate, while a temperature of 24 °C produced a slight inhibition. The rooting of Toledo cultures was more satisfactory at 21 °C than at 25 or 27 °C.

Lin and Monette (1987) reported that the quality of rhizome tip cultures was superior when these were subjected to a 2-week dark treatment upon culture initiation, as compared to being maintained under 30 μmol/s/m^2 (400–700 nm)

Table 6. Influence of IBA concentration and type of tissue on the mean number of roots produced in rhizome tip cultures of *Alstroemeria* Alsaan. [a](Lin and Monette 1987)

Type of tissue	Mean number of roots per culture			
	IBA concentration (μM)			
	0	0.15	1.48	14.75
Rhizome tip	0.25	0.25	1	1
Large shoot[b]	0	0.25	0	0.25
Small shoot[b]	0.25	0	0	0.25

SE of each mean = 0.28

Source of variation	Level of significance[c]
IBA	NS
Type	*
IBA × type	NS

[a] Four replicates per treatment combination.

[b] Large shoots > 5 cm; small shoots < 5 cm. Both types of shoots were borne on rhizome segments lacking a rhizome apical meristem.

[c] NS, *; NS = not significant, * = significant at 5% level.

with a 16-h photoperiod. The quality of the cultures deteriorated when the dark treatment was longer than 2 weeks. These authors used an illumination of 48 μmol/s/m^2 (400–700 nm) with a 16-h photoperiod for rhizome multiplication and shoot development. Shoot or rhizome development has also been conducted with either 420 or 2000 lx (Ziv et al. 1973), 1200 lx with a 12-h photoperiod (Hakkaart and Versluijs 1985), and 1500–2000 lx with a 16-h photoperiod (Gabryszewska and Hempel 1985). Pierik et al. (1988) reported that the rhizome multiplication rate was not significantly different whether the photoperiod was 8 or 16 h. Irradiance within the range of 1.5–9.7 W/m^2 had no significant influence on the rhizome multiplication rate, although shoot length was decreased at 6 W/m^2, in comparison with 1.5 W/m^2. These authors also reported that the rooting of Toledo was promoted by a photoperiod of 16 h, in comparison to 8 h. The optimal irradiance for rooting Toledo was 7 W/m^2, with a strong inhibition evident at 1.2–2.2 or 15.4 W/m^2. Pierik et al. (1988) reported that a 1–3-week dark treatment inhibited the rooting of Toledo. Monette and Lin (1991) incorporated a 2-week dark treatment at 8 °C for the rooting of Alsaan, but did not determine whether the dark treatment was essential for rooting success with this cultivar.

Other Factors. Lin and Monette (1987) used a liquid medium for the rhizome tip culture of Alsaan. Cultures tended to become vitrified (unpubl.), but this did not pose a problem either during micropropagation in vitro or during the acclimatization of regenerated plantlets. Pierik et al. (1988), by contrast, reported that the vitrification of Toledo obtained upon repeated subculture in liquid was very detrimental. Although the use of a liquid medium resulted in a higher rhizome multiplication rate than that obtained with a solid medium, they recommended the addition of 0.7% agar to the medium. These authors also reported that a reculture interval of 3 weeks was preferable to 3 3/4, 5, or 7 1/2 weeks. Monette and Lin (1991) transferred their rhizome tip cultures to fresh media every 2 weeks. Pierik et al. (1988) found that the number of buds/shoots per explant had a strong influence on the rhizome multiplication rate and that the best results were obtained with explants with only one bud/shoot. These authors routinely decapitated the shoots on their explants to a length of 0.5 cm upon subculture. This practice was also followed in the author's laboratory, as it appeared to stimulate rhizome multiplication (unpubl.).

2.2.3 Transfer of Plantlets from in Vitro Conditions

Hakkaart and Versluijs (1985) reported that rooted plantlets regenerated from rhizome axillary meristems and transferred to steam-sterilized soil survived and developed into normal plantlets only if they formed rhizomes. In plants which did not form rhizomes no new stems were produced; the existing stem elongated, ceased growth, and eventually died. These authors could not explain why some of their rhizome axillary meristem cultures survived transfer to soil while others did not, but commented on a possible relationship between survival and the position of the axillary meristem on the rhizome. Lin and Monette (1987) reported that 90% (45 out of 50) of their rooted rhizome tip cultures survived the transfer from in vitro conditions (Fig. 4D, E). They transferred cultures with one or two roots 1–2 cm long into 5-cm square plastic pots filled with peat:vermiculite:perlite (2:1:1 by vol) and placed them under intermittent mist in the greenhouse. Two to 4 weeks later the plantlets were moved onto a bench in the greenhouse (16/13 °C, day/night). Liquid fertilizer (20-20-20 at 300 mg/l N) was applied when irrigation was needed. Pierik et al. (1988) transplanted their rooted explants to soil and transferred them to a greenhouse where they were shaded and grown under plastic cover. During the first weeks in soil, the plants were sprayed periodically with 0.2% Previcur-N (Schering).

2.2.4 Genetic Variability

Lin and Monette (1987) reported that the 45 Alsaan plantlets regenerated from in vitro rhizome tip cultures and successfully transferred to the greenhouse all developed into true-to-type flowering plants. Pierik et al. (1988) reported that all their in vitro produced plants flowered normally. Spontaneous mutations are a rare event with *Alstroemeria* (Verboom 1980) and the information available to date on micropropagation by direct organ regeneration from stem explants or rhizome tip cultures suggests that these procedures do not *per se* stimulate the production of mutants for directly visible characteristics.

3 Prospects

New cultivars are now being developed through somatic embryogenesis (Mark Bridgen, pers. commun.). An even greater diversity of phenotypes should be achieved by combining either somatic embryogenesis or in vitro rhizome tip culture with a mutagenic treatment, such as irradiation of the cultures or exposure in vitro to chemical mutagens. The use of colchicine to produce polyploids is particularly appealing, as we know that several commercially valuable cultivars are polyploid. Alternatively, chemicals such as 5-bromouracil, 2-aminopurine, nitrous acid, hydroxylamine, or the acridine dyes could be added to the medium in which the *Alstroemeria* cultures are growing. Evaluation of the mutants for flower size and color and a number of other traits would necessitate regenerating plantlets and bringing a crop to harvest. It should be possible, however, to evaluate certain characteristics in vitro.

4 Protocol

The following protocol is suggested for *Alstroemeria* micropropagation by rhizome tip culture: *Initiation.* Rhizome tips 5–10 mm long are cut from actively growing rhizomes in the spring. They are washed free of soil in running tap water for 30 min and then stirred vigorously for 20 min in a 0.6% solution of sodium hypochlorite containing 0.1% Tween 20. They are rinsed three times in sterile distilled water and 1 mm is trimmed from the cut end of the rhizome tips. These are then inserted into an initiation medium consisting of full strength Murashige and Skoog (MS; 1962) salts (except iron),

25 mg/l NaFeEDTA, 3% sucrose, 2 mg/l BA, 0.4 mg/l vitamin B_1, and 0.7% agar, at a pH of 5.7. The rhizome tip cultures are placed in the dark at 8 °C for 2 weeks, after which they are transferred to a 16-h photoperiod under 30 μmol/s/m² (400–700 nm) and a temperature of 21 °C.

Rhizome Multiplication. Four to 6 weeks after initiation, the cultures are transferred to a liquid medium, pH 5, containing the same components as above. The cultures are placed on a tilter apparatus and maintained at 21 °C with a 16-h photoperiod under 48 μmol/s/m². They are transferred to fresh medium every 2 weeks and they are divided as the lateral rhizomes start growing, each division containing one growing rhizome tip. Shoots are trimmed to 5 mm at the time of transfer.

Rooting. The cultures are transferred to a liquid medium, pH 6.0, consisting of three-quarter-strength MS salts, 1.48 μM vitamin B_1, and 21.5 μM NAA and kept at 21 °C under 30 μmol/s/m² with a 16-h photoperiod. One to 2 weeks later, the cultures are transferred to agar-solidified growth regulator-free medium and kept at 8 °C in the dark for 2 weeks, after which they are returned to a 16-h photoperiod at 21 °C.

Transfer to Soil. When roots 10 to 20 mm long are present in the cultures, these are transferred to a potting mix of peat:vermiculite:perlite (2:1:1 by vol) and placed under intermittent mist in the greenhouse (16/13 °C, day/night) and fertilized with 20:20:20 (300 ppm N) as they become dry.

Acknowledgements. I am very grateful to W.C. Lin for helpful discussions and to D. Elvedahl for typing this manuscript.

References

Anonymous (1983) The culture of Vasa Alstroemeria. Van Staaveren/Alsmeer Bull No. 633, 16 pp

Baker JG (1888) *Alstroemeria.* In: Baker JG (ed) Handbook of Amaryllidaceae. George Bell & Sons, London, pp 132–142

Bridgen MP (1986) Somatic embryogenesis and organogenesis of *Alstroemeria.* HortScience 21(3): 858 (Abstr 1448)

Broertjes C, Verboom H (1974) Mutation breeding of *Alstroemeria.* Euphytica 23: 39–44

Brunt AA, Phillips S (1981) *Alstroemeria.* Rep Glasshouse Crops Res Inst for 1979, p 151

Coats AM (1969) The quest for plants. Studio Vista Limited, London

Darlington CD, Wylie AP (1955) Chromosome atlas of flowering plants. Allen and Unwin, London

Gabryszewska E, Hempel M (1985) The influence of cytokinins and auxins on *Alstroemeria* in tissue culture. Acta Hortic 167: 295–300

Goemans JAM (1962) Breeding of alstroemerias. J R Hortic Soc 87: 282–284

Hakkaart FA, Versluijs JMA (1985) Viruses of *Alstroemeria* and preliminary results of meristem tip culture. Acta Hortic 164: 71–75

Healy WE, Wilkins HF (1981a) Alstroemerias show promise as energy-efficient crop. Flor Rev 169: 16, 40–45

Healy WE, Wilkins HF (1981b) *Alstroemeria* culture. Minn State Flor Bull 30(5): 5–7

Healy WE, Wilkins HF (1986) Relationship between rhizome temperatures and shoot temperatures for floral initiation and cut flower production of *Alstroemeria* Regina. J Am Soc Hortic Sci 111(1):94–97

Healy WE, Wilkins HF, Celusta M (1982) The role of light quality, photoperiod, and high-intensity supplemental lighting on flowering of *Alstroemeria* Regina. J Am Soc Hortic Sci 107(6): 1046–1049

Heins RD, Wilkins HF (1979) Effect of soil temperature and photoperiod on vegetative and reproductive growth of *Alstroemeria* Regina. J Am Soc Hortic Sci 104(3): 359–365

Hughes J, Tsujita MJ, Blom TJ, Brown WW (1986) Alstroemeria. Canadian Florist, Greenhouse & Nursery 81(7): 4, 38–41

Hussey G, Hilton J, Lumsden P (1979) In vitro propagation of *Alstroemeria.* John Innes Inst 70th Annu Rep, pp 56–57

King JJ, Bridgen MP (1987) In vitro induction of axillary and adventitious shoots in *Alstroemeria.* HortScience 22(5): 1149 (Abstr 799)

Lin WC (1985) Influence of soil cooling and high intensity lighting on the growth and flowering of *Alstroemeria* Regina. HortScience 20(3): 378–380
Lin WC, Molnar JM (1983) *Alstroemeria* Alsaan. Can J Plant Sci 63: 565–566
Lin WC, Monette PL (1987) In vitro propagation of *Alstroemeria* Alsaan. Plant Cell Tissue Organ Cult 9: 29–35
Linsmaier EM, Skoog F (1965) Organic growth factor requirements of tobacco tissue cultures. Physiol Plant 18: 100–127
Maat DZ (1977) *Alstroemeria.* Annu Rep Res Inst Plant Prot, Wageningen 1976: 63
Maat DZ (1983) *Alstroemeria.* Annu Rep Res Inst Plant Prot, Wageningen 1982: 80
Monette PL, Lin WC (1991) Rooting of *Alstroemeria* Alsaan rhizome tip cultures. North Am Plant Prop 3(2): 15–16.
Murashige T, Skoog F (1962) A revised medium for rapid growth and bioassays with tobacco tissue cultures. Physiol Plant 15: 473–497
Oosthoek H (1967) *Alstroemeria*-hybriden aantrekkelijk in meer dan een opzicht. Vakbl Bloem 22: 1572–1573
Perry F (1972) Flowers of the world. The Hamlyn Publishing Group Ltd, Toronto
Phillips S, Brunt AA (1983) *Alstroemeria.* Rep Glasshouse Crops Res Inst for 1981, p 142
Phillips S, Brunt AA (1986) Four viruses of *Alstroemeria* in Britain. Acta Hortic 177: 227–233
Pierik RLM, van Voorst A, Booy G, van Acker CAM, Lelivelt CLC, de Wit JC (1988) Vegetative propagation of *Alstroemeria* hybrids in vitro. Acta Hortic 226: 81–89
Powell MC, Bunt AC (1986) The effects of propagation date on flower production in *Alstroemeria* Campfire and Red Sunset. Sci Hortic 28: 147–157
Quak F (1974) *Alstroemeria.* Annu Rep Res Inst Plant Prot, Wageningen 1973: 52
Robinson GW (1963) *Alstroemeria.* J R Hortic Soc 88: 490–494
Tsuchiya T, Hang A (1987) Chromosome studies in genus *Alstroemeria.* Acta Hortic 205: 281–287
Uphof OC (1952) A review of the genus *Alstroemeria.* Plant Life (Herbertia Edition) 8: 36–53
Van Hoof H (1971) Alstroemeria. Rep Res Inst Plant Prot for 1970, p 82
Verboom H (1980) Alstroemerias and some other flower crops for the future. Sci Hortic 31: 33–42
Vonk Noordegraaf C (1975) Temperature and daylength requirements of *Alstroemeria.* Acta Hortic 51: 267–274
White PR (1963) The cultivation of animal and plant cells, 2nd edn. Roland Press, New York
Wilkins HF (1985) The history of *Alstroemeria* production in North America. Herbertia 41: 15–19
Ziv M, Kanterovitz R, Halevy AH (1973) Vegetative propagation of *Alstroemeria* in vitro. Sci Hortic 1: 271–277

II Micropropagation of Snapdragon (*Antirrhinum majus* L.)

H.J. Newbury, E.A.B. Aitken, N.J. Atkinson, and B.V. Ford-Lloyd[1]

1 General Account

1.1 Botanical Aspects

The genus *Antirrhinum* L. (Scrophulariaceae) includes 23 species with the majority occurring in the Old World, and often being narrowly endemic to the Iberian Peninsula. The wild species grow as dwarf shrubs or perennial herbs, although the cultivated forms, confined mainly to the species *A. majus* (the garden snapdragon) are grown as annuals (Fig. 1). Of particular economic importance are the flowers (Fig. 1B) which are zygomorphic, occurring in terminal racemes or solitary in the leaf axils, and are conspicuous because of their variable colours. Little is known of the cytology except that *A. majus* has a somatic (diploid) chromosome number of 16. The distinction of some of the Mediterranean endemics has become difficult because of hybridisation and introgression with a few wider-ranging species, some of these having been dispersed outside their original range of cultivation for ornament and subsequent naturalisation (Rothmaler 1956). *A. majus* suffers from a variety of microbial pathogens. Seeds can be infected internally with several fungi, such as *Alternaria alternata*, *Phyllosticta antirrhini* and *Stemphyllium botryosum* (Rogers 1980). Seedlings and older plants are susceptible to pathogens such as botrytis blight (Rogers 1980), mildews (*Perenospora antirrhini*; Paulus et al. 1984) and rust (*Puccinia antirrhini*; Gawthrop and Jones, 1981). There have also been reports of infection by *Alternaria* and *Helminthosporium* species (Engelhard 1971), *Verticillium albo-atrum* (Dutta 1982; Junge 1982), *Pythium ultimum* (Paulus et al. 1984; Pittis and Colhoun 1984; Popple and Rigby 1985), *Rhizoctonia solani* (Yang and Leu 1981), snapdragon leafspot (*Cercospora antirrhina*, Porter and Aycock 1967) and diseases caused by the bacteria *Xanthomonas campestris* pv. *vitians* (Tsuchiya et al. 1981) and *Pseudomonas antirrhini* (Simpson et al. 1971). A review of *Antirrhinum* diseases has been prepared by Judd (1975).

1.2 Importance of this Species

A. majus has found widespread use as an annual bedding plant, although it can survive moderate winter conditions and flower again in the following season. It has

[1] School of Biological Sciences, University of Birmingham, P.O. Box 363, Birmingham B15 2TT, United Kingdom

Biotechnology in Agriculture and Forestry, Vol. 20
High-Tech and Micropropagation IV (ed. by Y.P.S. Bajaj)

Fig. 1. A A commercial variety of *Antirrhinum majus* grown as a bedding plant. Although varieties may vary in their habit, most possess several upright stems that bear spikes of up to a dozen flowers. **B** Flowers of *A. majus*

no special requirements for its growth, although it will not thrive in shaded locations; it is one of the bedding plants recommended for coastal gardens since it is reasonably tolerant to salt. In Britain, the flowering season is quite extended, running from July to October; individuals will continue to produce flowering spikes if seed heads are removed. The prolific production of white, yellow, pink or red flowers makes *Antirrhinum* an attractive plant, although it does not now appear to be as fashionable as was once the case. Recent developments have led to the introduction of some varieties with altered floral morphology. Some varieties are less obviously bilaterally symmetrical, with a more open corolla tube; others possess an extra whorl of petalloid appendages between the corolla and the stamens. Attempts have recently been made to use *A. majus* as an ornamental pot plant; several groups have succeeded in stunting the growth of this species (normal height 25–120 cm) using growth retardants (Munsi et al. 1980; Shanks 1982; Murray et al. 1986; Shedeed et al. 1986; Wainwright and Irwin 1987) and using appropriate treatments more compact plants with increased numbers of inflorescences and numbers of florets per inflorescence, have been reported.

In some parts of the world, including the USA, *A. majus* represents an important source of cut flowers. The cultivars first grown in the USA flowered during long days in the field or glasshouse, but in the mid-1920s a grower in Ohio introduced a cultivar, Cheviot Maid, that flowered during the winter (Lindstrom 1966). Later, F_1 hybrid greenhouse-forcing types were introduced, combining winter flowering characteristics with the strong vegetative growth of F_1 hybrid plants (Ball 1952; Rogers 1980). Greenhouse-forcing types are now classified into four response groups based upon growth and flowering response in relation to temperature and daylength (Rogers 1980). Supplementary lighting has been shown to increase flower production in glasshouses in winter by 63% (Stefanis and Langhans 1982). Freshly cut flower spikes usually have a vase life of about 1 week, but substantially longer life, greater number of florets opening and greater increase in spike length can be achieved (Marousky and Raulston 1970; Johnson 1972; Nowak 1981).

Quite apart from its importance in the horticultural industry, *A. majus* is of particular significance in the field of classical, and more recently molecular, genetics. The characterisation and isolation of transposons from this species provides an opportunity to isolate genetic loci of commercial interest. This species is therefore of great significance to the plant biotechnologist. *A. majus* has been used in many classical genetic experiments. Much of the genetical research has been on the biosynthetic pathway of the pigment anthocyanin (Fincham 1987; Martin ct al. 1987a). The study of the anthocyanin pathway in *A. majus* is simpler than for many other species. In *A. majus* relatively few steps, about seven are involved in the production of the red and pink flower colours; in many other plants various modifications of anthocyanin by additional hydroxylation, glycosylation and methylation are required for the production of blue and purple flowers (Martin et al. 1987a). The study of genes involved in the biosynthesis of anthocyanin has increased our understanding of the mechanism by which the genes regulate development (Martin et al. 1987a).

In some lines of *A. majus* the genes do not always appear to segregate in a normal Mendelian fashion. The earliest references to such unstable genes were by de Vries (1910). He noted that self-pollination of variegated forms of *A. majus*, which had flowers with red spots and stripes on an ivory background, produced progeny which were mostly of the variegated form, except for 2–30%, which were of the full-red flower type. Self-pollination of the full-red type produced progeny with three full-red types to one variegated type, as reviewed by Coen and Carpenter (1986). It was not until after the pioneering work by McClintock (1952) on maize that it was established that transposable elements were responsible for the instability of such genes in variegated forms.

The characteristics of transposable elements have been reviewed by several authors including Coen and Carpenter (1986), Saedler et al. (1987) and Peterson (1987). In brief, transposable elements are relatively short segments of DNA (3.5 to 17 kbp in *A. majus*) which can move (transpose) around the genome both in germinal and somatic tissue. Their insertion into and excision from genes is dependent on the presence of transposase, which in autonomous elements is coded for by the element itself. The insertion of a transposon into a gene will usually inactivate that gene, but following excision of the element, the gene may revert to its original active form. If, however, the excision of the element is not precise, the gene may be altered permanently. In somatic tissue, insertion into and excision out of pigment genes gives rise to the variegated flower colour types noted in some of the lines of *A. majus*. In germinal tissue, transposition into or out of a gene leads to apparent deviations from the expected Mendelian ratios of flower colour in progeny populations.

The characterisation and isolation of the *Antirrhinum* transposon Tam3 have allowed this element to be used in a tagging protocol designed to clone genetic loci that cannot be isolated using more conventional technology. The only other plant species that contains transposons sufficiently well characterised to allow this approach to be used at present is maize, where, for example, the bronze locus has been isolated (Fedoroff et al. 1984). In *A. majus*, the transposon tagging technique is now available for the isolation and characterisation of genes of commercial interest (Coen and Carpenter 1986); this method has already been used for the mutagenesis of a range of loci concerned with floral and foliar development including pallida, floricaula, olive, deficiens and delila (Martin et al. 1985; Luo et al. 1991; Carpenter and Coen 1990; Sommer et al. 1990) and is now being applied to a rust resistance locus in our laboratory (Newbury 1992).

Although, *A. majus* may not be considered of sufficient commercial interest to warrant the effort involved in gene isolation using transposons, such work is taking place for two reasons. First, techniques pioneered with this species may well become available for major crops in the future. Secondly, the genes isolated from *A. majus* should possess similar structures (i.e. have similar base sequences) to homologous genes in crop species. This will allow the isolation of these genes from the latter using well established DNA hybridisation techniques.

1.3 Conventional Propagation and Breeding

Antirrhinum majus is normally sold as seed, or as small established plants in trays ready for planting for bedding purposes. It is possible to produce rooted plantlets simply by immersing the bases of small (~ 5 cm) shoots in water. Roots usually form after a few weeks, but this practice is not normally adopted. *A. majus* possesses a number of characteristics which make it amenable to both genetic research and breeding programmes (Harte 1974). It has a short life cycle (two generations per year under controlled conditions), and a low number of relatively large floral parts. It can be readily self-pollinated or bud-emasculated prior to cross-pollination, and the organisation of the flowers into spikes is convenient for bagging. Each fruit (capsule) can contain 200 seeds, so that high progeny numbers are possible. One slight disadvantage is the small size of the seeds; these produce rather fragile seedlings which require transplanting from the original seed tray.

Commercial breeders have understandably concentrated upon the production of plants with attractive floral characteristics (colour, number and, recently, shape) along with an acceptable growth habit. Attempts have been made to incorporate rust resistance into some commercial varieties. Results in our laboratory (Aitken et al. 1989) have revealed that *Antirrhinum* rust currently comprises two races, and that a single dominant nuclear gene found in some varieties confers resistance to one race, whilst no resistance has been discovered for the latter.

1.4 Need for Micropropagation

Because of its relatively short generation time, and because each capsule on a plant can contain about 200 seeds, the advantage offered by vegetative multiplication techniques, in terms of speed, will never be as great as those encountered with many slow growing perennial plants. However, vegetative cloning avoids the possibility of rearrangement of genetic material consequent upon sexual reproduction, so that in vitro micropropagation techniques do offer the opportunity of rapidly cloning a selected genotype. Such a genotype might carry an advantageous horticultural character, such as a floral feature, disease resistance etc. In relation to the molecular genetic studies outlined above, it may be important to propagate a candidate plant in which a transposon has inactivated a gene of interest.

2 Previous Tissue Culture Work

Most previous tissue culture work carried out on *A. majus* has not been directed towards clonal propagation, but has involved a phase of dedifferentiation of the explant tissue. Plants produced by methods involving a callus phase may exhibit somaclonal variation so that one cannot ensure authentic clone production; however, such variation can be extremely useful in the production of differing forms of a plant which may be of commercial significance. The following review has been organised into sections related to the culture of different cell and organ types.

2.1 Protoplast Isolation and Culture

A number of studies have been carried out on both the isolation and culture of *A. majus* protoplasts. Aidid and Thain (1980) explored the effects of protoplast isolation techniques on the ionic relations of the mesophyll cells of *A. majus* cv. White Rocket. They found that much of the K^+ and Cl^- was lost during extraction, and suggested that an analysis of the ion content of intact *A. majus* cells would be a worthwhile guide in determining a protoplast culture medium with the correct level of salts to maintain a relatively normal ionic environment within protoplasts. Morris et al. (1981) examined the effects of the enzymes used in protoplast production on *A. majus* membranes and found that they caused depolarisation of the membrane potential even when they had been desalted or dialysed. The effects, however, were reversible if the enzymes were used in low concentrations, or at high concentrations for short periods.

Two independent accounts of the culture of isolated *A. majus* protoplasts have been published; however, regeneration of whole plants from protoplasts has not been reported. Poirier-Hammon et al. (1974) isolated protoplasts from mature leaves of *A. majus* cv. Kimosy Blanc. The protoplasts regenerated cell walls after 3 days and divided by the 5th day in a suitable hormonal environment, and by 2 weeks small colonies had formed. A separate study by Binding et al. (1981) on *A. majus* (cv. not stated) yielded different results and indicated that the performance of protoplasts in culture is dependent on the source material used. Experiments were carried out on apex and leaf material of outdoor-grown, glasshouse-grown or tissue cultured plants. Protoplasts of mature leaves did not regenerate cell walls and protoplasts of leaves of culture-grown shoots regenerated a thin cell wall and exhibited frequent budding. Protoplasts derived from the apices of culture-grown shoots, however, produced a rigid cell wall and proliferated to give highly differentiated cells. Further culture on low osmotic media resulted in sustained proliferation to give callus which later yielded embryoid-like structures. Differences between these reports may be at least partly attributable to a difference in the cultivars of *A. majus* used; however, the data of Binding et al. (1981) show that *A. majus* protoplasts isolated from apical regions and from young tissue cultured material were superior to those produced from the mesophyll region of mature leaves.

Sub-protoplasts have been reported in a study of cultured *A. majus* pollen tubes (Cheng et al. 1984). It was found that nucleate and enucleate sub-protoplasts behaved in the same way, suggesting the presence of presynthesised mRNA in the germinating tubes.

2.2 Root Culture

Limited work has been carried out on both the nutritional requirements for the growth of differentiated *A. majus* roots in culture and on the culture of roots for callus production and plant regeneration. Harte et al. (1977) isolated roots of *A. majus* line Sippe 50 and determined that for maintenance in culture, the inorganic nutrient solution has to be supplemented with sucrose as the preferred sugar. The addition of a small amount of glucose and casein was also necessary. Plants have been successfully regenerated from *A. majus* root cultures. Seedlings of cv. Kimosy White were germinated in sterile conditions and after 24 days roots were excised. Culture on a range of different liquid and solidified media induced callus formation, and treatment with a number of hormonal combinations led to shoot production (Pfister and Widholm 1984).

2.3 Leaf Disc Culture

Two varieties have been used in separate studies of *A. majus* leaf disc culture. Rao and Harada (1973) used cv. Kimosy White, whilst Rao et al. (1976) used cv. Red Majestic Chief. Similar results were obtained for both varieties. Limited "embryo" and "embryoid" production was reported in appropriate hormonal environments.

2.4 Embryo Sac Isolation

Techniques for the isolation of viable *A. majus* embryo sacs have been developed using a combination of physical and enzymatic separation methods (Zhou 1985; Zhou and Yang 1985). Successful culture of these embryo sac has not been reported, but if this can be achieved it could prove useful for the genetic manipulation of this species.

2.5 Stem and Hypocotyl Segment Culture

The formation of callus from stem sections of *A. majus* was first reported by Melchers and Bergman (1959). Subsequent reports on the culture of mature stem segments of cv. Kimosy White on agar-solidified media (Rao and Harada 1973;

Poirier-Hamon et al. 1974; Sangwan and Harada 1975) showed callus production, moderate or profuse root production, and the formation of embryoids. Similar results were obtained by Rao et al. (1976) using mature stem segments of cv. Red Majestic Chief; these authors also found that mature stem segment-derived *A. majus* callus possessed a decreased ability to grow, or to produce either roots or embryoids, after it had been subjected to 2–8 kR gamma radiation. Ten to 20 kR treatments entirely suspended growth, whilst above 20 kR was lethal. Both Harada et al. (1975) and Rao et al. (1976) have reported that callus growth and embryogenesis occurred in appropriate conditions in liquid media.

Initially, attempts to regenerate plants from these embryoids failed. However, the production of whole plants from *A. majus* mature stem segment callus was finally achieved using cv. Kimosy Blanc when small globular embryos produced on either solid or liquid media were transferred to a sugar-rich, agar-solidified medium supplemented with 10% coconut milk (Sangwan and Harada 1975). As a consequence of these results, Rao et al. (1976) attempted to regenerate plants from Red Majestic Chief embryos produced in liquid culture in the presence of IAA, NAA or 2,4-D by transferring them to agar-solidified medium supplemented with coconut milk, casein hydrolysate or malt extract; in no case was plant regeneration achieved. A later attempt to regenerate plants from mature stem segments of cv. Kimosy White also failed (Pfister and Widholm 1984); neither could regeneration be achieved using callus of cvs. Rocket Golden or Potomac Yellow.

A key factor in the regeneration of plants from *A. majus* stem-derived callus appears to be the age of the stem tissue used as explant material. When 24 day-old hypocotyl segments of culture-grown seedlings of cv. Kimosy White were cultured, callus was produced profusely and shoots were regenerated at high frequencies on appropriate media. In our laboratory we have found that regeneration from mature stem-derived callus is extremely poor. Callus derived from seedling shoot tips can be induced to form new shoots in six of seven varieties tested (see Sect. 3.1 below). However, regeneration was achieved in all seven varieties when callus was produced from hypocotyl segments. In this case, five of the varieties produced shoots directly, one (Yellow Freedom) produced leaves from the petioles of which new shoots emerged, and one (White Monarch) regenerated plants chiefly through the production of embryoids.

Recently, attempts have been made to genetically engineer *A. majus* using *Agrobacterium tumefaciens* as a gene vector. Initial experiments have assessed the efficiencies of different strains of *A. tumefaciens*, and the effect of added acetosyringone, on the transfer of wild-type T-DNA to hypocotyl explants (Godwin et al. 1991). Further work is now being carried out in our laboratory to develop a transformation system using disarmed forms of *A. tumefaciens* to transfer genes to *A. majus* hypocotyl cells; these will then be induced to form callus from which shoots will be regenerated.

3 Micropropagation

Three sets of experiments have been carried out to investigate the ease with which cultivars of *A. majus* can be micropropagated. The major differences between these experiments were in the ages of the shoot tip explants; considerable differences were noted in the reactions of shoot tips of different ages to culture conditions.

3.1 Culture of Seedling Shoot Tips

In order to determine the rate of shoot production using very young shoot tips as explants, *A. majus* seeds of different varieties were germinated in vitro and the seedling shoot tips excised. Seeds were first surface sterilised using 70% (v/v) ethanol for 15 s, followed by treatment with 2.5% (w/v) hypochlorite for 5 min, and finally washed three times with sterile distilled water. Batches of ten seeds were then placed in 25 × 150 mm Pyrex test tubes in which were placed 15-ml aliquots of

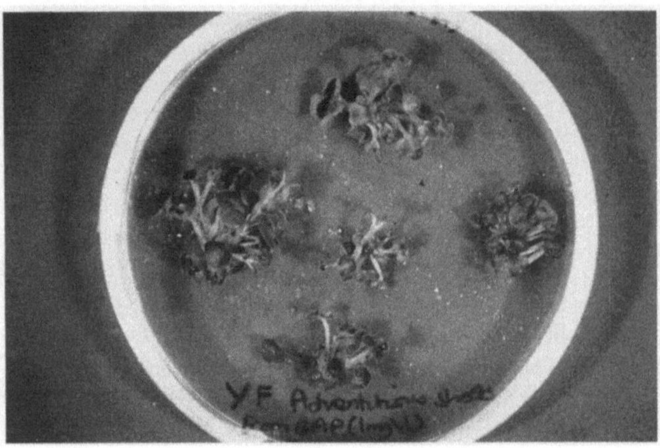

Fig. 2. Shoot tip cultures of the *A. majus* variety Yellow Freedom during culture on MS medium containing 3% sucrose, 0.7% agar and 1 mg/l BAP. Callus growth has occurred at the bases of the shoot tips and some of the new shoots are emerging from this.

Table 1. Fortyeight-day-old tips of seven varieties were cultured for 6 weeks on medium containing 1 mg/l BAP (MS + 3% sucrose, 0.7% agar, pH 5.7). Results are the means of eight replicates for each variety. (Atkinson et al. 1989)

Variety	Mean no. of shoots/explant	Diam. (mm) of basal callus
Line 75	3.0 ± 1.1	8.5
Yellow Freedom	14.0 ± 2.3	8.2
Victory	2.1 ± 1.6	7.8
Black Prince	1.6 ± 0.6	5.3
Little Darling	4.4 ± 2.6	3.5
Coronet	5.0 ± 2.3	7.0
White Monarch	0	5.5

Murashige and Skoog (1962) medium (MS) containing 0.7% (w/v) agar and 1% (w/v) sucrose. The tubes were incubated at 27 °C with a 16-h daylength, and after 48 days the shoot tip was removed. The tips were then placed further in Pyrex tubes containing 15 ml of MS medium with 0.7% (w/v) agar, 3% (w/v) sucrose and 1 mg/l BAP so that 5 mm of the base of the high shoot penetrated the medium. Following 6 weeks' incubation at 25 °C in a 16-h daylength, significant callus growth had occurred at the bases of the shoot tips, and in six of the seven varieties tested adventitious shoots were emerging from this basal callus (Table 1, Fig. 2).

Although continued production of shoots could be maintained by subdivision of the shooting clump and transfer to fresh medium in separate tubes, this method may be regarded as generally unsuitable for clonal plant production, since shoots were produced from basal callus and could be subject to somaclonal variation.

3.2 Culture of Shoot Tips from in Vitro-Grown Plantlets

Seeds of a number of varieties of *A. majus* were germinated on agar as above, and the seedlings maintained in culture for several months. Shoot tips were then removed and placed on one of two media, incubated for 5 weeks, and the size of callus formed and the numbers of axillary and adventitious shoots were recorded. Callus grew at the shoot tip bases on both media and the results (Table 2) show that large numbers of shoots were formed on medium 2; however, these were almost all derived from the callus rather than directly from the organised tissue of the initial explant.

It can be seen that material which is several months old produces much less basal callus and fewer adventitious shoots than 48-day-old material cultured on the same medium, but the level is still unacceptable compared to the number of axillary shoots produced, from the point of view of potential genetic instability of the regenerants.

Table 2. Shoot tips were removed from plants grown for 6–7 months in culture and placed on one of two media. Medium 1 consists of MS medium containing 3% sucrose and 1 mg/l BAP, solidified with 0.7% agar; medium 2 was Murashige Shoot Multiplication Medium (MSMM) obtained from Imperial Laboratories, solidified with 0.7% agar. Each result is the mean of four replicates. Standard errors are given (Atkinson 1990).

Variety	Medium 1			Medium 2		
	No. of shoots		Callus	No. of shoots		Callus
	Axillary	Advent.	diam. mm	Axillary	Advent.	diam. mm
Yellow Freedom	0.4 ± 0.2	1.2 ± 1.1	2.0	0.0	10.25 ± 2.6	1.0
Line 75	0.0	0.0	2.5	0.0	8.0 ± 2.9	2.0
Victory	0.0	0.0	9.0	1.0 ± 0.8	14.67 ± 1.5	1.8
Coronet	0.67 ± 0.3	0.17 ± 0.15	3.8	0.0	4.0 ± 2.9	1.9
Black Prince	1.5 ± 1.0	0.0	7.0	0.0	13.5 ± 5.5	1.2
Little Darling	0.0	0.0	8.5	0.0	40.3 ± 3.3	2.0

3.3 Culture of Shoot Tips from Glasshouse-Grown Plants

Plants of several varieties of *A. majus* were raised under glasshouse conditions and shoot tips were removed when the plants were 70–150 mm tall. Tips (2–3 mm long) were surface sterilised for 15 min in 10% (w/v) hypochlorite containing 0.1% Tween 20, washed three times in sterile distilled water, and placed in MS medium containing 3% (w/v) sucrose, 0.8% (w/v) agar and growth regulators as specified, in 180 × 21 mm tubes with closures. The cultures were maintained at 28 °C under white fluorescent tubes with a light intensity of 130 μmol/m/s and a 16-h daylength.

An initial experiment using the single variety Trumpet Serenade indicated that shoot production could be induced using a range of different plant growth regulator regimes within 22 days (Table 3; Newbury 1986) The multiplication rate was significantly higher ($P = 0.05$) in the two treatments that employed BAP alone than in the two treatments using kinetin. The majority of new shoots formed in each case were clearly from pre-existing axillary buds.

In another experiment (Table 3; Newbury 1986) shoot tips of other varieties of *A. majus* were cultured on the three media that had shown the fastest rate of shoot production with variety Trumpet Serenade (i.e., not including media containing kinetin). Analysis of the results (Table 3) showed that variation in the numbers of shoots produced due to the use of different plant growth regulator concentrations were not significant. However, explants of variety Yellow Freedom produced

Table 3. The effect of different media on the multiplication rate of shoots of different varieties of *Antirrhinum majus*. Shoot tips (at least 12 replicates of each) were cultured for 22 days in media containing: 1, 0.5 mg/l BAP; 2, 1 mg/l BAP; 3, 0.5 mg/l kinetin; 4, 1 mg/l kinetin; 5, 0.5 mg/l BAP and 0.1 mg/l IBA. (Newbury 1986)

Variety	Numbers of transferable shoots on media				
	1	2	3	4	5
Trumpet Serenade	5.45 ± 2.62	5.91 ± 2.21	3.27 ± 1.85	3.91 ± 1.51	5.00 ± 1.79
Victory	1.71 ± 0.76	2.14 ± 0.69	–	–	2.71 ± 1.38
Wisley Golden Fleece	2.20 ± 1.30	3.83 ± 0.75	–	–	3.83 ± 0.98
Yellow Freedom	4.13 ± 0.83	4.86 ± 2.79	–	–	4.14 ± 1.95

Table 4. Numbers of transferable shoots produced by a further range of *A. majus* varieties cultured on medium 2 (containing 1 mg/l BAP) for 22 days. (Newbury 1986)

Variety	Mean number of shoots per explant
Orange Glow	3.83 ± 2.32
Scarlet Monarch	4.00 ± 1.54
White Monarch	2.50 ± 1.29
Crimson Monarch	4.33 ± 3.88
Coral Monarch	4.50 ± 4.18

significantly higher ($P = 0.01$) shoot numbers than those of variety Victory. Since medium 2 had generally proved to be the most efficient for micropropagation, shoot tips of five other varieties of *A. majus* were cultured on this medium (Table 4; Newbury 1986). In all cases reasonable numbers of transferable shoots were obtained within 22 days.

3.4 Transfer of Cultured Shoots to Glasshouse Conditions

Rooting could be achieved in regenerants from all sources using a double filter paper bridge made from two strips of 3-MM Whatman filter paper 11 cm long folded into an M shape. The bridge was then placed in a glass tube (16 mm × 150 mm) into which was placed 7.5 ml of MS plus 3 g/l sucrose and 0.05 mg/l IAA, pH 5.7, and this was then autoclaved. An excised shoot was then placed into the tube so that the shoot base was in the V of the bridge. Adequate roots to allow transfer to soil occurred within 3 weeks (Fig. 3). This transfer was most effectively brought about firstly by using autoclaved potting compost and water for maintenance of the plants. Prior to planting, the roots of the plants were washed in sterile water, and immediately after planting the shoots were trimmed to a height of between 2.5 and 4 cm (to roughly the first node). The use of fully enclosed propagators with transparent lids gave a high level of plant recovery in a soil based medium.

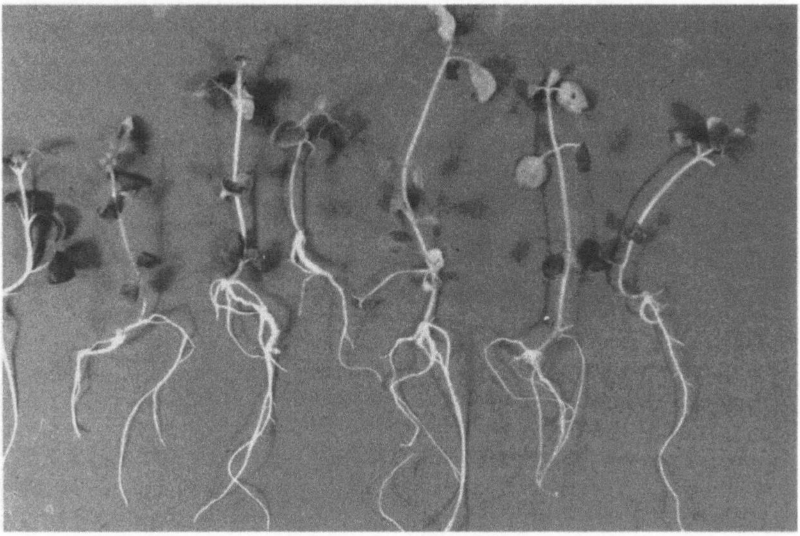

Fig. 3. Root production on micropropagated shoots of seven different varieties of *A. majus* following culture on a filter paper bridge in contact with MS medium containing 3% sucrose and 0.05 mg/l IAA

3.5 Commercial Aspects

Snapdragons were propagated almost exclusively from cuttings in the early part of this century, but are now largely seed propagated. However, the ability to rapidly clone specific genotypes remains important. There is an obvious need to multiply mutant forms that show improved or novel characteristics in a way that does not involve the genetic recombination events associated with sexual reproduction. Micropropagation of this species offers a rapid means of achieving this and can provide a very large number of rooted clones of a particular individual within a few months.

The economic importance of *A. majus* appears to have declined during the last 30 years. In the USA, it was ranked as the seventh most valuable cut flower plant in 1959, with a wholesale value of $4.5 million out of a total value of $142.5 million (i.e. 3.2%). By 1977 the wholesale value of *Antirrhinum* cut flowers had fallen to $2.8 million against a background of increasing sales of other species (Rogers 1980). Sales of *Antirrhinum* as a bedding plant also appear to have declined in the USA. In 1973, the Michigan Wholesale Growers reported that *Antirrhinum* represented 2.7% of their sales of bedding plants, whereas in 1978 this had fallen to 1%. In the UK, *Antirrhinum* was ranked as the tenth most popular bedding plant in 1989; it represented 3% of wholesale bedding plant value, amounting to a market of about $1.5 million (Andrew Eames, British Bedding Plant Association; pers. commun.).

4 Summary and Conclusions

Research in our laboratory has been aimed at different aspects of tissue culture of a range of *Antirrhinum* varieties. On the one hand we have been interested in determining the effect of tissue culture on genetic stability and somaclonal variation. It is clear that this can best be achieved using juvenile tissues as explants. In this way it is possible to maximise the rapid production of callus followed by de novo shoot formation and hence the possible production of somaclonal variants. This can be achieved even using young shoot tips where callus followed by adventitious shoots form on a range of media, including ones normally used for true micropropagation. Such a response in culture may be due to a greater developmental plasticity of juvenile tissue.

If, on the other hand, we wish to stabilise our tissue culture system genetically, so that we can be most certain that true clonal propagation of parental material is effected, then mature tissue explants are chosen, since callus production is reduced to a minimum, and new shoots are produced from axillary buds ensuring a duplication of the parental genotype.

5 Protocol

The optimum protocol for cloning *A. majus* plants in vitro is slightly variety-dependent, but, in general, is as follows. Lateral or apical shoot tips, 2–3 mm long, are surface sterilised in 10% (w/v) sodium hypochlorite containing 0.1% Tween 20 for 15 min. The tips are then washed three times in sterile distilled water and are placed, cut end down, onto agar-solidified (0.8%, w/v; Oxoid No. 3) MS medium (Murashige and Skoog 1962) containing 3% (w/v) sucrose and 1 mg/l BAP. The samples are then incubated at 28 °C under white fluorescent tubes with a light intensity of 130 μmol/m^2/s. After 3–4 weeks the shoots produced are transferred onto double filter paper bridges in boiling tubes containing 7.5 ml of MS medium with 3% (w/v) sucrose and 0.05 mg/l IAA. Three weeks after transfer, the shoot tips are removed and placed onto a fresh double filter paper bridge (as above), and within another 3 weeks the rooted shoots are ready for transfer to autoclaved standard mix compost.

Plants which have been exposed to exogenous BAP may require several cycles of culture on hormone-free medium, roots will form unaccompanied by callus. The most efficient way of rooting these plants is to place the shoot on MS for 4–5 weeks, and then to remove the shoot tip and place on a filter paper bridge in contact with liquid rooting medium.

After transfer to compost, autoclaved water is used for watering. Initially the plants are held in conditions of high humidity but this is decreased during a 2-week period until the plants are growing under standard glasshouse conditions.

References

Aidid SB and Thain JF (1980) Comparative studies of leaf tissue and isolated mesophyll cells of *Antirrhinum majus* L. II Ion relationships. J Exp Bot 31: 1043–1049

Aitken EAB, Newbury HJ, Callow JA (1989) Races of rust (*Puccinina antirrhini*) of *Antirrhinum majus* and the inheritance of host resistance. Plant Pathol 38: 169–175

Atkinson NJ (1990) A study of transposable element activity in somaclones of *Antirrhinum majus*. PhD thesis, University of Birmingham, UK

Atkinson NJ, Ford-Lloyd BV, Newbury HJ (1989) Regeneration of plants from *Antirrhinum majus*. L. callus. Plant Cell Tissue Organ Cult 17: 59–70

Ball V (1952) The Ball Red Book, 8th edn. GJ Ball Inc, West Chicago, Illinois

Binding H, Nehls R, Kock R, Finger J, Mordhorst G (1981) Comparative studies on protoplast regeneration in herbaceous species of the Dicotyledoneae class. Z Pflanzenphysiol 101: 119–130

Carpenter R, Coen ES (1990) Floral homeotic mutations produced by transposon-mutagenesis in *Antirrhinum majus*. Genes Dev. 4: 1483–1493

Cheng C, Xie Y-M, Hu S-Y (1984) Isolation and cultural behaviour of pollen tube subprotoplasts in *Antirrhinum majus* L. Acta Bot Sin 26: 459–465

Coen ES, Carpenter R (1986) Transposable elements in *Antirrhinum majus*: generators of genetic diversity. Trends Genet 2: 292–296

Coen ES, Romero JM, Doyle S, Elliot R, Murphy G, Carpenter R (1990) floricaula: a homeotic gene required for flower development in *Antirrhinum majus*. Cell 63: 1311–1322

de Vries H (1910) *Antirrhinum majus striatum*. In: Framer JB Darbishire AD (eds) The mutation theory, The Open Court Publ Co, New York pp 120–136

Dutta BK (1982) Effect of the chemical and physical condition of the soil on *Verticillium* wilt of *Antirrhinum*. Plant Soil 63: 217–226

Engelhard AW (1971) Botrytis-like diseases of rose, chrysanthemum, carnation, snapdragon and King aster caused by *Alternaria* and *Helminthosporium*. Proc Fla State Hortic Soc 83: 455–457

Fedoroff NV, Furtek DB, Nelson OE (1984) Cloning of the bronze locus in maize by a simple and generalizable procedure using the transposable controlling element Activator (Ac). Proc Natl Acad Sci USA 81: 3825–3829

Fincham JRS (1987) Transposable elements; patterns of flower pigmentation. Nature 325: 390–391

Gawthrop FM, Jones BMG (1981) Evidence for genetic change in the *Antirrhinum* rust (*Puccinia antirrhini*). Ann Bot 47: 197–202

Godwin I, Todd G, Ford-Lloyd BV, Newbury HJ (1991) The effects of acetosyringone and pH on *Agrobacterium*- mediated transformation vary according to plant species. Plant Cell Rep 9: 671–675

Harrison BJ, Fincham JRS (1968) Instability at the pal locus in *Antirrhinum majus*. 3. A gene controlling mutation frequency. Heredity 23: 67–77

Harte C (1974) *Antirrhinum majus* L. In: King RC (ed) Handbook of genetics, vol 2. Plants, viruses and protists. Plenum Press, New York, pp 315–332

Harte C, Stich B, Beissbarth R (1977) Der Nahrstoffbedarf isolierter Wurzeln von *Antirrhinum majus*. Biol Zentralbl 96: 665–667

Johnson CR (1972) Effectiveness of floral preservatives on increasing the vase-life of snapdragons. Flor Rev 149: 95–97

Judd RW Jr (1975) Snapdragon diseases. Conn Greenhouse Newslett 65: 6–12

Junge H (1982) Combined use of various plant protectants in ornamental plant production. Gartenbau 29: 27–28

Lindstrom RS (1966) Snapdragons – 60 °F and CO_2. Flor Rev 139: 51–54

Luo D, Coen ES, Doyle S, Carpenter R (1991) Pigmentation mutants produced by transposon mutagenesis in *Antirrhinum majus*. The Plant J 1: 59–69

Marousky FJ, Raulston JC (1970) Enhancement of snapdragon floret colour with light and floral preservatives. HortScience 5: 355 (Abstr.)

Martin C, Carpenter R, Sommer H, Saedler H, Coen ES (1985) Molecular analysis of instability in flower pigmentation of *Antirrhinum majus*, following isolation of the pallida locus by transposon tagging. EMBO J 4: 1625–1630

Martin C, Coen ES, Robbins T, Bartlett J, Almeida J, Carpenter R (1987a) The control of floral pigmentation in *Antirrhinum*. Biochem Soc Trans, 15: 14–17

Martin C, Carpenter R, Robbins T, Hudson A, Almeida J, Bartlet J, Mackay, Coen ES (1987b) Transposable elements in *Antirrhinum majus*. J Cell Sci (Suppl) 7: 109–122

McClintock B (1952) Chromosome organisation and genic expression. Cold Spring Harbor Symp Quant Biol 16: 13–47

Melchers G, Bergman L (1959) Untersuchungen an Kulturen von haploiden Geweben von *Antirrhinum majus*. Berl Dtsch Bot Ges 71: 459–473

Morris P, Linstaed P, Thain JF (1981) Comparative studies of leaf tissue and isolated protoplasts. III Effects of wall-degrading enzymes and osmotic stress. J Exp Bot 32: 801–811

Munsi PS, Nayak KC, Mukherjee D (1981) Effects of ethrel on growth and flowering of some ornamental plants. Indian Agric 24: 153–159

Murashige T, Skoog F (1962) A revised method for rapid growth and bioassays with tobacco tissue cultures. Physiol Plant 15: 473–497

Murray GE, Sanderson KC, Williams JC (1986) Application methods and rates of ancymidol on plant height and seed germination of bedding plants. HortScience 21: 120–122

Newbury HJ (1986) Multiplication of *Antirrhinum majus* L. by shoot tip culture. Plant Cell Tissue Organ Cult, 7: 39–42

Newbury HJ (1992) Fungal resistance: the isolation of a plant R gene by transposon tagging. In: Gatehouse A, Hilder V, Boulter D (eds) Genetic manipulation of pest and pathogen resistance in plants. CAB International, Oxford, in press

Nowak J (1981) Chemical pretreatment of snapdragon (*Antirrhinum majus*) spikes to increase cut flower longevity. Sci Hortic 15: 255–262

Paulus A, Besemer S, Nelson J, Munoz F, Otto H (1984) Fungicides for the control of Perenosporales and Erysiphales fungi. Meded Fac Landbouwwet, Rijksuniv Gent 49: 311–318

Peterson PA (1987) Mobile elements in plants. CRC Crit Rev Plant Sci 6: 105–208

Pfister JM, Widholm JM (1984) Plant regeneration from snapdragon tissue cultures. HortScience 19: 852–854

Pittis JE, Colhoun J (1984) Isolation and identification of pythiaceous fungi from irrigation water and their pathogenicity to *Antirrhinum*, tomato and *Chamaecyparis lawsoniana*. Phytopathol Z 110: 301–318

Poirier-Hamon S, Rao PS, Harada H (1974) Culture of mesophyll cell protoplasts and stem segments of *Antirrhinum majus* (snapdragon). Growth and organisation of embryoids. J Exp Bot 25: 752–760

Popple SC, Rigby H (1985) Evaluation of fungicides for the control of *Pythium* root rot of *Antirrhinum*. Tests Agrochem Cultiv 6: 34–35

Porter DM, Aycock R (1967) Snapdragon leafspot caused by *Cercospora antirrhina*. NC Agric Exp Stn Tech Bull 179: 1–31

Rao PS, Harada H (1973) Hormonal regulation of morphogenesis in organ cultures of *Petunia inflata*, *Antirrhinum majus* and *Pharbitis nil*. In: Plant growth substances (Proc. 8th Intern. Conf. on Plant Growth Subst.) Hirokowa Publ Co, Tokyo, pp 1113–1120

Rao PS, Bapat VA, Harada H (1976) Gamma radiation and hormonal factors controlling morphogenesis in organ cultures of *Antirrhinum majus* L. cv. Red Majestic Chief. Z Pflanzenphysiol 80: 144–152

Rogers MN (1980) Snapdragons. In: Larson RA (ed) Introduction to floriculture. Acad Press, New York, pp 108–131

Rothmaler W (1956) Taxonomischer Monographie der Gattung *Antirrhinum*. Feddes Repert 136: 1–124

Saedler H, Gierl A, Sommer H, Schwartz-Sommer Z (1987) Plant transposable elements and their role in the evolution of regulatory units and proteins. J Cell Sci (Suppl) 7: 139–144

Sangwan RS, Harada H (1975) Chemical regulation of callus growth, organogenesis, plant regeneration and somatic embryogenesis in *Antirrhinum majus* tissue and cell cultures. J Exp Bot 26: 868–881

Shanks JB (1982) Growth regulation of several ornamental crops with a triazoline derivative of norbornenodiazetine (BAS 106). Proc 9th Conf Plant Growth Regulation Soc Am, pp 60–67

Shedeed MR, El-Gammasy KM, Hashim ME, Almulla AMN (1986) Effect of some growth regulators on the growth, flowering and seed production of some winter annuals. Ann Agric Sci (Cairo) 31: 691–706

Simpson CJ, Jones GE, Taylor JD (1971) A seedling blight of *Antirrhinum* caused by *Pseudomonas antirrhini*. Plant Pathol 20: 127–130

Sommer H, Carpenter R, Harrison BJ, Saedler H (1985) The transposable element Tam3 of *Antirrhinum majus* generates a novel type of sequence alteration upon excision. Mol Gen Genet 199: 225–231

Sommer H, Beltran J-P, Huijser P, Pape H, Lonnig W-E, Saedler H, Schwartz-Sommer Z. (1990) Deficiens, a homeotic gene involved in the control of flower morphogenesis in *Antirrhinum majus*: the protein shows homology to transcription factors. EMBO J 9: 605–613

Stefanis JP, Langhans RW (1982) Snapdragon (*Antirrhinum majus*) production with supplemental irradiation from high-pressure sodium lamps. HortScience 17: 601–603

Tsuchiya Y, Ohata K-I, Azekami K, Matsuzaki M (1981) Pathogenicity of *Xanthomonas campestris* var. *vitians* to various crops and weeds. Bull Nat Inst Agric Sci Ser C, 0(35): 57–72

Wainwright H, Irwin HL (1987) The effects of paclobutrazol and pinching on *Antirrhinum* flowering pot plants. J Hort Sci 62: 401–404

Yang CH, Leu LS (1981) The occurrence of snapdragon (*Antirrhinum majus*) wilt disease in Taiwan. Plant Prot Bull 23: 55–57

Zhou C (1985) Isolation and identification of viable embryo sacs in several angiosperm species. Acta Bot Sin 27: 258–262

Zhou C, Yang HY (1985) Observations of enzymatically isolated, living and fixed embryo sacs in several angiosperm species. Planta 165: 225–231

III Micropropagation of *Begonia* spp.

J. SIMMONDS[1]

1 General Account

1.1 Distribution and Importance

The genus *Begonia* L. (family Begoniaceae) comprises more than 1000 species native to tropical regions of Africa, Central and South America, and Asia, as well as a great number of horticultural hybrids and cultivars. Many are grown commercially as flowering potted plants with tremendous variability in flower size, form, and color. Some are also grown as foliage plants because of the diversity of leaf shape and markings available. Begonias also are among the most popular bedding plants, and cultivars have been developed for either sunny or shaded locations.

1.2 Morphology and Cytology

Begonias are mostly herbaceous, or rarely shrubby perennial plants. The growth habit is monopodal and indeterminate. Leaves are arranged in an alternate phypllotaxy. All begonias are monoecious, usually bearing male and female flowers in the same inflorescence. Male flowers are initiated first and develop acropetally. The inflorescence branches are ternated by female flowers with three-winged inferior ovaries (Matzke 1938).

Begonias of major commercial interest can be classified into five horticultural groups. Each of these groups are cytologically diverse:

Fibrous Rooted. These are used as bedding plants, e.g., *B. semperflorens* Link and Otto (2n = 33 – 66) (Darlington and Wylie 1955).
Rhizomatous. These are used as foliage plants, e.g., *B. rex* Putz (2n = 32 – 34) (Sharma and Bhattacharyya 1957).
Tuberous Begonias. These are flowering pot plants and garden specimens, e.g., *B.* × *tuberhybrida* Voss (2n = 27 – 56). The origin of these hybrids is not well documented, but at least four Peruvian and two Bolivian species form the genetic background of this group. The diploid numbers of the wild ancestors are 26 and 28 (Legro and Doorenbos 1969). An analysis of chromosome numbers of the hybrid tuberous begonias showed variation of the diploid number between 27 and 28, the

[1]Agriculture Canada Research Branch, Plant Research Centre, Ottawa, Ontario, K1A OC6, Canada

Biotechnology in Agriculture and Forestry, Vol. 20
High-Tech and Micropropagation IV (ed. by Y.P.S. Bajaj)

triploid number between 41 and 42 and the tetraploid number between 52 and 56 (Legro and Haegeman 1971).

Christmas Begonia. *B. cheimantha* Everett ($2n = 27 - 54$) is used as a potted plant. This hybrid was the result of a cross between *B. socotrana* Hook ($2n = 28$) and *B. dregei* Otto and Dietr ($2n = 26$) (Doorenbos and Legro 1968).

Elatior Begonias. *B.* × *chiemalis* Fotsch ($2n = 35 - 63$) are used as potted plants. These are the result of crossing *B. socotrana* with various diploid begonia tuberhybrida species. Twenty-one of 23 cultivars analyzed were found to be triploids having 26, 27, or 28 tuberhybrida, and a constant number of 14 socotrana chromosomes. The remaining two triploids had a reduced number of socotrana chromosomes. Two hybrids having tetraploid tuberhybrida background were also detected (Arends 1970).

1.3 Breeding and Conventional Propagation

Most fibrous rooted begonias used are F_1 hybrids and are propagated from seed although additional plants can be obtained by taking stem cuttings from stock plants.

Rhizomatous begonias are grown for their foliage characteristics. Environmental conditions that favor flowering inhibit vegetative growth so it is most convenient to propagate these plants vegetatively by leaf vein cuttings.

The diploid ancestors of *B. tuberhybrida* cross readily to form fertile hybrids, which explains the rapid expansion of this group. The diploid hybrids are fertile but the triploids tend to be sterile and the tetraploids have varying degrees of fertility. Seed propagation is therefore possible with some cultivars, but all can be propagated vegetatively by tubers. Stem cuttings are taken early in the season so that there is ample time for small tubers to form. The resulting tubers are stored and started into growth in the following season.

Development of Christmas begonias relies on other *B. socotrana* crosses that have characteristics similar to the original *B. socotrana* × *B. dregei.* Propagation is by leaf petiole and stem cuttings; some of the original cheimantha begonias are still being propagated vegetatively.

Elatior hybrids can be produced by crossing various tuberous species with *B. socotrana,* which must be used as the male parent because of incompatability in the reciprocal cross (Gleed 1961). Elatior begonias are sterile and are propagated by leaf petiole cuttings and terminal cuttings.

1.4 Need for in Vitro Culture

Conventional methods of vegetative propagation are applied to all the groups of commercially important begonias in order to maintain desirable hybrid traits. In the case of elatior and cheimantha types and many of the tuberhybridas which are sterile, these techniques are mandatory for maintaining the lines. In vitro micropropagation procedures offer an efficient means of maintaining hybrid lines and allow for very rapid multiplication of new hybrid cultivars for early introductions.

Micropropagation by the induction of adventitious buds can also be used in mutation breeding programs (Roest et al. 1981). This technique is based on the phenomenon that adventitious shoots arising on petiole and leaf explants can originate from single epidermal cells. Hence, after mutation induction by irradiation or chemical mutagens and subsequent adventitious shoot formation, nonchimeric mutants can be produced. This technique provides an extra source of variation for floriculturists. Genotypes such as the sterile elatior begonias cannot be improved through cross-breeding, which makes mutation breeding a promising option. Several groups have produced commercially acceptable mutant begonias by this means (Brown and Harney 1974; Doorenbos and Karper 1975; Mikkelsen et al. 1975).

When vegetatively propagated plants are systemically infected with pathogens, they are readily passed from one vegetative generation to the next, and it is possible that the entire population of a particular clonal variety can be lost. Bacterial leaf spot and blight (*Xanthomonas begoniae*) can cause significant problems in greenhouse production of elatior begonias, but this can be eliminated by meristem culture of adventitiously produced buds (Hakkaart and Versluijs 1983; Westerhof et al. 1984). Similarly, foliar nematodes (*Aphelenchoides fragariae*) have been eliminated from stock plants of elatior begonias in commercial greenhouse in Canada. Virus infestations are common and may be symptomless, but nevertheless will cause loss of vigor and yield. Thermotherapy has been used to eliminate viruses from vegetatively propagated tuberhybrida begonias and these plants provided more flowers and larger tubers than the infected stock (Welvaert et al. 1980). By using thermotherapy in combination with micropropagation procedures it was possible to produce commercial size tubers of virus-indexed plants in one growing season (Samyn et al. 1984).

2 Tissue Culture Work

Growth and differentiation of tissue explants of begonia have been reported by many workers using *B. semperflorens* (Thakur 1973; Sehgal 1975). *B. rex* (Wirth 1960; Bigot and Nitsch 1968; Arora et al. 1970; Chlyah-Arnason and Tran Than Van 1975; Cassells and Morrish 1987a), *B.* × *tuberhybrida* (Peck and Cumming 1984; Samyn et al. 1984), *B.* × *cheimantha* (Heide 1965; Ringe and Nitsch 1968; Fonnesbech 1974a, b) and *B.* × *hiemalis* (Appelgren 1976; Hilding and Welander 1976; Mikkelsen and Sink 1978).

Petiole and leaf segments are the most common explants used for studying nutritional and growth substance requirements for adventitious bud production. However, other explants have been used; stem, flower stalk, and apices were also included in a study of vegetative propagation of elatior begonias (Khoder et al. 1981) and leaf axillary buds were used by Hakkaart and Versluijs (1984). Similarly, apical and axillary buds have been used for propagation of *B.* × *tuberhybrida* (Samyn et al. 1984). Petals have provided a source of somatic embryos with *B.* × *hiemalis* (Margara and Phelouzat 1984), and various inflorescence explants were used for the vegetative propagation of *B. venosa* Skan (Pierik and Tetteroo 1987).

Attempts to obtain haploid plants by anther culture have also been successful. *B. semperflorens* anthers containing microspores at the late uninucleate state were cultured, and monoploid plants recovered (Zang et al. 1982). Anthers of an elatior begonia cultivar (2n = 42) provided callus, and subsequently plants with 21 chromosomes were recovered (Khoder et al. 1984).

Begonia tissue culture systems have been used to study a number of fundamental aspects of developmental biology. *B. rex* cultures were used to investigate the significance of cell ontogeny, callus aging, and frequency of subculture on karyological changes in the callus and the regenerants (Cassells and Morrish, 1987b). In vitro responses can also be influenced by the physiological status of the donor tissue. Adventitious bud production of *B. x hiemalis* petiole explants was similar when taken from long-day or short-day-grown plants only during the period of vegetative growth. As flowering intensified, the explants lost the capacity to produce buds and consequently the short-day donors produced ineffective explants (Simmonds and Nelson 1988). *B. rex* leaf explants composed of only a single epidermal layer and three to five subepidermal collenchyma layers is a very responsive system for studying mechanisms controlling organogenesis. Three developmental pathways leading either to unicellular hairs, buds, or roots can be precisely manipulated by using three defined culture media (Chlyah and Tran Than Van 1984). The sex-regulating mechanism in *B. franconis* was explored by culturing male and female bud primorida on chemically defined media. Cytokinin was essential for bud growth, but none of the plant growth regulators used could alter the development of the floral organs, which were therefore strongly determined at this stage (Berghoef and Bruinsma 1979a). Similarly, when undifferentiated primordia were cultured, sex expression was not influenced by the plant growth regulators. It was concluded that the sex of the bud was endogenously regulated by already established primordia (Berghoef and Bruinsma 1979b).

3 Micropropagation

The use of in vitro techniques for asexual propagation is the area of plant tissue culture which currently has the most practical application. Advantages for the cultivation of begonias include:

1. Propagation of genetically uniform plants in large numbers for commercial production and maintenance of clonal lines.
2. Year-round propagation schemes can be developed.
3. Elimination of systemic pathogens.
4. Rapid multiplication of new hybrid cultivars for evaluation and eventual commercial production.

3.1 Micropropagation Systems

In vitro asexual propagation can be achieved by either enhanced precocious axillary shoot formation, the induction of adventitious shoots directly on organ explants or explant-derived callus, or by somatic embryogenesis. Regardless of the cloning system employed, the progeny must be true to type and have acceptable growth habit and developmental behavior.

3.2 Genetic (Phenotypic) Stability

Phenotypic variation in plants regenerated by adventitious organogenesis may arise from variant somatic cells within the explant (D'Amato 1977) or arise de novo as a consequence of genome instability in vitro (Bayliss 1980).

Adventitious regeneration of *B. rex* plants occurs predominantly from the epidermal cells with no formation of intervening callus, but in vitro systems for plant regeneration from callus cultures have also been developed (Cassells and Morrish 1985). In this study, phenotypic analyses were made of plants regenerated from leaf cuttings in vivo and in vitro from petiole, axenic leaf, multiple recycled axenic leaf, and callus cultures. No significant variation in plant height, plant diameter, leaf number, leaf area, petiole height, and rhizome development was found in any of the progenies. It was concluded that the high stability of the regenerants of *B. rex* Lucille Closon indicated the potential for cloning this plant by in vitro procedures (Cassells and Morrish 1985). The results of this progeny phenotype analysis suggest that the genome is relatively stable in vitro. However, karyological analysis of the callus refuted this. Callus subcultured once showed evidence of aneuploidy, and this trend increased with aging of the callus and subculture generation (Cassells and Morrish 1987b). It was suggested that the low frequency of variation in the adventitious regenerants may be a consequence of a loss of competence in the hypodiploid callus cells.

Micropropagation of *B.* × *tuberhybrida* from apical and axillary buds provide material for a large-scale field analysis of the regenerants (Samyn et al. 1984). The demonstration of a very high rate of recovery of plants which were all true to type and the fact that commercial-size tubers could be produced in one season, provided good support for commercialization of the in vitro technique. In a smaller study concerning the production of *B.* × *tuberhybrida* plants by adventitious organogenesis from leaf explants, it was also concluded that the regenerants were similar to the donor plant, and this system has good potential for large-scale propagation (Peck and Cumming 1984).

The problem of phenotypic variation of *B.* × *hiemalis* clones regenerated by in vitro propagation was thoroughly investigated by Westerhof et al. (1984). Many workers have reported successful regeneration by adventitious bud production, but few have provided information on the quality of plants produced, and even then the information is not substantiated. Some notable exceptions are Mikkelsen and Sink (1978), who reported that from 54 regenerants of Schwabenland Red, six plants had a flower color deviation and one of these also had an altered leaf form. Roest et al. (1981) obtained six mutations out of 82 plants after two cycles of in vitro shoot formation in Schwabenland Red; with Tiara they noted 29 mutations among 68 plants. Bigot (1981) obtained 8% variation in 340 Mixed regenerants. Westerhof et al. (1984) screened several thousand regenerants from two cultivars for 18 types of variations of leaf, stem, and flower characteristics. The initial cycle of regenerants from leaf explants gave nearly uniform offspring, but when this was followed by one or two further cycles of in vitro propagation, the variation increased. Most of the variants occurred in the small slow-growing plantlets. Thus, an important source of variation can be avoided by eliminating the smallest in vitro plantlets. The nature of the variations was not easy to ascertain. Some of the deviating plants returned to the normal type and were therefore phenotypic changes, others were more stable. Whether true mutations or physiological effects are involved is not easily determined. For the commercial propagator this distinction is not important; neither type of variation is acceptable.

3.3 Tissue Culture Procedure

Almost any plant tissue or organ may be used as an explant for micropropagation. It has been found that, for begonias, petiole or leaf explants are the most responsive tissues for adventitious organogenesis. The following methods have been developed for *B.* × *hiemalis*, but similar procedures are suitable the other classes of begonias.

3.3.1 Source of Explant

Selection of appropriate donor tissue is the most critical aspect affecting the performance of micropropagation systems. The physiological age of the explant influences the type and extent of morphogenesis. In general, young tissues are more responsive than older tissues (e.g., Croes et al. 1985). The influence of temperature and photoperiod on the physiological status of the donor tissue can also affect subsequent in vitro responses (Cassells 1979; Paterson and Rost 1979; Rashid and Reinert 1983). In vitro response of peduncle segments of *B.* × *hiemalis* increased concomitantly with an increase in total irradiation (Applegren 1985) and the percentage response of petiole explants from long-day-treated plants was higher than that obtained from explants of short-day-treated donor plants (Hilding and Welander 1976). The photoperiod does not directly influence the explant competence, but regenerative capacity is affected by the changing status of the donor plant in response to photoperiodic treatment. Adventitious bud production on petiole explants from long-day and short-day-treated donor plants was similar for the first 28 days of treatment when the plants were vegetative. Loss of regenerative capacity occurred later in the growth cycle of short-day-treated plants and was best correlated with increasing reproductive weight (Simmonds and Nelson 1988). Petiole or leaf explants should be taken from young expanding leaves before the plants are flowering strongly. Maintaining donor plants in long days increases bud production and extends the duration in which the donor plants can be used effectively in commercial micropropagation programs (Simmonds and Nelson 1989).

3.3.2 Explant Sterilization

Surface contaminants such as bacteria and fungi must be removed prior to culturing. The following procedure is recommended:

1. Excise leaves (Fig. 1A) and wash in running tap water for 1–2 h. This is not always necessary, but has been found to be an advantage with *B. × tuberhybrida*.
2. Cut leaves into segments (2 × 2 cm) or petioles into lengths of 2–5 cm and rinse in ethanol (95%) for 30 s.
3. Immerse in 7% (w/v) calcium hypochlorite solution containing a few drops of Tween 20 for 15 min.
4. Rinse three times in sterile distilled water.

3.3.3 Culture Medium for Bud Induction

Before planting the explants onto culture medium, cut away the bleached edges. Leaf segments (1 × 1 cm) should include a section of a major vein (Fig. 1B). Petiole segments (5–10 mm long) provide responsive explants (Fig. 1C). Murashige and Skoog (1962) mineral salts and sucrose 30 g/l provide the nutritional base for in vitro propagation of begonias. Both an auxin, e.g., naphthalene acetic acid (NAA) and a cytokinin, e.g., 6-benzylaminospurine (BAP) are required for bud induction. The pH is adjusted to 5.7 and the medium solidified with 0.8% agar. The cultures are grown in a 16-h photoperiod, provided by fluorescent light, $\mu E/m^2/s$ at shelf level, at 25 °C.

3.3.4 Growth Regulator Requirements for Bud Induction

Adventitious shoots were induced on petiole or leaf explants of *B. × hiemalis* cv. Schwabenland Pink in response to various combinations of NAA and BAP (Fig. 1C, D). Shoot initiation on petiole explants was earliest on medium with NAA:BAP (0.1:0.5 mg/l) and this also produced the most shoots per explant (Fig. 2). Shoot initiation was delayed on medium with lower BAP (NAA:BAP, 0.1:0.1 mg/l) and fewer shoots were formed (Fig. 2) but subsequent shoot growth was better on this low BAP medium. After 6 weeks on medium with BAP 0.1 mg/l, each explant produced ten shoots longer than 20 mm, while with BAP 0.5 mg/l only five such shoots were produced (Simmonds 1984a).

3.3.5 Culture Conditions for Shoot Growth

B. × hiemalis explants can produce large numbers of adventitious buds, and several studies have indicated the potential for producing vast numbers of plants by this means (Khoder et al. 1981; Takayama and Misawa 1981). In practice, this potential is rarely realized because many of the adventitious buds remain too small to be rooted, and considerable plant loss also occurs at the rooting and acclimatization

Fig. 1A–I. Micropropagation system for elatior begonias. **A** Young expanding leaf (3–6 cm diam.) as source of explants. **B** Leaf segment including a major vein. **C** Responding petiole explants. **D** Adventitious bud production on leaf explant. **E** Adventitious bud production on petiole explant. **F** In vitro-rooted regenerant after 3 weeks on agar medium with NAA (0.1 mg/l). **G** In vivo-rooted regenerant after 3 weeks in mist frame. **H** Greenhouse-established plant. **I** Multistemmed habit of in vitro plants

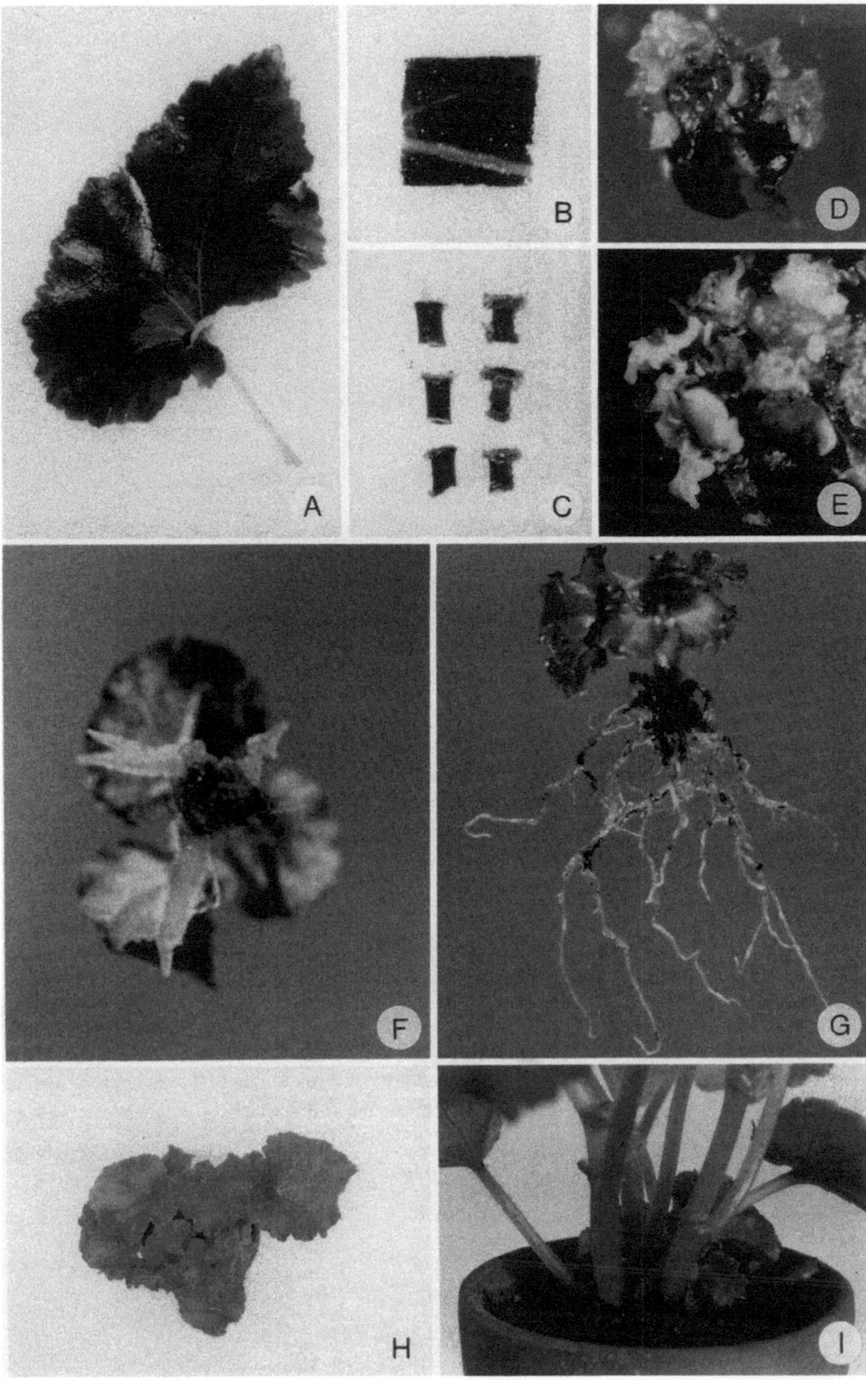
A
B
C
D
E
F
G
H
I

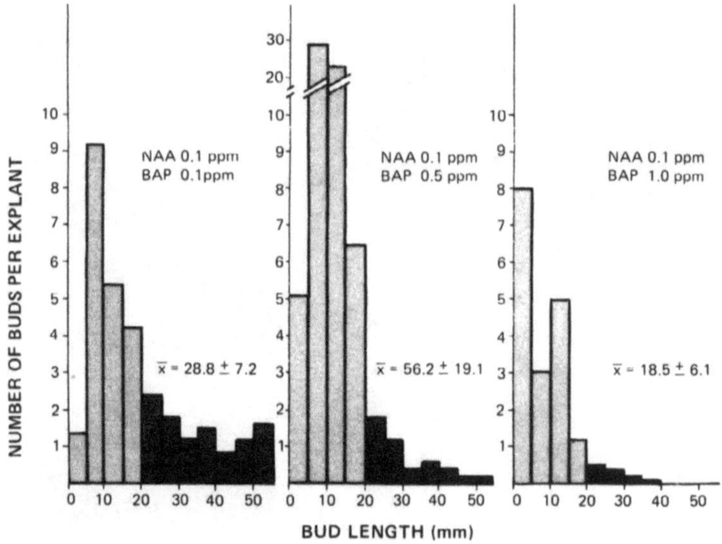

Fig. 2. Influence of BAP on induction and growth of adventitious buds of *B.* × *hiemalis*. Incubation at 6 weeks on agar medium. (Simmonds 1984a)

stages (Mikkelsen and Sink 1978). It has also been reported that a large source of phenotypic variation is associated with small slow-growing regenerants (Westerhof et al. 1984). For these reasons it is desirable to develop culture systems which produce the greatest number of large explants. When cultured for 9 weeks on agar medium, petiole explants produced about 50 adventitious buds, but only a few were large enough (> 1 cm) to be selected for rooting procedures suitable for commercial application (Table 1). If the explants were transferred to liquid-shake culture after 3 or 6 weeks on agar, bud growth was enhanced and significantly more shoots were produced which were large enough for easy handling, rooting and establishment (Table 1).

Table 1. The effect of a period of liquid-shake culture on adventitious bud development and on subsequent rooting and plant establishment. (Simmonds and Werry 1987)

Culture period (weeks)	Adventitious shoots/ explant[a]	Percent rooting	No. trans-planted	Percent established	Established plants/ explant
Agar (9) + liquid (0)	0.4[b] ab	75	3	67	0.2 a
Agar (3) + liquid (6)	19.0 b	86	163	75	12.2 b
Agar (6) + liquid (3)	26.9 c	93	250	79	19.7 c

[a]Petioles > 1 cm long.
[b]Mean separation by Duncan's multiple range test 0.05% level. Mean of ten replicates.

Table 2. In vitro and mist-frame rooting of adventitious shoots of *Begonia* × *hiemalis* cv. Schwabenland Pink (rooting period 3 weeks). (Simmonds 1984b)

Response	In vitro Agar NAA	Mist frame Peat IBA	
	0.1 mg l	0.1%	0.4%
Explants with roots (%)	91	100	100
No. of roots/explant	4.1 a[a]	7.4 b	9.0 c
Root length (cm)	0.5 a	3.2 b	4.1 b

[a] Values within rows followed by a common letter are not significantly different according to Duncan's multiple range test ($P = 0.05$).

3.3.6 Rooting and Plant Establishment

Maximum in vitro rooting of adventitious shoots was obtained after 3 weeks on medium with NAA 0.05 or 0.1 mg/l. Each explant initiated approximately four roots, but these were relatively short and poorly developed (Fig. 1f, Table 2). Acclimatization losses of in vitro-rooted plants were variable, but as high at 30%, depending on ambient greenhouse conditions.

Rooting adventitious shoots directly in a mist frame resulted in the development of a much larger root system (Fig. 1G, Table 2). For direct or in vivo rooting the adventitious shoots were removed from the culture vessel, the basal region was dipped in commercial rooting powder containing 0.4% indolebutyric acid (IBA), and then planted in cell packs containing sphagnum peat. These were transferred to mist frames with bottom heat (21 °C). Timed misting was controlled by an electronic leaf moisture senser. During the 1st week the plants were misted 24 h/day and then for 16 h/day for another 2 weeks. The plants were then transplanted to soil mix, fed with N:P:K (10:52:10) fertilizer (0.9 g/l) and grown in a shaded greenhouse. These plants were established without any losses and subsequently developed into superior multistemmed plants (Fig. 1H, I). Plant growth was vigorous, resulting in heavily blossoming, commercially acceptable plants (Fig. 3).

3.4 Commercial Aspects of Micropropagation

Although in vitro multiplication at one time was viewed sceptically as a "high technology" phenomenon, it has now gained commercial acceptance and is considered as part of normal greenhouse procedures. The technology is relatively easy to incorporate into the greenhouse operation, and once established becomes an essential component.

The potential of using micropropagation technology for the commercial production of *B.* × *hiemalis* was demonstrated by Takayama and Misawa (1981). Leaf explants provide a source of plantlets after 30 days in agar culture which could be further multiplied in liquid shake culture. After 20 days on the shaker, the plantlets

Fig. 3. Commercially superior micropropagated *B.* × *hiemalis* cv. Schwabenland Pink in full bloom

were transferred to agar media for rooting and then transplanted into soil and established in the greenhouse. By harvesting plantlets at 50-day intervals and recycling the explants it was estimated that theoretically more than 10^{14} plantlets could be produced in a year from a single 7 × 7 mm leaf segment. Such rates of multiplication have not yet been used in commercial greenhouses. In Canada, 60 greenhouses, each with more than 7500 square feet of growing area, are involved in *Begonia* production. These have been able to accommodate market needs by conventional shoot tip or leaf cutting propagation from stock plants. However, imported stock plants, while initially appearing healthy, can, after a few cycles of vegetative propagation, develop severe disease symptoms resulting in considerable losses. Tissue culture-derived stock plants can provide a unique source of pathogen-free clones, protecting the grower against systemic pathogens (see Sect. 1.4). The ease of maintaining a wide range of valuable stock plants for long periods, giving market flexibility, and the enhanced quality control resulting from tissue culture-derived stock plants has been successfully exploited commercially (Fig. 4).

Using leaf section explants plated in large culture vessels (Fig. 4A), tens of thousands of adventitiously derived plants can be easily generated. These are readily rooted in vivo (Fig. 4B) and established in the greenhouse (Fig. 4C). These

Fig. 4A–D. An example of commercial application of micropropagation of elatior begonias: **A** Multiple leaf explants in large vessels on bud induction medium. **B** In vivo rooting and plant establishment of regenerants. **C** Micropropagated stock plants to be used for shoot-tip propagation. **D** Large scale production of commercial plants. Courtesy of Planski Gardens, Campbellville, Ontario, Canada

pathogen-indexed plants then provide an excellent supply of material for further propagation by shoot tip cuttings to generate the large numbers of plants required for commercial production (Fig. 4D). Stock plants are regularly passed through this one-cycle micropropagation process to ensure that systemic pathogens are kept under control. The rapidity of the micropropagation system also allows easier accommodation to changes in market requirements.

4 Summary and Conclusions

Micropropagation by adventitious bud production can effectively eliminate bacterial, fungi, and nematode pathogens. It is also possible to identify virus-free meristems and establish virus-indexed clones by these procedures. Regenerants produced from in vitro systems develop strong multi-stemmed plants which are commercially more acceptable than single stemmed cuttings.

The potential for cloning *B. rex* by multiple cycling axenic cultures has been demonstrated and these systems are used to propagate foliage begonias. Phenotypic variation is a problem when multiple cycles of adventitious bud production are used with elatior begonias and until larger studies with *B.* × *tuberhybrida* have been reported, this group should also be considered suspect. Single cycle micropropagation in combination with traditional shoot-tip vegetative propagation provides an efficient means of commercially propagating pathogen free clones of elatior begonias.

5 Protocol

Best Explants. Petiole or leaf segment explants taken from young expanding leaves: The donor stock plants should be pathogen-indexed and should be allowed to begin flowering in order to check for phenotypic uniformity.

Best Medium. MS salts, 3% sucrose, NAA 0.1 mg/l, BAP 0.1 mg/l.

Optimal Conditions. 21–25 °C, 16-h photoperiod (40 $\mu E/m^2/s$). Culture the explants on agar for 6 weeks and then transfer to liquid-shake culture for 3 weeks. Root in vivo in mist frame and establish plants in a shaded greenhouse.

References

Appelgren M (1976) Regeneration of *Begonia* × *hiemalis* in vitro. Acta Hortic 64: 31–38

Appelgren M (1985) Effects of supplementary light to mother plants on adventitious shoot formation in flower peduncle segments of *Begonia* × *hiemalis* Fotsch in vitro. Sci Hortic 25: 77–83

Arends JC (1970) Somatic chromosome numbers in *elatior*-begonias. Meded Landbouwhogedch Wageningen 70(20): 1–18

Arora YK, Nakao S, Nakajima T (1970) Perpetuation of *Begonia rex* by aseptic culture with micro-leaf cuttings under various conditions of auxin and cytokinin. Jpn J Breed 20: 257–281

Bayliss MW (1980) Chromosomal variation in plant tissues in culture. Intl Rev Cytol (11A). Academic Press, New York, pp 113–144

Berghoef J, Bruinsma J (1979a) Flower development of *Begonia franconis* Liebm. II. Effects of nutrition and growth-regulating substances on the growth of flower buds in vitro. Z Pflanzenphysiol 93: 345–357

Berghoef J, Bruinsma J (1979b) Flower development of *Begonia franconis* Liebm. III. Effects on growth-regulating substances on organ initiation in flower buds in vitro. Z Pflanzenphysiol 93: 377–386

Bigot C (1981) Multiplication végétive in vitro de *Begonia* × *hiemalis* (Rieger et Schwabenland) 2 Conformité des plantes élevées en serre. Agronomie 1: 441–447

Bigot C, Nitsch JP (1968) Effect du moment de l'application d'acide gibbérelique ou de 2,6-diaminopurine sur la néoformation des bourgeons chez *Begonia rex* Putz. CR Acad Sci Paris D 267: 619–621

Brown JL, Harney PM (1974) Induction of non-chimaeral mutants in *Begonia*. HortScience 9: 276

Cassells AC (1979) The effect of 2,3,5-triiodobenzoic acid on caulogenesis in callus cultures of tomato and pelargonium. Physiol Plant 46: 159–164

Cassells AC, Morrish FM (1985) Growth measurement of *Begonia rex* Putz plants regenerated from leaf cuttings and in vitro from leaf petioles, axenic leaves, re-cycled axenic leaves and callus. Sci Hortic 27: 113–121

Cassells AC, Morrish FM (1987a) Genetic, ontogenetic and physiological factors and the development of a broad-spectrum medium for *Begonia rex* Putz cultivars. Sci Hortic 31: 295–302

Cassells AC, Morrish FM (1987b) Variation in adventitious regenerants of *Begonia rex* Putz Lucille Closon as a consequence of cell ontogeny, callus ageing and frequency of callus sub-culture. Sci Hortic 32: 135–143

Chlyah A, Tran Than Van K (1984) Histological changes in epidermal and subepidermal cell layers of *Begonia rex* induced to from de novo unicellular hairs, buds and roots. Bot Gaz 145: 55–59

Chlyah-Arnason A, Tran Than Van K (1985) Differential reactivity in epidermal cells of *Begonia rex* excised and grown in vitro. Physiol Plant 35: 16–20

Croes AF, Creemers-Molenaar J, van den Dende G, Kemp A, Barendse GWM (1985) Tissue age as an endogenous factor controlling in vitro bud formation on explants from the inflorescence of *Nicotiana tabacum* L. Exp Bot 30: 1771–1779

D'Amato F (1977) Cytogenetics of differentiation in tissue and cell cultures. In: Reinert J, Bajaj YPS (eds) Applied and fundamental aspects of plant cell, tissue, and organ culture. Springer, Berlin, Heidelberg, New York, pp 343–357

Darlington CD, Wylie AP (1955) Chromosome atlas of flowering plants. Allen and Unwin, London

Doorenbos J, Karper JJ (1975) X-ray-induced mutations in *Begonia* × *Liemalis*. Euphytica 24: 13–19

Doorenbos J, Legro RAH (1968) Breeding Gloire de Lorraine begonias. Meded Landbouwhogedch Wageningen 68(19): 1–14

Fonnesbech M (1974a) The influence of NAA, BA and temperature on shoot and root development from *Begonia* × *cheimantha* petiole segments grown in vitro. Physiol Plant 32: 49–54

Fonnesbech M (1974b) Temperature effects on shoot and root development from *Begonia* × *cheimantha* petiole segments grown in vitro. Physiol Plant 32: 282–286

Gleed CJ (1961) Hybrid winter flowering begonias. The story of raising some early varieties. J R Hortic Soc 86: 319–322

Hakkaart FA, Versluijs JMA (1983) Control of leaf curl and *Xanthomanas begoniae* in begonia *elatior* by meristem culture and an isolation test. Acta Hortic 131: 299–301

Heide OM (1965) Interaction of temperature, auxins and kinins in the regeneration ability of *Begonia* leaf cuttings. Physiol Plant 18: 891–920

Hilding A, Welander T (1976) Effects of some factors on propagation of *Begonia* × *Liemalis* in vitro Swed J Agri Res 6: 191–920

Khoder M, Villemur P, Jonard R (1981) La multiplication végétative de l'éspèce florale Begonia elatoir (cultivar Rieger) à partir de différents organes cultivés in vitro. CR Acad SC Paris Ser III 293: 403–408

Khoder M, Villemur P, Jonard R (1984) The production of monoploid and triploid plants through in vitro androgenesis in *Begonia* × *Liemalis* Fotsch. Bull Soc France Let Bot 131: 43–48

Legro RAH, Doorenbos J (1969) Chromosome numbers in *Begonia*. Neth J Agric Sci 17: 189–202

Legro RAH, Haegeman JFV (1971) Chromosome numbers of hybrid tuberous begonias. Euphytica 20: 1–13

Margara J, Phelouzat R (1984) Structure and óntogeny of neoformations observed in vitro on *Begonia* × *elatior* petals. Can J Bot 62: 2798–2803

Matzke EB (1938) Inflorescence patterns and sexual patterns in *Begonia semperflorens*. Am J Bot 25: 465–478

Mikkelsen EP, Sink KC (1978) In vitro propagation of Rieger elatior begonias. HortScience 13: 242–244

Mikkelsen JC, Ryan J, Constantin MJC (1975) Mutation breeding of Rieger elatior begonias. Am Hortic 54: 18–21

Murashige T, Skoog F (1962) A revised medium for rapid growth and bioassays with tobacco tissue cultures. Physiol Plant 15: 473–497

Paterson KE, Rost TL (1979) Effects of light and hormones on regeneration of *Crassula argentea* from leaves. Am J Bot 66: 463–470
Peck DE, Cumming BG (1984) In vitro propagation of *Begonia* × *tuberhybrida* from leaf sections. HortScience 19: 395–397
Pierik RLM, Tetteroo FAA (1987) Vegetative propagation of *Begonia venosa* Skan in vitro from inflorescence explants. Plant Cell Tissue Organ Cult 10: 135–142
Rashid A, Reinert J (1983) Factors affecting high-frequency embryo formation in ab initio pollen cultures of *Nicotiana*. Protoplasma 116: 155–160
Ringe F, Nitsch JP (1968) Conditions leading to flower formation on excised *Begonia* fragments cultures in vitro. Plant Cell Physiol 9: 639–652
Roest S, van Berkel MAE, Boeelmann GS, Broertjes C (1981) The use of an in vitro adventitious bud technique for mutation breeding of *Begonia* × *hiemalis*. Euphytica 30: 381–388
Samyn GL, Debergh PC, Vermaerke D (1984) Field performance and phenotypic stability of virus-free tissue-cultured *Begonia* × *tuberhybrida* multiflora. Sci Hortic 24: 185–191
Sehgal CB (1975) In vitro differentiation of foliar embryo and adventitious buds from leaves of *Begonia semperflorens* Link & Otto. Indian J Exp Biol 13: 486–488
Sharma AK, Bhattacharyya YC (1957) Cytological studies in *Begonia*. II. *Begonia*. Carylogia 14: 279–301
Simmonds J (1984a) Induction of growth and direct rooting of adventitious shoots of *Begonia* × *hiemalis*. Plant Cell Tissue Organ Cult 3: 283–289
Simmonds J (1984b) Improved rooting of micropropagated *Begonia* × *hiemalis* (Riegers) shoots in a mist frame. Canadex 286.24 Agric Can Publ, Ottawa Canada
Simmonds J, Nelson SD (1988) Adventitious bud production on explants of *Begonia* × *hiemalis* depends on the developmental state of the donor plant. Physiol Plant 73: 360–367
Simmonds J, Nelson SD (1989) Improved micropropagation of *Begonia* × *hiemalis* by maintaining donor plants in long-day conditions. HortScience 24: 831–832
Simmonds J, Werry T (1987) Liquid-shake culture for improved micropropagation of *Begonia* × *hiemalis*. HortScience 22: 122–124
Takayama S, Misawa M (1981) Mass propagation of *Begonia* × *hiemalis* plantlets by shake culture. Sci Hortic 16: 65–75
Thakur S (1973) In vitro foliar shoot formation in *Begonia semperflorens*. Curr Sci 42: 430–432
Welvaert W, Samyn G, Van Wymersch E (1980) On the production of virus-free *Begonia tuberhybrida* Voss cv. multiflora varieties. Acta Hortic 110: 253–257
Westerhof J, Hakkaart FA, Versluijs JMA (1984) Variation in two *Begonia* × *hiemalis* clones after in vitro propagation. Sci Hortic 24: 67–74
Wirth K (1960) Experimentelle Beeinflüssung der Organbildung on in vitro kultivierten Stücken von *Begonia rex*. Planta 54: 265–293
Zang S, Fan Y, Zhao J (1982) Studies on anther culture of *Begonia semperflorens* Link & Otto. Acta Hortic Sin 9: 61–64

IV Micropropagation of Cacti (*Cactaceae*)

J.F. Hubstenberger, P.W. Clayton, and G.C. Phillips[1]

1 Introduction

The Cactaceae is a large and diverse family, containing many ornamental species. To avoid confusion resulting from myriad synonyms, scientific nomenclature cited by the authors conforms to Hunt (1967). Benson (1982) and Britton and Rose (1963) provide useful botanical details for many of the species. The following introductory sections present brief highlights about the family.

1.1 Botanical Aspects

Cacti are dicotyledonous perennial flowering plants adapted for survival in arid climates. Generally, members of this family have a specialized cuticle rich in waxes to prevent water loss, and a highly modified plant body. Leaves have been reduced to non-transpiring spines or are extremely ephemeral (e.g., *Opuntia*, see Fig. 1). The result is a fleshy stem that may be elaborated into tree- or shrub-like forms. Stem tissues function as the primary photosynthetic factory (using Crassulacean acid metabolism), and support many reduced, often dormant, lateral meristems. Lateral meristems usually are composed of two buds, one that produces the spine whorls and one that may produce axillary branches or flower buds. These buds are contained in the areole, a structure unique to cacti. Areoles occur along the ribs of some species, and may be isolated on projections of the ribs called tubercles. Specialized tubercles, as in *Leuchtenbergia*, contain the rudimentary axillary meristem. Flowers may be regular (sometimes zygomorphic) and perfect (although sometimes self-incompatible). The fruit is a one-celled berry containing many seeds. Seed may exhibit various degrees of dormancy and germinate over many years, ensuring that viable seed remains in the habitat over a long period of time. The basic chromosome number within the family is $x = 11$. Most species studied to date are diploid, but polyploids also occur.

1.2 Distribution and Importance

Cacti are native to North and South America, from Canada to Chile. Of the many arid regions of these two continents, Mexico has the richest diversity of cacti,

[1] Department of Agronomy and Horticulture, Plant Genetic Engineering Laboratory, New Mexico State University, Las Cruces, NM 88003-0003, USA

Biotechnology in Agriculture and Forestry, Vol. 20
High-Tech and Micropropagation IV (ed. by Y.P.S. Bajaj)

Fig. 1. Mature specimen of *Opuntia erinacea* in habitat

having more genera and species than any other country (Marshall and Bock 1977). The cacti are interesting candidates for the study of evolution and plant biochemistry. The religious significance of *Lophophora* to certain Native American tribes is well known. In addition, *Opuntia* has been used as a host plant for cochineal insects for the production of dyes. Certain species of *Opuntia* are cultivated as food crops in Mexico. *Opuntia* and other genera are utilized by local villages for emergency food and feed. The chief economic importance of cacti lies in worldwide horticultural trade as ornamental and collectible plants. It is difficult to determine the value of the cactus trade. An estimate of the rare and endangered cactus trade places the value at 5 to 7.5 million dollars annually in the USA, with 2.5 to 5 million units solid (David Kraenzel, NMSU, pers. commun.). The ornamental trade in cacti must be considerably larger than the specialized trade in rare and endangered cacti. Major areas for production of ornamental cacti include Holland, West Germany, Great Britain, Japan, and the USA.

1.3 Conventional Propagation Practices

Cacti are usually propagated by seed and cuttings. Commercial propagation by cuttings is most effective with species that produce numerous lateral branches. Species that never, or seldom, produce lateral branches traditionally rely on seed propagation or grafting. These methods are inadequate when germination and growth rates are low. Extremely slow-growing cacti, rare species, or unusual mutations such as chlorophyll-deficient forms, may by grafted. Fortunately, cacti can be grafted to unrelated genera such as *Trichocereus* or *Myrtillocactus*, chosen as rootstocks because of their ease of propagation. However, commercially grafted plants are especially susceptible to fungal infection, and plants grafted to tropical zone species may not enter into natural dormancy cycles. Although certain cactus species have disease and pest problems in nature, the economic importance of such problems is not clear for commercial situations. Breeding programs have been developed for only a few species. *Epiphyllum* and related genera have been extensively studied as ornamentals, while *Opuntia amyclaea* has been improved as a food crop. Further development of exotic hardy types would increase the use of cacti as garden plants. Many beautiful interspecific hybrids are popular as ornamentals.

1.4 Need for in Vitro Culture

The conventional methods of propagation are often inadequate to meet the commercial demands for those cacti that exhibit low germination rates, slow rates of growth, or low rates of lateral branching. In vitro propagation is a potential alternative for production of these plants. In addition to commercially important species, many wild species would benefit from tissue culture production. About 25% of all cacti are listed as endangered or potentially threatened with extinction because of habitat loss and collection from the wild. Although the US Endangered Species Act has made field collections illegal for targeted species, and seeks to preserve the native habitat, wild populations are the principal source of rare, difficult-to-propagate cacti. A reliable source of vegetatively propagated plants for the commercial cactus trade would reduce collection pressure on vulnerable wild populations. Micropropagation by proliferation of axillary branches will maintain genetic fidelity of the discrete donor clones, while efficiently producing large number of plants. The potential for development of cactus micropropagation techniques makes extinction of any species unjustifiable.

2 Review of Tissue Culture Work

Early studies of the in vitro culture of cactus focused on callus formation and proliferation (Nitsch 1951; King 1957; Sachar and Iyer 1959; Colomas 1971; Minocha and Mehra 1974). Kolar et al. (1976) were the first to obtain cactus shoots from in vitro callus cultures of *Mammillaria woodsii*. Shoots were subsequently recovered from callus of *Mammillaria elongata* (Johnson and Emino 1979a). Mauseth studied the hormonal control of cactus differentiation (Mauseth 1977b, 1978, 1979a), and demonstrated that axillary shoots could be forced to grow on intact seedlings (Mauseth and Halperin 1975; Mauseth 1976). Based on these results, Mauseth (1977a) proposed that tissue culture could be used for axillary shoot propagation of cacti. Micropropagation of cacti by axillary shoot proliferation was successfully demonstrated by Mauseth (1979b) and Johnson and Emino (1979b). Following these pioneering efforts micropropagation systems have been developed for 36 species of cactus representing 21 genera (Table 1). Axillary shoots proliferate with a four- to tenfold increase in shoot number every month (Sluis and Wochok 1980; Clayton et al. 1990). Shoots are rooted and plantlets are re-established with 80 to 100% efficiency (Vyskot and Jara 1984; Ault and Blackmon 1987).

Table 1. Species of cacti successfully micropropagated by axillary shoot proliferation

Species	References
Astrophytum myriostigma Lem.	Vyskot and Jara (1984)
Chamaecereus sylvestrii (Speg.) Br. & R.	Mauseth (1979b)
Coryphantha macromeris (Engelm.) Lem.	Smith et al. (1991)
Echinocereus engelmanii (Parry) Lem.	Ault and Blackmon (1985)
E. pectinatus (Scheidw.) Engelm.	Ault and Blackmon (1985)
Epiphyllum hybrid	Mauseth (1979b)
Escobaria missouriensis (Sweet) D. Hunt	Clayton et al. (1990)
E. robbinsorum (Earle) D. Hunt	Clayton et al. (1990)
Ferocactus acanthodes (Lem.) Br. & R.	Ault and Blackmon (1987)
F. covillei Br. & R.	Ault and Blackmon (1985)
F. wislizenii Br. & R.	Ault and Blackmon (1985)
Hatiora salicornioides (Haw.) Br. & R.	Mauseth (1979b)
Hylocereus calcaratus (Web.) Br. & R	Johnson and Emino (1979b)
Leuchtenbergia principis Hooker	Starling (1985)
Lobivia binghamiana Bckbg.	Mauseth (1979b)
Mammillaria carmenae M. Castenada	Vyskot and Jara (1984)
M. elongata D.C.	Johnson and Emino (1979b); Mauseth (1979b)
M. prolifera (Miller) Haw.	Vyskot and Jara (1984)
M. sanangelensis [= *M. haageana* (Peiff.) sensu Hunt]	Rubluo et al. (1986)
M. wrightii Engelm.	Clayton et al. (1990)
Opuntia amyclaeae Tenore	Escobar et al. (1986)
O. basilaris Engelm. & Big.	Mauseth (1979b)
O. polyacantha Haw.	Johnson and Emino (1979b); Mauseth (1979b)
Pachycereus pringlei (S. Wats.) Br. & R.	Mauseth (1979b)
Pediocactus bradyi L. Benson	Clayton et al. (1990)
P. despainii Welsh & Goodrich	Clayton et al. (1990)
P. knowltonii L. Benson	Clayton et al. (1990)
P. paradinei B. W. Benson	Clayton et al. (1990)
P. winkleri Heil	Clayton et al. (1990)
Pereskia aculeata Miller	Mauseth (1979b)
Rhipsalis teres (Vellozo) Steudel	Johnson and Emino (1979b)
Sclerocactus mesae-verdae (Boissevain) L. Benson	Clayton et al. (1990)
S. spinosior (Engelm.) Woodruff & Benson	Clayton et al. (1990)
Selenicereus grandiflorus (L.) Br. & R.	Mauseth (1979b)
Toumeya papyracantha Br. & R.	Clayton et al. (1990)
Trichocereus spachianus (Lem.) Riccob.	Vyskot and Jara (1984)

3 Micropropagation Procedures

Micropropagation of cacti by axillary shoot proliferation has resulted in a high degree of phenotypic uniformity, and presumed genetic stability, among the propagules thus obtained (Mauseth 1979b; Ault and Blackmon 1987). With the development of micropropagation procedures, cactus researchers have increasingly avoided callus proliferation because of the increased mutation rates generally associated with callus or cell cultures (Lee and Phillips 1988). The following sections outline the current procedures for cactus micropropagation by axillary shoot proliferation.

3.1 Choice of Explants for Micropropagation

Successful initiation of cactus cultures can begin with one or all of the following explants (Fig. 2). These include excised tubercles or areoles (axillary meristems), immature stem tips of greenhouse grown seedlings, mature stem tips from field-collected material, or epicotyls from seedlings germinated in vitro (Clayton et al. 1990).

The use of epicotyls from aseptically germinated seedlings has the advantage of eliminating microbial contamination. This explant is especially useful for rare species where field collection is impossible, or when lateral meristems are not available. Although germination rates of some cacti are low, and the small size of the explants generally results in a slow transition to the shoot multiplication stage, epicotyl explants may represent the surest way of initiating rare species or the progeny of intergeneric crosses.

Stems tips from either greenhouse-grown seedlings or mature field-collected specimens make satisfactory explants. Provided surface sterilization is successful, the large size of these explants ensures a rapid transition to shoot multiplication, in as little as one passage. Nondestructive sampling is especially vital for rare plants. Havel and Kolar (1983) developed a method for nondestructive sampling of donor plants by removing small amounts of areolar tissue with a sterile syringe; however, only callus growth was obtained in this case. Excision of entire individual tubercles or areoles is a more effective method of nondestructive sampling of donor plants, when the lateral meristems are already released from apical dominance (Clayton et al. 1990).

Whatever source of explant is used, material should be excised from vernalized, nondormant donor plants. Etiolation is sometimes used to induce new growth for explant collection of summer dormant species. For many nonclustering types of cacti, basipetal areoles may be highly recalcitrant unless the apical meristem is first excised.

3.2 In Vitro Culture Initiation

Surface sterilization of cactus plants is extremely important because spine clusters and hairs harbor large microbial populations. Surface sterilization agents have included 3% chloramine B solution (Vyskot and Jara 1984) or 5% calcium

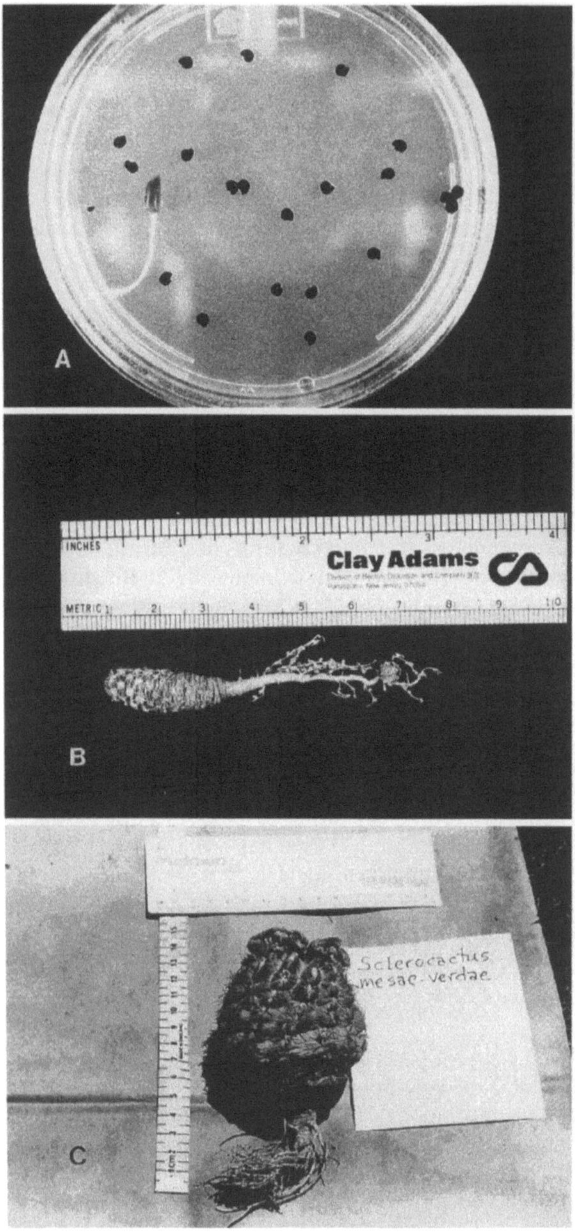

Fig. 2A–C. Materials used to provide cactus explants. **A** Aseptically germinated seedling of *Sclerocactus polyancistrus*. **B** Young seedling of *S. mesae-verdae* grown in the greenhouse. **C** Mature field-collected specimen of *S. mesae-verdae*

hypochlorite (Escobar et al. 1986) for 10 min. Sodium hypochlorite (commercial bleach) is also effective (Johnson and Emino 1979a, b). Intact stem apices are immersed in 95% ethanol for 1 min, followed by 2% sodium hypochlorite for 7 min, then rinsed three times in sterile distilled water (Clayton et al. 1990).

Heavily contaminated field collected materials are soaked in water containing detergent or other surfactant, then rinsed under running tap water for 5 min. Roots and spines are removed and the stem is placed into 95% ethanol for 1 min, followed by immersion in acidified 0.001% mercuric chloride for 2 min with constant agitation. Explants are then rinsed in three changes of sterile distilled water.

All vessel types can be used for cactus tissue culture. Large 100 × 25 mm Petri dishes are effective because they allow easy access to the plants, and adequate room for shoot multiplication. Transfers to fresh media are typically performed monthly.

Incubation environments have been reported to range from 24 to 27 °C, with a 16-h photoperiod. Clayton et al. (1990) reported incubation temperatures of 29 °C with uninterrupted fluorescent and incandescent light in excess of 10000 lx (150 $\mu mol/m^2/s$). Under these conditions the problem of vitreous shoots is virtually eliminated. As might be expected from the natural environment of these species, lower light intensities cause considerable etiolation and encourage callus proliferation. Continuous high light eliminates programming of midsummer dormancy. For example, *Echinocereus viridiflorus* var. *davisii* and *Escobaria minima* continue to flower in vitro as they do in early summer (Clayton et al., unpubl.).

3.3 Culture Media for Axillary Shoot Proliferation

The Murashige and Skoog (1962) basal nutrient medium has been used frequently for cactus micropropagation (Johnson and Emino 1979b; Sluis and Wochok 1980; Ault and Blackmon 1985, 1987; Escobar et al. 1986; Rubluo et al. 1986). Clayton et al. (1990) systematically compared five different basal media for cactus axillary shoot proliferation. Media formulations evaluated were Murashige and Skoog (1962) (MS), Schenk and Hildebrandt (1972) (SH), Gamborg et al. (1968) (B5) as modified by Dunstan and Short (1977), Phillips and Collins (1979) (L2), and a modified MS with half-strength major salts as described by Gladfelter and Phillips (1987) (MMS). In that study, the L2 basal medium was the superior basal nutrient formulation. MS, SH, and B5 were comparable for supporting axillary shoot proliferation. The low-salts MMS performed less satisfactorily than any of the other four media, all categorized as high-salts media.

Clayton et al. (1990) noted that the superior basal medium, L2, contains high total major and minor salts compared to the other media. L2 also contains high levels of magnesium and calcium ions, which may be important for cactus nutrition. The choice of any of the high-salts basal medium formulations may not be critical to establish cactus micropropagation cultures; however, this choice may become more critical as commercial interests seek to optimize the cost per unit in cactus micropropagation systems.

Early work in cactus micropropagation suggested that the growth regulator combinations required for axillary shoot proliferation are unique to each species (Johnson and Emino 1979a, b). Subsequent research has confirmed that the sources of auxin and cytokinin needed for optimal production of axillary shoots interact strongly with species (Clayton et al. 1990). In general, low levels of auxin or no auxin are required, in combination with moderate to high levels of cytokinin, for cactus axillary shoot proliferation (Table 2). Many cacti appear to have the capacity to produce excess auxin in vitro. The excess auxin often stimulates callus

Table 2. Growth regulator treatments used successfully to micropropagate various cacti (Clayton et al., unpubl.)

Taxa	Conservation status[a], if any	Auxin source[b] (mg/l)	Cytokinin source[b] (mg/l)
Ancistrocactus tobuschii W.T. Marshall	E	None	2-ip (5)
Ariocarpus agavoides (Castenada) Anderson		None or IAA (2)	2-ip (5) or Kin (4)
Arrojadoa penicillata (Gurke) Br. & R. var. *spinosior* Buin.		IAA (2)	Kin (4)
Astrophytum asterias (Zucc.) Lem.	C	NAA (0.02)	Zea (5)
Aztekium ritteri Boed.	P	None	BA (5) or 2-ip (5)
Coryphantha scheeri (Muehl.) Lem. var. *valida* (Engelm.) L. Benson		NAA (0.02)	Zea (5)
Echinocactus horizonthalonius Lem.	P	IAA (2)	Kin (4)
Echinocereus chloranthus Engelm.	P	IAA (2)	Kin (4)
E. fendleri (Engelm.) Ruempler var. *kuenzleri* (Castetter et al.) L. Benson	E	None	2-ip (5)
E. poselgeri Lem.		IAA (2)	Kin (4)
E. schmollii (Wein.) N.P. Taylor		IAA (2)	Kin (4)
E. viridiflorus (Engelm.) var. *davisii* (A.D. Houghton) W.T. Marshall	E	None	2-ip (5)
Echinofossulocactus violaciflorus (Quehl) Br. & R.		IAA (2)	Kin (4)
Echinomastus intertextus (Engelm.) Br. & R var. *dasyacanthus* (Engelm.) Bckbg.		None	2-ip (5)
Epithelantha micromeris (Engelm.) Web. var. *bokei* L. Benson		None	2-ip (5)
Escobaria hesteri (Y. Wright) F. Buxb.	P	NAA (0.05)	BA (1)
E. minima (Baird) D. Hunt	E	NAA (0.05)	BA (1)
E. missouriensis (Sweet) D. Hunt		IAA (2)	Kin (4)
E. organensis (D. Zimmerman) Castetter et al.		IAA (2)	Kin (4)
E. robbinsorum (Earle) D. Hunt	T	None or IAA (2)	2-ip (5) or kin (4)
E. sneedii Br. & R	E	None or IAA (2)	2-ip (5) or kin (4)
E. sneedii (Br. & R) var. *leei* (Boed.) D. Hunt	T	None	2-ip (5)
Leuchtenbergia principis Hooker		None	2-ip (5)

Mammillaria grahamii Engelm.		None	BA (5)
M. pectinifera (Web.) Hunt	P	None or IAA (0.02)	2-ip (5) or kin (4)
M. pottsii Scheer		None	2-ip (5)
M. thornberi Orcutt	P	NAA (0.05)	BA (1)
M. wrightii Engelm.		None	2-ip (5)
Melocactus matanzanus Leon		None	2-ip (5)
Neoevansia striata (Brand.) Sanchez-Mejorada		NAA (0.02)	2-ip (5)
Opuntia arenaria Engelm.	C	IAA (2)	Kin (4)
O. basilaris (Engelm. & Big.) var. *brachyclada* (Griffiths) Munz	C	IAA (2)	Kin (4)
O. erinacea Engelm. & Big.		None	2-ip (5)
Pediocactus bradyi (Earle) L. Benson	E	None	2-ip (5)
P. despainii Welsh & Goodrich	C	IAA (2)	Kin (4)
P. knowltonii L. Benson	E	NAA (0.05)	BA (1)
P. paradinei B. W. Benson	C	IBA (1)	2-ip (10)
P. peeblesianus (Croizat) L. Benson	E	IAA (2)	Kin (4)
P. Sileri (Engelm.) L. Benson	E	IAA (2)	Kin (4)
P. simpsonii (Engelm.) Br. & R		None	2-ip (10)
P. simpsonii (Engelm) Br. & R. var. *robustior* (Coult.) L. Benson		None	2-ip (5)
P. winkleri Heil	C	None	Kin (4)
Peniocereus greggii (Engelm.) Br. & R.	C	None	2-ip (5)
Sclerocactus glaucus (J.A. Purpus) Bckbg.	T	NAA (0.02)	Kin (5)
S. mesae-verdae (Boissevain) L. Benson	T	IBA (1)	2-ip (10)
S. polyancistrus (Engelm. & Big.) Br. & R.		None	2-ip (5)
S. spinosior (Engelm.) Woodruff & Benson	P	IAA (2)	Kin (4)
Thelocactus bicolor (Gal.) Br. & R.		IAA (2)	Kin (4)
Toumeya papyracantha (Engelm.) Br. & R.	C	NAA (0.02)	Zea (5)

[a]E = endangered species (US Federal List); T = threatened species (US Federal List); C = candidate for threatened status (US Federal List); and P = Proposed as threatened species (US or Mexico).
[b]IAA = indole-3-acetic acid; Kin = 6-furfurylaminopurine; NAA = 1-naphthaleneacetic acid; BA = 6-benzylaminopurine; IBA = indole-3-butyric acid; 2-ip = 6-dimethylallylaminopurine; zea = 6-[4-hydroxy-3-methylbut-2-enylamino]-purine.

production, which can be problematic in micropropagation systems. Species interact more strongly with cytokinin source than with auxin source (Clayton et al. 1990). Zeatin was the best cytokinin source for most species, but it is also the most expensive. This study concluded that growth regulator combinations may need to be empirically determined for each species.

The information summarized in Tables 1 and 2 represent excellent starting points for the micropropagation of a large number of cacti. The potential for axillary shoot production in several cacti are illustrated in Fig. 3. Some species also produce flowers in vitro during the axillary shoot proliferation stage (Fig. 4). Based on the results of Clayton et al. (1990), a set of three media could be used to screen the micropropagation potential of previously untried species of cacti. These screening media could be based on L2 or MS basal media, and supplemented with 5 to 10 mg/l zeatin as cytokinin, in combination with 0 to 0.02 mg/l NAA as auxin. In addition, we have found combinations of 2-ip or kinetin as cytokinin, with picloram or IAA as auxin, to be useful for several cacti. Most species should respond to one or more of these combinations.

3.4 Rooting and Re-Establishment

After the shoot multiplication phase, individual plantlets are placed on MS media for rooting (Fig. 5). This passage is the last stage where shoot quality, size, and vigor can be manipulated before re-establishment. Some species benefit from a specific auxin/cytokinin treatment to enhance shoot size and quality before root induction.

Ault and Blackmon (1987) report spontaneous rooting on auxin-free media with *Ferocactus acanthodes*. *Opuntia amyclaea* also roots without exogenous auxin (Escobar et al. 1986). Generally, there is a correlation between rooting in vivo and treatments required in vitro (Table 3). Species that root freely in vivo, such as certain species of *Opuntia* and *Chamaecereus*, also root spontaneously on auxin-free media. Some clustering types, such as certain species of *Escobaria* and *Mammillaria*, also are easily rooted on auxin-free media. Species that do not root easily on auxin-free media may root effectively after an initial auxin treatment (Table 3). Transfer to hormone-free media immediately after auxin treatments improves root development.

Growth regulator concentrations used for shoot multiplication stages can also have dramatic effects on subsequent rooting. The high cytokinin used for shoot proliferation adversely affects root induction. This problem can be eliminated by a passage on hormone-free medium to help reduce cytokinin carry over, before application of specialized rooting treatments. Root development on micropropagated shoots is illustrated in Fig. 6.

Ault and Blackmon (1987) report that nonrooted shoots of *Ferocactus acanthodes* were re-established with 90% success. With hard-to-root species, such as *Sclerocactus mesae-verdae*, unrooted shoots do not re-establish readily or uniformly (Clayton et al., unpubl.). Rooted plantlets must be protected from intense sunlight and desiccation. This can be done with polyethylene plastic (Ault and Blackmon 1987), or by providing plantlets with shade and humidity control. Adaptable species, such as *Opuntia*, will tolerate a wide variety of re-establishment

Fig. 3A–D. Axillary shoot proliferation of cacti cultured in vitro for 3–12 months. **A** *Epithelantha micromeris* var. *bokei* on MS medium + 5 mg/l 2-ip. **B** *Escobaria minima* on MS medium + 0.05 mg/l NAA + 1 mg/l BA. **C** *Mammillaria pectinifera* on MS medium + 5 mg/l 2-ip. **D** *Pediocactus paradinei* on MS medium + 1 mg/l IBA + 10 mg/l 2-ip

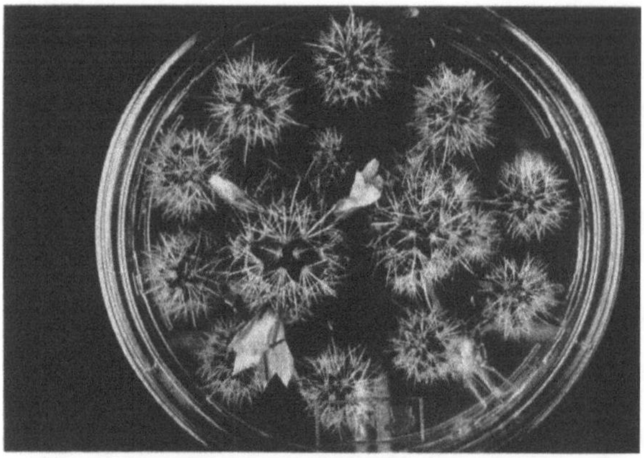

Fig. 4. Cultures of *Mammillaria thornberi* (10 months of age) showing in vitro flower formation on MS medium + 0.05 mg/l NAA + 1 mg/l BA

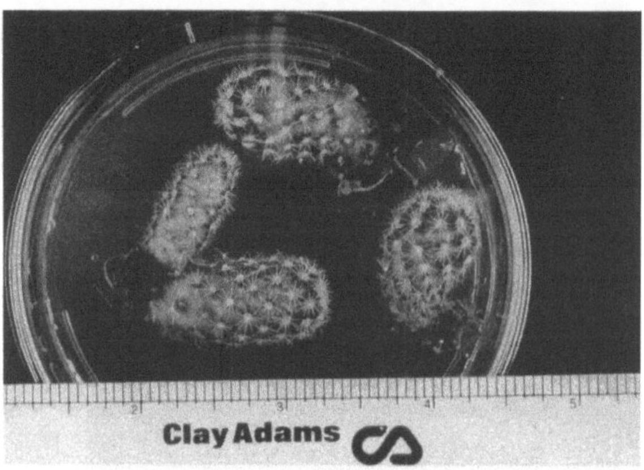

Fig. 5. Individual shoots of *Pediocactus bradyi* (9 months of age) derived from axillary shoot proliferation cultures, illustrating the size and quality desired for root induction (grown on hormone-free MS medium)

conditions. Species that require a specialized environment for cultivation will also benefit from these same or similar conditions for re-establishment. The genus *Pediocactus*, for example, requires cooler summer temperatures, low moisture, and excellent drainage during re-establishment.

Although cacti show fast growth rates during the axillary shoot proliferation stages, rooted propagules of cactus revert to normal, slower growth rates. Re-established plants of some genera and species develop a quality and size comparable to much older seedlings, a distinct advantage for slow-growing species (Figs. 7, 8).

3.5 Commercialization Using Biotechnology

The large number of cactus species that can be micropropagated indicates the potential for successful future commercialization. Optimization of multiplication and rooting media, as well as refinement of the re-establishment transition, is necessary for each species of interest. In addition, economic success dictates that

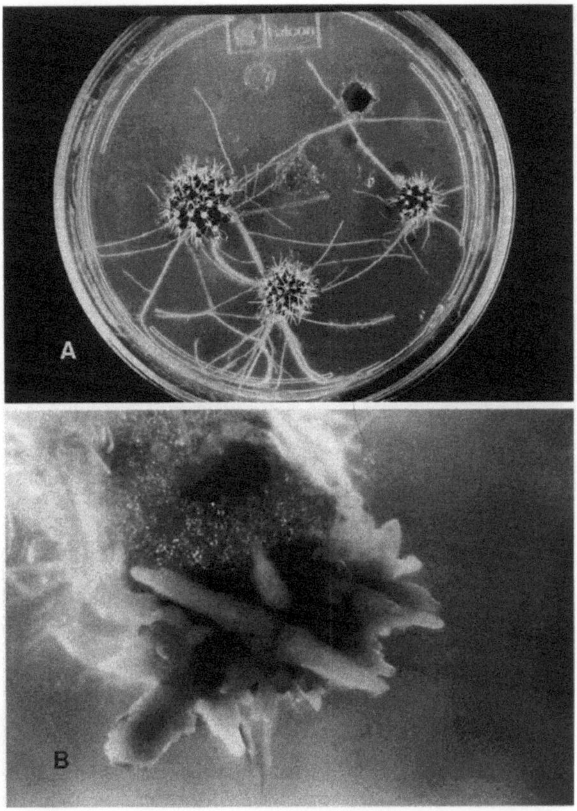

Fig. 6A, B. Root development on micropropagated shoots of cacti (9–15 months of age). **A** *Echinocereus fendleri* var. *kuenzleri* on hormone-free MS medium. **B** *Pediocactus bradyi* on MS medium + 5 mg/l NAA

more efficient production methods reduce the cost per unit. The greatest factor in bringing these techniques into the business sector will be the ability to produce plants at a price competitive with current market values. Demand for a single species may not reach the levels necessary for economy of scale. Overhead and labor costs will make some plants too expensive to market. Some species or cultivars have high enough intrinsic values to make these techniques economical. New hybrids, extremely rare species, and species that are difficult to cultivate will benefit most from these techniques. Commercial micropropagation would help utilize the rich, natural resource of endemic cacti for export in the world market.

Biotechnology could facilitate the commercialization of these underutilized family of unique plants in several ways. For example, only a few genera of cacti (such as *Epiphyllum* and *Schlumbergia*) have been intensively developed as horticultural crops. New cultivars and hybrids continually appear as novelties, keeping the trade active. Most cacti species have exacting cultural needs, and bloom in cultivation only if natural dormancy requirements are met. The production of exotic horticultural varieties that tolerate cultivation and bloom repeatedly would greatly stimulate the trade. The development of new cultivars may be facilitated by exploiting somaclonal variation in plants regenerated from callus or cell cultures.

Table 3. Rooting and re-establishment of 41 species of cacti

Species	Rooting without auxin	Rooting + auxin (mg/l)	Re-establishment in vivo	Reference
Arrojodoa penicillata var. *spinosior*	Yes	Unnecessary	Yes	[a]
Astrophytum asterias	Yes	?	Yes	[a]
A. myriostigma	?	Yes	Yes	Vyskot and Jara (1984)
Aztekium ritteri	Yes	?	No	[a]
Chamaecereus silvestrii	Yes	Unnecessary	Yes	[a]
Echinocereus fendleri var. *kuenzleri*	Yes	?	Yes	[a]
E. pectinatus var. *neomexicanus*	No	Yes (1) NAA	Yes	Ault and Blackmon (1985)
E. poselgeri	Yes	Unncessary	Yes	[a]
E. schmollii	Yes	?	Yes	[a]
E. viridiflorus var. *davisii*	Yes	Unnecessary	Yes	[a]
Echinomastus intertextus var. *dasyacanthus*	Yes	?	No	[a]
Epithelantha micromeris var. *bokei*	Yes	Unnecessary	Yes	[a]
Escobaria missouriensis	Yes	Yes (2–5) NAA or (5–10) IBA	Yes	Clayton et al. (1990)
E. robbinsorum	Yes	Yes (2–5) NAA or (5–10) IBA	Yes	Clayton et al. (1990)
E. sneedii	Yes	Unnecessary	Yes	[a]
E. sneedii var. *leei*	Yes	Unnecessary	Yes	[a]
Leuchtenbergia principis	Yes	Yes (5–10) IBA	Yes	Starling (1985)
Mammillaria carmenae	No	Yes (1) IAA	Yes	Vyskot and Jara (1984)
M. grahamii	Yes	?	Yes	[a]
M. pectinifera	Yes	?	Yes	[a]
M. pottsii	Yes	?	No	[a]
M. prolifera	No	Yes (1) IAA	Yes	Vyskot and Jara (1984)
M. thornberi	Yes	Unnecessary	Yes	[a]
M. sanangelensis	Yes	?	Yes	Rubluo et al. (1986)
M. wrightii	Yes	Yes (5–10) IBA	Yes	Clayton et al. (1990)
Neoevansia striata	Yes	Unnecessary	Yes	[a]

Opuntia amyclaea Copena-5	Yes	Yes (0.5) IBA	Yes	Escobar et al. (1986)
O. arenaria	Yes	Unnecessary	Yes	[a]
O. basilaris var. *brachyclada*	Yes	Unnecessary	Yes	[a]
O. erinacea	Yes	Unnecessary	Yes	[a]
Pediocactus bradyi	Yes	Yes (2–5) NAA or IBA or IAA	Yes	Clayton et al. (1990)
P. despainii	Yes	Yes (5–10) IBA or (5–10) NAA + (5–10) IAA	Yes	Clayton et al. (1990)
P. knowltonii	Yes	Yes (2–5) NAA	Yes	Clayton et al. (1990)
P. paradinei	Yes	Yes (2–10) NAA or IBA or IAA	Yes	[a]
P. simpsonii	Yes	?	Yes	[a]
P. winkleri	Yes	Yes (2–5) NAA or IBA or IAA	Yes	Clayton et al. (1990)
Peniocereus greggii	Yes	?	Yes	[a]
Sclerocactus spinosior	Yes	Yes (5–10) IBA or (5–10) NAA + (10) IAA	Yes	Clayton et al. (1990)
S. mesae-verdae	No	Yes (5–10) IBA or (5–10) NAA + (10) IAA	Yes	Clayton et al. (1990)
Toumeya papyracantha	Yes	Yes (5–10) IBA or (5–10) NAA + (10) IAA	Yes	Clayton et al. (1990)
Trichocereus spachianus	Yes	Yes (1) IAA	Yes	Vyskot and Jara (1984)

[a]Clayton et al., unpubl.

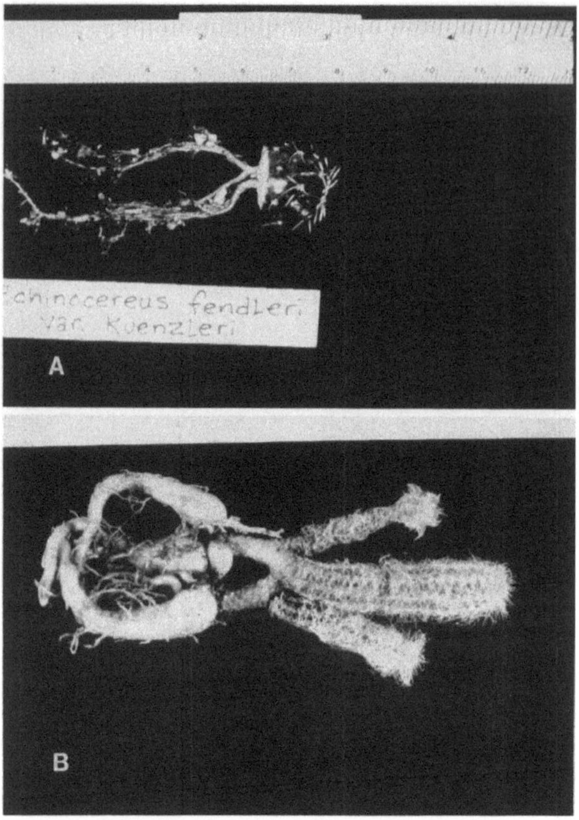

Fig. 7A, B. Micropropagated plants of cacti after 6 to 9 months of growth in soil mixes under greenhouse conditions. **A** *Echinocereus fendleri* var. *kuenzleri*. **B** *E. schmollii*

Fig. 8. Mature plant of micropropagated *Leuchtenbergia principis* after 2 years growth under greenhouse conditions

Embryo rescue and protoplast fusion techniques could ultimately succeed in the hybridization of selected species when sexual crosses prove impossible. Shoot meristem-tip cultures could be used to provide disease-free stocks for commercial development or export without quarantine. Cacti also make excellent candidates for the study of secondary plant products (Steinhart 1962). Any of these techniques combined with micropropagation could revolutionize the improvement, production, and utilization of cacti.

3.6 Considerations for Conservation and Preservation

A more immediate benefit of micropropagation, however, is in the areas of conservation and preservation. Reduction of collection pressure on wild populations may depend on the successful commercialization of these species. Given the current economic constraints, this is likely to take time. However, continued development of micropropagation techniques will provide direct aid for endangered species. It is feasible to produce plants to distribute to the scientific community so that further collection of plants from the wild is unnecessary. For some species, it is possible to adequately represent existing genetic variation by sampling a population of seedlings from a diverse locale. For other species, an appropriate sampling of a natural population can be done in a nondestructive manner. Thus, an entire gene pool can be adequately represented in a set of discrete clones. In addition to reducing collection pressure and providing cloned plants for study, micropropagation can be utilized for long-term germplasm storage. All of these approaches represent practical ways to immediately preserve rare plant material.

Direct benefits in the area of conservation may be more difficult to achieve. The most urgent priority of conservation is the protection of natural habitat. Natural ecosystems are vulnerable to disturbance, which can place species at risk of extinction. Certain species, such as *Pediocactus knowltonii*, exist in small type localities. A natural disaster could effectively eliminate the total existing population. Therefore, habitat preservation alone may be inadequate to ensure the survival of a species.

Some conservationists advocate the use of outplanting or transplanting of endangered species. The naturalization of such species into new favorable habitats has the potential to save a species. Micropropagated plants from balanced sets of discrete clones could be used to facilitate this approach, thus avoiding the risks of monoculture. Some evidence exists that short-term outplantings of *Mammillaria sanangelensis* are successful (Rubluo et al. 1986). However, Fahselt (1988) points out that persistence of transplants for more than a few seasons is unusual, and that transplantation is prohibitively expensive. Furthermore, there is a danger in manipulating the sensitive ecological balance of a specialized natural community (Fahselt 1988). Desert habitats are extremely fragile, and extinction of species therefore remains a danger.

One remaining alternative use of micropropagated cacti may be for outplanting into botanical gardens. However, even if the botanical collection is in an area that approximates natural conditions, such collections are likely to be ephemeral. In cases such as *Pediocactus knowltonii*, where large numbers of micropropagated plants are available, conservation efforts should be facilitated. If such propagules are not suitable for outplanting, a reliable population base exists for cultivation. Ultimately, cultivation is preferable to extinction.

4 Summary and Conclusions

Successful micropropagation of many cactus species has already been accomplished. Although additional species will require the identification of individualized growth regulator treatments, data exist that provide an excellent starting point for many cacti. Continued optimization of micropropagation techniques promises to clarify the conditions required for commercial production. Opportunities exist to expand the horticultural value of cacti, both through micropropagation and through genetic improvement. However, to preserve endangered cacti effectively it is necessary to understand the biology, physiology, and reproductive capacity of these plants in the wild. Micropropagated plants retain the genetic constitution of donor plants with the greatest fidelity. Clearly, the present work provides a basis for a rational approach to conservation of members of the Cactaceae. Fragile ecosystems in the arid areas of the Americas make this family, as a whole, vulnerable to loss. It is important that the conservation of plant species exploits technology that guarantees the survival of species that have long generation times and low reproduction rates. Micropropagation of rare species can accomplish these goals in the Cactaceae. In addition, cacti possess unique chemical components and have evolved genetic strategies to survive in extreme environments, which merit further basic study.

5 Protocol

The best explants for micropropagation of cacti are stem apices of mature plants, stem apices of immature seedlings, and excised epicotyls from aseptically germinated seedlings. Non-destructive sampling of mature donor plants is feasible by excision of individual areoles or lateral meristems.

The best basal medium for axillary shoot proliferation is the L2 formulation, although MS is also appropriate. Combinations of 5 to 10 mg/l zeatin as cytokinin, with 0 to 0.02 mg/l NAA as auxin, are effective for many species of cactus. Many cacti root spontaneously on hormone-free medium, or following treatment with 5 to 10 mg/l auxin (NAA or IBA).

Optimal incubation conditions include high light intensity ($>$10000 lx) and temperatures from 26 to 29 °C.

Acknowledgements. The authors express their appreciation to the many individuals who have supported their research. New Mexico Agricultural Experiment Station Special Report No. 83.

References

Ault JR, Blackmon WJ (1985) In vitro propagation of selected native cacti species. Hort Science 20: 541 (Abstr)
Ault JR, Blackmon WJ (1987) In vitro propagation of *Ferocactus acanthodes* (Cactaceae). Hort Science 22: 126–127
Benson L (1982) The cacti of the United States and Canada. Stanford Univ Press, Stanford, CA
Britton NL, Rose JN (1963) The Cactaceae. Dover, New York
Clayton PW, Hubstenberger JF, Phillips GC, Butler-Nance SA (1990) Micropropagation of members of the Cactaceae subtribe Cactinae. J Am Soc Hortic Sci 115: 337–343
Colomas J (1971) Obtention de cultures de tissues à partir de fragments de tiges de *Pachycereus pringlei* (S. Wats.) Br. et R. CR Acad Sci Paris Ser D 272: 1380–1382
Dunstan DI, Short KC (1977) Improved growth of tissue cultures of the onion, *Allium cepa*. Physiol Plant 41: 70–72
Escobar HA, Villalobos VM, Villegas A (1986) *Opuntia* micropropagation by axillary proliferation. Plant Cell Tissue Organ Cult 7: 269–277
Fahselt D (1988) The dangers of transplantation as a conservation technique. Nat Areas J 8: 238–244
Gamborg OL, Miller RA, Ojima K (1968) Nutrient requirements of suspension cultures of soybean root cells. Exp Cell Res 50: 151–158
Gladfelter HJ, Phillips GC (1987) De novo shoot organogenesis of *Pinus eldarica* Medw. in vitro. I. Reproducible regeneration from long-term callus cultures. Plant Cell Rep 6: 163–166
Havel L, Kolar Z (1983) Microexplant isolation from Cactaceae. Plant Cell Tissue Organ Cult 2: 349–353
Hunt DR (1967) *Cactaceae*. In: Hutchinson GE (ed) Genera of flowering plants. vol II. Oxford Univ Press, London, pp 427–467
Johnson JL, Emino ER (1979a) In vitro propagation of *Mammillaria elongata*. Hort Science 14: 605–606
Johnson JL, Emino ER (1979b) Tissue culture in the Cactaceae. Cactus Succulent J (US) 51: 275–277
King MR (1957) Studies in the tissue culture of cacti. Cactus Succulent J (US) 29: 102–104
Kolar Z, Bartek J, Vyskot B (1976) Vegetative propagation of the cactus *Mammillaria woodsii* through tissue cultures. Experientia 32: 668–669
Lee M, Phillips RL (1988) The chromosomal basis of somaclonal variation. Annu Rev Plant Physiol Mol Biol 39: 413–437
Marshall WT, Bock TM (1977) Cactaceae. Abbey Garden Press, Pasadena, CA
Mauseth JD (1976) Cytokinin and gibberellic acid induced effects on the structure and metabolism of shoot apical meristems in *Opuntia polyacantha* (Cactaceae). Am J Bot 63: 1295–1301
Mauseth JD (1977a) Cactus tissue culture: a potential method of propagation. Cactus Succulent J (US) 49: 80–81
Mauseth JD (1977b) Cytokinin and gibberellic acid induced effects on the determination and morphogenesis of leaf primordia in *Opuntia polyacantha* (Cactaceae). Am J Bot 64: 337–346
Mauseth JD (1978) An investigation of the phylogenetic and ontogenetic variability of shoot apical meristems in the Cactaceae. Am J Bot 65: 325–333
Mauseth JD (1979a) Cytokinin elicited formation of the pithrib meristem and other effects of growth regulators on the morphogenesis of *Echinocereus* (Cactaceae). Am J Bot 66: 446–451
Mauseth JD (1979b) A new method for the propagation of cacti: sterile culture of axillary buds. Cactus Succulent J (US) 51: 186–187
Mauseth JD, Halperin W (1975) Hormonal control of organogensis in *Opuntia polyacantha* (Cactaceae). Am J Bot 62: 869–877
Minocha SC, Mehra PN (1974) Nutritional and morphogenetic investigations of callus cultures of *Neomammillaria prolifera* Miller (Cactaceae). Am J Bot 61: 168–173
Murashige T, Skoog F (1962) A revised medium for rapid growth and bioassays with tobacco tissue cultures. Physiol Plant 15: 473–497
Nitsch JP (1951) Action du jus de tomate sur la croissance de tissue de crown-gall cultivés in vitro. CR Acad Sci Paris 233: 1676–1677
Phillips GC, Collins GB (1979) In vitro tissue culture of selected legumes and plant regeneration from callus cultures of red clover. Crop Sci 19: 59–64
Rubluo A, Chavez V, Martinez O, Martinez A (1986) The recovery of endangered Mexican plant

population through tissue culture. VI Int Congr on Plant tissue and Cell Cult Abstr, Univ Minnesota, Minneapolis, MN, p 428

Sachar RC, Iyer RD (1959) Effect of auxin, kinetin and gibberellin on the placental tissue of *Opuntia dillenii* Haw. cultured in vitro. Phytomorphology 9: 1–3

Schenk RU, Hildebrandt AC (1972) Medium and techniques for induction and growth of monocotyledonous and dicotyledonous plant cell cultures. Can J Bot 50 :199–204

Sluis CJ, Wochok ZS (1980) In vitro propagation of an endangered *Pediocactus* species. Plant Physiol 65: S-36 (Abstr)

Smith RH, Burdick PJ, Anthony J, Reilley AA (1991) In vitro propagation of *Coryphantha macromeris*. Hort Science 26: 315

Starling RJ (1985) In vitro propagation of *Leuchtenbergia principis*. Cactus Succulent J (US) 57:114–115

Steinhart CE (1962) Tissue culture of a cactus. Science 137: 545–546

Vyskot B, Jara Z (1984) Clonal propagation of cacti through axillary buds in vitro. J Hortic Sci 59: 449–452

V Micropropagation of *Chrysanthemum*

Y.P.S. BAJAJ[1], M.M.S. SIDHU[2], and A.P.S. GILL[3]

1 General Account

Chrysanthemums (familty Compositeae) are valued for their cut flowers and potted plants. The common commercially available chrysanthemum is *Chrysanthemum morifolium* Ramat (*Dendranthema morifolium, D. grandiflora*). They are available around the world however, in terms of stem production per year Japan is the leading most country (2 billion), followed by the Netherlands (800 m), Colombia (600 m), Italy (500 m), and the USA (300 m) (Anon 1987). The plants are vegetatively propagated but the production is highly affected by the many viral diseases (Horst 1989).

To obtain virus-free plants, meristem culture has been successfully applied and micropropagation through in vitro culture as a means of large-scale production of disease-free chrysanthemum is now being widely practised. The plants can be regenerated from segments of the stem, leaf, capitulum, and from meristems, callus, protoplasts, etc., and numerous publications have appeared on the subject (see Table 1). The plants show a wide range of chromosome numbers (Miyazaki and Tashiro 1978), and somaclonal variants for salt tolerance (Dalsou and Short 1985), plants with different flower color, size, shape, etc. have been obtained (Khalid et al. 1989). In this chapter our work on various aspects of the in vitro culture of *Chrysanthemum morifolium* and the factors affecting regeneration is summarized.

2 In Vitro Culture and Micropropagation

The literature on various aspects of in vitro culture of chrysanthemums is given in Table 1, and the results are shown in Figs. 1–5.

2.1 Establishment of Cultures and Media

Segments (1–1.5 cm) of stem, leaves, peduncle, inflorescence, or entire small capitulum and shoot tips of greenhouse-grown *Chrysanthemum morifolium* Ramat

[1] Former Professor of Tissue Culture, Punjab Agricultural University, Ludhiana, India. (Present address: A-137 New Friends Colony, New Delhi 110065, India)
[2] Carewel Nursery, Pakhowal Road, Ludhiana (Punjab), India
[3] Punjab Agricultural University, Ludhiana, India

Biotechnology in Agriculture and Forestry, Vol. 20
High-Tech and Micropropagation IV (ed. by Y.P.S. Bajaj)

Table 1. Various explants and tissues used for micropropagation of *Chrysanthemum morifolium*

Explant/tissue	Reference
Flower buds	Tsukamoto and Fujime (1971)
Meristems	Marani (1972), Ben-Jacov and Langhans (1972)
Shoot tips	Earl and Langhans (1974a, b), Bush et al. (1976) Bajaj (1976), Prasad and Chaturvedi (1988)
Pedicel segments	Roest and Bokelmann (1975)
Petal segments	Bush et al. (1976)
Leaf segments	Kaul et al. (1990)
Stem segments	Lu et al. (1990)
Inflorescence segments	Bajaj et al. (1992)
Callus	Arunyanart and McComb (1986)
Protoplasts	Sauvadet and Brochard (1989)

cv. Fredyule were surface sterilized with freshly prepared chlorine water for 5–10 min.The explants were then washed twice with sterile distilled water, cut to appropriate sized segments, and cultured aseptically on various media.

The basal media of MS (Murashige and Skoog 1962) and B5 (Gamborg et al. 1968) with the following combinations of growth regulators were employed:

MS1 (MS + IAA 2 mg/l + kin 0.5 mg/l)
MS2 (MS + IAA 4 mg/l + kin 2 mg/l)
MS3 (MS + NAA 0.2 mg/l + kin 2 mg/l)
B5 + 2,4-D 2 mg/l

The media contained 2% sucrose, and were gelled with 0.8% agar. The pH of the media was adjusted to 5.6–5.8, and 12 ml of each was poured into test tubes, and autoclaved. The cultures were maintained in a well-illuminated room at 24–27 °C.

Five- to 6-week-old in vitro-grown plants, when they had developed some roots, were taken out of the test tubes, and the agar around the roots was gently removed by washing under running tap water. They were then transferred to various soil mixtures in small plastic pots, covered with polythene bags or glass beakers for a week. During the next 2–3 weeks when the plants had developed sufficient roots and established well, they were transplanted into larger pots and maintained in them. For further multiplication, the terminal cuttings were planted in the pots containing sand.

2.2 Factors Affecting Regeneration/Micropropagation

2.2.1 *Effect of Different Explants and Media*

Stem tips, capitulum, leaves, segments of the stem, peduncle, and inflorescence were cultured on various media and their growth response is shown in Table 2 and Fig. 3.

Table 2. Differential growth responses of various explants of *Chrysanthemum morifolium* cv. Fredyule on different media

Explants	MS2			MS3			B5		
	Callus	Roots	Shoots	Callus	Roots	Shoots	Callus	Roots	Shoots
Stem tip	+	+ + +	+ +	+	–	+ + +	+ + +	–	–
Stem segments	+	+ +	+ +	+	–	+ +	+ +	–	–
Leaf	+ +	+ +	–	+	–	–	+ +	–	–
Inflorescence segments	+	+	–	+ +	–	+ +	ND	ND	ND
Capitulum	+	+ +	–	+ +	–	–	ND	ND	ND
Florets	+	+ +	–	+	–	–	+	–	–

– : No response, + : poor response, + + : good response, + + + : best response, ND: not done

1. Culture of Capitulum and Segments of the Peduncle Within a week of culturing, the peduncle elongated, and the cut end proliferated to form callus (Fig. 1A). The capitulum increased in size and in some cases produced profuse rooting (Fig. 1B). The callus produced buds and multiple shoots in 3–5 weeks (Fig. 1C, D). The segments of the peduncle from young capitulum grown on MS + IAA (4 mg/l) + kin (2 mg/l) developed shoots from the basal cut end, and all along the sides. The number of shoots produced was much higher than that produced by other explants. These results are in conformity with those obtained by Roest and Bokelmann (1975).

2. Leaf Culture. Young leaves and leaf segments, within 2 weeks of culturing on MS + 2,4-D (2 mg/l), showed localized initiation of small patches of white callus on their lower surface, and at the cut end of the petiole. The callus continued to grow, and spread all round (Fig. 1F). On MS2 and MS3 media, the leaves developed roots at the petiolar end (Fig. 1E).

3. Culture of Meristem Tip and Stem Segment. The meristem tip (1–5 mm) and segments of stem (1–2 cm) on MS1, started to elongate within 3 days, and produced shoots and plants in 3–7 weeks either directly (Fig. 4A), or via callus formation (Fig. 1G–I). The shoots which did not develop roots on MS1, when transferred to the basal medium containing NAA (2 mg/l) started rhizogenesis within a week; the root initials appeared at the basal end of such shoots, and plantlets were established.

The maximum response has been shown by capitulum, segments of the peduncle and stem, shoot tips, florets, and leaves in descending order. Medium MS3 was found to be better than MS2 and B5 (Fig. 3). The growth response of the explants on B5 in terms of regeneration was poor although callus formation was initiated on this medium.

2.2.2 Effect of Size and Orientation of the Explants on Growth Response

Stem tips and segments of sizes varying from 0.25–1.5 cm were cultured on MS2, MS3, and B5 media. The explants of less than 0.5 cm size mostly formed hard and green callus (Table 2), whereas the explants longer than 1 cm resulted in plantlet formation on MS2 and MS3 media. However, on B5 medium there was callus induction irrespective of size, although the callus proliferation was observed more frequently from the stem tip cultures than the other explants.

Experiments were also conducted on stem tips and stem segments by culturing them in two different orientations, i.e., flat and upward position. When the explants

Fig. 1A–I. The in vitro culture of various explants of *Chrysanthemum*. **A, B** Two- and 4-week-old cultures of capitulum on MS2 showing callus formation, and rooting at the basal cut end respectively. **C, D** Two- and 3-week-old cultures on MS3 showing the regeneration of multiple shoots at the basal end of the peduncle, and anthesis of the capitulum. **E, F** Culture of excised leaves on MS2 and B5 media, regenerating the roots and callus respectively. **G, H** The segment of a stem on MS1, 2 and 4 weeks after culturing, showing the regeneration of a plant. **I** An entire plant obtained from a meristem tip 7 weeks after culture. (Bajaj et al. 1992)

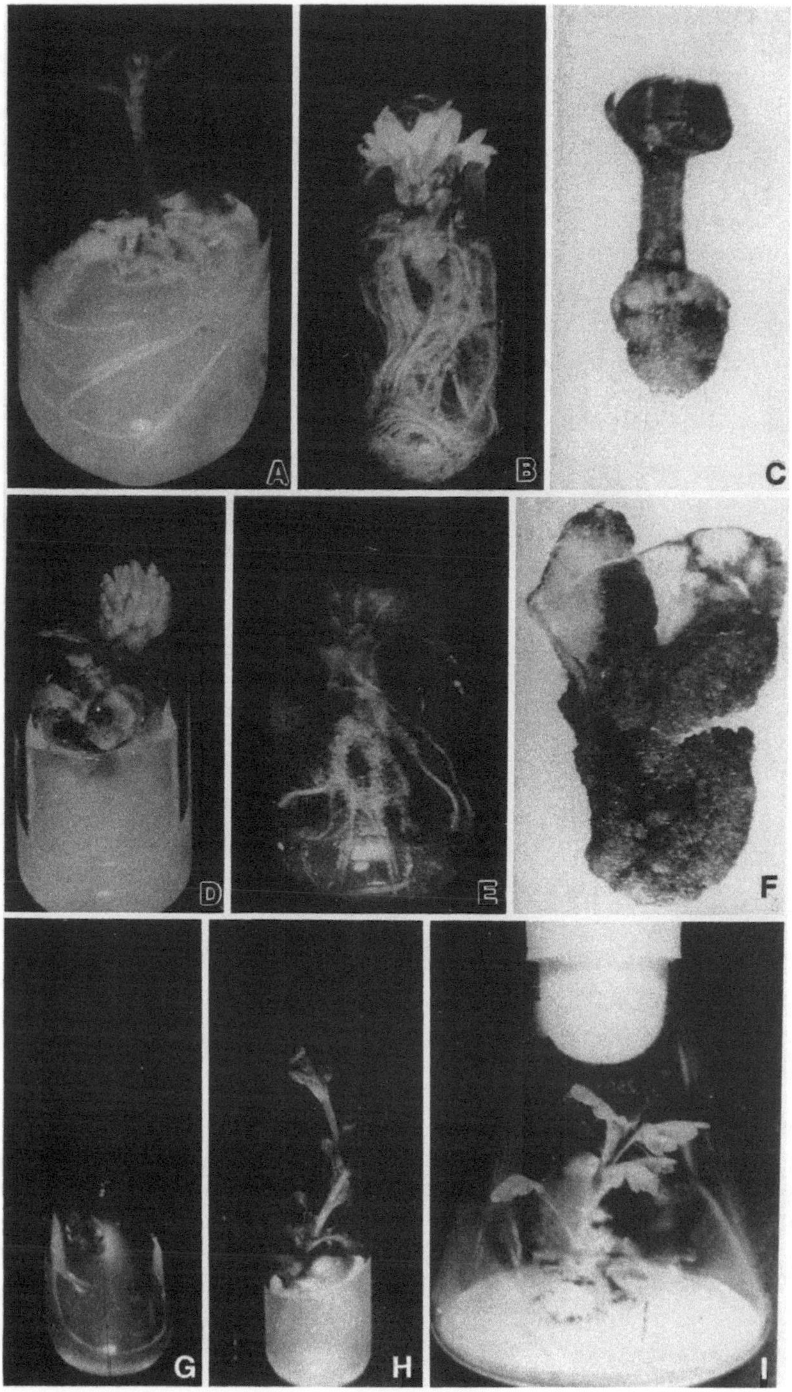
A
B
C
D
E
F
G
H
I

Fig. 2A–D. Multiple shoots and plants (about 50) obtained from a single capitulum 6 weeks after culture, and the transferring of plants to pots

were cultured flat on the surface of media MS2 and MS3, instead of regeneration of plantlets they frequently formed callus, whereas rooting was observed from callus on MS2 medium. The callus induction on B5 medium was also more from the flat oriented explants.

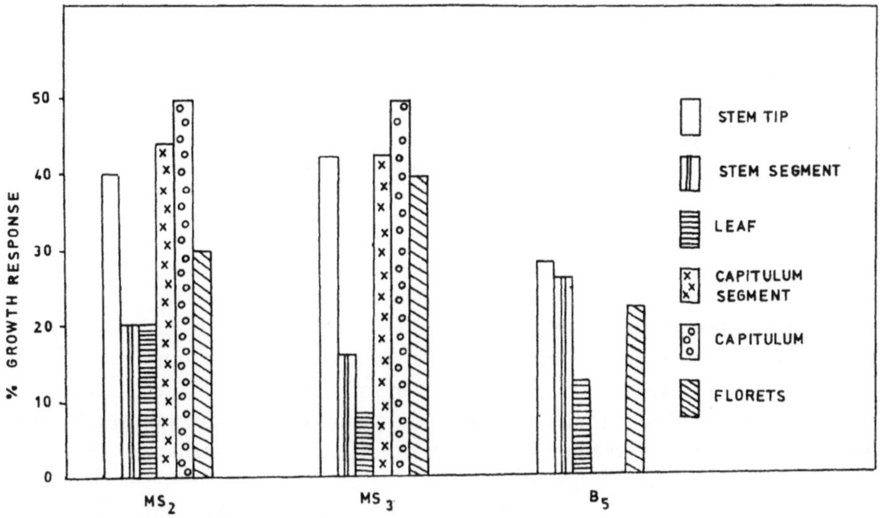

Fig. 3. Comparison of the in vitro response of different explants of *Chrysanthemum morifolium* cv. Fredyule on various media

2.2.3 Effect of Growth Regulators on Organogenesis in Cultured Explants

The effect of IAA, kinetin, and 2,4-D in various concentrations and combinations on different explants was studied. Table 2 shows the differential growth responses of various explants cultured on MS2, MS3 and B5 media. On MS2, callus formation and rooting were observed in almost all the explants, whereas shoot formation occurred only from the stem tip and stem segment. On MS3, none of the explants showed rooting; however, induction of hard and compact green callus was observed in almost all the explants. The shoot formation occurred only from stem tip, stem segment, and segments of the flower bud. On B5, rooting and shoot formation were completely absent, but callus formation was observed in all the cultured explants.

2.2.4 Differentiation of Plants from Callus

The segments of inflorescence cultured on MS3 medium showed swelling at the cut end after 7 days, there was also growth of the petals and formation of callus at their base within 20–25 days. This callus was mostly white, slow-growing, compact, and hard, showing green pigmentation in places when periodically subcultured on MS + 2,4-D for over 1 year, and on transferring to MS3, this callus differentiated to form multiple shoots and plantlets in 3 weeks. Regeneration into plantlets took about 35 days.

For differentiation studies the calli obtained from various explants on different media were subcultured on MS3. The calli derived from stem tip and stem segments on B5 medium, when transferred to MS3, lost their original pale yellow color and turned green. Callus derived from the stem segments cultured on MS3 medium took about 50 days for differentiation (Fig. 1G–I). Multiple shoot formation was observed 60 days after culture. The shoots obtained via callus differentiation were also cut into segments and put on to fresh MS2 medium to obtain complete plantlets. As many as 51 plantlets and multiple shoots were formed from a single callus.

2.2.5 Large-Scale Propagation of Plants

The explants for mass multiplication in vitro were cultured on MS2 and MS3. Emphasis was placed on culturing excised stem segments and shoot tips of 1.5 cm in size (Fig. 4A) to generate more plantlets in minimum time. The stem segments and shoot tips on MS3 started to elongate and produced new leaves during the first week. The shoots that emerged from the shoot tips and stem segments doubled in length after 20 days of culturing (Fig. 4A).

The excised stem segments (1.5 cm) and shoot tips (0.5 cm) exhibited differential response to plant regeneration. In the case of stem segment, the regeneration was 20% on MS2, and 16% on MS3, whereas plant formation from stem tips was observed in 40% cases on MS2, and 42.5% on MS3 (Fig. 3). The segments took a longer time (20 days) to produce 5–6-cm-long shoots, each having seven or eight leaves, on MS3 medium. Delayed rooting as well as shoot formation was observed on MS2 medium. However, segments having one bud each from the in vitro-grown plantlets showed better growth response (80%) both on MS2 and MS3 media. In

Fig. 4A–G. Mass multiplication of *Chrysanthemum morifolium* **A** Various stages in the formation of plants from shoot tips cultured for 8, 20, 30, and 40 days on MS + IAA (2 mg/l) + kin (0.05 mg/l). **B** Large-scale incubation of such cultures at 25 °C. **C** Storage of cultures on a glass trolley for easy transport. **D, E** In vitro-obtained plantlets immediately after transfer to pots covered with glass beakers and polyethylene bags to avoid desiccation. **F, G** Propagation of plants through terminal cuttings of an in vitro-obtained plant; **F** a plant suitable for taking terminal cuttings; **G** the terminal cuttings planted in sand for the induction of roots. (Bajaj 1986)

the case of segments cultured on MS2, profuse rooting was observed within 15–20 days, and complete plantlet formation took about 40 days.

Nineteen hundred cultures were raised (Fig. 4) from different types of explants, and after discarding the infected ones, nearly 400 complete plants were obtained in a period of 9 weeks. Furthermore, segments from 3-week-old plants which had originally been obtained in vitro upon subsequent culturing onto the fresh medium provided the necessary boost in enhancing multiplication rate. Likewise, starting with one stem segment (1.5 cm long), it was possible to produce 50 complete plants in 9 weeks.

2.2.6 Transfer of Plants to Soil

The in vitro-regenerated plants were transferred to two types of soil mixtures, and covered with glass beakers or polythene bags to avoid excessive desiccation.

1. Effect of Soil. A comparison of the two soil mixtures was made and the plants were transferred to small earthen pots (Fig. 4D) containing Mixture I (sand 40% + leaf mold 40% + farmland manure 20%) and Mixture II (sand 50% + leaf mold 25% + sawdust 25%). There was a mortality rate of 80% and 25% in Mixture I and Mixture II respectively. This may be attributed to the differences in the texture, aeration, and drainage of the two soils. The sawdust in Mixture II is responsible for better aeration and drainage.

2. Effect of Covering. The young in vitro-obtained plantlets transferred to the soil in pots, if left uncovered, have a tendency to undergo quick desiccation and wither within a day. Thus the plants were covered either with an inverted glass beaker or by a transparent polythene bag (Fig. 4D, E) for a week. In general, covering the plants with polythene bags was better than the glass beakers. Moreover, polythene bags with small holes in them yielded healthy-looking and sturdier plants, the holes helped better gaseous exchange and removal of excessive moisture around the plants.

2.2.7 Propagation of in Vitro-Obtained Plants Through Terminal Cuttings

Plants were further propagated by taking the terminal cuttings (5 cm) from the in vitro-obtained plants established in the pots (Fig. 4F), and transferring them in sand (Fig. 4G). The rooting occurred within 2 weeks. The advantage of this procedure lies in the fact that profuse rooting occurred in the terminal cuttings, which is more or less impossible in the cuttings from the old flowering plants. This method proved of great help in the multiplication process.

3 Somaclonal Variation

Induction of somaclonal variants is of special significance in floriculture for the production of novel plants. A number of examples of variant ornamental plants showing different types of flower size, color, shape, and other aesthetic traits are now known (Bajaj 1990). In chrysanthemum, Broertjes et al. (1976) induced different types of mutants by X-ray irradiation of the explants. The mutants were almost exclusively of a solid (nonchimeral) nature. Likewise, variation in chromosome numbers (Miyazaki and Tashiro 1978), salt-tolerance (Dalsou and Short 1985) and other changes in leaf shape, etc. were observed (Sutter and Langhans 1981). Recently, assessment of somaclonal variants was made and plants of potential commercial value obtained (Khalid et al. 1989).

4 Cold Storage of Germplasm

The in vitro storage of germplasm has several advantages over the conventional practice, such as maintenance under disease-free environments, and requiring less space. Various methods to reduce the rate of growth have been employed (Bajaj 1986), however, low-temperature storage has proved quite useful.

Cold acclimation and freeze preservation studies (Bannier and Steponkus 1972, 1976) on chrysanthemum showed that callus stored at $-3.5\,^{\circ}C$ in MS + 10%

Fig. 5A–C. Regeneration of plants from cultures of *Chrysanthemum morifolium* stored up to 6 years (1978–1984) in a refrigerator (4 °C). **A** Cultures stored for 3 years. **B** Stocks of germplasm from **A** cold-stored for 6 years. **C** Plant obtained from a stored culture capable of flowering. (Bajaj 1991)

sucrose for 28 days survived. Unacclimated callus was able to survive freezing down to only − 6.6 °C, whereas callus acclimated for 6 weeks at 4.5 °C exhibited complete survival down to − 16.1 °C (Bannier and Steponkus 1976). Moreover, in vitro-grown explants, proliferating shoot tips and plantlets of chrysanthemum stored for 6 years in a refrigerator (4 °C) survived. Low temperature reduced the growth drastically. The plants which had started to turn yellow, however, with brief exposure to light in the laboratory at room temperature turned green. The plants showed normal flowering on transfer to soil (Fig. 5). This method of storage avoids the need for periodical transfer, resulting in reduced maintenance cost in commercial micropropagation laboratory (see Bajaj 1991).

5 Summary and Conclusions

Complete plants can be regenerated from different types of tissues/explants, i.e., stem, segments, petals, inflorescence, shoot tips, callus, isolated protoplasts, etc., and mass propagated. Of the different explants, the best response was obtained from excised segments of the inflorescence grown on MS medium supplemented with IAA and kinetin. The explants and plantlets cold-stored (4 °C) for 6 years underwent normal flowering. Somaclonal variants for salt tolerance, differences in flower size, color, and shape have been obtained, and this is an area which holds promise for commercial exploitation of elite and novel variants. A combination of tissue culture and the conventional cuttings method has been successfully practised for clonal propagation.

6 Protocol

1. Segments (1–1.5 cm) of stem, leaf, inflorescence, peduncle, or entire small capitulums are sterilized with chlorine water for 5–10 min, and washed twice with sterilized water.
2. Explants are cultured on Murashige and Skoog's basal medium supplemented with indole acetic acid and kinetin.
3 Roots are induced on basal MS containing NAA
4. Plants are transferred to soil mixture of sand (50%) + leaf mold (25%) + sawdust (25%).
5. For further multiplication the terminal cuttings from these plants are transferred to moist sand; rooting occurs within 2 weeks.

References

Annonymous (1987) The worldwi.le trade in chrysanthemums. World Flower Trade Mag 4: 32–33
Arunyanart S, McComb JA (1986) *Chrysanthemum* plants regenerated from petal and bud callus. Proc VI Int Congr Plant tissue and cell culture. Minnesota Univ, Minneapolis, Abstr, p 237

Bajaj YPS (1986) In vitro preservation of genetic resources. Proc Int Symp Nuclear techniques and in vitro culture for plant improvement. IAEA, Vienna, pp 43–57

Bajaj YPS (1990) Somaclonal variation – origin, induction, cryopreservation and implications in plant breeding. In: Bajaj YPS (ed) Biotechnology in agriculture and forestry, vol 11. Somaclonal variation in crop improvement I. Springer, Berlin Heidelberg New York Tokyo, pp 3–48

Bajaj YPS (1991) Storage and cryopreservation of in vitro cultures. In: Bajaj YPS (ed) Biotechnology in agriculture and forestry, vol. 17. High-Tech and Micropropagation I. Springer, Berlin Hiedelberg New York Tokyo pp 361–381

Bajaj YPS, Sidhu MMS, Gill APS (1992) Some factors affecting micropropagation of *Chrysanthemum morifolium* Ramat. (submitted)

Bannier LJ, Steponkus PL (1972) Freeze preservation of callus cultures of *Chrysanthemum morifolium* Ramat. HortSci 7: 194

Bannier LJ, Steponkus PL (1976) Cold acclimation of *Chrysanthemum* callus cultures. J Am Soc Hort Sci 101: 409–412

Ben-Jaacov J, Langhans RW (1972) Rapid multiplication of *Chrysanthemum* plants by stem-tip proliferation. HortSci 7: 289–290

Broertjes C, Roest S, Bokelmann GS (1976) Mutation breeding of *Chrysanthemum morifolium* Ram. using in vivo and in vitro adventitious bud techniques. Euphytica 25: 11–19

Bush SR, Earle ED, Langhans RW (1976) Plantlets from petal segments, petal epidermis, and shoot tips of the periclinal chimera *Chrysanthemum morifolium* Indianapolis. Am J Bot 63: 729–737

Dalsou V, Short KC (1985) Selection for sodium chloride tolerance in chrysanthemums. Acta Hortic 212: 737–740

Earle ED, Langhans RW (1974a) Propagation of *Chrysanthemum* in vitro I. Multiple plantlets from shoot tips and the establishment of tissue cultures. J Am Soc Hort Sci 99: 128–131

Earle ED, Langhans RW (1974b) Propagation of *Chrysanthemum* in vitro. II. Production, growth, and flowering of plantlets from tissue cultures. J Am Soc Hort Sci 99: 352–358

Gamborg OL, Miller RA, Ojima K (1968) Nutrient requirements of suspension cultures of soybean root cells. Exp Cell Res 50: 151–158

Horst RK (1989) Chrysanthmum. In: Ammirato PV, Evans DA, Sharp WR, Bajaj YPS (ed) Handbook of plant cell culture, vol 5, Ornamental species. McGraw Hill, New York, pp 319–336

Kaul V, Miller RM, Hutchinson JF, Richards D (1990) Shoot regeneration from stem and leaf explants of *Dendranthema grandiflora* Tzvelev (syn. *Chrysanthemum morifolium* Ramat). Plant Cell Tissue Organ Cult 21: 21–30

Khalid N, Davey MR, Power JB (1989) An assessment of somaclonal variation in *Chrysanthemum morifolium*: the generation of plants of potential commercial value. Sci Hortic 38: 287–294

Lu CY, Nugent G, Wardley T (1990) Efficient direct plant regeneration from stem segments of Chrysanthemum (*Chrysanthemum morifolium* Ramat cv. Royal Purple). Plant Cell Rep 8: 733–736

Marani F (1972) Studies on meristem culture for producing virus-free *Chrysanthemum*. Informatore Fitopatologico 22: 3–8

Miyazaki S, Tashiro Y (1978) Tissue culture of *Chrysanthemum morifolium* Ramat III. Variation in chromosome number and flower colour of plants regenerated from different parts of shoots in vitro. Bull Shizuoka Agric Exp Stn 44: 13–31

Murashige T, Skoog F (1962) A revised medium for rapid growth and bioassays with tobacco tissue cultures. Physiol Plant 15: 473–497

Prasad RN, Chaturvedi HC (1988) Effect of season of collection of explants on micropropagation of *Chrysanthemum morifolium*. Biol Plantarum 30: 20–24

Roest S, Bokelmann GS (1975) Vegetative propagation of *Chrysanthemum morifolium* Ram. in vitro. Sci Hortic 3: 317–330

Sauvadet MA, Brochard P (1989) Regeneration of plants from protoplasts of *Chrysanthemum morifolium*. Abstr Int Symp Implication of biotechnology in agriculture, Amiens

Sutter E, Langhans RW (1981) Abnormalities in *Chrysanthemum* regenerated from long term cultures. Ann Bot 48: 559–568

Tsukamoto Y, Fujime Y (1971) The in vitro culture of *Chrysanthemum* flower buds. J Jpn Soc Hort Sci 40: 262–267

VI Micropropagation of Dogwoods (*Cornus* spp.)

R.N. Trigiano, R.M. Beaty, and K.W. Lowe[1]

1 Introduction

The genus *Cornus* L. (dogwood) (Cornaceae) is a morphologically diverse group of plants comprised of approximately 65 species (Eyde 1988) primarily distributed in the northern hemisphere (Brinkman 1974). The majority are either small trees or shrubs; two species have a herbaceous habit. Dogwoods can be separated into two main groups based on bracts and fruit color. In the first group, flowers are subtended by showy bracts and fruits are red at maturity, whereas in the second group, bracts are either reduced or absent and fruits are blue or occasionally white. Although species separation in the red fruit group is relatively straight-forward, species distinction in the larger blue fruit group is ambiguous (see review by Eyde 1988 for details).

Dogwoods are predominantly cultivated for their ornamental characteristics: showy bracts (Fig. 1), attractive seasonal foliage, and fruit and twig color, but also have other minor uses. The fruits of *C. mas* L. can be used to make jelly and other species provide hardwood for handicrafts. The bark of some species yields a compound that can be substituted for quinine (Brinkman 1974), and the fruit of *C. officinalis* Sieb. and Zucc. has been utilized in Chinese medicine as an astringent, in a tonic and as a hemostatic agent (Yazaki and Okuda 1989). Dogwood species also provide habitat and food for wildlife (Eyde 1988).

Many dogwoods are cultivated but the most common commercially grown species include the following: *C. florida* L. (flowering dogwood), *C. kousa* Hance (Chinese dogwood), *C. nuttalli* Audubon (Pacific dogwood), *C. canadensis* L. (bunchberry), *C. mas* (cornelian cherry) and *C. stolonifera* Michx. (red-osier dogwood). The most important North American species is *C. florida*, a native understory tree of the eastern United States. It produces a head of small inconspicuous flowers subtended by large white, pink, or red bracts in the spring. Green foliage appears after flowering, and red berries and leaves provide fall color to the landscape. Native dogwoods are propagated from seeds and can exhibit a wide range of growth habits and bract characteristics. There are over 100 named cultivars of flowering dogwood that have been selected for various characteristics including white, pink, or red bracts, size of bracts, variegated foliage, or dwarf habit (Santamour and McArdle 1985). Cultivars are vegetatively propagated by grafting buds onto native rootstock or by rooting cuttings (Dirr and Heuser 1987).

[1] Department of Ornamental Horticulture and Landscape Design, Institute of Agriculture, University of Tennessee, Knoxville, TN 37901-1071, USA

Biotechnology in Agriculture and Forestry, Vol. 20
High-Tech and Micropropagation IV (ed. by Y.P.S. Bajaj)

Fig. 1. A large, white bract cultivar (Cherokee Princess) of *Cornus florida* (flowering dogwood). The tree is approximately 1.5 m in height

Production of plants using either of these methods can be difficult and expensive and yield unpredictable results (Badenhop et al. 1985).

Flowering dogwood is susceptible to a number of fungal diseases. Common leaf spots, caused by several genera of fungi and spot anthracnose, incited by *Elsinoe corni* Jenkins and Bitanc., generally produce only cosmetic damage and seldom require control measures. Root rot caused by *Phytophthora cinnamomi* Rands can be avoided by selecting a well-drained planting site. However, dogwood anthracnose, a recently described disease caused by *Discula destructiva* sp. nov. (Redlin 1991), is decimating native populations of flowering dogwood (Hibben and Daughtery, 1988) and threatening to limit the use of this species as an ornamental. There is no known natural resistance and a mature tree may die within 3 or 4 years after the initial infection (Santamour et al. 1989); seedlings are more severely affected and may be killed within 1 year. The fungus also attacks *C. nuttalli*, but usually does not destroy the tree (K. Britton pers. commn.) and although leaves of *C. kousa* may be infected, little or no permanent damage is sustained (Santamour et al. 1989). The interspecific hybrid of *C. florida* × *C. kousa* (Orton 1969) is purported to have some resistance to the disease (Block 1991).

An in vitro propagation system for flowering dogwood has the potential to reliably and efficiently produce clones of dogwood cultivars that are difficult to propagate using conventional methods. Furthermore, tissue culture techniques may provide useful tools for selecting, engineering and regenerating plants resistant to dogwood anthracnose.

2 In Vitro Approaches

Cornus stolonifera callus has been used to study ultrastructural changes associated with chilling injury in plant cells (Niki et al. 1978, 1979) and alteration of respiratory pathways of cold-stressed cells (Yoshida and Tagawa 1979). *Cornus officinalis* suspension and callus cultures have been used to produce secondary products (tannins) (Yazaki and Okuda 1989).

Rooted shoots of *C. kousa*, *C. nuttalli* (D. McCown pers. commun.) and *C. canadensis* (Pennell 1983) produced via tissue culture methods are commercially available; however, these production procedures have not been published. There has been limited success in micropropagating shoots from apical buds of *C. florida* (Coker 1982). In this study, nondormant buds proliferated best on Knop's medium supplemented with either 1 or 2 mg/l 6-benzylaminopurine (BA). One to four microshoots could be produced per explant, but rooting was not achieved. In another study (Trigiano unpubl), axillary buds from mature trees and seedlings with three pairs of leaves were surfaced sterilized with 95% ethanol and 2.62% NaOCl and placed on MS medium with either 0.5, 1, 3, or 5 mg/l BA. Material collected from mature trees was very difficult to establish in culture; after 1 month of culture, only 40% of the buds were free of contaminating organisms. Most of these cultures became necrotic after 3 months and only 14 of the original sample of 400 exhibited proliferation (Fig. 2). Seedling explants (nodes with both axillary

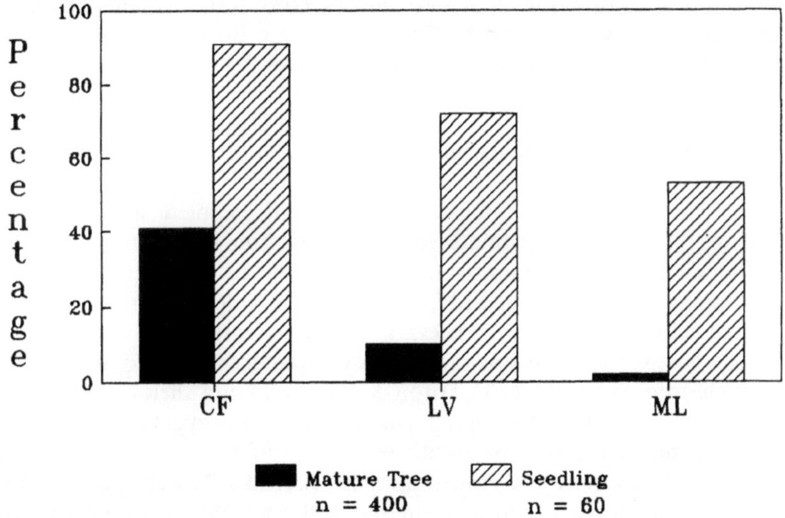

Fig. 2. Axillary bud culture of *Cornus florida*. *CF* contaminant free cultures; *LV* explants living after 3 months in culture; *ML* shoot multiplication

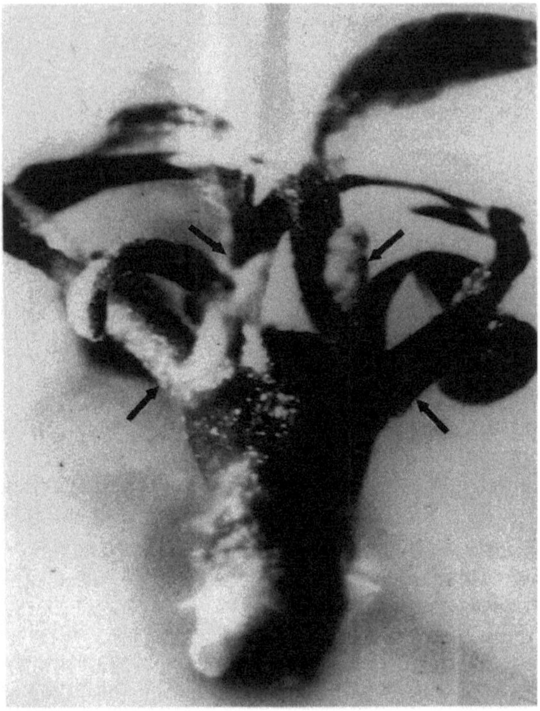

Fig. 3. Shoot multiplication from an axillary bud excised from a seedling of *Cornus florida* cultured on MS medium supplemented with 1 mg/l BA. *Arrows* indicate shoots

buds) were more easily established in culture and most were green and healthy after 3 months in culture (Fig. 2). Approximately 60% of the explants produced one to four microshoots per month (Fig. 3); the best treatment was 1 mg/l BA. All attempts to root the microshoots were unsuccessful.

2.1 Direct Somatic Embryogenesis

Although there are no published accounts of complete in vitro proliferation systems for *Cornus* spp., somatic embryogenesis from immature zygotic embryos of *C. florida* has been demonstrated (Trigiano et al. 1989). In this investigation, developing fruits were collected from four native (wild-type) trees and one tree of a named cultivar, Cherokee Princess, at weekly intervals from approximately 4–17 weeks post-anthesis (WPA). Drupes were surface sterilized in 2.62% sodium hypochlorite plus 0.1% (v/v) Triton X-100 for 15 min then rinsed three times in sterile distilled water. Developing ovules or immature zygotic embryos, depending on the sample date, were aseptically removed from the stony endocarp and placed on either SH (Schenk and Hildebrandt 1972) or MS (Murashige and Skoog 1962) basal media supplemented 30 g sucrose and 8 g/l phytagar (Gibco). Treatments consisted of factorial combinations of either 1, 2, 3, or 5 mg/l 2,4-dichlorophenoxyacetic acid (2,4-D) and either 0 or 1 mg/l kinetin. Cultures were initially incubated at 22 °C in the dark for 7 weeks, then transferred to SH or MS basal medium lacking growth regulators. These cultures were incubated at 24 °C with 25 μmol/s/m^2 of light for 16 h/day for 4 weeks.

Ovules and globular zygotic embryos collected from all five trees prior to 9 WPA produced callus after 3 or 4 weeks on all media used. These cultures exhibited little morphogenic potential except for the formation of a few roots and root hairs. Young expanding bracts and leaves, pistils, anthers, and flower receptacles produced callus but were not morphogenic. Zygotic embryos collected from 16 and 17 WPA fruits produced numerous adventitious root and scanty amounts of callus, but no somatic embryos.

The ability to produce somatic embryos was dependent on both genotype and the developmental stage of the zygotic embryo explant. Direct embryogenesis was restricted to two of the native trees and limited to within the 12–15 WPA sampling periods when explants were 3.6 ± 0.2 to 5.3 ± 0.5 mm in length respectively. Fruit characteristics corresponding to these developmental stages of the embryo were: red color on the distal end of the ovary and red coloration on approximately 50% of the fruit surface, respectively (Dietrich et al. 1990). Explants from the other two native trees and Cherokee Princess failed to differentiate embryos and only produced callus. Both SH and MS basal media amended with either 3 mg/l, 2,4-D or 5 mg/l 2,4-D + 1 mg/l kinetin supported direct embryogenesis. Approximately 4% of the explants produced embryos at a rate of 3.5 embryos per explant.

Somatic embryos were observed 3 weeks after initiation of cultures as small regular protuberances on the surface of cotyledons that had been separated from the embryo axes. Initiation and/or the developmental rate of embryos were usually asynchronous; 7 weeks after culture, groups of white-opaque globular and early cotyledonary-stage embryos forming on the same explant were recognizable. Occasionally, synchronous development was observed (Fig. 4). Many of the globular somatic embryos appeared to be developmentally arrested and had callused at the distal end. Most of the directly formed somatic embryos that had advanced beyond the globular stage were fused in pairs. Each of these embryos possessed well-developed closed vascular systems and root meristems, but lacked

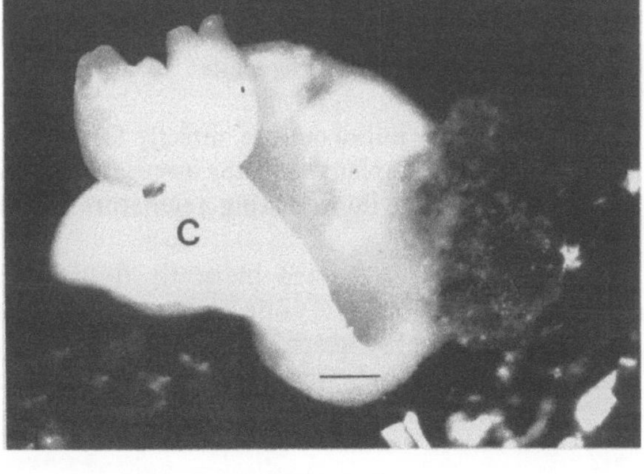

Fig. 4. Early cotyledonary-stage somatic embryos developing directly from a zygotic cotyledon (*C*) of *Cornus florida* 7 weeks after culture on MS medium supplemented with 3 mg/l 2,4-D. *Bar* 1.5 mm

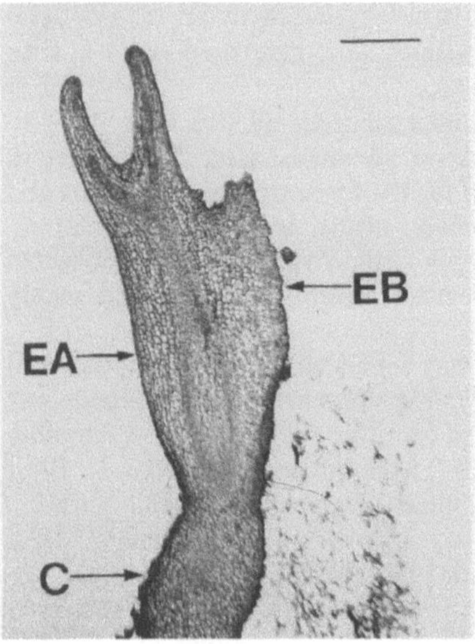

Fig. 5

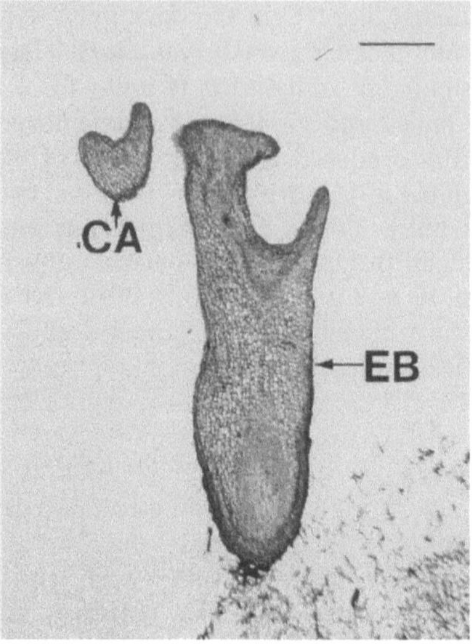

Fig. 6

Figs. 5 and 6. Serial longitudinal sections through fused somatic embryos (EA and EB) of *Cornus florida* formed 7 weeks after culture on MS medium supplemented with 3 mg/l 2,4-D. *C* zygotic embryo cotyledon; *CA* cotyledon of somatic embryo *EA*. *Bars* 360 μm. (Trigiano et al. 1989)

any well-defined or organized shoot apices (Figs. 5, 6). Maturation was promoted by incubation in the light on medium lacking regulators; occasionally, a single morphologically mature embryo was formed.

In another investigation, somatic embryos were stimulated to form directly from immature zygotic embryos of *C. florida* without the use of growth regulators. In these experiments, the capacity to form embryos was maintained for over 18 months. Conversion of embryos to plantlets was enhanced by desiccation (Pierson 1990).

2.2 Indirect Somatic Embryogenesis

Eleven weeks after initial culture, directly formed somatic embryos were removed from the explants. Explants and any associated callus from all sampling times were transferred from medium lacking regulators through the protocol depicted in Fig. 7.

Somatic embryos formed indirectly from approximately 2.5% of the calli derived from 9–15 WPA zygotic embryos of the same two trees that had formed direct embryos. Embryogenic calli varied from hard compact (Fig. 8) to friable and milky white (Fig. 9) or watery (Fig. 10). Somatic embryos were only initiated after incubation on medium containing 3 mg/l 2,4-D (see Fig. 7) and globular (Fig. 8)

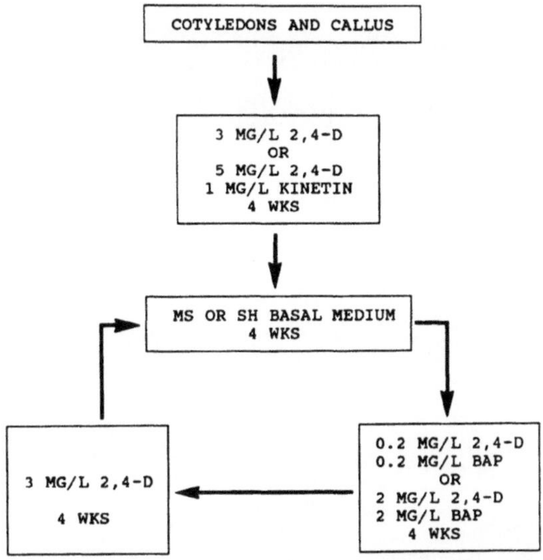

Fig. 7. Protocol for initiation and maintenance of embryogenic cultures and induction of indirect somatic embryos of *Cornus florida*. Unless otherwise noted the basal medium was SH. (Trigiano et al. 1989)

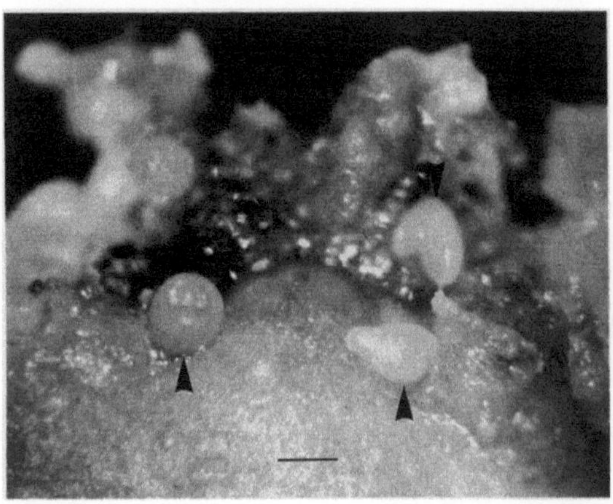

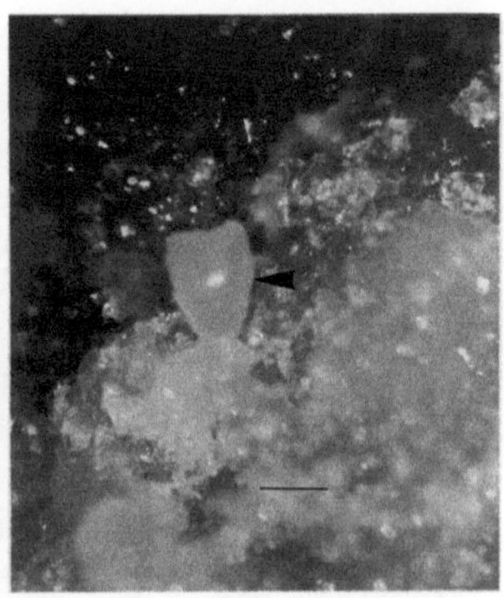

Fig. 8 **Fig. 9**

Fig. 8. Indirect somatic embryos (*arrowheads*) of *Cornus florida* formed on hard compact embryogenic callus 4 weeks after culture on SH medium supplemented with 3 mg/l 2,4-D. *Bar* 1 mm

Fig. 9. A cordate somatic embryo (*arrowhead*) of *Cornus florida* formed from friable embryogenic callus 4 weeks after culture on SH medium supplemented with 3 mg/l 2,4-D. *Bar* 0.8 mm. (Trigiano et al. 1989)

through cordate stage embryos (Fig. 9) formed during this 4-week period. Maturation of embryos occurred only after the cultures were transferred to medium without growth regulators. Fused embryos (Fig. 10), similar to those formed in direct embryogenesis, were common. Production of embryos decreased with each culture cycle (see Fig. 7) over a 16-month period.

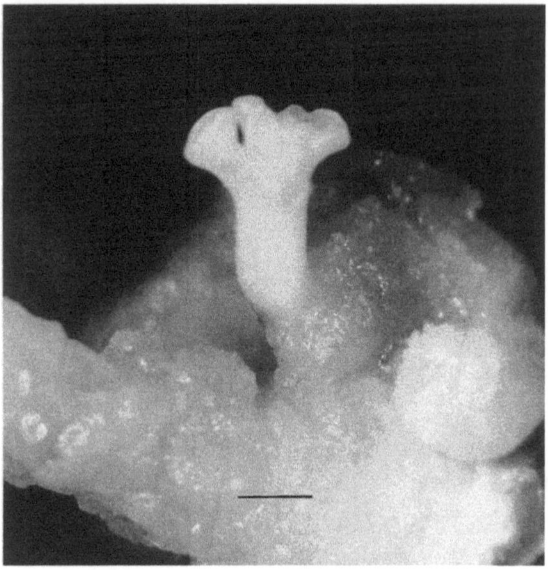

Fig. 10. Fused somatic embryos of *Cornus florida* formed indirectly from watery embryogenic callus 4 weeks after culture on SH medium supplemented with 3 mg/l 2,4-D. *Bar* 1.3 mm. (Trigiano et al. 1989)

2.3 Embryo Conversion

Approximately 50% of the morphologically mature somatic embryos placed on SH medium without growth regulators produced a primary root, but only 12% were capable of conversion into plantlets. In all cases, shoot development was limited to the first pair of true leaves (Fig. 11) and no additional growth occurred.

Fig. 11. *Cornus florida* plantlet regenerated from a somatic embryo and grown for 5 weeks on SH medium lacking growth regulators. *Bar* 1.2 cm. (Trigiano et al. 1989)

However, these leaves were similar in shape, venation pattern and arrangement (opposite) to those that were formed from zygotic embryos.

3 Summary and Conclusions

The frequency of somatic embryogenesis from zygotic embryo cotyledons and derived calli was low and limited to native or wild type trees; Cherokee Princess and some red bract cultivars did not produce embryos. Embryos converted to plantlets, but were incapable of growth beyond the first true leaf stage. Moreover, any plant that originated from a zygotic embryo explant would be a clone of a sexually recombinant genotype and not the donor tree. These problems would preclude the use of this technique for commercial production. However, it may be useful in a breeding program for resistance to dogwood anthracnose. If the rate of embryogenesis can be substantially improved and the number of embryogenic genotypes increased, then production of multiple clones from unique recombinant genotypes of interspecific hybrids, e.g., *C. florida* × *C. kousa*, may be possible.

4 Protocol

For *direct somatic embryogenesis*, excise torpedo and cotyledonary-stage zygotic embryos, separate cotyledons from the hypocotyl-radicle axis and place on MS medium augmented with either 3 mg/l 2,4-D or 5 mg/l 2,4-D + 1 mg/l kinetin. Incubate the cultures in the dark at 22 °C for 7 weeks. Somatic embryos will continue to develop when transferred to MS medium without growth regulators and incubated in the light at 24 °C for 4 weeks. For *indirect somatic embryogenesis*, culture the original explant and any associated callus according to the scheme presented in Fig. 7. Maintain cultures in the dark at 22 °C. Somatic embryos will form on SH medium supplemented with 3 mg/l 2,4-D and mature on SH medium without growth regulators.

References

Badenhop MB, Witte WT, Glasgow TE (1985) Production systems and costs for producing balled and burlapped trees of dogwood cultivars, Tennessee, 1984. Tenn Agric Exp Stn Bull 637: 29

Block L (1991) Hope springs eternal. Amer Nurseryman July 1: 30–41

Brinkman KA (1974) Cornus L. Dogwood. In: Seeds of woody plants in the United States, USDA Agric Handbook No 450, USDA, Forest Service, pp 336–342

Coker AL (1982) In vitro culture of flowering dogwood, *Cornus florida* L. MS Thesis, Univ of Tennessee, 28 pp

Dietrich JT, Trigiano RN, Graham ET (1990) Developmental anatomy of the zygotic embryo in *Cornus florida* (Cornaceae). Phytomorphology 40: 369–375

Dirr MA, Heuser CW (1987) The reference manual of woody plant propagation. Varsity Press, Athens, Georgia, USA, 239 pp

Eyde RH (1988) Comprehending *Cornus*: puzzles and progress in the systematics of the dogwoods. Bot Rev 54: 233–251

Hibben CR, Daughtrey ML (1988) Dogwood anthracnose in Northeastern United States. Plant Dis 72: 199–203

Murashige T, Skoog F (1962) A revised medium for rapid growth and bioassays with tobacco tissue cultures. Physiol Plant 15: 473–497

Niki T, Yoshida S, Sakai A (1978) Studies on chilling injury in plant cells I. Ultrastructural changes associated with chilling injury in callus tissues of *Cornus stolonifera*. Plant Cell Physiol 19: 139–148

Niki T, Yoshida S, Sakai A (1979) Studies on chilling injury in plant cells II. Ultrastructural changes in cells rewarmed at 26 °C after chilling treatment. Plant Cell Physiol 20: 899–908

Orton ER Jr (1969) Hybridizing woody ornamentals. Proc Int Plant Prop Soc 19: 371–375

Pennell D (1983) The future of micropropagation in the United Kingdom. Proc Int Plant Prop Soc 33: 249–253

Pierson PE (1990) Direct somatic embryogenesis of *Cornus florida* in the absence of growth regulators. In vitro 26: 45A

Redlin SC (1991) *Discula destructiva* sp nov., cause of dogwood anthracnose. Mycologia 83: 633–642.

Santamour FS Jr, McArdle AJ (1985) Cultivar checklists of the large-bracted dogwoods: *Cornus florida*, *C. kousa*, and *C. nuttallii*. J. Arboric 11: 29–36

Santamour FS Jr, McArdle AJ, Strider PV (1989) Susceptibility of flowering dogwood of various provenances to dogwood anthracnose. Plant Dis 73: 590–591

Schenk RU, Hildebrandt AC (1972) Medium and techniques for induction and growth of monocotyledonous and dicotyledonous plant cell cultures. Can J Bot 50: 199–204.

Trigiano RN, Beaty RM, Dietrich JT (1989) Somatic embryogenesis and plantlet regeneration in *Cornus florida*. Plant Cell Rep 8: 270–273

Yazaki K, Okuda T (1989) Gallotannin production in cell cultures of *Cornus officinalis* Sieb. et Zucc. Plant Cell Rep 8: 346–349

Yoshida S, Tagawa F (1979) Alteration of the respiratory function in chill-sensitive callus due to low temperature stress III. Involvement of the alternate pathway. Plant Cell Physiol 20: 1243–1250

VII In Vitro Culture and Micropropagation of *Euphorbia* spp.

J.E. Preece and K.P. Ripley[1]

1 Introduction

There are more than 1600 species in the genus *Euphorbia* (Bailey and Bailey 1976). These herbaceous plants, shrubs, and trees all have a milky white latex (Fig. 1) and are often spiny and resemble cacti. Some species have ornamental value, some may have utilization as bioenergy crops, and some produce valuable secondary products, including medicinals. The white latex in many of the euphorbias produces a dermatitis in some people and exudates from broken roots have been fatal to fish; in fact, Bailey and Bailey (1976) report that arrow poisons for hunting fish have been extracted from some *Euphorbia* species.

The latex from *Euphorbia* has a molecular weight of about 20 000 and remains a liquid upon dehydration (Calvin 1979). Members of this genus have great potential for the production of liquid fuels, and the fact that they will grow in arid areas

Fig. 1. Milky white latex on the cut end of a shoot of *E. lathyris*

[1]Department of Plant and Soil Science, Southern Illinois University, Carbondale. IL 62901-4415, USA

Biotechnology in Agriculture and Forestry, Vol. 20
High-Tech and Micropropagation IV (ed. by Y.P.S. Bajaj)

could result in a transformation of rather unproductive areas into energy plantations (Calvin 1979, 1983). Roth et al. (1982) analyzed 508 plants for hydrocarbon, oil, polyphenol, and protein contents, and for use as renewable energy resources. They listed the following *Euphorbia* species as having potential as hydrocarbon or oil-producing crops: *E. antisyphilitica, E. corollata, E. dentata, E. esula, E. fulva, E. geniculata, E. heterophylla, E. lathyris,* and *E. pulcherrima.*

Various parts of plants and especially the juice and latex of *Euphorbia* species have been used against cancer, tumors, warts, callous, corns, and other human disorders (Hartwell 1969). In fact, Hartwell (1969) listed 74 species and a list where only the *Euphorbia* genus was mentioned that have been used worldwide as treatments for human diseases. Interestingly, members of the genus also produce tumor-promoting (cocarcinogenic) factors. Irritants have been extracted from some euphorbs and have been used as insecticides (Kinghorn 1979). Determination of the structures of various toxic compounds from *Euphorbia* spp. has been described (Hirata 1975). Because of these varied properties, there is interest in secondary product production from in vitro cultures of members of this genus.

2 In Vitro Approaches

Although this is a large genus, in vitro work has focused on only a few species. These studies have been on micropropagation, production of callus, chimera separation, elimination of viruses, induction of mutations, protoplast cultures, secondary product production, in vitro selection, and laticifer growth (Table 1). Because members of the family Euphorbiaceae are naturally polyembryonic (Litz 1985), it is surprising that there have not been more studies published on somatic embryogenesis of *Euphorbia* spp. In addition, adventitious shoots form naturally on the hypocotyls of seedlings of *E. esula* (Raju 1975) and *E. Lathyris* (Fig. 2). Because of this natural phenomenon, caulogenesis might be achieved with in vitro cultures of euphorbs without great difficulty.

2.1 Micropropagation

2.1.1 Axillary Shoot Proliferation

Members of the genus *Euphorbia* produce numerous toxic substances (Hirata 1975), and unless they are handled properly, shoot explants of at least *E. lathyris* produce a dark exudate that stains the culture medium (Fig. 3). This exudate is toxic to the explant (Preece and Ripley 1984; Ripley and Preece 1986) and has been shown to inhibit the growth of tobacco callus (Compton and Preece 1988). If explant wounding is minimized, however, exudation can be reduced or eliminated. No other authors in the references cited here have reported inhibitory exudation from *Euphorbia* spp. explants. If it is noticed, it can cause serious problems and must be dealt with either through rapid transfers to fresh medium or by minimizing explant wounding.

Table 1. Summary of in vitro studies with *Euphorbia* spp.

Plant species	Explant	Objectives	Medium	PGRs and other additives	Results	Reference
Euphorbia antisyphilitica	Apical shoot tips	Micropropagation	MS, half-strength MS for rooting	NAA (mg/l) 0.009–0.25 IBA (mg/l) 0.1–3.0 BA (mg/l) 0.1–7.0 sucrose 30 g/l	20–30 axillary shoots with 0.02 mg/l NAA and 1.0 mg/l BA in 5–6 weeks, 100% rooting in 12 days with 0.1 mg/l IBA	Jakobek et al. (1986)
E. characias	Stem without cortex, veins	Obtain photoheterotrophic suspension cultures	MS	NAA 1 mg/l BA 1 mg/l sucrose 30 g/l, 15 g/l for sub-cultures	Early stationary phase exhibited good photosynthetic ability	Hardy et al. (1987a)
E. characias	Stem without cortex, veins	Study transition to photoheterotrophy	MS	NAA 1 mg/l BA 1 mg/l sucrose 30 g/l, 15 g/l for sub-cultures	Photoautotrophic cultures grow 6 times more slowly than photoheterotrophic cultures	Hardy et al. (1987b)
E. characias	Cell suspension	Study growth, physiol, photomixotrophic cells	MS	NAA 1 mg/l BA 1 mg/l sucrose 15 g/l	Photosyn. increased during exponential and mid-stages, decreased late-stationary phases	Chagvardieff et al. (1988)
E. cyparissias	Root tissue	Learn requirements for continuous production of callus	MS LS	IAA NAA 2,4-D Picloram each at 0.02–22 mg/l kinetin 0.17 mg/l, GA_3	Callus light brown and declined after 18 months in vitro, good growth with 0.2 mg/l NAA and 0.01 mg/l kinetin	Lee and Starratt (1972)
E. escula	Root tissue	Learn requirements for continuous production of callus, and major lipid constituents in callus	MS LS	IAA NAA 2,4-D Picloram each at 0.02–22 mg/l kinetin 0.0002–0.2 mg/l GA_3, sucrose 5–30 g/l	Auxin but not cytokinin required for callus growth, green callus still required sucrose, lipids found in callus	Lee and Starratt (1972)

Table 1. (*Continued*)

Plant species	Explant	Objectives	Medium	PGRs and other additives	Results	Reference
E. flanaganii	Stem segments	Micropropagation	MS	IAA NAA 2,4-D each at 0.1,1,10,100, 1000 mg/l kinetin 0.1 mg/l	Most callus with NAA 10 mg/l and kinetin 0.1 mg/l, adventitious shoots formed from callus, shoots rooted without auxin	Silberstein et al. (1983)
E. fulgens	Shoot tip and node	Micropropagation	MS	IBA 0.5–2.0 mg/l NAA 0.5–2.0 mg/l zeatin 1.0–5.0 mg/l BA 1.0 mg/l 2-iP 10.0 mg/l kinetin 5.0 mg/l sucrose 25 g/l	Axillary shoots grew with 2.0 mg/l zeatin, adventitious shoots initiated with 5.0 mg/l zeatin, microshoots were rooted	Zhang et al. (1987)
E. hirta	Leaf	Micropropagation	MS	IAA 2 mg/l NAA 0.5–1 mg/l BA 0.5–3 mg/l GA 2 mg/l sucrose 20 g/l	Shoots formed from callus, needed to be elongated with GA prior to rooting	Baburaj et al. (1987)
E. hirta	Leaf, stem	Study anthocyanin production and morphogenesis from callus	MS	IAA NAA 2,4-D BA	Shoots and roots formed from callus, cyanidin-3-glucoside was isolated from cyanic not acyanic callus	Bahadur et al. (1987)
E. lathyris	Shoot tip	Micropropagation	Modified MS, half-strength major salts, ten times normal vitamins	2-iP 0.1 mg/l	Only one or two shoots from many primordia grew every 5 weeks, shoots rooted in soil	Lee et al. (1982)

E. lathyris	Node Internode leaf	Micropropagation	MS half-strength MS	NAA 0.1–1.0 mg/l BA 0.6–2.3 mg/l	Callus formed from all explants, shoots, that rooted slowly formed from nodes	Tideman and Hawker (1982)
E. lathyris	Shoot tip	Micropropagation	Not reported	Cytokinin gibberellin	Dark explant exudate reported, shoots responded to cytokinin	Preece and Ripley (1984)
E. lathyris	Shoot tip	Micropropagation	MS	NAA 1 mg/l IBA BA 5–20 mg/l sucrose 20 g/l	BA necessary for callus, adventitious not axillary shoots, rooting In vitro with NAA	Ripley and Preece (1986)
E. lathyris	Shoot tip	Study effects of explant exudate on tobacco callus	MS	IBA 0.2–2.0 mg/l BA 0.2–2.3 mg/l sucrose 30 g/l	Exudate inhibitory to tobacco callus	Compton and Preece (1988)
E. lathyris	Stem	Study chemical constituents of callus	MS	2,4-D NAA each at 0.1–10 mg/l sucrose 20 g/l either malt or yeast extract or CH 2 g/l	Callus formed with malt extract and 2,4-D, and contained sitosterol, stigmasterol, campesterol, palmatic acid, and linoleic acid	Yamamoto (1988)
E. marginata	Seeds	Study development of nonarticulated laticifers	See Mahlberg (1959)	NAA 0.2 mg/l 2,4-D 0.2 mg/l sucrose 40 g/l yeast extract 1 g/l	Long, branched laticifers grew apically from the zygotic embryo into callus that formed from explant	Mahlberg (1959)
E. marginata	Embryonic axis	Study laticifer and shoot apex growth	Mahlberg (1967)	CW 2.5–15% yeast extract 5–15% IAA 0.002–17.5 mg/l kinetin 0.002–22.5 mg/l GA_3 0.003–3.5 mg/l sucrose 30 g/l	0.2 mg/l IAA best for shoot growth, meristem activity stimulated laticifer growth	Mahlberg (1967)
E. millii	Leaf	Investigate chemical constituents of callus	MS	NAA 2,4-D each at 0.1–10 mg/l malt or yeast extract or CH at 2 g/l	Callus contained sitosterol, stigmasterol, campesterol, palmitic acid, linoleic acid, cyanidin glycoside	Yamamoto et al. (1981)

Table 1. (*Continued*)

Plant species	Explant	Objectives	Medium	PGRs and other additives	Results	Reference
E. millii	Leaf	Selection for high and stable red pigmented cells	MS	2,4-D 0.2 mg/l malt extract 2 g/l sucrose 20 g/l	After 24 selections, the red pigment was 7 × that of original cells, pigment productivity in cells was stable	Yamamoto et al. (1982)
E. millii	Leaf	Selection for red pigmented callus	MS	2,4-D 0.2 mg/l malt extract 2 g/l sucrose 20 g/l	After 24 subcultures of cell aggregates, they became stable for red pigment, this was superior to single cell cloning	Yamada and Fujita (1983)
E. millii	Leaf	Investigate chemical constituents of callus	MS LS Gamborg et al. (1968) Nitsch and Nitsch (1967) Heller (1953) White's	NAA 2,4-D each at 0.1–10 mg/l NAA 0.2–18.6 mg/l 2,4-D 0.02–2.2 mg/l BA 0.0002–0.002 mg/l malt or yeast extract or CH 2 g/l	Callus contained sitosterol, stigmasterol, campesterol, palmitic acid, linoleic acid, and a selected strain produced 7 × more of the red pigment cyanidin glycoside	Yamamoto (1988)
E. peplus	Shoot tip node internode leaf	Micropropagation	MS half-strength MS	NAA 0.1–0.9 mg/l BA 0.6–2.3 mg/l	Shoot tips and leaves were unsuccessful, callus and shoots formed from nodes and internodes	Tideman and Hawker (1982)
E. pulcherrima	Decoated seed	Micropropagation	White's	IAA 2,4-D kinetin GA_3 each at 1 mg/l CW 10%	Seeds germinated, adventitious shoots formed from hypocotyl cortex	Nataraja et al. (1973)

E. pulcherrima	Petiole, internode	Micropropagation	Full and half-strength MS	2,4-D NAA IBA kinetin BA zeatin 2-iP, each at 0.01–1 mg/l sucrose 20 g/l adenine 3×10^{-4} M	Best callus with NAA and 2-iP adventitious shoots with adenine and 2-iP in vitro rooting on MS	DeLange et al. (1974)
E. pulcherrima	Decoated seed	Obtain somatic embryogenesis	White's	IAA 1 mg/l 2,4-D 1.2 mg/l kinetin 1.5 mg/l GA_3 1 mg/l CW 10% sucrose 20 g/l	Embryoids formed from callus and their shoots and roots elongated	Nataraja (1975)
E. pulcherrima	Shoot tip	Eliminate two viruses using cell suspensions	MS	IAA 2 mg/l BA 0.2 mg/l sucrose 30 g/l	163 out of 166 plants regenerated from cell suspensions were free from PoiMV and PoiCV	Preil et al. (1982)
E. pulcherrima	Shoot tip	Analyze and separate chimeras	MS	IAA 2 mg/l BA 0.2 mg/l sucrose 30 g/l	From two cultivars, cell suspensions yielded plants with two distinctly different phenotypes, offering proof that stock plants were chimeras	Preil and Engelhardt (1982)
E. pulcherrima	Cell suspension	Obtain somatic embryogenesis	MS	IAA 1–2 mg/l CPA 0.1–2 mg/l BA 0.2 mg/l	Somatic embryogenesis with CPA and BA both at 0.2 mg/l in 20% of cultures, embryos grew into complete plantlets	Kleffel and Preil (1986)
E. pulcherrima	Shoot tips, stem segments	Induce mutations in embryogenic cell suspensions with X-rays	MS	IAA 1–2 mg/l BA 0.2 mg/l CPA 0.1–0.2 mg/l sucrose 30 g/l	Somatic embryogenesis with CPA and BA both at 0.2 mg/l, highest mutation rate with 60 Gy X-ray dose	Kleffel et al. (1986)

Table 1. (*Continued*)

Plant species	Explant	Objectives	Medium	PGRs and other additives	Results	Reference
E. pulcherrima	Stem segments	Studies on mass propagation using bioreactors	MS	IAA 2 mg/l CPA 0.2 mg/l BA 0.05–0.2 mg/l	Somatic embryogenesis with CPA and BA, cell doubling time decreased by increasing O_2 to 60% maximum air saturation	Preil et al. (1988)
E. tannensis ssp. *eremophila* var. *eremophila*	Seed node internode leaf	Micropropagation	MS half-strength MS	NAA 0.1–0.9 mg/l BA 0.6–2.3 mg/l	Callus formed from nodes, leaves, internodes, shoots formed from nodes	Tideman and Hawker (1982)
E. tirucalli	Stem	Determine effects of the medium on sterol production by callus	MS, White, Nitsch and Nitsch (1956) Eriksson (1965) Heller, B5	IAA NAA IBA 2,4-D each at 0.02–2 mg/l kinetin BA zeatin 2-iP each at 0.02–20 mg/l sucrose 20 g/l CH 1 g/l squalene 1–10 mg/l	Callus produced 4,4′-dimethyl sterols, euphol, and tirucallol, IAA and squalene promoted sterol production, 0.2 mg/l 2,4-D and 2.2 mg/l kinetin were the most stimulatory to callus production	Biesboer and Mahlberg (1979b)
E. tirucalli	Stem	Compare euphol/tirucallol ratios in stem explants latex, and callus	Not mentioned	IAA DL-mevalonic acid 1–10 mg/l lanosterol 1–10 mg/l	More euphol in latex, more tirucallol in stem and callus	Biesboer and Mahlberg (1979a)

E. tirucalli	Stem	Establish cell lines with unique easy to I.D. metabolic markers (sterols)	MS	IAA NAA IBA 2,4-D each at 0.02–2 mg/l kinetin BA zeatin 2-iP each at 0.02–20 mg/l	Callus age and sterol concentration were not correlated, IAA stimulated, BA inhibited sterol biosynthesis, squalene in the medium enhanced sterol production	Biesboer and Mahlberg (1980)
E. tirucalli	Shoots	Investigate chemical constituents of callus	MS	NAA 2,4-D each at 0.1–10 mg/l malt or yeast extract or CH at 2 g/l	Callus contained sitosterol, stigmasterol, campesterol, palmitic acid, linoleic acid	Yamamoto et al. (1981)
E. tirucalli	Stem	Isolate and culture protoplasts from callus and cell suspensions	MS Kao and Michayluk (1975)	2,4-D 1 mg/l NAA 1,2 mg/l kinetin 0.25–0.5 mg/l casamino acid 0.2%	More normal protoplasts from suspensions, oil body formation from protoplasts from callus	Misawa et al. (1982)
E. tirucalli	Stems	Induce callus and cell suspensions, isolate and successfully culture protoplasts	MS 1-B5	NAA 1,2 mg/l 2,4-D 0.2, 1 mg/l kinetin 0.5 mg/l zeatin 0.5 mg/l casamino acids 1 mg/l	Callus and suspensions were initiated and grew, protoplasts were isolated and grew into large cell clusters	Ohyama et al. (1984a)
E. tirucalli	Cell suspension	To observe and analyze oil bodies in cultured cells	1-B5	See Ohyama et al. (1984b)	Oil bodies were seen, major constituents were steroids diterpenoids triterpenoids	Ohyama et al. (1984b)
E. tirucalli	Young shoot	Study chemical constituents of callus	MS	NAA 2,4-D each at 0.1–10 mg/l sucrose 20 g/l either malt or yeast extract or CH 2 g/l	Callus formed with malt extract and 2,4-D, and contained sitosterol, stigmasterol, campersterol, palmatic acid, and linoleic acid	Yamamoto (1988)

Fig. 2. Adventitious shoot formation on the hypocotyl of a seedling of *E. lathyris*. This is a natural phenomenon, and this plant had not received any special treatment

Tideman and Hawker (1982) were two of the first to report axillary shoot proliferation from any *Euphorbia* spp. They were unsuccessful in obtaining shoot proliferation from fragmented shoot apices of *E. peplus*. However, they achieved multiple axillary shoots from node explants of *E. lathyris*, *E. peplus*, and *E. tannensis* on both full and half-strength MS (1962) medium with 0.6–2.3 mg/l benzyladenine (BA), with or without 0.93 mg/l naphthalene acetic acid (NAA). The most shoots formed from *E. peplus* and the fewest from *E. lathyris*; *E. tannensis* was intermediate. Shortly after shoots of *E. peplus* formed, they flowered and produced fruit in vitro. When shoots from *E. peplus* and *E. tannensis* were excised for rooting studies, latex was evident on the cut ends. All rooting attempts were in vitro on half-strength MS medium with 0.93 mg/l NAA. Rooting and transplanting into the greenhouse were successful for all three species; however, *E. lathyris* shoots were very slow and roots formed on only a few of the microshoots. Plantlets of *E. tannensis* died when transplanted into peat blocks in the greenhouse, but responded well in a 6:3:1 mix of coarse sand: peat:perlite.

During 1982, Lee et al. also reported shoot proliferation from 5-mm-long apical shoot tips of *E. lathyris*. They used half-strength MS salts and 10 × vitamins. The most shoots formed with 0.1 mg/l 2-ip. Most of these shoots remained as buds and

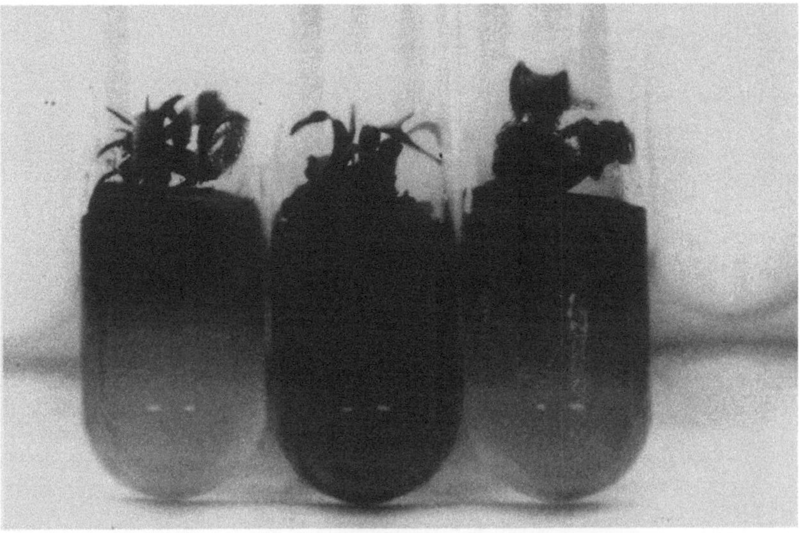

Fig. 3. Unless shoot tip explants of *E. lathyris* are handled properly, they produce a dark inhibitory exudate that stains the medium. Explants had been on MS medium for above 1 week

only one or two grew from each culture every 5 weeks. The long shoots were excised and rooted in soil.

The next published reports of studies on axillary shoot proliferation were with *E. lathyris* (Preece and Ripley 1984, Ripley and Preece 1986). Two-cm shoot tip explants with the two visible, vertically oriented leaves surrounding the apex were cultured on MS medium. Unless the explants were oriented vertically with the lower 1 cm of stem inserted into the agar solidified medium, survival was poor, and the stems appeared white and bleached out. It was shown that BA was not necessary for axillary shoot proliferation. Axillary shoots were visible after 3 months in vitro on medium without plant growth regulators (Fig. 4). In vitro rooting was accomplished using shoots harvested from cultures that had been in darkness for 1 month. When IBA was used in vitro or as a dip prior to placing microshoots in a greenhouse medium, rooting was poor. When NAA (1 mg/l) was used in vitro with MS medium containing half-strength $CaCl_2$, NH_4NO_3, and KNO_3, rooting was good (Table 2). These plantlets were acclimatized to the greenhouse.

In 1986, Jakobek et al. published a paper on micropropagation of *E. antisyphilitica.* Their stock plants had been clonally propagated by division and were grown in the greenhouse. Although they achieved axillary shoot proliferation with all combinations of BA and NAA that they tried on MS medium, the optimum shoot multiplication was with 1.0 mg/l BA and 0.02 mg/l NAA. At higher concentrations of BA, the shoots were short and yellow. They had 100% rooting after 12 days when 0.1 mg/l IBA was incorporated into the in vitro medium. These plantlets could be directly transplanted to greenhouse medium and acclimatization was not

Fig. 4. Axillary shoot proliferation from a shoot tip explant of *E. lathyris* after 3 months in vitro with no plant growth regulators in the medium

Table 2. The effect of NAA on root number, root length, callus formation, shoot length, and percent rooting of *E. lathyris*[a]

NAA (mg/l)	Total root number	Length of longest root (mm)	Callus diameter (mm)	Shoot length (mm)	Percent rooting
0	0	0	0	47.7	0
1	5.2*	7.9**	3.6**	52.1 NS	85.7[b]
10	0 NS	0 NS	11.3**	35.6 NS	0
LSD 5%	4.02	4.08	2.19	18.12	
LSD 1%	5.51	5.60	3.01	24.81	

[a] Each value represents the mean of seven replications.
[b] Two weeks later 100% of the shoots treated with 1 mg/l NAA rooted.
*,**,[NS] Significantly different from the control at the 5% (*) or 1% (**) level or non-significant (NS) according to the t-test.

necessary. Shoot proliferation and rooting remained constant, even with cultures more than 2 years old.

Zhang et al. (1987) found that node explants of *E. fulgens* produced more shoots and fewer buds than shoot tips on MS medium with zeatin. Microshoots rooted well in vitro without the addition of vitamins, glycine, or auxin. Best acclimatization was 6 days under intermittent mist followed by 6 days under shade than placing the plantlets in full sunlight.

2.1.2 Organogenesis

Seeds of *E. pulcherrima* germinated in vitro on White's (1963) medium with 2% sucrose (Nataraja et al. 1973). Adventitious shoot buds were observed to form from the cortical cells of the hypocotyls of the in vitro seedlings. The same phenomenon has been observed on greenhouse-grown seedlings of *E. esula* (Raju 1975) and *E. lathyris* (Fig. 2). Nataraja et al. (1973) reported that while vascular differentiation occurred in the adventitious shoots, it was not connected to the vascular tissue in the parental hypocotyl. Similar results were reported by Nataraja (1975) where shoot buds were observed on hypocotyls on White's medium without plant growth substances, or with 10% coconut water, and/or 1 mg/l IAA, or 1 mg/l kinetin. Coconut milk (CM) was necessary for shoot bud production when 1 mg/l GA_3 or 1 mg/l 2,4-D were used.

When petiole explants were excised from the uppermost leaves of greenhouse-grown *E. pulcherrima* plants and cultured in vitro on half-strength MS medium with 5 g/l meso-inositol, 2 to 10.2 mg/l 2-iP, and 0.93 mg/l NAA, adventitious buds formed mainly in the transition zone between the callus and explant (De Langhe et al. 1974). These buds did not grow out well. If cultures were incubated in darkness, the auxin that was required when the cultures were incubated in a 12-h photoperiod was eliminated. When internode explants were placed onto full strength MS, different qualities of callus could be induced in the darkness depending on the plant growth regulators in the medium. If the callus was red, adventitious shoots grew when the medium contained 10.5 mg/l IAA and 1.1 mg/l kinetin. A grayish green callus produced buds even without plant growth regulators in the medium. Considerably more buds formed when the same callus was cultured on medium with 3.5 mg/l IAA and 4.1 mg/l 2-iP plus 0.2 mg/l BA. There was a requirement of 10% coconut water for the buds to grow out into shoots. The adventitious shoots could be rooted in vitro on full strength MS, without vitamins, meso-inositol, casein hydrolysate, or plant growth regulators.

Tideman and Hawker (1982) reported that adventitious shoots formed from internode explants of *E. peplus*. They observed no organogenesis from this species on full-strength MS medium, but when half-strength MS medium containing 0.6 mg/l or 2.3 mg/l BA was used, adventitious shoots formed. *E. lathyris* and *E. tannensis* explants did not form adventitious shoots. Leaf explants from all three species were not organogenic on any of the media that were tried.

Callus that formed from stem explants of *E. flanaganii* produced adventitious shoots when subcultured on MS (1962) medium with 1 to 10 mg/l kinetin and 0.01 to 1 mg/l IAA (Silberstein et al. 1983). Best rooting of the microshoots in vitro was on MS medium with 0.01 mg/l NAA. Plantlets were transplanted into soil and grown in the greenhouse.

When BA (5 to 20 mg/l) was incorporated into MS medium, adventitious buds formed from callus on the basal internode of shoot tip explants of *E. lathyris* (Ripley and Preece 1986). Of these buds, only one or two at a time would elongate, the remainder were quiescent.

A chlorophyllous callus that would produce adventitious shoots developed from leaf explants of *E. hirta* on MS medium with NAA and BA, both at 1 mg/l (Baburaj et al. 1987). The greatest number of shoots formed when the callus was transferred to MS medium with 3 mg/l BA and 0.5 mg/l NAA. The shoots were transferred to medium with 2 mg/l GA for elongation. These shoots were rooted in 1 week on half-strength MS medium with 2 mg/l IAA.

Bahadur and Reddy (1987) obtained both adventitious roots and shoots from leaf and stem callus of *E. hirta* on MS. It was necessary to include both BA and either NAA or IAA in the medium for shoot production, but only auxin was necessary for rhizogenesis.

2.1.3 Somatic Embryogenesis

Nataraja et al. (1973) sowed seeds of *E. pulcherrima* in vitro. They briefly reported that embryoids formed from callus in several cultures. They used White's medium, but did not give details about the culture conditions for induction or growth of the embryoids. In a similar study, Nataraja (1975) reported somatic embryogenesis from callus that formed from seed explants of the same species. This callus initiated from only the hypocotyl region. The somatic embryos did not produce well developed cotyledons and did not develop synchronously. The radicles from the somatic embryos grew down into the medium. Conversion of embryos into plantlets was not reported.

Kleffel and Preil (1986) obtained somatic embryogenesis from cell suspensions of *E. pulcherrima* when the cells were subcultured onto MS liquid medium with 0.1 mg/l 4-chlorophenoxyacetic acid (CPA) and 0.2 mg/l BA. The somatic embryos would develop into whole plantlets unless they were plated onto agar solidified medium with 2 mg/l IAA and 0.2 mg/l BA. This combination caused the embryos to produce callus followed by embryogenesis and organogenesis.

Preil et al. (1988) used a bioreactor to automate mass propagation using embryogenic suspensions of *E. pulcherrima*. Callus was initiated from stem segments and cell aggregates were retransferred for development into embryogenic callus on agar solidified MS medium with 0.2 mg/l BA and 2 mg/l IAA. From cell suspensions, somatic embryogenesis was induced with 0.2 mg/l BA and 0.2 mg/l CPA. These embryos grew into plantlets when they were plated onto medium with 0.2 mg/l BA and 2 mg/l IAA. Callus that developed from these embryos was inhibited on medium with no auxin and 0.05 mg/l BA. In their embryogenic suspensions, there were approximately 100 000 somatic embryos per liter of medium.

2.2 Chimera Separation

Preil and Engelhardt (1982) initiated cell suspensions from *E. pulcherrima* Annemie-pink sport, Annette Hegg, and Brillant. Callus was initiated from stem explants on MS medium with 0.2 mg/l BA and 2 mg/l IAA. Cell suspensions were obtained by transferring the callus to the same medium, but without the agar. When cells and cell clusters were again transferred to the same agar-solidified medium, organogenesis occurred. These shoots were rooted on the same medium without the BA. Plants regenerated from Annemie-pink sport, had either only red or only white bracts, indicating that the pink sport was a chimera. Plants regenerated from the Annette Hegg cell line either had short internodes and bracts typical of the cultivar, or had long internodes with narrow bracts. This was interpreted as evidence of a chimera. Plants regenerated from Brillant cells all had one phenotype, they had long internodes and oakleaf-form leaves, unlike the cultivar. All of these plants were uniform. It was believed that this plant was also a chimera. The stock plant was infected with virus and it was suggested that one type of cell may have been killed off by the infection, and only the cell type that gave rise to this abnormal but uniform phenotype survived.

2.3 Elimination of Viruses

Cell suspensions of *E. pulcherrima* Preduza and Brillant were established for the purpose of eliminating *Poinsettia* mosaic virus (PoiMV) and *Poinsettia* cryptic virus (PoiCV) (Preil et al. 1982). Shoot tip explants were excised from infected stock plants and stimulated to form callus on MS medium with 0.2 mg/l BA and 2 mg/l IAA. This callus was transferred to liquid medium where cell suspensions were grown for 20 (Preduza) and 60 (Brillant) days. Cells and aggregates were plated onto the same agar-solidified medium as above and organogenesis occurred in the callus. These shoots were rooted on the same medium, but without the BA. They tested 166 plants for virus and 163 were free from both viruses. Of the remaining three plants, one was infected with PoiMV and two were infected with PoiCV. This method therefore proved to be an effective method for escaping these viruses. However there was a problem with Brillant which proved to be a chimera and resulted in regenerated plants with a different phenotype than the parent plant.

2.4 Induction of Mutations

To better understand mutability of *E. pulcherrima,* a cultivar (Angelika), that is heterozygous for anthocyanin synthesis (WHwh) was studied (Kleffel et al. 1986). Callus was induced to form on stem explants on MS medium with 0.2 mg/l BA and 2 mg/l IAA. Cell suspensions were then created from this callus. When 0.2 mg/l BA, 1 mg/l IAA, and 0.1 mg/l CPA were used, a high percentage of single cells was obtained. When the subculture period was changed from every 2 to every 4 weeks and 0.2 mg/l BA and 0.2 mg/l CPA were used in the medium, somatic embryogenic suspensions were obtained. These suspensions were plated at a concentration of 10^3 microclusters/ml and immobilized by mixing with agar. The immobilized cell clusters were irradiated with 10 to 60 Gy doses of X-rays. As the X-ray dosage increased, the mutation rate increased, as measured by number of plants that were homozygous (whwh), but the survival rate of embryoids decreased down to 11% (controls were considered to have 100% survival). Therefore the dosages were split and periods of time were allotted so the cells could recover from the X-ray treatment. It was found that survival rate increased if the irradiation dose was split with a time interval between. This was measured by the relative effectivity that was calculated by the number of mutants per number of irradiated units. The highest relative effectivity was with an X-ray dose of 40 Gy applied as two 20 Gy exposures separated by an interval of 1 hour.

2.5 Protoplast Cultures

Protoplasts were isolated from fast-growing callus and cell suspensions of *E. tirucalli* (Misawa et al. 1982). The callus was induced from stem explants on MS medium with 1 mg/l 2,4-D and 2 mg/l NAA. It was then maintained on the same medium supplemented with 0.5 mg/l kinetin. They then established cell suspensions. Protoplasts were isolated from callus and cell suspensions by digesting the cell walls with 1% Dricelase and 0.1% Pectolyase Y23. The resulting protoplasts were cultured on KM medium 8P (Kao and Michayluk 1975) supplemented with 1 mg/l 2,4-D in a thin layer of 1 ml protoplast suspension per 35 × 15 mm plastic dish. Protoplasts isolated from callus underwent abnormal divisions, although oil bodies were formed after 40 days in vitro. However, protoplasts from the suspension cultures had their first normal cell divisions after 50 h in vitro. When they were transferred back to agar solidified medium, callus grew rapidly. Within 20 days, visible callus also formed when suspensions of isolated protoplasts were cultured directly on agar medium with 0.34 M glucose. No oil bodies there were observed in the protoplasts from cell suspensions.

Ohyama et al. (1984a) used the same methods and species described in the above paragraph, except that they only isolated protoplasts from cell suspensions and during the isolation also used 3 mM MES buffer and 0.6 M mannitol described by Kao and Michayluk (1975). They found that although cell suspensions grew faster in Gamborg et al. (1968) 1-B5C medium with 1 mg/l casamino acids than in 1-B5 medium, it was easier to isolate protoplasts from cells grown in 1-B5 medium. They found that casamino acids were inhibitory to protoplast isolation and culture. The protoplasts divided rapidly on medium with 0.5 mg/l zeatin, 1 mg/l NAA, and 0.2 or 1 mg/l 2,4-D.

2.6 Metabolism of Callus

Photoautotrophic Callus. Green friable calli that were produced from stem explants of *E. characias* were selected for suspension cultures (Hardy et al. 1987a). As the suspension cultures depleted the sucrose in the medium, growth slowed and chlorophyll content decreased. Chlorophyll level then increased from day 14 to day 32 (early stationary phase of growth). At this point, the supply of any exogenous carbon source or CO_2 was the limiting factor for growth. Without any added carbon, cultures senesced after 2 months. If CO_2 was the only source of carbon, the authors were able to maintain viable photoautotrophic suspension cultures for at least 2 years. The chloroplasts from the photoautotrophic cells appeared to possess normal chloroplasts with stacked grana and stromal lamellae.

In a second study with suspensions of *E. characias*, Hardy et al. (1987b) found that as the sucrose in the medium was utilized by photomixotrophic cells, the Compensation Ratio (C.R.) decreased to zero, then as the cells became photoautotrophic, the C.R. stabilized at approximately 1.5. When photoautotrophic suspensions were transferred to medium with 5 g/l sugar, there was a return to photomixotrophy within 4 days, and then a transition to photoautotrophy after 15 days in the new medium. The transition to photoautotrophy is therefore a physiological adaptation of the cells to altered culture conditions. The cell doubling time was nearly six times longer for photoautotrophic than photomixotrophic suspensions.

Chagvardieff et al. (1988) also found that the chlorophyll content of photomixotrophic suspensions of *E. characias* decreased during the first week in medium with sucrose, then increased during the mid-exponential phase of growth. Growth ceased as the sugar was depleted. Under conditions of saturating light and CO_2 levels, it was shown that photosynthesis increased linearly after 2 weeks of the stationary phase of growth. After day 25, gross photosynthesis, on a chlorophyll basis, decreased to 50%, even though the chlorophyll content still increased in the suspensions. This decrease was thought to be from a factor, e.g., an inorganic nutrient, that became limiting in the culture medium. During the exponential and mid-stationary growth phases, cell respiration was mainly confined to the mitochondria. However, in the late stationary phase, chloroplast activities, such as photoreduction, accounted for about 15% of the O_2 consumption. Evidence was presented that the cells photorespired after 3 weeks in culture.

2.7 Secondary Product Production

Green and non-green calli of *E. esula* developed from roots of young seedlings on MS medium and were analyzed and compared for their lipid constituents (Lee and Starratt 1972). The composition of both types of callus was similar. The least polar fraction that was isolated was shown to be triglycerides. They also found β-sitosterol, and fatty acids that were primarily palmitic acid. Using thin layer chromatography, the authors could detect no long chain aldehydes, alcohols, or triterpenoids in the callus tissues.

Callus tissue derived from stem explants of *E. tirucalli* was analyzed for its sterol content and the effects of modifications of the culture medium on sterol production (Biesboer and Mahlberg 1979a, b, 1980). The 4,4'-dimethyl sterols, euphol, and tirucallol were shown by gas-liquid chromatography to be produced by callus. Latex extracted from the whole plant contained more euphol than tirucallol; however, this ratio was reversed in stem explants and callus. The cultured cells produced four times less sterol than the parent, greenhouse-grown plants. By adding [2-^{14}C] mevalonate to the culture medium, the authors were able to demonstrate that sterol biosynthesis occurred in the callus. Callus on media of White and MS produced the highest amounts of sterols compared to other media that were tested. Of the plant growth regulators tested, IAA stimulated the highest cell and sterol yields, whereas BA suppressed both. Optimum cell yield (as measured by dry weight) appeared correlated to sterol production and levels of both euphol and tirucallol followed this trend. Of the precursors of sterol production that were added to the medium, only squalene at 1 mg/l enhanced sterol yield (by 24%), whereas mevalonic acid, lactone, and lanosterol had no effect; high concentrations (> 2.5 mg/l) of lanosterol were inhibitory to both callus growth and sterol yield. No laticifers were evident in the callus; this may indicate that in whole plants, sterols are synthesized in tissues and transported to the laticifers.

Calli were induced to grow from shoot explants of *E. tirucalli* and leaves of *E. millii* on MS medium with either 2,4-D or NAA plus either malt or yeast extract or casein hydrolysate (Yamamoto et al. 1981). From media containing 2,4-D, they selected a yellowish-white callus of *E. tirucalli*, and a yellow and a varied colored callus from *E. millii*. All calli produced approximately the same amounts of the same sterols: sitosterol (72%), stigmasterol (25.5%), campesterol (2.5%), and a trace of cholesterol. The fatty acid fractions contained palmitic and linoleic acids. The colored callus of *E. millii* produced a red pigment that was identified as cyanidin glycoside. They suggested the apparent lack of agreement between their results and those of Biesboer and Mahlberg (1979a, b, 1980) on the sterols in *E. tirucalli* may be because Yamamoto et al. used 2,4-D in their medium.

Using an OSO_4 stain, Ohyama et al. (1984b) detected oil bodies in the cytoplasms of cells from fine suspension cultures of *E. tirucalli*. There were more oil bodies in cells of later stages of growth than earlier stages. The major oil components of these cells were shown to be the sterols β-sitosterol, stigmasterol, and either euphol or tirucallol; and both di- and triterpenoids.

Bahadur and Reddy (1987) placed leaf and internode explants in vitro from *E. hirta* plants that produced either only green (acyanic) or only purplish red (cyanic) leaves. The explants from the purplish red plants produced callus with both green and red pigmented areas. Using paper chromatography, they identified the red pigment to be cyanidin-3-glucoside. They detected no anthocyanin formation in calli from the all green plants.

Yamamoto (1988) grew callus from shoot explants of *E. tirucalli*, leaves of *E. millii*, and stems of *E. lathyris* on MS medium with 0.1 to 10 mg/l 2,4-D or NAA plus malt or yeast extract or casein hydrolysate. Calli from all three species were shown to contain sitosterol, stigmasterol, campesterol, palmitic acid, and linoleic acid. The callus from *E. millii* also contained the anthocyanin, cyanidin glycoside.

2.8 In Vitro Selection

Yamamoto et al. (1982) and Yamada and Fujita (1983) cultured leaf-derived calli of *E. millii* on MS medium with 0.2 mg/l 2,4-D. These calli were mottled red and white and the red pigment was cyanidin monoglucoside. Mottled calli were cut into sections and transferred to fresh medium. As these grew, each was cut, and one-half was analyzed for red pigmentation. The remaining reddest cell aggregates were then selected and cut into several sections. Such selections were made whenever the calli were transferred to fresh medium; this was done 28 times. Between the 16th and 28th transfers, the pigment content was measured and reported in the article. The red pigment increased nearly three times between the 16th and 22nd transfers. After the 23rd transfer, the content of the pigment was seven times higher than that of the original callus. The pigment content then stabilized after the 24th transfer and was at the same high level for the remainder of the study. The authors reported that cloning and selecting cell aggregates was superior to single cell selection techniques, because of a failure to obtain high red pigment producing clones with the single cell method.

Using the same cell aggregate selection and cloning technique, Yamamoto (1988) selected cell aggregates of *E. millii* that contained the most red pigment. It was shown that the calli produced the most anthocyanin on Gamborg et al. (1968) medium than on Linsmaier and Skoog (1965), MS (1962), Nitsch and Nitsch (1967), Heller (1953), or White (1963) media. There was 1.3 times more anthocyanin from calli B5 medium (Gamborg et al. 1968) than on MS (the medium that promoted second highest anthocyanin production from calli). When testing a range from one to nine percent sucrose in the medium, the calli with the most anthocyanin were on medium with 5 to 7%. Yamamoto (1988) showed that the calli produced more anthocyanin when 2,4-D was in the medium compared to NAA. The combination of 0.2 mg/l 2,4-D and 0.002 mg/l BA was optimum for callus and anthocyanin production. If the concentration of BA was ≥ 0.2 mg/l, callus growth was inhibited. The author also reported that anthocyanin production was promoted when there was 210 mg/l nitrogen in the medium in the form of NH_4^+ and NO_3^-, and the NH_4^+/NO_3^- ratio was 1/16 to 1/4. When cell suspensions were cultured in a jar fermenter, the optimum rotation for anthocyanin production was 50 to 100 rpm when a small paddle was used in the vessel. A rotatory shaker resulted in four times more anthocyanin production because the *E. millii* were so fragile that the agitation with the paddle caused damage.

2.9 Laticifer Growth

The nonarticulated laticifers that were in the cultured zygotic embryos of *E. marginata* were elongated and branched cells with several, usually spindle-shaped nuclei (Mahlberg 1959). These laticifers grew intrusively into the callus that formed from the embryos. Laticifers grew from both their tips and by cell elongation. The elongated laticifers occupied only intercellular spaces in the callus.

Short segments of injured laticifers grew in cultured apical meristems from zygotic embryo explants of *E. marginata* (Mahlberg 1967). The growth of the injured laticifer segments appeared to be stimulated by apical meristem activity. Latex was evident when the developing plants were injured.

3 Conclusions and Prospects

Considering the number of species within the genus and both the current and potential uses of whole plants and the varied activities of the chemical constituents of these plants, only a very few in vitro studies have yet been published on *Euphorbia* spp. Those that are present in the literature indicate that in vitro techniques ranging from micropropagation to in vitro selection have been successful. In the future, as fossil fuels become depleted, plants, such as many euphorbs, that will grow on lands that are currently unproductive will have increasing economic importance. Additionally, euphorbs produce various secondary plant products, such as those with activities as pesticides and compounds that aid in prevention, curing, or may cause human diseases. In general these types of compounds are receiving increasing attention. In vitro studies with various species of *Euphorbia* will be of major importance as we attempt to meet important economic challenges and solve various other problems of humanity.

4 Protocol

A protocol for micropropagation of *Euphorbia* spp., based on our experience with *E. lathyris*, is as follows.

Shoot tip explants should be up to 5 cm long when excised from stock plants. Following surface sterilization in 0.5% NaOCl with wetting agent (Tween 20) for 20 min, then three 5-min rinses, shoot tip explants may be cut to a length of 2.0 cm. Explant preparation is critical. Carefully, leaves should be cut near the stem, rather than removed by tearing or pulling. Cutting minimizes the wounded surface area, visible dark exudation is reduced or eliminated, and rapid transfers to fresh medium become unnecessary. Explants oriented vertically have better survival rates than those placed horizontally onto the surface of the medium. Axillary shoot proliferation occurs on MS medium with no cytokinin or 5 mg/l benzyladenine (BA). Those on medium with the BA produce microshoots about 1 month sooner than if no cytokinin is present. Adventitious shoots form from callus on the cut ends of the shoot tip explants if BA is in the medium. Adventitious shoots are especially evident after three months in vitro on medium with 5 or 10 mg/l BA. Microshoots root better if they are excised and placed on modified MS with one-half strength calcium chloride, ammonium nitrate, potassium nitrate, and 1 mg/l NAA. Once rooted, acclimitization can be accomplished by removing the lids on culture vessels for about 1 week prior to planting into pots with a greenhouse medium. These plants should be in a high relative humidity for one or more weeks before being placed onto a greenhouse bench without supplemental humidity control.

References

Baburaj S, Damotharan R, Santhaguru K (1987) Regeneration in leaf callus cultures of *Euphorbia hirta* Linn. Curr Sci 56: 194

Bahadur B, Reddy KRK (1987) Anthocyanin biosynthesis in tissue cultures of *Euphorbia hirta* (Euphorbiaceae). Cell Chromosome Res 10: 110–115

Bailey LH, Bailey EZ (eds) (1976) Hortus Third, Macmillan, New York

Biesboer DD, Mahlberg PG (1979a) Production of sterols by callus cultures of *Euphorbia tirucalli* L. Botan Soc America. Abstr Pap Presented at the Meetings of the Botanical Soc Am and Certain Affiliated Groups, Am J Bot 66: 40, ms 157

Biesboer DD, Mahlberg PG (1979b) The effect of medium modification and selected precursors on sterol production by short-term callus cultures of *Euphorbia tirucalli*. J Nat Prod 42: 648–657

Biesboer DD, Mahlberg PG (1980) Sterol synthesis and identification in cultures of *Euphorbia tirucalli* L. In: Sala F, Parisi B, Cella R, Ciferri O (eds) Plant Cell Cultures: Results and Perspectives. Elsevier/North-Holland Biomedical Press, Amsterdam, pp 351–357

Calvin M (1979) Petroleum plantations for fuel and materials. BioScience 29: 533–538

Calvin M (1983) New sources for fuel and materials. Science 219: 24–26

Chagvardieff P, Pean M, Carrier P, Dimon B (1988) Oxygen exchange during growth of photomixotrophic cell suspensions of *Euphorbia characias* L. Plant Cell Tissue Organ Cul 12: 243–251

Compton ME, Preece JE (1988) Response of tobacco callus to shoot tip exudation from five species. HortScience 23: 208–210

De Langhe E, Debergh P, van Rijk R (1974) In vitro culture as a method for vegetative propagation of *Euphorbia pulcherrima*. Z Pflanzenphysiol 71: 271–274

Eriksson T (1965) Studies of the growth requirements and growth measurements of cell cultures of *Haplopappus gracilis*. Physiol Plant 18: 976–993

Gamborg OL, Miller RA, Ojima K (1968) Nutrient requirements of suspension cultures of soybean roots. Exp Cell Res 50: 151–158

Hardy T, Chaumont D, Brunel L, Gudin C (1987a) Photoautotrophic suspension cultures I—Obtention of photoautotrophic cultures from *Euphorbia characias* L. J Plant Physiol 128: 11–19

Hardy T, Chaumont D, Wessinger ME, Bournat P (1987b) Photoautotrophic suspension cultures II – Transition from photoheterotrophic to photoautotrophic growth. J Plant Physiol 130: 351–361

Hartwell JL (1969) Plants used against cancer. A survey. Lloydia 32: 153–205

Hecker E (1977) New toxic, irritant and cocarcinogenic diterpene esters from Euphorbiaceae and from Thymelaeaceae. Pure Appl Chem 49: 1423–1431

Heller R (1953) Researches on the mineral nutrition of plant tissues. Ann Sci Nat Bot Biol Veg 11th Ser 14: 1–223

Hirata Y (1975) Toxic substances of the Euphorbiaceae. Pure Appl Chem 41: 175–199

Jakobek JL, Backhaus RA, Herman K (1986) Micropropagation of candelilla, *Euphorbia antisyphilitica* Zucc. Plant Cell Tissue Organ Cult 7: 145–148

Kao KN, Michayluk MR (1975) Nutritional requirements for growth of *Vicia hajastana* cells and protoplasts at a very low population density in liquid media. Planta 126: 105–110

Kinghorn AD (1979) Cocarcinogenic irritant diterpenoids of the Euphorbiaceae. In: Kinghorn AD (ed) Toxic Plants. Columbia Univ Press, New York, pp 138–159

Kleffel B, Preil W (1986) Studies on embryogenesis in suspension culture of poinsettia (*Euphorbia pulcherrima*). In: Somers DA, Gengenbach BG, Biesboer DD, Hackett WP, Green CE (eds) Abstr 6th Int Cong Plant Tissue Cell Cult. Univ Minnesota, p 294

Kleffel B, Walther F, Preil E (1986) X-ray-induced mutability in embryogenic suspension cultures of *Euphorbia pulcherrima*. In: Nuclear Techniques and In-Vitro Culture for Plant Improvement. IAEA, Vienna, pp 113–120.

Lee CW, Yeckes J, Thomas JC (1982) Tissue culture propagation of *Euphorbia lathyris* and *Asclepias erosa*. HortScience 17: 533 (Abstr)

Lee TT, Starratt AN (1972) Growth substance requirements and major lipid constituents of tissue cultures of *Euphorbia esula* and *E. cyparissias*. Can J Bot 50: 723–726

Linsmaier EM, Skoog F (1965) Organic growth factor requirements of tobacco tissue cultures. Physiol Plant 18: 100–127

Litz RE (1985) Somatic embryogenesis in tropical fruit trees. In: Henke RR, Hughes KW, Constantin MJ, Hollaender A, Wilson CM (eds) Tissue culture in forestry and agriculture. Plenum Press. New York, pp 179–193

Mahlberg PG (1959) Development of the non-articulated laticifer in proliferated embryos of *Euphorbia marginata* Pursh. Phytomorphology 9: 156–162

Mahlberg PG (1967) Growth response of the laticifer and the juvenile shoot apices of *Euphorbia marginata* Pursh. Phytomorphology 17: 429–437

Misawa N, Yamano Y, Ohyama K, Komano T (1982) Protoplast culture of *Euphorbia tirucalli* and oil body formation in the cells. Proc. Int. Cong Plant Tissue Cell Cult: 607–608

Murashige T, Skoog F (1962) A revised medium for rapid growth and bioassays with tobacco tissue culture. Physiol Plant 15: 473–497

Nataraja K (1975) Morphogenesis in embryonal callus of *Euphorbia pulcherrima* in vitro. Curr Sci 44: 136–137

Nataraja K, Chennaveeraiah MS, Girigowda P (1973) In vitro production of shoot buds in *Euphorbia pulcherrima*. Curr Sci 42: 577–578

Nissen S, Foley ME (1987) *Euphorbia escula* L. root and shoot bud indole-3-acetic acid levels at three phenologic stages. Plant Physiol 84: 287–290

Nitsch JP, Nitsch C (1956) Auxin-dependent growth of excised *Helianthus* tissues. Am J Bot 43: 839–851

Nitsch C, Nitsch JP (1967) Induction of flowering in vitro in stem segments of *Plumbago indica*. I. The production of vegetative buds. Planta 72: 353–370

Ohyama K, Misawa N, Yamano Y, Komano T (1984a) Protoplast isolation from *Euphorbia tirucalli* L. cell suspension cultures and sustained cell division. Z Pflanzenphysiol 113: 367–370

Ohyama K, Uchida Y, Misawa N, Komano T, Fujita M, Ueno T (1984b) Oil body formation in *Euphorbia tirucalli* L. cell suspension cultures. Plant Cell Rep 3: 21–22

Preece JE, Ripley KP (1984) Clonal propagation and field culture of *Euphorbia lathyris* In: Harvesting the sun. International Bio-Energy Directory and Handbook-1984, pp 109–110

Preil W, Engelhardt M (1982) In vitro-Entmischung von Chimärenstrukturen durch Suspensions-Kulturen bei *Euphorbia pulcherrima* Willd. (In vitro separation of chimeras by suspension cultures of *Euphorbia pulcherrima* Willd.) Gartenbauwissenschaft 47: 241–244

Preil W, Koenig R, Engelhardt M, Meier-Dinkel A (1982) Eliminierung von poinsettia mosaic virus (PoiMV) und poinsettia cryptic virus (PoiCV) aus *Euphorbia pulcherrima* Willd. durch Zellsuspensionskultur. Phytopathol Z. 105: 193–197

Preil W, Florek P, Wix U, Beck A (1988) Towards mass propagation by use of bioreactors. Acta Hortic 226: 99–106

Raju MVS (1975) Experimental studies of leafy spurge (*Euphorbia esula* L.). I. Ontogeny and distribution of buds and shoots on the hypocotyl. Bot Gaz 136: 254–261

Ripley KP, Preece JE (1986) Micropropagation of *Euphorbia lathyris* L. Plant Cell Tissue Organ Cult 5: 213–218

Roth WB, Cull IM, Buchanan RA, Bagby MO (1982) Whole plants as renewable energy resources: checklist of 508 species analyzed for hydrocarbon oil, polyphenol, and protein. Trans Ill State Acad Sci 75: 217–231

Silberstein SP, Huang FH, Klingaman GL (1983) In vitro propagation of succulent *Euphorbia flanaganii* N.E.Br. Cactus Succulent J 55: 80–83

Tideman J, Hawker JS (1982) In vitro propagation of latex-producing plants. Ann Bot 49: 273–279

White PR (1963) The cultivation of animal and plant cells, 2nd edn. Ronald Press, New York

Yamada Y, Fujita Y (1983) Production of useful compounds in culture. In: Evans DA, Sharp WR, Ammirato PV, Yamada Y (eds) Handbook of plant cell culture vol 1. Macmillan, New York, pp 717–728

Yamamoto Y (1988) Tissue cultures of *Euphorbia* species In: Hanover JW, Keathley DE (eds): Genetic manipulation of woody plants. Plenum Press, New York, pp 365–376

Yamamoto Y, Mizuguchi R, Yamada Y (1981) Chemical constituents of cultured cells of *Euphorbia tirucalli* and *E. millii*. Plant Cell Rep 1: 29–30

Yamamoto Y, Mizuguchi R, Yamada Y (1982) Selection of a high and stable pigment-producing strain in cultured *Euphorbia millii* cells. Theor Appl Genet 61: 113–116

Zhang B, Stoltz LP, Snyder JC (1987) In vitro propagation of *Euphorbia fulgens*. HortScience 22: 486–488

VIII Micropropagation of Candelilla (*Euphorbia antisyphilitica* Zucc.)

R.A. BACKHAUS[1] and J. HO[1,2]

1 Introduction

1.1 Botany

Candelilla, *Euphorbia antisyphilitica* Zucc. (*E. cerifera* Alocer) is a succulent, desert shrub with numerous erect pale green, leafless stems, 3–4 mm thick and 2–10 dm tall. It is a cespitose perennial with fleshy rhizomes 3–4 mm thick (Correll and Johnston 1970). Although leafless most of the year, it produces tiny, sessile leaves, 2–3 mm long, at the tips of new growth in spring (Fig. 1). Minute white to lavender staminate or pistillate flowers are borne in cyathia in the upper axils of the stem during spring and fall. A member of the Euphorbiaceae, this native of the Chihuahuan desert of North America is noted because its stems secrete a high quality wax which has been harvested from plants growing in the wild. For this reason, it has attracted attention as a possible new crop suitable for cultivation on marginal arid lands (Anonymous 1975).

Candelilla is Spanish for "little candle" and probably owes its name to the growth habit of the plants, which have the shape of small candles, or perhaps to its wax, which has been used in Mexico to manufacture candles. The botanical name, given by Zuccarini in 1829, referred to the medicinal value of the plant's milky latex which was used as a treatment for venereal disease in Mexico in the early 1800s. The value of the plant's wax was not recognized until much later when commercial wax production began in 1912 (Villa Salas 1981).

1.2 Distribution and Importance

Candelilla is abundant throughout the Coahuilan desert in Mexico, extending from Hildago north to Coahuila and eastern Chihuahua including the Big Bend area of western Texas. It survives on only 100–600 mm of annual rainfall, at an elevation range of between 550–2400 m. It grows best on well-drained slopes and appears rarely on low-lying, heavy, clay soils. The median temperature range for candelilla is 18–22 °C, with a maximum of 47 °C and a minimum of − 14 °C (Villa Salas 1981).

[1] Department of Botany, Arizona State University, Tempe, AZ 85287-1601, USA
[2] Shihezi Agricultural College, Shihezi, Xinjiang, P.R. China

Biotechnology in Agriculture and Forestry, Vol. 20
High-Tech and Micropropagation IV (ed. by Y.P.S. Bajaj)

Fig. 1. A Mature candelilla plant showing its growth habit 6 years after being propagated by division. **B** Close-up of candelilla shoot tip showing developing inflorescences. *Bar* 1 cm

Candelilla plants are not usually cultivated, but are gathered from wild stands primarily in the deserts of northern Mexico. The plants are uprooted and brought to a makeshift "fabrica" or extraction camp, where they are boiled in large vats (Fig. 2). The wax floating to the surface is skimmed off and collected for further processing and packing. Once an area has become depleted of candelilla, the extraction plant is moved to a new location where harvesting resumes. This

Fig. 2. A candelilla "fabrica" showing workers processing candelilla plants gathered from wild stands growing in the Chihuahuan desert of Mexico. Large bundles of candelilla (*background*) are hauled by burro to the site where they are boiled in a large vat. The wax floats to the surface and is skimmed off and collected. Following the extraction, spent candelilla stalks are dried and later burned in the pit beneath the vat to provide heat for the boiling process. Much of the wax produced in this manner is smuggled out of Mexico to the USA. (Photo D.J. Keil)

destructive method of harvest has led to the near eradication of candelilla in certain regions and has warranted investigations into the cultivation of this species (Anonymous 1975; Villa Salas 1981).

The harvest and exportation of candelilla wax has declined somewhat, over the years, probably due to the depletion of easily accessible stands. During the early 1950s, the USA, the world's largest consumer of candelilla, imported 3.6 million kg of wax annually (Hodge and Sineath 1956). Accurate data regarding the amount of candelilla wax currently produced by Mexico and exported to the USA is not available, because much of the wax is illegally smuggled out of Mexico north to the Big Bend area of Texas. There it is legally sold to merchants, who then resell it to American manufacturers. Data for 1966 and 1987 (Table 1) shows the comparison of USA imports of candelilla with various other plant waxes. Although candelilla is one of the most expensive of plant waxes, annual imports to the USA in 1987 were only one-fifth of that in the 1950s.

Candelilla wax has been used as a substitute for carnauba and beeswax. Compared to other plant waxes, the melting point of candelilla wax varies widely (Table 2), probably due to the time of year and locale where plants are harvested and the heterogeneous nature of the plant material gathered. An additional species, *Pedilanthus pavonis*, is also referred to and harvested as candelilla in Mexico, and wax prepared from mixtures of the two species often occur. This undoubtedly

Table 1. United States Imports of Plant Waxes (US Dept of Commerce 1987; US Tariff Commission 1968)

Wax	1966	1987	1987
	Thousands of kg		Price US $ per kg
Candelilla	1383	776	0.93
Carnauba	6549	4024	0.84
Japan and Ouricari	142	541	0.69

Table 2. Properties of some commercial plant waxes. (Hodge and Sineath 1956; Tulloch 1976)

Wax	Melting point (°C)
Candelilla	67–80
Carnauba	83–86
Ouricari	80–83

results in altered chemical and physical properties of the wax, which have been duly noted (Ashraf-Khorassani and Taylor 1990). Some of the chemical constituents and physical properties of candelilla wax have been characterized (Hodge and Sineath 1956; Tulloch 1976). The major components of crude candelilla wax have been characterized and consist of 49–50% hydrocarbons, 20–29% wax esters, 12–14% alcohols, 7–9% free acids, and the remainder as moisture and inorganics (Ashraf-Khorassani and Taylor 1990). The major constituent of the hydrocarbon fraction is n–$C_{31}H_{64}$ with lesser quantities of C_{25}–C_{35} also present.

Candelilla wax has been used as hardener and extender for softer paraffin (synthetic) waxes, in chewing gum, lip gloss, polishes, varnishes, paper sizing, sealing waxes, electrical insulation, leather processing, cosmetics, printing ink, typewriter ribbons, and to facilitate the extrusion of munition powder (Hodge and Sineath 1956; Anonymous 1975; Tulloch 1976; Ashraf-Khorassani and Taylor 1990).

As with most plants, epicuticular wax is thought to reduce transpiration, especially under drought conditions, and to increase disease and frost resistance. These properties are related to the amount, composition, and physical form of the wax deposited (Tulloch 1976). Unlike other plant species, however, candelilla can produce relatively large quantities of wax. Candelilla wax can account for 1.5 to 3% of the weight of the entire plant, and can exceed 7% when measured as a percentage of stem weight (Villa Salas 1981). The time of year for greatest wax production is thought to be in winter, and it is harvested mostly between April and May (Villa Salas 1981). Scanning electron micrographs reveal that the wax is deposited as plates or encrustations by exudation through the cuticle onto the surface of the stem (Fig. 3A). When these tissues are treated with organic solvents, such as chloroform, it removes the wax to reveal the de-waxed cuticle of the stem (Fig. 3B).

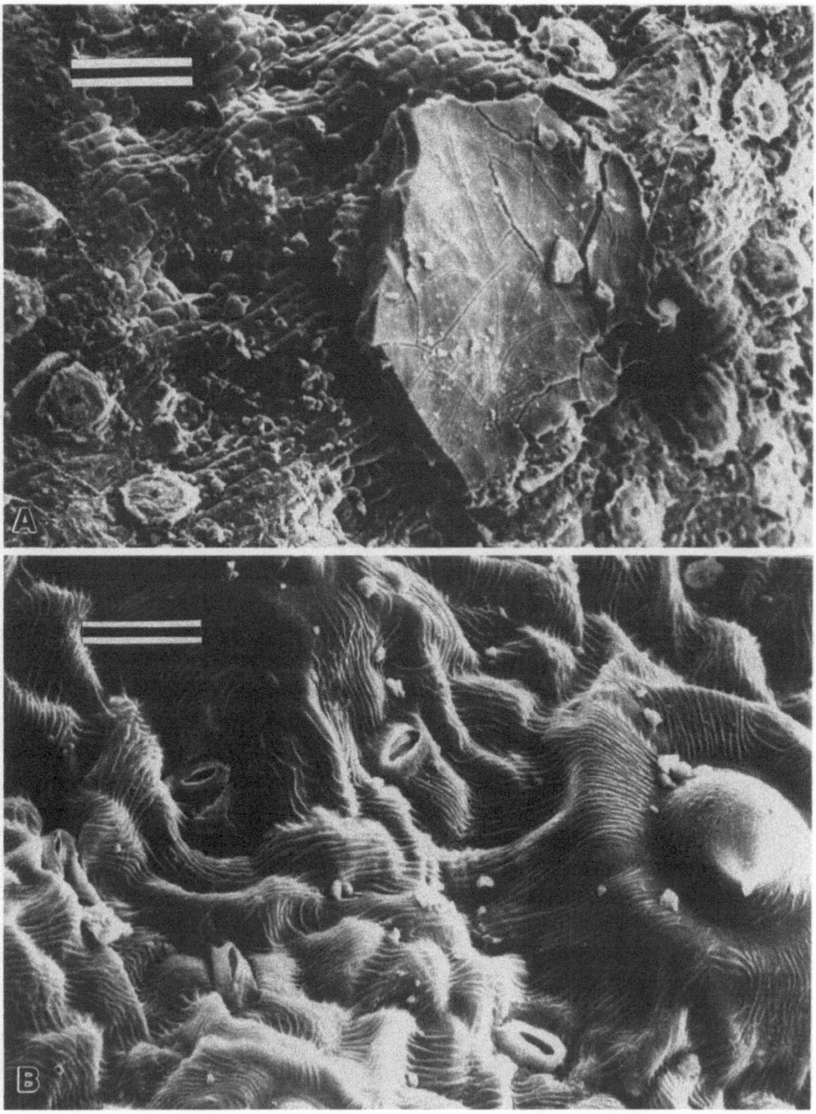

Fig. 3. A Scanning electron micrograph of candelilla stem surface showing wax encrustations being secreted from the cuticle. Circular encrustations are those surrounding stomata. *Bar* 200 µm. **B** Same tissue after being extracted with chloroform reveals the de-waxed surface. *Bar* 200 µm. (Micrographs G. Hendricks)

1.3 Conventional Propagation and Breeding

Candelilla can be asexually propagated by stem cuttings or by the division of shoots and rhizomes which form dense clumps. Rooting of stem cuttings is enhanced by the use of auxin-type rooting compounds at a concentration of 500 to 5000 ppm (Villa Salas 1981). Propagation by vegetative means is preferred for commercial cultivation if superior wax-producing clones are ever identified. It is known that, once harvested from an area, candelilla is slow to regenerate new stands. This has been a factor in its over-exploitation in Mexico. Unfortunately, little is known about the environmental and cultural factors which control wax formation and yield. Even less is known about the genetics or breeding characteristics of candelilla. So little is known about its reproductive behavior that, for example, it is not known whether native candelilla stands are made up of genetically distinct individuals or if they are derived from vegetative clones which arose via rhizomes.

1.4 Need for in Vitro Culture

Since the inherent knowledge of the reproductive behavior of candelilla is so limited, it seems most feasible that individual clones or isolated lines collected from diverse, wide-ranging stands should be the starting material for developing candelilla as a new crop. Tissue culture of candelilla for the purpose of micropropagation is based on its potential economic value as a cultivated source of wax, and its evaluation in field trials will require substantial numbers of cloned plants. The fastest way to achieve this is through tissue culture, because conventional propagation by division or rooting of stem cuttings is slow. A second reason for developing tissue culture protocols for candelilla is as a means of conserving native plant material which may be subject to eradication by over-exploitation. Finally, as a long-term goal, it is not impossible that culture methods could be developed for in vitro wax production using candelilla tissue as the starting material.

2 Micropropagation

2.1 Explanting

Selected or promising plants growing in wild stands can be dug and transported, bare-root, for up to 1 week before transplanting into a greenhouse. Specimens can be easily transplanted into 15-cm pots containing a coarse, well-drained commercial potting mix (i.e., a 1:1, peat:sand) and kept in a greenhouse until new axillary shoots appear. It is best to grow plants to be explanted in a greenhouse on a raised bench to avoid contamination by splashing water and soil. After about 2 months, newly emerged shoot tips are excised when they become 3–4 cm long and 0.2–0.5 mm in diameter. These are surface sterilized in a 0.5% solution of sodium hypochlorite (commercial bleach diluted 1:9, v/v, with 0.1% Tween 20 as a surfactant) by immersing the tissue for 10 min under a partial vacuum of − 40 to

– 50 kPa followed by three washes in sterile, distilled water. The bottom 1 cm of basal tissue is removed under sterile conditions, and the remaining shoot tip is planted in a semi-solid culture medium.

2.2 Media

The basal medium consists of Murashige and Skoog (1962) salts supplemented with 87.6 mM sucrose, 555 μM inositol, 1.2 μM thiamine-HCl and TC agar (Hazelton-KC, Lenexa, KS). To this medium, supplements are added (Table 3) to produce a Stage I/Stage II shoot explanting/shoot proliferation medium and a Stage III rooting medium. The pH of each medium is adjusted to 5.7 prior to autoclaving at 121 °C, 103.4 kPa, for 15 min. All cultures are grown in 25 ml of media in 25 × 150 mm tubes capped with plastic Kaput closures (Bellco, Vineland, NJ). Cultures are maintained at 25–28 °C under a photon flux of 25 $\mu mol/s/m^2$ for all culture stages. Irradiance is provided by cool white fluorescent tubes at 16 h per day.

Axillary shoots in Stage II media are first observed after 1–2 weeks. New axillary shoots elongate over the next 2–6 weeks and grow to 1–3 cm long (Fig. 4A), when they should be subcultured to a fresh medium of the same composition or can be transferred to a rooting medium. Cultures should not be kept in the same culture vessel any longer than 10 weeks, or shoots begin to deteriorate. The usual symptom is severe chlorosis leading to shoot death from the apex down. Long-term cultures also develop excess callus at the base of the shoots. In the case of cultures which are kept too long, it is possible to retrieve green shoots which might remain in a cluster of dead, yellow shoots. These can simply be subcultured to Stage II medium with no ill effects, except that they are slow to resume growth.

Root initiation occurs on up to 100% of the shoots transferred to the rooting medium containing 0.49 μM IBA (Fig. 4B). Roots develop on shoots transferred to

Table 3. Amounts of chemical supplements added to basal medium to give Stage I, II, and III media. (Jakobek et al. 1986)

Chemical	Stage I explanting and Stage II shoot proliferation media	Stage III rooting medium
MS salts	1X	0.5×
Thiamine-HCl	1.2 μM	1.2 μM
Inositol	555 μM	555 μM
NaH_2PO_4–H_2O	1.23 mM	—
Sucrose	87.6 mM	87.6 mM
6-Benzylaminopurine (BA)	4.4 μM	—
Naphthaleneacetic acid (NAA)	0.13 μM	—
Indole-3-butyric acid (IBA)	—	0.49 μM
TC Agar	8 g/l	6 g/l

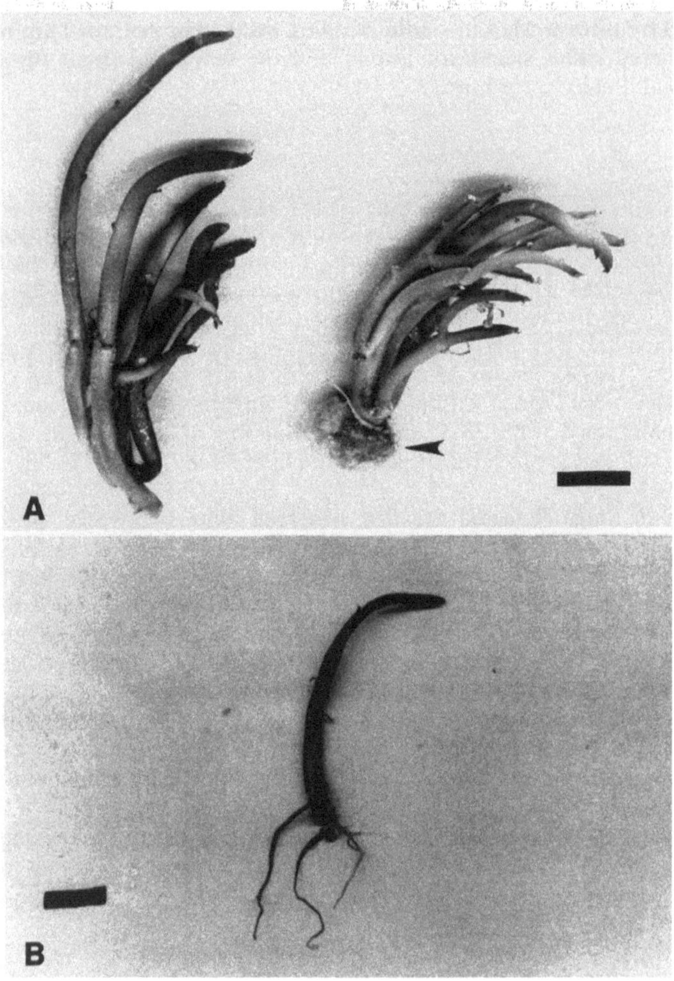

Fig. 4. A Candelilla following tissue culture after 4 weeks in Stage II shoot proliferation medium (Table 3). Excess callus can develop on the base of shoots, especially if not subcultured after 6 weeks. **B** Rooting of shoot tips following 4 weeks of culture in Stage III rooting medium. *Bars* 1 cm

rooting medium after 2–5 weeks and rooted plants can be transferred almost immediately to 8 cm diameter pots containing a commercial potting mix (Jiffy Mix, JPA, West Chicago, IL). Because candelilla is a succulent, it does not become desiccated as easily as other species when removed from the culture vessel. Also, the fact that it has the capacity to produce a thick epicuticular wax layer may help it in preventing excess water loss during acclimatization. Plants can be placed in a shaded greenhouse until shoot growth resumes and can be placed in full sun for eventual transplanting to the field.

2.3 Controlling Callus and Shoot Morphogenesis

One of the inevitable byproducts of this micropropagation procedure is the production of large quantities of callus, which form at the base of the shoots during the axillary shoot proliferation step. Excess callus production is undesirable because it suppresses shoot proliferation and causes existing shoots to deteriorate and die as the callus increases in mass. That is why candelilla must be subcultured at 3–5-week intervals. Attempts at suppressing callus production by increasing the levels of cytokinin up to 44 μM had the effect of severely suppressing shoot elongation and resulted in numerous, abnormal, stubby shoots which could not be easily subcultured in Stage II or rooted in Stage III. Reducing the cytokinin concentration or eliminating the auxin in Stage II cultures caused a reduction in the shoot proliferation rate with little concomitant effect on reducing callus production. Thus, after numerous trials, the protocol outlined in Table 3 for micropropagation appears to be the best so far developed for candelilla. No serious attempt has been made to exploit the callus-producing abilities of candelilla for either somatic embryogenesis or direct, *in vitro* wax production. These techniques could be developed further.

3 Conclusions

As with any potential new crop, commercial wax production from cultivated candelilla will require a considerable research effort. Superior clones have to be identified, optimum cultural practices must be developed and knowledge of the genetics and reproductive characteristics of the plant must be better understood before improvements in wax yield and quality can take place. Despite this, candelilla wax from unimproved, wild plants continues to be used for industrial purposes. Genetic improvement of the wax composition could make it more competitive with some of the higher-melting point plant waxes, such as carnauba. Cultivated candelilla will probably have its best chance of success in countries where land costs are inexpensive. Although it is currently harvested and processed by hand, these tasks could easily be mechanized if high labor costs are a consideration.

Candelilla has proved to be relatively simple to grow by tissue culture techniques. The rate of plant proliferation and cloning is certainly much faster than the traditional technique of division by rhizomes. Such rapid proliferation could lead to the possible commercialization of prospective high-yielding clones if cultivation of this plant is attempted.

4 Protocol

1. Excise a 3–4-cm-long portion of shoot tip from greenhouse-grown plants (Jakobek et al. 1986).
2. Surface sterilize with a freshly prepared solution of 0.5% sodium hypochlorite and 0.1% Tween 20 by immersing tissue for 10 min under a partial vacuum and washing three times in sterile water.

3. Excise the top 2–3 cm of shoot tip and place explant vertically on a solid medium containing MS salts, supplemented with 87.6 mM sucrose, 1.23 mM NaH_2PO_4, 555 μM inositol, 1.2 uM thiamine-HCl, 4.4 uM 6-benzylaminopurine, 0.13 uM naphthaleneacetic acid and 8 g/l TC Agar at pH 5.7.
4. Cultures in 25 × 150 mm tubes are incubated in a 16-h photoperiod at a photon flux of 25 $\mu mol/s/m^2$ at 25–28 °C for 4–6 weeks.
5. Axillary shoots which are 1 cm long are subcultured on the above medium at 3–6-week intervals.
6. Rooting of axillary shoots which are at least 1 cm long can be accomplished in a medium containing 0.5 × MS salts, 87.6 mM sucrose, 555 uM inositol, 1.2 uM thiamine-HCl, 0.49 uM indole-3-butyric acid and 6 g/l TC Agar.
7. Shoots are rooted after 2–5 weeks in 25 × 150 mm tubes incubated in a 16-h photoperiod at a photon flux of 25 $umol/s/m^2$ at 25–28 °C.
8. Rooted shoots are transplanted to 8 × 8 cm pots containing a commercial potting mix such as Jiffy Mix (JPA, West Chicago, IL) or coarse vermiculite, and are placed under 50% shade in a greenhouse for 2–6 weeks until they are hardened off.

Acknowledgments. I would like to thank J. Jakobek, C. Odgers, and P. Giemsoe, who participated in this research, G. Hendricks, who provided the electron micrographs, and D.J. Keil, who furnished the photograph of the candelilla fabrica in Mexico.

References

Anonymous (1975) Candelilla. In: National Academy of Sciences (ed) Underexploited tropical plants with promising economic value. Washington DC, pp 141–144

Ashraf-Khorassani M, Taylor LT (1990) Analysis of crude, purified, and synthetic candelilla wax using supercritical fluids. LC GC 8: 314–320

Correll DS, Johnston MC (1970) Manual of the vascular plants of Texas. Texas Research Foundation, Renner TX, p. 965

Hodge WH, Sineath HH (1956) The Mexican candelilla plant and its wax. Econ Bot 10(2): 134–154

Jakobek JL, Backhaus RA, Herman K (1986) Micropropagation of candelilla, *Euphorbia antisyphilitica* Zucc. Plant Cell Tissue Organ Cult 7: 145–148

Murashige T, Skoog F (1962) A revised medium for rapid growth and bioassays with tobacco tissue cultures. Physiol Plant 15: 473–497

Tulloch AP (1976) Chemistry of waxes of higher plants. In: Kolattukudy PE (ed) Chemistry and biochemistry of natural waxes. Elsevier, New York, pp 235–287

US Dept of Commerce (1987) United States Imports and Imports for Consumption, Schedule A, Commodity by Country. FT 135/December 1987, Bureau of the Census, Washington DC, p 79

US Tariff Commission (1968) Summaries of Trade and Tariff Information, Schedule 4, vol 12, TC Publication 239, Washington DC, p 121

Villa Salas AB (1981) Investigacion sobre candelilla (*Euphorbia antisyphilitica*) en el area del CIFNE. Cienc Forestal 6(32): 3–18

IX Micropropagation of Gardenia (*Gardenia jasminoides* Ellis)

A.S. Economou[1]

1 Introduction

Gardenia jasminoides Ellis (also *G. grandiflora* Lour, *G. florida* L., *G. radicans* Thunb.) known as Cape jasmine, is a common and popular plant species grown in mild climates in Europe and North America as a garden bush and small tree, or under protection as a pot plant (Fig. 1). The double-flowered type is used as an ornamental plant, while the single-flowered type is used medicinally (Mizukami 1988).

1.1 Botanical and Morphological Characteristics

The genus *Gardenia* Ellis (n = 11) belongs to the family Rubiaceae and includes some 200 species. They are nearly or quite evergreen shrubs or small trees native to China, Indochina, Japan, South Africa, and western tropical Africa. The plant was introduced for cultivation in England in 1754 and was named after A. Garden. *Gardenia* has mostly opposite leaves, and rarely a whorl of three leaves. The leaves are thick, leathery, and shiny of dark green color and with lanceolate to obovate shape up to 8 cm long. The flowers are solitary, multipetaled, white or cream, and very fragrant. They are terminal or appear on short stems in the leaf axils and are 5 to 10 cm in diameter. The calyx is tubular with five long teeth, while the corolla is short and tubular with five to 11 waxy and petal-like lobes.

1.2 Distribution and Importance

Gardenia is not a major flowering potted plant. Its popularity has decreased since 1940s and 1950s (Wilkins 1986), but recently there has been a renewed interest in gardenia plants, with steadily increasing demand for well-grown, flowering plants. In Greece *Gardenia* is a flowering pot plant with sales of more than 1 million annually. The individual flowers are also used as cut flowers for wedding bouquets, corsages, and various decorations. More than 1 million cut flowers are harvested per year in Greece, with a wholesale value of 0.5 million US dollars.

Gardenia is also a medicinal plant. The dried fruits of the single-flowered types have been used for curing a number of diseases. Important secondary metabolites,

[1] Department of Horticulture, Aristotle University, P.O. Box 281, 54006 Thessaloniki, Greece

Biotechnology in Agriculture and Forestry, Vol. 20
High-Tech and Micropropagation IV (ed. by Y.P.S. Bajaj)

Fig. 1. *Gardenia jasminoides* plants in bloom

such as crocin and various iridoid glucosides, have been isolated from the fruits, and cell suspension cultures of *Gardenia* can be applied to produce some of the iridoid glucosides for pharmaceutical purposes (Mizukami 1988).

1.3 Conventional Propagation

Gardenia is propagated by terminal cuttings taken from growing shoots of the year. The cuttings are removed from mother stock plants grown in protective shelters during any time of the year, the preferable period being from November to March if mature cuttings are needed. Mother stock plants are grown in large containers or in beds, using a well-drained peat-based soil mixture with a pH 5.0–5.5, and the cuttings are harvested when the shoots have been developed sufficiently.

The growing conditions of the stock plants can influence the rootability of the cuttings. Proper nutrition of the mother stock plants is achieved with a fertilizer of the formulation 15-30-15 and an adequate amount of micro-nutrients to avoid iron deficiency. It is necessary to apply iron occasionally, in a chelate form, to prevent yellowing of the plants. Excessive nitrogen content in the soil mixture may produce succulent shoots which are susceptible to disease infection in the propagation bench. In winter the mother stock plants need full sunlight, while in summer a shading material is applied over the plants. Cuttings should be avoided from plants suffering from diseases and particularly from the canker caused by the fungus *Phomopsis gardeniae*.

Gardenia cuttings, 7–10 cm long, are cut above a pair of leaves. The lower leaves are not taken off, in order to prevent entry of disease spores through the wounds.

Application of a rooting hormone (in water solution or powder) promotes or accelerates rooting. The cuttings are rooted under intermittent mist or a polyethylene tent, where high humidity is maintained. Uniformity of rooting is obtained by under-bench heating. A temperature of 22–24°C in the rooting mixture, which consists of 1:1 (v/v) sand:peat, is necessary for fast rooting. Some shading over the propagation bench is desirable during periods of high light intensities. Watering or misting the cuttings with alkaline water may cause problems in rooting or may inhibit further growth. Roots are developed after 4–6 weeks in the rooting medium. Propagation by cleft grafting has also been reported for *Gardenia* (Mustard and Lynch 1977). This technique is more suitable to avoid nematodes infection by using *Gardenia thunbergia* L. rootstocks, which are nematode-resistant species.

2 In Vitro Culture Studies

Although *Gardenia* is propagated readily by conventional methods, cuttings are very susceptible to disease in the rooting mixture, resulting in many losses. Moreover, several years are required to produce adequate stock plants using cuttings of a newly released cultivar. Rapid propagation through tissue culture can alleviate these problems. However, little work has been done, so far, on in vitro propagation of *Gardenia*, although considerable progress has been made in secondary metabolite production from *Gardenia* in vitro cultures (Mizukami 1988). This chapter will review the tissue culture work on *Gardenia*, focusing mainly on clonal propagation.

2.1 Callus Formation

Callus formation has been reported on various explants such as stem internodes (Dumanois et al. 1984), leaf segments (Mizukami et al. 1983; Dumanois et al. 1984), seedlings (Ueda et al. 1981), and floral parts (Dumanois et al. 1984; Tabata et al. 1984). In our work callus was formed on shoot internodes as well as on leaf sections from in vitro-derived shoots.

Auxins used for callus formation were 2,4-D, NAA, or IAA with the presence or absence of cytokinins. Auxin type and concentration varied with explant source. Callus cultures facilitated, so far, cell suspension cultures for secondary metabolite production (Ueda et al. 1981; Mizukami et al. 1983, 1985), and have not been used for in vitro propagation of *Gardenia.*

2.2 Shoot Formation

For micropropagation, shoot tips were used as explants for initiating aseptic cultures (Pontikis 1983; Economou and Spanoudaki 1985, 1986), as well as apical and axillary meristems (Scaramuzzi and D'Elia 1984) and nodal segments (Dumanois et al. 1984). Shoot tip cultures were established on Anderson's medium (1975) supplemented with 0.1 mg/l GA_3 and 0.5 or 2.5 mg/l BA (Pontikis 1983) or on a low salt medium (Economou and Read 1984) with the addition of 5–20 mg/l 2-iP (Economou and Spanoudaki 1985). Shoots (approximately two or three per explant) arose from the existing lateral buds of the explants after 4–10 weeks in culture (Table 1, Fig. 2). In experiments of shoot tip explants on media with the cytokinins 2-iP, BA, or kinetin, various effects were found on lateral bud sprouting; more shoots were formed with BA than with 2-iP or kinetin, whereas the longest and highest quality rating shoots were obtained on medium supplemented with 2-iP (Table 2). However, when the original shoot tip explants (basal mass), after removal of the proliferated shoots, were subcultured undivided on fresh medium containing 2-iP, BA, or kinetin similar numbers of shoots (four or five per explant) of the same length and quality rating were produced (Economou and Spanoudaki

Fig. 2. Shoot formation by *Gardenia* shoot tip explants on a low salt medium after 8 weeks in culture

Table 1. Effect of 2-iP on shoot formation in vitro from *Gardenia* shoot tip explants. (Economou and Spanoudaki 1985)

2-iP (mg/l)	Shoot No.	Shoot Length (cm)
5	1.6a[a]	1.9b
10	3.1b	0.9b
20	2.1a	0.5a

[a] Mean separation in columns by HSD, 5% level.

Table 2. Effect of 2-iP, BA, and kinetin on shoot formation in vitro from *Gardenia* shoot tip explants. (Economou and Spanoudaki 1986)

Cytokinin treatment	No. of shoots Concentration (mg/l)				Shoot length (mm) Concentration (mg/l)				Shoot quality rating Concentration (mg/l)			
	1	5	25	Mean[a]	1	5	25	Mean	1	5	25	Mean
2-iP	1.4	2.4	2.1	2.0a	6.1	6.0	6.2	6.1a	3.1	3.3	2.9	3.1b
BA	2.8	5.0	2.6	3.5b	2.6	2.6	1.3	2.2a	2.4	2.0	1.6	2.0a
Kinetin	1.2	2.0	3.0	2.0a	3.3	3.0	1.5	2.6a	2.8	2.2	1.5	2.2a

[a] Separation in the mean columns by HSD, 5% level.

1986). In vitro-derived shoots have also been used for establishing new aseptic cultures of *Gardenia*. BA, which was superior to 2-iP and kinetin for promotion of shoot formation from in vitro-derived shoot explants, retarded shoot elongation and reduced shoot quality rating (Economou and Spanoudaki 1986).

Apical and axillary meristems (3–5 mm long) with one or two leaf primordia were used by Scaramuzzi and D'Elia (1984) for *Gardenia* in vitro propagation on MS medium (Murashige and Skoog 1962) with the addition of IAA, or IBA and BA, or kinetin. Shoot formation was enhanced on medium supplemented with 1–5 mg/l IAA and 10 mg/l BA in a 6- to 7-week culture period. In our research, shoot apices, containing two or three leaf primordia, were used as explants to overcome contamination problems due to bacterial infections. Shoot apex growth and new shoot formation were induced with the presence of 10 mg/l BA or 2-iP in the medium.

Table 3. Effect of cytokinins (BA, 2-iP) in the presence of auxin (IAA) on axillary budding of *Gardenia* stem nodes after 6 weeks. (Dumanois et al. 1984)

Plant growth regulators (mg/l)	No. of shoots per ten explants	Mean shoot length (mm)
BA 0.3	18.7 ± 3.0[a]	10.8 ± 1.2
BA 0.3 + IAA 1	15.4 ± 2.0	12.5 ± 2.9
BA 1	95.8 ± 25.0	8.7 ± 0.8
BA 1 + IAA 1	85.4 ± 16.0	8.7 ± 0.8
2-iP 0.3	16.6 ± 2.0	16.6 ± 4.2
2-iP 0.3 + IAA 1	12.9 ± 2.9	11.6 ± 1.2
2-iP 1	15.4 ± 2.9	20.0 ± 3.7
2-iP 1 + IAA 1	10.0 ± 2.0	20.0 ± 5.0
IAA 1	15.8 ± 2.5	16.6 ± 2.9
None	9.1	4.5

[a] All statistical results at 5% level.

Nodal explants (approximately 1 cm long) from greenhouse grown *Gardenia* plants were cultured on medium consisting of MS macronutrients (half strength) and micronutrients and vitamins of Morel and Wetmore's medium (1951) for 6 weeks (Dumanois et al. 1984). Axillary bud sprouting was induced with the addition of 1 mg/l BA and 0 or 1 mg/l IAA (Table 3). The position of the nodal segment affected the rate of axillary shoot formation. Nodal explants from the apical zone responded less actively than those from lower zones in the shoot. In addition, nodal explants excised from in vitro-derived shoots produced new shoots in monthly subcultures providing, thus, an efficient system for *Gardenia* micropropagation (Dumanois et al. 1984).

Table 4. Effect of 2-iP plus GA_3 on shoot formation in vitro from *Gardenia* in vitro-derived shoot and basal mass explants. (Economou and Spanoudaki 1986)

Treatment (mg/l)	Explant source					
	In vitro-derived shoot			Basal mass		
	No. of shoots	Shoot length (mm)	Shoot quality rating	No. of shoots	Shoot length (mm)	Shoot quality rating
2-iP (5)	3.2a[a]	6.9a	2.2a	4.6a	7.1a	2.1ab
2-iP (5) + GA_3 (0.2)	3.7ab	13.5b	3.4c	5.5ab	16.1b	3.1c
2-iP (5) + GA_3 (1)	5.5c	15.3bc	3.1bc	8.0c	26.9c	2.8bc
2-iP (5) + GA_3 (5)	4.5bc	16.1c	2.0a	6.9bc	29.5c	1.8a

[a] Separation in columns by HSD, 5% level.

Fig. 3. Shoot formation by *Gardenia* basal mass explant after 6 weeks on a medium containing 5 mg/l 2-iP and 1 mg/l GA_3. (Economou and Spanoudaki 1986)

The addition of GA_3 (0.2 or 1 mg/l) to the medium before autoclaving in combination with 5 mg/l 2-iP increased the number of shoots produced from in vitro-derived shoot explants of *Gardenia* (Economou and Spanoudaki 1986). Furthermore, GA_3 in the medium increased shoot length while at concentrations 0.2 or 1 mg/l improved shoot quality rating. With basal mass cultures, GA_3 addition to the medium also affected the number of shoots produced, shoot length and shoot quality rating (Table 4, Fig. 3).

2.3 Root Formation

Excised shoots from in vitro-proliferated *Gardenia* explants were transferred vertically onto agar-based medium for rooting. Roots were developed in the presence of IBA at concentrations from 1 mg/l (Pontikis 1983) to 10 mg/l (Scaramuzzi and D'Elia 1984) in the medium. Rooting lasted from about 2 to 4 weeks. Dumanois et al. (1984) achieved in vitro *Gardenia* shoot rooting in medium containing 1 mg/l IAA and 2 g/l activated charcoal in 3 weeks. In all cases a high percentage of rooting occurred.

Rooting in vivo of *Gardenia* shoots has also been achieved. Dumanois et al. (1984) reported successful rooting in 3–5 weeks by dipping the bases of the in vitro-produced shoots in solutions of IAA (4–6 mg/l) or IBA (1–2 mg/l). The application of growth regulators enhanced rooting without, however, any effect on both the number of roots which were produced and the percentage of rooting. In our work, *Gardenia* shoots rooted exclusively ex vitro in soilless medium (Economou and Spanoudaki 1985, 1986). They were inserted into sphagnum moss peat or

Fig. 4. *Gardenia* in vitro-produced shoots during rooting on a 2:1 (v/v) peat:perlite mixture. (Economou and Spanoudaki 1986)

Fig. 5. In vitro-produced shoot of *Gardenia* which rooted ex vitro after 5 weeks. (Economou and Spanoudaki 1985)

Fig. 6. *Gardenia* plant, 4 months old, derived from tissue culture

sphagnum moss peat and perlite mixture (2:1 v/v) in plastic containers without any rooting-hormone treatment (Fig. 4). Cover with clear polyethylene film and thorough misting assured successful rooting in 5 weeks (Figs. 5, 6).

2.4 Plantlet Acclimatization

Plantlets derived from rooted shoots were transferred from the agar medium to pots containing a mixture (1:1:1 v/v) of peat, vermiculite, and soil (Pontikis 1983) or straight sphagnum moss peat (Dumanois et al. 1984) for further growth. Acclimatization of the plantlets took place in a shaded greenhouse on a bench covered with a polyethylene sheet for maintaining high humidity (Pontikis 1983). Dumanois et al. (1984) hardened plantlets in vermiculite at 18 °C for 3 weeks and then switched to 24 °C by using mist watering. On the other hand, ex vitro-rooted *Gardenia* shoots, after planting in a mixture of sphagnum moss peat and perlite (1:1 v/v), were placed for acclimatization in the greenhouse on a mist-propagation bench under 50% shade for 2 weeks, followed by placement under 20% shade on a nonmisted bench (Economou and Spanoudaki 1986).

3 Micropropagation

3.1 Selection of Plant Material

The source of explant material is essential to the success of *Gardenia* micropropagation. Mother stock plants should be maintained clean and disease-free. Generally, greenhouse-grown plants are less contaminated than field-grown plants. Spraying periodically with chemicals to control contaminating microorganisms is advisable. Plants usually suffer from systemic infections due mainly to bacteria rather than to fungi.

Shoot tip explants (1–1.5 cm long) from actively growing shoots or nodal explants (0.5–0.8 cm long) bearing axillary buds from lower parts of the shoot should be chosen for the cultures. When severe contamination problems appear, shoot apices with two or three leaf primordia are used as explants.

3.2 Explant Disinfestation

The shoot tips with expanded leaves removed and the nodal segments are soaked for 10–12 min in 0.1–0.2% Captan solution (w/v). Further surface sterilization of the explants is accomplished with a solution of 1–2% sodium hypochlorite (often Chlorox or other commercial bleach containing 1–2% NaOCl) in dips for 15 min. The highest sodium hypochlorite concentrations are applied mostly to nodal explants. The surfactant Tween 20 at 15 drops/l is added to the disinfestation solution. When bacteria problems exist, soak for 10 min in mercuric chloride ($HgCl_2$) solution at 200–300 mg/l preceding sodium hypochlorite treatment. Following surface sterilization the explants are rinsed three times for 5 min each with sterile water. Before transferring, the cut ends of the explants are trimmed aseptically. Once cultures have been initiated, shoots from in vitro cultures can provide clean and much more productive explants than shoot tips or nodal segments taken from plants.

3.3 Media

A medium containing a low salt concentration appears to be suitable for *Gardenia* in vitro culture. The azalea medium (Economou and Read 1984), Anderson's (1975) rhododendron medium, as well as Lloyd and McCown's (1980) woody plant medium (WPM) have been used successfully, so far, for *Gardenia* explant establishment and shoot proliferation. Murashige and Skoog's (1962) medium has also been applied by other workers, whereas in our studies a browning appearance of the explants was noticed on MS medium shortly after isolation.

For explant establishment inclusion of BA or 2-iP (2–10 mg/l) in the medium is essential, while addition of IAA (0.2–4 mg/l) is of minor importance. GA_3 (0.2–1 mg/l) is added to the medium before autoclaving, and only in cultures of proliferated explants to increase shoot number and length and also to improve shoot quality. The pH of the medium can be adjusted to a level from 5.0 to 5.5 with 0.1 N KOH prior to agar (0.6–0.8%) and after sucrose (3%) addition. Medium autoclaving requires 18 min at 121 °C.

3.4 Establishment of Cultures

Explants after disinfestation are placed horizontally onto the medium and are pushed half-way into the agar. The cultures are incubated in the growth room at 25 ± 1 °C under a 16-h photoperiod, provided by cool white fluorescent lamps at about 30–40 μmol/s/m^2 (2400–3200 lx).

3.5 Regeneration

Approximately after 6 and 10 weeks in culture of nodal and shoot tip explants, respectively, a few new shoots (0.5–1 cm long) are developed from the lateral buds. These shoots are cut off and subcultured for new shoot formation. The remaining tissues (basal masses) of the original explants without any division are also subcultured for additional shoot production. In both explant types of in vitro-derived shoots and basal masses, subculture and shoot harvesting are repeated routinely every 6 weeks for five or six times. Then the initial explants are discarded because the shoot formation rate decreases. The number of harvested shoots is about four to eight per explant during subculturing. In all subcultures the basal medium is similar to that of the establishment stage except for the auxin, which is omitted, and GA_3, which is added at concentrations of 0.2–1 mg/l. Culture conditions are the same as in the establishment stage.

Harvested shoots are either subcultured or transferred to the rooting medium. Shoots for rooting, 1–1.5 cm long, are inserted directly into a peat-perlite mixture (2:1 v/v), without any rooting-hormone treatment, in plastic packs or plugs. After insertion of shoots, the pack or plugs are covered with a polyethylene sheet to maintain high humidity, and placed on shelves of a growth room at 24 ± 1 °C under 50–60 μmol/s/m^2 (4000–4800 lx) illumination from cool white fluorescent lamps provided for 16 h daily. During rooting, shoots are misted with water on a regular basis to avoid desiccation. Roots usually start to emerge during the 3rd week. After 4 weeks in the rooting room, packs or plugs with the rooted shoots are moved to the greenhouse for further growth and subsequent hardening.

3.6 Acclimatization

Rooted shoots are planted then to 5 × 5 cm pots containing a mixture of peat:perlite (1:1 v/v) and placed in the greenhouse for 2 weeks under 50% shade. A suitable environment is created by covering young plantlets with tents of polyethylene film and subjecting them to intermittent water misting. During the following 3 weeks the humidity is gradually reduced, and the shade is changed from 50 to 20%. At the end of the 3rd week the polyethylene cover is removed and plantlets are self-supporting. A few weeks later they are shipped to growers.

4 Summary and Conclusions

Gardenia, a woody ornamental species with shiny dark green leaves and white fragrant flowers, is propagated by terminal cuttings. In vitro techniques have been employed to build up stock plants of newly released cultivars or to free *Gardenia* plants from pathogens. Even though in vitro culture increases the propagation rate, micropropagation has not yet become a widely used propagation practice in *Gardenia*. However, it shows promise for large-scale commercial application in the near future.

5 Protocol

Explant Shoot tips (1–1.5 cm long) and nodal segments (0.5–0.8 cm long).

Medium A low salt medium (Economou and Read 1984) containing BA or 2-iP (2–5 mg/l) in the establishment stage plus GA_3 (0.2–1 mg/l) in the multiplication stage is suitable for shoot formation. Shoots harvested from in vitro cultures are rooted ex vitro in a peat:perlite mixture (2:1 v/v).

Culture Conditions Aseptic cultures are maintained at 25 ± 1 °C in a growth room furnished with cool white lamps at about 2400–3200 lx for 16 h daily. The rooting of shoots takes place in the growth room at 24 ± 1°C with a 16-h photoperiod of 4000–4800 lx from cool white lamps. Plantlet acclimatization is achieved in the greenhouse by gradually reducing both humidity and shading.

References

Anderson WC (1975) Propagation of rhododendrons by tissue culture, part 1. Development of a culture medium for multiplication of shoots. Proc Int Prop Soc 25: 129–135

Dumanois G, Godin B, Bigot C (1984) In vitro vegetative multiplication of *Gardenia jasminoides* Ellis. J Plant Physiol 116: 389–407

Economou AS, Read PE (1984) In vitro shoot proliferation of Minnesota deciduous azaleas. HortScience 19: 60–61

Economou AS, Spanoudaki MJ (1985) In vitro propagation of *Gardenia*. HortScience 20: 213

Economou AS, Spanoudaki MJ (1986) The influence of cytokinins and gibberellic acid on *Gardenia* tissue cultures. Sci Hortic 29: 155–161
Lloyd GB, McCown BH (1980) Commercially feasible micropropagation of mountain laurel, *Kalmia latifolia*, by use of shoot-tip culture. Proc Int Prop Soc 30: 421–437
Mizukami H (1988) *Gardenia jasminoides* Ellis: in vitro propagation and the formation of iridoid glucosides. In: Bajaj YPS (ed) Biotechnology in agriculture and forestry, vol 7, Medicinal and aromatic plants II. Springer, Berlin Heidelberg New York, pp 213–226
Mizukami H, Terao T, Miura H, Ohashi H (1983) Glucosylation of salicyl alcohol in cultured plant cells. Phytochemistry 22: 679–680
Mizukami H, Terao T, Osashi H (1985) Partial purification and characterization of UDP-glucose: salicyl alcohol glucosyltransferase from *Gardenia jasminoides* cell cultures. Planta Med 2: 104–105
Morel G, Wetmore RH (1951) Tissue culture of monocotyledons. Am J Bot 38: 141–143
Murashige T, Skoog F (1962) A revised medium for rapid growth and bioassays with tobacco tissue cultures. Physiol plant 15: 473–497
Mustard MJ, Lynch SJ (1977) Nursery propagation and anatomical union of cleft grafted gardenias. Proc Fla State Hortic Soc 90: 347–349
Pontikis CA (1983) In vitro propagation of *Gardenia jasminoides*. Plant Prop 29: 13–14
Scaramuzzi F, D'Elia C (1984) Clonal multiplication of *Gardenia grandiflora* Lour. CR Acad Sci Paris 299: 657–662
Tabata M, Umetani Y, Shima K, Tanaka S (1984) Glucosylation of esculetin by plant cell suspension cultures. Plant Cell Tissue Organ Cult 3: 3–9
Ueda S, Kobayashi K, Muramatsu T, Inouye H (1981) Studies on monoterpene glucosides and related natural products. Part XL. Iridoid glucocides of cultured cells of *Gardenia jasminoides* f. *grandifolia*. Plant Med 41: 186–191
Wilkins HF (1986) *Gardenia jasminoides*. In: Halevy AH (ed) Handbook of flowering, vol V. CRC Press, Boca Raton, pp 127–131

X Micropropagation of *Gladiolus*

Y.P.S. Bajaj[1], M.M.S. Sidhu,[2] and A.P.S. Gill[3]

1 Introduction

The genus *Gladiolus* (family Iridaceae) comprises about 180 species, and numerous cultivars and hybrids which are of ornamental value as cut flowers. The commercially available large flower plants are *G.* × *grandiflorus* or *G.* × *hortulans*. The plants are vegetatively propagated through corms and cormels, and develop annually from axillary buds on the corm (Fig. 1), or a new corm forms annually by the swelling of the basal internodes of the inflorescence stalk. Numerous branched stolons, bearing cormels at the tips, develop from the basal node of the daughter corm during flowering.

Because of the immense commercial importance of gladiolus, it is highly desirable to speed up its rate of multiplication for mass propagation. By using conventional practices, the number of daughter corms produced by one mother

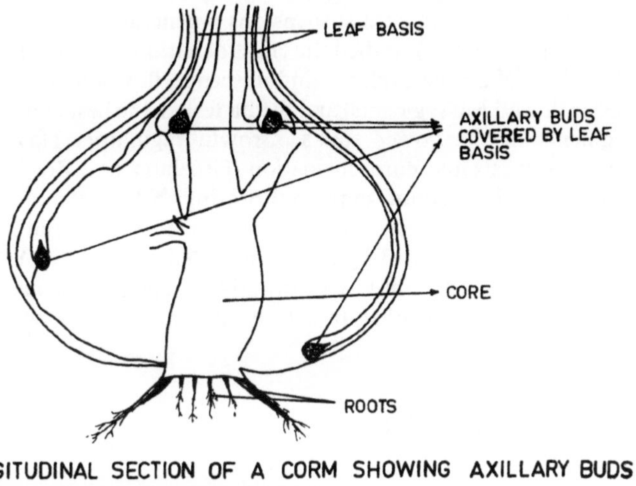

Fig. 1. Longitudinal section of a corm showing axillary buds

[1] Former Professor of Tissue Culture, Punjab Agricultural University, Ludhiana, India. (Present address; A-137 New Friends Colony, New Delhi 110065, India)
[2] Carewel Nursery, Pakhowal Road, Ludhiana (Punjab), India
[3] Punjab Agricultural University, Ludhiana, India

Biotechnology in Agriculture and Forestry, Vol. 20
High-Tech and Micropropagation IV (ed. by Y.P.S. Bajaj)

corm in 1 year is small, and it takes 8–10 years to produce a clone which is large enough for commercial purposes. However, by using various in vitro methods, this time lapse can be reduced to 2 years. Although work on gladiolus was started two decades ago (Ziv et al. 1970) and considerable progress has been made since then, yet the reproducible regeneration of plants from callus cultures, and also increasing the number of plants per corm are still problems that prevent the application of tissue culture methods for mass propagation and exploitation by the flower industry. More basic work on various factors influencing regeneration is needed before the method is developed to provide sufficient material for field planting and mass propagation (Bajaj et al. 1982).

2 Review of in Vitro Culture Studies

The subject has been recently reviewed (see Ziv and Lilien-Kipnis 1989 for details). The work on in vitro culture of gladiolus was initiated by Ziv et al. in 1970, who employed explants from inflorescence stalk and obtained callus, buds, and cormlets.[4] The best combination for organ initiation was 10 mg/l NAA and 0.5 mg/l of kinetin (kin). Later, Simonsen and Hildebrandt (1971) reported virus-symptomless plants of cv. Hit Parade. Likewise, plants free from BYMV and CMV were obtained (Logan and Zettler 1985).

The callus established from various segments was shown to be capable of rhizogenesis or regenerating complete plants (Wilfret 1971; Hussey 1977; Bajaj et al. 1982). Recently, cell suspensions regenerated plants (Kamo et al. 1990).

Bajaj et al. (1982) studied the effect of various factors on in vitro propagation of gladiolus. Of all the explants (inflorescence, flower stalks, denuded flower, bracts, perianth, and leaf segments) and media tested, the best callus was obtained from the segments of the flower stalks. Growth regulators play an important role in morphogenesis and corm formation (Ginzburg and Ziv 1973). Omission of cytokinin from the medium improved rooting. NAA promoted growth and the total yield of corm and cormels obtained by Lilien-Kipnis and Kochba (1987) from ten in vitro plantlets was 13.2 g for those rooted without NAA, 19.8 and 22 g for those rooted on 0.3 and 4 μm NAA respectively. The precocious formation of corm and cormels can be controlled in cultures (Steinitz and Lilien-Kipnis 1989):

3 Micropropagation

Gladiolus and its hybrids (Sutter 1986) can be regenerated from callus cultures, axillary buds, meristems, and from the segments of the cormels. The plantlets left to grow in cultures produce cormlets, which can be further multiplied and plants mass propagated.

[4] Corm is an underground bulb with one to several buds. Cormel is a single budded bulblet attached firmly to the mother corm and appearing as a sword above ground. Cormlets are the in vitro-produced microcorms.

3.1 Culture of Explants

The segments of the young inflorescence (3–5 mm), flower stalks (3–5 mm), denuded flower, bract, perianth, and leaf segments (0.5–1.5 cm) were cultured on MS1, G1 or G5 media, and their in vitro response in terms of callusing is summarized in Table 1 and shown in Fig. 2.

Of the various media tested, G1 induced the proliferation of all the explants, although the degree of callus formation varied a great deal. On MS1, none of the

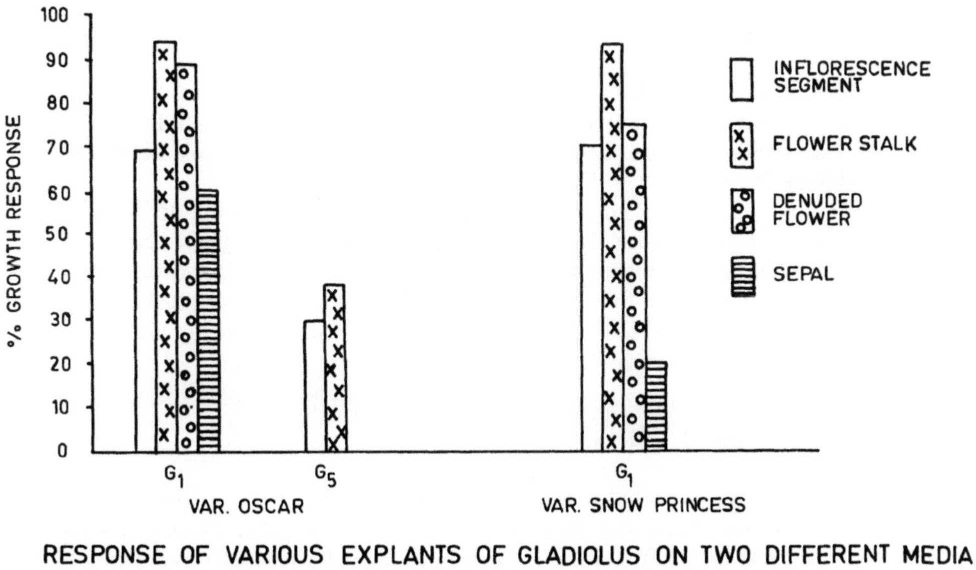

Fig. 2. Response of various explants on two different media (From Bajaj et al. 1982)

Table 1. Differential growth response of various explants of two cultivars of gladiolus on various media. (Bajaj et al. 1982)

Cultivar and medium	Inflor. segment (3–5 mm)	Flower stalk (3–5 mm)	Denuded flower	Bract
Oscar G1				
No. of cultures raised	33	17	28	5
No. of cultures response	23	16	25	3
% Response	70	95	89	60
Oscar G5				
No. of cultures raised	10	8	Not done	
No. of cultures response	3	3	—	—
% Response	30	37	—	—
Snow Princess G1				
No. of cultures raised	20	15	12	10
No. of cultures response	14	14	9	2
% Response	70	93	75	20

segments formed a sustained callus, and only a limited callus was induced by G5. The maximum proliferation response was shown by the segments of the flower stalk, and there was no marked difference between Oscar and Snow Princess (Bajaj et al. 1982).

After 1 week of culturing, there was swelling at both the cut ends of the explants, except with leaf explants. Later, a thin layer of callus was formed, with a few root initials which eventually developed into roots (Fig. 3A–D). The callus was mostly hard, yellow, and slow-growing.

3.2 Differentiation of Callus

The callus was periodically subcultured and maintained on G6 medium. The regeneration potential was studied by lowering the amount of 2,4-D from 2 to 0.1 mg/l, and adding various combinations and concentrations of IAA, NAA, and kin (Table 2).

The callus induced on G6 continued to grow without showing any sign of differentiation, whereas with that on MS2 medium there was differentiation of roots within 2 weeks. Rhizogenesis was also observed on G1 and G5 media, but the number of roots formed on these media was less than on MS2. Occasionally (in 5% cases), the formation of shoots was observed on G2 medium, but in none of the media were complete plants obtained.

3.3 Culture of Excised Anthers

The in vitro growth response of the anthers varied with stage of development. The young and medium-sized anthers cultured on G7, MS + 2,4-D (0.5 mg/l) + coconut milk (5%) and kin (0.1 mg/l), developed leaf-like (Fig. 3E) and petaloid (Fig. 3F) structures in 6–8 weeks in some cases. The maturing anthers elongated considerably and developed callus and roots (Fig. 3G). Occasionally, the callus showed some organized growth. Microscopic examination revealed multinucleate and multicellular pollen. This in vitro response of the anthers is similar to that of freesia (Bajaj and Pierik 1974).

3.4 Culture of Axillary buds and Cormels

The growth responses of the axillary bud and the cormel tip cultured on media G1 and G5 are summarized in Table 3. The in vitro response, in terms of both growth and shoot formation, was somewhat better on G5 medium, although complete plants with normal shoots and well-developed roots were obtained on both media.

When whole cormels were cultured, they formed complete plants (Fig. 3H–M). Such plantlets left to grow in the culture tubes produced cormlets. In some cases,

→

Fig. 3A–M. In vitro propagation of *Gladiolus grandiflorus* Oscar and Snow Princess. **A** Segment of a young inflorescence stalk of Oscar, 15 days after culture on G1 medium (MS + NAA 10 mg/l + kin 0.5 mg/l) showing the initiation of callus. **B** Mass of an inflorescence-derived callus 35 days after subculture on G6 (MS + 2,4-D 2 mg/l). **C** Callus transferred to MS2 (MS + IAA 2 mg/l); note the differentiation of roots after 35 days. **D** An excised bract showing callusing and multiple root formation

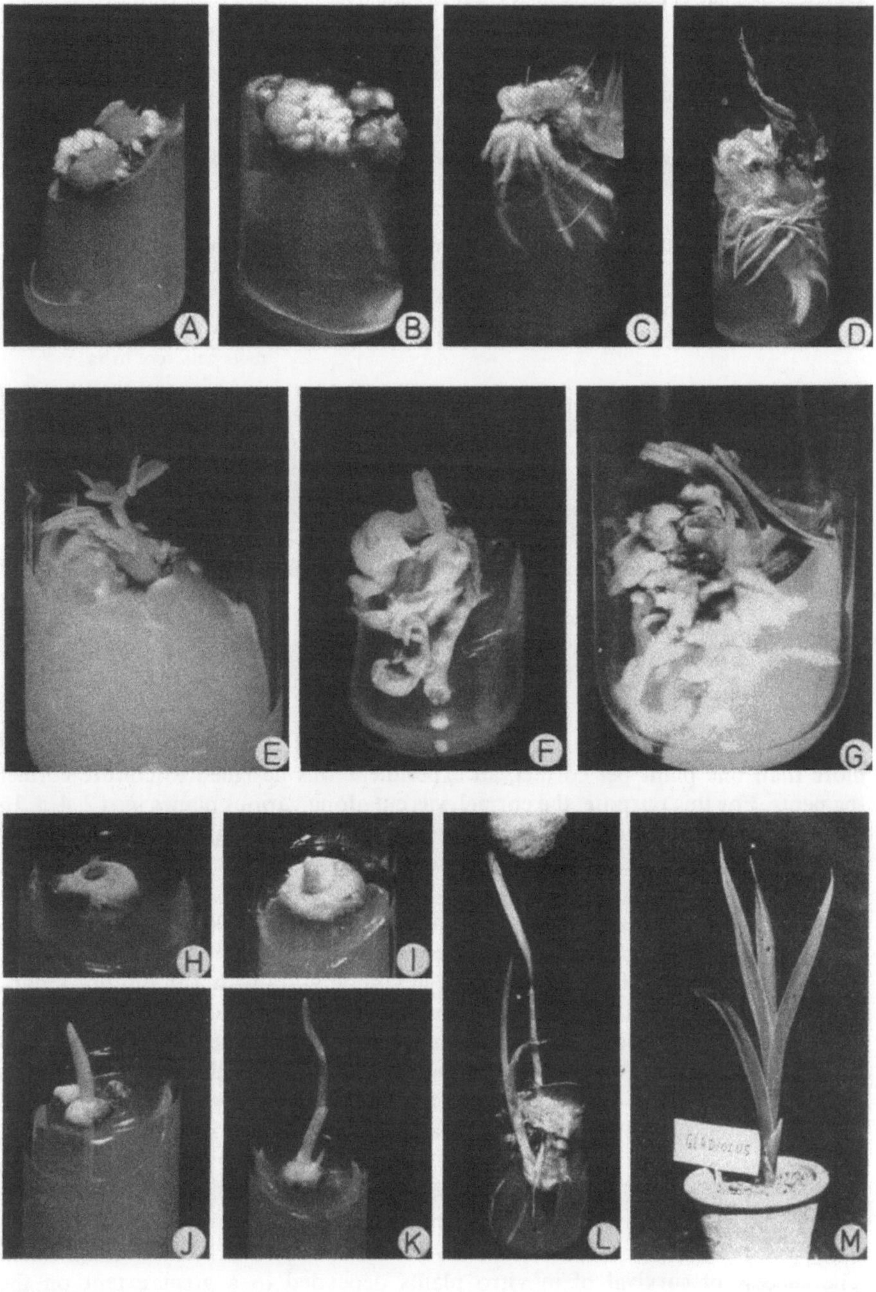

35 days after culture. **E-G.** Cultures of young (**E**), medium (**F**), and maturing (**G**) anthers showing the formation of leaf, petaloid structures, and roots, respectively. **H-L.** Various stages in the formation of plants from cormel of Snow Princess on G1 medium, 7, 15, 25, and 45 days after culture. **M** An in vitro plant transferred to the soil. (Bajaj et al. 1982)

Table 2. Subculture of callus of gladiolus Oscar on various media.[a] (Bajaj et al. 1982)

Medium	1–2 weeks	3–4 weeks	5–6 weeks
MS2	Rhizogenesis	Complete roots (0.5–1 cm)	Increase in root length (1–2 cm)
G1	Formation of root primordia	Rhizogenesis occurred, but it was less than on MS2 medium	Well-developed roots with root hairs
G2	Proliferation of callus	Formation of the green islets resembling shoot primordia	Formation of shoot, but rhizogenesis did not occur
G5	Rhizogenesis	Well-developed roots	Well-developed multiple roots with root hairs
G6	Yellowish callus, started turning white	Good proliferation without rhizogenesis	Callus continued to proliferate without differentiation

[a]The following combinations with the basal medium (MS) of Murashige and Skoog (1962) were used:
MS1 = MS + indole-3-acetic acid (IAA, 2 mg/l) + kinetin (kin, 0.5 mg/l)
MS2 = MS + IAA (2 mg/l) + kin (0.05 mg/l)
G1 = MS + naphthalene acetic acid (NAA, 10 mg/l) + kin (0.5 mg/l)
G2 = MS + NAA (0.1 mg/l) + kin (2 mg/l)
G5 = MS + NAA (0.1 mg/l) + kin (0.5 mg/l)
G6 = MS + dichlorophenoxyacetic acid (2,4-D, 2 mg/l)
G7 = MS + 2,4-D (0.5 mg/l) + coconut milk (5%) + kin (0.1 mg/l).

multiple shoots were also formed (Fig. 3L). To explore the possibility of producing more than one plant per cormel, an experiment was designed to culture cormel segments. For this purpose, the cormel was cut along various planes into 2, 4, 8, 16 or 32 segments for culture (Fig. 4). Complete plants were obtained from up to four segments. however, as the number of segments was increased, the regeneration potential decreased.

There was a marked interaction between NAA and kin. The high-auxin medium G1 was better than the G5 medium, containing a low level of NAA. A lower concentration of NAA (0.5 mg/l), in combination with kin, did not induce callus in any of the cultured explants, and a high concentration of NAA (10 mg/l) was required for callusing and the differentiation of roots. These results agree with those of Ziv et al. (1970). However, in our studies (Bajaj et al. 1982), the explants from the distal end of the inflorescence showed a better response than proximal ones. These results are attributed to differences in endogenous growth regulators.

3.5 Microcorm Production and Mass Propagation

The success of survival of in vitro plants depended to a great extent on the induction of a good rooting system. Activated charcoal (0.3%) added to the pretransplant medium improved root growth (Ziv 1979) and after transplanting, survival rate was enhanced in plantlets rooted on charcoal medium. Horticultural grade vermiculite (instead of agar) enriched with half-strength liquid MS supple-

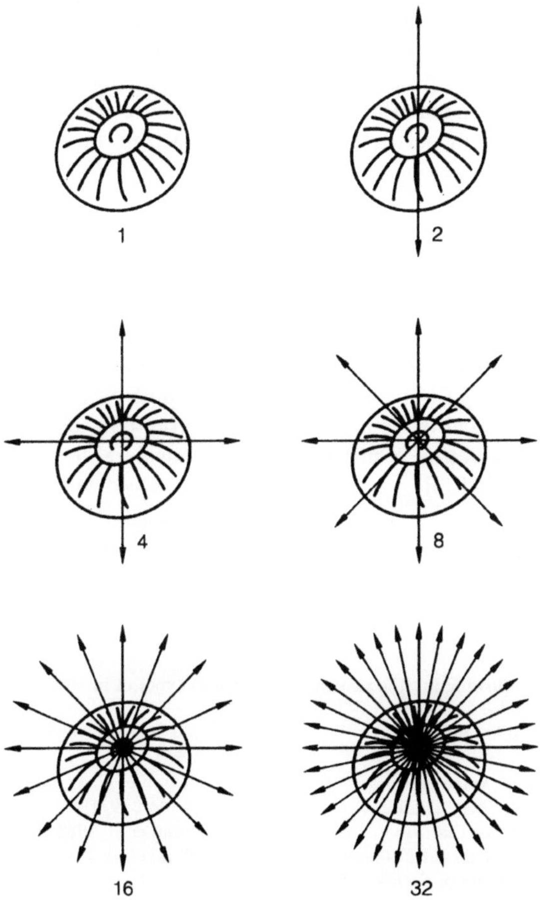

Fig. 4. Plan showing the division of cormel into segments

Table 3. In vitro response of axillary buds and cormel-tips cultured on media G1 and G5. (Bajaj et al. 1982)

Medium		Axillary bud	Cormel tip
G1	No. of cultures raised	15	9
	No. of cultures response	9	6
	% Response	60.0	66.7
	Average no. of shoots per culture	1	2
G5	No. of cultures raised	12	10
	No. of cultures response	9	8
	% Response	75.0	80.0
	Average no. of shoots per culture	1	1

mented with NAA and activated charcoal ensured a good root system (Logan and Zettler 1985).

With the refinement in in vitro technology and the manipulation of media, explants (Bajaj et al. 1982), and culture conditions it has been possible to mass propagate gladioli (Lilien-Kipnis and Kochba 1987), and to enhance cormlet production (Ziv 1989). Utilizing jar fermenters, Takayama et al. (1987) reported harvesting about 300 corms from 2-l bioreactors. Likewise, Kamo et al. (1990) established cell suspension cultures and regenerated plants. Microcorm production through bioreactors is likely to attract more attention in future for commercial exploitation.

4 Summary

Gladioli are plants of considerable ornamental value. In vitro culture and micropropagation studies have yielded positive response, and plants can be regenerated from callus cultures, as well as from cormel segments and axillary buds. In vitro production of microcorms has considerable potential for commercial propagation of gladioli.

5 Protocol

Following are the references for various types of in vitro cultures of gladiolus.

Callus cultures	(Ziv et al. 1970; Simonsen and Hildebrandt 1971; Bajaj et al. 1982)
Cell Suspension	(Kamo et al. 1990)
Meristem Culture	(Wilfret 1971)
Anther Culture	(Bajaj et al. 1982)
Corm/cormel Culture	(Bajaj et al. 1982; Ziv 1989)
Transfer of Plants to Soil	(Ziv 1979)
Mass Propagation	(Takayama et al. 1987; Lilien-Kipnis and Kochba 1987)

References

Bajaj YPS, Pierik RLM (1974) Vegetative propagation of *Freesia* through callus cultures. Neth J Agric Sci 22: 153–159

Bajaj YPS, Sidhu MMS, Gill APS (1982) Some factors affecting the in vitro propagation of *Gladiolus*. Sci Hortic 18: 269–275

Ginzburg C, Ziv M (1973) Hormonal regulation of cormel formation in *Gladiolus* stolons grown in vitro. Ann Bot 37: 219–224

Hussey G (1977) In vitro propagation of *Gladiolus* by precocious axillary shoot formation. Sci Hortic 6: 287–296

Kamo S, Chen J, Lawson R (1990) The establishment of cell suspension cultures of *Gladiolus* that regenerate plants. In Vitro 26: 425–430

Lilien-Kipnis H, Kochba M (1987) Mass propagation of new *Gladiolus* hybrids. Acta Hortic 212: 631–638

Logan AE, Zettler FW (1985) Rapid in vitro propagation of virus indexed gladioli. Acta Hortic 164: 169–180

Murashige T, Skoog F (1962) A revised medium for rapid growth and bioassays with tobacco tissue cultures. Physiol Plant 15: 473–497
Simonsen J, Hildebrandt AC (1971) In vitro growth and differentiation of *Gladiolus* plants from callus cultures. Can J Bot 49: 1817–1819
Steinitz B, Lilien-Kipnis H (1989) Control of precocious gladiolus corm and cormel formation in tissue culture. J Plant Physiol 135: 495–500
Sutter EG (1986) Micropropagation of *Ixia viridifolia* and a *Gladiolus* × *Homoglossum* hybrid. Sci Hortic 29: 181–189
Takayama S, Amo T, Fukano M, Akita M (1987) Mass propagation of gladiolus by liquid culture methods. Abstr 10th Meet Jap Assoc Plant Tiss Cult, p 88
Wilfret GJ (1971) Shoot tip culture of *Gladiolus*: an evaluation of nutrient media for callus tissue development. Proc Fla State Hortic Soc 84: 389–393
Ziv M (1979) Transplanting *Gladiolus* plants propagated in vitro. Sci Hortic 11: 257–260
Ziv M (1989) Enhanced shoot and cormlet proliferation in liquid cultured *Gladiolus* buds by growth retardants. Plant Cell Tissue Organ Cult 17: 101–110
Ziv M, Lilien-Kipnis H (1989) Gladiolus. In: Ammirato PV, Evans DA, Sharp WR, Bajaj YPS (eds) Handbook of plant cell culture, vol 5. Ornamental species. McGraw-Hill, New York 461–478
Ziv M, Halevy AH, Shilo R (1970) Organs and plantlet regeneration of Gladiolus through tissue culture. Ann Bot 34: 671–676

XI Micropropagation of Hyacinths (*Hyacinthus orientalis* L.)

A. Bach[1]

1 General

The hyacinth is native to Greece, Asia Minor, and the Balkans, and was introduced to cultivation through the Orto Botanico of Padua before 1568. Few species of *Hyacinthus* are in cultivation and there is only one, *H. orientalis*, which is of horticultural importance. It is from this species that the familiar hyacinth has been developed, which is widely cultivated in open fields as well as being forced in the greenhouse during the winter (Rees 1972).

1.1 Botany

Hyacinth is a perennial bulb ornamental and belongs to the family Liliaceae. The mature 4–6-year-old hyacinth bulb has been described by Blaauw (cf. Rees 1972). It consists of the dead inflorescence stalk inside, which is surrounded by the bases of six foliage leaves whose aerial parts have recently died and, outside these, two fleshy scales. Surrounding these are the bases of the foliage leaves of the previous year, and outside these again the two scales of the corresponding annual cycle. All these scales (15–18 in number) and leaf bases are fleshy and swollen with storage material. The outer scales are thin and pigmented either purple, brown, or white. The new axillary bud is near the base of the inflorescence stalk.

The flowers are in racemes. They are long and dense, composed of a large number of flowers. The racemes produced by strong bulbs measure 20–25 cm in length and 18–23 cm in circumference. The perianth is petaloid, occupying the two outer whorls, followed by two whorls of three stamens, with a superior ovary of three carpels in the centre of the flowers. The fruit is a capsule.

Originally blue, the hyacinth can now be white, pink, yellow, mauve, and various shades of blue. Most modern cultivated varieties are single-flowered (Fig. 1).

Cytological studies have shown that cultivars exist with various numbers of chromosomes. The diploid form has 16 chromosomes of five different types, and there are triploids ($3n = 24$) and a large number of heteroploids, e.g., cv. L'Innocence (27), cv. Ostara (25), and cv. City of Haarlem (23); in fact, all possible chromosome numbers between 16 and 31 have been found in hyacinths, except 18. No true tetraploid ($4n = 32$) has been discovered (Darlington et al. 1951; Rees 1972).

[1] Department of Horticulture, University of Agriculture of Kraków, al. 29 Listopada 54, Kraków, Poland

Biotechnology in Agriculture and Forestry, Vol. 20
High-Tech and Micropropagation IV (ed. by Y.P.S. Bajaj)

Fig. 1. Flowers of *Hyacinthus orientalis* L.

1.2 Distribution

The most important and largest producer of hyacinth nowadays is The Netherlands, which covers 95% of the total world production of the bulbs. The cultivation is spread over an area of ca. 1000 ha. (Table 1), which is 6.5% of total bulb production, ranking fifth after tulip 46.8%, lilies 11.8%, gladiolus 11.7%, and

Table 1. Hyacinth bulb production and area of cultivation. (Jaarverslag-Statistiek, 1989)

Distribution	Year of production	
	1976/77	1987/88
Area of production (ha)	866.0	1057.9
Total bulb production (mill.)	174.1	180.0
Export of bulbs (mill.)	135.4	158.9

narcissus 10.4% (Jaarverslag Statistiek 1989). The most popular cultivated varieties are pink 46.8% (e.g., cv. Anna Marie, cv. Pink Pearl), blue 26.8% (e.g., cv. Delft's Blue, cv. Ostara), white 15.3% (e.g., cv. Carnegie, cv. White Pearl), and others (orange, yellow, and purple).

1.3 Conventional Propagation

Generative hyacinth propagation is applicable in breeding new cultivars only because the plants obtained in this way are variable; for this reason the principal horticultural cultivars are mainly complex hybrids which originated as a single plant selection. Hyacinth production is therefore dependent on vegetative propagation.

Under garden conditions, a hyacinth bulb may produce one or more bulblets, but this rate of multiplication is too low for commercial production. For this reason, in horticultural practice the hyacinth bulbs are usually encouraged to increase the number of daughter bulbs by cuttage to accelerate propagation.

The main two artificial methods commercially used to enhance the rate of propagation are: scooping and cross-cutting (or scoring) (Hartman and Kester 1976). Only mature bulbs which have been dug after the foliage has died down, and are 17–18 cm or more in circumference, may be used in these techniques of propagation. In the former method, the basal plate of the bulb is scooped out, exposing the scales, in the latter three straight knife cuts are made across the base of the bulb, each deep enough to go through the basal plate and growing point.

In the third basal cuttage method, called coring, which seems to be most commonly used in the USA, the center of the bulb is removed entirely by means of a core borer 1–1.8 cm in diameter.

All these operations lead to the removal of apical dominance, and result in callusing of the exposed scale tissue. Bulbs treated in this way are then forced at a high temperature to form small adventitious bulbs. Depending on the cultivar and the size of the parent bulb, a scooped bulb will produce 40–60 bulblets, but 4–5 years will be required to produce flowering; a scored bulb will produce 12–24, requiring 3–4 years; and a cored bulb will produce 10, requiring 2–3 years to develop flowers (Hartman and Kester 1976).

Hyacinth can be propagated also from leaf cutting and flower stems as well as from scales of the old bulb: the adventitious bulbs develop from epidermal and subepidermal cells (Naylor 1940; Marston 1973; Krause 1980).

There are also vegetative methods of propagation of hyacinth in which the parent bulb is preserved undamaged. Amano and Tsutsui (1980) showed that high temperature treatment of the hyacinth mother bulb induced the formation of new bulblets at the proximal end of the bulb scale on the abaxial side.

1.4 Need for in Vitro Culture

With traditional methods it takes a long time to multiply bulbs and to introduce the new hyacinth cultivars commercially. The application of plant tissue culture techniques to the propagation of flower bulb crops has expanded rapidly over the past 15 years, and hyacinths have certainly shared in this development.

The in vitro method, in combination with other propagation techniques, e.g., scoring or scooping, can be a valuable means to build up commercial stocks of selected plants with desired characteristics, namely new genotypes or disease-free plants. Application of such methods would increase breeding and selection efforts, which in turn improves the quality of the bulbs produced (Aartrijk and Linde 1986).

2 Brief Review of Tissue Culture Work

A close look at the publications concerned with the techniques of hyacinth micropropagation shows that they concentrate mainly on defining optimal conditions for the initial stage of culture. Less attention is paid to the other stages in propagation, e.g., the multiplication stage, transfer to soil, and commercial feasibility.

The majority of the in vitro methods described are based on adventitious formation of bulblets or plantlets originating directly from one or a few cells within an explant or indirectly, i.e., buds develop within previously formed callus.

The many factors that accelerate the regeneration process of hyacinths may be divided into tissue factors and growing condition factors. Independently of these, the rate of propagation still differs between cultivars.

The main factors related to plant tissue are: origin and size of explants, and their polarity and pretreatment. Many sources of tissue explants are used to establish the initial step of hyacinth culture, including bulb scale, leaf, stem, and various parts of the inflorescence, ovary, anther, and flower bud. With the aim of obtaining virus-free hyacinth plants, shoot tips were used (Asjes et al. 1974; Matsumoto 1983). Inflorescence stem sections and foliage tissues produce more and faster bulblets and shoots than bulb scale tissue (Hussey 1975; Chung et al. 1983; 1984; Paek et al. 1987). Pierik and Post (1975) have reported that the number and total fresh weight of bulblets increased almost linearly with the size of primary explants. Thus, Amaki et al. (1984) describe an original method for producing numerous large bulblets from one hyacinth bulb by desiccation with silica gel.

Organogenesis of hyacinth tissue depends also on its position, both in the bulb and in the media. The yields of bulblets increased more in the basal part of bulb scale and leaf segments than in the distal parts (basipetal polarity) and also in the abaxial part of the scale in comparison to the adaxial layer (Pierik and Post 1975; Paek 1982; Chung et al. 1984). In contrast to bulb scale and leaf, regeneration ability in the distal part of the inflorescence stem was higher than in the basal part (Paek 1982). In vitro studies on the polarity of adventitious root and bulblet formation in etiolated leaves and in floral buds of hyacinth revealed that root formation depends on the presence of auxin only in the medium,whereas formation of bulblets depends on the orientation of segments and takes place only at the morphologically basal end of the explant in the presence of cytokinin and auxin (Saniewski 1975). The fresh weight of bulblets originating on the youngest scales was lower than that of older ones (Paek 1982).

The regeneration of hyacinths depends also on the pretreatment of the plant tissue. The storage time of the parent bulb and temperature are the main factors in this method of bulblet induction. Successful results, i.e., faster regeneration of plantlets and larger bulblets, have been obtained after 2 months' storage at low temperature at 2 °C (Aartrijk and Bolm-Barnhoorn 1983).

The optimal stage for culture initiation of bulb scales was just before planting, but for leaves or flower stalk tissue it was the flowering stage (Chung et al. 1984).

Nutrient media, plant growth regulators, and environmental factors must be considered in hyacinth propagation. Nutrient media for adventitious shoots and bulb culture of hyacinth have not been widely studied. The medium most frequently used in the initiation stage was MS (Murashige and Skoog 1962). Some authors used a mixture of Knop's macroelements and Heller's microelements (Pierik and Steegmans 1975; Chun et al. 1982). The optimal concentration of sucrose is ca. 3–5% and depended on the kind of tissue, e.g., higher concentrations were required for the growth of bulblets (Chung et al. 1984). The optimal pH value was usually 5.5–6.0.

To save manual labor and to induce quick neoformation, a technique combining liquid and agar-solidified layers of media (double-phase media) has been applied to the micropropagation of hyacinth cv. Carnegie in the multiplication stage (Bach and Cecot 1988, 1989). The fresh weight of explants grown on the double-phase media was greater (190–270 mg) in comparison to those cultivated on solid media (90–190 mg), and the rate of propagation proved greater on the double-phase media than on the single-phase (Table 2). The yield of new shoots was higher on media supplemented with NAA in comparison with IAA and IBA, regardless of the type of basal medium. The induction of bulblets was poor (primary explants were one-leaf shoot microcuttings) on both types of medium (Bach and Cecot 1989).

Plant growth regulators play the role of trigger compounds in the media for increasing the potential number of bulblets in the initial stage. Pierik and coauthors (1971, 1973, 1975b), in a series of single-factor experiments, reported that IAA and IBA stimulated initiation of adventitious bulblets from bulb scale explants. Hussey (1975) reported also that the stems, leaves, and ovaries responded well to concentrations of IAA and NAA (2.0–8.0 mg/l and 0.03–012 mg/l respectively). Auxins generally act according to an optimum curve; however, the effect was dependent on the kind of auxin used. Cytokinins do not seem to be necessary in all cases, especially when explants from bulbs are used (Pierik and Steegmans 1975); this is contradicted by Kim (1981), who reported that BA (1–3 mg/l) was obligatory

Table 2. Growth of hyacinth cv. Carnegie on two types of culture medium. (Bach and Cecot 1989)

Type of medium	Rate of propagation	Mean no. of: Leaves (per shoot)	Roots	Bulbs	Mean % of: Rooted explants	Bulbing explants
Single-phase	1.3a[a]	1.26a	0.6a	0.2a	13.9a	3.5a
Double-phase	1.8b	1.42b	0.6a	0.1a	14.5a	3.8a

[a] Means followed by the same letter are not different according to the 5% level of Student's t-test.

Table 3. Effect of different auxins on growth of hyacinths cv. Carnegie cultivated on single- and double-phase media. (Bach and Cecot 1989)

		Mean no. of:			Mean % of:	
Growth reg.	Rate of propagation	Leaves (per shoot)	Roots	Bulbs	Rooted explants	Bulbing explants
IAA[a]	1.6a	1.19a	0.4a	0.4a	7.9a	7.8a
IBA[a]	1.6a	1.11a	0.3a	0.4a	2.4a	14.1b
NAA	2.2b	1.33b	1.2b	0.1a	32.4b	1.6c

[a] The basic growth medium was supplemented with BA (1 mg/dm^3).

in obtaining proper growth of adventitious bulblets. Higher concentrations of auxins, particularly of NAA, as well as cytokinins, stimulated callus formation.

On the other hand, Saniewski (1975) proved that the application of the mixture of auxin with cytokinin in the proportion 1:10 stimulated differentiation of bulblets. Auxin applied alone resulted in root differentiation, whereas cytokinin (BA) alone did not affect the differentiation of either bulblets or roots.

The formation of bulblets and roots in vitro was reported to be strongly inhibited by gibberelins, ethylene, and abscisic acid (Pierik and Steegmans 1975).

The results of Lu et al. (1988) showed that it is possible to control the regeneration of tepals, stamens, and ovules in explants from the perianth of hyacinth. Regeneration could be controlled by means of exogenous hormones, BAP and 2,4-D.

Few data on the effects of environmental factors are available. According to Pierik and Ruibing (1973), the optimal temperature for differentiation of bulblets in vitro is 13 °C, whereas for further growth it ranges between 21 and 25 °C. An increase in culture temperature in the range of ca. 20–30 °C results in decrease in the number of bulblets per explant (Paek et al. 1983). No difference in number of bulblets regenerated was observed between light or dark conditions, but bulblet growth was greatly promoted in dark conditions (Paek et al. 1983). The growth of stem tips was stimulated at high intensities of light (cool white fluorescent light) and inhibited in dark (Matsumoto 1983). Callus induction was inhibited by light (Kim et al. 1981). In vitro studies on hyacinth have been recently reviewed by Paek and Thorpe (1989).

3 Micropropagation

In vitro plant propagation usually includes three procedural stages, the initial stage, multiplication, and/or rooting. The data concerning multiplication and rooting of hyacinth indicate a different subculture technique in comparison with other plant subcultures which follow the initial stage. The two main methods of multiplication are listed below:

1. multiplication of split adventitious plantlets (Hussey 1975) and culture of bulb scale or leaf tissue of adventitious plantlets (Chung et al. 1983),

2. The subculture technique, which involves division of a clump of adventitious shoots to single shoots of hyacinths plant (microcuttings) (Bach and Cecot 1987) or reculturing of regenerated young basal shoots explants (Paek et al. 1987).

Both of these schemes entail some disadvantages which limit their use for commercial propagation. The mean proliferation rate is usually low and varies between 0.1 and 2 adventitious bulblets or shoots per explant. The rate of propagation is limited also by the number of subcultures, particularly in method (1). The explants subcultured three or four times at 20–23 °C tend to fall into the dormant state.

In the search for a commercial method of hyacinth propagation, a new method has been worked out in our laboratory, according to which, the initial stage is similar to that already described, but the multiplication stage and rooting and/or bulbing are different. Sections 3.1, 3.2, 3.3, 3.4, and 3.5 provide a detailed description of the proposed method, which was developed and tested at the University of Agriculture in Kraków, Poland, in 1986–88.

3.1 Culture Initiation

To achieve proper primary explants for further experiments, optimal hyacinth plants were selected, and the bulbs were stored at low temperature (2 °C) for 3–4 months. As initial explants the basal parts of immature leaves were removed from inside the nonsterile hyacinth bulbs and sterilized by soaking in Chloramina B solution (7% w/v or 3 min). These explants are generally (95–97%) free from contamination, in contrast to our previous experiments with bulb scale tissue (60–70% respectively).

Fig. 2. Initiation stage: regeneration of primary explants (MS, IBA 1 mg/l)

The explants were cultured on MS medium with the addition of IBA (1 mg/l). The medium contained 0.7% Difco Bacto Agar, 30 g sucrose/l, and the pH value was adjusted to 5.8. The cultures were incubated under 16-h illumination provided by a cool white fluorescent light at 23 ± 1 °C.

The adventitious buds of the initial explants began to develop after 2 weeks in culture. In general, buds produced bulblet primordia and then white, single, closed scale bulblets with green shoot tips. After a short time (4–6 weeks), these tips grew into single shoots, which produced one or two juvenile, cylindrical leaves (Fig. 2). The mean percentage of the regenerated explants was 95–98%.

3.2 Multiplication Stage

Research conducted in our laboratory revealed a high degree of regeneration of epidermal and subepidermal cell layers peeled from the excised leaves or bulblet scales. The procedure was continued in two different ways.

In the first method the leaf or bulblet epidermal explants were taken directly from regenerated in vitro shoot explants (microcuttings), while in the other, such shoots were chilled at 4 °C for 8 weeks before isolation of the epidermal explants. As a result, in the first case the adventitious shoots were obtained, and in the second bulblets (Bach 1989) (Fig. 3).

The explants taken from unchilled adventitious shoots started to regenerate new shoots on the wounded edges of the explants within 8 weeks when cultivated on a medium containing, e.g., BA (1 mg/l) and NAA or IAA (0.01 and 0.1 mg/l, respectively), at 23 °C. The rate of propagation was six to ten adventitious shoots per explant, depending on the size of the explant and the cultivar. Regardless of the origin of the explant (bulblet or leaf), the epidermal tissue developed adventitious shoots only, but those obtained from the leaves were more numerous (Table 4).

A B

Fig 3A, B. Multiplication stage: adventitious shoots and bulblets regenerating on epidermal explants excised of unchilled (**A**) or chilled (**B**) microcuttings (MS, BA 1 mg/l, NAA 0.01 mg/l). (Bach 1989)

Table 4. Effect of different tissue explants on multiplication of hyacinths cv. Delft's Blue cultivated in vitro

Explants	Mean no. of: Shoots (per explant)	Leaves	Mean length of new leaves (mm)
Leaf epidermal layer	9.0a[a]	11.3a	8.8a
Leaf mesophyll	1.2c	2.8c	8.2a
Bulblet epidermal layer	4.2b	4.4b	9.0a
Bulblet storage tissue	1.3c	2.7c	5.9b

[a] Means followed by the same letter within a column are not different according to the 5% level of Student's t-test.

Newly formed shoots occurred as a tuft of single, cylindrical green leaves and could be used again in year-round propagation or for both bulbing and rooting (Fig. 4).

In the second method the microcuttings (i.e., regenerated explant shoots), removed from primary explants and treated at low temperature (4 °C), formed adventitious roots after 4–6 weeks, but it was necessary to keep them for 8 weeks at 4 °C before peeling epidermal explants in order to obtain bulblets as a result of next 8 weeks' treatment at 23 °C. Chilling microcuttings for 4 weeks or less results only in adventitious shoots; chilling over a period of 8–12 weeks produces shoots and bulblets (Table 5). Like the shoots, the yield of bulblets was greater on epidermal explants originating from leaves of chilled microcuttings in comparison with those taken from bulblets. The majority of adventitious bulbs were formed on the edges of epidermal explants and could be either rooted or planted into soil. The first white nodules were visible 10–15 days after excising epidermal explants. Bulb primordia appeared as soon as 25–30 days after transplanting to the medium. The growth regulators necessary for developing the bulblets were similar to those for shoot regeneration. However, we observed the highest formation of bulblets on MS medium with the addition of kinetin and IAA (2 and 2 mg/l, respectively).

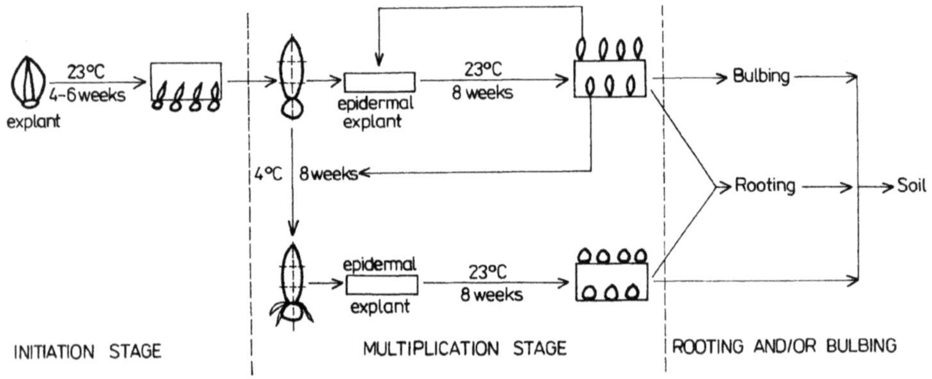

Fig. 4. Scheme for mass propagation of hyacinth. (Bach 1989)

Table 5. Effect of temperature and duration of incubation on the morphogenesis of hyacinth microcuttings cv. Carnegie cultivated in vitro (Bach 1990)

Duration of incubation (weeks)	Temperature of incubation: 23 °C			4 °C			23 °C following 4 °C		
	Shoots	Roots	Bulbs	Shoots	Roots	Bulbs	Shoots	Roots	Bulbs
0	0[a]	0	0	0	0	0	0	0	0
2	0	0	0						
4	5	0	0	0	0	0			
6	51	0	0						
8	52	40	0	0	0	0			
10	56	40	20						
12	67	60	30	5	2	0	65	21	17
14	85	40	30						
16	78	40	30	5	2	0	74	52	64
18									
20				10	16	0	50	67	86
22									
24							0	72	93
26									
28							0	80	98

[a] The percentage of microcuttings forming adventitious shoots, roots, or bulblets.

3.3 Bulbing and Rooting

To obtain bulblets by this method of propagation, clumps of adventitious shoots were divided and transplanted to fresh media. Various factors were tested to stimulate bulbing: high sucrose level, treatment with GA_3 and paclobutrazol added to the media, different concentrations of auxins and cytokinins applied to one- or double-phase media, and cold treatment.

Suitable duration of the cold treatment at 4 °C in combination with a high concentration of sucrose proved to be the most successful combination (Bach 1990). A bulblet induced in vitro at the basal region of hyacinth microcuttings usually developed only one close, fleshy scale (Fig. 5). It was observed that microcuttings planted on the media with high sucrose level (5% w/v) and chilled, developed well-formed adventitious bulblets with thin, papery outer scale, purple or white depending on the cultivar (Fig. 6).

The adventitious bulblets began to grow after ca. 8 weeks' cold treatment, so the period from the isolation of explants to the establishment of bulb multiplication would take about 16–20 weeks (Table 6).

The following in vitro method of hyacinth rooting has been developed: planting bulblets, explants of leaves or ovaries on the rooting media without cytokinin and/or low concentrations of NAA (0–0.01–0.1 mg/l) (Saniewski 1975; Kim et al. 1981; Paek 1982) and IBA (Chung et al. 1983; Paek et al. 1987).

In our laboratory, three additional alternative procedures of rooting are applied:

1. Adventitious bulblets of hyacinths or one-leaf microcuttings are stored at 4 °C for 4 weeks. Root formation starts from the vegetative apex of bulblets or the basal end of microcuttings. Their occurrence depends on the length of the storage period and the concentration of the plant growth regulators in the nutrient media
2. The double-phase system of Bach and Cecot (1989), in which a liquid medium containing low concentrations of NAA is added to a solid multiplication medium is also recommended. In this case, the promotive influence of NAA on the formation of adventitious roots was significant.

Table 6. Effect of cold treatment and concentration of sucrose on bulb development in hyacinth microcuttings cv. Delft's Blue cultured in vitro

Duration of cold treatment before incubation for 8 weeks at 23 °C (weeks)	Concentration of sucrose in media			
	30 g/l		60 g/l	
	Mean no. of bulbs (per shoot)	Mean width of bulbs (mm)	Mean no. of bulbs (per shoot)	Mean width of bulbs (mm)
0	0	0	0	0
4	1.1	1.8	0.3	1.0
8	1.6	2.2	0.7	1.3
12	3.9	2.9	3.5	2.3
16	4.6	3.4	5.4	3.0
20	5.0	3.6	5.6	4.1
24	5.2	2.7	5.2	4.1

Fig. 5. Bulbing and rooting of one-leaf microcuttings of hyacinth (MS, BA 1 mg/l, NAA 0.1 mg/l, 30 g/l of sucrose, 12–16 weeks of incubation at 4 °C followed by 8 weeks at 23 °C)

Fig. 6. Adventitious bulblets formed on high sucrose level media (MS, BA 1 mg/l, NAA 0.1 mg/l, 60 g/l sucrose)

3. The rooting stage for microcuttings of hyacinth can be carried out not only under aseptic conditions but also in a greenhouse nursery, where the percentage of rooted microcuttings was 90–95% (Bach, unpubl.).

3.4 Transfer to Soil

Plantlets formed on agar media or rooted in vivo were planted directly into soil in plastic greenhouses. Generally, 90–95% of them survived and began to grow. Hyacinth bulblet behavior in soil was similar to that of bulblets obtained by scooping or scoring. Bulblets should be planted in the original in vitro clumps and not divided (Fig. 7).

Fig. 7. Adventitious bulblets ready to plant into soil

Fig. 8. Plants transferred to the greenhouse

Hyacinth plantlets, especially when they have been subcultured several times at 23 °C, tend to fall into dormancy, i.e., they develop new leaves and cannot form bulblets. A short cold treatment (2–4 weeks at 2–4 °C) breaks the dormancy (Fig. 8).

Bulblets can also be chilled to break dormancy and obtain continuous growth (Paek 1982).

3.5 Commercial Aspects

The in vitro method described here guarantees an efficient crop and enables optimal planning of the annual cycle of production in tissue culture laboratories.

From our results the production of bulblets and microcuttings was calculated as follow: a few parent bulbs (even one) are sufficient to initiate a culture. A 2-month multiplication rate of at least three to five bulblets or microcuttings results in 10000–18000 pieces of bulblet or one-leaf microcuttings.

In the epidermal explant method, it is possible to obtain yearly ca. 10000 plantlets or bulblets per 1–2 m^2 of culture shelf.

Five varieties of hyacinth (cvs. Carnegie, Delft's Blue, Anna Marie, Pink Pearl, City of Haarlem) were propagated by this method for a period of 2 years. They showed similar type of growth; the rate of propagation varied, however, depending on the cultivar.

Another advantage of this method is that no special laboratory equipment is required, only a growth- and a cool chamber.

Irrespective of their developmental stage, hyacinth plantlets may be stored in darkness and at 0–2 °C for 4–6 months. Thus, the possibility of long-term storage of hyacinth tissue includes this method in a whole-year cycle of in vitro production, as in other plant species.

4 Protocol

Primary Explant. Basal part of young leaves removed from bulbs. *Multiplication Explant.* An epidermal and subepidermal cell layer peeled from chilled or unchilled leaves of adventitious shoots

Medium. Murashige and Skoog's with addition to initial medium 1 mg/l IBA; to multiplication media BA 1 mg/l; 0.01 mg/l NAA, or 0.1 IAA; to rooting medium 0.01 mg/l NAA.

Optimal Conditions. 23 ± 1 °C, 16-h day for multiplication of adventitious shoots, and 4 ± 1 °C for 8–12 weeks before incubation at 23 °C, to obtain adventitious bulblets.

References

Aartrijk J van, Blom-Barnhoorn GJ (1983) Tissue culture of ornamental bulb plants (in Dutch: Weefselcultuur van bolgewassen). Ann Bulb Res Centre Lisse: 62–63

Aartrijk J van, Linde PCG van der (1986) In vitro propagation of flower-bulb crops. In: Zimmermann RH et al. (eds) Conf Tissue culture as a plant production system for horticultural crops, 1985, Beltsville pp 317–331 Martinus Nijhoff Publishers, Dordrecht

Amaki W, Shinohara Y, Hayata Y, Sano H, Suzuki Y (1984) Effects of bulb desiccation and storage on the in vitro propagation of hyacinth. Sci Hortic 23: 353–360

Amano M, Tsutsui K (1980) Propagation of hyacinth by hot temperature treatment. Acta Hortic 109: 279–287

Asjes CJ, Bunt MH, Slogteren DHM (1974) Production of hyacinth mosaic virus-free hyacinths and lily symptomless virus-free lilies by meristem tip culture. Acta Hortic 36: 223–228

Bach A (1989) High efficiency of plant regeneration in epidermal cell layer explant of *Hyacinthus orientalis* L. Folia Hortic I/1: 45–53

Bach A (1990) Shoot multiplication and bulblet production of hyacinth (*Hyacinthus orientalis* L.) in vitro. Zeszyty Naukowe AR Krakow, Ser. Rozprawy habilitacyjne, 150: 1–82

Bach A, Cecot A (1987) Micropropagation of *Hyacinthus orientalis* cv. Carnegie in a double-phase culture medium. In: Grueber G, Hass-Tschirschke J, Hentig WU (eds) Abstr Int Symp Propagation of Ornamental plants-, Geisenheim, p 88

Bach A, Cecot A (1988) Micropropagation of *Hyacinthus orientalis* cv. Carnegie in a double-phase culture medium. I. Effect of cytokinins on growth and development. Acta Hortic 226: 607–610

Bach A, Cecot A (1989) The application of double-phase culture media in micropropagation of *Hyacinthus orientalis* cv. Carnegie. II. Effect of auxins on growth and development. Acta Hortic 251: 215–221

Chun CK, Suh YK, Chung JD (1982) In vitro propagation of *Hyacinthus orientalis* L. V. Effect of basal media, auxin concentrations, inoculated time on organogenesis from bulb-scale tissues. Res Rev Kyungpook Natl Univ 34: 437–446

Chung JD, Chun CK, Suh YK, Rhee EM (1981) In vitro propagation of *Hyacinthus orientalis* L. Effect of auxin on callus formation and bulblet differentiation from flower organ and flower stem tissue. J Kor Soc Hortic Sci 22: 146–151

Chung JD, Chun CK, Suh YK, Rhee EM (1983) In vitro propagation of *Hyacinthus orientalis* L. Effect of auxin concentrations and light or dark incubation on organogenesis from flower stalk tissue and comparison of totipotency between bulb-scale and flower stalk tissue. J Kor Soc Hortic Sci 24: 76–85

Chung JD, Chun CK, Choi SO (1984) Variation of totipotency from individual organ parts of *Hyacinthus orientalis* cultured in vitro at different growth stage. J Kor Soc Hortic Sci 25: 297–304

Darlington CD, Hair JB, Hurcombe R (1951) The history of the garden hyacinths. Heredity 5: 233–252

Hartman HT, Kester DE (1976) Plant propagation: principles and practices. Prentice Hall, Englewood Cliffs, New Jersey

Hussey G (1975) Propagation of hyacinths by tissue culture. Sci Hortic 3: 21–28

Jaarverslag-Statistiek (1989) Production of ornamentals in the Netherlands. PVC, Den Haage, pp. 34–58

Kim J, Hasegawa PM, Bressan RA: (1981) In vitro propagation of hyacinth. Hort Science 16: 645–647

Krause J (1980) Propagation of hyacinth by leaf cuttings. Histological changes in leaf cuttings of *Hyacinthus orientalis*. Acta Hortic 109: 271–277

Lu W, Enomoto K, Fukunaga Y, Kuo C (1988) Regeneration of tepals, stamens and ovules in explants from perianth of *Hyacinthus orientalis* L. Importance of explant age and exogenous hormones. Plant 175: 478–484

Marston ME (1973) Propagation of hyacinths by scaling. Hortic Res 13: 11–16

Matsumoto M (1983) Propagation of virus free hyacinths by tissue culture. Bull Toyama Veg Ornam Crops Res Sta 1: 9–22

Murashige T, Skoog F (1962) A revised medium for rapid growth and bioassays with tobacco tissue culture. Physiol Plant 15: 473–497

Naylor EE (1940) Propagation of *Hyacinthus* by leaf cuttings. Bull Torrey Bot Club 67: 602–606

Paek KY (1982) Studies on the clonal propagation by the culture of bulb scale segments, inflorescence stems and flower-buds in hyacinth. Proc 5th Int Congr Plant Tissue and Cell Culture, Maruzen, Tokyo pp 713–714

Paek KY, Thorpe TA (1989) Hyacinth. In: Ammirato PV, Evans DA, Sharp WR, Bajaj YPS (eds) Handbook of plant cell culture, vol 5. Ornamental crops. McGraw-Hill, New York, pp 479–508

Paek KY, Chun CK, Chung JD (1983) Factors influencing regeneration of bulblets and effects of chilling treatment of parent bulblets on scale segments of hyacinth in vitro. J Kor Soc Hortic Sci 24/1: 68–75

Paek KY, Lee CW, Choi JK, Hong YP (1987) Effect of explant source, breaking of apical dominance and dormancy on shoot proliferation and bulbing of hyacinth, narcissus and lily in vitro. J Kor Soc Hortic Sci 28/1: 88–98

Pierik RLM, Post AJM (1975) Rapid vegetative propagation of *Hyacinthus orientalis* L. in vitro. Sci Hortic 3: 293–297

Pierik RLM, Ruibing MA (1973) Regeneration of bulblets on bulb scale segments of hyacinth in vitro. Neth J Agric Sci 21: 129–138

Pierik RLM, Steagmans HHM (1975) Effect of auxins, cytokinis, gibberellins, abscisic acid and ethephon on regeneration and growth of bulblets on excised bulb scale segments of hyacinth. Physiol Plant 34: 14–17

Pierik RLM, Woets J (1971) Regeneration of isolated bulb scale segments of hyacinth. Acta Hortic 23, 423–427

Rees AR (1972) The growth of bulbs. Acad Press London

Saniewski M (1975) Root and bulblet regeneration by flower buds of *Hyacinthus orientalis* L. in tissue culture. Bull Acad Pol Sci Ser Sci Biol 23: 121–123

Saniewski M (1977) Polarity of adventitious root and bulblet formation in etiolated leaves and flower buds of *Hyacinthus orientalis* L. Bull Acad Pol Sci Ser Sci Biol 25: 63–66

Saniewski M, Nowak J, Rudnicki RM (1974) Studies on the physiology of hyacinth bulbs. IV Hormonal regulation of induction of roots and bulblets in *Hyacinthus orientalis* L. grown in culture. Plant Sci Lett 2: 373–376

XII Micropropagation of *Impatiens* spp.[1]

L.C. Stephens, J.L. Weigle, S.L. Krell, and K. Han[2]

1 General Background

1.1 Commercial Importance of the Genus

Impatiens rank at or near the top for sales of bedding plants in the USA, according to Census Bureau and grower's association surveys (Anonymous 1982). Total bedding plant sales topped 500 million dollars in the USA in 1988, of which 2–3% are estimated to be from sales of *Impatiens* (pers. commun., Dr. David Koranski, Iowa state University 1988). The main reasons for the increasing popularity of *Impatiens* are the ability of plants to flower in the shade, to flower over many months, and the neat appearance of the plants because of their self-cleaning habit (Winters 1973).

There are several important species under cultivation. *Impatiens wallerana* Hook., originally native to Africa, is purchased as seeds by commercial growers and sold as transplants to the public or as seeds sold directly to the public in colorful packets. *I. wallerana* is also sometimes referred to as *I. holstii* or *I. sultanii.* Another species, *I. balsamina* L., garden balsam, native to tropical and subtropical Asia, is sold as a retail packet-seed item to be sown directly in the garden.

A group of species that has been cultivated much more recently is collectively known as New Guinea Hybrid *Impatiens*, and whereas most cultivars are native to New Guinea, some are hybrids among species from Celebes, New Guinea, and Java. The cultivated Java species is *I. platypetala* Lindl., the Celebes species is *I. aurantiaca* Teysm., and the species native to New Guinea have been identified as *I. hawkeri* W. Bull., *I. schlecteri* Warb., *I. mooreana* Schltr., *I. linearifolia* Warb., and *I. herzogii* K. Schum by Winters (1973). Grey-Wilson (1979) has suggested that the scientific name for all New Guineas be *I. hawkeri* Bull. Crossing data at Iowa State University (ISU) have indicated that one species designation would be appropriate, especially for cultivated varieties (Pasutti 1977). Indeed, the ISU breeding program is a result of intercrossing all five species of the original US Department of Agriculture (USDA) introductions. The New Guinea and New Guinea hybrid *Impatiens* are useful as bedding and houseplants, and are available

[1] Journal paper no. J-13132 of the Iowa Agriculture and Home Economics Experiment Station, Ames, Iowa. Project No. 2495.

[2] Department of Horticulture, Iowa State University, Ames, IA 50011, USA

Biotechnology in Agriculture and Forestry, Vol. 20
High-Tech and Micropropagation IV (ed. by Y.P.S. Bajaj)

as transplants and finished flowering plants from retail florists and greenhouses via specialized propagators. The advantages of these over the seed-propagated *I. wallerana* cultivars include larger flowers and variegated foliage and the ability to bloom in sunny and light-shaded areas (Winters 1973). However, New Guinea *Impatiens* are not as tolerant to heat and drought as *I. wallerana* cultivars (Anonymous 1982).

1.2 Propagation Methods

All cultivars of *I. wallerana* are propagated by seeds. Many F_1 hybrids have been developed by major American and European flower seed companies. The cultivars of *I. balsamina* are mostly open-pollinated lines that are grown as mixtures of pastel flower colors. Cultivars derived from *I. balsamina* and *I. wallerana* are not propagated by cuttings because of the vigor, uniformity, and cheaper production costs of seeds. Certain of the New Guinea and Indonesian *Impatiens* are also propagated by seed, but because Celebes, New Guinea, and Java *Impatiens* all have different chromosome numbers, the interspecific hybrids are usually sterile unless amphidiploids are produced. Sweet Sue is a New Guinea × Celebes hybrid amphidiploid that is uniform and produces seed consistently (Anonymous 1982). Tangeglow is another interspecific hybrid involving Celebes as one parent and either New Guinea or Java as the other (Pasutti 1977). This cultivar is also a probable amphidiploid that exhibits good fertility. The only diploid from the Indonesian group that is propagated from seed is Tangerine, a cultivar developed from Celebes *Impatiens*. The most recent introduction of a seed-propagated *Impatiens* is F_1 Tango, derived from New Guinea *Impatiens* (All-America Selections).

All other New Guinea and New Guinea Hybrid *Impatiens* are propagated asexually, mainly because of their heterozygous nature, probable polyploid status, and general tendency to produce variable seedling progenies. Also, leaf variegation is an important trait in *Impatiens* that does not come true from seed, and the genetics of this trait in *Impatiens* is poorly understood. Propagation is usually by shoot-tip cuttings. Most cultivars produce numerous axillary branches, which can also be propagated as shoot-tip cuttings, with the notable exceptions of genetic dwarf selections, for which growth of axillary branches is inhibited. These can be cut into nodal sections, but the adventitious rooting and growth of the shoot is very slow, and plants produced in this way are variable in growth, plant size, and first flowering. Some propagators are now using micropropagation to be certain of having clean stock because of recent invasions of tomato spotted wilt virus (TSWV) and it's vector, the western flower thrips, *Frankliniella occidentalis* (Lyndon Drewlow, pers. commun. 1988, Mikkelsen's Inc., Ashtabula, Ohio). In recent years, TSWV virus has posed a serious threat to the greenhouse and floral crops industry, causing production losses in numerous operations in the USA and Canada (Matteoni et al. 1988). All diseased crops were infected with the vector and in some cases, TSWV outbreaks were associated with the introduction of infected New Guinea *Impatiens* (Matteoni et al. 1988). Thus, the micropropagation of virus-indexed plants has become crucial to the survival of New Guinea *Impatiens* as a floral crop.

1.3 Interspecific Hybridization and Improvement Programs

The improvement of cultivars of *I. wallerana* and *I. balsamina* has been carried out largely by private flower-seed companies and the improved cultivars released as F_1 hybrids. Many of the first *I. wallerana* F_1 hybrids were developed by Claude Hope of Linda Vista SA, Costa Rica. The improvement of Indonesian and New Guinea *Impatiens* initially involved a collaborative collecting expedition in 1970 underwritten by Longwood Gardens, Kennett Square, Pennsylvania (USA), and the USDA. With the introduction of 23 *Impatiens* selections, breeding work was begun at Longwood Gardens, with the first ten clonal cultivars released in 1972 (Armstrong 1974, Martin 1984). Both public and private breeders soon followed with more introductions (Martin 1984). Breeding programs have continued at the USDA, Iowa State University (ISU), and several private US companies, including Mikkelsens Inc., Ashtabula, Ohio, and Pan American Plant Co, West Chicago, Illinois.

The USDA research was carried out primarily by Toru Arisumi. Initially, he treated *Impatiens* interspecific hybrids with colchicine to obtain fertile polyploids among the New Guinea, Java, and Celebes crosses (Arisumi 1973). Certain of these amphidiploids were fertile, and seedlings within progenies were uniform enough to use as seed-propagated cultivars (Arisumi 1975). Arisumi (1980a) found that, of interspecific crosses amongst African, Indian, Indonesian, and New Guinea *Impatiens*, many produced embryos to the globular, heart-shaped, or torpedo stage. That is, the crossing barrier evidently was post-zygotic rather than pre-zygotic. Crosses among African, Indian, and New Guinea *Impatiens* included successful crosses with *I. wallerana* Elfin White, suggesting that these hybrids may be useful for incorporating yellow flowers into *I. wallerana* seed-propagated cultivars and New Guinea *Impatiens* clonal cultivars (Arisumi 1987). Other useful traits that may be incorporated are drought tolerance and double flowers from *I. flaccida* into either seed-propagated *I. wallerana* or New Guinea *Impatiens* clonal cultivars (Arisumi 1987).

The early research at ISU produced crosses amongst Java, New Guinea, and Celebes *Impatiens* that yielded interspecific hybrids in all combinations, but crosses of these with *I. wallerana* were unsuccessful (Beck et al. 1974). Subsequently, the breeding objectives have been to develop New Guinea *Impatiens* plants that will withstand or grow in low-light conditions such as office building or home setting, and to develop New Guinea hybrids that will grow under adverse conditions outdoors. An example of work on improving *Impatiens* for the indoor environment are the recent selections of genetic dwarf plants, 7915-3 (to be released as Tinted Angel) and 8350-3. Both have short internodes, and both produce many fewer and smaller axillary branches, making conventional shoot-tip propagation time-consuming. Both are relatively slow-growing compared with normal-internode plants. Interspecific hybrids among Java, New Guinea, and Celebes *Impatiens* have shown desirable horticultural traits, including low plant habit, rapid growth, and prolific summer flowering (Weigle and Pasutti 1976). Both Celebes and Java *Impatiens* have early and year-round flowering, a low spreading habit, and heat and wind tolerance, but variation for other important bedding-plant traits is missing, so hybrids with New Guinea *Impatiens* are needed to obtain variegated foliage, glossy leaves, and other flower colors (Pasutti et al. 1977). The major drawback of these hybrids has been lack of variation in flower color because the orange color of the Celebes and the magenta color of the Java species are dominant to the range of colors found in the New Guinea species (Weigle and Pasutti 1976, Pasutti and Weigle 1980b). Recently, successful backcrossing has demonstrated that the orange Celebes pigment can be deleted from inbreds that show the range of colors expected from the New Guinea species (Stephens et al. 1988). Two New Guinea × Java interspecific hybrid cultivars released by ISU are Blue Moon and Burgundy (Weigle and Pasutti 1979, Weigle et al. 1979).

2 In Vitro Culture Studies

In vitro cultural techniques that have been useful with *Impatiens* to date include: embryo and ovule culture, protoplast isolation and culture, callus and suspension culture, and shoot-tip culture (Table 1).

2.1 Embryo and Ovule Culture

Initial attempts at ovule and embryo culture of interspecific hybrids of *I. wallerana* with several Indian and African species were unsuccessful, but several hybrids involving Indian and African crosses were successfully embryo-cultured (Arisumi 1980b). Embryos from 15 to 26 days and ovules from 5 to 20 days after pollination were cultured. Ovules and older embryos were successfully cultured to the seedling stage (Arisumi 1980b). The media used were MS (Murashige and Skoog 1962), B5 (Gamborg et al. 1968), or Nitsch (1969). However, no conclusions were drawn as to which medium was best because the experiments were not designed to detect differences among media (Arisumi 1980b). Further research using ovule culture was used to obtain 25 progenies of interspecific hybrids, including eight from among African species, one between two Indian species, four between the African and Indian species, and twelve between the African and New Guinea-Indonesian species (Arisumi 1985). The culture medium used was Nitsch and Nitsch (1969) modified with half the ammonium and potassium nitrate concentration and supplemented with 100 ml/l deproteinized coconut water, 1 mg/l indole-3-acetic acid, 1 mg/l kinetin or zeatin, and 20 g/l sucrose (Arisumi 1985). Ovule culture

Table 1. Response of different explants of *Impatiens* cultured in vitro

Type of explant	In vitro response	Reference
Embryo	Growth, development to seedling stage	Arisumi (1980b)
Ovule	Growth, development to seedling stage	Arisumi (1980b, 1985, 1987)
Cotyledon	Callus, root initiation	Stephens (1985)
Hypocotyl	Callus, root initiation	Stephens (1985)
Leaf protoplast	None	Delafield and Grant (1986)
Shoot tip	Axillary shoot proliferation	Han and Stephens (1987), Stephens et al. (1985)

generated only 0–16.5% rescued seedlings (Arisumi 1985). There was great variation in success rate by cross, and the overall low success rate indicated that cultural factors may need to be improved (Arisumi 1985).

2.2 Protoplast Isolation and Culture

The expected merger of New Guinea, Java, and Celebes germplasms into the domestic *I. wallerana* has not occurred because of seemingly incompatibility barriers (Arisumi 1973, Beck et al. 1974). These barriers have been shown to be post-zygotic rather than pre-zygotic (Arisumi 1980a). Thus, emphasis has been on rescue of embryos and ovules of interspecific hybrids. If this area of research is not fruitful, somatic hybridization via protoplast fusion may be an attractive alternative, although regeneration from protoplasts has yet to be demonstrated. Somatic hybridization may also allow cytoplasmic hybrids (cybrids) that contain cytoplasmic traits of one parent with the nuclear genome of another, with a much better chance of obtaining fertile plants. Procedures for protoplast isolation of *I. wallerana* have been developed by Delafield and Grant (1986). Mature axenic shoots growing on a basal B5 medium proved a superior source of protoplasts. Leaf mesophyll protoplasts were isolated from young fully expanded leaves by using low enzyme concentrations and a short incubation time. However, protoplasts were unsuccessfully cultured in liquid drops of MS medium with 0.5 mg/l NAA (Delafield and Grant 1986).

2.3 Callus and Suspension Culture

Several attempts have been made to obtain regeneration from callus and suspension cultures, none of which have been successful to date, although callus and suspension cultures can be maintained by using routine culturing procedures. Cotyledon and hypocotyl explants of *I. wallerana* F1 Futura Mix and *I. balsamina* L. Tall Double Mix were plated on combinations of cytokinin and auxin concentrations included in an MS medium (Stephens 1985). Concentrations of BA or kinetin from 0–30 mg/l with IBA or NAA from 0–0.1 mg/l did not cause shoot regeneration, but shoot-like buds formed on one cotyledon explant of *I. balsamina* at 3 mg/l BA (Stephens 1985). Arisumi (1983) reported shoots from callus of interspecific *Impatiens* embryos. Callus cultures of New Guinea × Java interspecific hybrids were successfully initiated and maintained from greenhouse-grown leaf and stem explants by culturing on MS medium with 0.1 mg/l 2,4-D (Stephens unpubl.). Suspension cultures were also maintained and subcultured on the same medium. Transfer of callus to MS medium without growth regulators or to MS medium with a wide range of cytokinin and auxin concentrations failed to induce shoot proliferation on the callus (Stephens unpubl.). Regeneration of adventitious shoots was also unsuccessful with sections of young internodes of the New Guinea *Impatiens* Pikinini, although these internodes rooted extensively on a wide range of media (Koenigsberg and Langhans 1976). A systematic effort is needed to determine the regeneration potential of *Impatiens*.

2.4 Micropropagation

Micropropagation is suitable when conventional asexual propagation is less effective, slower, or more expensive per propagule. More importantly in the case of New Guinea *Impatiens* cultivars, micropropagation provides a means of maintaining virus-indexed shoot cultures which can be propagated to provide clean stock plants. Micropropagation of asexually propagated *Impatiens* provides other important advantages. For breeding programs, selections can be more rapidly introduced to the commercial trade. Certain genotypes, such as New Guinea *Impatiens* genetic dwarfs, may be difficult to propagate by conventional vegetative means. Micropropagation also produces a continuing source of vegetative shoots, and more shoots are produced per unit time in less space than by a conventional shoot-tip propagation method.

2.4.1 Establishment

Axenic cultures of Indonesian and New Guinea *Impatiens* are best obtained by harvesting shoot tips from fast-growing, vegetative shoots, and from plants that are not watered from overhead. As described previously for the establishment of aseptic shoot cultures (Stephens et al. 1985), vegetative stem tips are cut from plants in the glasshouse and rinsed in cold tap water. After the expanded leaves are cut off, the tips are trimmed to 2-3 cm long and put in a side-arm flask containing 0.5% NaOCl (10% v/v commercial laundry bleach). The flask is aspirated for 10 min and swirled continuously, after which all further steps are performed aseptically in a laminar-flow transfer hood. After four rinses in sterile, deionized water, all leaves and visible leaf primordia are trimmed from each shoot, and it is cut to a length of 2–3 mm from cut base to apical meristem. The prepared explant is plated onto a semi-solid medium, cut base down in a suitable culture vessel. These methods have been successful in obtaining axenic shoot cultures, but initial contamination rates are usually high. The most serious contamination seems to be from bacteria present within older tissues of the stem. Thus, small shoot tips 2–3 mm long were required to escape this seemingly internal contamination. From small numbers of clean cultures, any *Impatiens* genotype can be rapidly multiplied because growth in culture is relatively rapid. Our basal culture medium is composed of inorganic MS salts, with the addition of 100 mg/l m-inositol, 0.5 mg/l nicotinic acid, 0.5 mg/l pyridoxine HCL, 0.1 mg/l thiamine HCl, 2 mg/l glycine, and 30 g/l sucrose (Stephens et al. 1985; Han and Stephens 1987). The pH is adjusted to 5.8, 8 g/l Difco Bacto agar is added, and the medium autoclaved 15 min at 1.1 kg/cm^2 and 121 °C. Cytokinins must be added in small amounts (1–2 mg/l) to obtain growth of a single shoot. Initiation of shoot tips on MS basal medium without cytokinins or auxins results in very slow growth initially. Eventually, roots will form unless these shoot cultures are transferred to higher concentrations of cytokinins. Cultures are incubated in a growth chamber at 26 °C under 16-h illumination of ca. 45 μmol/s/m^2. Twenty-four-hour illumination causes bleaching of chlorophyll in leaves of newly initiated shoots.

2.4.2 Multiplication

For the production of multiple shoot proliferation, some cytokinins are more effective than others. Kinetin at 30 mg/l is the greatest concentration that we have tested, and the most shoots were produced in this concentration (Stephens et al. 1985). A comparison of cytokinins showed that both 2-ip and BA were more effective than kinetin (Han and Stephens 1987). In general, fairly high concentrations of cytokinin are required for inducing multiple shoots. For genetic dwarf genotypes, BA at 10 mg/l is required for optimal response (Table 2). Beyond this level, shoot numbers continue to increase, but if cytokinin concentration is too high, shoots become so short that transfer to soil becomes difficult. However, cytokinin concentration must be high enough to keep roots from forming. Once this happens, one or a few shoots exert apical dominance, and more axillary shoots will not be produced. Cytokinin concentrations that are effective for inducing multiple shoots need to be two to five times greater than for most other crops, but these New Guinea *Impatiens* have also shown resistance to growth-retarding chemicals in vivo (Pasutti and Weigle 1980a), so there seems to be a precedent for the results that we have obtained in vitro (Stephens et al. 1985).

NAA, the only auxin tested, was not clearly effective in increasing number or length of shoots. In combination with kinetin, NAA at 0.5 mg/l was ineffective in stimulating more shoots (Stephens et al. 1985). At concentrations from 0.4 to 1.0 mg/l, NAA had no effect on shoot number for one genotype and clearly inhibited shoot number increase of Star Fire (Han and Stephens 1987).

Stephens et al. (1985) showed, and Han and Stephens (1987) confirmed that genotype response varied over a range of growth-regulator concentrations. It is probably best to test each genotype over a range of cytokinin and auxin concentrations. In general, higher concentrations of cytokinins may be required than are customary for other species. This may explain why, in the first report of shoot-tip

Table 2. Mean number of shoots per explant of dwarf New Guinea *Impatiens* 7915-3 after 4 weeks' culture

	Cytokinin		
mg/l	BA	BA	2-iP
0	1.1c	1.0d	1.4c
1	1.8b		
2	2.2b		3.0b
3	3.1a		
4	3.1a	3.4c	3.3b
6			3.8ab
8		5.2b	4.6a
10			3.2b
12		7.1a	
16		7.8a	

Means in columns with the same letter are not significantly different by Duncan's multiple range test. Each column is a separate experiment, so rows are not statistically comparable.

culture, Koenigsberg and Langhans (1976) were unsuccessful in obtaining more than a single shoot per culture. We have found that concentrations of 2 mg/l and less of a cytokinin will only cause cultures to form single shoots, and usually within 3 weeks, adventitious roots begin to form.

The New Guinea genetic dwarf selections behave similarly to previously tested Indonesian *Impatiens* and interspecific hybrids. Multiplication rates were stable over three successive subcultures for the New Guinea *Impatiens* dwarf genotype 7915–3 (Table 3). These cultures were divided by cutting into equal-sized pieces containing one or more shoots and transferring to a 25 × 150 mm culture tube. The pattern of shoot proliferation is remarkably similar to normal genotypes (Fig. 1A, Table 2); that is, high concentrations of cytokinin are required. As with other genotypes, kinetin proved relatively ineffective (Stephpens et al. 1985; Han and Stephens 1987). The highest concentration of BA produced small shoots and clumps of very small shoots that were difficult to transfer successfully to soil (Table 4). When proliferating shoots were grown in 25 × 150 mm culture tubes, each culture was easily divided into several new cultures. Twenty healthy clumps (of one or more shoots) were transferred at each subculturing date. The highest multiplication rate, over three subcultures, was achieved with 12 mg/l BA (Table 3). Shoots were shortest in the highest BA concentration, and progressively longer as BA concentration was decreased (Table 4). When shoots were smaller than 5 mm, it was easier to transfer a clump of tissue containing one or more shoots.

Table 3. Multiplication rate of dwarf New Guinea *Impatiens* 7915-3. Number of divisions into new cultures on three successive subcultures from the initiation date. (Each number is the mean of 20 cultures)

BA (mg/l)	Date initiated	Dates of subcultures		
	8/18	9/21	10/31	12/2
0	1	1.3	1.0	1.0
4	1	1.9	1.4	2.0
8	1	2.8	2.5	3.0
12	1	3.4	3.3	3.8
16	1	2.8	3.1	4.4

Table 4. Percentages of each category of shoot sizes of cuttings transferred to soil after the fifth subculture of dwarf New Guinea *Impatiens* 7915-3

BA (mg/l)	Single shoots longer than 5 mm	Single shoots equal to or shorter than 5 mm	Clumps of very small shoots
0	95	5	0
4	70	30	0
8	23	48	29
12	43	29	29
16	4	71	25

For incubating the cultures, 25 °C provides rapid growth, requiring transfer every 3–4 weeks. Thus, commercial culture rooms with banks of lights hung under shelving probably will provide acceptable growth rates without extra heating provided. For lighting conditions, cultures should be placed within 15 cm of cool white fluorescent light tubes providing ca. 45 µmol/s/m^2 during a 16-h photoperiod. As with cultures in the establishment phase, we have found that continuous light causes chlorosis in shoot cultures during a 2- to 4-week culture period.

2.4.3 Rooting/Transfer to Soil

In medium containing low concentrations of cytokinin (up to 2 mg/l), and no auxin, production of roots will occur, usually within several weeks of culture transfer. Initial results showed that rooted shoots 2 cm long survived when transferred to a mist propagation bed for 1–2 weeks before transfer to a growing bench (Stephens et al. 1985). Transfer of proliferating cultures from high to low concentrations of cytokinin will also produce sizable plantlets for transfer to soil, but this in vitro rooting requires an extra step in culture and seems not to be necessary. Later tests have shown that cuttings without roots in mist-propagation beds survive at high rates so that rooted shoots are not necessary for successful transfer to soil (Fig. 1B,C). Before transfer to soil of genetic dwarf genotypes, sizes of shoots were categorized by size. Not surprisingly, the lowest percentage of longest shoots (greater than 5 mm) occurred in the multiplication medium with the highest BA concentration (Table 4). For this genotype, 10 mg/l BA in multiplication medium was a good compromise between the most shoots that were transferrable to soil and the largest.

For the transfer of cultured shoots to soil, we use a soil mix of equal volumes of sphagnum peat moss and vermiculite with the addition of a granulated fertilizer and ground limestone to adjust the soil pH to ca. 6.5. The survival of shoots is related to the size of the shoot from culture, so that unrooted shoots longer than 5 mm had a 95% survival after 6 weeks in small 6.4 to 7.6-cm pots (Table 5). Shoots less than or equal to 5 mm and clumps of very small shoots had a much poorer survival rate (Table 5).

First flowering occurred ca. 12–16 weeks after transfer to soil, with one transplanting to 10-cm pots (Fig. 1D). No obvious variants were found in 275 plants that have been grown to first flowering. All closely resembled the original genotype used for cloning. That no variants were detected may be regarded as a priori evidence that shoot proliferation is axillary in origin rather than adventitious because of the high rates of variants that usually occur with adventitious

Table 5. Effect of shoot length of dwarf New Guinea *Impatiens* 7915-3 and 8350-3 on survival after transfer to soil in a mist-propagation bed for 2 weeks

Length of cultured shoot	Percent survival after 6 weeks
Longer than 5 mm	95
Shorter than or equal to 5 mm	57
Clumps (small, multiple shoots)	57

Fig. 1A–D. Stages of micropropagation of New Guinea *Impatiens* genetic dwarf 8350-3. **A** Multiple shoot production after 4 weeks in culture on MS medium with 10 mg/l BA. **B** Plantlets surviving after transfer of unrooted shoots to soil in cell packs. **C** One plantlet ready for transfer to 10 cm pot. **D** First flowering of plantlet 12 weeks after transfer to soil

shoot regeneration. Sample testing of micropropagated plants would still be wise to assure maintenance of clonal integrity.

2.4.4 Commercial Aspects

Mikkelsen's Inc., a commercial breeder and propagator of New Guinea *Impatiens*, uses micropropagation to produce rooted cuttings of their cultivars (Fig. 2) (Lyndon Drewlow, pers. commun., 1990, Mikkelsen's Inc., Ashtabula, Ohio). They initiate shoot-tip cultures from TSWV-indexed stock plants. Shoots from multiplying shoot cultures are rooted in soil in an isolation house and the rooted plantlets

Fig. 2. Commercial micropropagation of New Guinea *Impatiens*. Micropropagated and TSWV-indexed *Impatiens* plants growing in a production greenhouse of Mikkelsen's Inc., Ashtabula, Ohio

are used as stock plants from which cuttings are taken. These cuttings are rooted in a thrips-free screen house and the virus-indexed rooted cuttings are sold to wholesale and retail plant growers. By this method, New Guinea *Impatiens* cultivars can be grown in the presence of other floral crops without carrying the TSWV virus to other host plants.

3 Conclusions and Prospects

For successful micropropagation of New Guinea *Impatiens*, shoot tips from fast-growing, vegetative shoots provide the best explant for the establishment of aseptic cultures. A Murashige and Skoog (1962) culture medium with the addition of 2–10 mg/l of a cytokinin such as BA produces rapidly multiplying shoot cultures. The addition of an auxin has not improved the multiplication response for the genotypes we have tested. Additionally, each genotype should be tested for its response to cytokinin type and concentration. Cultures will maintain a rapid growth rate under a constant 25 °C and a 16-h photoperiod of 45 $\mu mol/s/m^2$. Under these conditions, cultures should be divided and transferred monthly. Shoots 5 mm or longer, harvested from multiplying shoot cultures, can be easily rooted in mist-propagation beds so that rooted shoots are not necessary for successful transfer to soil. Many factors that could affect the efficiency of the micropropagation process remain unexplored. The culture medium and incubation conditions are two areas with potential for increased production efficiency.

Another area of in vitro culture that warrants future research is assessing systematically the regeneration potential of the important *Impatiens* species. Still unexplored is the potential for immature embryos to exhibit regeneration that older tissues seem not to exhibit. Such a regeneration method would open the way for genetic transformation of *Impatiens*, providing a powerful new means of improving *Impatiens* cultivars. More effort also seems warranted in the area of protoplast fusion and generation of somatic hybrids of incompatible crosses, especially involving the commercial seed-propagated *I. wallerana*. In the meantime, because there seem to be few pre-zygotic barriers to interspecific hybridization, improving embryo-culture techniques to integrate important traits may allow even greater successes than have been achieved so far.

Acknowledgments. We thank the Society of Iowa Florists and the Ohio Floriculture Foundation for supporting the original research reported herein.

References

Anonymous (1982) Petunia lovers beware, new *Impatiens* hybrids may be taking over the market. Flor Rev 170 (4416): 12–13

Arisumi T (1973) Morphology and breeding behavior of colchicine-induced polyploid *Impatiens* ssp. L. J Am Soc Hortic Sci 98: 559–601

Arisumi T (1975) Phenotypic analysis of progenies of artificial and natural amphiploid cultivars of New Guinea and Indonesian species of *Impatiens* L. J Am Soc Hortic Sci 100: 381–383

Arisumi T (1980a) Chromosome numbers and comparative breeding behavior of certain *Impatiens* from Africa, India, and New Guinea. J Am Soc Hortic Sci 105: 99–102

Arisumi T (1980b) In vitro culture of embryos and ovules of certain incompatible selfs and crosses among *Impatiens* species. J Am Soc Hortic Sci 105: 629–631

Arisumi T (1983) Post-germination abnormalities among ovule-cultured *Impatiens*. HortScience 18: 617 (Abstr)

Arisumi T (1985) Rescuing abortive *Impatiens* hybrids through aseptic culture of ovules. J Am Soc Hortic Sci 110: 273–276

Arisumi T (1987) Cytology and morphology of ovule culture-derived interspecific *Impatiens* hybrids. J Am Soc Hortic Sci 112: 1026–1031

Armstrong RJ (1974) An *Impatiens* circus, the Longwood New Guinea hybrid Impatiens. Am Hortic 53(1): 14–18

Beck AR, Weigle JL, Kruger EW (1974) Breeding behavior and chromosome numbers among New Guinea and Java *Impatiens* species, cultivated varieties, and their interspecific hybrids. Can J Bot 52: 923–925

Delafield SJ, Grant WF (1986) Protoplast isolation from *Impatiens wallerana*. HortScience 21: 803 (Abstr)

Gamborg OL, Miller RA, Ojima K (1968) Nutrient requirements of suspension cultures of soybean root cells. Exp Cell Res 50: 151–158

Grey-Wilson C (1979) *Impatiens* in Papuasia. Studies in Balsaminaceae: I. Kew Bull 34: 661–688

Han K, Stephens LC (1987) Growth regulators affect in vitro propagation of two interspecific *Impatiens* hybrids. Sci Hortic 32: 307–313

Koenigsberg S, Langhans R (1976) Tissue culture studies with the New Guinea-Java hybrid *Impatiens*. HortScience 11: 321 (Abstr)

Martin T (1984) New Guinea *Impatiens*. Horticulture 63 (8): 32–37

Matteoni JA, Allen WR, Broadbent AB (1988) Tomato spotted wilt virus in greenhouse crops in Ontario. Plant Dis 72: 801

Murashige T, Skoog F (1962) A revised medium for rapid growth and bioassays with tobacco tissue cultures. Physiol Plant 15: 473–497
Nitsch JP (1969) Experimental androgenesis in *Nicotiana*. Phytomorphology 19: 389–404
Nitsch JP, Nitsch C (1969) Haploid plants from pollen grains. Science 163: 85–87
Pasutti DW (1977) Cytological study of some New Guinea, Java, and Celebes *Impatiens* interspecific hybrids. MS Thesis. Library, Iowa State University, Ames IA
Pasutti DW, Weigle JL (1980a) Growth-regulator effect on New Guinea *Impatiens* hybrids. Sci Hortic 12: 293–298
Pasutti DW, Weigle JL (1980b) Pollen fertility in Java × New Guinea *Impatiens* interspecific hybrids. Can J Bot 58: 384–387
Pasutti DW, Weigle JL, Beck AR (1977) Cytology and breeding behavior of some *Impatiens* hybrids and their backcross progeny. Can J Bot 55: 296–300
Stephens LC (1985) In vitro organogenesis on seedling explants of *Impatiens*. HortScience 20: 539 (Abstr)
Stephens LC, Krell SL, Weigle JL (1985) In vitro propagation of Java, New Guinea, and Java × New Guinea *Impatiens*. HortScience 20: 362–363
Stephens LC, Weigle JL, Krell SL (1988) Flower color inheritance in inbred progenies of *Impatiens* interspecific hybrids. J Hered 79: 136–137
Weigle JL, Pasutti DW (1976) Approaches to transfer of characteristics between ploidy levels in New Guinea *Impatiens*. Acta Hortic 63: 109–112
Weigle JL, Pasutti DW (1979) Blue Moon *Impatiens*. HortScience 14: 766
Weigle JL, Pasutti DW, Beck AR (1979) Burgundy *Impatiens*. HortScience 14: 768
Winters HF (1973) New *Impatiens* from New Guinea. Am Hortic 52(3): 16–22

XIII Micropropagation of Iris with Special Reference to *Iris* × *hollandica* Tub.

P.C.G. VAN DER LINDE[1,2] and J.A. SCHIPPER[1]

1 General Account

1.1 Botany and Classification

Irises have a rootstock which is rhizomatous, bulbous, or cormous. The leaves are distichous and the large spathe-valves are either membranous or herbaceous. The flowers are very variable in shape and color, and many combinations of colors are found within one flower. The petals are joined in a perianth tube; the outer segments (falls) are usually reflexed and the inner segments (standards) are more or less erect, giving the flower the well-known and characteristic habitus of an iris. The stamens are inserted at the base of the falls and have short filaments and basifixed anthers. The ovary is trilocular and contains many ovules. The styles are petaloid and crested, often covering the stamens and part of the falls. The capsule is oblong and angled, and contains many globule seeds (Grey 1937).

Irises are found throughout the temperate and subtropical zones of the Northern Hemisphere. More than 300 species are known. This is only a small number compared to the one obtained when all subspecies, varieties, natural hybrids, and cultivars are added. The classification of irises was subjected to revisions by Baker (1892), Dykes (1913), Lawrence (1953), Rodionenko (1961), and Mathew (1981). Rodionenko's classification (see Table 1) has proved to be the most practical (Köhlein 1987).

Rodionenko classified the rhizomatous irises in the genus *Iris*. The bulbous and cormous irises are classified as four separate but related genera, viz., *Xiphium*, *Iridodictyum*, *Gynandriris*, and *Juno*. Irises belonging to the genus *Xiphium* are characterized by a large bulb without fleshy roots during dry storage, a tall habitus, and a late flowering period (May-June). The genus *Iridodictyum* contains irises having a small bulb with a reticulate skin and rush-like leaves with four to eight ribs. The plants remain small and usually produce only one flower on a short scope in the early spring. Irises having a tuber surrounded by reticular sheets of death leaves, are grouped in the genus *Gynandriris*. The last genus, *Juno*, contains irises having a bulb with fleshy roots during dry storage, deeply furrowed or ridged stem-clasping leaves, and very small standards.

[1] Bulb Research Centre, Central Research Laboratory for Tissue Culture of Horticultural Crops, P.O. Box 85, 2160 AB Lisse, The Netherlands
[2] Present address: RZ Biofleur C.V., P.O. Box 2, 9123 ZR Metslawier, The Netherlands

Biotechnology in Agriculture and Forestry, Vol. 20
High-Tech and Micropropagation IV (ed. by Y.P.S. Bajaj)

Table 1. Classification of the genus *Iris* L. and related genera according to Rodionenko (1961)

Genus	Subgenus	Section	Subsection
Iris	Limniris	Limniris	Apogon
			Ensatae
			Tenuifoliae
			Syriacea
		Unguiculares	
		Ioniris	
	Xyridion	Xyridion	
		Spathula	
	Nepalensis		
	Pardanthopsis		
	Crossiris	Crossiris	
		Lophiris	
		Monospatha	
	Iris	Iris	
		Hexapogon	Regelia
	Pseudoregelia		
		Oncocyclus	
Xiphium			
Iridodictyum		Iridodictyum	
	Monolepis		
Gynandriris			
Juno	Juno		
	Physocaulon		
	Acanthospora		

1.2 Distribution and Importance

In horticultural practice, a simple classification of commercially important irises is used. Two large groups are distinguished: the rhizomatous and the bulbous irises. The rhizomatous irises are divided into bearded (eupogon, e.g., *Iris germanica*) and nonbearded (apogon, e.g., *Iris sibirica*) species, in proportion to the presence or absence of hairs on the falls. The chromosome number varies between 20 and 108. The plants are found throughout the temperate and subtropical zones of the Northern Hemisphere. They are used as garden plants.

Although the number of bulbous species is smaller, their economic value is much higher. The bulbous irises are divided into several groups; viz., Reticulata, Juno, and Xiphium. The Reticulata group (genus *Iridodictyum*) originates from eastern Turkey, the Middle East, the Caucasus and Iran. The chromosome number varies between 16 and 28. They are used as garden plants and, on a limited scale, as pot plants. *Iridodictyum reticulatum*, *I. danfordiae*, and *I. histrioides* are the best known cultivars.

The Juno group (genus *Juno*) originated from the Caucasus and spread to Central Asia, Sicily, Spain, Northern Africa, Eastern Turkey, Asia Minor, and Afghanistan. The chromosome number varies between 18 and 50. Juno irises grow in stony, lime-rich soil, which determines their main commercial use as a rock-garden plant. For the bulb industry they are of minor importance, because they are not hardy. *Juno bucharica*, *J. magnifica*, and *J. willmottiae* are less cold-sensitive and have been grown successfully outdoors.

The Xiphium group (genus *Xiphium*) is the most important for the bulb industry. Xiphium irises originate from the western Mediterranean (Spain, Portugal, southern France, Morocco, Algeria, Tunisia). The chromosome number varies between 28 and 56. The most important species are the English iris (*X. latifolium*) from the Pyrenees, the Spanish iris (*X. vulgare*) from the Iberian peninsula, and the Moroccan iris (*X. tingitanum*) from Morocco. These irises are used for (year-round) cut-flower production.

1.3 Conventional Propagation and Breeding

From crosses between several *Xiphium* species and varieties (*Xiphium vulgare* varieties *praecox* and *lusitanicum*, *X. tingitanum*, *X. boisseiri*, *X. filifolium*, and the Spanish cultivar Cajanus), a group of hybrids was obtained presently known as Dutch irises; *Iris* × *hollandica* Tub. (Fig. 1A). They are characterized by a taller habitus and an earlier flowering period than the original *Xiphium* species. An enormous assortment has been built up in the last century. Dutch irises are grown in The Netherlands (more than 2000 acres), the United Kingdom (ca. 100 acres), France (ca. 100 acres), Israel, the United States of America (ca. 300 acres), and Japan.

The existing assortment of Dutch irises is divided into small- and large-bulbing varieties. The small-bulbing varieties flower at a circumference of the bulb of 6

Fig. 1. A Habitus of Dutch iris: blooming plant (*right*) and non-blooming, three-leaved plant (*left*). **B** Virus-infected (*left*) and virus-free (*right*) flower of bulbous iris cv. Professor Blaauw

(sometimes 5) cm or more and originate from crossings between different varieties of the diploid species *X. vulgare* (2n = 34). In most cases they are easily intercrossed. Some known cultivars are White Excelsior, White Emperor, Sapphire Beauty, and Purple Sensation. The large-bulbing varieties (about 90% of the cultivated area in The Netherlands) flower at a circumference of 8 cm or more and originate from crosses between small-bulbing varieties and *X. tingitanum* (2n = 28). Consequently, they have 31 chromosomes and are sterile. The best-known cultivars for year-round cut-flower production are Wedgwood, Ideal, Prof. Blaauw, and Blue Magic. Breeding research now concentrates on early flowering, a long vase life, and chromosome duplication in large-bulbing varieties (Eikelboom and van Eijk 1987).

Dutch irises cannot be propagated by seed. The propagation fully depends on vegetative daughter-bulb formation in the field. Since only two to four new bulbs are formed annually, it takes many years for the breeder to raise a stock of sufficient size to test bulb productivity, disease resistance, and forcing abilities of a new hybrid or cultivar. Subsequently, even more time is needed to bulk up for commercial purposes. During these years of culture in the field, the new variety may devaluate, and the crop is exposed to many pathogens. The bulb grower has to cope with *Ditylenchus destructor*, a nematode causing eelworm disease (Bergman et al. 1978), with several fungi, e.g., *Fusarium oxysporum* f. gladioli (Bruhn 1955), *Penicillium corymbiferum* (Saaltink 1968), *Pythium* spp. (Koster and de Rooij 1979), *Sclerotium* sp. and *Botrytis cinerea* (Bergman et al. 1978), and *Rhizoctonia solani* (Sonderman and MacLean 1949), causing several types of soft and dry rots either in the field or during dry storage, and with *Erwinia carotovora* pv. carotovora, causing bacterial soft rot (Kamerman 1986).

Virus diseases are also observed (Derks et al. 1985). Almost every bulbous iris is infected with iris mild mosaic virus (IMMV), causing mild mosaic disease. In addition, several cultivars are infected to a great extent with a second virus, narcissus latent virus (NLV), causing a mild yellow mosaic disease. Severe mosaic disease, caused by an additional infection with iris severe mosaic virus (ISMV), is controlled in The Netherlands by visual selection in the field (Asjes 1979). Bean yellow mosaic virus (BYMV) was found to invade the crop in Israel without causing any visible symptoms (Alper and Loebenstein, 1981). IMMV, BYMV, and ISMV are potyviruses with filamentous particles of ca. 750 nm. NLV is a carlavirus with particles of ca. 650 nm. All these viruses are transmitted by aphids in a nonpersistent manner (Hammond et al. 1985).

1.4 Need for in Vitro Culture

To improve the quality of bulbous irises, efforts were and are being made to produce virus-free plants. As every commercial stock is infected with virus, only tissue culture can be used to eradicate these pathogens. Baruch and Quak (1966) were the first to use this technique to obtain virus-free plants of the cultivar Wedgwood. Since their work was published, many other varieties have been made virus-free. Field tests were performed to determine the quality of the virus-free stocks. They were found to produce a 30–50% higher bulb yield than virus-infected ones (Hirata and Kunishige 1982; Asjes 1988). In forcing experiments, bulbs obtained from meristem-tip cultures show a longer flower stem and a better quality cut flower (Fig. 1B) than the original commercial stock (Hirata and Kunishige 1982). Unfortunately, reinfection with the virus occurs rapidly (Hirata and Kunishige 1982; Anonymous 1987), which necessitates spraying of the crop with mineral oil and pesticides to limit virus spread by aphids (Swenson 1971; Asjes 1989a,b).

With regard to the slow propagation rate in the field and the rapid reinfection of virus-free stocks, a fast and safe propagation method is highly desirable. One possible method for enhanced propagation is chipping. The development of a chipping method for iris is in progress (Alkema 1971; Schipper and Van der Weijden 1984; Jones and Hanks 1988). However, chipping must be regarded as a second step in bulking up, because it needs a reasonable amount of bulbs to start with and it cannot be used year after year on the same stock. Consequently, the need for propagation in vitro as a first step in bulking up is evident.

2 Review of Literature

Aseptic culture of iris tissue is described in more than 40 publications (Table 2). The objectives of the reported studies may be grouped into five categories: propagation,

Table 2. Compilation of literature on aseptic culture of iris tissue

Culture	Result	Reference
Embryo	Germination	Werckmeister (1936)
Embryo	Germination	Randolph and Cox (1943)
Embryo	Germination	Randolph (1945)
Embryo	Germination	Werckmeister (1952)
Embryo	Germination	Lenz (1954)
Embryo	Germination	Lenz (1955)
Embryo	Germination	Randolph and Randolph (1955)
Embryo	Germination	Werckmeister (1956)
Embryo	Germination	Shockey (1959)
Embryo	Germination	Randolph and Khan (1960)
Embryo	Germination	Farrington (1961)
Shoot apex	Flower initiation	Rodrigues Pereira (1962)
Embryo	Germination	Holden (1962)
Shoot apex	Flower initiation	Rodrigues Pereira (1962)
Embryo	Germination	Werckmeister (1962)
Shoot apex	Flower initiation	Rodrigues Pereira (1964, 1965)
Embryo	Germination	Kidd (1966)
Apical meristem	Virus eradication	Baruch and Quak (1966)
Shoot apex	Flower initiation	Rodrigues Pereira (1966)
Embryo	Germination	Stolz (1967)
Embryo	Germination	Stolz (1968)
Embryo	Germination	Farrington (1969)
Apical meristem	Virus eradication	Mori et al. (1969)
Embryo	Development	Stolz (1971)
Axillary bud	Shoot regeneration	Fujino et al. (1972, 1973)
Stem, leaf, root	Callus	Marsavina et al. (1973)
Inflorescence stem	Shoot regeneration from callus	Meyer et al. (1975)

Table 2. (*Continued*)

Culture	Result	Reference
Embryo, shoot apex	Embryogenesis from callus	Reuther (1975)
Cells	Callus, suspension	Sevruk et al. (1975)
Axillary bud, scale, apical meristem, inflorescence stem	Shoot regeneration Bulblet growth	Fujino (1976)
Inflorescence stem	Shoot regeneration	Hussey (1976a)
Adventitious shoot	Propagation	Hussey (1976b)
Shoot	Propagation	Hussey (1977)
Apical meristem	Shoot regeneration from callus	Lunjova (1977)
Embryo	Embryogenesis from callus	Reuther (1977a)
Apical meristem	Shoot regeneration and embryogenesis from callus	Reuther (1977b)
Embryo	Development	Stolz (1977)
Embryo	Germination	Weiler and Emershad (1977)
Shoot apex	Shoot regeneration	Doss (1979)
Shoot apex	Flower initiation	Doss and Christian (1979)
Embryo	Germination	Tamberg (1979)
Axillary bud	Virus eradication	Allen and Anderson (1980)
Apical meristem, axillary bud	Virus eradication	Hirata and Kunishige (1982)
Embryo	Development	Yabuya and Yamagata (1982)
Apical meristem	Virus eradication	Brunt (1983)
Shoot apex	Flower initiation	Doss et al. (1983)
Shoot apex	Shoot regeneration, embryogenesis from callus	Laneri and Lucretti (1983)
Flower stalk	Shoot regeneration	Meyer (1984)
Developing ovule	Embryo development	Yabuya et al. (1984)
Embryo	Amphidiploidy	Yabuya (1985)
Propagated shoot	Bulblet growth	Mielke and Anderson (1986)
Scale	Shoot regeneration, bulblet growth	Van der Linde et al. (1986, 1987)
Adventitious meristem	Virus eradication	Blom-Barnhoorn and Van der Linde (1987)
Embryo	Embryogenesis from callus	Radojevic et al. (1987)
Suspension	Essential oil production	Aslanyants et al. (1988)
Flower stalk	Shoot regeneration, propagation, bulblet growth	Bach (1988)
Rhizome	Shoot regeneration	Kromer (1988)
Scale	Shoot regeneration, bulblet growth	Van der Linde et al. (1988)
Adventitious shoot	Propagation, bulblet growth	Van der Linde and Hol (1988)
Pollen	protoplast isolation and culture	Zhou (1988, 1989)
Suspension	Terpenoid bioconversion	Bagdasarova et al. (1989)

virus eradication, embryo culture, physiology of flower initiation, and other purposes.

2.1 Propagation

2.1.1 Shoot Regeneration on Explants

Many types of explants regenerate shoots in tissue culture. For bulbous irises adventitious shoots are formed on explanted axillary buds (Fujino et al. 1972, 1973; Fujino 1976), pieces of scale and/or basal plate (Fujino 1976; Hussey 1977; Van der Linde et al. 1987, 1988; Van der Linde and Hol 1988), sections from flower stems (Fujino 1976; Hussey 1976a, 1977; Bach 1988) and damaged apical meristems on basal plate, containing a few leaf primordia (Doss 1979). For rhizomatous irises adventitious shoot regeneration on flower-stem sections (Meyer 1984) and axillary shoot regeneration from shoot apices (Laneri and Lucretti 1983; Kromer 1988) have been reported.

There is consensus in the literature on the type of medium to be used for shoot regeneration. With the exception of some early reports (Fujino et al. 1972, 1973; Doss 1979), the formulation of Murashige and Skoog (1962) medium (MS) either at half or at full strength, a sucrose concentration of 2–4%, and a pH of 5–6 have been reported. The addition of activated charcoal enhances shoot regeneration from immature flower-stem sections (Bach 1988). The reported culture conditions vary. A culture temperature of 15–25 °C has been reported and the cultures were placed either in the light (500–4000 lx, 2–24 h/day) or in the dark.

The addition of 0.03–2 mg/l NAA (Hussey 1976a) or equal amounts of NAA and BA or kinetin (0.5–1 mg/l) is required (Fujino 1976; Hussey 1977; Meyer 1984) for optimal regeneration of shoots on axillary buds, scale pieces and flower-stem sections from bulbous iris. No growth regulators are necessary for regeneration on scale explants containing a piece of basal plate (Van der Linde et al. 1988). Fujino and co-workers (Fujino et al. 1973; Fujino 1976) found an increase in the number of shoots per explant on media containing a relatively high salt concentration or TIBA. Addition of GA_3 to the medium increased the size of the regenerated shoots. When GA_3 was the only regulator in the medium, the adventitious primordia developed into bulblets, which produced shoots thereafter. Regeneration of axillary shoots from explanted shoot apices of rhizomatous iris required the addition of 2.4 μM NAA and 30 μM BA (Laneri and Lucretti 1983).

2.1.2 Propagation of Shoots

For bulbous irises, Fujino et al. (1972) briefly mention the continuation of shoot development in sliced groups of adventitious shoots cultured on modified Kano medium containing BA. Hussey (1976b, 1977) describes the development of axillary shoots from isolated shoots, cultured at 20 °C and 16 h/day light on MS medium, containing 0.03–0.12 mg/l BA. He was able to obtain a propagation factor of 4 with a cycle of 6–8 weeks. Increasing the BA concentration to 2 mg/l or more resulted in greatly distorted and highly branched shoots. Van der Linde and co-workers induced axillary shoots in halved basal parts of shoots, cultured at 15 or 20 °C and 16 h/day light on MS medium containing 0.1 mg/l NAA and 0.3–0.5 mg/l BA. Quite similarly to Hussey's results, they obtained a propagation factor of 4.4 with a cycle time of 7 weeks (Van der Linde et al. 1987; Van der Linde and Hol 1988). Shoot propagation was slightly enhanced after the addition of activated charcoal to the medium (Bach 1988). Shoots of rhizomatous iris were propagated on Linsmaier and Skoog (1965) medium (LS), containing 0.5 mg/l of both NAA and kinetin, to which BA was added (Meyer 1984). No data on the propagation factor or culture conditions were given.

2.1.3 Propagation Through Callus

Callus cultures are established from apical meristems and excised axillary buds of bulbous iris on MS medium containing high auxin concentrations (2–5 mg/l NAA). These calli regenerate shoots spontaneously (Fujino 1976; Hirata and Kunishige 1982).

Shoot apices and zygotic embryos of rhizomatous irises develop embryogenic callus on LS and MS medium containing 2,4-D alone or in combination with kinetin (Reuther 1975, 1977a,b; Radojevic et al. 1987), 2,4-D and BA (Laneri and Lucretti 1983) or IAA and kinetin (Radojevic et al. 1987). Callus can be propagated in liquid medium containing IAA and kinetin (Reuther 1975). The somatic embryos germinate readily on solid LS or Knudson (1946) medium without growth regulators and develop into normal plantlets (Reuther 1975, 1977a,b; Laneri and Lucretti 1983). Radojevic et al. (1987) observed embryogenesis after the second subculture of the callus. The embryos were allowed to develop into plantlets in liquid medium containing GA_3, IAA, and IBA.

Apical meristems of rhizomatous iris develop organogenic callus on MS medium containing kinetin and IAA (Lunjova 1977). Callus is also obtained from stem, leaf, and root tissue of rhizomatous iris on medium containing kinetin, NAA, and 2,4-D (Marsavina et al. 1973). Meyer et al. (1975) describe the growth of callus from the peduncles of young inflorescences of rhizomatous iris in the dark at 20 °C on MS medium containing a high concentration of phosphate and organic nutrients, NAA, and kinetin. Rooted plantlets were obtained after cutting the callus, transferring it to medium without growth regulators, and subculturing in the light.

2.1.4 Bulblet Growth in Shoots

Various methods have been reported for induction of bulb growth in shoots. Bulblet growth in ca. 50% of the shoots on scale explants was obtained by interrupting the culture at 20 °C with a cold treatment of 4 weeks at 5 °C (Van der Linde et al. 1986, 1988). In shoots propagated at 15 °C, efficient bulblet growth was obtained after transfer to a low-auxin medium at 15 °C in the light for 6 weeks, followed by a cold treatment of 8 weeks 5 °C and another 3.5 months of culture at 20 °C in complete darkness. In this way, each shoot produced at least one bulblet with an average weight of 200 mg (Van der Linde and Hol 1988). In contrast,

Mielke and Anderson (1986) found that the best way to obtain bulblets from shoots is to culture them for 6 months at 25 °C in darkness without a pre-bulbing chilling period. Fujino (1976) was able to obtain bulbs from elongated shoots within 7 weeks after transferring them to medium containing 10 mg/l ancymidol. A cold treatment of 4 weeks at 3 °C "seemed to promote bulbing only slightly". Bach (1988) found that both the induction and the growth of bulblets in propagated shoots was stimulated by the addition of activated charcoal to the medium.

2.1.5 Transfer to Soil and Soil Performance

Either rooted shoots or bulblets obtained in vitro may be transferred to soil. Rooting of shoots is easily obtained on medium containing auxin and no cytokinin (Meyer et al. 1975; Fujino 1976; Hussey 1976a; Meyer 1984). Hirata and Kunishige (1982) transferred rooted shoots of bulbous iris to vermiculite and found 41–70% survival. All surviving shoots produced a bulb. The largest bulbs (13.5 mm diameter) were obtained from shoots planted in autumn. Hussey (1976a) mentions the transfer of rooted shoots to compost, but does not give details.

Rooted shoots of rhizomatous iris were transferred to a mixture of soil, peat, and perlite (1:1:1) and hardened under mist before transfer to a greenhouse. Although no data on the success of this method are given, no abnormalities were observed in the flowering plants (Meyer et al. 1975). In a later report, the transfer is described as "easy", and again all plants appeared to be true to type (Meyer 1984). The latter was already observed by Reuther (1977a,b) in flowering plants obtained from somatic embryos of rhizomatous iris and confirmed by Radojevic et al. (1987), who observed neither chromosome abnormalities nor changes in isoenzyme patterns of plants obtained by somatic embryogenesis from callus cultures of *I. pumila*.

Van der Linde and co-workers transferred in vitro-grown bulblets to soil. Prior to planting the dormancy of the bulblets was broken by dry storage at 30 °C for 6–10 weeks. The bulblets were planted in potting compost and cultured at 10–18 °C. After 9–14 days, 85–100% sprouting was observed (Van der Linde et al. 1986, 1987).

2.2 Virus Eradication

Reports on virus eradication by tissue culture deal with bulbous irises only. Several culture techniques have been reported for the production of virus-free plants. Baruch and Quak (1966), Mori et al. (1969), and Brunt (1983) used direct shoot regeneration from excised apical meristems. The same process was used by Blom-Barnhoorn and Van der Linde (1987) with excised adventitiously formed meristems on bulb-scale explants and by Allen and Anderson (1980) with basal-plate explants. Hirata and Kunishige (1982) used shoot regeneration from callus obtained from excised apical meristems.

It is difficult to give an indication of the percentage of surviving and growing meristems obtained by the various techniques, since most reports do not give details on this subject. Baruch and Quak (1966), in search for a good medium, obtained only 20 plantlets from 700 excised meristems and Mori et al. (1969) only one out of nine. On the other hand, Blom-Barnhoorn and Van der Linde (1987) obtained many bulblets by using MS medium with 0.1 or 1 mg/l NAA, a culture temperature of 15 or 20 °C and illumination for 16 h/day. The survival was 90–100% (data not shown).

The efficiency of virus eradication varies a lot. Baruch and Quak (1966), using an undefined serological method, reported that 80% of the plantlets were virus-free. A comparable efficiency (65%) was found by Brunt (1983), after selection with electron microscopy (EM) and enzyme-linked immunosorbent assay (ELISA). Only 40% virus-free plants were obtained after visual selection by Hirata and Kunishige (1982). Even lower values are reported by Allen Anderson (1980), who found 9–17% virus-free plants after selection with EM, ELISA, indicator hosts, and immunosorbent electron microscopy (ISEM). A similar low efficiency was reported by Blom-Barnhoorn and Van der Linde (1987), who found 10–26% virus-free plants after selection with EM and on symptoms during 3 years of culture in soil. These discrepancies may be explained by the different methods of tissue culture and selection, and by differences in the years of testing after transfer of the plants to soil.

2.3 Embryo Culture and Development

The early reports on embryo culture have been reviewed by Kidd (1966). Seeds from rhizomatous irises show a delayed germination (Randolph and Cox 1943; see Table 3) due to inhibitors present in the endosperm (Randolph and Cox 1943; Lenz 1954; Stolz 1968) and/or to embryo dormancy (Werckmeister 1952; Lenz 1955). To enhance germination Werckmeister (1936) experimented with the culture in vitro of mature embryos. He accelerated germination and also obtained germination of embryos from crossings between different groups of irises, which produced abnormal seed. In later reports, the technique was improved by applying a cold treatment to the embryos (Lenz 1955), and adjusting the nutrient composition (Randolph and Khan 1960; Shockey 1967; Yabuya and Yamagata 1981) and the agar concentration (Stolz 1971; Yabuya and Yamagata 1981) of the medium. To improve the transfer to soil, the effects of many growth regulators (Stolz 1977; Yabuya and Yamagata 1981; Yabuya et al. 1984) and the presence of the placenta (Yabuya et al. 1984) on the development of the embryos was studied.

Table 3. Germination percentage in time of seeds and excised mature embryos of rhizomatous iris in vitro. (Randolph and Cox 1943)

Time (days)	Seeds (%)	Embryos in contact with 1/2 endosperm (%)	Embryos (%)
3	0	21.7	57.2
5	0	21.7	84.0
6	0	41.3	94.6
41	18.7	45.7	94.6

2.4 Flower Induction in Excised Meristems

Large-bulbing irises flower when the bulbs have reached a certain size. However, when such bulbs are stored at 25–30 °C they remain vegetative (retarded) for a long time. They can be transferred to the generative state by a subsequent cold storage of several weeks at 13 °C. Due to this characteristic, apical meristems of large-bulbing iris constitute an ideal model system to study the transition of apices from the vegetative to the generative state. In a number of publications Rodrigues Pereira (1961, 1962, 1964, 1965, 1966) describes the culture and the development of shoot apices from retarded bulbs and the influence of temperature pretreatments, cocultivation with cold-treated scales, plant hormones and presence of primordial leaves and scales on the transition to the generative state. He concludes that during the cold storage chemicals are produced in the scale tissue, which are transported to the apical meristem and induce the transition to the generative state. He also showed evidence that these chemicals are probably gibberellins. His results were confirmed by Doss and Christian (1979). The promotive effect of gibberellin on the transition is inhibited by ABA and decanoic acid (Doss et al. 1983).

2.5 Other in Vitro Studies

Yabuya (1985) describes the regeneration of amphidiploid plants from excised embryos of *Iris laevigata* × *I. ensata* treated with colchicine. Werckmeister (1962) already obtained amphidiploid plants by this method but no details were given. Sevruk et al. (1975) describe the growth in correlation with the mitotic index of callus and suspension cultured cells of *Iris sibirica*. Aslanyants et al. (1988) describe the production of essential oil from cell cultures of *Iris sibirica*. They reduced the costs of production by 70% by applying a hydrolysate of protein-vitamin concentrate instead of casein hydrolysate. Recently, the same group also reported on the bioconversion of terpenoids in the same cultures (Bagdasarova et al. 1989) Zhou (1988, 1989) reports on the isolation and culture of pollen protoplasts from *Iris tectorum*.

3 Micropropagation of Bulbous Iris

3.1 General

Micropropation of bulbous iris consists of four steps:

1. Regeneration of adventitious shoots on bulb-scale explants.
2. Propagation of shoots by axillary bud formation.
3. Induction and growth of bulblets in propagated shoots.
4. Breaking of dormancy and transfer to soil.

For a working protocol the reader is referred to Section 5 of this chapter. In the following paragraphs important features of each step in the protocol are discussed.

3.2 Regeneration of Adventitious Shoots on Bulb-Scale Explants

Many factors influence the regeneration of shoots on bulb-scale explants, viz., the decontamination period, explant size and preparation, salt concentration of the nutrient medium, culture temperature, and light. A hypochlorite treatment of halved bulbs (Figs. 3 and 4A) is necessary to decontaminate the tissue from fungi, e.g., *Penicillium, Aspergillus*, and *Rhizopus*. When halved bulbs are immersed in 1% (w/v) sodium hypochlorite for 30 min and subsequently rinsed with sterile water, contamination is reduced to 15% without loss in regeneration capacity. Extending the decontamination period to 60 min does not further reduce contamination and affects the regeneration capacity (Van der Linde et al. 1988; Fig. 4A). Before cutting the explants, it is important to remove the hypochlorite-infiltrated tissue along the cut surfaces, i.e., the basal plate and the longitudinal cut (Fig. 2). Explants containing hypochlorite-infiltrated tissue show a strongly reduced regeneration capacity (Fig. 4B).

After removal of the hypochlorite-infiltrated tissue, single outer-scale and double inner-scale explants are cut (Figs. 2 and 3). From a contamination point of view, small explants are preferred, but for iris only large explants show a high regeneration capacity. Cutting halved bulbs into more than three sections or reducing the height of the scales reduces the capacity of the explants to regenerate shoots (Van der Linde et al. 1988; Fig. 4C). Therefore, not more than 12 explants should be cut from a single bulb with a circumference of 9 cm or more.

Care should be taken not to break the explants apart but to cut them carefully and to leave a piece of the basal, abaxial site of the second scale attached to the outer-scale explant (Fig. 2). Breaking the scales apart instead of cutting reduced the

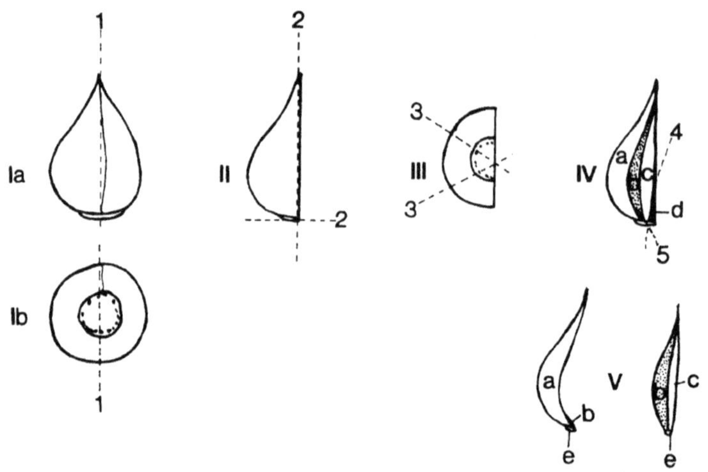

Fig. 2. Preparation of scale explants. Bulb after removal of dry tunics and roots showing mark where outer scale grew together (*Ia*); bottom view (*Ib*); lateral view of halved bulb (*II*); bottom view of halved bulb (*III*); lateral view of bulb section (*IV*); lateral view of single outer- and double inner-scale explant (*V*). *1–5* (*broken lines*) represent successive cuts (for explanation see text). *a* outer scale; *b* milddle scale; *c* inner scale; *d* remainder of shoot apex; *e* basal plate

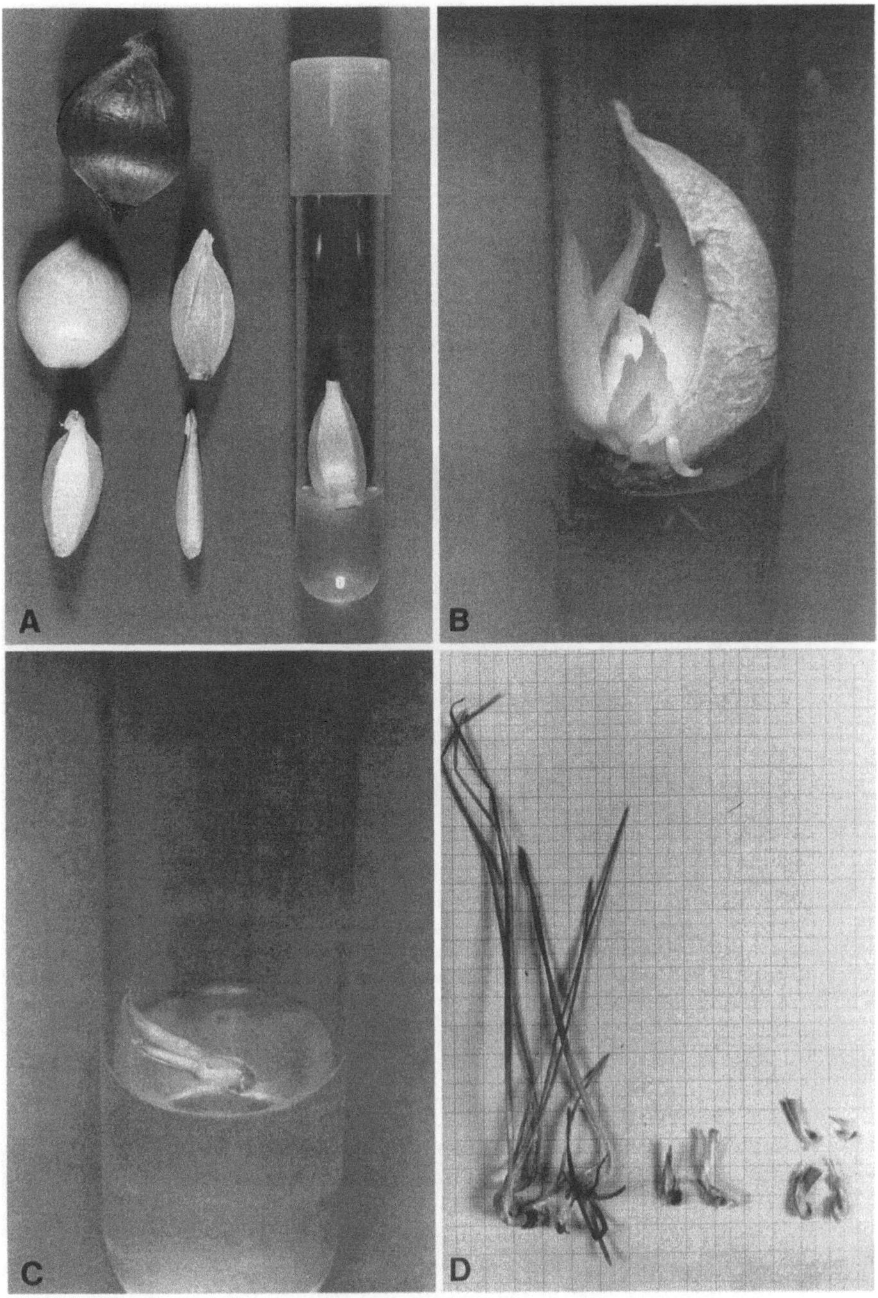

Fig. 3A–D. Caption see p. 186

Fig. 3A–F. Successive stages of propagation in vitro of bulbous iris cv. Prof. Blaauw. Explant preparation (**A**), adventitiously formed shoots on single-outer scale explant after 9 weeks of culture at 20 °C in darkness (**B**), halved basal part of shoot from **B** (**C**), shoot propagation by axillary bud regeneration: 7-week-old shoots from **C** (**D**, *left*), basal parts of shoots (**D** , *middle*), halved basal parts (**D**, *right*), bulblets from shoot propagated at 15 (**E**, *left*) and 20 °C (**E**, *right*) after 14 weeks of culture at 20 °C in darkness, soil performance: bulbs planted in March (**F**, *left*) and bulbs harvested in November (**F**, *right*)

regeneration capacity of single outer-scale explants from 96 to 78%. When this kind of explant was cut without a piece of the second scale, its regeneration capacity was reduced from 96 to 84%.

The scale explants are cultured at 15 or 20 °C in the dark. At 25 °C no regeneration occurs (Fig. 5A). Illumination of the explants for 16 h/day with 30 $\mu E/m^2/s$ from the start of the culture also reduces the regeneration capacity (Van der Linde et al. 1988). However, when the light was switched on after 2 weeks of culture, no decrease in regeneration capacity was observed (Fig. 5B) and the shoots developed more homogeneously.

The number of shoots formed per scale explant is affected by the salt concentration in the medium, by the presence of axillary buds or a part of the apical meristem, and by the addition of plant growth regulators to the medium. When a remainder of the apical meristem or an axillary bud was present on the scale explant, only one large shoot developed and no adventitious shoots were formed (data not shown). Removal of the buds and the apical meristem allows the

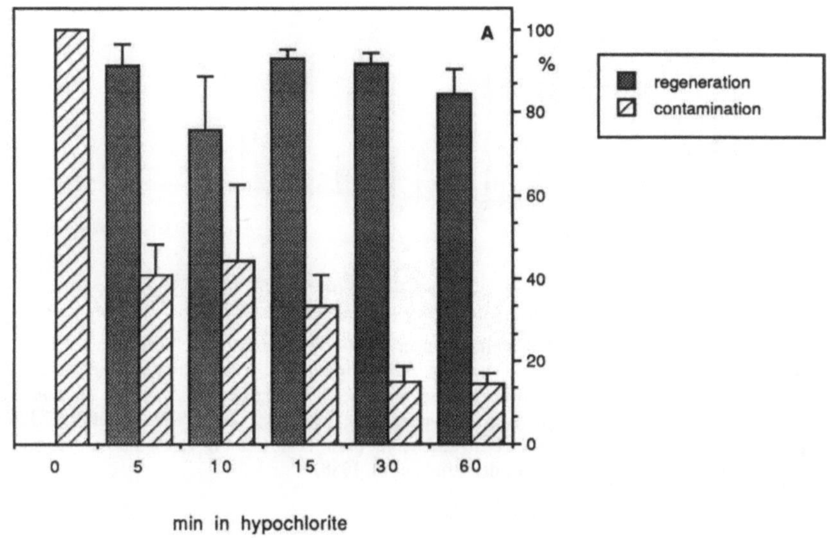

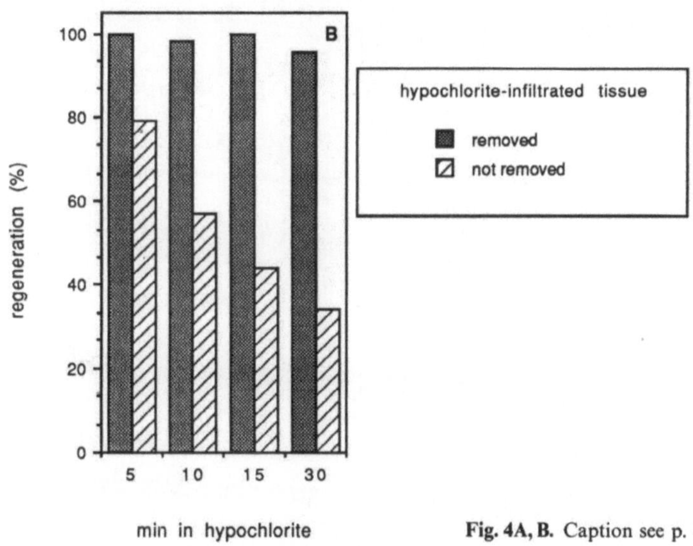

Fig. 4A, B. Caption see p. 188

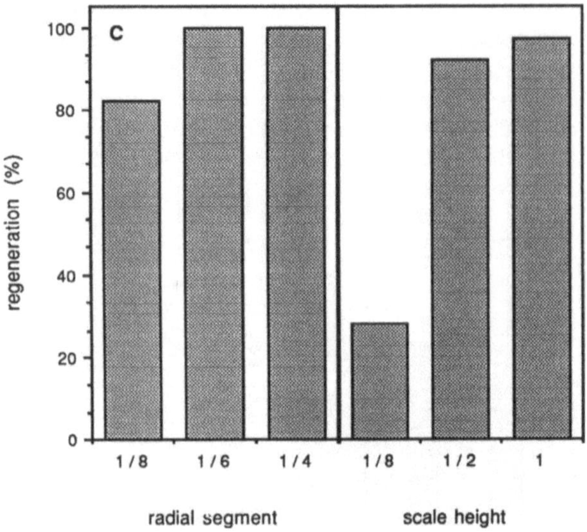

Fig. 4A–C. **A** Effect of duration of immersion in 1% (w/v) sodium hypochlorite on regeneration of shoots and on contamination in scale explants. *Bars* represent SE. (After Van der Linde et al. 1988). **B** Effect of removal of hypochlorite-infiltrated tissue on the shoot regeneration capacity of scale explants as influenced by the time of immersion in hypochlorite. (Van der Linde et al., unpubl). **C** Effects of explant size, viz., radial segment of the bulb and height of the scale (see Fig. 2), on the shoot regeneration capacity of scale explants. (After Van der Linde et al. 1988)

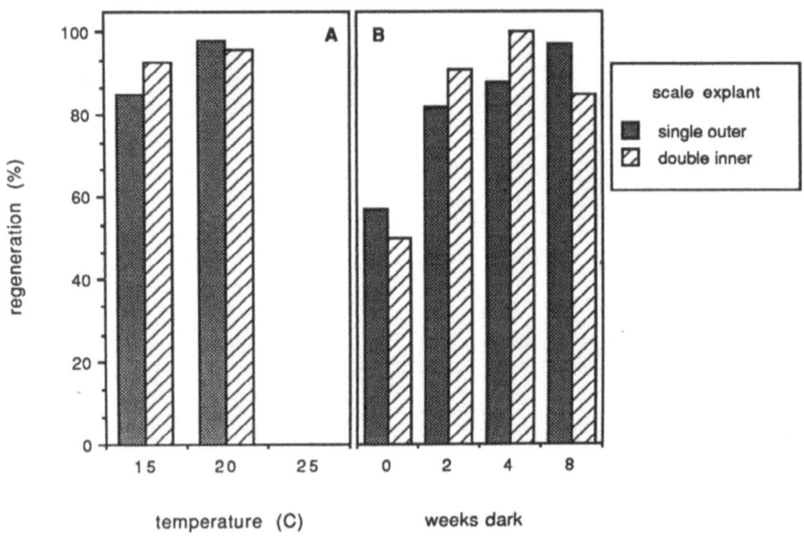

Fig. 5. **A** Effect of culture temperature on the shoot-regeneration capacity of scale explants. (After Van der Linde et al. 1988). **B** Effect of light 30 $\mu E/m^2/s$) given at different times after starting the culture on the shoot-regeneration capacity of scale explants. (After Van der Linde et al. 1988)

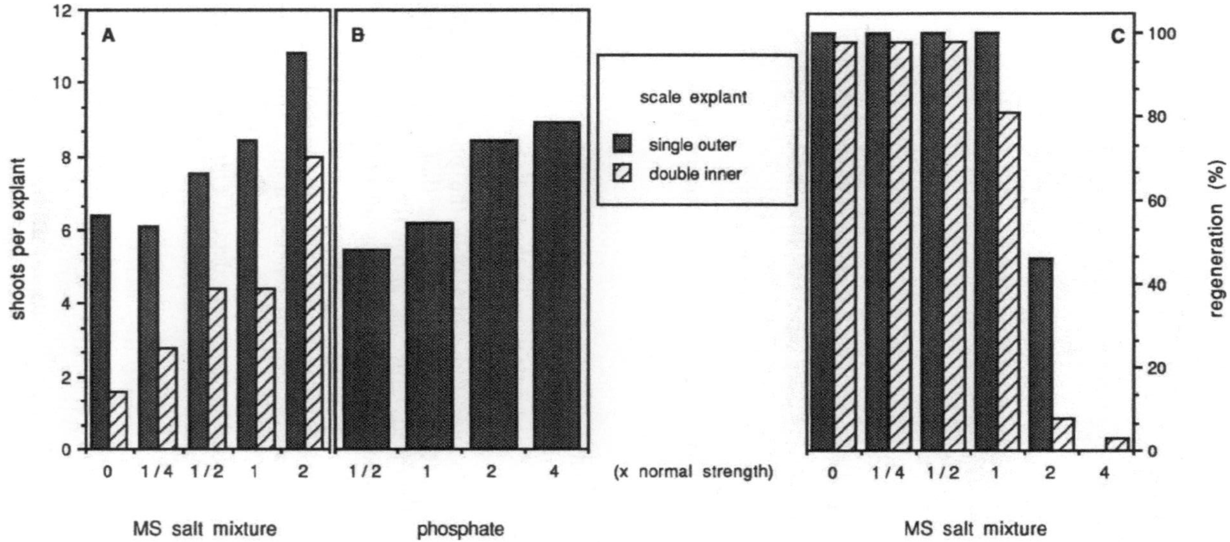

Fig. 6A–C. **A** Effect of the concentration of Murashige and Skoog salt mixture on the number of adventitious shoots per scale explant (Van der Linde et al., unpubl.). **B** Effect of the phosphate concentration (KH_2PO_4) on the number of adventitious shoots per scale explant. (Van der Linde et al., unpubl). **C** Effect of the concentration of Murashige and Skoog salt mixture on the shoot-regeneration capacity of scale explants. (After Van der Linde et al. 1988)

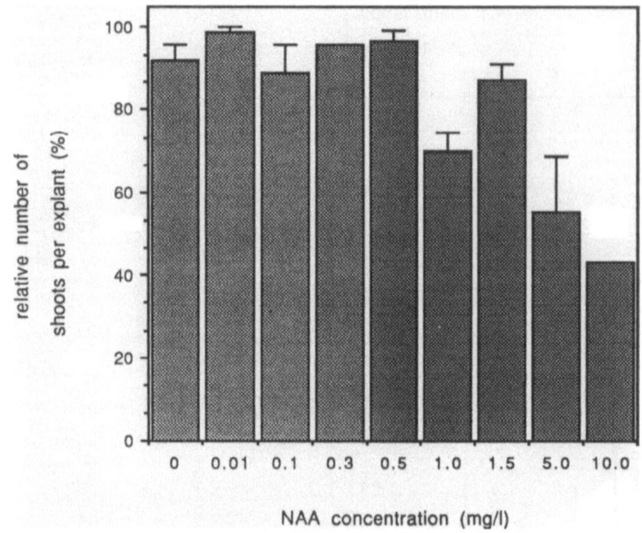

Fig. 7. Effect of the NAA concentration on the relative number of shoots per explant. *Bars* represent SE. (Van Aartrijk and Van der Linde, unpubl.)

regeneration of adventitious shoots from the junction of the scale and the basal plate (Fig. 3B) even on medium containing only agar and water.

Increasing the salt concentration in the medium also increased the number of adventitious shoots per scale explant, especially on double-inner scales (Fig. 6A). The same effect was observed when only the phosphate concentration of the medium was increased (Fig. 6B), indicating that the buffering capacity of the medium may be involved. However, above the normal concentration of MS salts the regeneration capacity of the scales is affected (Van der Linde et al. 1988; Fig. 6C).

The addition of a low NAA concentration to the medium did not influence the mean number of shoots per explant, whereas a high NAA concentration reduced this number (Fig. 7). Addition of BA had no effect on the mean number of shoots per explant, but inhibited the negative effect of a high NAA concentration (data not shown).

3.3 Propagation of Shoots by Axillary Bud Formation

After ca. 9 weeks of culture the shoots are large enough to be used for propagation. The shoots are cut from the explants and from each other, and the basal part of each shoot (1.5–2.5 cm) is collected. This part contains the meristem and one or several leaves in successive stages of development. The basal part is cut longitudinally and both pieces, containing a part of the meristem, are grafted on 5 ml MS medium (Fig. 3C), containing 0.1 mg/l NAA and 0.5 mg/l BA, and cultured at 15 or 20 °C with 30 $\mu E/m^2/s$ illumination for 16 h/day. After 7 weeks ca. two new shoots have developed from each halved shoot, which can be subcultured according to the same procedure (Fig. 3D). In this way a propagation factor of 4.4 was obtained over a period of six subcultures. No change of the propagation factor was observed during a period of 1.5 year (Van der Linde and Hol 1988).

3.4 Induction and Growth of Bulblets in Propagated Shoots

Van der Linde and Van Aartrijk (1987) list conditions for optimal growth of bulblets in vitro. These include a relatively low salt, high sucrose, and low auxin concentration, a relatively low pH and high temperature, and darkness. When 7-week-old shoots propagated at 15 or 20 °C were transferred to bulb-induction medium containing half-strength MS-salt mixture, 30 g/l sucrose, 0.03 mg/l NAA, pH 5.5 and cultured at 20 °C in darkness, bulblets developed within 3 months of culture. However, only for shoots propagated at 15 °C was a high percentage of bulbing observed (100% versus 72% for shoots propagated at 20 °C). Furthermore, the bulblets had the tendency to sprout in vitro before they were fully developed.

To prevent sprouting in vitro, a cold treatment of 8 weeks 5 °C is given to the shoots after 6 weeks of culture on bulb-induction medium at the propagation temperature. After the cold treatment they are cultured at 20 °C in the dark. In this way dormancy is induced in the bulblets, which is seen by the development of a brown tunic after ca. 3.5 months of culture at 20 °C in darkness. This brown tunic is only developed by bulblets from shoots propagated at 15 °C (Fig. 3E). With this procedure more than one bulblet with an average weight of ca. 200 mg is produced from each shoot (Van der Linde and Hol 1988).

3.5 Breaking of Dormancy and Transfer to Soil

After 3.5 months of culture at 20 °C in darkness, the bulblets were harvested. Bulblets with a weight exceeding 100 mg (ca. 80%) were stored in a gauze sack at 30 °C for 6–8 weeks to enhance sprouting. After this treatment, they were planted in potting compost at a depth of ca. 5 cm and cultured at 10–18 °C. The best planting time was the end of March and beginning of April. After 9 weeks ca. 90% of the bulblets had sprouted (Fig. 8). Large bulblets sprouted earlier than small ones. Bulblets exceeding a weight of 750 mg developed two leaves. At the end of the culture, bulbs with an average weight of ca. 3 g were harvested (Fig. 3F). The ratio of number of planted tissue-cultured bulbs over number of harvested bulbs was ca. 1.7.

Due to the time-consuming period of bulb induction, bulb growth, and breaking of dormancy, it is tempting to transfer rooted shoots to soil. However, in our experience, the number of surviving shoots varied from 50 to 85%. Many of the shoots produced more than two leaves, which is quite remarkable for young bulbous irises. Even 5 to 11 leaves were observed (Fig. 9A). By the time of harvest in November the cultures contained normal dormant bulbs, but also shoots still producing new leaves and without a bulb at the base, sprouted bulbs already starting a new growing period, and bulbs containing shoots at the base, which drained the bulbs resources (Fig. 9B-D). These abnormalities, which were never

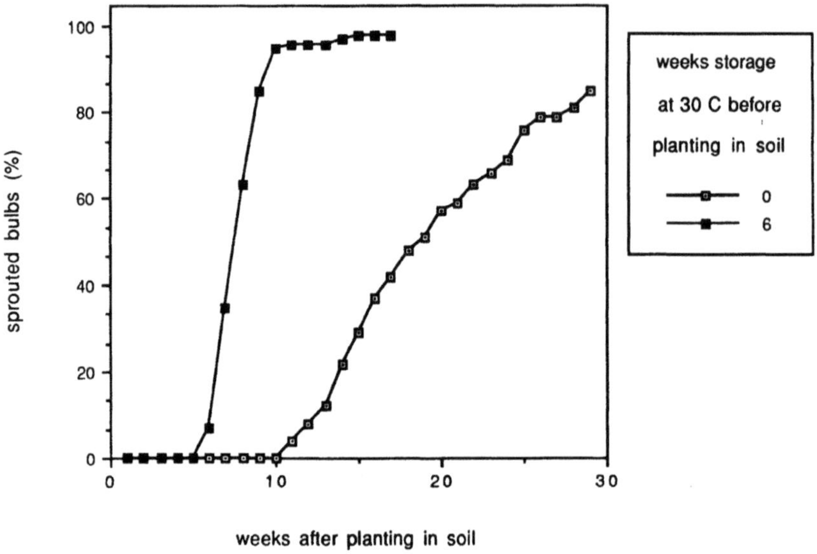

Fig. 8. Influence of 6 week's dry storage at 30 °C on the sprouting in soil of bulblets produced in vitro. (Van der Linde and Van Aartrijk, unpubl.)

observed when bulbs were planted, reduced the ratio of number of planted rooted shoots over number of harvested bulbs to 0.6.

4 Summary and Conclusions

Aseptic culture of iris tissue has been described since 1936. The main application described by the early reports was the acceleration of germination in seeds of rhizomatous irises. A new impulse to tissue-culture research was given, when virus-free bulbous irises were obtained from meristem-tip cultures. The virus-free stocks proved to be of superior quality to the virus-infected ones. Since reinfection with viruses in the field occurs rapidly and the natural propagation is very slow, the need for propagation in vitro became clear.

This chapter describes data from literature and from our own research on shoot regeneration and propagation in vitro, as well as on the transfer of propagules to soil. A protocol is presented for micropropagation of bulbous iris, which has been adopted by Dutch commercial tissue culture laboratories in recent years. It is expected that micropropagation of bulbous iris, as a first step in bulking up, will increase in proportion to the increase in cultivated virus-free stocks. However, this still leaves the need for a chipping method as a second step in bulking up.

Fig. 9A–D. Abnormalities observed after planting rooted shoots of bulbous iris produced in vitro: plants with more than three leaves (**A**); bulbs with shoots at the base (**B**); bulbs sprouting at harvest (**C**); Plants without a bulb and still producing new leaves at harvest (**D**). (Van der Linde et al., unpubl.)

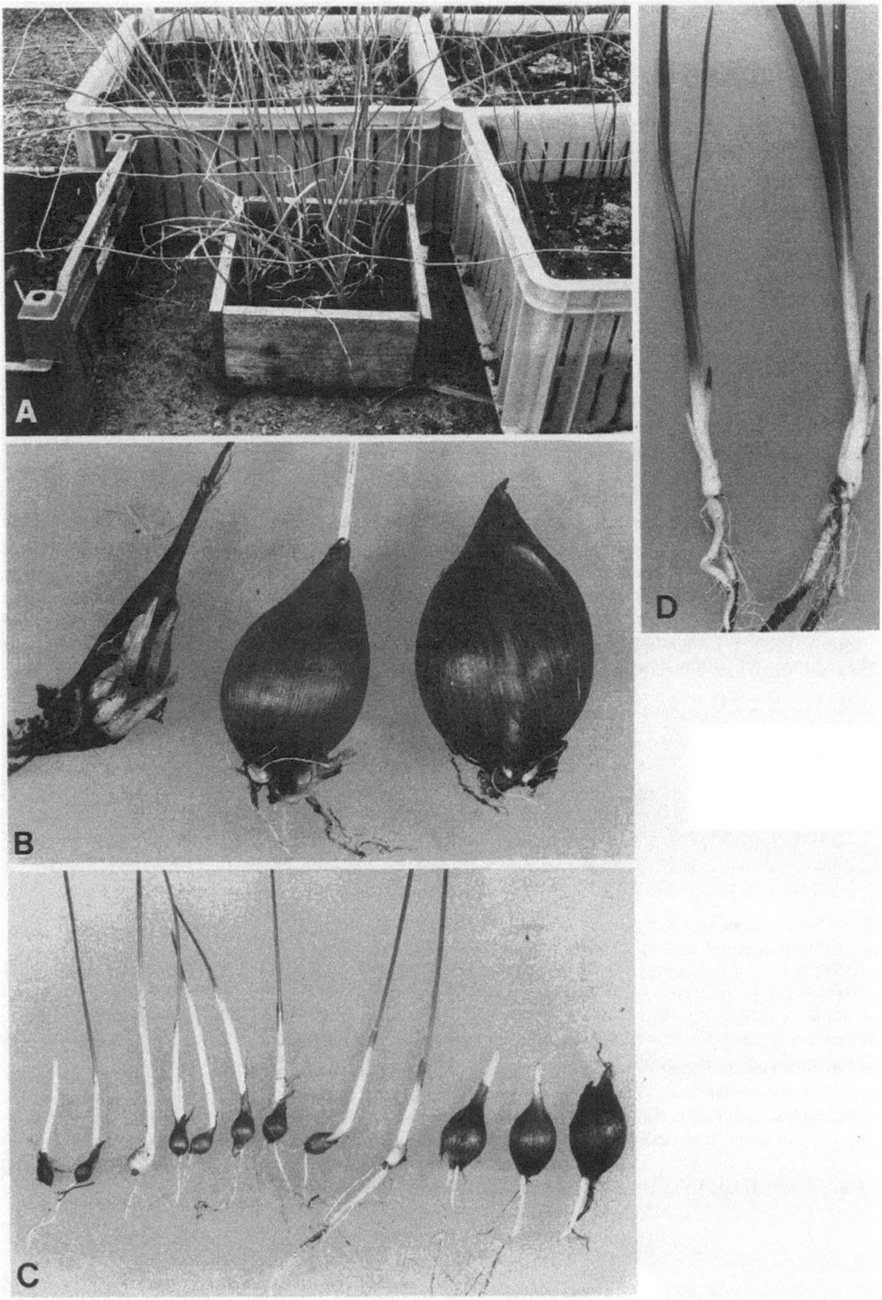
A
B
C
D

5 Protocol

Plant Material. Field-grown bulbs are harvested, dried, and stored at 30 °C. They can be stored for up to 1 year without any loss in regenerative capacity of the explants.

Explant Preparation and Culture. Bulbs with a circumference of 9 cm or more are selected. The dry tunics and root primordia are removed. Damaged bulbs are discarded. The bulbs are sliced longitudinally to give two equal parts (Fig. 2, cut 1). The half bulbs are decontaminated in 1% sodium hypochlorite in water for 30 min and rinsed three times with sterile water. All subsequent actions are performed in a laminar air-flow cabinet. The half bulbs are allowed to dry on sterilized filter paper. The hypochlorite-infiltrated tissue along the longitudinal cut and along the bulb base is removed (Fig. 2, cuts 2). The half bulb is cut into three equal-sized sections (Fig. 2, cut 3). The sections contain three scales and a piece of the apical meristem held together by the basal plate. The remainder of the apical meristem is removed (Fig. 2, cut 4). Subsequently, an outer-scale and a double-inner scale explant are cut (Fig. 2, cut 5). In this way 12 explants are cut from the single bulb. The explants are grafted in upright position on 14 ml medium Murashige and Skoog (1962) macro- and micronutrients, 0.4 mg/l thiamine, 100 mg/l myo-inositol, 30 g/l sucrose, 10 g/l Bacto-agar (Difco), pH = 6) in glass culture tubes (25 × 150 mm) and cultured at 15 °C in the dark. After 4 weeks the cultures are illuminated for 16 h/day with 30 $\mu E/m^2/s$ cool fluorescent white light (Philips TL-33). After 9 weeks of culture ca. ten shoots per explant are formed.

Propagation of Shoots. The shoots are cut from the explants and from each other. The basal part of each shoot (1.5–2.5 cm) is collected and carefully cut longitudinally in order to slice the meristem of the shoot into two equal parts. Each half shoot is grafted in upright position on 5 ml medium (Murashige and Skoog macro- and micronutrients, 0.4mg/l thiamine, 100 mg/l myo-inositol, 30 g/l sucrose, 6 g/l agar, 0.1 mg/l 1-naphthaleneacetic acid, 0.5 mg/l benzylaminopurine, pH = 6) in glass culture tubes and cultured at 15 °C under 30 $\mu E/m^2/s$ for 16 h/day. After 7 weeks each half shoot has formed one to four new shoots, which are subcultured again. This procedure can be repeated at least ten times without change in shoot production.

Bulblet Growth in Propagated Shoots. For the production of bulblets from propagated shoots. 7-week-old shoots are transferred to 14 ml medium (half-strength concentration of Murashige and Skoog macro- and micronutrients, 0.4 mg/l thiamine, 100 mg/l myo-inositol, 30 g/l sucrose, 0.03 mg/l 1-naphthaleneacetic acid, 6 g/l agar, pH = 5.5) in glass culture tubes and cultured for 6 weeks at 15 °C under 30 $\mu E/m^2/s$ for 16 h/day. After this period the shoots are placed at 5 °C in the dark for 8 weeks and subsequently cultured at 20 °C in complete darkness for 3–4 months. By that time the bulblets have developed a yellow-brown tunic. Each shoot gives rise to one to two bulblets with an average weight of 250 mg.

Breaking of Dormancy and Transfer to Soil. Tunicated bulblets of 100 mg or more are collected. They are stored in a guaze sack at 30 °C for 6–8 weeks. The bulblets are planted in potting compost in beds at a depth of 5 cm. For outside culture the best planting time is the end of March to the beginning of April. After 4–8 weeks 95–100% of the bulblets have sprouted. Bulblets with a weight exceeding 750 mg may produce more than one leaf.

Acknowledgments. The authors owe gratitude to ir. A.F.L.M. Derks, ir. M. Benschop, dr. R.J. Bogers, dr. G.J. de Klerk, and drs. A.H.T. te Boekhorst for critically reading and improving the manuscript, to dr. Z. Kulescha for providing summaries of the papers in Russian, to dr. W.J. de Munk for providing Fig. 1A and to H. Dijkman for preparing the photographs.

References

Alkema HY (1971) Nieuwe vermeerderingsmethoden bij bolgewassen. Bloembollencultuur 82: 1211–1212

Allen TC, Anderson WC (1980) Production of virus-free ornamental plants in tissue culture. Acta Hortic 51: 245–249

Alper M, Loebenstein G (1981) Bean yellow mosaic virus in bulbous irises in Israel. Plant Dis 65: 694–695

Anonymous (1987) Virusvrije iris gevoelig voor herbesmetting. Bloembollencultuur 98(33): 6

Asjes CJ (1979) Viruses and virus diseases in Dutch bulbous iris (*Iris hollandica*) in the Netherlands. Neth J Plant Pathol 85: 269–279

Asjes CJ (1988) Still having to live with viruses. Grower 109(6): 28–31

Asjes CJ (1989a) Bestrijding virusverspreiding in lelies. 1. Gevoeligheid voor virus varieert sterk. Bloembollencultuur 100(7): 34–35

Asjes CJ (1989b) Bestrijding virusverspreiding in lelies. 2. Veel factoren spelen een rol. Bloembollencultuur 100(8): 30–31

Aslanyants LK, Marsavina ZV, Kazaryan AG (1988) Productivity of *Iris sibirica* L. cell culture grown on simplified nutrient medium. Rastit Resur 24: 107–110

Bach A (1988) Effect of activated charcoal on in vitro propagation of *Iris* × *hollandica*. Bull Pol Acad Sci Biol Sci 36: 107–111

Bagdasarova RM, Aslanyants LK, Uzunyan LV (1989) Bioconversion of terpenoids by iris (*Iris sibirica*) cell cultures. Biochem Microbiol 24: 638–642

Baker JG (1892) Handbook of the Irideae, George Bell & Sons, London, New York

Baruch ER, Quak F (1966) Virus-free plants of iris Wedgewood obtained by meristem culture. Neth J Plant Pathol 72: 270–273

Bergman BHH, Eijkman AJ, Muller PJ, Van Slogteren DHM, Weststeijn G (1978) Ziekten en afwijkingen bij bloembolgewassen. 2. Amaryllidaceae, Iridaceae, e.a., Laboratorium voor Bloembollenonderzoek, Lisse

Blom-Barnhoorn GJ, Van der Linde PCG (1987) Iris virusvrij vermeerderd. Bloembollencultuur 98:14–15

Bruhn C (1955) Untersuchungen über die *Fusarium*-Krankheit der Gladiolen. Phytopathol Z 25: 1–38

Brunt AA (1983) Recent progress in the production, propagation and distribution of virus-free flower bulbs. Annu Rep Glasshouse Crops Res Inst Littlehampton, 109–116

Derks AFLM, Hollinger TC, Vink-van den Abeele JL (1985) Identification and symptom expression of four elongated viruses infecting bulbous irises. Acta Hortic 164: 309–317

Doss RP (1979) Some aspects of daughter bulb growth and development and apical dominance in bulbous *Iris*. Plant Cell Physiol 20: 387–394

Doss RP, Christian JK (1979) Relationships between bulb size, apex size and flowering in bulbous *Iris* cv. Ideal. Physiol Plant 54: 215–218

Doss RP, Kimura Y, Christian JK (1983) Endogenous levels of abscisic acid and decanoic acid in Dutch iris bulbs and the influence of abscisic acid and decanoic acid on iris meristems cultured in vitro. Plant Physiol 72: 713–716

Dykes WR (1913) The genus *Iris*. Cambridge Univ Press, Cambridge

Eikelboom W, Van Eijk JP (1987) Veredelingsonderzoek aan boliris, gericht op forceerbaarheid en kleur, 1964–1988. Inst voor Veredeling van Tuinbouwgewassen, Rapp 239, Wageningen, The Netherlands

Farrington BB (1961) Simplified plate culture of iris seed. Bull Am Iris Soc 160: 43–44

Farrington BB (1969) Laboratory culture of iris seeds and their embryos. Bull Am Iris Soc 195: 45–50

Fujino M (1976) Multiplication of Dutch iris by organ culture. Thesis Hyogo Pref Agric Exp Stn 84 pp

Fujino M, Fujimara T, Hamada K (1972) Multiplication of Dutch iris (*Iris hollandica*) by organ culture. 1. Effect of growth regulators, culture media, pH, peptone, and agar on the growth and differentiation of excised lateral buds. J Jpn Soc Hortic Sci 41: 66–71

Fujino M, Fujimara T, Hamada K (1973) Multiplication of Dutch iris (*Iris hallandica*) by organ culture. 2. Effect of growth regulators on the organ formation of excised lateral buds. *Bull* Hyogo Pref Agric Exp Stn 20: 119–124

Grey CH (1937) Hardy bulbs vol. 1. Iridaceae. Williams and Morgate Ltd, London

Hammond J, Brunt AA, Derks AFLM, Inouye N, Barnett OW, Allen TC, Lawson RH (1985) Viruses infecting bulbous iris: a clarification of nomenclature. Acta Hortic 164: 395–397

Hirata Y, Kunishige M (1982) Studies on the virus disease of bulbous iris (*Iris hollandica* H.): obtaining and multiplication of virus-free bulbs through tissue culture, and bulb production by using it. Bull Veg Ornamental Crops Res Stn Ser C (Kurume) 6: 83–99
Holden J (1962) Embryo and seedling culture of bearded iris. Aril Soc Int, YB: 16–20
Hussey G (1976a) Propagation of Dutch iris by tissue culture. Sci Hortic 4: 163–165
Hussey G (1976b) In-vitro release of auxillary shoots from apical dominance in monocotyledonous plantlets. Ann Bot 40: 1323–1325
Hussey G (1977) In-vitro propagation in some members of the Liliaceae, Iridaceae, and Amaryllidaceae. Acta Hortic 78: 303–309
Jones SK, Hanks GR (1988) Bulking up bulbous iris. Plantsman 9: 247–251
Kamerman W (1986) Some factors affecting the incidence of bacterial soft rot in bulbous iris. Acta Hortic 177: 509–512
Kidd KK (1966) A review of factors in iris seed germination. Bull Am Iris Soc 181: 14–22
Knudson L (1946) A new nutrient solution for the germination of orchid seeds. Bull Am Orchid Soc 15: 214–217
Köhlein F (1987) Iris. Timber Press, Portland, Oregon
Koster ATJ, de Rooij M (1979) Chemical control of root rot (*Pythium* spp.) in bulb cultivation by soil disinfection. Med Fac Landbouww Rijksuniv Gent 44: 537–543
Kromer KD (1988) Regeneration of some monocotyledonous plants from subterranean organs in vitro. Acta Agrobot 38: 65–87
Laneri U, Lucretti S (1983) Tissue culture of tall bearded iris: results and perspectives. Genet Agrar 37(1–2): 178–179
Lawrence GHM (1953) A reclassification of the genus *Iris*. Gentes Herb 8: 346–371
Lenz LW (1954) The endosperm as a barrier to intersectional hybridization in iris. Aliso 3: 57–58
Lenz LW (1955) Studies on iris embryo culture. Aliso 3: 173–182
Linsmaier E, Skoog F (1965) Organic growth factor requirements of tobacco tissue cultures. Physiol Plant 18: 100–127
Lunjova LS (1977) Propagation of iris (Iridaceae) through apical meristem culture in vitro. Bot Zh 62(3): 416–421
Marsavina ZV, Sevruk OG, Serdakova EH (1973) Cultivation of isolated tissue of *Pelargonium roseum* and *Iris sibirica*. Iswestija Akad Nauk 26(6): 13–18
Mathew B (1981) The Iris. BT Badsford Ltd, London
Meyer MM (1984) In-vitro propagation of siberian iris from flower stalks. HortScience 19: 575
Meyer MM, Fuchigama LH, Roberts AN (1975) Propagation of tall bearded irises by tissue culture. HortScience 10: 479–480
Mielke KA, Anderson WC (1986) In-vitro bulbing of bulbous iris. HortScience 21: 774
Mori K, Hamaya E, Shimomura T, Ikegami Y (1969) Production of virus-free plants by means of meristem culture. J Cent Agric Exp Stn 13: 45–111
Murashige T, Skoog F (1962) A revised medium for rapid growth and bioassays with tobacco tissue cultures. Physiol Plant 15: 473–497
Radojevic L, Sokic O, Tucic B (1987) Somatic embryogenesis in tissue culture of iris (*Iris pumila* L.). Acta Hortic 212: 719–723
Randolph LF (1945) Embryo seed culture. Bull Am Iris Soc 97: 33–45
Randolph LF, Cox LG (1943) Factors influencing the germination of iris seed and the relation of inhibiting substances to embryo dormancy. Proc Am Soc Hortic Sci 43: 284–300
Randolph LF, Khan R (1960) Growth response of excised mature embryos of iris and wheat to different culture media. Phytomorphology 10: 43–49
Randolph LF, Randolph FR (1955) Embryo culture of iris seed. Bull Am Iris Soc 139: 7–17
Reuther G (1975) Induktion der Embryogenese in Kalluskulturen. Planta Med: 42–58 (Suppl)
Reuther G (1977a) Embryoide Differenzierungsmuster im Kallus der Gattungen *Iris* und *Asparagus*. Ber Dtsch Bot Ges 90: 417–437
Reuther G (1977b) Adventitious organ formation and somatic embryogenesis in callus of *Asparagus* and *Iris* and its possible application. Acta Hortic 78: 217–224
Rodionenko GI (1961) Rod. Iris-*Iris* L., Woprosy morfologu, biologu, ewoljuou i sistematiki. Moskwa, Leningrad (Transl by the Br Iris Soc London, 1987)
Rodrigues Pereira AS (1961) Flower initiation in excised stem disks of Wedgwood iris. Science 134: 2044–2045

Rodrigues Pereira AS (1962) Physiological experiments in connection with flower formation in Wedgwood iris (*Iris* cv. Wedgwood). Acta Bot Neerl 11: 97–138
Rodrigues Pereira AS (1964) Endogenous growth factors and flower formation in Wedgwood iris bulbs. Acta Bot Neerl 13: 302–321
Rodrigues Pereira AS (1965) Physiological analysis of flower formation in Wedgwood iris. J Exp Bot 16: 405–410
Rodrigues Pereira AS (1966) Effect of 2-thiouracil and gibberellic acid on flower formation in Wedgwood iris. Acta Bot Neerl 15: 215–223
Saaltink GJ (1968) *Penicillium corymbiferum* entering bulbous iris through wounds. Neth J Plant Pathol 74: 85–93
Schipper JA, Van der Weijden JA (1984) Investigations on the technique for growing and forcing of irises. Jaarversl Lab Bloembollenonderz, Lisse
Sevruk OG, Marsavina ZV, Pogos AZ, Gasparon AV (1975) Some cytological peculiarities of callus and suspended tissue culture of iris (*Iris sibirica*). Biol Zh Arm 28(8): 11–15
Shockey HG (1959) Culture of Regelia embryos. Bull Am Iris Soc 153:25–26
Sonderman CH, MacLean NA (1949) Rhizoctonia neck and bulb rot of iris in the Pacific Northwest. Phytopathology 39: 174–175
Stolz LP (1967) Simplified iris embryo culture. Bull Am Iris soc 187: 90–98
Stolz LP (1968) Iris seed dormancy. Physiol Plant 21: 1328–1331
Stolz LP (1971) Agar restriction of the growth of excised mature iris embryos J Am Soc Hortic Sci 96: 681–684
Stolz LP (1977) Growth regulator effects on growth and development of excised mature iris embryos in vitro. HortScience 12: 495–496
Swenson KG (1971) Oil sprays prevent virus spread. N AM Lily Soc Yb 24: 7–10
Tamberg T (1979) The embryo cutting method for seed germination. Iris Yb (Br Iris Soc): 93–95
Van der Linde PCG, Hol GMGM (1988) Weefselkweek iris: vermeerdering sterk verbeterd. Vakwerk 62(50): 11, 15
Van der Linde PCG, Van Aartrijk J (1987) Micropropagation of flower-bulb crops. In Alderson PG, Dullforce WM (eds) Micropropagation in horticulture. Practice and commercial problems. Univ Nottingham, School of Agriculture, pp 123–134
Van der Linde PCG, Blom-Barnhoorn GJ, Van Aartrijk J (1986) Towards in-vitro propagation of bulbous iris. Acta Hortic 177: 569
Van der Linde PCG, Blom-Barnhoorn GJ, Hol GMGM (1987) Aanzienlijke verbetering methode. Een irisbol kan 500 tot 1500 bollen produceren. Bloembollencultuur 98(11): 14–15
Van der Linde PCG, Hol GMGM, Blom-Barnhoorn GJ, Van Aartrijk J, De Klerk GJ (1988) In-vitro propagation of *Iris hollandica* Tub. cv. Prof. Blaauw. 1. Regeneration on bulb-scale explants. Acta Hortic 226: 121–128
Weiler J, Emershad R (1977) Tissue culture for iris hybridizers. Bull Am Iris Soc 58: 36–39
Werckmeister P (1936) Über die Herstellung und künstliche Aufzucht von Bastarden Gattung *Iris*. Gartenbauwissenschaft 10: 500–520
Werckmeister P (1952) Embryokulturversuch zur Frage des Keimversuches von Irissamen aus der Sektion Regelia Forster. Ber Dtsch Bot Ges 65: 321–325
Werckmeister P (1956) Practical methods and problems of iris embryo culture. Iris Yb (Br Iris Soc): 94–102
Werckmeister P (1962) Studien über die künstliche Aufzucht von Embryonen ausserhalb des samens in der Gattung *Iris* L. In: Garnaud J-C (ed) Advances in Horticultural Science and their Practical Applications, vol 2. Pergamon Press, Oxford London New York Paris pp 351–359
Yabuya T (1985) Amphidiploids between *Iris laevigata* Fisch. and *I. ensata* Thunb. induced through culture of embryos treated with colchicine. Jpn J Breed 35: 136–144
Yabuya T, Yamagata H (1982) Embryo growth and cultural conditions ln *Iris ensata*. Jpn J Breed 31: 377–382
Yabuya T, Mima H, Ideguchi N, Adachi T, Nagatomo T (1984) Studies on ovule culture of *Iris ensata* Thunb. 1. The effect of plant hormones and placentas on the growth of globular proembryos. Bull Fac Agric Miyazaki Univ 31: 69–77
Zhou C (1988) Mass isolation and culture of pollen protoplasts from three plant species. Acta Bot Sin 30: 362–367
Zhou C (1989) A study on isolation and culture of pollen protoplasts. Plant Sci 59: 101–108

XIV Micropropagation of Surprise Lily (*Lycoris squamigera*)

F.H. HUANG[1], G.L. KLINGAMAN[1], and H.H. CHEN[2]

1 Introduction

1.1 Botany of the Plant

The genus *Lycoris* (family Amaryllidaceae) consists of a dozen or more bulbous geophytic herbs native to China, Japan, Taiwan, and Burma. The bulbous habit of growth primarily occurs between 23–45° north and south latitudes (Rees 1966). It is most commonly found in species which have evolved under a Mediterranean climatic rhythm of hot, dry summers and mild, moist winters. *Lycoris* species are ideally evolved to avoid droughts by producing their foliage during the fall or spring and then going dormant during the summer dry period. Blooming coincides with the end of the drought period, occurring in late summer or early fall before the foliage appears. This blooming pattern has led to such fanciful common names as Magic Lily, Surprise Lily, Naked Lady, and Hurricane Lily (Fig. 1).

Blooms are borne terminally in an umbel on a solid scape in late summer or early fall. Perianth segments form a short tube or segments are separate with colors ranging from the usual pink red to purplish, red-orange, and white. The perianth segments are often recurved, exposing the long, curved stamen. The ovary is inferior with few, black seeds borne in the dehiscent capsule.

Lycoris have a sympodial branching habit with each meristem producing from five (for *L. radiata*) to ten (for *L. squamigera*) leaf primordia before terminating with the inflorescence. At the time of flower bud initiation, lateral growing points have formed in the axils of the upper two or three leaves. The uppermost bud, when growth resumes, will form the next unit of the sympodium. The lower buds will form daughter bulbs. In Japan, floral initiation occurs, as indicated by a flattening and broadening of the apical meristem, by mid April in all species studied (Mori and Sakanishi 1977).

Members of the *Lycoris* genus are botanically separated into two subgenera based on flower form, the trumpet-shaped flower as found in *L. squamigera* being the type for one subgenus and the recurved perianth group with exposed stamens as found in *L. radiata* serving as type for the second subgenus. In addition to the floral differences, there is also significant difference with regard to the time of year foliage is produced. *Lycoris africana*, *L. albiflora*, and *L. radiata* produce foliage in

[1] Department of Horticulture and Forestry, University of Arkansas, Fayetteville, AR 72701, USA
[2] Former graduate student

Biotechnology in Agriculture and Forestry, Vol. 20
High-Tech and Micropropagation IV (ed. by Y.P.S. Bajaj)

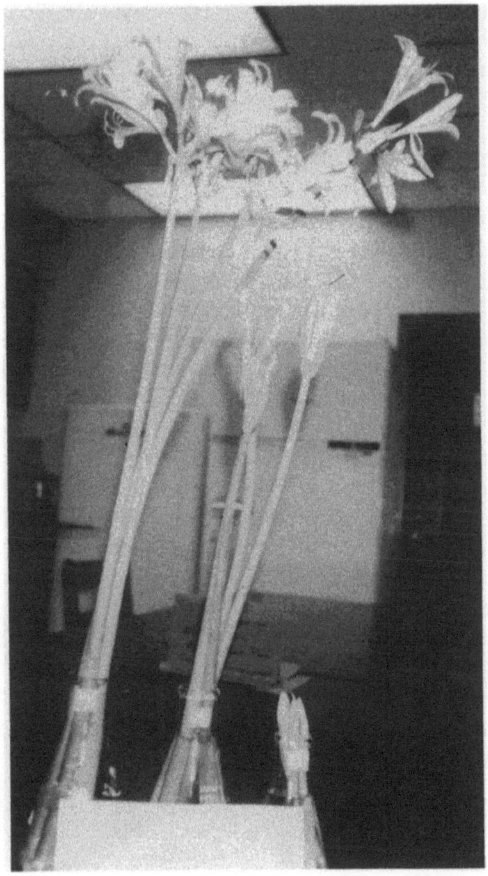

Fig. 1. Flowers of *Lycoris squamigera* (Huang et al. 1985)

the autumn with senescence occurring by late April. *Lycoris incarnata*, *L. sanguinea*, *L. sprengeri*, and *L. squamigera* produce foliage in late winter (February) with senescence occurring by late May (Mori and Sakanishi 1977).

1.2 Conventional Propagation

Propagation of *Lycoris* by conventional means is slow, explaining the rarity of some of the choicest selections in American gardens. In general, members of the amaryllis family increase slowly by natural means with 10 years or longer required to build sufficient stock volume for commercial release using natural means. During this long production time new cultivars often become contaminated with virus diseases, thus reducing their landscape performance before even being released (Hussey 1980). Natural bulb increase usually results in the production of one to three daughter bulbs per year.

Attempts to increase the yield of daughter bulbs by bulb cuttage techniques such as scoring or scooping (Caldwell 1962; Hartman and Kester 1983) have worked on an experimental basis but have not been adopted by commercial bulb growers in the USA. Caldwell reports that in trials with *L. houdysheli* only three daughter bulbs were produced per mature bulb and they required 3 years to reach blooming size. This relatively low return and long-term production time has dampened the enthusiasm for the use of conventional bulb cuttage techniques.

The use of seed to increase plant numbers has not been important with *Lycoris*, primarily because many selections are sterile. Additionally, all clonally propagated progenies are self-incompatible. Artificial pollination has produced seeds of *L. radiata*, and these have flowered true to type (Caldwell 1962). The inherent variability which could creep into a slow-to-flower plant such as *Lycoris* grown from seed, makes the wide scale use of seedlings impractical.

Unlike most important garden bulb crops, *Lycoris* has not received the attention given to other species. Limited breeding attempts have been reported by Caldwell (1962). He indicated that, if fertile parents are used, there is wide compatibility in the small genus. Crosses within the two subgenera are always compatible, and from limited trials, crosses between subgenera have been successful. A limited number of improved cultivars have been selected, but their availability is extremely limited. Because most *Lycoris* bulbs used in this country are imported from Japan, there has not been an adequate mechanism available to get improved cultivars into the propagation channels of that country.

Tissue culture propagation has been used successfully on other members of the amaryllis family, including the notoriously slow to propagate *Hippeastrum* × *hybridum*. Yields of propagules have been sufficiently high to make this an important means of propagation for newly selected tulip, amaryllis, hyacinth, and other flowering bulb crops (Hussey 1980).

2 In Vitro Approaches

2.1 Literature Review

2.1.1 Types of Explants Used for Micropropagation

A variety of plant organs have been used as explant sources in micropropagation. Bapat and Narayanaswamy (1976) reported that amaryllis bulb scales, floral peduncles, or immature anthers could be cultured in vitro to induce callus or bud formation. Seabrook and Cumming (1977) used inner bulb scales, immature leaves, scapes, peduncles, or ovaries of amaryllis as explant sources, and they all responded well. In a survey of explants used for induction of organs, however, bulb scales were the most widely used and usually showed the most successful results in various bulbous species (Takayama and Misawa 1979; Hussey 1982a).

In tissue culture, the mother plant materials (whether from greenhouses or field-grown), their developmental stages, their ages, or even the seasons of the year all play important roles in initiating the callus or organ formation and development

(Evans and Sharp 1981; Flick et al. 1983; Hussey 1983). Fujieda et al. (1979) concluded that explants obtained from post-dormant bulbs of the onion (*Allium cepa* L.) were more active in forming adventitious shoots than those taken from dormant bulbs. In a study of the morphogenetic potential of different age tulip bulbs, Nishiuchi (1979) pointed out that only in certain growing stages could the tulip bulb scale cultured in vitro exert the regeneration potential. This morphogenetic potential was species-dependent even within the same genus. The cause of these phenomena, as Nishiuchi inferred, might be due to the changes of endogenous hormones in plants as time progressed.

Auxiliary shoots might also be cultured in vitro, but this type of regeneration is not feasible in most bulbous species with strong apical dominance (Hussey 1982b).

2.1.2 *Effect of Various Cultural Conditions on Explants*

Among various cultural conditions, temperature was thought to be the most important factor in explant regeneration of *Lilium* bulb scales (Takayama and Misawa 1979). The suitable temperature was set at about 25 °C. As reported by Leshem et al. (1982), light stimulated *Lilium* bulblet growth and leaf emergence, but had no significant influence on its regenerative capacity. They observed the effects of explant orientation, in which the "upright-inverted" positioning was less desirable than the "dorsal-ventral" position (Leshen et al. 1982). They concluded that there were polarity effects on regenerative ability as far as proximal or distal explant scale orientation was concerned.

2.1.3 *Rate of in Vitro Propagation*

Seabrook and Cumming (1977) reported that a total of 450 plantlets could be produced from one amaryllis bulb cultured in vitro in 8 weeks. Theoretically, 1.2×10^{10} *Lilium speciosum* and 3.2×10^{12} *Lilium auratum* plants could be obtained from one medium sized bulb in 1 year under bulb scale shake culture (Takayama and Misawa 1983).

2.1.4 *Genetic Stability*

The genetic stability of adventitious shoots, which were induced either from explant or callus tissue, was determined by chance, i.e., a single or very small group of totipotent cell(s) which gave rise to adventitious shoots was responsible for the genotype of the proliferating plantlet. If the cell(s) carried a mutant gene, the risk of obtaining an aberrant plantlet was high (Hussey and Falavigna 1980). The chances, however, were rare in many instances. For example, only one tetraploid was observed out of 226 plantlets of *Allium cepa* (Hussey and Falavigna 1980). Stimart et al. (1980) reported that 19% of the hybrid *Lilium* Black Beauty regenerated from 1-year-old callus culture bore variegated foliage, but all of those that flowered, including some variegated ones, were true to mother floral type. Only a long-term callus culture, up to 30 months, was unacceptable in which the variegated foliage was prevailing and plants showed slow and weak growth (Stimart et al. 1982). In another experiment, Stimart and Ascher (1980) reported a true-to-type plantlet regeneration from *Freesia* callus after 27 months of sub-culturing in liquid medium.

3 Micropropagation

3.1 General

3.1.1 Explant and Sterilization

Lycoris squamigera bulbs were obtained from a commercial nursery: Wayside Gardens, South Carolina, and local plantings. Figure 2 shows explant production from a mother bulb. After this stage, the separated scales were wrapped together in cheese cloth, dipped in diluted liquid Ivory soap water (1 : 60 v/v) for 15 min, sterilized in 0.5% sodium hypochlorite (Chlorox) for 15 min, rinsed in sterile distilled water several times after each step, and finally cut and trimmed to explants of about 10 mm × 20 mm size.

An additional sterilization procedure was introduced in the second part of the study. It was the use of Incyte, a tissue culture decontaminant manufactured by Alcide Corporation. A 1 : 1 :50 v/v/v concentration of Incyte solution in water was used as a disinfectant after the 15-min Chlorox treatment. Sterilization time in the Incyte solution lasted for 30 min. Explants then were rinsed twice in sterile distilled water before being placed on the first culture medium.

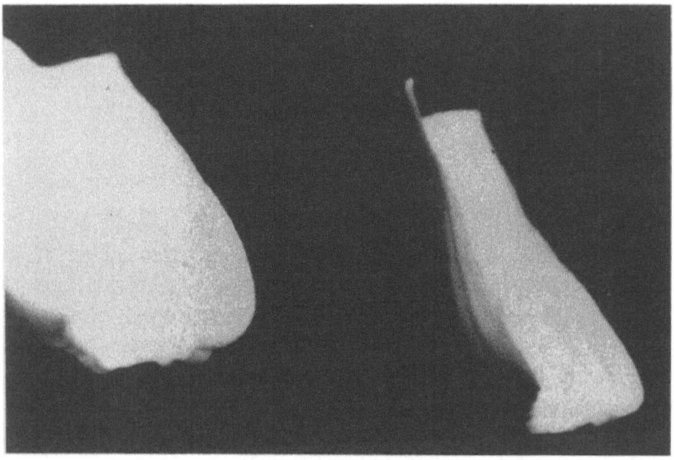

Fig. 2. Explants, *Lycoris squamigera* inner bulbs scales (0 day). (Huang et al. 1985)

3.1.2 Medium Preparation

Throughout the experiments, a modified Murashige and Skoog (MS) culture medium was used (Seabrook and Cumming 1977). For each culture tube, 20 ml of medium with or without growth regulator(s) was added with 3% sucrose and 0.7% agar. Culture media then were adjusted to pH 5.5 ± 0.1, and autoclaved at 15 psi, 121 °C for 15 min.

3.1.3 Culture Condition

One sterilized explant was inserted into a solidified culture medium tube. All explants were incubated on the shelf under 50 μE/m^2/s white fluorescent tubes for 16 h daily at constant temperature of 25 ± 2 °C.

3.1.4 Experimental Design for Root Induction

A) A factorial design of cytokinin: benzylaminopurine (BAP) and auxins: 1-naphthylacetic acid (NAA) or 2,4-dichlorophenoxyacetic acid (2,4-D) was adopted in the first experiments. The selected concentration gradients of each growth regulator were set up as follows: 0, 0.1, 1, and 4 mg/l. Each concentration of auxin was incorporated into the medium at each concentration of BAP, so that there were 32 different treatments in total, with five replicates each. The experiments were repeated twice.

B) A second experiment was conducted following the completion of auxin and cytokinin treatments for root induction. Four concentrations of NAA: 0, 4, 8, and 12 mg/l were tested in order to determine the optimum NAA concentration for *Lycoris squamigera* rooting morphogenesis. There were four treatments with ten replicates, and the experiment was repeated twice.

3.1.5 Shoot Induction

At the end of the 4th cultural week, a change of the medium to 4 mg/l BAP and 1 mg/l NAA was required to induce shooting from rooted explants. Once the adventitious shoots were induced, they were separated from each other with the root(s) attached, and transferred into fresh MS medium for further elongation of the adventitious organs.

3.1.6 Plantlet Hardening

After 1 to 2 weeks growth in MS medium, the plantlets were subcultured and transferred into half-strength MS first pretransplant medium. Then, after another 2 weeks, they were subcultured (if more shoots were formed), and transferred into one-fifth strength pretransplant MS medium. After another 2 weeks, the growing

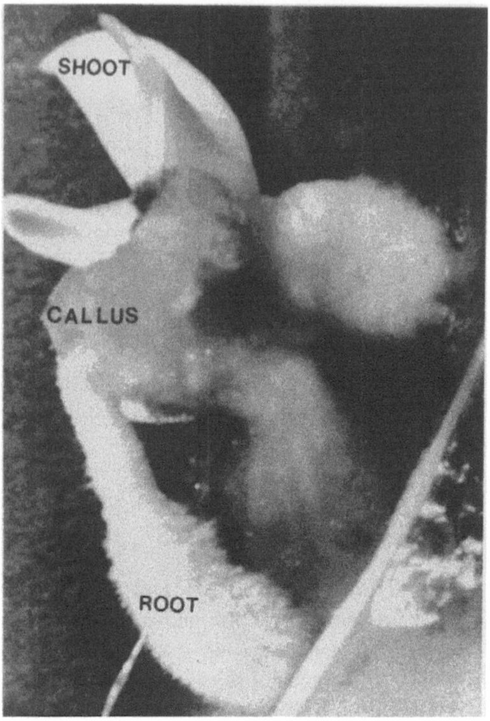

Fig. 3. Organogenesis of *Lycoris squamigera* after 60 days of culture (60 days, MS medium). (Huang et al. 1985)

Fig. 4. Tissue cultured plantlets in the pot (126 days, Promix). (Huang et al. 1985)

plantlets were put into 5-cm pots containing Pro-Mix and covered with transparent polyethylene bags (Fig. 7). After a 2-week hardening period, all plantlets were removed from bags and placed in the greenhouse (Fig. 8).

3.1.7 Microscopic Study

A microscopic study of the callus was made during the period of initiation of roots from the calli to substantiate organogenesis of *Lycoris squamigera.* Samples of callus were fixed in formalin-acetic acid-alcohol, dehydrated in a series of water, ethanol, and tertiary butyl alcohol mixture, then embedded in paraplast. Tissue selections were cut at 15 μm and stained with toluidine blue.

3.2 Results

3.2.1 Explant Establishment

Explants of *Lycoris squamigera* inner bulb scales responded to the culture medium within 2 to 3 weeks after insertion. The response was revealed when white scale tissue became swollen and turned gradually yellowish green in color.

Contamination Rate. During the first 2 weeks of culture, the contamination of explants reached the highest point. After that, it subsided. Table 1 shows the total explant contamination rates in the experiments.

The Use of Incyte. Most of the attempts at bulbous plant tissue culture use sodium hypochlorite solution as the explant disinfectant. This technique was adopted in the first part of this study and resulted in a contamination rate of 25–30% (Table 2). In the second experiment, however the contamination rate was reduced to 0–15% using Incyte. This sharp decrease in explant contamination rate seems to suggest that the use of Incyte is a means of improving the sterilization technique.

Explant Size. The size of explants might be critical in organ regeneration. From all *Lycoris* bulb scale experiments, explants less than 2 cm in length and 1 cm in width were often incapable of regenerating adventitious roots. Those responding well were those with thick scales (approximately 3 mm in thickness).

Table 1. Explant contamination and survival rates (Huang et al. 1990)

Experiment trial		Beginning explant	Sterilant used	Rate of contamination	Rate of survival
Part A	trial I	160	Chlorox	25%	75%
	trial II	160	Chlorox	30%	70%
Part B	trial I	40	Chlorox/Incyte	0%	100%
	trial II	20	Chlorox/Incyte	15%	85%

Table 2. Means of the interaction of auxin type and level on callus formation, adventitious rooting, and adventitious shooting in auxin/cytokinin treatments (Huang et al. 1985)

Auxin type	Level mg/l	Total treatment number	Mean of callus formation	Mean of adv. rooting	Mean of adv. shooting
NAA	0	4	0.00b[a]	0.00b	0.00c
	0.1	4	0.00b	0.00b	0.00c
	1	4	0.69a	0.43b	0.16c
	4	4	1.00a	0.88a	0.50a
2,4-D	0	4	0.00b	0.00b	0.00c
	0.1	4	0.10b	0.10b	0.10c
	1	4	0.44ab	0.44b	0.25bc
	4	4	0.05b	0.05b	0.00c
LSD			0.45	0.50	0.21

[a] Values within a column followed by the same letter are not significantly different by LSD at 1% significance level.

3.2.2 Callus Formation

In the auxin/cytokinin treatments (Part A), callus tissue formed first on explant parts exposed to the air. It occurred following the expansion of explant tissue in treatments with 1 and 4 mg/l NAA or 0.1, 1, and 4 mg/l 2,4-D levels.

Auxin Effects. It is generally believed that auxin promotes callus formation (Hussey, 1980; Gamborg and Shyluk 1981). The addition of auxin, either in the form of NAA or 2,4-D, did induce callus. At 1 mg/l concentration, the promoting effect between the two auxins was not significantly different. When the concentration was increased to 4 mg/l, the difference became significant, with NAA producing more callus than 2,4-D. Although it was a potent auxin in its activity, 2,4-D at 4 mg/l concentration level imposed an adverse effect which eventually caused browning in *Lycoris* explant tissue.

As far as the BAP concentration and its interaction effect with the auxin on callus formation is concerned, there is no significant effect shown in the experiment. This fact is also true for adventitious root, adventitious shoot, and total plantlet formation.

3.2.3 Adventitious Root Formation

Toward the end of the first month of culture, adventitious roots were formed in auxin/cytokinin treatments (Part A) from auxin levels of 1 mg/l to 4 mg/l in NAA, and 0.1 mg/l to 4 mg/l in 2,4-D, but the regenerating degree varied. In the second NAA concentration treatment (Part B), rooting occurred between 4 and 12 mg/l with most effective rooting in medium containing 4 mg/l NAA.

Auxin Effect. It is evident that auxins promote adventitious rooting in *Lycoris squamigera.* Rooting increased as NAA concentration was increased from 0.1 mg/l

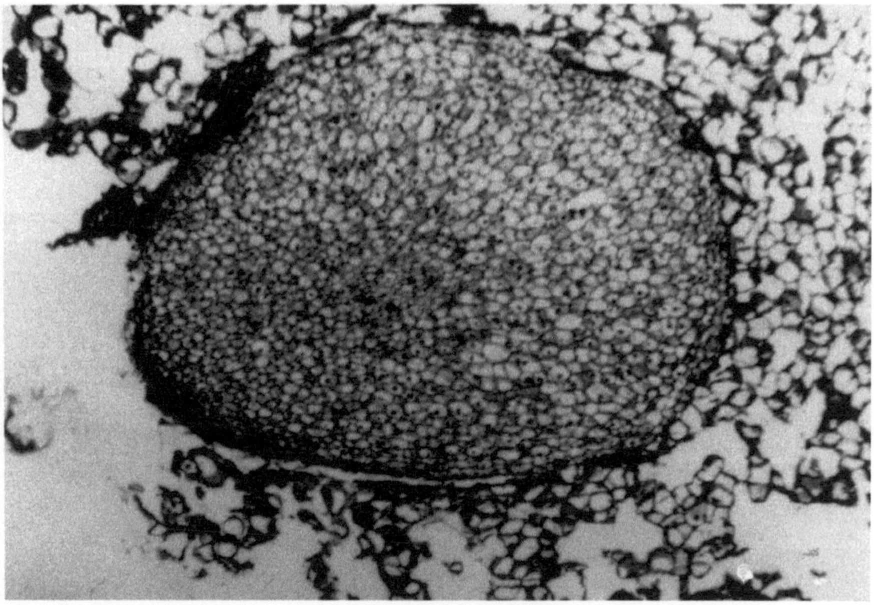

Fig. 5. Cross-section of an adventitious root initiating among callus tissue (15 days, MS medium). (Huang et al. 1985)

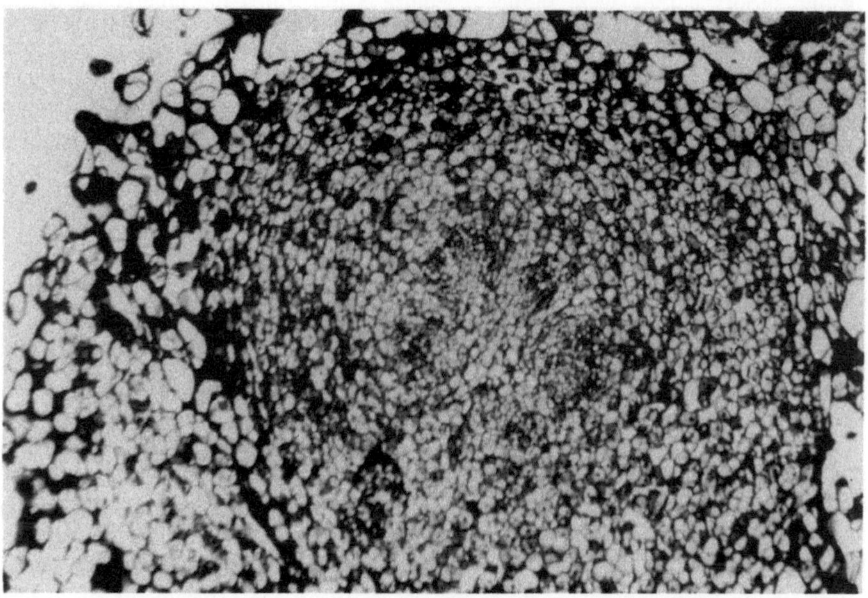

Fig. 6. Further organization, adventitious root showing a developing polyarch stele (22 days, MS medium)

to 4 mg/l. The effect of 2,4-D dropped when the concentration was increased from 1 mg/l to 4 mg/l. The mean difference (0.825) between NAA and 2,4-D at 4 mg/l concentration level was greater than the LSD of 0.498 (at 1% significance level). This big difference can be interpreted as an antagonistic effect of a high 2,4-D concentration in culture medium, resulting in explant tissue browning and death.

Microscopic Study. In auxin/cytokinin treatments (Part A) during the stage of adventitious root formation, callus tissue with root primordia was studied under the microscope. It was found that adventitious roots were formed at random among the callus cells. Figures 5 and 6 reveal two different stages of adventitious root development in the callus of *Lycoris squamigera* explants. In the second NAA concentration treatments (Part B), adventitious roots appeared prior to the callus formation. This observation agrees with the common belief that in vitro adventitious meristems can be induced either from organ explants or from associated callus (Bapat and Narayanaswamy 1976; Hussey 1980; Gamborg and Shyluk 1981). Adventitious root formation always precedes shoot formation in *Lycoris squamigera* bulb scale culture.

3.2.4 Adventitious Shoot Formation

As early as 3 weeks after changing the first culture medium to shoot induction medium, adventitious shoots began to form in some rooted explants. The shoots sometimes multiplied when subdivided. On average, rooted explants had 3 adventitious shoots if they were induced. Figure 3 illustrates the emergence of the adventitious roots and shoots from callus tissue. The mean difference between NAA and 2,4-D at 4 mg/l concentration level was 0.5, which was significantly greater than its LSD value (0.206, at 1% level). In the second study, shoots formed with 4 to 8 mg/l NAA, with most shooting at the lower level. Since shoots were induced after roots formed, this difference could also be traced back to the auxin effect on adventitious rooting.

3.2.5 Plantlet Formation

Plantlets formed as soon as the adventitious shoots were induced on rooted explants, and continued to grow into whole plants (Fig. 8). A few aberrant plantlets were found (e.g., albinos), but their growth was stunted and produced no survivors. In the second part of the investigation, the maximum number of plantlets was achieved using 4 mg/l NAA.

Regeneration Rate. As described by Murashige (1974), a meristemoid is a cluster of undifferentiated cells which receive stimuli and form new organ(s). If the callus or explant tissue fails to transform itself into meristemoids, there will be no regenerative ability from explant or its callus tissue. In all the experiments of *Lycoris squamigera* bulb scale culture, a regeneration rate as high as 60% was achieved when 4 mg/l concentration of NAA was used in the culture medium.

Plantlet survival rate in the greenhouse averaged to 76.6% (Table 3).

Fig. 7. Plantlet hardening in polyethylene bag (112 days, Promix). (Huang et al. 1985)

Fig. 8. Regeneration of plantlets after 11 weeks of culture (77 days, MS medium). (Huang et al. 1985)

Table 3. Plantlet survival rates in the greenhouse for auxin/cytokinin treatments of *Lycoris squamigera* (Part A experiment). (Huang et al. 1985)

Experiment trial	Total plantlets at beginning	Plantlets survived	Survival rate (%)
I	44	36	81.8
II	7	5	71.4
		Average	76.6

Commercial Aspects of Micropropagation. The reproductive potential of *Lycoris squamigera* by inner bulb scale culture can be estimated as follows:

- explant/layer × four layer/bulb = 16 explant/bulb.
- 16 explant/bulb × 90% of explant survival rate = 14.4 explant survived/bulb.
- 14.4 explant survived/bulb × 60% of regeneration rate × 3* plantlet/explant survived = 25.9 plantlet/bulb.
- 25.9 plantlet/bulb × 77% of plantlet survival rate = 20 plantlet survived/bulb.

*average plantlets obtained from one shooted explant.

The conventional method of propagating this species is slow, with only two to three bulblets obtained from one mother bulb in 3 years. With this approach, 20 plantlets can be produced from one bulb in just 4 months. It would be even more economically productive for the micropropagation of *Lycoris squamigera* if a cell suspension culture technique (Takayama and Misawa 1983) could be adopted, or the route of embryogenesis could be explored in the future.

4 Summary and Conclusion

In *Lycoris squamigera* in vitro inner bulb scale culture, an auxin, such as NAA at a concentration of 4 mg/l, is needed to regenerate adventitious roots. There is no great difference in hormone type except at concentrations higher than 4 mg/l.

A high cytokinin to low auxin ratio (BAP:NAA = 4:1 mg/l) in the culture medium is used to induce adventitious shoots from rooted explants. Explant sterilization can be enhanced by the addition of Incyte, in which the inoculates can be maintained nearly as high as 90% contamination-free.

The regeneration pattern is either indirectly via callus cells or directly through explant tissue to induce adventitious roots, followed by the formation of adventitious shoots.

As an explant, inner bulb scales prove to be a satisfactory source. An average of 20 plantlets can be obtained from one mother bulb in about 4 months. This rate of reproduction would greatly exceed the rate through conventional bulblet propagation, which produces only two or three bulblets within 3 years. Therefore, the use of organ culture in the propagation of *Lycoris squamigera* seems quite promising.

The offspring of bulb scale organogenesis may not be disease-free unless some pathogen detection methods are used, and may not be true to type unless the horticultural characteristics are checked.

5 Protocol

Best Explant. The size and thickness of explants are critical in plant regeneration, explants larger than 10 × 20 mm in size and 3 mm in thickness of inner bulb scales respond well in plant regeneration.

Best Medium. Four mg/l NAA is needed to induce adventitious roots in a modified MS medium; 4 mg/l BAP to 1 mg/l NAA is used for the formation of adventitious shoot from rooted explants.

Optimal Conditions. Fifty $\mu E/m^2/s$ cool white fluorescent lamps for 16 h daily at a constant temperature of 25 ± 2 °C are considered as optimal conditions.

References

Bapat VA, Narayanaswamy S (1976) Growth and organogenesis in explanted tissues of amaryllis in culture. Bull Torrey Bot Club 103(2): 53–56

Caldwell S (1962) *Lycoris*—a progress report. Am Hortic Mag 41: 61–93

Evans DA, Sharp WR (1981) Plant regeneration from cell cultures. In: Janick J (ed) Hortic Rev vol III, Avi Pub Co, Inc Westport, CN pp 214–314

Flick CE, Evans DA, Sharp WR (1983) Organogenesis. In: Evans DA et al (eds) Handbook of plant cell culture vol I Techniques for propagating and breeding. Macmillan, New York, pp 13–81

Fujieda K, Matsouka N, Fujita Y (1979) Vegetative multiplication of onion, *Allium cepa* L., through tissue culture. J. Jpn Soc Hortic Sci 48(2): 186–194

Gamborg OL, Shyluk JP (1981) Nutrition, media and characteristics of plant cell and tissue cultures. In: Thorpe TA (ed) Plant tissue culture, methods and applications in agriculture. Acad Press, New York, pp 21–44

Hartmann HT, Kester DE (1983) Plant propagation: principles and practices, 4th edn. Prentice-Hall, New Jersey, 727 pp

Huang FH, Chen HH, Klingaman GL (1985) Micropropagation of *Lycoris squamigera* through tissue culture as a plant production system for horticultural crops. Oct 20–23 1985, USDA-ARS Bellsville, MD

Hussey G (1980) Propagation of some members of the Liliaceae, Iridaceae and Amaryllidaceae by tissue culture. In: Brickell CD (ed) Petaloid monocotyledons, Acad Press, London, pp 33–42

Hussey G (1982a) In vitro propagation of monocotyledonous bulbs and corms. In: Proc 5th Int Congr Plant Tissue and Cell Cult. Maruzen, Tokyo, pp 677–680

Hussey G (1982b) In vitro propagation of *Narcissus*. Ann Bot 49: 707–719

Hussey G (1983) In vitro propagation of horticultural and agricultural crops. In: Mantell SH (ed) Plant Biotechnology. Cambridge Univ, MA, pp 111–138

Hussey G, Falavigna A (1980) Origin and production of in vitro adventitious shoots in the onion, *Allium cepa* L. J Exp Bot 31: 1675–1686

Leshem B, Hannah LK, Steinitz B (1982) The effect of light and of explant orientation on the regeneration and subsequent growth of bulblets on *Lilium longiflorum* Thunb. bulb scale sections cultured in vitro. Sci Hortic 17: 129–396

Mori G., Sakanishi Y (1977) Studies on the growth and flowering of bulbous plants of Amaryllidaceae. I. Growth and flowering of *Lycoris* plants grown in the field. J Jpn Soc Hortic Sci 45(4): 389–396

Murashige T (1974) Plant propagation through tissue cultures. Annu Rev Plant Physiol 25: 135–116

Nishiuchi Y (1979) Studies on vegetative propagation of Tulip. II. Formation and development of adventitious buds in the excised bulb scale cultured in vitro. J. Jpn Soc Hort Sci 48(1): 99–105
Rees AR (1966) The physiology of ornamental bulbous plants. Bot. Rev 32: 1–23
Seabrook JA, Cumming BG (1977) The in vitro propagation of *Amaryllis* (*Hippeastrum* spp. *Hybridus*). In Vitro 13: 831–836
Stimart DP, Ascher PD (1982) Plantlet regeneration and stability from callus cultures of *Freesia* × *Hybrida* Bailey cultivar Royal. Sci Hortic 17: 153–157
Stimart DP, Ascher PD, Zagorski JS (1980) Plants from callus of the interspecific hybrid *Lilium* Black Beauty. HortScience 15(3): 313–315
Takayama S, Misawa M (1979) Differentiation in *Lilium* bulb scales grown in vitro. Effect of various cultural conditions. Physiol Plant 46: 184–190
Takayama S, Misawa M (1983) A scheme for mass propagation of *Lilium* in vitro. Sci Hortic 18(4): 353–362

XV Micropropagation of *Nematanthus*

C.L. Lê[1]

1 Introduction

Nematanthus Schrader, a member of the family Gesneriaceae, resulted from the cross of *Hypocyrta* and *Nematanthus* (Moore 1973), is classified into 26 neotropical species (Chautems 1985). Wiehler (1979, 1983), based on the morphology of flowers and leaves, and ecological distribution, has clearly defined the genus *Nematanthus* as epiphytic, ornithophilous, perennial herbs and subshrubs, restricted to southeastern Brazil. The basic chromosome number for all *Nematanthus* species is n = 8. The leaves of these species are characterized by the presence of hypodermis. The corollas, which might be orange, red, or yellow, are usually pouched or enlarged at the bottom and more or less laterally compressed. The fruit is a displayed capsule type.

The genus *Nematanthus*, an endemic group closely related to the coastal rainforest of Brazil, grows between 14 and 29° latitude south (Chautems 1985). Most of them are found along the coast of Rio de Janeiro, Sao Paulo, Pasana, and Santa Catarina, at an altitude between 500 and 1200 m. It has been noted that each species settled an area more or less gregariously under these natural growth conditions (Chautems 1985).

In horticultural practice, *Nematanthus* is propagated by seed or vegetatively from shoot cuttings. These techniques can be employed at any time of the year (Reist 1979).

Seed propagation is relatively easy, but not advisable due to the fact that seed progeny is extremely heterogeneous. The growth rate of plants obtained from seed germination is low, and flowering will take place about 18 months after the seedlings have been sown (Reist 1978b, 1979). However, this procedure is now used only for genetic improvement programs (Chautems 1985). On the other hand, cutting is the preferred method for traditional mass propagation. Reist (1978b) showed that tip cuttings of *N. gregarius* with four to eight leaf pairs, removed from mother plants and placed on a rooting medium without growth regulators and under mist environment, root with 100% success after 1 month. After being potted and pinched back to two or three pairs of leaves, the cultures are continued from 8 to 10 months more – depending on the period of propagation – to reach the flowering stage.

[1] Station fédérale de Recherches Agronomiques de Changins (RAC), 1260 Nyon, Switzerland

Biotechnology in Agriculture and Forestry, Vol. 20
High-Tech and Micropropagation IV (ed. by Y.P.S. Bajaj)

At present, *N. gregarius*, with its highly attractive foliage and bloom, is almost exclusively produced for the pot plant market. Recently, some of the existing species, e.g., *N. fissus*, *N. fritschii*, *N. hirtellus*, *N. longipes*, *N. nervosus*, *N. perianthomegus*, *N. strilligosus*, and *N. wettsteinii*, have given rise to a number of valuable hybrids which could be of interest for indoor plant production (Reist 1978a).

Despite the ease with which most gesneriads, including the genus *Nematanthus*, can be propagated by conventional methods, in vitro culture techniques could offer a valuable alternative (Lê 1984) for the rapid clonal multiplication of new cultivars if rapid breeding is to be done.

2 In Vitro Studies

The successful application of tissue culture methods to mass propagation of a number of gesneriads has been widely demonstrated during the last decade (Start and Cumming 1976; Johnson 1978a, b; Hughes and Barni 1978; Bilkey and McCown 1979; Lê 1980; Lê and Collet 1981). However, little is known about in vitro clonal multiplication of *Nematanthus*, except for our preliminary work (Lê 1984, 1988). The method used to regenerate complete plantlets of *Nematanthus* through in vitro tissue culture is reported here.

2.1 Establishment of Shoot Cultures

Actively growing shoots (2–4 cm long) can be excised from the floral branches of mother plants cultivated under greenhouse conditions at any time of the year, but plant material should preferably be collected from February till June if to obtain high rate of explant growth during the establishment stage is desired (unpubl.).

Leaves were removed and washed for 30 min under running tap water. The washed material was then surface sterilized by dipping in 70% ethanol for a few seconds to break surface tension, and soaked twice for 10 min in 0.5% (vol/vol) sodium hypochlorite containing 2 drops Teepol (per 100 ml) as a wetting agent. The shoot pieces were then rinsed three times with sterile distilled water. Finally, 5-mm-long apices (Fig. 1) were excised from the surface sterilized shoot tips, using a scalpel and a pair of forceps, and transferred to the culture medium.

2.2 Media and Culture Conditions

The basal nutrient medium consisted of Murashige and Skoog's (MS) (1962) inorganic salts supplemented with 0.4 mg/l thiamine-HCl, 0.5 mg/l nicotinic acid, 0.5 mg/l pyridoxine-HCl, 100 mg/l Myo-inositol and 3% sucrose. Various combinations of auxin and cytokinin were added to the basal medium in order to induce shoot proliferation. For further growth into complete plantlets, regenerated shoots (1–2 cm long) were excised from shoot-proliferating explants and transferred to the

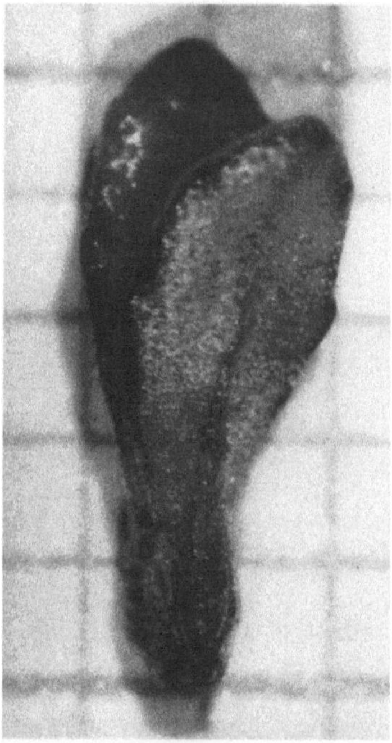

Fig. 1. An initial explant using shoot tip (0.5 cm in length) at the time of culture. *Bar* 1 mm

basal medium containing various concentrations of indole-3-butyric acid (IBA), which was the sole growth regulator supplement, in the range of 0.01 to 5.0 mg/l. Sucrose was reduced to 2% in all treatments of rhizogenesis. The media were solidified with 0.7% Difco-Baco Agar and the pH was adjusted to 5.7 prior to autoclaving at 121 °C (1.1 kg/cm^2) for 15 min.

All cultures were incubated in a temperature-controlled environmental chamber at 23 °C in complete darkness for 3–6 days; afterwards, they were placed in a 16-h photoperiod and 23 °C day/18 °C night temperature regime. The irradiance was 850 μW/cm^2 (measured over the range of wavelengths 400–700 nm) provided by cool white fluorescent lamps (Mazda Aviva TF/65AVI).

2.3 Mineral Salts

Experiments carried out in our laboratory using shoot tip explants showed that the *Nematanthus* spp. required a high concentration of MS macronutrients for rapid proliferation of shoot primordia (Fig. 2). At low concentrations (one-quarter or one half) of the basal MS medium, the frequency of shoot explants forming new shoots was relatively reduced; shoots thus obtained were dwarf and etiolated in appearance, and most of them were unavailable for further subculturing operations. While increasing the concentration from one half to four-fifths of the basal MS medium, a significant increase in the number of shoot tip explants giving rise to vigorous shoots was found. At full strength of the MS macronutrients, more than 95% of the initial explants developed shoots.

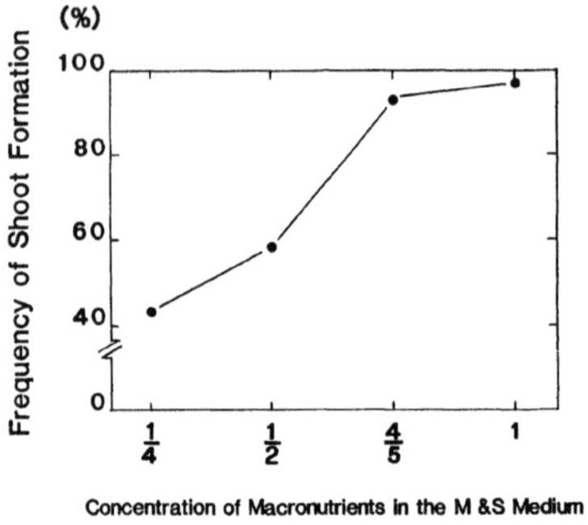

Fig. 2. Effect of various concentrations of the macronutrients in the MS basal medium on frequency of shoot formation from shoot explants of *Nematanthus* sp

2.4 Effect of Growth Regulators

2.4.1 Effect of NAA

The incorporation of naphthalene-acetic acid (NAA) into the basal medium at 0.01 mg/l did not show any apparent effect on shoot proliferation (Fig. 3). At low concentration (0.01 to 0.01 mg/l), NAA stimulated root development after several weeks of culture, whereas increasing NAA level from 1 to 2 mg/l resulted in stimulation of callus formation, which had an inhibitory effect on the initiation of shoot primordia on shoot tip explants.

2.4.2 Effect of BA

Shoot tip cultures on the basal medium containing 0.1 to 1.0 mg/l 6-benzyl-aminopurine (BA) as the sole growth regulator source led to moderately developed shoots (Fig. 3). Most of them were deformed and appeared to be injured. At higher concentrations of BA (2.0 mg/l)), shoot growth was severely inhibited and profuse friable callus formation took place in the basal cut ends.

2.4.3 NAA/BA-Interaction

As shown in Fig. 3, the incubation of shoot tip explants on the medium supplemented with BA and NAA resulted in higher shoot proliferation rates, as opposed to those where BA or NAA was added singly to the medium. In cultures grown in the presence of 1.0 mg/l BA and 0.01 or 0.1 mg/l NAA, a maximum development of normal shoots was observed within 6 weeks of transfer from basal medium (Fig. 4). The addition of NAA at the concentration of 0.1 mg/l or more yielded limited shoot proliferation or did not promote shoot initiation when 0.1 mg/l BA was supplied in the basal medium.

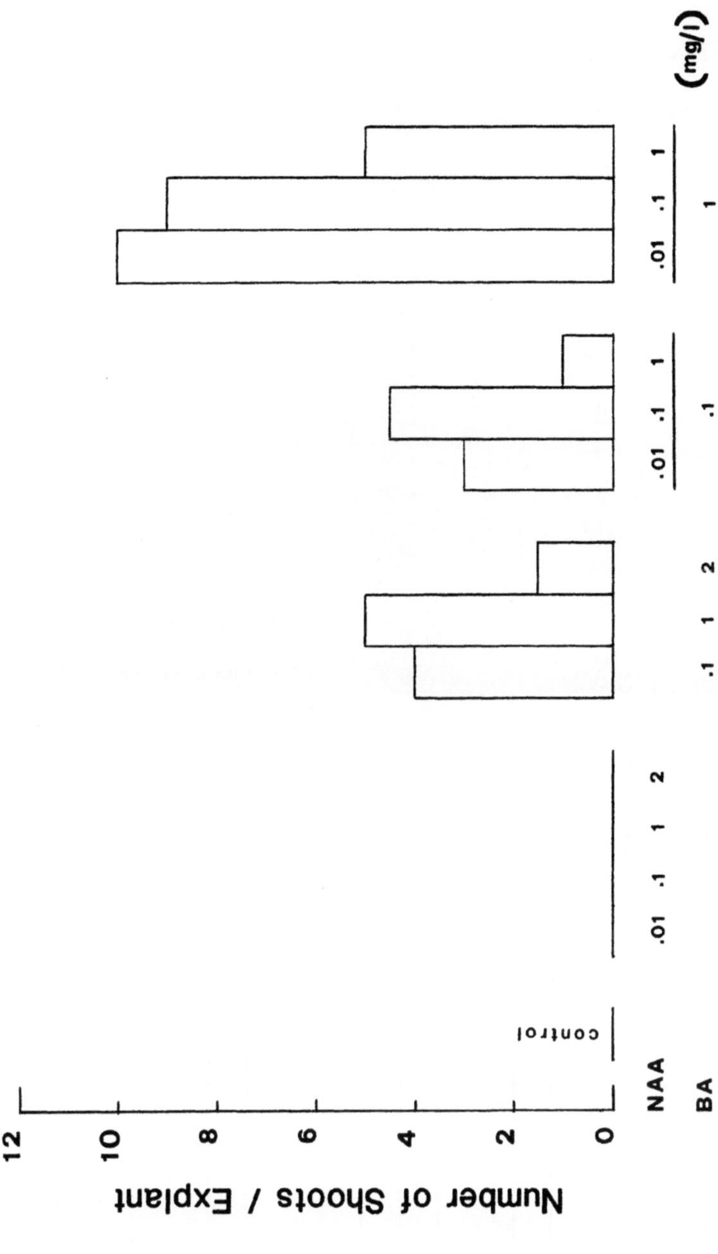

Fig. 3. Effect of BA and NAA on shoot proliferation of *Nematanthus* shoot explants

Fig. 4. Shoot development from shoot tip explant on MS medium supplemented with 1.0 mg/l BA and 0.01 mg/l NAA. Bar 1 cm

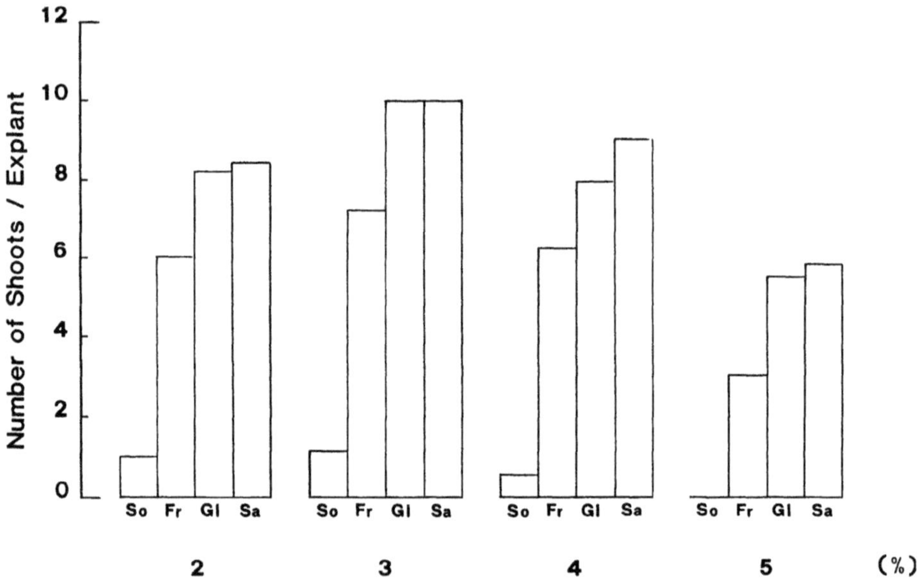

Fig. 5. Effect of different carbon sources on shoot formation from shoot tip explants *So* sorbitol; *Fr* fructose; *Gl* glucose; *Sa* sucrose

2.5 Carbon Sources

The carbon source in the medium seems to play a role for in vitro growth of *Nematanthus*. The effect on in vitro development of different carbon sources such as sucrose, glucose, fructose, and sorbitol ranging from 2 to 5% have been investigated. As shown in Fig. 5, glucose and sucrose at a concentration of 3% were found to be optimal for new shoot development, while fructose at all concentrations tested yielded growth rate less than glucose and sucrose treatments. If sorbital is added in the medium as the sole source of carbon, the growth rate of initial explants was severely reduced and the formation of new shoots failed to occur at higher concentrations (up to 4%).

2.6 pH

The optimum pH for in vitro growth of *Nematanthus* was in the range of 5.5 to 5.8. pH lower than 5.5 could retard the induction of shoot development on initial explants, whereas increasing pH to higher than 6 significantly decreased the multiplication rate per subculture.

2.7 Subculture of Shoots

For repeated subculturing, shoots 0.5–1 cm in length excised from growing shoot clusters were transferred to the basal medium supplemented with reduced BA concentration at the level of 0.5 mg/l, in combination with 0.1 mg/l NAA. This procedure has been now maintained for more than 30 subcultures without any loss of morphogenetic responses. The multiplication rate was ten fold per subculture at 4-6-week intervals.

2.8 Rooting

Adventitious root formation could be induced on indole-3-acetic acid (IAA) as well as on indole-3-butyric acid (IBA), but the rooting frequency with IAA treatment was very low in comparison with IBA treatment. Moreover, if the medium was supplied with naphthalene-acetic acid (NAA), the risk of forming profuse callus at the basal cut end of the cuttings was extremely high during the rooting stage. For these reasons, all rooting treatments were performed with various concentrations of IBA, which led to satisfactory results. The most effective concentrations of IBA for root induction was 0.05 mg/l under our conditions of experiments (Fig. 6). Roots occured, without callus formation, in a great number of 3 weeks of culture and were concentrated on the proximal part of the cuttings. Complete inhibition of rooting accompanied by callus development occurred at higher concentration (up to 2.0 mg/l) (Lê 1988). It was also noted that the percentage of shoots forming roots improved with increasing subcultures.

Fig. 6. Root initiation on shoot explant at 3 weeks after transfer to the medium containing 0.05 mg/l IBA. *Bar* 1 cm

Fig. 7. Regenerated *Nematanthus* plantlets growing in the greenhouse

2.9 Transfer to the Greenhouse

In the greenhouse, plantlets regenerated in vitro were removed from the culture containers and rinsed in tap water to discard excess agar from roots before transferring to soil mixture using peat:perlite (5:1). They were then maintained under intermittent mist conditions for hardening during the first 10 days, and were allowed to grow on their own after the mist period (Fig. 7). More than 90% of plants transferred to greenhouse conditions continued to grow normally and were identical to the parents.

3 Summary and Conclusion

The present results demonstrate that, with appropriate media and environmental conditions, the culture and regeneration of *Nematanthus* can be achieved by using shoot tip explants from mature plant material. This finding can offer an adequate technique as potential for the rapid clonal propagation of valuable cultivars of *Nematanthus* for commercial production in addition to the conventional methods of multiplication.

4 Protocol

1. Establishment of surface sterilized shoot tips (0.5 cm long) on basal MS medium containing 10 mg/l BA and 0.1 mg/l NAA.
2. Shoot multiplication can be induced on the basal medium supplemented with 1.0 mg/l BA and 0.01 mg/l NAA. Repeated subculturing can be achieved at 4- to 6-week intervals onto fresh medium with reduced BA concentration (0.5 mg/l).
3. Rapid rooting is accomplished by culturing isolated shoots on the basal medium supplied with 2% sucrose and 0.05 mg/l IBA as the sole inducing agent.
4. Plantlets with well-developed root system are transferred to a peat:perlite (5:1) mixture and gradually acclimatized to greenhouse conditions.

References

Bilkey PC, McCown BH (1979) In vitro culture and propagation of *Episcia* sp. (Flame violet). HortScience 14: 109–114

Chautems A (1985) Révision taxonomique et possibilités d'hybridation de *Nematanthus* Schrader (Gesneriaceae), genre endémique de la forêt côtière brésilienne. PhD Thesis, Université de Genève (Switzerland)

Harney PM (1982) Tissue culture propagation of some herbaceous horticultural plants. In: Application of plant cell and tissue culture to agriculture and industry. Univ. Guelph, Ontario, Canada, pp 187–208

Hughes KW, Barni B (1978) Regeneration of *Kohleria amabilis* (Gesneriaceae) from leaf and petiole explants. Symp Prop high plants through Tiss Cult, Univ Tennessee, Knoxville, April 16–19
Johnson BB (1978a) In vitro propagation of *Gloxinia* from leaf explants. HortScience 13: 149–150
Johnson BB (1978b) In vitro propagation of *Episcia cupreata*. HortScience 13: 596
Lê CL (1980) Microbouturage in vitro de l'*Aeschynanthus hildebrandii* Hemsl. Rev Suisse Vitic, Arboric, Hortic 12(5): 245–248
Lê CL (1984) Multiplication végétative in vitro de *Nematanthus* hybride. Rev Suisse Vitic, Arboric, Hortic 16(4): 185–188
Lê CL (1988) Rapid propagation of *Nematanthus* hybrid in vitro. Acta Hortic 226: 635–638
Lê CL, Collet GF (1981) Advantages et contraintes de la multiplication in vitro de *Saintpaulia ionantha* Wendl. Rev Suisse Vitic, Arboric, Hortic 13(5): 265–270
Moore HE (1973) Comments on cultivated Gesneriaceae. Baileya 19: 35–41
Murashige T, Skoog F (1962) A revised medium for rapid growth and bioassays with tobacco tissue cultures. Physiol Plant 15: 473–497
Reist A (1978a) Il n'y a plus d'*Hypocyrta*. Hort Mar Rom 12: 5–7.
Reist A (1978b) *Nematanthus gregarius*, multiplication et culture. Hort Mar Rom 12: 7–13
Reist A (1979) Growing *Nematanthus gregarius*. Gloxinian 29: 19–23
Start ND, Cumming BG (1976) In vitro propagation of *Saintpaulia ionantha* Wendl. HortScience 11: 204–206
Wiehler H (1979) Generic delimitation in a new classification of the neotropical Gesneriaceae. PhD Thesis, Univ Miami, Fl
Wiehler H (1983) A synopsis of the neotropical Gesneriaceae. Selbyana 6 (1–4): 1–219

XVI Micropropagation of Orchids – *Spathoglottis plicata* and *Epidendrum radicans*

F. SINGH[1]

1 Introduction

Orchids, the doyens among ornamentals, are very distinctive plants. Taxonomically they represent the most highly evolved family (Orchidaceae) among monocotyledons, and the largest of all flowering plant families with 600–800 genera and 25 000–35 000 species (Garay 1960; Schultes and Pease 1963; Sheehan 1983). They account for 7% of the total species of flowering plants of the world (Van der Pijl and Dodson 1966).

Orchids are perennial herbs exhibiting an incredible range of diversity in size, shape, and color of flowers. They are commercially grown in many countries, such

Fig. 1A,B. Morphological features of the two species. **A** *Spathoglottis plicata*. **B** *Epidendrum radicans*

[1] Orchid Laboratory, Indian Institute of Horticultural Research, Post Box No. 8025, Bangalore 560 080, India

Biotechnology in Agriculture and Forestry, Vol. 20
High-Tech and Micropropagation IV (ed. by Y.P.S. Bajaj)

as Thailand, Singapore, Japan, France, Australia, and the USA, both for cut flower production or as pot plants (Rao 1977).

Breeding of orchids is often a slow and costly undertaking. The conventional methods of propagation through symbiotic seed germination and division of back bulbs, offshoots, and keikis (plantlets) are very slow. The use of micropropagation techniques have revolutionized the commercial orchid industry. Orchids are one of the few flowering plants of commercial value to be propagated in vitro both through seed and tissue culture.

The present chapter deals with the status of micropropagation of two most important orchid species, *Spathoglottis plicata.* Bl. and *Epidendrum radicans* Pav Ex Lindl.

1.1 *Spathoglottis plicata* Bl.

(*Bletia angustifolia* Gaud, *Paxtonia rosea* Ldl., *Phaius rumphii* Bl., *Spathoglottis lilicina* Griff., *S. plicata* Ldl.).

The genus *Spathoglottis* was created by Blume in 1825, and the name refers to the spathe-like lip in the species. It is terrestrial, with pseudobulbs tightly clustered, ovoid to conical, bright or dull green. Leaves deep green, sheathing linear, lanceolate acute, ribbed, 50–60 cm long, 8–10 cm wide. Inflorescence rigidly erect 65–70 cm tall, 15–20 flowers with pinkish persistent bract. Flowers 3–4 cm in diameter, sepals and petals violet red or pink purple, lip three angled, crisp and attractive.

Spathoglottis plicata is one of the most important horticultural species cultivated for its attractive and ever-blooming flowers (Fig. 1A). There are many horticultural variants, differing largely in color and size of flowers. Flower color in the species is very extensive and ranges from brilliant yellow, purple, magenta, crimson pink, to white. Holtum 1957 in *Flora of Malaya* (Teuscher 1978) mentions seven varieties of the species along with variety Vieillardii which he considered the best, he also mentions about 19 hybrids of *S. plicata* which have been produced in Malaya. A natural hybrid between *S. plicata* × *S. vanoverberghii* has been reported from the Philippines (Hawkes 1969).

The species is native to India, Thailand, Malaysia, Indonesia, and the Phillippines (Kamemoto and Sagarik 1975), and has acclimatized in NE India, Malaya, Japan, Taiwan, Thailand, the Phillippines, and Hawaii (USA). The profusion of flowers and length of blooming period makes the species among the most charming of terrestrial orchids.

1.2 *Epidendrum radicans* M.S. Exc Lindl.

(Gen et sp. Orch 104, 1831, *E. rhizophorum.* Batem in Bot. Reg. *E. ibaguense* HBK, *E. pratense* Rchb. f. *E. radicans* Pav Ex Ldl.)

Epidendrum is the largest genus of Orchidaceae, the total number of taxa exceeding 1000. Eu *Epidendrum* is epiphytic, stem scandent, long, often branched near the base, roots hanging in the air, 20–30 cm long. Leaves ovate oblong 5–7 cm long, emarginate, peduncle slender sheathed by imbricating adherent bracts and terminating in many-flowered corymbiform raceme (Fig. 1B). Flowers 4–6 cm in diameter, varying in colors from orange, red, yellow, scarlet, vermilion, or red. Sepals and petals waxy, spreading, elliptic, oblong, lip with three fringed lobes, of which the middle one is bi-partite, bright orange yellow, deeper towards the margin.

Epidendrum radicans is the most brilliant of the bright-colored epidendrums. They are easy to grow and can be trained on trellis or other contrivance to give a very pleasing and effective appearance. They are grown exclusively in Hawaii and other tropical areas and have been introduced into all gardens and nurseries.

Epidendrum radicans is exclusively American, ranging from North Carolina and Mexico throughout the West Indies, Central America, and Argentina. The species is now acclimatized in Mexico, Colombia, Brazil, and Hawaii.

1.3 Conventional Methods of Propagation

Both the species are propagated through vegetative means. The seeds, although produced in large number, cannot be germinated through conventional methods, they need in vitro methods for germination due to their complex requirements, which will be discussed later in the chapter.

1.3.1 Spathoglottis plicata

This species bears prominent pseudobulbs which can be separated and potted separately; the process, however, is very slow, and only a few plants can be raised through this method; further, pseudobulbs cannot be separated very frequently, as it may affect overall growth and flowering in the mother plant.

Being a terrestrial species, the plants do not produce any offshoots or keikis, which are otherwise found in all other epiphytic monopodial orchids.

1.3.2 Epidendrum radicans

As there are no pseudobulbs, the vegetative propagation is restricted to cuttings. A 10–12-cm nodal cutting with two or three aerial roots is most ideal for repotting and propagation; even if roots are not visible, a 10-cm cutting produces them from dormant buds which lie underneath the outer skin covering the stem.

Offshoots and keikis are frequently produced on the mother plant, and can be separated and treated as plantlets. However, the rate of conventional vegetative multiplication is very slow with a maximum of four or five new canes and eight to ten keikis (plantlets) being derived from one old cane in any growing season.

2 Need for Micropropagation

Orchids may well be the first flowering plants of commercial value to be propagated in vitro through both seed and tissue culture. Aseptic germination of seeds is essential, as orchid seeds lack any metabolic machinery and have no endosperm, no cotyledons, and no root initials. They do not have the capacity for directly utilizing the substrate available in nature (Savina 1974; Sheehan 1983; Singh 1981, 1988). Thus, in spite of a very large number of seeds produced only 2–3% of the seeds germinate in nature with the help of mycorrhizal fungus.

The major drawback in the orchid industry has been the length of time taken for clonal propagation of selected species and hybrids. According to Blowers (1964), it takes about 10 years to cultivate about a dozen good-sized propagations. Further, most of the cultivated orchids are complex hybrids and are highly heterozygous, and there is very little hope of breeding pure lines out of them.

Cut flower production for the industry has to be programed in such a manner that the right flower is ready on the day it is required; only a clone or pure line can be used for this purpose to respond to the culture treatment.

The incidence of genetic variability in orchid mericlones is very low as compared to chrysanthemum and carnation, which mutate very readily. Orchids are highly stable and mutate rarely and most of the mericlones are true to type. Mericloning of orchid has also helped to some extent in the elimination of virus and other diseases. It is, however, advisable to use only indexed material for mericloning (Bertsch 1966).

3 Brief Review of in Vitro Approaches in the Two Species

3.1 Embryo Culture

After the successful asymbiotic germination of orchid seeds on nutrient media (Knudson 1922, 1946), many commercial growers and hobbyists tried to improve their media and to germinate other species (for review, see Arditti 1967, 1982, 1984; Stoutmire 1974; Sheehan 1983).

In addition to seed germination, various factors affecting seed germination and subsequent development were also studied, and techniques such as green pod or embryo culture (Valmayor and Sagawa 1967; Lucke 1971) and suspension culture (Singh 1987) were developed during the course of time.

During seed germination (embryo culture) a protocorm-like structure is formed from the clump of embryo cells and on this a shoot meristem can be distinguished. Later, rhizoids are seen on the opposite end. In light the protocorms become green, autotrophic, and more leaves are produced.

3.2 Tissue Culture

Since the pioneering work of Morel in 1960 (Morel 1964), tissue culture has been the most actively researched area in orchid propagation. Orchid hobbyists and growers were the first to experiment with different nutrient media, growth regulators, explant and other factors to bring the technique to commercial level. At present, more than 56 genera are being mericloned.

Table 1. Survey of micropropagation work on *Spathoglottis plicata* and *Epidendrum radicans*

Explant	Medium	Result	Reference
		Spathoglottis plicata	
Seed	White's	Direct differentiation into root and shoot	Chennaveeraiah and Patil (1975)
Hypocotyl	White's	Plantlets	Chennaveeraiah and Patil (1975)
Callus	Murashige and Skoog 1962	Rhizogenesis (95%) callus	Bapat and Narayanaswamy (1977)
Young seedling rhizome	Murashige and Skoog 1962	Callus formation	Bapat and Narayanaswamy (1977)
Seeds	Vacin and Went 1949	Germination and plant formation	Beechey (1970)
Roots	Vacin and Went 1949	Callus and plantlet formation	Beechey (1970)
Seeds	Knudson C 1946	Plantlet formation	Sangama (1981)
		Epidendrum radicans Obrienianum	
Seedling leaf tips	Churchill et al. 1970	Callus formation	Churchill et al. (1970, 1973)
Callus	Churchill et al. 1970	Callus, protocorm and abnormal seedlings	Churchill et al. (1970, 1973)
Protocorms, plantlets	Harrison and Arditti, 1970	Plant formation and growth	Churchill et al. (1970)
Excised aerial roots	Ojima and Fujiwara 1962	Root culture	Churchill et al. (1972)
Seeds	Knudson C 1946	Seedling germination and growth	Vacin and Went (1949)
	Vacin and Went 1949	Seedling growth	Vacin and Went (1949)
Root tips from in vitro-raised plants	Ojima and Fujiwara	Callus formation	Stewart and Button (1978)
Stem, single node sections	Knudson 1946 Heller 1953 Lindermann et al. 1970	Plantlet formation	Stewart and Button (1976)
Flower stalk nodal section	Lindermann et al. 1970	Plantlets and multiple plantlets	Stewart and Button (1976)
Seeds	Vacin and Went 1949	Germination and plantlet formation	Nair (1982)
Flower stalk cuttings	Vacin and Went 1949	Protocorm-like bodies, plantlets	Singh and Prakash (1982)

New techniques and data continue to be added. However, the development of tissue culture procedures in orchids is still highly empirical (Vajrabhaya 1977, Pierik 1987, Withner 1988, Singh 1988).

A perusal of earlier literature (Table 1) reveals that not much work has been done on the two species *Spathoglottis plicata* and *Epidendrum radicans* with respect to embryo culture or tissue culture except for standardization of media, etc.

4 Micropropagation and in Vitro Approaches in *Spathoglottis plicata* Bl.

4.1 Seed Culture

Seed setting in the species is very high and it takes 62–64 days for the capsules to dehisce. However, for in vitro germination the capsules were harvested after 30–35 days (Fig. 2A) and inoculated.

The seeds are dustlike, light brown in color, and contain an undifferentiated embryo surrounded by seed coat (Fig. 2B). Ultrastructure studies conducted in the laboratory (Sangama and Singh 1989) with the help of JEOL-10C × II TEM have revealed that at the protocorm stage all the embryo cells are packed with food reserve in the form of lipid bodies and proteins, (Fig. 2C). Throughout the embryo the cytoplasm is dense and stains heavily for protein. No starch or other carbohydrates were seen except very rarely within proplastids. The mitochondria and endoplasmic reticulum were not observed in the protocorm cells. During the protocorm development stage, lipid and protein bodies were utilized for differentiation.

4.1.1 Sterilization of Seeds

It is always convenient to use the green pod culture or ovule culture technique to raise the seedlings; however, if the seeds are collected from the dehisced capsule they can be sterilized by shaking them with 2% sodium hypochlorite for 10–15 min, followed by a thorough rinse with sterile water so that all traces of chlorine are washed away.

In ovule culture the green capsule was thoroughly cleaned with sodium hypochlorite and sterile water and finally surface sterilized by flaming with alcohol. The capsule was split open longitudinally and ovules scooped and sown on nutrient medium under aseptic conditions.

→

Fig. 2A–C. Capsule and seed structure in *Spathoglottis plicata*. **A** Capsule. **B** Seeds showing embryo, × 52 *e* embryo; *sc* seed coat. **C** Electron micrograph of *Spathoglottis plicata* seeds showing embryo cells with nucleus (*N*) lipid (*L*) protein bodies (*PB*) and cell wall (*CW*), × 2700

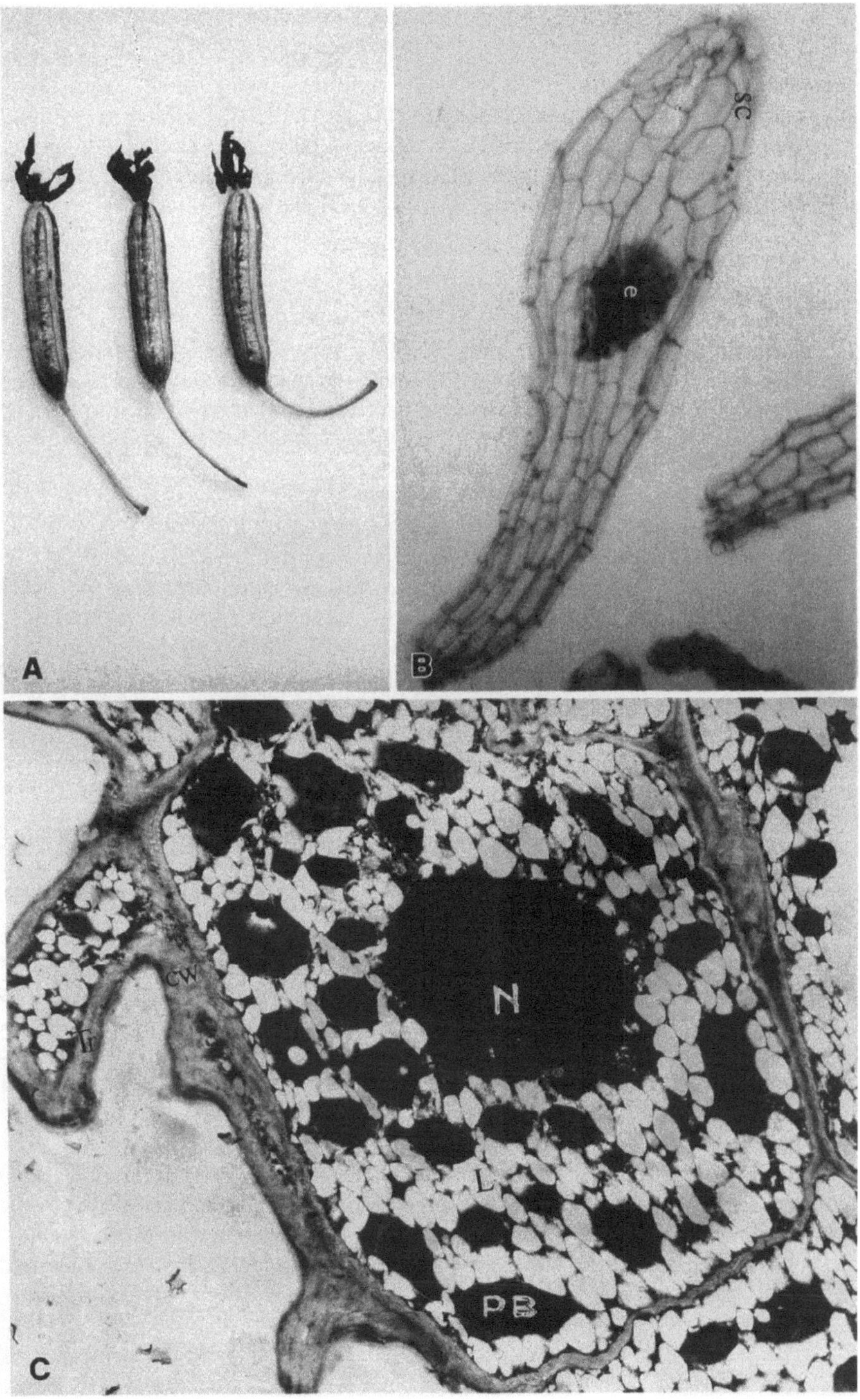
A
sc
e
B
cw
Tr
N
L
PB
C

4.1.2 Viability Testing of Seeds

It is preferred to check the viability of seeds on nutrient media before sowing them. The method standardized in this laboratory (Singh 1981) using differential staining of embryo with tetrazolium chloride and malachite green was followed to ascertain the percentage of viable seeds without having to run a germination test (Sheehan 1983).

4.1.3 Seed Storage

Spathoglottis plicata seeds lose their viability very fast, and the conventional methods of seed storage do not help. However, the seeds can be stored in organic solvents (Singh 1988) such as petroleum ether for more than 15 months to be used later without much loss in viability.

4.1.4 Embryo Culture

Embryo culture in orchids is defined as growing of ovules containing immature embryos excised from green fruits under aseptic conditions; it is also referred to as green pod culture, ovule culture etc. (Sagawa and Valmayor 1966; Sauleda 1976, Singh 1987). The technique has many advantages over the conventional method of sowing seeds from ripe capsules.

In *Spathoglottis plicata* the time interval between pollination and fertilization is between 15–18 days (Swamy 1949) and capsules can be safely harvested for embryo culture after 30–45 days and sown under aseptic conditions (Fig. 3a–c)

4.1.5 Nutrient Media

Out of the five different media used, (Table 2) it was found that germination of *Spathoglottis plicata* seeds was best on Knudson C medium (20 days) as compared to Burgeff N_3f (94 days). Subsequent growth of seedlings was also best on Knudson C medium (Table 2), which proved to be superior to other media. Chennaveeraiah and Patil (1975), however, have reported White's medium as suitable for this species, while Beechey (1970) reported Vacin and Went medium to be most suitable.

A perusal of earlier literature reveals that Knudson C medium has proved to be a better medium for other orchid species like *Cymbidium* (Torikata et al 1965) *Cattleya* (Boesmann 1962) and *Dendrobium* (Mowe 1973). This may be due to the low concentration of salt in Knudson C medium as compared to others.

The best carbon source for the germination of *Spathoglottis plicata* was found to be sucrose (15 g/l) followed by glucose (15 g/l), fructose (10 g/l), and myo-inositol (250 mg/l). It was also observed that organogenesis was promoted at sub-optimal

Fig. 3A–C. Germination of *Spathoglottis plicata*. **A** Protocorm formation. **B** Seedlings on Knudson C media. **C** Seedling transferred to pots

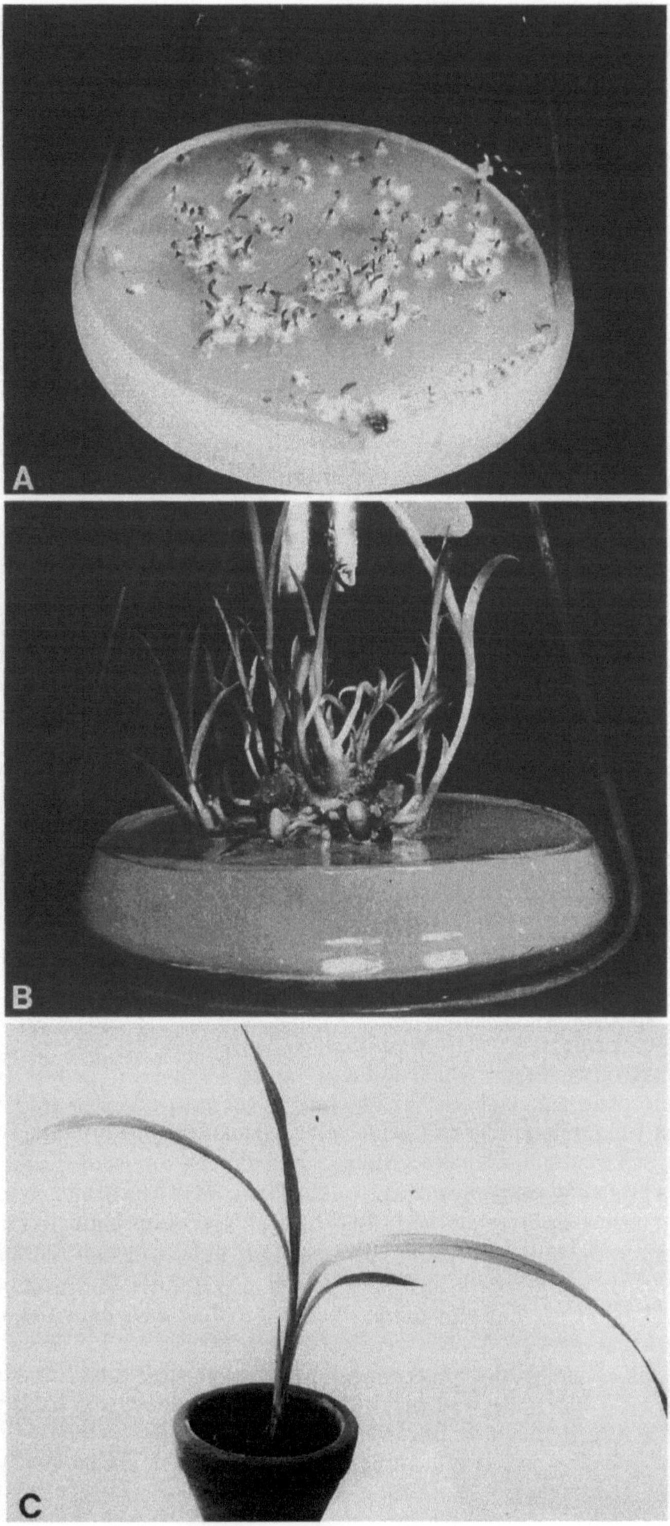

Table 2. Effect of different media on number of days taken for greening, first leaf, second leaf and fourth leaf stages in *Spathoglottis plicata*

	No. of days taken for:			
Medium	Greening	1st leaf	2nd leaf	4th leaf
Knudson C	20	45	80	132
Burgeff N_3f	94	130	158	213
Murashige and Skoog	33	59	87	149
Raghavan and Torrey	35	63	99	153
Vacin and Went	37	62	80	140

concentration while protocorm proliferation is enhanced by a supraoptimal level. After a certain stage of development (four leaves), the exogenous supply of sucrose was not required.

Experiments conducted to study the effect of different growth regulators and additives on seed germination in *Spathoglottis plicata* (Sangama and Singh 1989) revealed that initial germination was accelerated by 1 mg/l of NAA, IBA, and GA_3. However, further growth was inhibited by a higher concentration, which induced callusing, with a total loss of growth of 10 mg/l, The pronounced effect of NAA at 1 mg/l was noteworthy, as was earlier reported by Strauss and Reisinger (1976).

GA_3 at 0.1 mg/l enhanced seed germination; however, it resulted in small sized protocorms and elongated leaves.

Benzyl adenine (BA) at 0.1 mg/l concentration had no effect on seed germination. While 1 mg/l resulted in healthy protocorms and short and stout roots.

Treatment with 2,4-D at different concentration (1,2,3 mg/l) produced only callusing. This is in conformity with the earlier reports by Ueda and Torikata (1969) and Fonnesbech (1972). Chennaveeraiah and Patil (1975), however, reported rapid growth with 0.5 and 1 mg/l 2,4-D.

4.1.6 Encapsulation of Protocorms

Protocorms of *Spathoglottis plicata* were encapsulated in alginate to produce individual beads, which could be stored for more than 180 days at 4 °C without much loss of viability. The technique of encapsulation will provide an easy and novel propagation and delivery system (Singh 1991).

Twenty day-old green protocorms of *S. plicata* were selected for encapsulation. The protocorms were mixed with 20 ml of 2.5% sodium alginate (Sigma). For encapsulation the protocorms were dropped individually or two or three together with the help of a sterile bent tip forcep into 50 mM solution of calcium chloride in water (1.36 g/150 ml). Calcium alginate beads were formed around the protocorms within 10–15 min at 25 °C on a rotary shaker at 80 rpm.

The alginate-embedded protocorms (beads) were then rinsed in sterile water, dried on sterile filter paper and were either germinated directly on nutrient medium or stored at 4 °C in the dark. Encapsulated protocorms germinated well even after storage of 180 days (70%), indicating that this method could be used to preserve

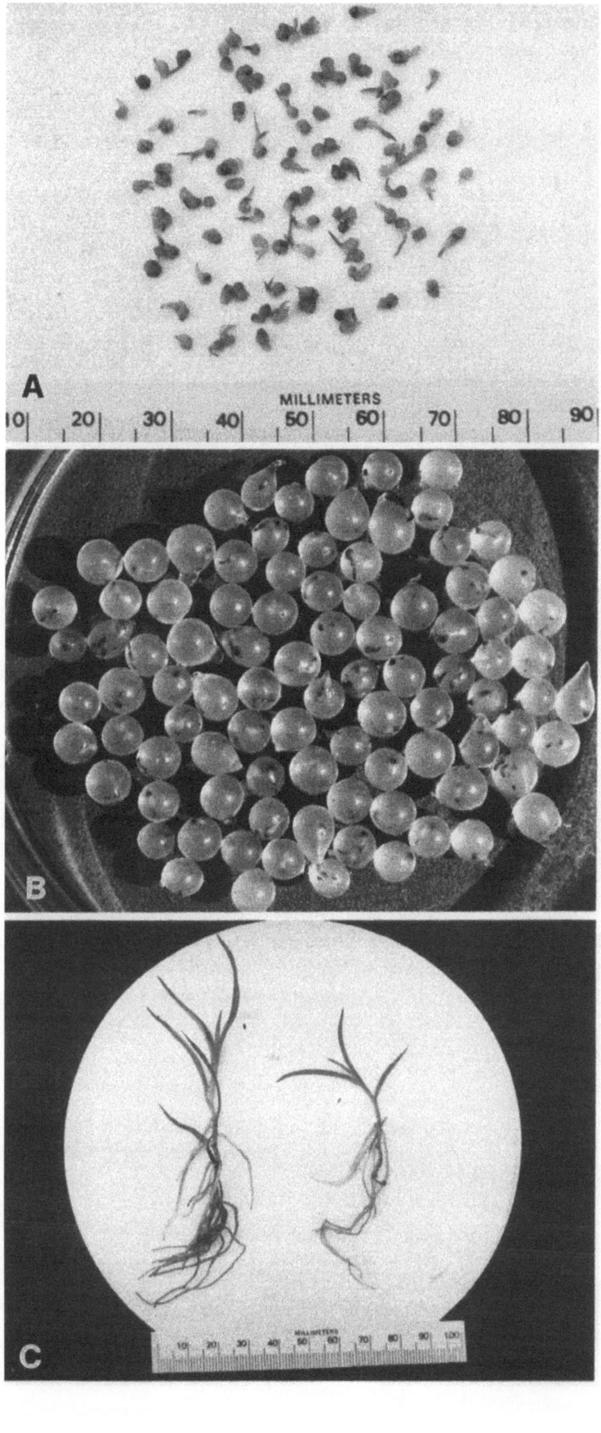

Fig. 4A–C. Encapsulation of protocorms of *Spathoglottis plicata*. **A** Protocorms. **B** Encapsulated protocorms (beads). **C** Seedlings raised from encapsulated protocorms (beads) on VW medium desirable genotypes for the future, and also as an efficient delivery system. The seedlings raised through the encapsulation method were successfully transferred to the field (Fig. 4A–C)

5 Micropropagation and in Vitro Approaches in *Epidendrum radicans*

5.1.1 Seed Germination and Embryo Culture

The seed set in the species is very high, the capsule takes 120–130 days for maturity, the seeds are dull white, comparatively large, weighing 6 μg (Arditti 1979). The seed coat is membranous and net-like, the embryo is undifferentiated and situated in the center.

The capsule can be harvested safely after 90–105 days and the embryo germinated on nutrient media. The embryo/green pod culture technique offers several advantages over the conventional method of sowing seeds from ripe capsules. (Sagawa and Valmayor 1966).

5.1.2 Nutrient Media

Out of the four different media tested, the best and earliest seed germination was observed on Vacin and Went medium followed by KC, MS, and Burgeff N_3f (Table 3).

Table 3 shows that all four nutrient media supported good germination and subsequent growth of seedlings. Being an epiphytic orchid, *E. radicans* prefers low salt concentration for optimal growth, which is in accordance with earlier work by Sagawa and Valmayor (1966) and Valmayor and Sagawa (1967). Experiments conducted by Nair (1982) have revealed that NAA 1 mg/l was very effective in enhancing germination and subsequent growth; however, at higher concentrations it stimulated callus formation and affected the growth of plantlets. The growth-promoting effects of NAA have been reported for a number of other orchid species, e.g., *Cypripedium* (Boesmann 1962) and *Vanda* (Arditti 1979; Mathew and Rao 1980).

Table 3. Effect of different media on seed germination in *Epidendrum radicans*

Nutrient media	Mean no. of days for				
	Greening	Ist leaf	2nd leaf	3rd leaf	4th leaf
Knudson C (KC)	19	48	83	115	114
Murashige and Skoog (MS)	21	54	82	110	142
Vacin and Went (VW)	17	39	65	03	125
Burgef N_3f (BN)	24	44	79	110	148

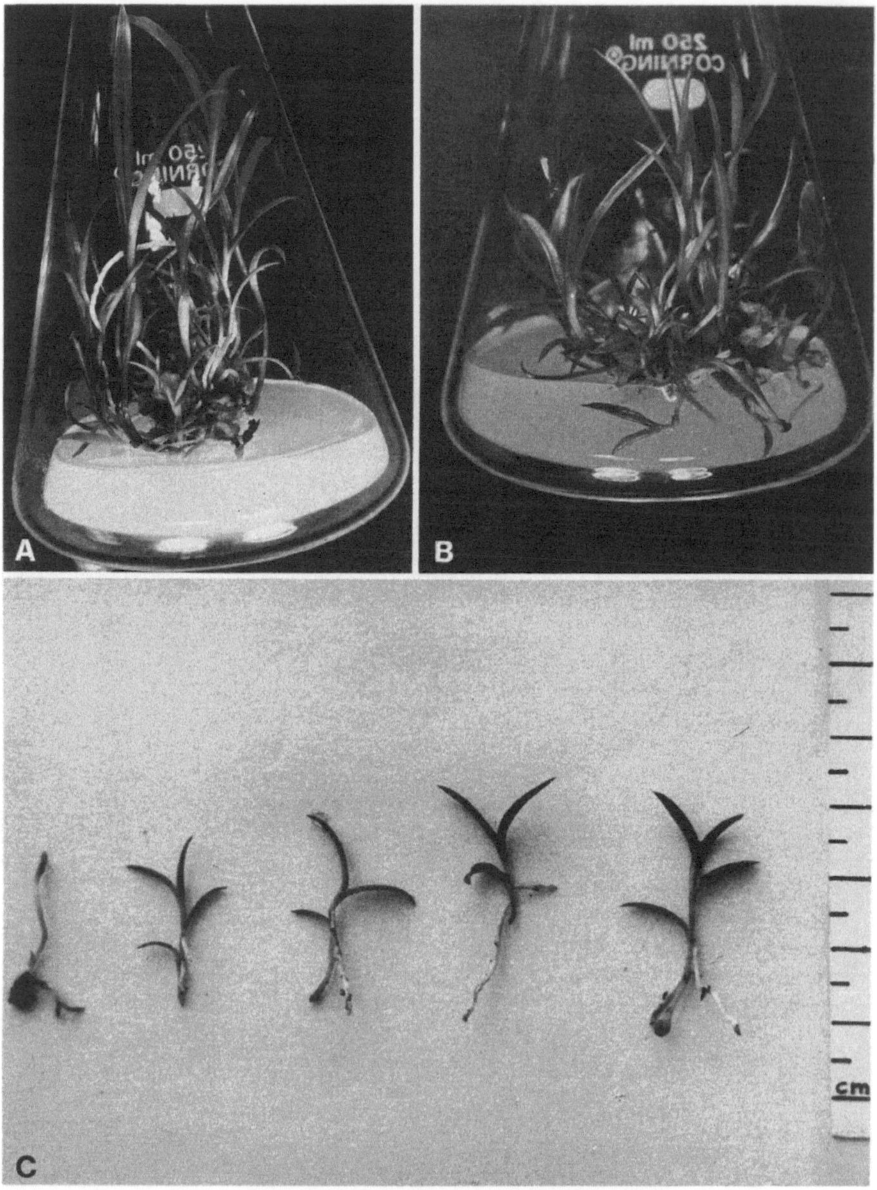

Fig. 5A–C. Germination of orchid seedlings on nutrient media. **A** VW medium. **B** VW medium supplemented with coconut milk 15% (v/v). **C** Young seedlings ready for transplanting to field

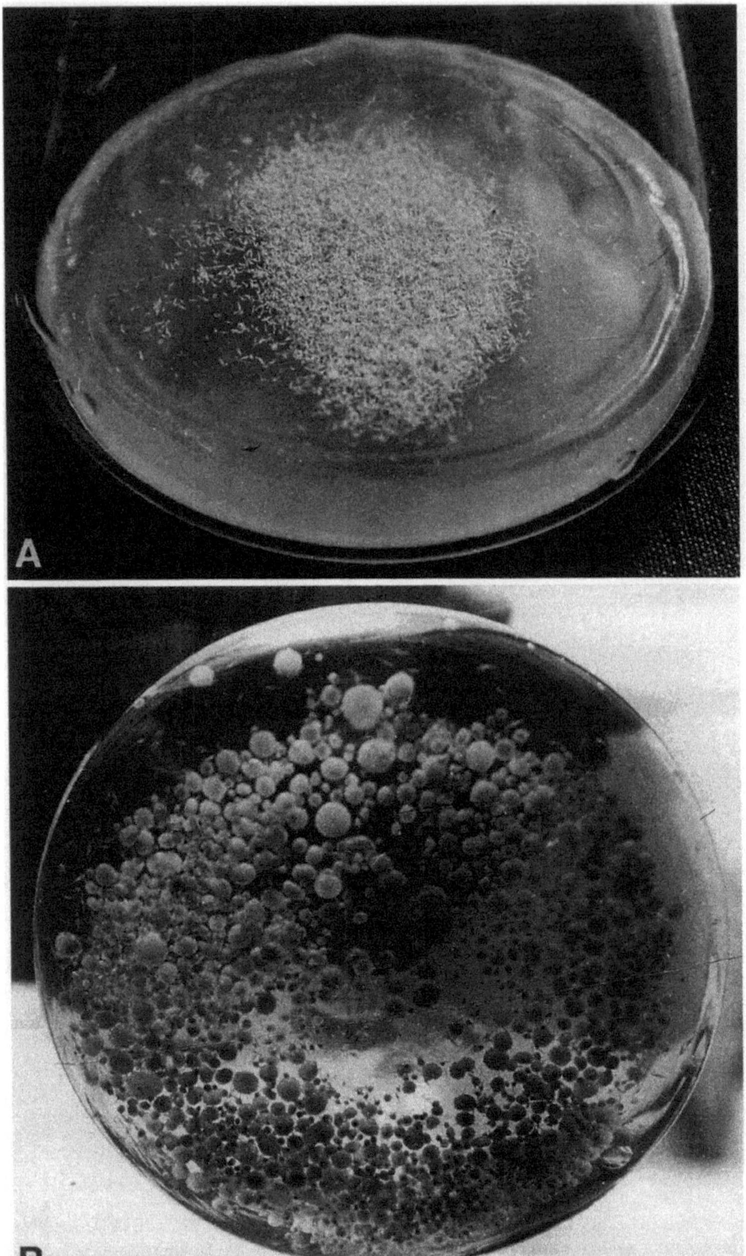

Fig. 6A,B. Suspension culture of seeds of *Epidendrum radicans*. **A** Solid medium. **B** Liquid medium

Treatment with 2,4-D at a lower level (1 mg/l) produced few seedlings, while a high concentration was inhibitory and induced callus formation.

Additives like coconut milk (15% v/v) accelerated the growth of seeds and resulted in very robust seedlings. A combination of NAA 1 mg/l. IBA 3 g/l and CM (15% v/v) was found to be the best for germination of *Epidendrum radicans.* seeds (Fig. 5A–C)

5.1.3 Suspension Culture

Most of the nutrient media used for the culture of orchid embryos are solid media (Valmayor and Sagawa 1967; Pierik and Steegman 1972; Arditti 1984) or liquid media with glass wool or cotton wool for the support of the seeds (Ernst 1975).

A technique was developed (Singh and Prakash 1985) to culture embryos of *Epidendrum radicans* in suspension, and the method has given much better results as compared to germination on solid medium (Fig. 6A–B). Growth of embryos in liquid VW medium was substantially faster than that on solid VW medium (Fig. 6A–B). After 31 days, the difference in growth in the two media was remarkable, and the seedlings orginating from liquid media formed much healthier plantlets.

The effect of liquid medium on orchid embryo germination may be due to

1. Better aeration provides optimum conditions for respiration, protein synthesis, and salt intake.
2. Increased surface area which is exposed to nutrients, resulting in healthy growth.
3. The liquid medium may help in the dilution of some specific inhibitors or stalling products produced in the process of growth and differentiation.

The liquid culture media not only accelerate the germination process and subsequent growth but also save 8–10 months of time interval between pollination and flowering plant production (Singh 1988).

5.2 Tissue Culture Propagation

As compared to other orchids, such as *Catlleya, Cymbidium,* or *Dendrobium,* clonal propagation in *Epidendrum* has been relatively less studied (Stewart and Button 1976; Arditti 1977; Singh and Prakash 1982). Root tips and leaf tips were also tried by Churchill et al. (1970, 1972) and Stewart and Button (1978) without much success, callus or plantlet formation was observed in rare cases.

5.2.1 Nodal Sections

Nodal sections from young healthy plants were selected as explant (Fig. 7A). After washing with sterile water for 2–3 min, the sections were sterilized for 10 min with 2% sodium hypochlorite to which a drop of Tween was added, and then thoroughly rinsed in sterile water to remove the last traces of chlorine. Bleached tissue at the cut ends of the sections was removed before inserting the sections into the

Table 4. Effect of different media combinations on organogenesis in *Epidendrum radicans* nodal sections[a]

	Differentiation of nodal section into:				
	No growth	Callus formation	Plantlet formation	multiple plantlet formation	Remarks
Knudson C	–	4	5	1	Plantlets healthy
Vacin and Went	7	–	3	–	Poor rooting
Murashige and Skoog	8	–	2	–	Poor rooting
Kc + NAA (1 mg/l)	3	7	–	–	Callus degenerated
Kc + NAA (2 mg/l)	8	2	–	–	Callus degenerated
Kc + CW (15% v/v)	–	–	6	4	Plantlets healthy

[a] Number of cultures in each treatment = 10.

media under aseptic conditions. All the sections which were cultured had undamaged intact buds.

All three media, (Knudson, Vacin and Went, Murashige and Skoog), supported the growth and buds developed spontaneously into plantlets; Knudson C was, however, superior to the other media in all growth characteristics. (Table 4). Knudson C medium also induced callus formation (7B) and multiple plantlet formation in some cases. No callusing was found on Vacin and Went medium or Murashige and Skoog medium (Table 4).

Addition of NAA (1 mg/l) suppressed the development of plantlets and induced callus formation. Additives like coconut milk (15% v/v) when incorporated into the Kc medium accelerated the growth of plantlets and also induced multiple plantlet formation; the stimulatory effect may be due to the presence of cytokinins in coconut milk. The plantlets formed in vitro developed normally, within 150–160 days they had developed roots; at this stage individual plants were separated from the original explant and transferred to community pots in the orchid house (Fig. 7a–d).

Formation of multiple plantlets, with the addition of CM from single node section is very useful in the micropropagation of orchids like *Epidendrum radicans.* Multiple plantlet formation was reported earlier in other orchids (Fonnesbech 1972; Pierik and Steegman 1972; Singh and Prakash 1984). It may be helpful in propagating cultivars which are available in limited quantity and for which other methods of propagation are not suitable.

5.2.2 Flower Stalk Cuttings

Epidendrum radicans is known to have lot of offshoots and keikis from the base of the flower canes; 70% of the nodes on the flower stalk carry a dormant bud which can be stimulated to develop into normal plantlets on an appropriate medium (Singh and Prakash 1982).

Fig. 7A–D. Tissue culture propagation of *Epidendrum radicans* nodal section as explant. **A** Nodal section. **B** Plantlet formation. **C** Fully grown plantlet. **D** Plantlet transferred to field

A
B
C
D

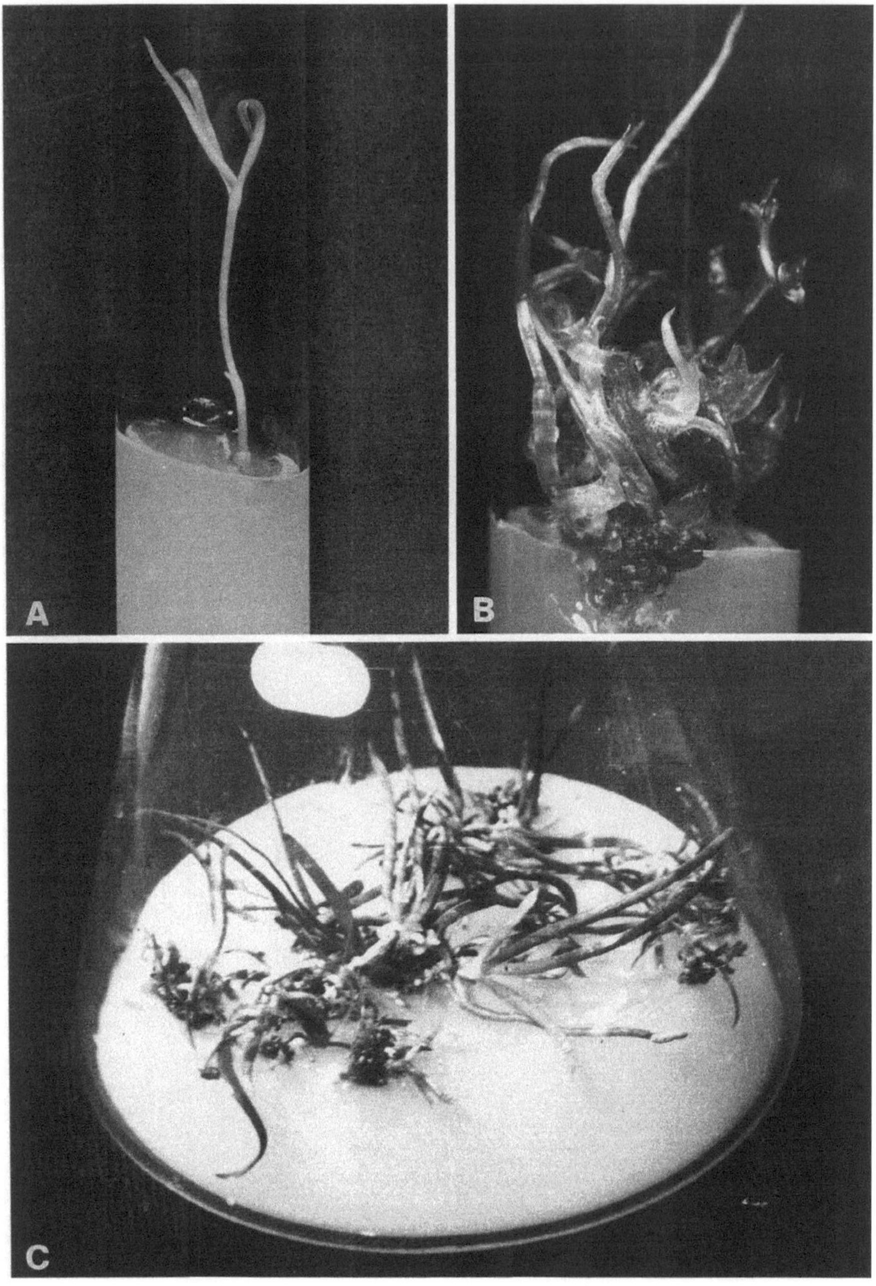

Fig. 8A–C. Tissue culture propagation of *Epidendrum radicans* flower stalk cuttings as explant. **A** Direct plantlet formation on VW medium. **B** Multiple plantlet formation. **C** Mericlones

Flower stalks were wiped clean with 95% ethyl alcohol; 4–6 cm sections, each with a dormant bud, were washed with sterile water and sterilized in 2% sodium hypochlorite, washed again with sterile water, and planted on VW medium.

The medium was supplemented with coconut milk (15%, v/v), 1 mg/l BA, 1 mg/l NAA and gelled with 9 g/l agar or 0.25% Gelrite; pH was adjusted to 5.2 before autoclaving.

After 35–40 days, when the buds had sprouted they were transferred to solid VW medium supplemented with 3 mg/l IBA. After 120–130 days, fully grown plantlets were ready to be taken out and transferred to community pots (Fig. 8).

The technique of using flower stalk cuttings as explant for propagation is well known and has been tried in other orchid species, *Thunia alba* (Singh and Prakash 1984), *Vanda* (Sagawa and Sehgal, 1967), and *Phalaenopsis* (Rotor 1949; Griesbach 1983). The technique is, however, not suitable for sympodial orchids.

5.3 Commercial Production

When an excised meristem measuring 0.25–1 mm in length with meristem dome is placed on a simple medium, instead of giving one plantlet, the meristem first develops into a mass of tissue morphologically similar to the protocorms which are often found when orchid seeds germinate in vitro. When these masses or protocorms are cut into pieces, each piece proliferates to give more masses which can be further subdivided or made to differentiate into plantlets; thus from one isolated

Fig. 9. Young plants in net pots ready for sale to hobbyist and growers

meristem many plantlets are obtained. It is estimated that if frequent cutting of protocorms could be successfully continued through a period of 1 year, 2 million plantlets could be produced from only one meristem.

For commercial production the saleable products are plantlets in net pots (Fig. 9) which have undergone hardening for a period of 2–3 months; the other items of sale include plantlets in flasks, or bottles, these plantlets are already differentiated from the protocorms and are at the transplanting stage. The use of tissue culture technique for commercial production of orchids is well known in countries like Singapore, Thailand, Hawaii (USA), etc.

6 Summary and Conclusion

Orchids, known to the horticulturists for the past 40 years (Garay 1974, Rao 1977), are one of the few flowering plants of commercial value to be propagated in vitro both through seed and tissue culture. Orchid seeds, although produced in a very large number, lack any functional endosperm and do not germinate on their own. Similarly, conventional methods of propagation are very slow and do not keep up with the fast-developing commercial industry.

Micropropagation techniques have been developed to propagate the two most important orchid species, *Spathoglottis plicata* (Terrestrial) and *Epidendrum radicans* (Epiphytic).

The embryo culture or green pod culture technique is used to raise seedlings, thus having a lot of time between pollination and flowering plant production. The use of liquid medium or the suspension culture technique to raise *Epidendrum radicans* seedlings has not only accelerated the growth of seedlings and their early development, but also saves the use of agar or other gelling agent in the medium. Experiments conducted to store the seeds of *S. plicata* have indicated that seeds can be successfully stored in organic solvents for later use (Singh 1988).

Protocorms of *S. plicata* were successfully encapsulated in sodium alginate to provide easy and novel propagation and delivery system.

New methods have been developed for the micropropagation of *Epidendrum radicans*. The explants used are nodal sections and flower stalk cuttings. Dormant buds on nodal section are successfully invigorated on Knudson C media. Multiple plantlet formation was observed when the media was supplemented with coconut milk (15% v/v). The formation of multiple plantlets from single node sections is very useful in propagating cultivars which are available in limited quantity and for which other methods of propagation are not suitable.

Use of flower stalk sections for the in vitro propagation of *Epidendrum radicans* has opened new avenues for orchid propagation, especially in species where acquisition of suitable explant is difficult.

The technique of micropropagation like, embryo culture, suspension culture, encapsulation of protocorms, use of flower stalk cuttings as explant and multiple plantlet formation will not only help in the propagation of these two species but also help in the propagation of other orchid hybrids and species which are difficult to propagate or where the explant is the limiting factor.

7 Protocols

1. Embryo Culture. Mature unripe capsules were harvested, surface sterilized, and ovules inoculated on Knudson C medium (*Spathoglottis plicata*) and Vacin and Went medium (*Epidendrum radicans*); the pH of the medium was adjusted to 5.2 before autoclaving.

2. Tissue Culture. Nodal sections and flower stalk cuttings of *Epidendrum radicans* were sterilized with 2% sodium hypochlorite and inserted into the medium under aseptic conditions, the media Kc (nodal sections) and VW (flower stalk cuttings) were gelled with 0.25% Gelrite; pH of the medium was adjusted to 5.5 before autoclaving.

3. Culture Conditions. Cultures were incubated at 25 $\pm$ 2 °C under 16-h photoperiod (1.5 W/m^2). The plantlets at the four-leaf stage were taken out of the flasks, washed thoroughly with water to remove all traces of gel, and transferred to community pots containing a mixture of three fern fiber and charcoal (1:1). The plantlets were not allowed to dry, and were kept moist.

References

Arditti J (1967) Factors affecting germination of orchid seeds. Bot Rev 33: 1–97

Arditti J (1977) (ed) Orchid biology 1. Cornell University, Ithaca, NY

Arditti J (1979) (ed) Aspects of the physiology of orchids in HW Woolhouse. Advances in botanical research, vol 7. Acad Press New York pp 422–697

Arditti J (1982) Orchid biology reviews and perspectives I. Cornell University, Ithaca, NY

Arditti J (1984) (ed) Orchid biology reviews and perspectives III. Cornell University, Ithaca, NY

Bapat VA, Narayanaswamy S (1977) Rhizogenesis in a tissue culture of the orchid *Spathoglottis*. Bull Torrey Bot Club 104: 2–4

Bernard N (1989) Sur la germination du *Neottianidus avis* CR Acad Sci Paris 128: 1253–1255

Beechey N (1970) Propagation of orchids from aerial roots. Am Orchid Soc Bull 39 1085–1088

Bertsch W (1966) A new frontier orchid propagation by meristem culture. Proc 5th World Orchid Conf Long Beach Am Arch Soc pp 225–229

Blowers JW (1964) Meristem propagation and possible effects. Orchid Rev 72: 407–408

Boesmann G (1962) Problèmes concernant le semis et l'amélioration des orchidées. Adv Hortic Sci 2 368–372

Chennaveeraiah MS, Patil SJ (1975) Morphogenesis in seed cultures of *Spathoglottis plicata*, Curr Sci 44: 68

Churchill ME, Ball EA, Arditti J (1970) Production of orchid plants from seedling leaf tips. Orchid Dig 34: 271–273

Churchill ME, Ball EA, Arditti J (1972) Tissue culture of orchids II, method for root tips. Am Orchid Soc Bull 41: 726–730

Churchill ME, Ball EA, Arditti J (1973) Tissue culture of orchids, method for leaf tips. New Phytologist 72 161–166

Ernst R (1975) Studies on asymbiotic culture of orchids Am Orch Soc Bull 44: 12–18

Fonnesbech M (1972) Growth hormones and propagation of *Cymbidium* in in vitro. Physiol Plant 27: 310–316

Garay LA (1960) On the origin of the Orchidaceae. Bot Mus Leaflets Harv Univ 19: 57–96

Garay LA (1974) The enigmatic power of orchids. Am Orchid Soc Bull 43: 384–393

Griesbach RJ (1983) The use of indoleacetylamino acids in the in vitro propagation of *Phalaenopsis* Orchids. Sci Hortic 19: 363–366

Hawkes AD (1969) Encyclopedia of cultivated orchids. Faber and Faber. London

Kamemoto H, Sagarik R (1975) Beautiful Thai orchid species. Orchid Soc of Thailand, Bangkok

Knudson L (1922) Non-symbiotic germination of orchid seed. Bot Gaz 73: 1–25

Knudson L (1946) A new nutrient solution for germination of orchid seed. Am Orchid Soc Bull 15: 214–217

Lucke E (1971) The effects of biotin on sowing of *Paphiopedilum*. Am Orchid Soc Bull 40: 24–26
Mathew VH, Rao PS (1980) In vitro multiplication of *Vanda* hybrid through tissue culture technique. Plant Sci Lett 17: 383–89
Morel G (1960) Producing virus free *Cymbidium*. Am Orchid Soc Bull 29: 495–497
Morel G (1964) A new means of clonal propagation of orchids. Am Orchid Soc Bull 31: 473–477
Morel G (1974) Clonal multiplication of orchids In: Withner CL (ed) The orchids scientific studies. John Wiley & Sons, New York pp 169–222
Mowe BL (1973) Germination and growth of *Dendrobium* in several culture media. Singapore J Primary Ind J: 20–30
Nair SR (1982) Seed germination and tissue culture studies in orchids PhD Thesis Univ Agric Sci Bangalore
Pierik RLM (1987) In vitro culture of higher plants. Martinus Nijhoff Publ Netherlands
Pierik RLM, Steegman HHM (1972) The effect of 6-benzylamino-purine on growth and development of cattleya seedlings grown from unripe seeds. S Pflanzen physiol 68: 2328–234
Rao AN (1977) Tissue culture in orchid industry In: Reinert J and Bajaj YPS (eds) Applied and fundamental aspects of plant cell, tissue, and organ culture. Springer, Berlin Heidelberg New York, pp 45–69
Rotor G (1949) A method of vegetative propagation of *Phalaenopsis* species and hybrids. Am Orch Soc Bull 18: 738–739
Sagawa Y, Sehgal OP (1967) Aseptic stem propagation of *Vanda* Miss Joaquime. Pac Orchid Soc Bull 25: 17–18
Sagawa Y, Valmayor HL (1966) Embryo culture of orchids. In: L. R. De Garmo (ed) Proc 5th World Orchid Conf, Long Beach, pp 99–101
Sangama (1981) Effect of growth regulators on the germination of orchid seds, MSc Thesis, Univ Agric Sci Bangalore
Sangama, Singh F (1989) Ultra structure changes during seed germination of *Spathoglottis plicata* Cytologia (submitted)
Sauleda RP (1976) Harvesting time of orchid seed capsule for the green pod culture process. Am Orchid Soc Bull 45: 416–419
Savina GI (1974) Fertilization in orchids. In: Linskens HF (ed) Fertilization in higher plants. American Elsevier Co, New York, pp 197–204
Schultes RE, Pease AS (1963) Generic names of orchids, their origin and meaning. Acad Press, New York
Sheehan TJ (1983) Recent advances in botany propagation and physiology of orchids. In: Janick J (ed) Hortic Rev 5: 279–315
Singh F (1981) Differential staining orchid seeds for viability testing. Am Orchid Soc Bull 50: 416–418
Singh F (1983) Orchid tissue culture, potential priorities and application. In: A Wadhwani (ed) Proc Natl Symp Tissue Culture Kasargod ICAR, New Delhi, pp 67–75
Singh F (1987) Application of tissue culture techniques in orchids. In: Reddy GM (ed) Proc Symp Plant Cell and Tissue culture of economically important plants. Osmania Univ Hyderabad pp 165–173
Singh F (1988) Storage of orchid seeds in organic solvents. Gartenbauwissenschaft 53: 122–124
Singh F (1991) Encapsulation of *Spathoglottis plicata* protocorms. Lindleyana 6: 61–63
Singh F, Prakash D (1982) In vitro propagation of *Epidendrum* through flower stalk cuttings. Proc Natl Semin Plant Prop Calcutta, pp 103–104
Singh F, Prakash D (1984) In vitro propagation oif *Thunia alba* (wall) Reichb F, through flower stalk cuttings. Sci Hortic 24: 385–390
Singh F, Prakash D (1985) Suspension culture technique for the culture of orchid embryos. Gartenbauwissenschaft 50: 236–238
Stewart J, Button J (1976) Tissue culture studies in *Paphiopedilum* In: Senghas K-H (ed) Proc VIIIth World Orchid Conf, Frankfurt, Germany pp 372–78
Stewart J, Button J (1978) Development of callus and plantlets from *Epidendrum* roots tips cultured in vitro. Am orchid Soc Bull 47: 607–612
Stoutmire W (1974) Terrestrial orchid sedlings. In: Withner CL (ed) The orchids. Scientific Studies. John Wiley, New York; pp 101–128
Strauss MS, Reisinger DM (1976) Effects of naphthaleneacetic acid on seed germination. Am Orchid Soc Bull 45: 722–723
Swamy BGL (1949) Embryological studies in the orchidaceae II Embryology. Am Midl Nat 41: 202–232

Teuscher H (1978) *Calanthe* and *Spathoglottis*. Am Orchid Soc Bull 47: 336–343
Torikata H, Sawa Y, Sisa M (1965) Non-symbiotic germination and growth of the orchid seeds. J Jpn Soc Hortic Sci 34: 63–70
Ueda H, Torikata H (1969) Organogensis in the meristem tissue culture of cymbidium. J Jpn Soc Hortic Sci 38: 188–193
Vacin E, Went F (1949) Some pH changes in nutrient solutions. Bot Gaz 110: 605–613
Vajrabhaya T (1977) Variation in clonal propagation In: Arditti J (ed) Orchid biology I, Cornell Univ Press, Ithaca, PP 177–202
Valmayor HL, Sagawa Y (1967) Ovule culture in some orchids. Am Orchid Soc Bull 36: 766–769
Van der Pijl L, Dodson CH (1966) Orchid flowers, their pollination and evolution. Univ of Miami Press, Miami
Withner CL (1988) Notes on the in vitro propagation of orchids. Orchid Dig 52: 173–177

XVII Micropropagation of *Phalaenopsis* spp.

M. TANAKA[1]

1 General Account

The genus *Phalaenopsis* (Orchidaceae) comprises approximately 40 species. They are distributed in the Himalayas, China, Tibet, Southeast Asia, Formosa, the Philippines, the Andaman Islands, Sumatra, Java, Borneo, Celebes, New Guinea, and northern Australia (Sweet 1980). The plant is an epiphyte with fleshy, copious, adventitious roots arising from the base and lower nodes of stem. The stem is short, leafy, and does not form pseudobulb. The leaf is distichous, persistent, and fleshy; varying in outline from elliptic to oblong-lanceolate and obovate-lanceolate. The inflorescence, usually a many-flowered raceme, arises laterally and is either straight or flexious (Fig. 1). Although the chromosome number of most of the species is $2n = 38$, that of the most hybrids varies from $2n = 38$ (diploid) to $2n = 114$ (hexaploid) (Woodard 1951; Kamemoto et al. 1961, 1964, 1967; Sagawa and Niimoto 1961; Sagawa 1962; Shindo and Kamemoto 1963a,b; Pancho 1966; Sharma and Chatterji 1966).

Utilizing mostly the large-flowered species, such as *Phalaenopsis amabilis*, *P. aphrodite*, *P. schillerana*, *P. sanderana* and *P. stuartiana*, modern hybrids with large-sized white flowers or pastel-colored flowers have been bred. *Phalaenopsis* is now one of the most popular orchid genera and is becoming an important ornamental plant. It is being grown commercially for cut flower production and as pot plant. Recently, hybridizers have been successful in producing various types of more colorful *Phalaenopsis* hybrids which became popular, and their production for the potted-plant market has also gradually increased. Most *Phalaenopsis* growers, however, still depend on seeds for their production. For this reason, a reliable method of clonal propagation has long been hoped for.

Since 1972 we have been working on leaf segment culture with the intention of establishing a convenient method of micropropagation in *Phalaenopsis*. This chapter reviews the various aspects of the micropropagation of *Phalaenopsis* based on our published works together with those of others, and also discusses its commercial aspects.

[1]Laboratory of Floriculture, Faculty of Agriculture, Kagawa University, Miki-cho, Kagawa 761-07, Japan

Biotechnology in Agriculture and Forestry, Vol. 20
High-Tech and Micropropagation IV (ed. by Y.P.S. Bajaj)

Fig. 1. A *Phalaenopsis* plant with a flower stalk at appropriate stage for micropropagation

2 Brief Review of Micropropagation

There are several nodes on the basal part of the *Phalaenopsis* flower stalk which have dormant buds. These buds could be induced to form plantlets in vitro for subsequent micropropagation. Plantlet induction can be done by culturing the bud, with a piece of flower stalk below and above it (Fig. 2) and was tried for the first time by Rotor in 1949. Later, Rotor's method was modified by many other workers, and was used as the first step to the micropropagation of *Phalaenopsis* (Sagawa and Niimoto 1960; Sagawa 1961; Urata and Iwanaga 1965; Scully 1966; Daerr 1967; Intuwong et al. 1972; Koch 1974a, b; Tews 1974; Reisinger et al. 1976; Arditti et al. 1977; Flamee and Boesman 1977; Bouriquet et al. 1982).

The lateral buds on the flower-stalk cutting cultured in vitro showed three growth patterns: they remain dormant, they grow to form vegetative shoots or they grow to form secondary flower stalks (Urata and Iwanaga 1965; Koch 1974a,b) (Fig. 3). Obtaining clean culture of the cuttings is the most serious problem that has been encountered. For this reason, this method of micropropagation yields, in practice, only a relatively small number of plantlets from an elite genotype.

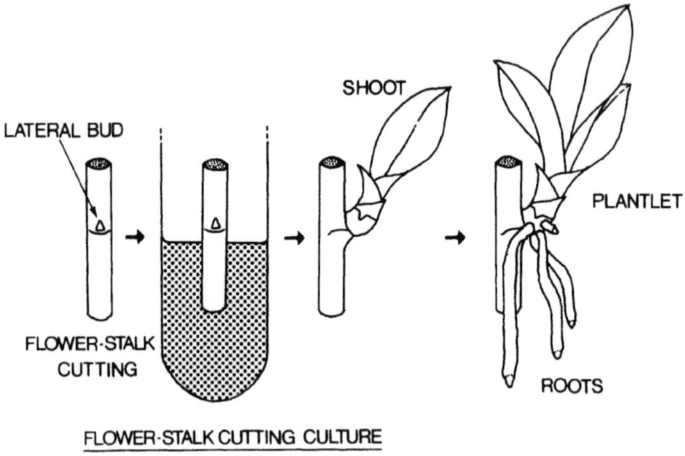

Fig. 2. Diagram illustrating the procedure for inducing plantlets on flower stalk cuttings of *Phalaenopsis*

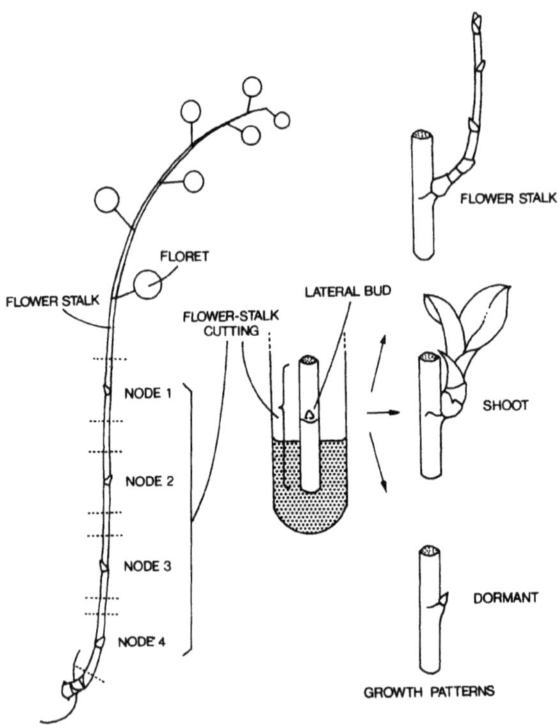

Fig. 3. Diagram illustrating the different patterns of growth of the lateral buds of *Phalaenopsis* flower stalk cuttings cultured in vitro. (Tanaka and Sakanishi 1978)

A faster method of multiplication would be the production of protocorm-like bodies (PLBs) from explants cultured in vitro, as in *Cymbidium* (Morel 1960, 1965; Wimber 1963) or other orchid genera. Hence, many studies have been reported on the micropropagation of *Phalaenopsis* using shoot tips (Intuwong and Sagawa 1974), root tips (Tanaka et al. 1976), axillary buds on the flower stalk (Tse et al. 1971; Koch 1974a,b; Fu 1978), segments of broken plantlets derived from the axillary bud (Pieper and Zimmer 1976; Zimmer and Pieper 1978, 1979), as well as internodal sections of the flower stalk (Homma and Asahira 1985; Lin 1986). To date, however, clonally propagated plants are still not available in large numbers to *Phalaenopsis* growers for production.

Excision of shoot tips or axillary buds from adult plants can kill or cause severe damage to the mother plant. In addition, the numbers of explants available is limited since *Phalaenopsis* is a monopodial orchid. Therefore, it is considered desirable to use leaf tissue as explant.

3 Micropropagation Through Leaf Segment Culture

Wimber (1965) was the first to report the formation of protocorm-like bodies (PLBs) from leaf tissues in orchids. The PLBs were formed on a leaf of *Cymbidium* plantlets that had developed from shoot tip maintained on a gyrotory shaker in liquid medium. Since then, the formation of PLB on cultured leaf segments has been reported in *Laeliocattleya* (Churchill et al. 1970, 1971, 1973; Ball et al. 1971), *Cattleya* (Champagnat et al. 1970), *Vanda* (Tanaka et al. 1975), *Aranda* (Fu 1979), *Renantanda* (Goh and Tan 1982), and *Rhynchostylis* (Vij et al. 1984).

3.1 Regenerative Capacity of *Phalaenopsis* Leaf Segments

Our preliminary work started by examining the capability of producing PLB in leaf segments derived from seedlings, adult plants, and shoots derived from flower stalk cuttings cultured in vitro. It was observed that half of the leaves derived from seedlings formed PLBs on the adaxial surface (Fig. 4A) when cultured on Murashige and Skoog's medium (1962) (MS) supplemented with 5 mg/l NAA, 10 mg/l kinetin, and 10%(v/v) coconut milk (CM). However, both leaf segments from adult plants and from shoots on flower stalk cuttings did not form PLB (Tanaka et al. 1975). These results suggest the possibility of regeneration from *Phalaenopsis* leaf segments, but at this stage it would be of no practical use if leaf segments from an adult plant or young shoot on flower stalk cuttings are not utilized.

In later experiments, the author was able to induce PLB formation on leaf segments excised from young shoots derived from flower stalk cuttings cultured in vitro. Among the media examined, MS supplemented with 1 mg/l NAA, 10 mg/l adenine, and 10 mg/l BA proved to be the most suitable for PLB formation on the leaf segments (Fig. 4B; Tanaka and Sakanishi 1977). This medium could also be useful for inducing PLB formation on the leaf segments excised from proximal, unexpanded parts of the youngest leaf (ca. 5 cm long) of adult plants, although only

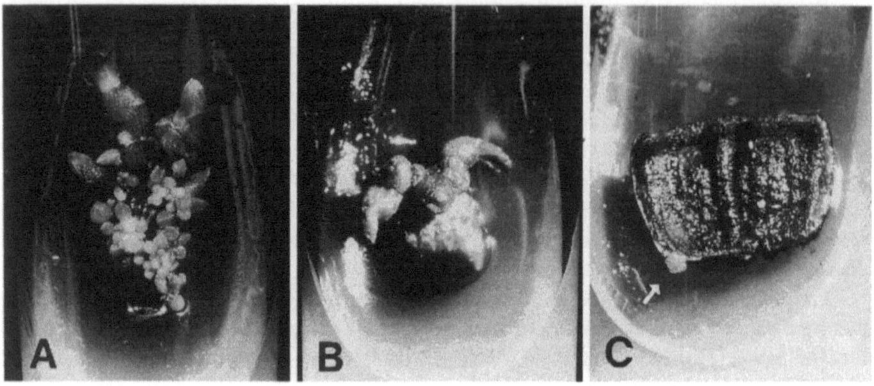

Fig. 4A–C. Protocorm-like bodies (PLBs) formed from the leaf segments from various sources of *Phalaenopsis amabilis* hybrid. **A** PLBs formed on adaxial surface of leaf excised from 120-day-old seedlings, 106 days after planting. **B** PLBs formed on adaxial surface of leaf segment of shoot derived from flower stalk cuttings, 125 days after planting. **C** PLB (*arrow*) formed on adaxial surface of leaf segment excised from newly expanded leaf (about 5 cm long) of mature plants, 239 days after planting. (Tanaka 1987)

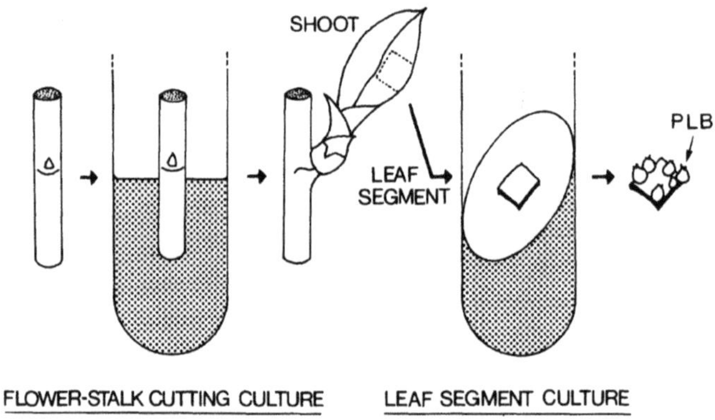

Fig. 5. Diagram illustrating the procedure for the micropropagation of *Phalaenopsis* through in vitro leaf segment culture (Tanaka 1987)

3.2% of the cultured leaf segments formed PLBs (Fig. 4C; Tanaka and Sakanishi 1985). Leaf segments derived from flower stalk cuttings, however, are preferable to those from adult plants because they have a higher rate of PLB formation and also their use does not damage the mother plants (Fig. 5).

3.2 Factors Affecting Shoot Production on Flower Stalk Cuttings

Since lateral buds on flower stalk cuttings cultured in vitro show three growth patterns as mentioned above, inducing as many buds to develop into shoots is necessary to obtain sufficient materials for propagation work from a selected mother plant. In the studies on flower stalk cutting culture of *Phalaenopsis*, the percentage of each growth pattern was different for different workers, and the effect of cultural conditions or culture media on the growth of the buds has not been previously studied in detail. Hence, experiments by this author were carried out to examine various factors that affect the production of shoots from flower stalk cuttings.

Flower stalks were collected from stock plants during the period when several florets had reached anthesis. They were wiped three times with gauze previously dipped in 70% ethanol, and then cut into 5-cm-long segments, each with 1 lateral bud in the middle. These cuttings were immersed in 70% ethanol for 10 s before surface sterilization for 10 min in a filtered suspension of 7% calcium hypochlorite containing 0.1% of Tween 20, and then were rinsed three times with sterile water. After surface sterilization, the bracts were removed and the exposed ends of the cuttings cut away. The 4-cm cuttings were inserted upright into the solid culture medium (Tanaka and Sakanishi 1978; Tanaka et al. 1988).

The medium used was Vacin and Went's (1949) supplemented with 20% (v/v) coconut milk (CM), 2% sucrose, and 1% agar. After planting, the tubes were closed with aluminum foil held firmly with a rubber band.

3.2.1 Temperature

We have previously reported the effects of the culture temperature on the patterns of bud growth in *Phalaenopsis amabilis* hybrids (Fig. 6). Cuttings obtained from different positions on the flower stalks were cultured at 20, 25, or 28 °C. At each temperature, some buds remained dormant. A higher rate of dormancy occurred in those buds excised from the highest position on the flower stalk. In sprouting buds

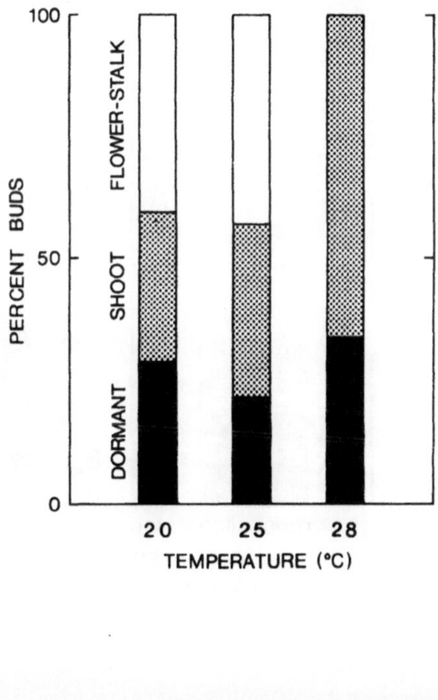

Fig. 6. Effects of temperature on the growth patterns of lateral buds on flower stalk cuttings cultured in vitro. (After data of Tanaka and Sakanishi 1978)

at 20 °C, the buds grew to become secondary flower stalks except for a few taken from a lower position on the flower stalk. The results obtained at 25 °C differed from those at 20 °C in that the proportion of buds growing to shoots increased slightly. At 28 °C, all those that developed formed shoots (Tanaka and Sakanishi 1978). According to Sakanishi et al. (1980), a temperature of 25 °C or lower was necessary for the emergence of a flower stalk from an intact plant of *P. amabilis* hybrid, and at 28 °C it did not occur. These results indicate that the effect of temperature on the in vitro growth of lateral buds of flower stalks is similar to that on the growth of lateral buds of the main axis in intact *Phalaenopsis*.

3.2.2 Light

The light conditions used by different workers for the culture of flower stalk cuttings differed, and their effect has not been reported in detail. When flower stalk cuttings were cultured under 8, 16 or 24 h daylength at 25 °C, the percentage of the buds failing to sprout decreased, but there was also a corresponding increase in the number of buds developing into secondary flower stalks. About 50% of the buds produced shoots at each daylength (Tanaka et al. 1988).

The effect of light intensity on the growth patterns of the bud was also examined. Flower stalk cuttings were cultured at 1600, 650, 400, or 170 lx. Some were also cultured in continuous darkness. The role of light intensity is clearly shown in Fig. 7. Shoot production increased with decreased light intensity from 1600 to 170 lx. On the other hand, the formation of secondary flower stalks was promoted by high light intensity. Maximum dormancy was reached in continuous darkness (Tanaka et al. 1988).

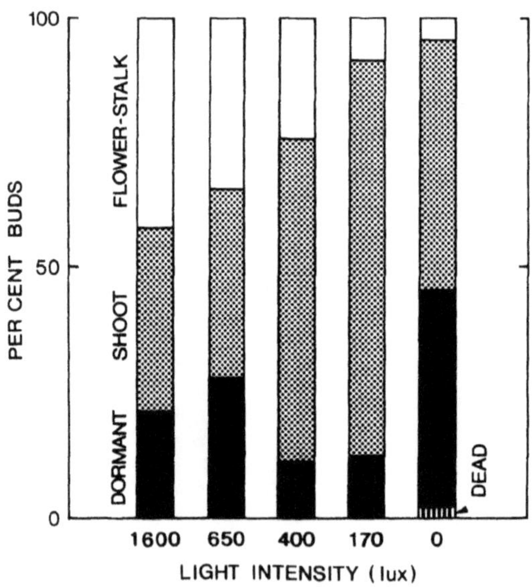

Fig. 7. Effects of light intensity on the growth patterns of lateral buds on flower stalk cuttings of *Phalaenopsis* White Falcon × *Phalaenopsis* Persistent cultured in vitro (Tanaka et al. 1988)

3.2.3 Additives to Culture Medium

As mentioned above, several cultured buds remained dormant in VW medium supplemented with 20% CM. The dormancy was especially frequent when the cuttings from the upper portion of the flower stalk were used. Therefore, we attempted to promote sprouting of these buds by adding BA to the basic VW medium without CM at 28 °C. With increased concentrations of BA, the sprouting rate of dormant buds increased progressively (Fig. 8). High concentrations of BA were necessary for the buds excised from a higher position on the flower stalk to sprout. At 5 mg/l BA or above, all buds developed shoots, but the leaves were malformed. At 28 °C, on VW medium with and without CM, the percentage of the cultured buds which remain dormant was 35 and 76%, respectively. Increased sprouting was caused by the presence of a cytokinin in the CM (Zwar and Bruce 1970). Buds released from dormancy by BA application were also influenced by the position on the flower stalk and by temperature. Thus, BA was effective principally in promoting the growth of dormant buds, while further bud development was determined by cultural temperature. Reisinger et al. (1976) cultured the flower stalk cuttings on modified Knop's medium supplemented with transcinamic acid, a known anti-auxin, to release them from apical dominance, and reported 59% plantlet formation.

When culturing the flower stalk cuttings, blackening of media has been observed (Kotomori and Murashige 1965; Scully 1966; Flamee and Boesman 1977). This blackening is due to the oxidation of phenolic compounds, which could possibly have a negative effect on growth and differentiation (Kotomori and Murashige 1965; Morel 1974; Flamee and Boesman 1977). Addition of charcoal and PVP (polyvinylpirrolidon) which are known to adsorb phenolic compounds

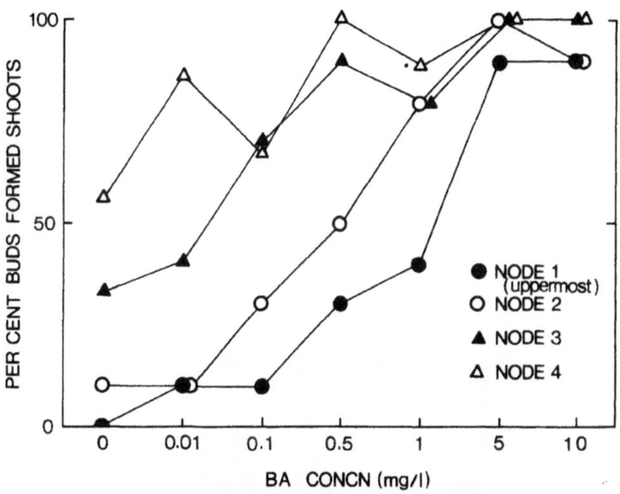

Fig. 8. Effects of BA concentration on shoot growth of lateral buds on flower stalk cuttings of *P. amabilis* hybrid cultured in vitro. Cuttings taken from ten flower stalks, each having four nodes, were used for each BA concentration. (After data of Tanaka and Sakanishi 1978)

showed that these substances in the medium inhibited shoot production on the flower stalk cuttings (Tanaka et al. 1988).

3.2.4 Length of Flower Stalk Cuttings

Generally, cuttings of 3–6 cm with one bud are used for culture. This author studied the effect of the length of flower stalk cuttings on the growth patterns of the explants. Flower stalk cuttings of 4, 2, or 1 cm in length, as well as excised buds, were cultured. Shoot production was enhanced when shorter cuttings were used. Maximum shoot formation was obtained with 1-cm cuttings. Also, the number of buds remaining dormant decreased with shorter cuttings (Fig. 9; Tanaka et al. 1988). This could be due to the flowering stimuli which may exist in the flower stalk tissue. This explanation is supported by the observation that secondary flower stalks usually develop from the upper buds, which may have a greater floral induction than buds on the lower end of the flower stalk (Urata and Iwanaga 1965; Koch 1974a; Tanaka and Sakanishi 1978). For the excised buds, all those that developed formed shoots, while others remained dormant or died.

3.2.5 Surface Sterilization of Flower Stalk Cuttings

When flower stalk cuttings were transferred onto nutrient medium, these explants were often contaminated with a wide range of microorganisms. This is so when the conventional method of sterilizing is adopted. Usually 20 to 25% of the cultures were contaminated when they were surface sterilized with calcium hypochlorite solution.

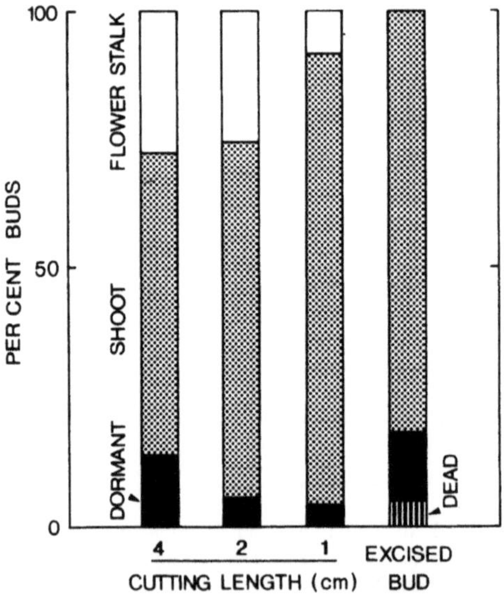

Fig. 9. Effects of length of flower stalk cuttings on the subsequent growth patterns of lateral buds of *P. amabilis* hybrid cultured in vitro. (Tanaka et al. 1988)

A new method has therefore been devised to overcome this contamination problem. The flower stalk cuttings were first immersed in 70% ethanol for 10 s, then rinsed three times in sterile water. Thereafter, they were immersed in antimicrobials solution for 30 min in the dark. The flower stalks were then transferred onto the nutrient medium and 1 ml of antimicrobial solution was added to the surface of the medium. The cultures were then placed in the dark for 10 days, after which they were transferred to light condition. Cultures were kept at 25 °C.

Among several combinations of antimicrobials examined, the solution containing 10 mg/l Benlate, 25 mg/l pentachloronitrobenzen (PCNB), 100 mg/l thiabendazole (TBZ), 10 mg/l rifampicin, 500 mg/l ampicillin, and 50 mg/l vancomycin brought about complete aseptic culture (Tanaka et al. 1983). In practical application of this novel surface sterilization method, the flower stalk cuttings should first be sterilized with conventional sterilizing chemicals, after which they can be treated with the antimicrobial solution. On the other hand, when the cuttings were cultured on the media in which antimicrobials were incorporated, many of them died, and the growth of the buds on aseptic cuttings was inhibited (Tanaka et al. 1983).

3.3 Factors Affecting PLB Formation in Leaf Segments

When the lateral buds on the cuttings developed into vegetative shoots with one to three leaves, leaf segments (6–8 × 6–8 mm) were excised from apical, medium, and basal parts of expanding uppermost leaves (ca. 2 cm in length) of the shoot (Fig. 10A, B). All segments were planted adaxial side up on slant agar medium in test tubes (18 × 180 mm) containing 8 ml of the medium.

3.3.1 Simplification of Medium

In order to simplify the culture medium for PLB formation, we adopted Kyoto solution (Tsukamoto et al. 1963) modified by the addition of 100 mg/l myo-inositol, 1 mg/l nicotinic acid, and 1 mg/l thiamine-HCl (modified Kyoto medium) to leaf

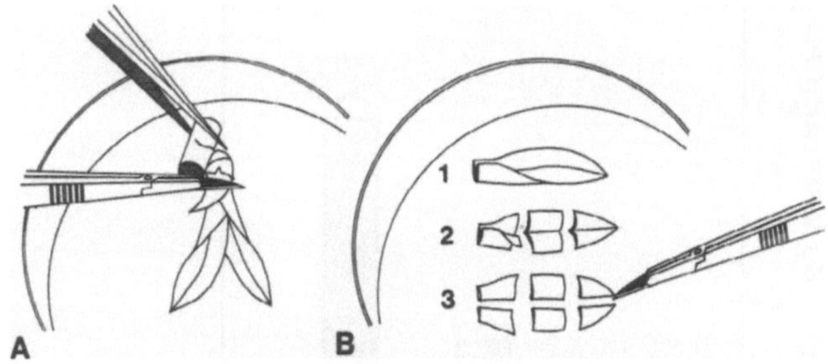

Fig. 10A, B. Diagram illustrating the procedure for preparation of leaf segments from flower stalk cuttings. **A** Excise a shoot at the base just above its apical meristem; so that the cuttings can be used again to obtain more shoot meterial. **B** Cut the uppermost leaf (about 2 cm long) into apical, middle and basal parts, then bisect them segment culture. The medium can be prepared more easily than MS medium, because mineral nutrients in modified Kyoto solution are based on 3.5 g/l Hyponex (Hyponex Japan, 6.5-6-19) alone, which is widely used by horticulturists as a plant food, and has proved to be applicable as mineral nutrient base of culture medium for asymbiotic germination of orchid seeds. As a result, 16.7% of leaf segments cultured on modified Kyoto medium supplemented with 1 mg/l NAA, 10 mg/l adenine, and 10 mg/l BA produced PLBs (Fig. 11), while 8.3% of those on MS medium with the same amount of the growth regulators did. The average number of PLBs per segment on modified Kyoto medium was 3.4, while that on the MS medium 1.4. Thus, modified Kyoto medium is recommended by reason of its efficacy for PLB formation in addition to its simplicity.

The efficacy of adenine for cytokinesis is generally admitted to be very low, but the rate of PLB formation was reduced when adenine was omitted from the above

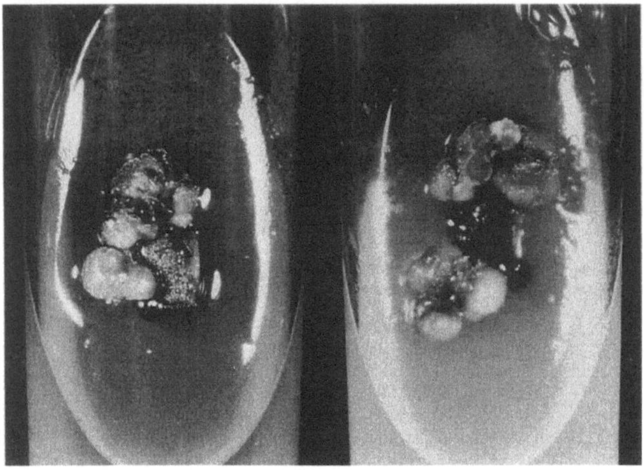

Fig. 11. PLBs formed on leaf segments cultured on modified Kyoto medium supplemented with 1 mg/l NAA, 10 mg/l adenine, and 10 mg/l BA. Hybrid used was *Phalaenopsis* (White Falcon × Persistent) × *Phalaenopsis* Jimmy Hall, 120 days after planting

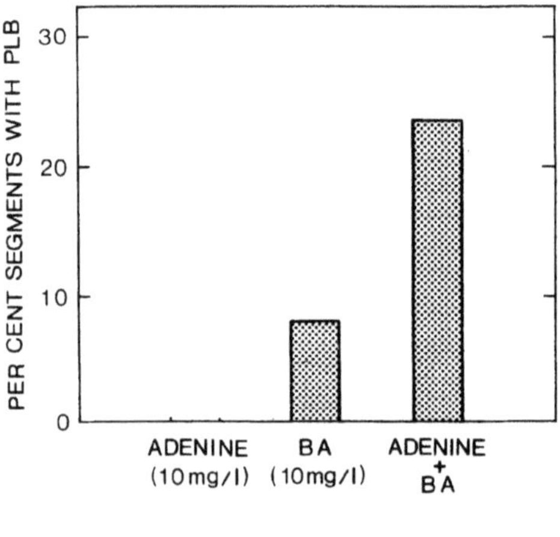

Fig. 12. Effects of adenine on the formation of PLB in the leaf segments of *P. amabilis* hybrid. (Tanaka and Sakanishi 1980)

medium, though adenine alone had no effect (Fig. 12; Tanaka and Sakanishi 1980). This indicates that adenine plays some role as a BA synergist for PLB formation. All of the vitamins in modified Kyoto medium were also found to be essential to induce higher rate of PLB formation in leaf segments (Tanaka 1987).

3.3.2 Replanting of Leaf Segment

Since the solid medium in which leaf segments were cultured became black, like the flower stalk cuttings mentioned above, it was expected that PLB formation in leaf segments could be affected by this exudate. Hence, leaf segments were transferred onto new medium at of 1-, 2-, and 4-week intervals until the end of the culture period. Leaf segments were also cultured on the same medium throughout the experiment (for 5 months) without being transferred to a new medium. Also some segments were cultured on the medium supplemented with 2 g/l activated charcoal.

PLB formation in leaf segments was enhanced when leaf segments were replanted on new medium, whereas those cultured on the same medium without replanting showed a lower percentage of PLB formation. The highest percentage of PLB formation was obtained in leaf segments when replanted every 2 weeks (Fig. 13; Tanaka 1987). Leaf segments replanted repeatedly produced higher mean number of PLBs than those without replanting. Therefore, in order to obtain good PLB formation, leaf segments need to be transplanted to fresh medium. This is also true with shoot tips (Intuwong and Sagawa 1974) or internodal sections of flower stalk (Homma and Asahira 1985). Although charcoal is known to absorb phenolic compounds that inhibit morphogenesis (Fridborg et al. 1978), all leaf segments died within 1 week when cultured on charcoal-enriched medium. However, the addition of PVP (MW.av. 160000, Tokyo Kasei, Japan) at the concentration of 800 ppm to the leaf segment culture medium with 0.2% Gelrite (Merck, USA) was found to enhance PLB formation in leaf segments (Tanaka et al. unpubl. data).

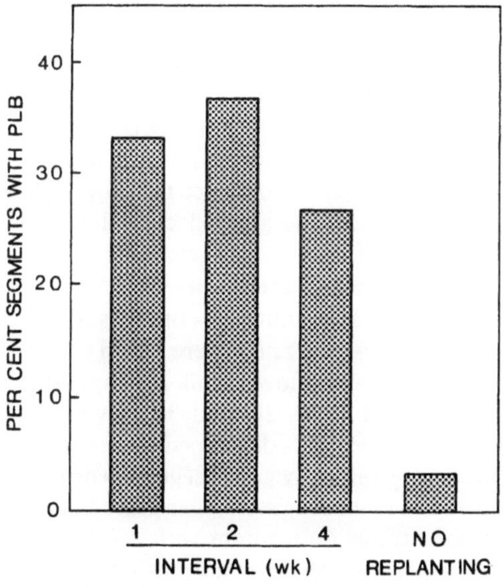

Fig. 13. Effects of replanting interval on the subsequent formation of PLB in leaf segments of *P. amabillis* hybrid. (Tanaka et al. 1990)

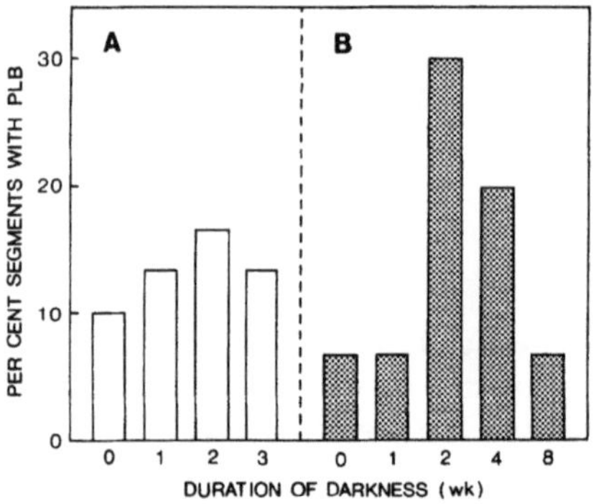

Fig. 14A,B. Effects of duration of darkness after commencement of culturing on the formation of PLB in leaf segments from two hybrids. **A** *Phalaenopsis* White Falcon × Persistent. **B** *Phalaenopsis* Enshyu. (Tanaka et al. 1990)

3.3.3 Darkness After Commencement of the Culture

In order to examine whether a period of darkness after the commencement of culture could enhance the formation of PLB or not, leaf segments were placed in darkness for 1 to 8 weeks at 25 °C after they were planted on the medium. After the durations of dark incubation, they were transferred to light condition (16-h daylength, 900 lx).

The highest percentage of PLB formation was obtained in the leaf segments placed in the dark for 2 weeks. Longer treatment of darkness (3, 4, and 8 weeks) inhibited PLB formation in leaf segments (Fig. 14; Tanaka 1987). Zimmer and Pieper (1978, 1979) reported that induction of PLB on the segments of broken *Phalaenopsis* plantlet increased when these segments were cultured in darkness for 2 weeks after commencement of culture, and also that this 2-week period in darkness seems to be necessary to suppress medium blackening. Our experimental results agree precisely with their observation.

3.3.4 Conditions During Flower Stalk Cutting Culture

We have examined the various factors that affect the production of shoots from flower stalk cuttings. The aim was to obtain a large number of leaf segments for culturing in order to enhance the propagation rate. At this stage, it is also necessary for us to check the after-effect of flower stalk cutting culture on the subsequent formation of PLB in leaf segments. Both the medium and environmental conditions under which the shoots were developed from flower stalk cuttings proved to have tremendous influences on the subsequent formation of PLB in leaf segments.

NAA, BA, and zeatin were added to the flower stalk cutting media at 0.01, 0.1, 1, and 10 mg/l. The flower stalk cuttings were also cultured on media containing 2% sucrose and 1, 10, 20, and 30% (v/v) coconut milk (CM). Leaf segments were excised from shoots developed on the cuttings and cultured on the modified Kyoto medium given above. Leaf segments originating from cuttings cultured on cytokinin-enriched media gave higher percentage of subsequent PLB formation than those from cuttings on medium with NAA or without growth regulator (Fig. 15). Leaf segments excised from cuttings on media supplemented with CM gave a higher percentage of PLB formation than those from cuttings without CM. The highest percentage was obtained when cultured on medium with 20% CM (Fig. 16). From these data, it seems that the regenerative capacity in leaf segments was enhanced by BA, zeatin, and zeatin-containing coconut milk (Zwar and Bruce 1970), which was

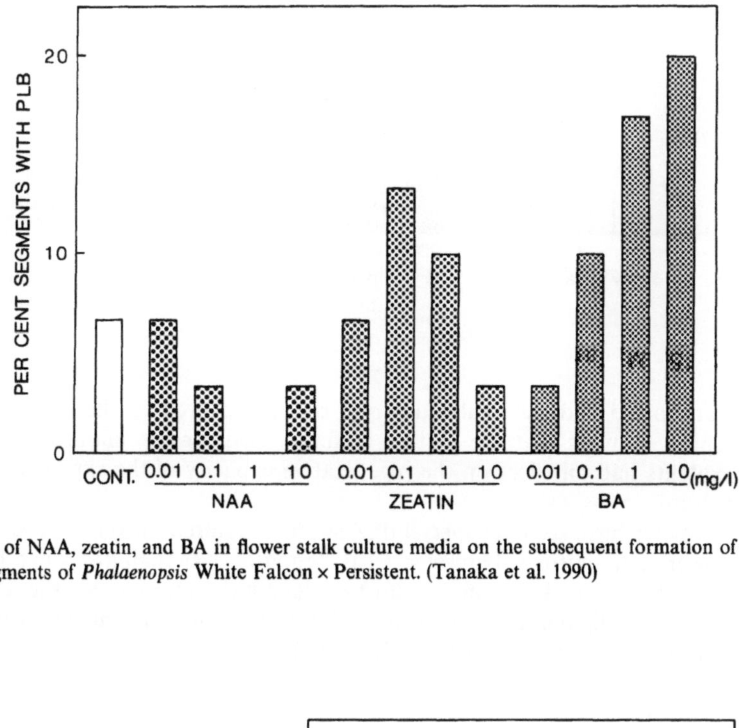

Fig. 15. Effects of NAA, zeatin, and BA in flower stalk culture media on the subsequent formation of PLB in leaf segments of *Phalaenopsis* White Falcon × Persistent. (Tanaka et al. 1990)

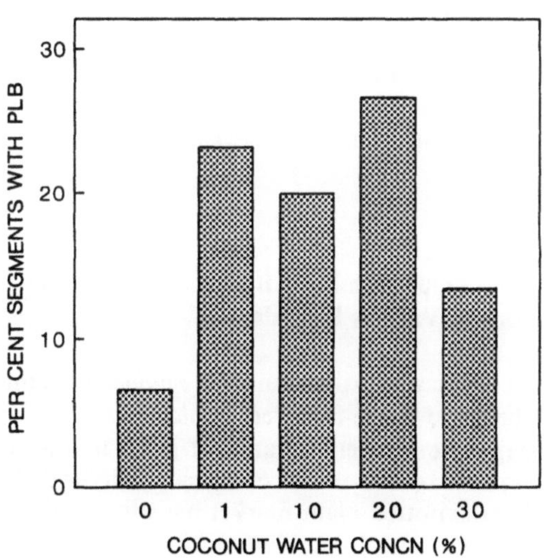

Fig. 16. Effects of coconut milk in flower stalk culture media on the subsequent formation of PLB in leaf segments of *Phalaenopsis* White Falcon × Persistent. (Tanaka et al. 1990)

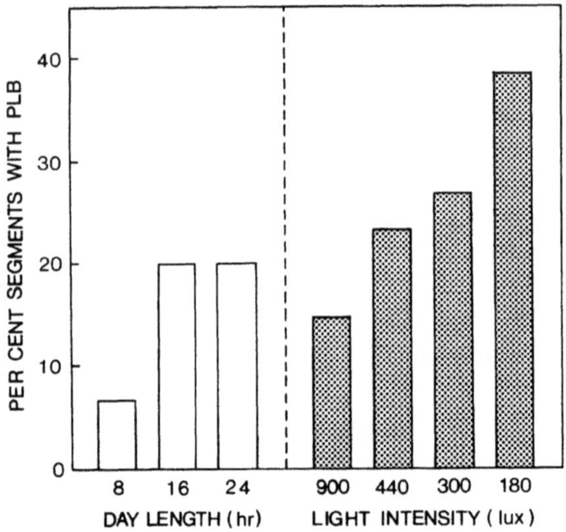

Fig. 17. Effects of light during flower stalk cutting culture on the subsequent formation of PLB in leaf segments of *Phalaenopsis* White Falcon × Persistent. (Tanaka et al. 1990)

transported and accumulated in the shoots. In orchids, there has never been a report showing that the culture medium used for the initial culture to obtain explants had an effect on the regenerative capacity of the explants in the second culture.

As we have seen, no special daylength is required for shoot production on the flower stalk cuttings. However, PLB formation was inhibited when leaf segments from shoots developed under short daylength (8 h) were used (Fig. 17). The after-effect of the light intensity is clear. PLB formation in leaf segments is accelerated by lowering the light intensity under which the flower stalk cuttings are cultured (Fig. 17). This may possibly be associated with the physiological state of the leaf segments, since shoots developed under low light intensity were somewhat etiolated or yellowish green in color. Shoot production in flower stalk cutting culture is also accelerated by a lowering of light intensity (Tanaka et al. 1988); this means that a large number of leaf segments can be obtained from a selected mother plant. This is also important for increased production.

3.4 Comparison in PLB Formation on Leaf Segments Among Various Hybrids

In a series of experiments on *Phalaenopsis* micropropagation, we have been using the large white-flowered *Phalaenopsis* hybrids, since these are important to the Japanese growers because of good market demand. Recently, however, various types of more colorful *Phalaenopsis* hybrids became popular, and their production for the potted plant market has also gradually increased.

Therefore, in our work many other hybrids were used for micropropagation. Twenty-five hybrids consisting of six general types: namely white, white with red lip, pink, yellow, spotted, and striped were used.

Flower stalks excised from these hybrids were cultured as described above, except that 4-cm cuttings were cultured under approximately 170 lx with 16-h photoperiod. The percentage of shoot production on the axenic cuttings differed with different hybrids 3 months after culture. Of the 25 hybrids, six produced no shoots from the flower stalk cuttings whose buds either developed secondary flower stalks or remained dormant. In three hybrids, however, shoots developed on the secondary flower stalk. In four hybrids, all cuttings produced shoots 3 months after culture. For the remaining 15 hybrids, the percentage of shoot formation varied from 13 to 89%. The number of leaves produced was also different, varying from one to three leaves per shoot.

All the leaf segments excised from these shoots were cultured under the conditions described above. Of the 22 hybrids, 19 were successfully propagated, while for the remaining three (hybrid code M, T, and V), the leaf segments did not produce PLBs. The rate of PLB formation in the leaf segments varied from 2.1 to 100% (Fig. 18). Figure 19 shows changes in PLB formation rate in leaf segment culture of 11 hybrids (A, C, D, E, H, J, O, Q, S, W, and X) selected among the 19 hybrids. PLB formation in most hybrids was observed 2 months after culture, but the rate of PLB formation was different with different hybrids. There were also differences between various hybrids in the number of PLBs formed per leaf segments, ranging from 1 to 15.6 PLBs per segment (Fig. 18).

Differences in the response of various species or hybrids to shoot-tip culture have been reported in *Cattleya* (Ichihashi and Kako 1973) and *Cymbidium* (Kano 1972). From our present results, it is clear that there is a great difference in the regenerative capacity between hybrids, which could possibly be due to differences of the parental species or hybrids used in their breeding. The leaf segments of the hybrids which gave a higher rate of PLB formation had a tendency to show a lower degree of media blackening, whereas those showing a higher degree of blackening showed a lower percentage of PLB formation. This suggests that the regenerative ability of in vitro-cultured leaf segments may be closely related to the amount of phenolic compounds in the leaves.

3.5 Proliferation of PLBs and Plantlet Development

Secondary proliferation of PLBs which were already formed on leaf segments is important for obtaining numerous plantlets. Several methods for PLB proliferation have been reported (Tanaka and Sakanishi 1980; Greisbach 1983; Tanaka 1987; Amaki and Higuchi 1989). When the PLBs were bisected transversally and the basal sections were placed upside down on VW medium supplemented with 20% coconut milk, it was found that each section produced an average of 13.7 PLBs (Table 1, Fig. 20; Tanaka 1987). The two third basal segments of *Doritaenopsis* PLB were cut longitudinally into four pieces, each piece, cultured on modified Kyoto medium supplemented with 2 g/l tryptone, produced 20.6 PLBs (Amaki and

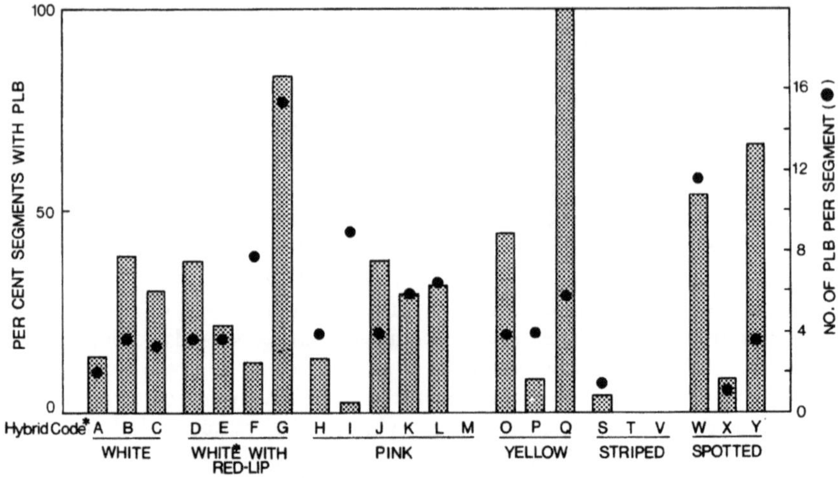

Fig. 18. Comparison in PLB formation on leaf segments among various hybrids. * *A* White Falcon × Persistent; *B* Enshyu; *C* (White Falcon × Persistent) × Jimmy Hall; *D* Pink Cheers × *mannii*; *E* George Woodward × (Makua Shore × Percy Porter); *F* unknown; *G* unknown; *H* Callie Flynn; *I* unknown; *J* Kyoto × Zauberrose; *K* Callie Flynn × [Lavender Lady × (Mount Kaala × Makua Shore)]; *L* Lavender Lady × [Lavender Lady × (Clarelen × Zada)]; *M* Zada; *N* Marion Fowler; *O mannii* × Mount Martina; *P* Golden Sands; *Q* Memoria Isolina Cestero; *R mannii* × Pat Darby; *S* Pink Cheers × [Lavender Lady × (Mount Kaala × Makua Shore)]; *T* Zada × [Lavender Lady × (Mount Kaala × Makua Shore)]; *U* [Lavender Lady × (Mount Kaala × Makua Shore)] × Promising Day; *V* unknown; *W* George Moler; *X* Lolita × *mannii*; *Y* Surfrider × *gigantea*. (Tanaka et al. 1990)

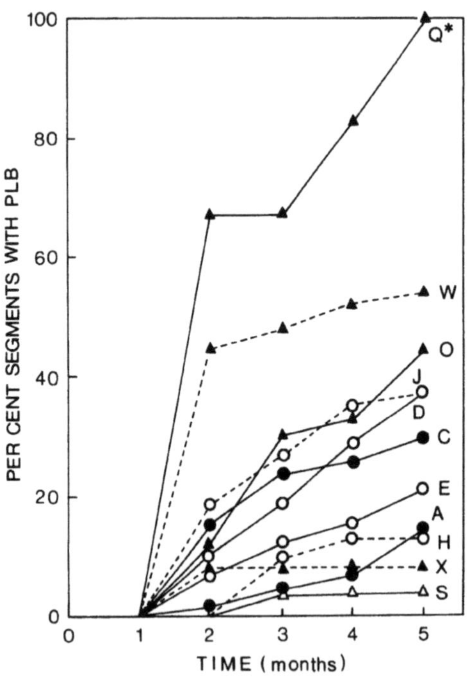

Fig. 19. Changes of rate of PLB formation in leaf segment culture of various hybrids. (Tanaka et al. 1990) *See legend to Fig. 18

Table 1. Comparison of daughter PLB formation and/or shoot development between upper and lower half of cut PLB. (Tanaka 1987)

Section	No. of cultures	Developed organ on culture				
		PLB and Shoot		PLB		Shoot
		%	No. of PLB/ culture	%	No. of PLB/ culture	%
Upper	16	31	1.0	38	5.3	31
Lower	16	0	–	100	13.7	0

Hybrid used: Lavender Lady × [Lavender Lady × (Clarelen × Zada)].
PLBs were transversally bisected, and cultured on VW + coconut milk 20% (v/v) medium, at 25 °C with a 16-h light (900 lx).
Data were recorded 55 days after culture.

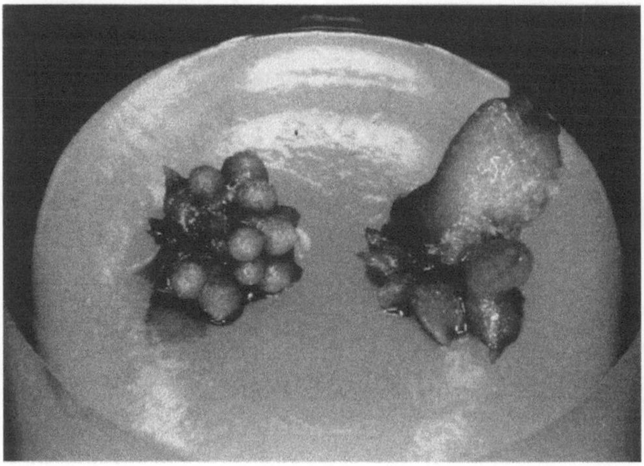

Fig. 20. Proliferation of PLBs from the upper and lower half of cut PLB derived from leaf segment culture. PLBs were transversally bisected, and cultured on Vacin and Went + 20% (v/v) coconut milk medium, 55 days after planting. *Left* basal half of transverse cut PLB; *right* apical half of transverse cut PLB. (Tanaka 1987)

Higuchi 1989). This cutting procedure can be repeated to obtain the number of PLBs desired. These results indicate that it is easier to proliferate PLBs than to induce PLB formation in leaf segment culture.

To obtain plantlets, PLBs were transferred to the medium containing 3 g/l Hyponex, Nitsch microelements (Nitsch and Nitsch 1967), 100 mg/l myo-inositol, 1 mg/l nicotinic acid, 2 g/l activated charcoal, 3% sucrose, and 1% agar (Tanaka 1987), or to the medium recommended for seed germination of *Phalaenopsis* (Tanaka and Sakanishi 1985). It took 5 to 7 months to obtain plantlets from the PLB stage in the flask (Fig. 21A). All these plantlets were planted in the conventional way in the greenhouse (Fig. 21B).

Fig. 21. A Plantlets derived from PLBs, 8 months after transferring to the medium containing 3.5 g/l Hyponex (6.5: 7:19), Nitsch microelements (1967), 100 mg/l inositol, 1 mg/l nicotinic acid, 2 g/l activated charcoal, 3% sucrose, and 1% agar. Hybrid used was *Phalaenopsis* (White Falcon × Persistent) × *Phalaenopsis* Jimmy Hall. **B** Clonal plants growing in a community pot. Hybrid used was *Phalaenopsis mannii* × Mount Martin. (After Tanaka 1987)

4 Conclusions

Phalaenopsis, being a monopodial, rarely produces vegetative offshoots. For this reason, a reliable method of micropropagation has long been expected. In Japan, where there is a great demand for large numbers of clonally propagated *Phalaenopsis* plantlets for cut flower and pot plant production, tissue culture laboratories are constantly seeking an efficient method to micropropagate this plant. The questions frequently asked are: which explant yields better results?, or how to induce PLB in explants and proliferate it? The method of micropropagation of *Phalaenopsis* as established in our laboratory and presented here would be an ideal method because: (1) the use of flower stalk neither damages nor endangers the selected mother plant, (2) only the flower stalk needs to be sent for micropropagation in the commercial tissue culture laboratory, (3) the propagation rate is relatively high since a larger number of leaf segments could be obtained from several shoots on the cuttings of a selected mother plant than the shoot tips or internodal sections of a flower stalk. After the leaves have been excised, the cuttings may be used again to obtain more shoot material, (4) the procedure is simple when compared to the shoot-tip culture method, and (5) dissecting microscope and gyrotory shaker are not required.

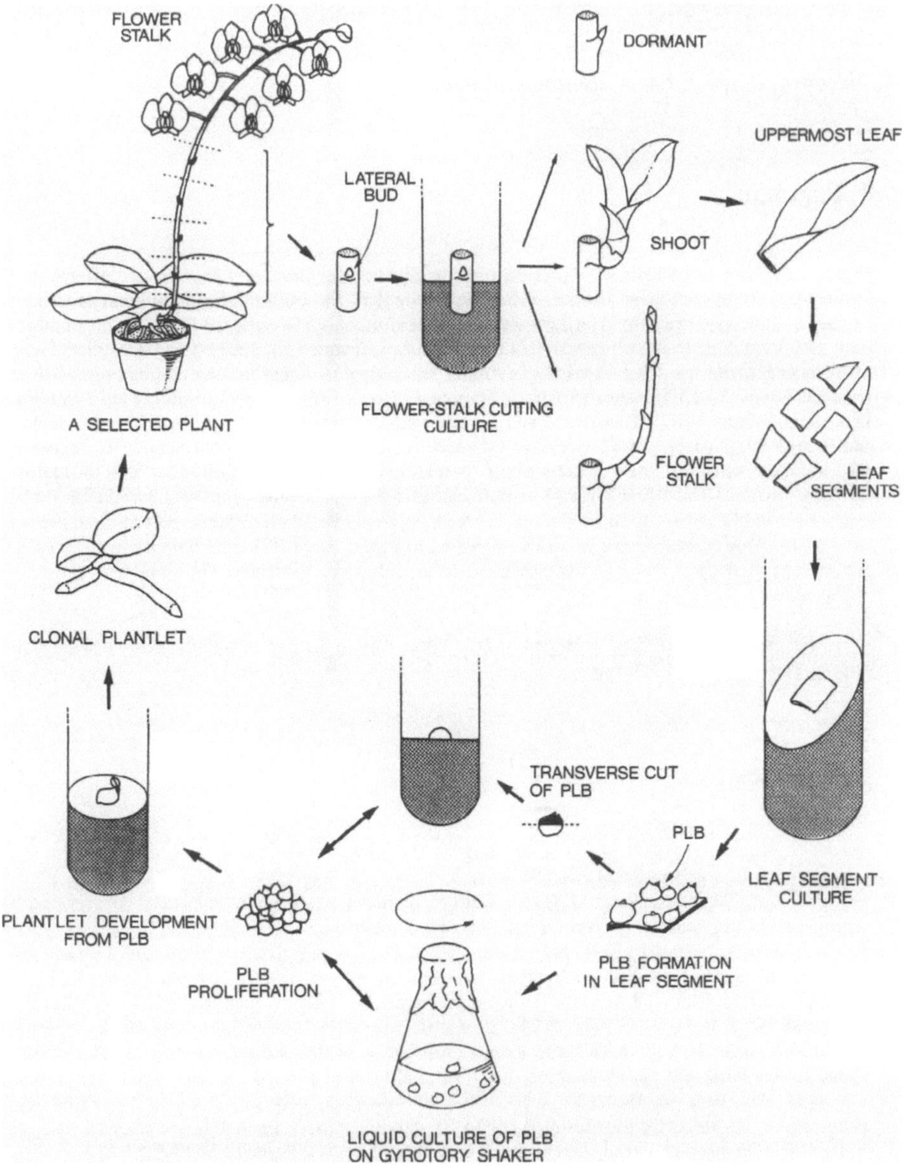

Fig. 22. Schematic representation of micropropagation of *Phalaenopsis*. (Tanaka 1987)

It is very important in practical application to determine whether the regenerated plants are true to type or otherwise. If micropropagation gives rise to a high frequency of variation, this method is undesirable. In 1986, a total of 420 plants from 30 different hybrids produced in vitro were examined and their flower color and shape compared with those of the mother plants. A total of 128 plants from 15 *Phalaenopsis* species were also examined. Only 2.9% of plants from three hybrids showed some variations in flower shape and color, while flowers of most plants from hybrids and those from all species were similar to their respective mother plants (Tanaka et al. unpubl. data). These results confirmed that this method of micropropagation has a commercial value.

5 Protocol

Figure 22 shows the schematic representation of micropropagation of *Phalaenopsis* through leaf segment culture. From the facts above, we may conclude that the best medium and optimal conditions for each culture stage are as follows. Leaf segments for culture should be obtained from shoots of flower stalk cuttings cultured on Vacin and Went's medium supplemented with 20% (v/v) CM at 25 °C with 16-h daylength under low light intensity of 170 lx. The segments should then be cultured on modified Kyoto medium (3.5 g/l Hyponex (6.5:6:19, Hyponex Japan), 100 mg/l myo-inositol, 1 mg/l nicotinic acid, 1 mg/l thiamine-HCl, 2% sucrose, and 1% agar) supplemented with 1 mg/l NAA, 10 mg/l adenine, and 10 mg/l BA in darkness for 2 weeks, and then transferred to a 16-h light (900 lx) at 25 °C. Segments need to be replanted onto new medium every 2 weeks, or they should be cultured on PVP (800 ppm)-enriched medium. The PLBs formed on leaf segments can be bisected transversally and the basal sections should be placed upside down on VW medium containing 20% coconut milk to obtain many more PLBs. This cutting procedure can be repeated to obtain more PLBs. To obtain plantlets, PLBs can be separated one by one and transplanted to a medium recommended for seed germination of *Phalaenopsis*.

Acknowledgments. Gratitude is expressed to Dr. C.K.H. Teo of Science University of Malaysia for his help in preparing the manuscript.

References

Amaki W, Higuchi H (1989) Effects of dividing on the growth and organogenesis of protocorm-like bodies in *Doritaenopsis*. Sci Hortic 39: 63–72

Arditti J, Ball EA, Reisinger DM (1977) Culture of flower stalk buds: a method for vegetative propagation of *Phalaenopsis*. Am Orchid Soc Bull 46: 236–240

Arditti J, Johnson JA, Perera RG (1981) Culture media which do not require sterilization: *Phalaenopsis* flower stalk nodes. Orchid Rev 89: 49–52

Ball EA, Arditti J, Churchill ME (1971) Clonal propagation of orchids from leaf tips. Orchid Rev 79: 281–288

Bouriquet R, Broly H, Legrand B (1982) Clonal propagation of *Phalaenopsis* (Orchidaceae) by in vitro culture. In: Earle ED and Y Demarly (eds) Variability in plants regenerated from tissue culture. Praeger, New York, pp 35–38

Champagnat M, Morel G, Mounetou B (1970) La multiplication végétative des *Cattleya* a parti de jeunes feuilles cueillies aseptiquement in vitro. Ann Sci Nat Bot, Paris, 12 Ser, XI: 97–114

Churchill ME, Ball EA, Arditti J (1970) Clonal propagation of orchid plants from seedling leaf tips. Orchid Dig 34: 271–273

Churchill ME, Arditti J, Ball EA (1971) Clonal propagation of orchids from leaf tips. Am Orchid Soc Bull 40: 109–113
Churchill ME, Ball EA, Arditti J (1973) Tissue culture of orchid. I. Methods for leaf tips. New Phytol 72: 161–166
Daeer M (1967) Vegetative Vermehrung bei *Phalaenopsis*. Die Orchidee 18: 322–324
Flamee M, Boesman G (1977) Clonal multiplication of *Phalaenopsis* hybrids by means of sections of the flower stalk. Med Fac Landbouww Rijksuniv Gent 42: 1865–1868
Fridborg G, Pedersen M, Landstrom L, Eriksson T (1978) The effect of activated charcoal on tissue culture: absorption of metabolites inhibiting morphogenesis. Physiol Plant 43: 104–106
Fu Fan May Lay (1978) Studies on the tissue culture of orchids. 1. Clonal propagation of *Phalaenopsis* by lateral buds from flower stems. Orchid Rev 86: 308–310
Fu Fan May Lay (1979) Studies on the tissue culture of orchids. 2. Clonal propagation of *Aranda, Ascocenda, Cattleya* by leaf tissue culture. Orchid Rev 87: 343–346
Goh CJ, Tan H (1982) Clonal propagation from leaf explants in *Renantanda* orchid hybrid. Orchid Rev 90: 295–296
Greisbach RJ (1983) The use of indoleacetylamino acids in the in vitro propagation of *Phalaenopsis* orchids. Sci Hortic 19: 363–366
Homma Y, Asahira T (1985) New means of *Phalaenopsis* propagation with internodal sections of flower stalk. J Jpn Soc Hortic Sci 54: 379–387
Ichihashi S, Kako S (1973) Studies on clonal propagation of *Cattleya* through tissue culture method. I. Factors affecting survival and growth of shoot meristem of *Cattleya* in vitro. J Jpn Soc Hortic Sci 42: 264–270
Intuwong O, Sagawa Y (1974) Clonal propagation of *Phalaenopsis* by shoot tip culture. Am Orchid Soc Bull 43: 893–895
Intuwong O, Kunisaki JT, Sagawa Y (1972) Vegetative propagation of *Phalaenopsis* by flower stalk cuttings. Na Okika O Hawaii (Hawaii Orchid J) 1: 13–18
Kamemoto HH, Tanaka R, Kosaki K (1961) Chromosome numbers of orchids in Hawaii. Univ Hawaii Agric Exp Stn Bull 127: 1–28
Kamemoto HH, Sagarik R, Kasemsap S (1964) Chromosome numbers of Sarcanthine orchid species of Thailand. Nat Hist Bull Siam Soc 20: 235–241
Kamemoto HH, Kosaki K, Shindo K (1967) Chromosome counts of orchids in Hawaii. 1960–64. Na Pua Okika O Hawaii Nei 17: 79–82
Kano K (1972) Seed germination of oriental *Cymbidium* and their shoot tip culture. In: Corrigan MJG (ed) Proc 6th World Orchid Conf, pp 133–142, Sydney
Koch L (1974a) Untersuchungen zur vegetativen Vermehrung bei *Phalaenopsis* in vitro. Thesis, TU Hannover, pp 1–169
Koch L (1974b) Erbgleiche Vermehrung von *Phalaenopsis* in vitro. Gartenwelt 74: 482–484
Kotomori S, Murashige T (1965) Some aspects of aseptic propagation of orchids. Am Orchid Soc Bull 34: 484–489
Lin CC (1986) In vitro culture of flower stalk internodes of *Phalaenopsis* and *Doritaenopsis*. Lindleyana 1: 158–163
Morel GM (1960) Producing virus-free cymbidiums. Am Orchid Soc Bull 29: 495–497
Morel GM (1965) Clonal propagation of orchids by meristem culture. Cymb Soc News 20: 3–10
Morel GM (1974) Clonal multiplication of orchids. In: Withner CL (ed) The orchids: scientific studies. Wiley-Interscience, New York, pp 169–222
Murashige T, Skoog F (1962) A revised medium for rapid growth and bioassays with tobacco tissue cultures. Physiol Plant 15: 473–497
Nitsch C, Nitsch JP (1967) The role of adenine in bud differentiation. Phytomorphology 17: 446–453
Pancho JV (1966) IOPB chromosome number reports III. Taxon 14: 50–57
Parliman BJ, Evans PT, Mazur AR (1982) Adventitious bud differentiation and development in leaf cuttings of *Dionaea muscipula* Ellis Ex. L. (Venus' fly-trap) cultured in vitro. J Am Soc Hortic Sci 107: 310–316
Pieper W, Zimmer K (1976) Clonal propagation of *Phalaenopsis* in vitro. Acta Hortic 64: 21–23
Reisinger DM, Ball EA, Arditti J (1976) Clonal propagation of *Phalaenopsis* by means of flower stalk node cultures. Orchid Rev 84: 45–52
Rotor G (1949) A method of vegetative propagation of *Phalaenopsis* species and hybrids. Am Orchid Soc Bull 18: 738–739

Sagawa Y (1961) Vegetative propagation of *Phalaenopsis* by stem cuttings. Am Orchid Soc Bull 30: 808–809

Sagawa Y (1962) Cytological studies of the genus *Phalaenopsis*. Am Orchid Soc Bull 31: 459–465

Sagawa Y, Niimoto D (1960) Vegetative propagation of *Phalaenopsis*. Fla Orchidist 3: 22

Sagawa Y, Niimoto D (1961) Cytological studies in the Orchidaceae. Am Orchid Soc Bull 30: 628–632

Sakanishi Y, Imanishi H, Ishida G (1980) Effect of temperature on growth and flowering of *Phalaenopsis amabilis*. Bull Univ Osaka Prefect Ser B 32: 1–9

Scully RM (1966) Stem propagation of *Phalaenopsis*. Am Orchid Soc Bull 35: 40–42

Sharma AK, Chatterji AK (1966) Cytological studies on orchids with respect to their evolution and affinities. Nucleus 9: 177–203

Shindo K, Kamemoto HH (1963a) Karyotype analysis of some Sarcanthine orchids. Am J Bot 50: 73–79

Shindo K, Kamemoto HH (1963b) Karyotype analysis of some species of *Phalaenopsis*. Cytologia 28: 390–398

Sweet HR (1980) The genus *Phalaenopsis*. California. The Orchid Dig Inc, Carmichael

Tanaka M (1987) Studies on the clonal propagation of *Phalaenopsis* through in vitro culture. Mem Fac Agric Kagawa Univ 49: 1–85

Tanaka M, Sakanishi Y (1977) Clonal propagation of *Phalaenopsis* by leaf tissue culture. Am Orchid Soc Bull 46: 733–737

Tanaka M, Sakanishi Y (1978) Factors affecting the growth of in vitro cultured lateral buds from *Phalaenopsis* flower stalks. Sci Hortic 8: 169–178

Tanaka M, Sakanishi Y (1980) Clonal propagation of *Phalaenopsis* through tissue culture. In: Kashemsanta MRS (ed) Proc 9th World Orchid Conf, pp 215–221, Bangkok

Tanaka M, Sakanishi Y (1985) Regenerative capacity of in vitro cultured leaf segments excised from mature *Phalaenopsis* plants. Bull Univ Osaka Prefect Ser B 37: 1–4

Tanaka M, Hasegawa A, Goi M (1975) Studies on the clonal propagation of monopodial orchids by tissue culture. I. Formation of protocorm-like bodies from leaf tissue in *Phalaenopsis* and *Vanda*. J Jpn Soc Hortic Sci 44: 47–58

Tanaka M, Senda Y, Hasegawa A (1976) Plantlet formation by root-tip culture in *Phalaenopsis*. Am Orchid Soc Bull 45: 1022–1024

Tanaka M, Kumura M, Goi M (1983) Surface-sterilization for in vitro culture of *Phalaenopsis* flower stalk cuttings using antimicrobials. Acta Hortic 131: 321–328

Tanaka M, Kumura M, Goi M (1988) Optimal conditions for shoot production from *Phalaenopsis* flower stalk cuttings cultured in vitro. Sci Hortic 35: 117–126

Tanaka M, Hayakawa J, Nishimura M, Goi M (1990) Micropropagation of *Phalaenopsis* by leaf segment culture. In: Shooter R (ed) Proc 3rd Asia Pacific Orchid Conf, pp 6–13, The Orchid Club of South Australia Inc., Adelaide

Tews G (1974) Einfache vegetative Phalaenopsisvermehrung. Orchideen (Greifswald, DDR) 3: 14–15

Tse AT, Smith RJ, Hackett WP (1971) Adventitious shoot formation on *Phalaenopsis* nodes. Am Orchid Soc Bull 40: 807–810

Tsukamoto Y, Kano K, Katsuura T (1963) Instant media for orchid seed germination. Am Orchid Soc Bull 32: 354–355

Urata U, Iwanaga ET (1965) The use of Ito-type vials for vegetative propagation of *Phalaenopsis*. Am Orchid Soc Bull 34: 410–413

Vacin EF, Went FW (1949) Some pH changes in nutrient solutions. Bot Gaz 110: 605–613

Vij SP, Sood A, Plaha KK (1984) Propagation of *Rhynchostylis retusa* Bl. (Orchidaceae) by direct organogenesis from leaf segment culture. Bot Gaz 145: 210–214

Wimber DE (1963) Clonal multiplication of *Cymbidium* through culture of shoot meristems. Am Orchid Soc Bull 32: 105–107

Wimber DE (1965) Additional observation on clonal propagation of *Cymbidium* through culture of shoot meristems. Cymb Soc News 19: 7–10

Woodard JW (1951) Some chromosome numbers in *Phalaenopsis*. Am Orchid Soc Bull 20: 356–358

Zimmer K, Pieper W (1978) Clonal propagation of *Phalaenopsis* by excised buds. Orchid Rev 86: 223–227

Zimmer K, Pieper W (1979) *Phalaenopsis*—Zur vegetativen Vermehrung. Gärtnerbörse und Gartenwelt 79: 258–260

Zwar JA, Bruce IM (1970) Cytokinins from apple extract and coconut milk. Aust J Biol Sci 23: 289–297

XVIII Micropropagation of Peony (*Paeonia* spp.)

J.A.T. Buchheim[1,2] and M.M. Meyer, Jr.[1]

1 Introduction

1.1 Importance and Distribution

The peony has been known for thousands of years both as a medicinal and landscape plant (Bailey 1914; Miller 1984). The herbaceous peony (*Paeonia lactiflora* Pall.) and tree peony (*P. suffruticosa* Andr.) are two landscape species found in the family Paeoniaceae. The family is native to north temperate regions, primarily southern Europe (where *P. lactiflora* originates), China (where *P. suffruticosa* originates), and northwestern America (where *P. californica* Nutt. ex Torr. & A. Gray originates) (Chant 1978). Early systematic studies placed the genus *Paeonia* in the family Ranunculaceae, where it was anomalous because of its persistent sepals, petals with a sepalar rather than a staminodial derivation, a perigynous disk, and seeds with arils (Chant 1978). Placement of the genus *Paeonia* in the Ranunculaceae was also anomalous in that *Paeonia* vessel members have scalariform perforation plates, whereas ranunculacean genera have simple perforation plates lacking cross walls (Trelease 1967). Recently, *Paeonia* has been placed in its own monotypic family, Paeoniaceae, which comprises 33 species of perennial, rhizomatous herbs and shrubs (Chant 1978). Most of the species are diploid ($2n = 10$), although several tetraploids have been described (Stebbins 1939).

1.2 Conventional Breeding and Propagation

The peony has several mechanisms to colonize and survive harsh environmental conditions, including vegetative bud dormancy and a complex sequential seed dormancy. Germination may take up to 2 years due to a requirement of warm temperatures (25 °C) for embryo development, then moderate temperatures (15 °C) for root growth, followed by epicotyl growth after cool temperatures (5 °C) (Krekler 1962; Hollingsworth 1979a, b). The germination time of peony seeds can vary considerably (Hollingsworth 1975). Peony growers and breeders have difficulty in obtaining controlled, uniform germination due to this variability and the complex sequential dormancy.

[1]Department of Horticulture, University of Illinois, Urbana, Il 61801, USA
[2]Present address: Department of Horticulture, Louisiana State University, Baton Rouge, LA 70803, USA

Biotechnology in Agriculture and Forestry, Vol. 20
High-Tech and Micropropagation IV (ed. by Y.P.S. Bajaj)

Fig. 1. **A** Dormant vegetative buds on a three year-old root clump of *Paeonia lactiflora* Emma Klehm. (Meyer, unpubl.). **B** Dormant vegetative bud from *Paeonia lactiflora* Emma Klehm root clump with cataphylls removed. Note the axillary buds (*arrows*). *Bar* 9 mm. (Thomas 1987)

Fig. 2. Globular somatic embryo developing on the adaxial cotyledon surface of a zygotic embryo cultured on LS medium for about three months. *Bar* 0.25 mm. (Thomas 1987)

Fig. 3. Zygotic embryo (*arrow*) excised from a portion of the endosperm after removal of the seed coat. (Thomas 1987)

Fig. 4. Thirteen-month-old somatic embryo, 1 month after topical application of 500 mg/l GA_3 and culture on LS medium. *Bar* 14 mm. (Thomas 1987)

Fig. 5. Adventitious shoot meristems formed on the periphery of an excised zygotic epicotyl that was cultured on LS medium with 2.202 mg/l thidiazuron for 2 months. *Bar* 5 mm. (Thomas 1987)

Fig. 6. Adventitious shoot derived from an epicotyl that was cultured on LS medium with 2.202 mg/l thidiazuron. The adventitious shoot was subcultured on cytokinin-free medium for 1 month. *Bar* 7 mm. (Thomas 1987)

Once germination is complete, seedlings must grow and develop for several years before flowering. The herbaceous peony requires up to 10 years to grow from a seed to a mature plant with typical bloom. Flowering plants from hybrid seed must be evaluated over several years before superior individuals can be identified and propagated (Krekler 1962).

Vegetative propagation is necessary to maintain the desirable characteristics of a particular peony plant. Herbaceous peonies are propagated by dividing tuberous root clumps into pieces containing three to five dormant vegetative buds (Fig. 1A).

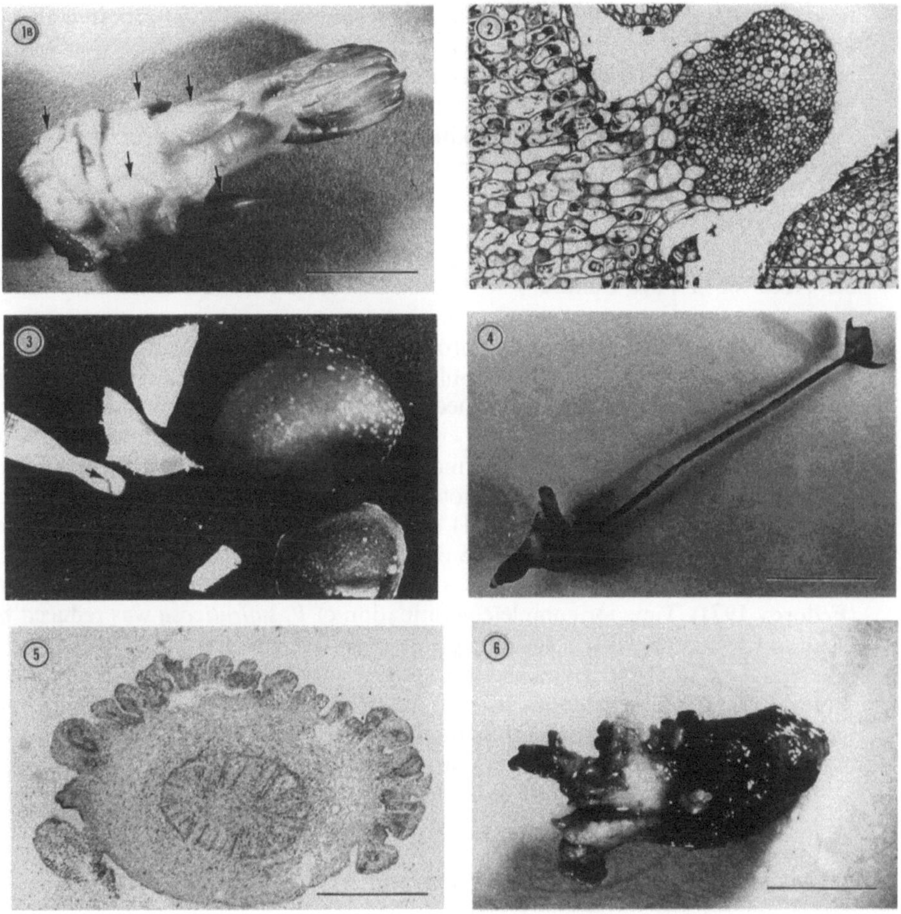

A grower may double his stock every 3 years by planting in a 3-year rotation (Shannon and Kamp 1959). Tree peonies are propagated by grafting a dormant shoot of a tree peony cultivar onto the rootstock of a herbaceous peony that has had its vegetative buds removed. The herbaceous peony rootstock is removed when the tree peony forms its own roots. Several years of growth are required before these plants reach a good flowering size. Thus, the release of a new peony cultivar may require many years of production before enough plants are obtained to sell (Wister 1962).

An important morphological difference between the herbaceous and tree peony is that the herbaceous peony grows from axillary buds located under bud scales (cataphylls) on large vegetative buds that develop underground on the root clump (Fig. 1A, B). When this vegetative bud expands, one to three lateral flower buds are typically visible in the axils of the upper leaves below the terminal flower (Vasil'eva 1976). Nodes below the flower buds typically do not develop vegetative or reproductive shoots in the axil of the leaves (Wilkins and Halevy 1985). Tree peonies form vegetative buds aboveground, in the axils of the basal two to four leaves (Wilkins and Halevy 1985). Shannon and Kamp (1959) reported that a vegetative bud is present in the leaf axil and lower stem-leaf portion of the herbaceous peony, which could be a source of cuttings. Although leaf-bud cuttings produced roots and developed callus tissue, the bud initials remained dormant, even when cold treatments, adenine, kinetin, or ethylene chlorohydrin were used to break dormancy. Unless this problem can be solved, leaf-bud cuttings cannot be used for propagation purposes.

1.3 Need for in Vitro Culture

Because conventional breeding and propagation are so difficult and time-consuming (see Sect. 1.2), peony culture could benefit greatly from micropropagation. Breeding programs could be hastened by reducing the time required for seed germination with the use of embryo culture. Certain growth regulators might be used to break epicotyl dormancy and replace the need for a cold treatment. Exogenous gibberellins have been shown to replace the need for moist-chilling of seeds of *P. tenuifolia* (Fedorov 1971) and *P. suffruticosa* (Barton and Chandler 1957). The application of GA_3 at 0.05 or 0.1% to seeds of *P. tenuifolia* successfully increased embryo growth during stratification and improved germination (Fedorov 1971). Time for complete germination of *P. suffruticosa* was reduced by treating hypocotyls with 1 to 100 μg GA_3 (Barton and Chandler 1957). Shoot differentiation in GA_3-treated tree peony hypocotyls after 9 days was about the same as untreated hypocotyls after 6 weeks at 5 °C. Besides embryo culture techniques, tissue culture techniques might also be employed to rapidly propagate clones of new cultivars, saving time, space, and labor.

2 Review of Tissue Culture Work and Micropropagation

Peony seedlings have been successfully grown in vitro (Meyer 1976a; Zilis and Meyer 1976; Lin 1980; Lin et al. 1987; Thomas 1987) by placing zygotic embryos on modified LS medium. The epicotyl dormancy that occurs in peony seeds in vivo was also observed in cultured embryos, and was overcome by exposure to 5 °C for 6 to 8 weeks (Meyer 1976a; Zilis and Meyer 1976). Epicotyl dormancy was not overcome when *P. lactiflora* seeds without seed coats were soaked in a 10 mg/l GA_3 solution prior to culture or when excised embryos were cultured on medium containing 10 or 25 mg/l GA_3 (Lin 1980).

Relatively little tissue culture research has been published on either the herbaceous or tree peony (see Table 1). Meyer (1976b), Lin (1980), Radtke (1983), and Hosoki et al. (1989) have attempted culturing herbaceous peony vegetative buds and buds formed under cataphylls. In most cases, only callus was obtained, and high contamination rates and excessive browning (assumed to be caused by phenolic compounds) of the tissues and culture medium presented major obstacles. Only Hosoki et al. (1989) have been successful in clonally propagating the herbaceous peony through shoot tip culture. Underground buds of the cultivars Takinoyosooi and Sarah Bernhardt were removed in the early spring, cataphylls removed, and sterilized with NaOCl (0.7% active chlorine). When main and axillary shoot tips (2–3 mm long) were removed and placed on modified MS with 0.5 mg/l BAP and 1 mg/l GA_3, axillary bud formation and development was promoted. Continuous shoot multiplication was achieved by vertically dividing the shoot axes and subculturing to fresh medium every 36 days. A high frequency of shoots were rooted when cultured on paper-bridges in liquid medium with 1 mg/l IBA. The authors estimate that 700 and 300 plants could be obtained from a single bud of Takinoyosooi and Sarah Bernhardt, respectively, in 1 year.

Radtke (1983) performed a series of experiments on several herbaceous peony cultivars and explants. Buds from leaf axils and from the axils of cataphylls were cultured on modified MS plus 0.3 mg/l IAA and 30 mg/l 2-iP. All explants from the

Table 1. Summary of in vitro studies conducted on several species of *Paeonia*

Plant species	Inoculum	Medium (mg/l)	Growth response	Reference
Hybrids	Microspore	MS	Embryos, plantlets	Roberts and Sunderland (1977)
Hybrids, *P. delavayi*	Microspore	MS	Embryos	Sunderland (1983a, b, c)
P. lactiflora	Microspore	MS + NAA(2) + CM(10%)	Callus	Ono and Harashima (1981)
P. lutea, *P. suffruticosa*	Anther	Mod. MS + kin(1) + IAA(1)	Embryos	Zenkteler et al. (1975)
Hybrids, *P. decora*, *P. triternata*	Anther	MS + low auxin MS + kinetin + high auxin	Callus, embryos	Sunderland et al. (1973)
Hybrids, *P. decora*, *P. triternata*, *P. emodi*	Anther	MS	Callus, roots, embryos, plantlets	Sunderland et al. (1975)
P. hybrida	Anther	MS + NOA(0.1)	Embryos, plantlets	Sunderland (1974), Sunderland and Dunwell (1974)
P. lactiflora	Flower bud, pedicel	Mod. MS + 2-iP(5) + NAA(2.5) or Picloram(1)	Callus	Redtke (1983)
P. lactiflora, *P. suffruticosa*	Flower bud	Mod. MS + kin(2.5) + NAA(2.5–10)	Callus, roots	Meyer (1976b)
P. lactiflora	Zygotic embryo	Mod. LS + NAA (2.5) + 2-iP(0.4)	Callus, roots	Lin (1980), Lin et al. (1987)
P. lactiflora	Zygotic embryo	Mod. LS + NAA (0.5)	Embryos	Lin (1980)
P. lactiflora	Zygotic embryo	Mod. LS	Embryos	Lin (1980), Lin et al. (1987), Thomas (1987)

Table 1. (*Continued*)

Plant species	Inoculum	Medium (mg/l)	Growth response	Reference
P. lactiflora	Zygotic + somatic embryo	Mod. LS + GA_3 (500)	Shoot expansion	Thomas (1987)
P. lactiflora	Zygotic embryo	Mod. LS + thidiazuron(2.2)	Callus, shoot meristems	Thomas (1987)
P. lactiflora, *P. suffruticosa*	Zygotic embryo	Mod. LS	Radicle expansion	Meyer (1976a)
P. suffruticosa	Zygotic embryo	Steeves (1955) + CM(150) + 2,4-D (0.2)	Callus	Demoise and Partanen (1969)
P. lactiflora, *P. suffruticosa*	Axillary vegetative bud	Mod. MS + kin (2.5) + NAA (2.5–10)	Callus, roots	Meyer (1976b)
P. lactiflora	Axillary veg. bud	Mod. MS + 2-iP (0.2) + NAA(1)	Callus	Lin (1980)
P. lactiflora	Axillary veg. bud	Mod. MS + 2-iP(15–30) + IAA(0.3) or NAA(2.5)	Callus, shoot expansion	Radtke (1983)
P. lactiflora	Axillary veg. bud	Mod. LS + thidiazuron (1.1)	Shoot meristems	Thomas (1987)
P. lactiflora	Axillary and main shoot tips	Mod. MS + BAP(0.5) + GA_3(1)	Multiplication and expansion of axillary buds	Hosoki et al. (1989)
P. suffruticosa	Axillary veg. bud	Mod. MS + kin(0.2–1) + BAP(0.5–1) + GA_3(0.1–0.5)	Callus, shoot meristems, shoots	Li et al. (1984)
P. suffruticosa	Leaf, petiole	Mod. MS + BAP(2) + NAA(0.1–0.5)	Callus, shoot meristems, plants	Li et al. (1984)
P. lactiflora	Stem, petiole	Mod. MS + 2-iP(5) + NAA(2.5)	Callus	Radtke (1983)
P. suffruticosa	Stem	Mod. SH or LB + NAA(10) or 2,4-D(0.2–2)	Callus, roots	Gildow and Mitchell (1977)
P. lactiflora, *P. suffruticosa*	Root	Mod. MS + kin (2.5) + NAA (2.5–10)	Callus, roots	Meyer (1976b)
P. lactiflora	Root	Mod. LS + 2,4-D (0.5–2)	Callus	Lin (1980)

Abbreviations: Basal media: LB Liau and Boll (1970); LS Linsmaier and Skoog (1965); MS Murashige and Skoog (1962); SH Schenk and Hildebrandt (1972); Supplements: IAA Indole-3-acetic acid; NAA Naphthaleneacetic acid; NOA Naphthoxyacetic acid; 2,4-D 2,4-Dichlorophenoxyacetic acid; Picloram 4-amino-3,5,6-trichloropicolinic acid; BAP Benzylamino purine; CM coconut milk; kin kinetin; 2-iP 6-g, g-(dimethylallyamino)purine; thidiazuron N-phenyl-N′-1,2,3-thiadiazol-5-ylurea; GA_3 gibberellic acid.

leaf axils died quickly due to excessive browning of the tissue. A dark treatment and the addition of 0.2% activated charcoal to the medium reduced production of phenolic compounds from axillary buds formed under the cataphylls. These explants elongated and formed leaf structures after 2 weeks in culture. Dark conditions favored callus production. One explant cultured in darkness without charcoal initiated callus, and 1 month later produced a leaf-like structure. When 2-iP concentration was reduced to 15 mg/l, buds elongated further. Larger buds were found to be better explant sources, as they were less susceptible to the deleterious effects of phenolic compounds. Differences in phenolic production and the ability to form callus in darkness were noted with the various cultivars.

Flower buds, pedicels, petioles, and stem segments were also tested (Radtke 1983). All stem and petiole tissue cultured on modified MS plus 2.5 mg/l NAA and 5 mg/l 2-iP without charcoal produced phenolic compounds within 2 days. If receptacle tissue was removed, flower bud explants produced significantly less phenolic compounds compared to those still intact. Despite frequent transfers, tissues grown on medium lacking charcoal eventually turned brown and died. More callus was obtained from flower buds and pedicels when cultured on MS plus 5 mg/l 2-iP and 1 mg/l Picloram instead of NAA. In an attempt to prevent browning, stem sections were also soaked in either citric acid (150 mg/l), ascorbic acid (100 mg/l), an equal volume of both, or sterile water. All four treatments were ineffective in preventing or reducing the browning of tissues. Neither organogenesis nor embryogenesis was obtained from any of these cultures.

Meyer (1976b) was also unsuccessful in regenerating plants from herbaceous and tree peony flower bud explants. Callus was obtained on MS plus 2.5 or 10 mg/l NAA and 2.5 mg/l kinetin after 2 months of culture in darkness at 27 °C. Callus cultures of one herbaceous peony cultivar produced roots, but never formed shoots. Meyer (1976b) also soaked root explants in ascorbic acid and citric acid, and added ascorbic acid to the medium, with little effect on reduction of phenolic compounds. A small amount of callus formed from root explants, and was subcultured for 2 years, but plants were never obtained from these cultures. Lin (1980) studied the effect of 0.5 mg/l 2,4-D on callus development from root segments. Callus was obtained from phloem pieces after 10 days and was subcultured on 1 or 2 mg/l 2,4-D. Callus production continued but differentiation did not occur.

Li et al. (1984) cultured axillary buds, and leaf and petiole segments of two cultivars of *P. suffruticosa*. Shoot tips (excised from axillary buds) callused and differentiated buds after about 7 weeks of culture on modified LS medium with 0.2–1 mg/l kinetin, 0.5–1 mg/l BAP, and 0.1–0.5 mg/l GA_3. Clusters of buds formed directly on shoot tips when they were cultured on MS medium with cytokinin but without GA_3. Callus formed on segments of young leaves and petioles when cultured on modified MS with 2 mg/l BAP and 0.1–0.5 mg/l NAA. Bud clusters differentiated from callus after subculture on MS plus 0.2–0.5 mg/l IAA and 2 mg/l BAP, with a higher frequency occurring on callus derived from petiole explants. Buds expanded up to 3–4 cm in height 25–35 days after subculture on MS media with 0.5–2 mg/l BAP and 0.2–0.5 mg/l GA_3. These shoots were then excised and dipped into a solution containing 50–100 mg/l IBA for 2–3 h prior to culture on modified MS medium with 0.5% activated charcoal for 25–35 days. After a period of acclimation, plantlets were successfully transplanted.

Li et al. (1984) estimate that 5000–10 000 plantlets could be produced yearly from each bud using these techniques.

Demoise and Partanen (1969) studied the effects of subculture and physical condition of the medium on nuclear behavior of *P. suffruticosa*. Embryos were excised and cultured on a modified medium developed by Steeves et al. (1955) with 0.8% Bacto agar. Callus cultures were obtained and subcultured on either liquid or solid medium with 150 mg/l coconut milk, and 0.2 mg/l 2,4-D. Studies of the mitotic index revealed 2.5 times more rapid callus growth in liquid medium compared to solid medium. Although the physical condition of the medium affected the mitotic index, the indices remained constant over a 153-day subculturing period. Neither organogenesis nor embryogenesis occurred in these callus cultures. No obvious changes were found in the morphology of chromosomes; however, the concept of homogeneity of cells in these cultures was rejected after evidence of varying or fluctuating heterogeneity, especially in ploidy number, was found.

Gildow and Mitchell (1977) also studied initiation, growth, and nuclear characteristics of *P. suffruticosa*. Attempts to initiate callus from light-grown stems, leaves, roots, and floral parts were unsuccessful due to excessive browning of the tissue and medium. This brown coloration was greatly reduced by using stem sections (3–5 mm long) from etiolated plants. Explants from stems etiolated at 15 °C produced greater callus growth than explants from stems etiolated at 25 °C. The addition of 0.2 mg/l 2,4-D or 10 mg/l NAA to either LB medium (Liau and Boll 1970) or SH medium plus 1650 mg/l NH_4NO_3 was equally effective in initiating callus. Callus was then transferred to various media supplemented with 0.2–2 mg/l 2,4-D and 0.1 mg/l kinetin and grown in the dark. Callus growth and protein content was greatest in callus cultured on SH medium with 1250 mg/l KNO_3. Only roots formed when callus was subcultured to liquid or solid medium lacking growth regulators. A histochemical comparison of tissue grown on the various media showed that the cells in all of the cultures remained diploid.

Induction of somatic embryogenesis in herbaceous peonies has been obtained in numerous instances (see Table 1). Lin (1980) and Lin et al. (1987) reported that somatic embryos formed on the adaxial cotyledon surface of zygotic embryos cultured on modified LS medium and on callus tissue derived from zygotic embryos. Maximum callus tissue developed from zygotic embryos cultured on modified LS medium plus 2.5 mg/l NAA and 0.4 mg/l 2-iP, whereas somatic embryo production from callus was greatest on modified LS plus 3800 mg/l KNO_3 and 0.5 mg/l NAA. Somatic embryos also arose directly on the adaxial cotyledon surface of zygotic embryos (Fig. 2) cultured on modified LS medium with 4% sucrose and 3300 mg/l NH_4NO_3 (Lin 1980; Lin et al. 1987; Thomas 1987; Thomas and Meyer 1987).

Structural development of somatic embryos derived from cotyledons appeared similar to that in zygotic embryos, with somatic embryos passing through globular, heart, and torpedo stages (Lin 1980; Thomas and Meyer 1986, 1987; Lin et al. 1987; Thomas 1987). Somatic embryogenesis on cotyledons and on callus was continuous and asynchronous. Somatic embryos arising from callus were connected by suspensor-like material. These embryos separated from the callus tissue as the cotyledons enlarged, but never formed true leaves. Somatic embryos remaining attached to the cotyledon surface developed roots but the cotyledons did not expand. Embryos isolated from the cotyledon surface and cultured on medium lacking growth regulators formed roots and expanded cotyledons, but never developed true leaves (Lin 1980; Thomas and Meyer 1986, 1987; Lin et al. 1987; Thomas 1987).

Embryogenesis has also been reported in both microspore and anther culture of several *Paeonia* species (see Table 1). Microspore culture of *P. lactiflora* was accomplished by harvesting flower buds when the calyx was intact and petals were not visible (Roberts and Sunderland 1977). Buds at this stage were easy to disinfest and anther walls produced less phenolic compounds. Anthers containing undivided microspores were floated on a liquid medium. Float cultures were necessary to allow phenolic compounds to diffuse away from the anthers, which produce these compounds rapidly after the first division of pollen. Repeated transfers to fresh medium were also necessary during the first 2 to 3 weeks to eliminate most of the phenolic compounds. Embryos were obtained by culturing pollen in a basal medium in darkness for 15 weeks. Embryos were then transferred to liquid medium on a rotary shaker in darkness for an additional 6 months before rooted plantlets were obtained. These plantlets were not successfully transplanted to soil (Roberts and Sunderland 1977).

Embryonic structures formed in peony microspore cultures failed to develop in an organized manner (Sunderland 1983c). A number of structures formed, some resembling globular- or heart-shaped embryos, while others were highly irregular and were probably multipolar in origin. On transfer to fresh medium, individual globular embryos proliferated slowly and gave rise to roots without shoots or shoots without roots, whereas multipolar structures formed highly complex shoot-like systems that lacked recognizable leaves. Normal bipolar development was only achieved with embryos from cultured anthers (Sunderland 1983c). The growth rate of anther-derived embryos was slow, with embryos emerging from anthers after 4 to 6 months of culture (Sunderland et al. 1975). Embryo emergence varied from as few as one to as many as hundreds per anther. Embryo clusters proliferated slowly and gave rise to more embryo clusters. Separation of these embryos and subculture to fresh medium resulted in reformation of clusters. Some embryos formed callus instead of cotyledons, which, upon subculture, resulted in root formation only. Sporadic cotyledon enlargement and chlorophyll production occurred, but normal shoots were not formed. Embryos that emerged individually developed more normal root and shoot poles, and continued to develop after several months of subculture; however, shoot development was not sufficiently advanced for transfer to the greenhouse (Sunderland et al. 1975). Histological studies showed that these individual embryos arose from an organized embryogenic body that resulted from predominantly periclinal cell divisions (Haccius et al. 1975). This elongated structure is thought to be analogous to a primitive suspensor; therefore, individual embryos could represent cases where only one embryo arose from a suspensor, and clusters could represent cases where multiple embryos arose from a suspensor (Haccius et al. 1975).

2.1 Zygotic Embryo Culture

Seeds were collected after the follicles were fully ripened. Seeds were sterilized in 0.26% NaOCl, rinsed with water, air-dried for 24 h at 26 °C in a sterile transfer hood, and stored in polyethylene bags at 5 °C prior to use. Seeds were cut in half with wire strippers, and seed coats were removed from the halves containing the rudimentary embryo. Several seed halves were then placed in a cheesecloth bag, sterilized in 0.53% NaOCl plus 0.1% Tween 20 (Sigma, St. Louis, MO) for 20 min, and rinsed three times in sterile, deionized water. Embryos were removed by slicing into the endosperm with a sharp scalpel and cracking the endosperm in half (Fig.3). Typically the endosperm cracks next to the embryo, facilitating its removal. Embryos were aseptically transferred to 150 × 25 mm test tubes containing 20 ml of modified LS medium (Table 2). The pH of all media was adjusted to 5.6 with 1 N KOH before the addition of Bacto agar (Difco, Detroit, MI). Growth regulators were incorporated into the media after melting the agar. Media were sterilized in an autoclave at 15 lb/in^2 pressure for 25 min. Cultures were grown at 26 °C under cool white fluorescent lights, 45–50 $\mu mol/s/m^2$, with an 18-h photoperiod.

Because gibberellins have been used successfully to overcome epicotyl dormancy in other species, including *P. suffruticosa* (Barton and Chandler 1957), herbaceous peony embryos were also exposed to exogenous GA_3. Embryos were excised and cultured on modified LS medium for 3 weeks. A stock of GA_3 was made by adding 2 mg of active GA_3 to each ml of double-deionized water. The stock was filter sterilized. Lower concentrations were made by first diluting the GA_3 stock with sterile water and then adding equal volumes of 1% agarose stock (Type VII, Sigma, St. Louis, MO). The agarose stock, made by dissolving the agarose in water, was melted and sterilized in an autoclave. Dilutions of GA_3 (for final concentrations of 100, 250, 500, 750 and 1000 mg/l GA_3 plus 0.5% agarose) were made in a sterile transfer hood after the agarose was cool but still liquid. The sterile GA_3-agarose solutions were stored at 5 °C and were melted in hot water (about 80 °C) prior to application. The melted GA_3-agarose solutions were aseptically applied to the epicotyledonary region of the embryo (after solutions had cooled to about 35 °C) with a sterile micropipet at the rate of 50 μl per embryo.

Table 2. Composition of modified Linsmaier and Skoog medium (Lin 1980)

Mineral salts[a]			
Major elements		Minor elements stock[b]	
Salts	g/l	Salts	g/l
NH_4NO_3	1.650	H_3BO_3	1.540
KNO_3	1.900	$MnSO_4 \cdot 4H_2O$	0.845
$CaCl_2 \cdot 2H_2O$	0.440	$ZnSO_4 \cdot 7H_2O$	0.575
$MgSO_4 \cdot 7H_2O$	0.370	KI	0.166
KH_2PO_4	0.170	$CuSO_4 \cdot 5H_2O$	0.125
KFeEDTA	0.005	NaCl	2.920
		85% Molybdic acid	0.017
Organic components			
Sucrose	40 g/l	*myo*-Inositol	100 mg/l
Bacto agar	6 g/l	Thiamine	0.4 mg/l
		Casein hydrolysate	500 mg/l

[a] pH adjusted to 5.6 with 1N KOH.
[b] One ml of minor element stock per liter of medium.

Embryos in the control were treated with a sterile solution of 0.5% agarose instead of GA_3.

Leaf number and petiole length were measured as an indicator of epicotyl growth because peony seedlings grown in vivo typically produce only a few leaves on a shortened stem that cannot be measured without using light microscopy. The application of GA_3-agarose resulted in a significant increase in epicotyl growth after four weeks. Both leaf number and petiole length increased as GA_3 concentration increased up to 500 mg/l, and decreased with the higher concentrations of GA_3 (Fig. 7). Spindly petioles, deformed leaves, and curled, elongated cotyledons developed after GA_3 application, especially at concentrations of 750 and 1000 mg/l. Seedlings did not transplant well to the greenhouse, even though steps were taken to slowly acclimatize them. The petioles were weak and desiccated quickly, and fungal pathogens killed many of the transplanted seedlings. By comparison, seedlings that were transferred to large jars containing 60 ml of modified LS medium exhibited typical leaf morphology after 6 to 8 weeks of culture. These leaves consisted of shorter, sturdier petioles, and larger, thicker, dark green blades.

Although there were some problems with establishing these germinated herbaceous peony embryos, the overall germination time was reduced from up to 2 years in vivo to only 6 weeks in vitro. Only about 5% of the embryos treated with the optimum concentration, 500 mg/l GA_3, did not respond to the treatment. The number of contaminated cultures was less than 2% of all of the embryos cultured. Thus, this GA_3 treatment shortened the germination time considerably, and resulted in a more uniform, reliable germination (Thomas and Meyer 1986;

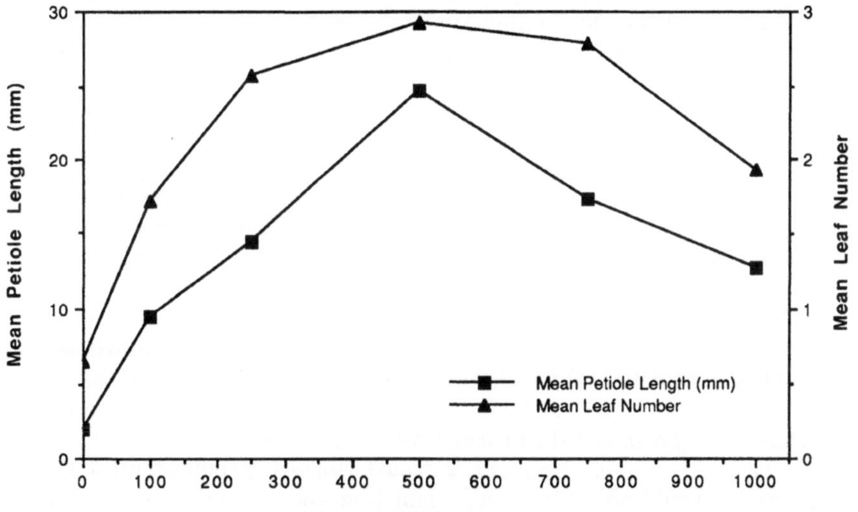

Fig. 7. Epicotyl growth of zygotic embryos, 4 weeks after topical application of GA_3-agarose solutions. Embryos were cultured on LS medium for 3 weeks prior to GA_3-agarose treatment. (Thomas 1987)

Thomas 1987). This could be of particular value to breeders who make crosses that result in only a few seeds.

2.2 Somatic Embryo Culture

Although embryos have been regenerated from herbaceous peony zygotic embryos, anthers, and pollen (see Sect. 2), shoot growth has never been sufficient for establishment in the greenhouse. True somatic embryos should resemble their zygotic counterparts in both morphology, anatomy, and physiology. It would follow that these somatic embryos might also possess an epicotyl dormancy. Therefore, somatic embryos were obtained by culturing zygotic embryos on 20 ml of modified LS medium (Table 2) per test tube (growth regulators were not necessary) until somatic embryos started appearing on the adaxial cotyledon surface (about 10 weeks). Somatic embryos were removed from the cotyledon surface and subcultured on modified LS medium every 2 months. Dome-shaped meristems, visible to the eye, formed after approximately 8 months of culture on 15% of these somatic embryos. Somatic embryos were cultured for an additional 4 months to observe growth, during which time more embryos formed visible epicotyls, but shoot growth did not occur. At this point, somatic embryos with and without visible epicotyls were treated with 50 µl of 500 mg/l GA_3-agarose or 0.5% agarose as a control (as described in Sect. 2.1). The embryos were in various stages of development, and some had secondary somatic embryos forming on the cotyledon surface and hypocotyl. Somatic embryos in the control treatment did not form true leaves during 4 weeks of culture. Cotyledons remained cream-colored and grew very little compared to GA_3-treated somatic embryos. Seventy-five percent of the embryos treated with 500 mg/l GA_3 produced enlarged, green cotyledons although root length was not significantly increased. True leaves formed on 29% of the somatic embryos treated with GA_3 (Fig. 4). These leaves first appeared about 3 weeks after treatment, although leaves were slower to emerge and develop compared to leaves on GA_3-treated zygotic embryos. Leaves on somatic embryos were also smaller, thinner, chlorotic, and had shorter petioles than leaves on zygotic embryos. Of the number of somatic embryos forming leaves, 33% had visible epicotyls prior to treatment, and 67% lacked visible epicotyls. Germinated somatic embryos did not transfer well to the greenhouse. They suffered from the same problems described (in Sect. 2.1) for zygotic embryos (Thomas 1987).

To our knowledge, this is the first instance of well-developed plantlets derived from somatic embryos of *Paeonia* (Thomas 1987). The propagation technique described above could be used in cases where only a few seeds are recovered from crosses between species. The embryos could be excised and cultured on basal medium until somatic embryos formed. The somatic embryos could be removed from the cotyledon surface and cultured further to produce more somatic embryos (secondary embryos), and then treated with GA_3 to induce epicotyl growth. The original zygotic embryos could also be treated with GA_3 to induce epicotyl growth. The breeder would not only have a plant from the original seed, but also have clones of this plant, increasing the stock considerably. This method could also be used to micropropagate clones of established peony cultivars, once a technique is found to obtain somatic embryos from vegetative tissue.

2.3 Tissue Culture

Herbaceous peony root clumps containing dormant vegetative buds (Fig. 1A) were removed from the ground, washed, air-dried, and stored in polyethylene bags at 5 °C until needed. The buds were removed and washed in deionized water with 0.26% NaOCl and 0.1% Tween 20. Because browning of the tissue occurred soon after the buds were cut, outer scales and browned tissue were removed. Buds were then sterilized in 0.53% NaOCl plus 0.1% Tween 20 for 20 min under a vacuum. The NaOCl was removed and buds were rinsed three times with sterile, deionized water. Buds were placed in test tubes containing 4 ml of liquid modified LS medium (Table 2) with approximately 50 mg insoluble polyvinylpyrrolidone (PVP; Polyclar, GAF Corp., NY) per test tube. The test tubes were placed on a rotating wheel for 24 h. This step was added to the disinfestation procedure to encourage growth of any fungal spores or bacteria that were not killed by the initial sterilization step, and to complex any phenolic compounds produced by the tissue with the PVP. The medium plus PVP was removed from each test tube and replaced with approximately 20 ml of 0.53% NaOCl plus 0.1% Tween 20. The buds were sonicated for 1 min in this solution and then exposed to the NaOCl for an additional 14 min. The NaOCl was then aspirated off and buds were rinsed three times in sterile, deionized water. The axillary buds (Fig. 1B), formed under the cataphylls, were aseptically removed and treated with 3% peroxide for approximately 15 sec, and given a final rinse with sterile, deionized water. Axillary buds were placed in test tubes containing 20 ml of modified LS medium (Table 2) plus either 2-iP, BAP, or thidiazuron. The concentrations tested were 0.2, 2.0, and 10.2 mg/l 2-iP, 0.2, 2.3, and 11.3 mg/l BAP, 0.002, 0.022, 0.220, and 1.101 mg/l thidiazuron (Nor-Am Agricultural Chemicals, Wilmington, DE), and a control containing no cytokinins. Thidiazuron, or Dropp, is commercially used as a cotton defoliant, but has been found to have strong cytokinin activity (Mok et al. 1982; Mok and Mok 1985). All cytokinin stocks were made in dimethyl sulfoxide. Axillary buds were subcultured after one week to the same medium with 1% soluble PVP-40 (Sigma, St. Louis, MO) to complex phenolic compounds.

Excessive browning of the bud tissue became evident immediately, especially on larger buds. Contamination was also a problem, with about 35% of the explants being infested. Variations in browning and contamination were noted with different cultivars. After 3 months of subculture, explants grown on medium with thidiazuron developed callus and remained greener than those on BAP or 2-iP. Adventitious shoots were only obtained from one cultivar, Emma Klehm, after culture on 1.101 mg/l thidiazuron. These adventitious shoots formed in a cluster (about five shoots per cluster) on the basal tissue of the expanded axillary bud. The shoots were removed and subcultured on modified LS medium with PVP-40 in the dark to reduce production of phenolic compounds. Unfortunately, after several months of subculture, all of the shoots turned brown and died.

Because contamination and browning of the vegetative bud explants were such problems, excised epicotyls from cultured zygotic embryos were also used as explants. The epicotyls are analogous to vegetative buds but have low rates of contamination and tissue browning. Embryos were treated with 500 mg/l GA_3-agarose for 4 weeks. Epicotyls were then excised by removing the cotyledons and root, but leaving intact the basal tissue that connects the epicotyl to the cotyledons and root. The concentrations tested were 0.2 and 2.0 mg/l 2-iP, 0.2 and 2.3 mg/l BAP, 0.002, 0.022, 0.220, 1.101, and 2.202 mg/l thidiazuron, and a control containing no cytokinins. After 3 weeks of culture, the basal tissue of epicotyls grown on thidiazuron swelled and often were tinged with a reddish pigment. Adventitious shoot meristems were visible around the periphery of the basal tissue, especially on explants grown on high concentrations of thidiazuron (Fig. 5). These meristems appeared as dark green, round structures protruding from the swollen basal tissue. Epicotyls cultured on 2.202 mg/l thidiazuron produced some callus along with these meristems. Swollen basal tissue and adventitious meristems were not observed on tissue grown on 2-iP, BAP, or the control, and more browning occurred on these explants compared to those on thidiazuron. Epicotyls cultured on 2.202 mg/l thidiazuron produced the most meristems and callus. These meristems expanded only one or two leaf-like structures while remaining attached to the basal epicotyl tissue. Adventitious shoot meristems that were excised and cultured on medium lacking growth regulators in either light or dark continued to develop and produced a few more leaf-like structures but never expanded a true shoot (Fig. 6). These meristems were then cultured on medium containing 0.20 or 1.02 mg/l IBA, 0.19 or 0.93 mg/l NAA, or a control lacking an auxin. After 3 months of culture, roots had not formed on any of the meristems, regardless of culture in light or dark (Thomas 1987).

Perhaps the cytokinin activity of thidiazuron was too strong and inhibited root formation. A similar response was seen in *Malus* shoots proliferated from shoot tips cultured on 2.202 mg/l thidiazuron (van Nieuwkerk et al. 1986). Although some problems exist with the use of thidiazuron, it has been used successfully to obtain shoots from *Acer* (Kerns and Meyer 1986), *Malus* (van Nieuwkerk et al. 1986; Fasolo et al. 1989; Sriskandarajah et al. 1990), *Syringa, Forsythia, Fraxinus, Ligustrum* (Einset and Alexander 1984), *Pyrus* (Chevreau et al. 1989), *Prunus* (Mante et al. 1989), *Rhododendron* (Briggs and McCulloch 1988; Briggs et al. 1988), *Ixia* (Meyer and van Staden 1988), *Oxydendrum* (Banko and Stefani 1989), and various species in the superorders Magnoliidae, Hamamelidae, Dilleniidae, Rosidae, and Asteridae (Einset 1985). Perhaps peonies can also be propagated in this way, but first a method is needed to reduce contamination and browning problems encountered with vegetative buds.

3 Summary and Conclusions

Herbaceous peony seeds may take up to two years to germinate in vivo. By culturing rudimentary embryos on modified LS medium and topically applying 500 mg/l GA_3-agarose to the epicotyl region, seedlings were obtained in only 6 weeks. Somatic embryos that spontaneously arose on the adaxial surface of cultured zygotic embryos possessed an epicotyl dormancy similar to that in zygotic embryos. This dormancy was also overcome by topical application of 500 mg/l GA_3-agarose. Unfortunately, seedlings derived from both zygotic and somatic embryos did not transfer well to the greenhouse. Embryos obtained from microspores and anthers (Sunderland et al. 1973; Sunderland 1974; Sunderland and

Dunwell 1974; Sunderland et al. 1975; Zenkteler et al. 1975; Roberts and Sunderland 1977; Ono and Harashima 1981; Sunderland 1983a, b, c) may also have this dormancy mechanism, and could benefit from GA_3 treatment.

Somatic embryos have not yet been obtained from vegetative material. One of the major obstacles with vegetative buds has been high contamination rates and excessive browning of the tissues. Once these problems are resolved, the cytokinin activity of thidiazuron might be used to propagate adventitious shoots. Excised epicotyls and axillary buds (from under cataphylls) formed adventitious shoots or shoot meristems when cultured on LS medium with 1.101 mg/l thidiazuron. Only two axillary bud explants developed adventitious shoots, and these shoots eventually turned brown and died after subculture. Adventitious shoot meristems from epicotyl explants expanded a few reduced, leaf-like structures, but did not develop roots even when subcultured on IBA or NAA. The strong cytokinin activity of thidiazuron may have inhibited root formation. Thidiazuron has been shown to be 100 times more effective than BAP or zeatin in radish callus and 10 000 times more effective in soybean callus (Thomas and Katterman 1986). The adventitious shoots and meristems derived from peony explants in vitro may also have a dormancy mechanism, as was seen in somatic embryos, that requires exposure to cold temperatures or GA_3.

Micropropagation of peonies could cut decades off the production time required to introduce a new cultivar on the market. Hosoki et al. (1989) have estimated that up to 700 plants could be obtained in one year by splitting and routinely subculturing a single vegetative bud. Breeders could use embryo culture to reduce the germination time to only a few weeks. Somatic embryos could be used to increase the amount of material breeders have to work with, especially in cases where a cross results in very few progeny. Although peonies are not yet routinely micropropagated, the advances described above move us closer to that goal.

References

Bailey LH (1914) Standard cyclopedia of horticulture. Macmillan, New York, pp 2431–2436

Banko TJ, Stefani MA (1989) In vitro propagation of *Oxydendrum arboreum* from mature trees. HortScience 24: 683–685

Barton LV, Chandler C (1957) Physiological and morphological effects of gibberellic acid on epicotyl dormancy of tree peony. Contrib Boyce Thompson Inst 19: 201–214

Briggs BA, McCulloch SM (1988) Micropropagation of azaleas using thidiazuron. Acta Hort 227: 330–333

Briggs BA, McCulloch SM, Edick LA (1988) Micropropagation of azaleas using thidiazuron. Acta Hort 226: 205–208

Chant SR (1978) Flowering plants of the world. Mayflower Books, New York, pp 80–81

Chevreau E, Skirvin RM, Abu-Qaoud HA, Korban SS, Sullivan JG (1989) Adventitious shoot regeneration from leaf tissue of three pear (*Pyrus* sp.) cultivars in vitro. Plant Cell Rep 7: 688–691

Demoise CF, Partanen CR (1969) Effects of subculturing and physical condition of medium on the nuclear behavior of a plant tissue culture. Am J Bot 56: 147–152

Einset JW (1985) Role of cytokinin in woody plant micropropagation. Proc Int Plant Prop Soc 35: 608–615

Einset JW, Alexander JH (1984) Multiplication of *Syringa* species and cultivars in tissue culture. Proc Int Plant Prop Soc 34: 628–636

Fasolo F, Zimmerman RH, Fordham I (1989) Adventitious shoot formation on excised leaves of in vitro-grown shoots of apple cultivars. Plant Cell Tissue Organ Cult 16: 75–87

Fedorov AK (1971) The effect of gibberellin on the development of plants overwintering with rudimentary inflorescences. Biologicheskaya 5: 776–779

Gildow FE, Mitchell JP (1977) Initiation, growth, and nuclear characteristics of tissue cultures of *Paeonia suffruticosa*. Physiol Plant 58: 790–795

Haccius B, Hausner G, Sunderland N (1975) Developmental history of *Paeonia* pollen embryoids. John Innes Inst Annu Rep 66: 60–61

Hollingsworth D (1975) Controlled germination of peony seeds indoors. Amer Peony Soc Bull 214: 27–30

Hollingsworth D (1979a) Techniques in producing peony seedlings. Am Peony Soc Bull 231: 39–42

Hollingsworth D (1979b) Peony seed germination illustrated. Am Peony Soc Bull 232: 18–22

Hosoki T, Ando M, Kubara T, Hamada M, Itami M (1989) In vitro propagation of herbaceous peony (*Paeonia lactiflora* Pall.) by a longitudinal shoot-split method. Plant Cell Rep 8: 243–246

Kerns HR, Meyer MM (1986) Tissue culture propagation of *Acer* × *freemanii* using thidiazuron to stimulate shoot tip proliferation. HortScience 21: 1209–1210

Krekler WH (1962) Peony culture, uses, and propagation. In: Wister JC (ed) The Peony Am Hortic Soc, Washington DC, pp 100–111

Li Y, Wu D, Pan S, Xu S, Wei Z, Xu Z, Li X (1984) In vitro propagation of *Paeonia suffruticosa*. Kexue Tongbao 29: 1675–1678

Liau DG, Boll WG (1970) Callus and cell suspension cultures of bush bean, *Phaseolus vulgaris*. Can J Bot 48: 1119–1130

Lin JJ (1980) In vitro embryo and tissue culture studies in the herbaceous peony (*Paeonia lactiflora* Pall.) MS Thesis, Univ Illinois, Urbana, IL, 84 pp

Lin JJ, Thomas JA, Meyer MM (1987) Tissue culture and embryoid formation in *Paeonia lactiflora* Pall. Am Peony Soc Bull 263: 24–30

Linsmaier EM, Skoog F (1965) Organic growth factor requirements of tobacco tissue. Physiol Plant 18: 100–127

Mante S, Scorza R, Cordts JM (1989) Plant regeneration from cotyledons of *Prunus persica*, *Prunus domestica* and *Prunus cerasus*. Plant Cell, Tissue Organ Cult 19: 1–11

Meyer HA, van Staden J (1988) In vitro multiplication of *Ixia flexuosa*. HortScience 23: 1070–1071

Meyer MM (1976a) Culture of *Paeonia* embryos by in vitro techniques. Am Peony Soc Bull 217: 32–35

Meyer MM (1976b) Culture of *Paeonia* callus by tissue culture techniques. Am Peony Soc Bull 218: 27–29

Miller JA (1984) Peonies: Greek portraits of a queen/heirs fit for a king. Sci News 126: 56–57

Mok MC, Mok DWS (1985) The metabolism of (^{14}C)-thidiazuron in callus tissues of *Phaseolus lunatus*. Physiol Plant 65: 427–432

Mok MC, Mok DWS, Armstrong DJ, Shudo K, Isogai Y, Okamoto T (1982) Cytokinin activity of N-phenyl-N-1,2,3-thiadiazol-5-ylurea (thidiazuron). Phytochem 21: 1509–1511

Murashige T, Skoog F (1962) A revised medium for rapid growth and bioassays with tobacco tissue cultures. Physiol Plant 15: 473–497

Ono K, Harashima S (1981) Induction of haploid callus from isolated microspores of peony in vitro. Plant Cell Physiol 22: 337–341.

Radtke GW (1983) Tissue culture of herbaceous peonies. Am Peony Soc Bull 246: 19–23

Roberts M, Sunderland N (1977) Pollen culture of *Paeonia*. John Innes Inst Annu Rep 68: 60–61

Schenk RV, Hildebrandt AC (1972) Medium and techniques for the induction and growth of monocotyledonous and dicotyledonous plant cell cultures. Can J Bot 50: 199–204

Shannon J, Kamp JR (1959) Trials of various possible propagation methods of herbaceous peonies. Ill State Flor Assoc Bull 197: 4–7

Sriskandarajah S, Skirvin RM, Abu-Qaoud H, Korban SS (1990) Factors involved in shoot elongation and growth of adventitious and axillary buds of three apple scion cultivars in vitro. J Hortic Sci 65: 113–121

Stebbins GL (1939) Notes on the systematic relationships in the genus *Paeonia*. Univ Calif Publ Bot 19: 245–266

Steeves TA, Sussex IM, Partanen CE (1955) In vitro studies on abnormal growth of prothalli of the bracken fern. Am J Bot 42: 232–245

Sunderland N (1974) Anther culture as a means of haploid induction. In: Kasha KJ (ed) Haploids and higher plants, advances and potentials. Univ Guelph, Ontario, pp 91–122

Sunderland N (1983a) Microspore culture. John Innes Inst Annu Rep 72: 142–144
Sunderland N (1983b) Comparison of microspore and anther culture. John Innes Inst Annu Rep 72: 144
Sunderland N (1983c) Products of microspore culture. John Innes Inst Annu Rep 72: 144–145
Sunderland N, Dunwell JM (1974) Pathways in pollen embryogenesis. In: Street HE (ed) Tissue culture and plant sciences. Acad Press, London, pp 141–167
Sunderland N, Dunwell JM, Roberts E (1973) Anther culture of paeony. John Innes Inst Annu Rep 64: 76–77
Sunderland N, Dunwell JM, Roberts E (1975) Anther culture in the genus *Paeonia*. John Innes Inst Annu Rep 66: 57–60
Thomas JA (1987) Hormonal regulation of epicotyl dormancy, anatomical studies, and tissue culture of *Paeonia lactiflora* Pall. MS Thesis, Univ Illinois, Urbana, IL, 111 pp
Thomas JA, Meyer MM (1986) Tissue culture and anatomical studies of *Paeonia lactiflora*. HortScience 21: 774 (Abstr)
Thomas JA, Meyer MM (1987) In vitro embryogenesis and development of *Paeonia lactiflora* Pall. In Vitro 23: 70A (Abstr)
Thomas JC, Katterman FR (1986) Cytokinin activity induced by thidiazuron. Plant Physiol 81: 681–683
Trelease W (1967) Winter botany. Dover Publ, New York, p 52
Van Nieuwkerk JP, Zimmerman RN, Fordham I (1986) Thidiazuron stimulation of apple shoot proliferation in vitro. HortScience 21: 516–518
Vasil'eva MY (1976) The formation of renewal organs in herbaceous peonies. Ref Zhurnal 55: 930 (Abstr)
Wilkins HF, Halevy AH (1985) *Paeonia*. In: Halevy HA (ed) CRC Handbook of flowering: IV. CRC Press, Boca Raton, FL, pp 2–4
Wister JC (1962) The peonies. Am Hortic Soc, Washington DC, 220 pp
Zenkteler M, Misiura E, Ponitka A (1975) Induction of androgenetic embryoids in the in vitro cultured anthers of several species. Experientia 31: 289–291
Zilis MR, Meyer MM (1976) Rapid in vitro germination of immature, dormant embryos. Proc Int Plant Prop Soc 26: 272–275

XIX Micropropagation of Commercial *Pelargonium* Species and Hybrids (Glasshouse Geraniums)

A.C. Cassells[1]

1 Introduction

Pelargoniums originated mainly in South Africa, where some 200 species have been described (Clifford 1970; van der Walt 1977; van der Walt and Vorster 1981, 1988). Commercial pelargoniums (geraniums) include four groups (Fig. 1). The zonal pelargoniums (*P.* × *hortorum* Bailey, syn. *P. zonale*) which are used as pot and bedding plants, the ivy-leaf pelargoniums [*P. peltatum* (Lin.) L'Her. ex Ait.],

Fig. 1. Leaf shape in commercial pelargoniums. *Top left* zonal *Pelargonium*; *top right* ivy-leaf *Pelargonium*; *bottom left* regal *Pelargonium*; *bottom right* scented leaf *Pelargonium*

[1] Department of Plant Science, University College, Cork, Ireland

Biotechnology in Agriculture and Forestry, Vol. 20
High-Tech and Micropropagation IV (ed. by Y.P.S. Bajaj)

used in window boxes and hanging baskets, the regal pelargoniums (*P.* × *domesticum* Bailey), used as pot plants, and the scented leaf or aromatic pelargoniums (*P. graveolens* L'Her. ex Ait. and others), used as pot and bedding plants but also for field production of essential oils (Clifford 1970; Swain 1974). The origins of the cultivars and hybrids have been discussed by Clifford (1970), Harney (1976), and Moore and Hyypio (1982).

Pelargoniums have increased their market share in the past decades – US sales alone in 1979 being in excess of $90 million (Voight 1982) and rank as one of the major world floriculture crops. Approximately 90% of the ornamental crop is propagated vegetatively (Reuther 1988), the remainder from seed. The ratio of zonal to ivy-leaf to regal sales in the USA is approximately 10:4:1 and in Europe 10:10:1.

1.1 Traditional Production Problems

Traditional production problems in *Pelargonium* have been associated with poor seed production and disease. Low fertility in commercial pelargoniums can be

Table 1. Principal characterised vascular diseases of *Pelargonium*

Common name	Pathogen	Reference
Agent		
Fungi		
Verticillium wilt	*Verticillium albo-atrum*	Dimock (1962)
Bacteria and bacteria-like		
Bacterial blight syn. bacterial stem rot and leaf spot	*Xanthomonas pelargonii*	Munnecke (1954), Knauss and Tammen (1967), Digat (1977)
Bacterial fasciation	*Corynebacterium fascians*	Dickey (1982)
Witches broom	MLO (unconfirmed)	Okuda (1978)
Southern bacterial wilt	*Pseudomonas solanacearum*	Strider et al. (1981)
Viruses		
Ringspot	Tomato ringspot virus Tobacco ringspot virus *Pelargonium* ringspot virus	Reinert et al. (1963) Kemp (1967) Hollings (1957)
Leaf curl syn. crinkle	*Pelargonium* leaf curl virus	Jones (1940) Hollings (1962)
Ring pattern	*Pelargonium* ring pattern virus	Stone (1980)
Line pattern	*Pelargonium* line pattern	Stone and Hollings (1977) Plese and Stefanac (1980)
Leaf break and mosaic	Cucumber mosaic virus	McWhorter (1957) Paludan (1976)
Yellow net-vein	*Pelargonium* net-vein virus	Russo et al. (1979)
Flower-break	*Pelargonium* flower-break virus	Stone and Hollings (1973) Stone (1980)

explained in cytological terms where euploidy, auto and allopolyploidy, and aneuploidy can exist in different cultivars of the same group (Daker 1969; Craig 1982). The consequence of this cytological diversity, along with the selection of chimeral types within the groups (Clifford 1970; Tilney-Bassett 1986), is that the crop is largely vegetatively propagated with the problems associated with stock maintenance, namely labor and space costs and disease risk.

Pelargoniums are affected by a range of fungal, bacterial, and viral vascular pathogens (Table 1) which attack the growing crop and cause significant losses in conventional propagation. The most serious losses are caused by *Xanthomonas pelargonii* (bacterial blight) and tomato ring spot virus (Reuther 1988). The latter reduces the vigor of stock plants (Stone and Hollings 1973).

Traditional strategies for the control of disease in stock involved roguing out symptomed individuals. The selected plants were then isolated and used as mother stock (Fig. 2a). With increased information on the causal agents of disease,

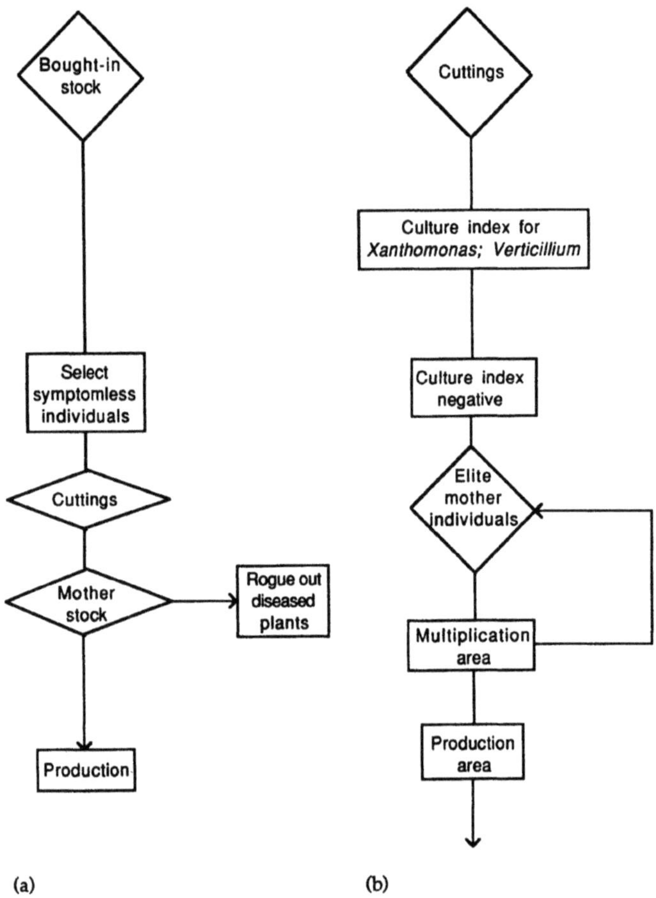

Fig. 2.a Strategy for production of healthy stock based on symptomatology and rogueing. **b** Strategy for production of *Xanthomonas*-free stock based on culture indexing

particularly *Xanthomonas*, in the crop, culture indexing was introduced (Dimock 1962; Fig. 2b). Heat therapy has been employed to eliminate viruses (Hollings and Stone 1968).

1.2 Production of Culture-Virus-Indexed Pelargoniums

Following the demonstration that meristem culture could be used to produce virus-free plants (Morel and Martin 1955), this technique was applied to *Pelargonium* (Holdings 1965; see also Table 2). A further important element in the virus disease control strategy in *Pelargonium* was the production of appropriate antisera for serodiagnosis (e.g., Newhart et al. 1980) which facilitated the development of reliable ELISA procedures (Horst 1982).

Thus by 1982, growers had available to them from the literature a protocol for the production of culture-virus-indexed (CVI) pelargoniums based on culture indexing for cultivable vascular pathogens (Table 1) – principally *Xanthomonas* followed by heat therapy and/or meristem tip culture to produce virus-free individuals (elites). Following confirmation by serodiagnostics that these individuals were indeed virus-free, they could be used as elite stock for subsequent production. Serodiagnosis and observation could be employed to monitor the elite stock and the cycle of CVI repeated to renew it. The scheme is outlined in Fig. 3.

1.3 The Role of Micropropagation in *Pelargonium* Production

Tissue culture offers techniques which can be used both to eliminate vascular pathogens – fungi, bacteria, and viruses (e.g., Cassells 1986, 1988) and to clone plants. While meristem culture has a contribution to offer in the production of CVI stock, such stock and the production from it can readily become reinfected and the

Table 2. The use of meristem culture to produce disease-free pelargoniums

Disease	Reference
Xanthomonas	Hamdorf (1976) Beauchesne et al (1977) Theiler (1977) Reuther (1988)
Virus (principally ringspot)	Pillai and Hildebrandt (1968) Gippert and Schmelzer (1973) Pena Iglesias et al (1974) Stone et al (1974) Hamdorf (1976) Horst et al (1976) Beauchesne et al (1977) Debergh and Maene (1977) Graifenberg and Giustiniani (1978) Hakkart and Hartel (1979)

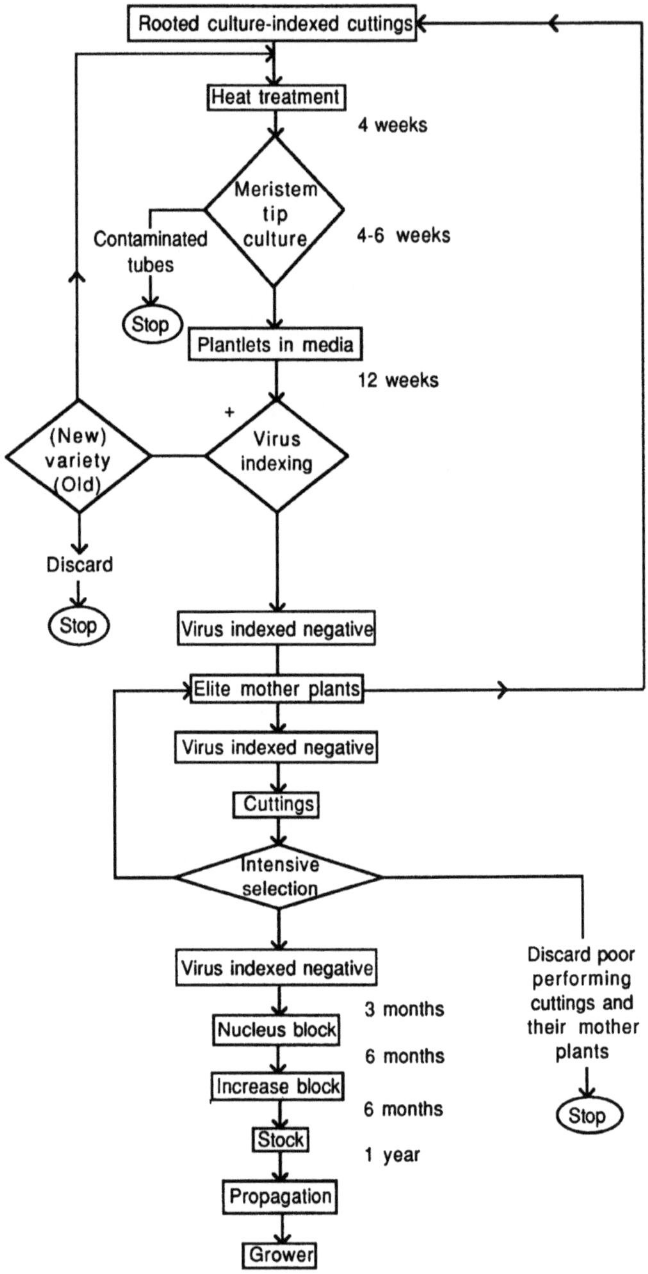

Fig. 3. Flow chart for production of CVI geraniums from CI geraniums. (After Oglevee-O'Donovan 1982)

CVI elites require regular renewal. Further, international trade in cuttings is subject to phytosanitary regulation and such cuttings may, albeit occasionally, harbor pests and diseases from the country of origin leading to rejection by inspectors and disruption to trade. Theoretically at least, micropropagated material may be axenic and pose no phytosanitary problems in international trade (Kahn 1986; Parliman 1986).

The aim of this review will be to assess the contributions tissue culture can make, with reference to micropropagation, to the production of the *Pelargonium* crop with emphasis on the production and storage of axenic stock and mass clonal propagation. For detailed background on all aspects of *Pelargonium* production the reader may consult Mastralerz and Holcomb (1982).

2 Application of Micropropagation

The operations in micropropagations have been divided into stages by previous authors for convenience (e.g., Murashige 1974; Debergh and Maene 1981). This approach will be adopted here but with an alteration in emphasis. The scheme to be followed is presented in Fig. 4. The first stage involves genetic selection; this is followed by disease/contaminant screening and elimination to produce axenic stock and stock storage. Then follows the choice of cloning strategy, monitoring for genetic stability and freedom from contamination. The final step is establishment, involving the rooting procedure and transfer to the environment. These steps will be discussed below.

2.1 Genetic Selection

Pelargonium contains many phenotypes which are inherently unstable, depending, in some cases, on the cloning pathway selected (Fig. 5). Mention has already been made of aneuploids (see above; Craig 1982) but also included are chimeras (Cassells and Minas 1983a; Tilney-Bassett 1986) and beneficially infected varieties (Cassells and Minas 1983a). Further, *Pelargonium* also contains putative transposable elements which may contribute to instability in vitro (Cassells and Carney 1987).

It is essential, if high levels of variation in the progeny are to be avoided (Skirvin and Janick 1976; Cassells and Carney 1987), that each variety is scrutinized before entry into micropropagation. Ideally, any variety showing genetic instability itself or in its antecedents should be rejected. There is an onus on the raiser to advise the micropropagator in this regard (Cassells 1989). If a proscribed type is accepted for micropropagation it should be with a caveat from the micropropagator and the appropriate cloning pathway should be employed. In general, the latter means meristem or nodal culture via precocious bud development. Meristem culture will not suffice for beneficially infected varieties (Cassells et al. 1980).

In genetic selection it should also be recognized that subclonal variation may exist and thus only individual plants which have been assessed should be used as tissue donors. Further, an adequate number of individuals should be used as donors to broaden the genetic base (Johansen et al. 1984).

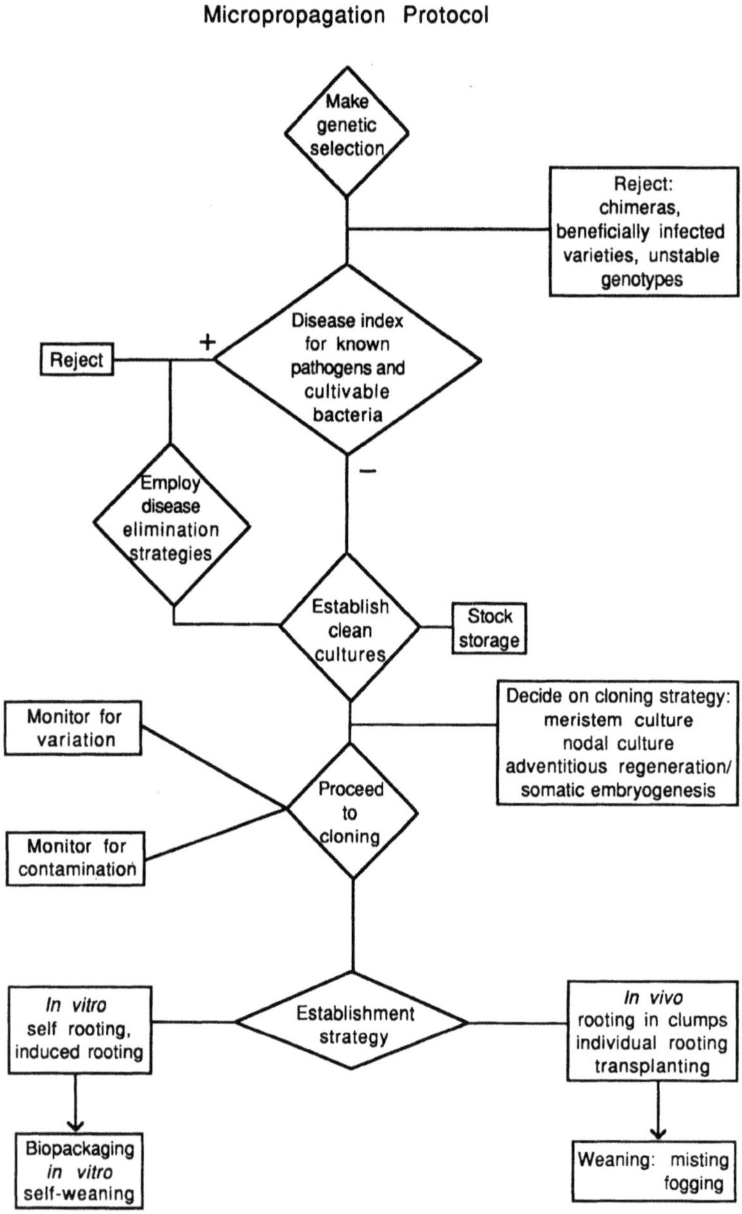

Fig. 4. Micropropagation protocol

Fig. 5. Phenotypically unstable pelargoniums. *Top left* flower chimera in zonal *Pelargonium* Mr Wren; *top right* leaf chimera in ivy-leaf *Pelargonium* L'Elégante; *bottom left* beneficial infection causing picotee effect in flower on *right*, in ivy-leaf *Pelargonium* Mexicana; *bottom right* unstable flower gene locus in zonal *Pelargonium* Speckles

2.2 Disease and Contaminant Screening and Elimination

Pelargoniums are known to be affected by a range of vascular pathogens (Table 1; Fig. 6). Many of these may be latent (e.g. *Xanthomonas* in regal pelargoniums) or express symptoms only in certain seasons (Horst 1982). Further, pelargoniums, like other species, may contain endophytic bacteria – either common soil inhabitants, e.g., *Erwinia* spp., *Pseudomonas* spp., *Corynebacterium* spp., or exotics (Cassells et al. 1988).

While it may be possible to defer disease/contamination monitoring, if using meristem culture, in anticipation of the fortuitous elimination of many or all such

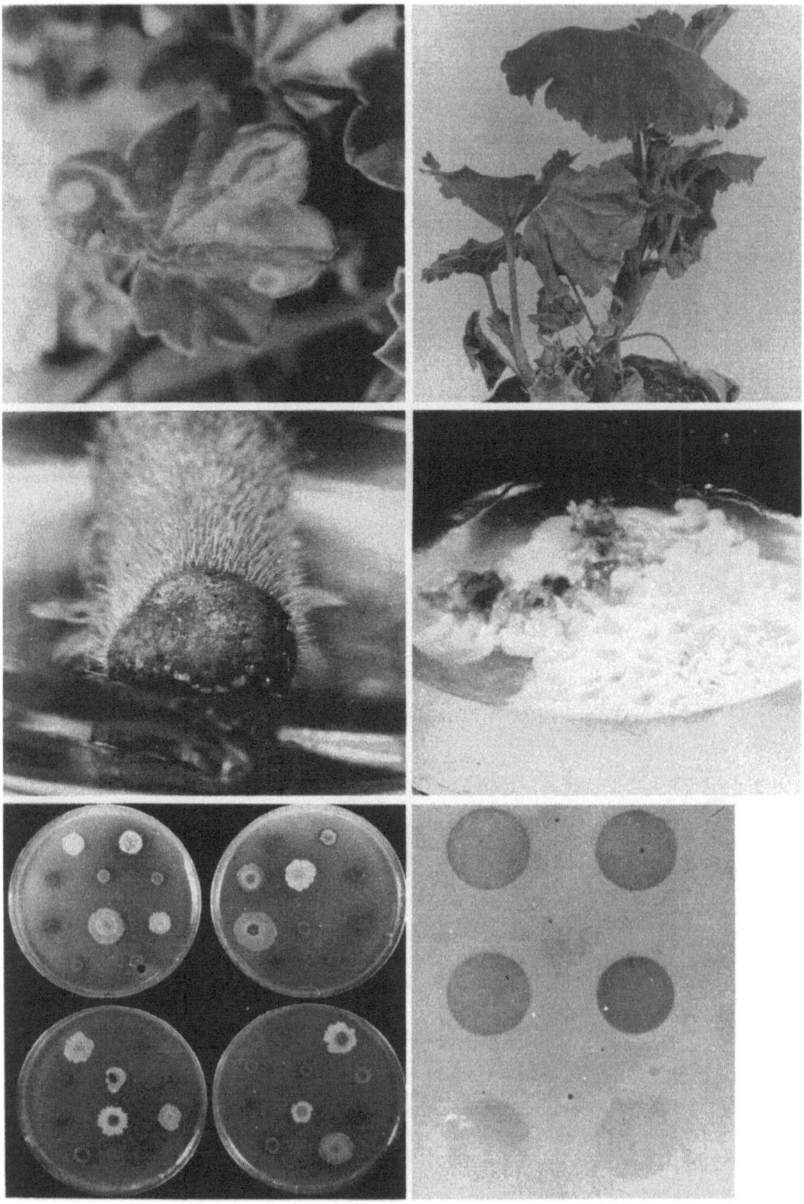

Fig. 6. Expression and detection of vascular pathogens of pelargoniums. *Top left* ringspot symptoms in ivy-leaf *Pelargonium*; *top right Xanthomonas* symptoms in zonal *Pelargonium*; *middle left Xanthomonas* exuding from vascular bundles in an explant culture; *middle right Xanthomonas* over-running culture on transfer of adventitious shoots to lower salt rooting medium; *bottom left* culture indexing of stem explants; *bottom right* detection of *Xanthomonas* by DNA probe

contaminants, it is known that many viruses or virus strains and *Xanthomonas* may pass through meristem culture (Reuther and Sonneborn 1983). Screening the donor plant, akin to culture indexing in Fig. 2, but including virus testing, may help to identify escapes and will provide an insight into what needs to be eliminated and thus facilitate screening of the in vitro material.

If nodal or explant culture is to be used, screening of the donor plant is essential to avoid clonal contamination. It should be stressed that virus symptoms may not be expressed in vitro and many cultivable contaminants can be suppressed by high salt culture media (Falkiner 1988). Virus symptoms may not be expressed until or after establishment, resulting in clonal loss while the bacteria may emerge on subculture, especially to lower salt media (Fig. 6; Cassells et al. 1988), affect growth rates in vitro (Long et al. 1988) or kill young plants at establishment (Cassells, unpubl.).

Screening involves employing methods which are adequately sensitive and reliable without being narrowly specific. It is essential to detect all strains of an organism. Normally this means using ELISA or DNA probes for known pathogens of the crop and culture indexing using a range of common bacterial and fungal media (Schaad 1980) (Fig. 6). The detection and identification of contaminants in plant tissue culture has recently been reviewed (Cassells 1991a,b; 1992)

For *Pelargonium*, in practice, symptomless donors are selected. These are then culture-indexed for *Xanthomonas* and other cultivable contaminants and tested using serology for tomato ringspot and occasionally, other viruses. If these tests are positive, the donor plants may be subjected to heat therapy for viruses (Nyland and Goheen 1969, see Fig. 3) and/or meristem culture (see below). To eliminate bacterial contaminants it has been recommended that the plants be grown under "hard" conditions (Debergh and Maene 1981) or treated with antibiotics, e.g., oxytetracycline (Mycoshield, Pfizer Chemical Corporation, NJ), to reduce bacterial contamination (Cassells et al. 1988). Procedures which facilitate bacterial spread, e.g., overhead watering, should be avoided, as should mechanical contact between plants and infestation with potential vectors. e.g., whiteflies and aphids.

Arguably, the best strategy to produce axenic cultures regardless of the ultimate cloning strategy, is meristem tip culture. Table 2 gives a list of workers who have used meristem culture primarily to eliminate specific diseases of *Pelargonium*. The aim of the work cited in Table 2 was the production of elite individuals for use as stock plants. Many of the authors cited reported varietal differences in response to their media. This aspect is discussed further below.

3 Development of a Mass Cloning Strategy

Relatively few studies have been published on the mass clonal propagation in vitro of *Pelargonium* (see Table 3). Most of the studies have involved meristem culture (Fig. 7) and have involved the production of relatively small numbers of elite stock plants. Only Pillai and Hildebrandt (1969), Stefaniak and Zenkteler (1982), and Cassells and Carney (1987) described the intentional use of adventitious regeneration (see also Holdgate 1977) (Fig. 7).

Table 3. Studies on the micropropagation of *Pelargonium* (see also Table 2)

Authors	
Mayer (1956)	Jelaska and Jelencic (1980)
Weise (1972)	Grunewaldt (1981)
Reuther (1975)	Hammerschlag and Bottino (1981)
Hamdorf (1976)	Stefaniak and Zenkteler (1982)
Horst et al. (1976)	Cassells and Minas (1983b)
Shirvin and Jarich (1976)	Menard et al. (1985)
Beauchesne et al. (1977)	Brown and Charlwood (1986)
Debergh and Maene (1977)	Noord and Hanniford (1988)
Moncousin (1977)	Dunbar and Stephens (1989)
Theiler (1977)	

3.1 Meristem Culture

Several workers have reported that seasonal or artificial conditions which stimulate bud break facilitate meristem establishment (e.g., Hamdorf 1976; Beauchesne et al. 1977; Cassells and Minas 1983b; Menard et al. 1985). Indirect evidence suggests that endogenous auxin critically influences the establishment and response of *Pelargonium* meristems in vitro. Cassells and Minas (1983b) found that dark treatment of donor plants, culture at low temperature, and dark incubation of the meristems favored establishment in vitro, while Menard et al. (1985) found that the auxin-transport inhibitor reduced apical dominance in the proliferating cultures.

Some authors favor the use of a hormone-free medium for meristem establishment (e.g., Debergh and Maene 1977). During early exposure to the culture medium, inhibitory phenolics may be produced by the meristems. This inhibition can be overcome by repositioning on the medium (Cassells and Minas 1983b) or by incorporation of polyvinylpyrrolidone (Menard et al. 1985).

In earlier work, various basal media were used, e.g., major elements of Murashige and Skoog (1962) with minor elements of Nitsch and Nitsch (1970), with additions (Debergh and Maene 1977). However, in a recent publication (Menard et al. 1985) Murashige and Skoog medium without additions is reported to be adequate for establishment and proliferation, although the addition of casein hydrolysate to facilitate expression of latent *Xanthomonas pelargonii* is recommended.

Development of precocious axillary meristems was stimulated by direct implantation (Reuther 1975; Cassells and Minas 1983b) or subculture onto hormone-containing media (Debergh and Maene 1977) (Fig. 7). Menard et al. (1985) recommend zeatin as the sole hormone addition, claiming it stimulates proliferation while suppressing callus growth. Other authors have used various auxins, cytokinins and some have included gibberellic acid (see Tables 2 and 3). As examples, the procedures of Cassells and Minas (1983b) and Menard et al. (1985) are shown in Table 4. Using these latter procedures, multiplications of two- to seven fold per month have been achieved (Cassells and Minas 1983b; Menard et al. 1985).

Menard et al. (1985) criticized early studies on the meristem culture of *Pelargonium* on the grounds that callusing of the explant occurred with the risk of

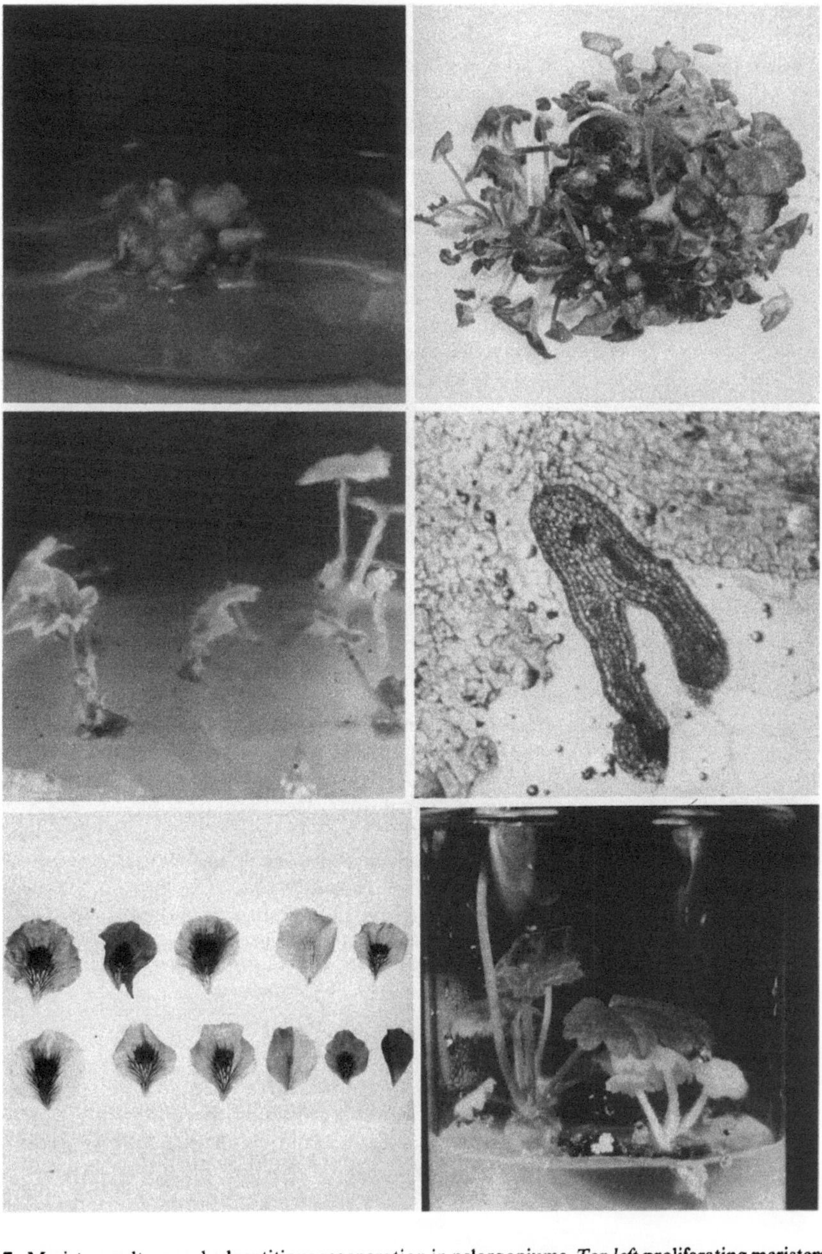

Fig. 7. Meristem culture and adventitious regeneration in pelargoniums. *Top left* proliferating meristem cluster; *top right* proliferating adventitious shoot culture; *middle left* leaf chimera perpetuated in meristem culture; *middle right* somatic embryo of zonal *Pelargonium*; *bottom left* variation in petal shape and pigmentation in adventitious regenerants of regal *Pelargonium* Grand slam; *bottom right* breakdown of leaf chimera in adventitious regeneration

Table 4. Representative meristem culture micropropagation procedures for *Pelargonium*

Cassells and Minas (1983b)	
Donor plant	Grow plants in short days or place in dark for 1–2 weeks. Grow at low temperature (10 °C)
Establishment and multiplication	Place on Murashige and Skoog (MS) basal medium supplemented with 0.825 g/l ammonium nitrate, 0.15 mg/l sodium dihydrogen phosphate, 1 g/l casein hydrolysate, 30 g/l sucrose, 2 mg/l IAA, 1 mg/l GA_3, 4 mg/l kinetin, 50 mg/l adenine sulfate, 100 mg/l meso-inositol, with 6 g/l Agar at pH 5.8. Allow 2 weeks for establishment, reposition if tannins develop around meristem. Subculture every 4–6 weeks depending on variety. For rooting, discontinue subculturing and allow shoots to develop in situ
Rooting	In vitro: place shoots ca. 1 cm in length on half-strength MS basal medium containing 15 g/l sucrose, 0.01 mg/l kinetin, and 0.1 mg/l IAA with 6 g/l agar, pH 5.8
	In vivo: Place developed shoot culture in seed compost (Cassells and Minas 1983b) in the double-membraned container described in Fig. 4 (Cassells, 1987). Allow to self-root
Comments	Cultures should be from CVI donor plants and should be monitored during production for cultivable bacteria. (Cassells 1990a,b,c)
Menard et al. (1985)	
Donor plant	Vigorous vegetative growth stage (March to August)
Establishment (2 weeks)	Basal MS medium with 500 mg/l polyvinylpyrrolidone, 250 mg/l casein hydrolysate, 250 mg/l yeast extract, 30 g/l sucrose, 8 g/l, pH 5.8
Differentiation of the meristem (4–6 weeks)	Basal MS medium with 250 mg/l polyvinylpyrrolidone, 2 mg/l zeatin, 0.02 mg.l triiodobenzoic acid, 30 g/l sucrose, 8 g/l agar, pH 5.8
Multiplication	Medium as differentiation stage, omit triiodobenzoic acid, reduce zeatin to 0.2 mg/l
Rooting (4 weeks)	Half-strength medium containing 0.1 mg/l NAA. Conventional establishment in seed propagators
Comments	The meristem establishment medium is enriched to stimulate expression of *Xanthomonas pelargonii* and *Corynebacterium fasciens*. Polyvinylpyrrolidone is incorporated to prevent tannin formation. Triiodobenzoic acid is used as antiauxin

somaclonal variation. In this laboratory we independently evaluated the media listed in Tables 2 and 3 and found the modified medium of Hamdorf (Cassells and Minas 1983b; Table 4) and that of Menard et al. (1985) (Table 4) satisfactory with respect to suppression of callusing of the explant and low variation in the progeny.

3.2 Adventitious Regeneration

Stefaniak and Zenkteler (1982) and Cassells and Carney (1988) reported cloning via adventitious regeneration in zonal and regal pelargoniums respectively (Fig. 7). As was discussed in Section 2.1, genetic considerations influence the pathway of cloning in *Pelargonium* (see Fig. 4). Indeed, the first report of somaclonal variation (Larkin and Scowcroft 1981) in tissue culture progeny plants came from the model study of Skirvin and Janick (1976) on aromatic pelargoniums.

To achieve high numbers of adventitious regenerants in regal *Pelargonium* cv. Grand Slam, Cassells and Carney (1988) found that exposure to a pulse of auxin was critical. Thereafter, transfer to a cytokinin-containing, auxin-free medium resulted in adventitious regeneration. Adventitious regeneration occurred directly in axenic explants when transferred to cytokinin-containing, auxin-free medium. Other authors, however, have used a conventional auxin:cytokinin factorial approach, e.g., Pillai and Hildebrandt (1969).

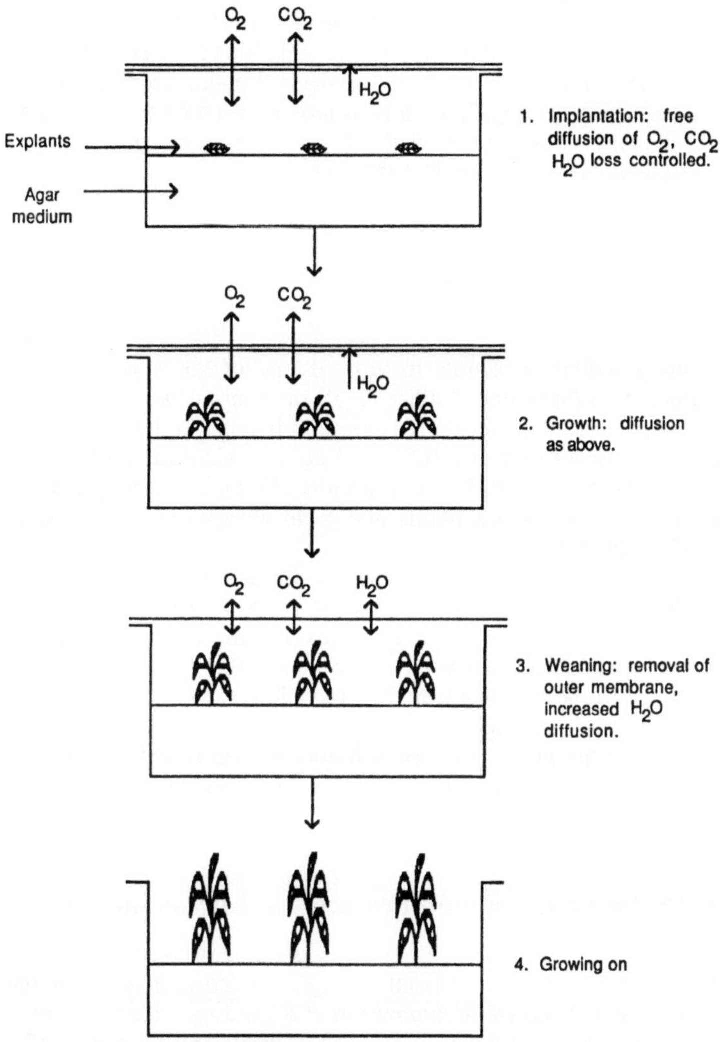

Fig. 8. A double membrane patented system which allows control of the gaseous culture environment during microplant development and weaning. (Cassells 1987)

Adventitious regeneration in *Pelargonium* has been associated with a high frequency of variation in the progeny (Skirvin and Janick 1976; Cassells and Carney 1987). However, this may be cultivar-specific for, as was mentioned in the introduction, chimerism, aneuploidy, and unstable loci may occur in ostensibly similar phenotypes (see Holdgate 1977). Variation in tissue culture progeny in *Pelargonium* is discussed further below.

3.3 Monitoring of Production

Cultures should be sampled routinely during production to ensure that the clean (or axenic) status is maintained. Further, production should be monitored to ensure that genetic stability has been maintained also. The range of variants which can be encountered in tissue culture-derived plants has been discussed previously (Cassells 1985; see Fig. 7). Finally, pelargoniums are prone to vitrification and the culture microclimate particularly the gaseous exchange in the vessels should be carefully controlled (Cassells 1989; Fig. 8).

4 Rooting and Establishment

Rooting individual shoots in vitro is one of the most costly steps in micropropagation (Sluis and Walker 1985). In *Pelargonium* it can also be a source of epigenetic variation, i.e., altered apical dominance in the progeny, if the root:shoot ratio is affected by the induction of an over-vigorous root system (Cassells and Minas 1983b; Cassells 1985). Hamdorf (1976), for example, has reported that micropropagated stock plants may yield 40% more cuttings than conventional stock from cuttings.

It may be more cost-effective for the mircopropagator to induce roots in proliferating shoot clumps in vitro or in vivo, albeit subsequent grading will be required. In this laboratory we induce rooting in shoot clumps and wean in a double-membraned container which allows the gaseous atmosphere, specifically humidity, to be altered during the transition from growth and rooting to weaning (Cassells 1987; see Fig. 8).

Unless the plants have been hardened during the in vitro stages, they may require misting or fogging to facilitate establishment (Scott 1986).

5 Problems and Future Prospects in Micropropagation

It is evident from the above that attempts to micropropagate *Pelargonium* have met with qualified success. Examination of a random selection of micropropagators' catalogs from around the world failed to locate any companies offering pelargoniums (Long, pers. commun.). In the studies reported here (see Tables 2 and 3), most authors have been concerned with the use of meristem culture to produce elite

Fig. 9. Micropropagation in *Pelargonium* production. *Top* random samples of production established in vivo for quality control; *middle* samples as above grown on to flowering; *bottom* progeny of meristem culture in large-scale production

stock in low volume. Many authors have raised the problem of varietal differences in response to a given medium (e.g., Hamdorf 1976; Theiler 1977). Marked genetic diversity may explain why this is so, and to date no attempt appears to have been made to develop a broad spectrum medium for each of the groups (see Cassells et al. 1986).

Two studies (Cassells and Minas 1983b; Menard et al. 1985) describe meristem culture procedures aimed at production of larger numbers of progeny (Fig. 9) giving multiplication rates comparable with other micropropagated crops (George and Sherrington 1984), while one study on regal pelargoniums reports an efficient method for cloning members of this group via an adventitious pathway (Cassells and Carney 1987). In the latter studies, as in other studies on the crop, high levels of genotype-dependent variation were reported in the progeny. Holdgate (1977) using an adventitious pathway of regeneration, claims the progeny were stable. This difference in behaviour may reflect the genetic diversity in the crop (see A; Craig 1982).

The question of genetic stability also has implications for the production, via somatic embryogenesis, of artificial seed (Ammirato 1986). Somatic embryogenesis has been demonstrated in *Pelargonium* (Cassells 1979; Fig. 7) and this potential could be exploited if genome stability can be maintained or if there is selection against unstable genotypes (Cassells 1989).

6 Conclusions

Meristem culture is now standard procedure in the production of elite mother stock for the *Pelargonium* industry. The high rate of reinfection, especially by *Xanthomonas*, requires that meristem culture-derived plants be used in the regular production (annual) cycle. The use of tissue culture, however, has not yet extended to the production of the crop per se. The reason for this may be historical, partly in that many popular varieties are intrinsically unstable, being derived via mutagenesis or arising as sports, and partly that the early meristem culture methods described involved callusing of the explant giving rise to somaclonal adventitious progeny as well as progeny from precocious axillary buds. Variation in the range 0–100% has been observed (see Cassells and Minas 1983a). With effective screening against unstable genotypes and the use of modern procedures, e.g., that of Menard et al. (1985) micropropagation may be as cost-effective as conventional production from elite mother stock.

Pelargonium is a major world floriculture and essential oil crop. It is amenable to nonconventional breeding strategies, e.g., somaclonal variation, to produce new economic variants (Skirvin and Janick 1976), somatic hybridization (Yarrow 1982) to exploit the wide diversity of species (Van der Walt 1977), and genetic engineering, as *Pelargonium* is susceptible to *Agrobacterium tumefaciens* and is capable of adventitious regeneration.

In the contexts of a biotechnological model of crop development, micropropagation enables genetic advances to be commercialized rapidly. In *Pelargonium* clean stock may be produced, but broad spectrum media and procedures need to be developed such that new introductions can be micropropagated without costly development work (e.g., see Cassells et al. 1986). Further studies are suggested aimed at exploiting the potential of artificial seed produced via somatic embryogenesis in bioreactors (Ammirato 1986).

7 Protocol

The recommended protocol (see also Fig. 4 and Table 4) is:

1. Carefully select stable genotypes, i.e., nonchimeral, without unstable loci and without subliminal infection or beneficial infection with virus or virus-like agents.
2. Grow the donor plant under conditions which suppress bacterial spread and colonization (Debergh and Maene 1981).
3. Where *Xanthomonas* is present in the stock plants, it may be advisable to treat with antibiotics (Cassells et al. 1988).
4. Remove the buds from actively growing plants, i.e., do not use dormant buds, dissect out the meristem ca. 0.3–0.5 mm in size. It may not be advisable to use cycocel-treated donor plants.
5. Culture using the procedure of Menard et al. (1985) or Cassells and Minas (1983b) as in Table 4. Following stages one and two of Menard et al. (1985) the explants may be transferred to the proliferation medium of Cassells and Minas (1983a). Proliferation rates in the range of two- to six fold per month are usual, but varietal differences are pronounced.
6. Cultures must be screened throughout for carry-over of contamination from the parent plant and for laboratory contamination arising during cloning (Cassells 1990a,b,c).
7. Progeny must be tested for trueness to type before release.

References

Ammirato PV (1986) Control and expression of morphogenesis in culture. In: Withers LA, Alderson PG (eds) Plant tissue culture and its agricultural applications Butterworths, London, pp. 23–46

Beauchesne G, Albouy J, Morand JC, Daguenet J (1977) Multiplication de clones de *Pelargonium hortorum* et *Pelargonium peltatum* a partir de culture de meristeme pour l'obtention de plantes sains. Acta Hortic 78: 397–402

Brown JT, Charlwood BV (1986) The control of callus formation and differentiation in scented pelargoniums–culture medium effect on propagation. J Plant Physiol 123: 409–417

Cassells AC (1979) The effect of 2,3,5-triiodobenzoic acid on caulogenesis in callus cultures of tomato and *Pelargonium*. Physiol Plant 46: 159–164

Cassells AC (1985) Genetic, epigenetic and non-genetic variation in tissue culture derived plants. In: Schafer-Menuhr A (ed). In vitro techniques: propagation and longterm storage Martinus Nijhoff/Dr W. Junk, Dordrecht, pp. 111–120

Cassells AC (1986) Production of healthy plants. In: Alderson PG, Dullforce WM (eds) Micropropagation in horticulture. Institute of Horticulture, London, pp. 53–70

Cassells AC (1987) Patent PCT/GB87/00218, Patent Office, London

Cassells AC (1988) (ed) Contamination of plant tissue cultures by bacterial and bacteria-like agents. Acta Hortic pp. 225

Cassells AC (1989) Pitfalls for nurserymen in micropropagation and how to avoid them. Proc Int Plant Prop Soc 33: 240–246

Cassells AC (1991a) Problems in tissue culture: Culture contamination. In: Zimmerman RH, Debergh PC (eds) Micropropagation of horticultural crops Kluwer, Dordrecht, pp 31–44
Cassells AC (1991b) Control of contamination in automated plant propagation. In: Vasil IK (ed) Cell Cult Somatic Cell Genet Vol 8 Academic Press Orlando. pp 179–192
Cassells AC (1992) Screening for pathogens and contaminating micro-organisms in micropropagation. In: Duncan JM, Torrance C (eds) Techniques for rapid detection and diagnosis in plant pathology. Blackwell, Oxford, pp 179–192
Cassells AC, Carney BF (1987) Adventitious regeneration in *Pelargonium* × *domesticum* Bailey. Acta Hortic 212: 419–425
Cassells AC, Minas G (1983a) Beneficially infected and chimeral pelargoniums: implications for micropropagation by meristem and explant culture. Acta Hortic 131: 287–297
Cassells AC, Minas G (1983b) Plant and in vitro factors influencing the micropropagation of *Pelargonium* cultivars by bud-tip culture. Sci Hortic 21: 53–65
Cassells AC, Minas G, Long RD (1980) Comparative tissue culture studies of pelargoniums using meristem and explant culture: chimeral and beneficially infected varieties. In: Ingram DS, Helgeson TP (eds) Tissue culture for plant propagators Blackwell, Oxford, pp 125–130
Cassells AC, Plunkett A, Kelleher D (1986) Screening of *Saintpaulia ionantha* Wendl. cultivars for caulogenetic potential based on the in vitro responses of young axenic leaves on auxin and cytokinin factorial media. Sci Hortic 30: 151–157
Cassells AC, Carney BF, McCarthy E, McHugh A, Harmey MA (1988) Problems caused by cultivable bacterial endophytes in the establishment of axenic cultures of *Pelargonium* × *domesticum*: the use of *Xanthomonas pelargonii*-specific ELISA, DNA probes and culture indexing in the screening of antibiotic treated and untreated donor plants. Acta Hortic 225: 153–161
Clifford D (1970) Pelargoniums including the popular geranium, 2nd edn. Blandford Press, London
Craig R (1982) Chromosomes, genes and cultivar improvement. In: Mastralerz JA, Holcomb EZ (eds) Geraniums, 3rd edn. Pennsylvania Flower Growers, Pennsylvania, pp 380–411
Daker MG (1969) Chromosome numbers of *Pelargonium* species and cultivars. J R Hortic Soc XCIV: 346–354
Debergh P, Maene LJ (1977) Rapid clonal propagation of pathogen-free pelargonium plantlets starting from shoot tips and apical meristems. Acta Hortic. 78: 449–454
Debergh PC, Maene LJ (1981) A scheme for commercial propagation of ornamental plants by tissue culture. Sci Hortic 14: 335–345
Dickey RS (1982) Bacterial fasciation. In: Mastralerz JA, Holcomb EJ (eds) Geraniums, 3rd edn Pennsylvania Flower Growers, Pennsylvania, pp 217–220
Digat B (1977) Aspects nouveaux matière de lutte contre les bactérioses du *Pélargonium*. Pépinieristes Hortic Maraîches 174: 17–23
Dimock AW (1962) Obtaining pathogen-free stock by cultured cutting techniques. Phytopathology 52: 1239–1241
Dunbar KB, Stephens CT (1989) Shoot regeneration of hybrid seed geranium (*Pelargonium* × *hortorum*) and regal geranium (*Pelargonium* × *domesticum*) from primary callus culture–plant regeneration from cotyledon culture and leaf culture. Plant Cell Tissue Organ Cult 19: 13–21
Falkiner FR (1988) Strategy for the selection of antibiotics for use against common bacterial pathogens and endophytes of plants. Acta Hortic 225: 53–56
George EF, Sherrington PD (1984) Plant propagation by tissue culture. Exegetics, Basingstoke
Gippert R, Schmelzer K (1973) Erfahrungen mit Spitzenmeristemkulturen von Pelargonien (*Pelargonium* zonale-Hybriden). Arch Phytopathol Pflanzensch 9: 353–362
Grafeberg A, Giustiniani L (1978) Propagazione del pelargonio attraverso coltura in vitro di apici vegetativi. Rev Ortoflorofrutt Ir 62: 141–146
Grunewaldt J (1981) Züchtungsforschung an Pelargonien-peltatum-Hybriden in vitro. 1. Regeneration von Pelargonien-peltatum-Hybriden in vitro. Gartenbauwiss enschaft 46: 250–252
Hakkaart FA, Hartel G (1979) Virus eradication from some *Pelargonium zonale* cultivars by meristem culture. Neth J Plant Pathol 85: 39–46
Hamdorf G (1976) Propagation of *Pelargonium* varieties by stem-tip culture. Acta Hortic 59: 143–151
Hammerschlag F, Bottino P (1981) Effect of plant age on callus growth, plant regeneration and anther culture of geranium. J Am Soc Hortic Sci 106: 114–116
Harney PM (1976) The origin, cytogenetics and reproductive morphology of the zonal geraniums: a review. HortScience 11: 189–194
Holdgate DP (1977) Propagation of ornamentals by tissue culture. In: Reinert J, Bajaj YPS (eds) Applied

and fundamental aspects of plant cell, tissue, and organ culture. Springer, Berlin, Heidelberg, New York, pp 18–43
Hollings M (1957) *Pelargonium* ringspot. Plant Pathol 6: 17–18
Hollings M (1962) Studies of *Pelargonium* leaf curl virus: host range, transmission and properties in vitro. Ann Appl Biol 50: 182–202
Hollings M (1965) Disease control through virus-free stock. Annu Rev Phytopathol 3: 367–396
Hollings M, Stone OM (1968) Technique and problems in the production of virus-tested planting material. Sci Hortic 20: 57–72
Horst RK (1982) Controlling virus diseases. In: Mastralerz JW, Holcomb EJ (eds) Geraniums, 3rd edn. Pennsylvania Flower Growers, Pennsylvania, pp 253–261
Horst RK, Smith SH, Horst HT, Oglevee WA (1976) In vitro regeneration of shoot and root growth from meristematic tips of *Pelargonium* × *hortorum* Bailey. Acta Hortic 59: 131–142
Jelaska S, Jelencic B (1980) Plantlet regeneration from shoot tip culture of *Pelargonium zonale* hybrid. Acta Bot Croat 39: 56–63
Johansen DG, Knutson KW, Slack SR, Jones ED (1984) Recommendations for standards and guidelines for acceptance of tissue cultured potato seed stocks into certification programmes. Am Pot J 61: 368–370
Jones LK (1940) Leaf curl and mosaic of geranium. Wash Agric Exp St Bull 390: 1–19
Kahn RP (1986) Plant quarantine and international shipment of tissue culture plants. In: Zimmerman RH, Griesbach RJ, Hammerschlag FA, Lawson RH (eds) Tissue culture as a plant production system for horticultural crops Nijhoff, Dordrecht, pp 147–164
Kemp WG (1967) Natural occurrence of tobacco ringspot virus in *Pelargonium* in Ontario. Can Plant Dis Surv 47: 295–300
Knauss JF, Tammen J (1967) Resistance of *Pelargonium* to *Xanthomonas pelargonii*. Phytopathology 52: 1178–1181
Larkin PJ, Scowcroft WR (1981) Somaclonal variation—a novel source of variability from cell cultures for plant improvement. Theor Appl Genet 60: 197–214
Long RD, Curtin TF, Cassells AC (1988) An investigation of the effects of bacterial contaminants on potato nodal cultures Acta Hortic 225: 83–92
Mastralertz JW, Holcomb EJ (1982) Geraniums, 3rd edn. Pennsylvania Flower Growers, Pennsylvania
Mayer L (1956) Wachstum und Organbildung an in vitro kultivierten segmenten von *Pelargonium zonale* und *Cyclamen persicum*. Planta 47: 401–406
McWhorter FP (1957) Virus diseases of geranium in the Pacific Northwest. Plant Dis Rep 41: 83–88
Menard D, Coumans M, Gaspar Th (1985) Micropropagation du *Pelargonium* a partir de meristèmes. Meded Fac Landbouwwet Rijksuniv Gent 50: 327–331
Moncousin C (1977) *Pelargonium* × *hortorum* et cultures tissulaires. Rev Hortic Suisse 50: 367–370
Moore HE, Hyypio PA (1982) Taxonomy of pelargoniums in cultivation. In: Mastralerz JW, Holcomb EJ (eds) Geraniums 3rd edn Pennsylvania Flower Growers, Pennsylvania, pp 341–379
Morel G, Martin C (1955) Guérison de pommes de terre atteintes de maladies à virus. C R Acad Sci Paris 41: 472–474
Munnecke DE (1954) Bacterial stem rot and leaf spot of *Pelargonium*. Phytopathology 44: 627–632
Murashige T (1974) Plant propagation through tissue cultures. Annu Rev Plant Physiol 25: 135–166
Murashige T, Skoog F (1962) A revised medium for rapid growth and bioassays with tobacco tissue cultures. Physiol Plant 15: 473–497
Newhart SR, Romaine CP, Craig R (1980) A rapid method for virus-indexing the florist's geranium. HortScience 15: 811–813
Nitsch JP, Nitsch C (1970) Néoformation de fleurs in vitro chez une espèce de jours courts: *Plumbago indica* L. Ann Physiol Veg 7: 251–258
Noord SM, Hanniford GG (1988) Effect of media on micropropagation of *Pelargonium* × *domesticum* – shoot tip culture medium optimization for geranium propagation. HortScience 23: 758
Nyland G, Goheen AC (1969) Heat therapy of virus diseases of perennial plants. Annu Rev Plant Physiol 7: 331–354
Oglevee-O'Donovan WA (1982) Culture indexing for vascular wilts. In: Mastralerz JW, Holcomb EJ (eds) Geraniums, 3rd edn. Pennsylvania Flower Growers, Pennsylvania, pp 233–252
Okuda S (1978) Studies on the causal agent of yellows or witches broom diseases of plants. Spec Bull Coll Agric Utsumomiga Univ 32: 70
Paludan N (1976) Virus diseases in *Pelargonium* × *hortorum* especially concerning tomato ringspot virus. Acta Hortic 59: 119–130

Parliman B (1986) Tissue culture techniques and plant introduction/quarantine procedures. In: Zimmerman RH, Griesbach RJ, Hammerschlag FA, Lawson RH (eds) Tissue culture as a plant production system for horticultural crops. Nijhoff, Dordrecht, pp 271–282
Pena Iglesias A, Ayuso P, Rubio Huertos M, Moreno R (1974) Research on *Pelargonium* viruses in Spain: 1. Preliminary work on virus identification: host range, serology, electron microscopy and tissue culture. Acta Hortic 36: 105–109
Pillai SK, Hildebrandt AC (1968) Geranium plants differentiated in vitro from stem tip and callus cultures. Plant Dis Rep 42: 600–601
Pillai SK, Hildebrandt AC (1969) Induced differentiation of Geranium plants from undifferentiated callus in vitro. Am J Bot 56: 52–58
Plese N, Stefanac Z (1980) Some properties of a distinctive isometric virus from *Pelargonium*. Acta Hortic 110: 183–190
Reinert RA, Hildebrandt AC, Beck GE (1963) Differentiation of viruses transmitted from *Pelargonium hortorum*. Phytopathology 53: 1292–1298
Reuther G (1975) Untersuchungen zur Vermehrung von Pelargonien-varietäten durch Gewebekultur. Gartenbauwiss 40: 181–187
Reuther G (1988) Problems of transmission and identification of bacteria in tissue culture propagated geraniums. Acta Hortic 225: 139–145
Reuther G, Sonneborn HH (1983) Untersuchungen zur Frühdiagnose von *Xanthomonas*. Infektion bei Pelargonien in Gewebekultur. Gartenbau Gartenwissenschafft 33: 854–857
Russo M, Castellano MA, Martelli GP (1979) Rhabdovirus-like particles in English ivy (*Hedera helix*) and ivy-leaf geranium (*Pelargonium peltatum*). Phytopathol Z 96: 122–131
Schaad NW (1980) Laboratory guide for identification of plant pathogenic bacteria. Am Phytopathol Soc Minnesota
Scott MA (1986) Weaning of cultured plants. In: Alderson PG, Dullforce WM (eds) Micropropagation in horticulture. Inst of Horticulture, London, pp 173–182
Skirvin RM, Janick J (1976) Tissue culture-induced variation in scented *Pelargonium* spp. J Am Soc Hort Sci 101: 281–290
Sluis CJ, Walker KA (1985) Commercialisation of plant tissue culture propagation. Int Assoc Plant Tissue Cult Newsl 47: 2–12
Stefaniak B, Zenkteler M (1982) Regeneration of whole plants of geranium from petioles cultured in vitro. Acta Soc Bot Pol 51: 167–172
Stone OM (1980) Nine viruses isolated from *Pelargonium* in the United Kingdom. Acta Hortic 110: 117–182
Stone OM, Hollings M (1973) Some properties of *Pelargonium* flower-break virus. Ann Appl Biol 75: 15–23
Stone OM, Hollings M (1977) Virus diseases. Annu Rep Glasshouse Crop Res Inst for 1976: 121–122
Stone OM, Hollings M, Pawley RR (1974) Rep Glasshouse Crops Res Inst 1973, p 125
Strider DL, Jones RK, Haywood RA (1981) Southern bacterial wilt of geranium caused by *Pseudomonas solanacearum*. Plant Dis 65: 52–53
Swain R (1974) Aromatic pelargoniums. Arnoldia 34: 97–124
Theiler R (1977) In vitro culture of shoot tips of *Pelargonium* species. Acta Hortic 78: 403–414
Tilney-Bassett RAE (1986) Plant chimeras. Edward Arnold, London, pp 199
Van der Walt JJA (1977) Pelargoniums of Southern Africa. Purnell, Capetown.
Van der Walt JJA, Vorster PJ (1981) Pelargoniums of Southern Africa vol 2. Juta, Kenwyn
Van der Walt JJA, Vorster PJ (1988) Pelargoniums of Southern Africa vol 3. National Botanic Gardens, Kirstenbosch
Voight AO (1982) Status of the industry. In: Mastralertz JW, Holcomb EJ (eds). Geraniums, 3rd edn. Pennsylvania Flower Growers, Penssylvania, pp 1–2
Weise KH (1972) Meristemkultur bei Pelargonien. Gartenwelt 72: 97–99
Yarrow SA (1982) An assessment of somatic hybridization in the genus *Pelargonium*. PhD Thesis, Univ Nottingham, UK

Note added in proof. Since the literature review for this paper was completed a report entitled 'Pelargonium Micropropagation and Pathogen Elimination' (M. Appelgren, C.S Hunter, N. Paludan, G. Reuther and R. Theiler-Hedrich, editors), pp. 127, has been published by the commission of the European Communities: BRIDGE, COST 87, programme March 1991.

XX Micropropagation of Azalea (*Rhododendron* spp.)

B.A. BRIGGS, S.M. McCULLOCH, and L. EDICK-CATON[1]

1 Introduction

Azaleas are an important woody ornamental throughout Europe and most of the United States, belonging to the Ericaceae or Heath family. The Ericaceae, a very primitive plant family, comprises 70 genera and approximately 1900 species. Azaleas belong to the genus *Rhododendron* which consists of over 800 species. Other important genera within the Ericaceae include: *Erica* (500 spp.), *Vaccinium* (400 spp.), *Gaultheria* (100 spp.), *Arctostaphylos* (55 spp.), *Gaylussacia* (40 spp.), and *Leucothoe* (35 spp.) (Bailey 1976).

Fig. 1. Flower of a mature azalea. (Photo S. McCulloch)

[1] Briggs Nurseries Inc., 4407 Henderson Blvd., Olympia, WA 98501, USA

Biotechnology in Agriculture and Forestry, Vol. 20
High-Tech and Micropropagation IV (ed. by Y.P.S. Bajaj)

The genus *Rhododendron* was first recognized by Linnaeus in his *Species Plantarum* (1753). He created a separate genus *Azalea* composed of six species. In 1834, George Don, author of *Dichlamydeous Plants*, subdivided *Rhododendron* into eight sections which are still retained today. Azalea was included under *Rhododendron* by Don.

Azaleas are morphologically distinct from rhododendrons by flower form. Most azaleas are diploid with a chromosome complement of 26; however, there are some tetraploids, and rarely, hexaploids. There are no consistant botanical differences between rhododendrons and azaleas. Rhododendrons are mostly evergreen and have companulate flowers, whereas azaleas are largely deciduous and have funnelform flowers, but there are exceptions to this loose description (Bailey 1976). Native throughout Europe, Asia, and the Americas, azaleas are most abundant in temperate regions. Azaleas are very sensitive to soil pH (requiring an acid soil), salinity, and winter injury. Azaleas are used in the landscape as a shrub border, mass planting, and foundation planting. There are an infinite number of cultivars available, providing a wide range of color, size, texture, and habit. Hybridizers are continually improving flower color, foliage quality, disease resistance, hardiness, and growth habit (Fig. 1).

1.1. Conventional Propagation

Propagation of azalea can be done by seed, stem cutting, layering, grafting, root cutting, division, and tissue culture (personal experience).

Seeds usually yield a variable crop which is valuable to the hybridizer, but not reliable for commercial production of named varieties. Seed capsules are collected in the late summer through early fall when still green and then allowed to air dry. Seed should be stored at room temperature. Seeds can be germinated in the house the following winter or outside in the spring. The potting soil should provide good moisture during germination and growth of first true leaves, at which time the seedling should be established.

Stem cuttings should be taken from new shoots of a healthy mature stock plant. Cutting size should normally be 8–10 cm or 2 cm for a dwarf azalea. Lower leaves should be stripped without tearing the bark. Prepared cuttings are placed in a well-drained rooting medium in a humid environment. Stem cuttings generally produce true-to-name plants, but bud sports or mutations may occur.

There are many methods of layering that can be used to propagate azaleas: mounding, air layering, and branch or stem layering. These methods are most successful when using 1-2-year-old wood during the late spring or summer.

Grafting is another technique that is primarily used to grow unusual forms or difficult-to-root azaleas.

All conventional methods of propagation mentioned are reliable but yield relatively small numbers of plants. To meet the needs for large commercial production of azaleas tissue culture is the most productive form of propagation. In the past two decades tissue culture has become a viable option for growers to produce large amounts of azaleas. It offers growers an opportunity to produce new introductions, older popular selections, difficult to root clones, and maintain azaleas free of certain pathogens.

2 In Vitro Propagation of Azaleas

Micropropagation of *Rhododendron* became possible with Anderson's (1975) development of a shoot tip culture technique. Anderson modified the Murashige and Skoog (1962) inorganic formula concentrations by using a bioassay system to reduce the rhododendron shoot toxicity response to Murashige and Skoog (MS) medium. In 1981, McCown and Lloyd presented their woody plant medium (WPM). This new formulation has shown to be particularly effective in culturing members of the Ericaceae. Economou and Read (1984) observed in vitro shoot necrosis on MS and on Andersons's medium and further modified MS medium for culture of deciduous azaleas. It has been established that *Rhododendron* spp. along with other woody plants have low mineral salt requirements in vitro.

2.1 Explant Source/Sterilization Procedure

The selection of the stock plant to be micropropagated is the most important step and should be based on superior form, health, and flowering characteristics. The ideal explant source is a greenhouse-grown stock plant which has been irrigated from below to reduce soil and waterborne contamination on the plant tissue. Stem cuttings are taken from new vegetative growth including the shoot tip and a few lateral buds. During the winter, dormant buds or flower buds may be used but are more difficult to surface sterilize.

Fig. 2. Preparation of explant material. (Photo S. McCulloch)

Sterilization is accomplished by first removing all excess surface area (leaves/petioles/bud scales). Nodal cuttings should be rinsed under cool tap water to remove all surface particulates, then washed with agitation in a surfactant such as Tween 20 (2 drops per 400 ml deionized water) for approximately 5–10 min. The cuttings should be thoroughly rinsed. The plant material is then disinfested with a 0.5% sodium hypochlorite solution for 20–40 min, depending on the quantity, cleanliness, and tenderness of the explant material. Follow this with a rinse of 0.05% sodium hypochlorite solution.

Explants are then ready for placement on fresh growth media. Basal ends are recut, shoot pieces are cut into tips or nodal sections, meristems or flower buds can be excised and the explant can be placed on the appropriate defined media. Cultures are then placed under cool white fluorescent light (50–70 μmol/s/m^2) with a 16-h photoperiod at approximately 25–30 °C (Fig. 2).

2.2 Media

Mineral salt formulations for micropropagation of azaleas vary slightly (Table 1). Experience has demonstrated that azaleas vary also and may grow on any or all of the defined media available, or may require individual optimization of certain inorganic minerals.

Cytokinin requirements of azaleas have been established in many studies. Cytokinins studies include: 6-benzylamino purine (BA), 6-(= *Y*, = *Y* dimethyally-

Table 1. Comparison of inorganic mineral salt concentrations of commonly used formulations for micropropagation of azalea. Expressed in mM. (After George et al. 1987)

Inorganic	Murashige and Skoog (1962)	Anderson (1974)	Economou and Read (1984)	Lloyd and McCown (1981)
NH_4NO_3	20.6	5.0	5.0	5.0
KNO_3	18.8	4.7	2.0	—
$(NH_4)_2SO_4$	—	—	1.0 ·	—
$CaCl_2 \cdot 2H_2O$	3.0	3.0	3.0	0.9
$MgSO_4 \cdot 7H_2O$	1.5	1.5	1.5	1.5
KH_2PO_4	1.25	—	3.0	1.25
$NaH_2PO_4 \cdot H_2O$	1.25	2.75	—	—
$Ca(NO_3)_2 \cdot 4H_2O$	—	—	—	2.3
K_2SO_4	—	—	—	0.6
H_3BO_3	0.1	0.1	0.1	0.1
$MnSO_4 \cdot H_2O$	—	0.1	0.1	0.15
$MnSO_4 \cdot 4H_2O$	0.1	—	—	—
$ZnSO_4 \cdot 7H_2O$	0.03	0.03	0.03	0.03
KI	0.005	0.0018	—	—
$Na_2MoO_4 \cdot 2H_2O$	0.0011	0.0011	0.001	0.001
$CuSO_4 \cdot 5H_2O$	0.0001	0.0001	0.0001	0.001
$CoCl_2 \cdot 6H_2O$	0.0001	0.0001	0.0001	—
$FeSO_4 \cdot 7H_2O$	0.1	0.2	—	0.1
$Na_2EDTA \cdot 2H_2O$	0.1	0.2	—	0.1
FeNaDTPA	—	—	0.1	—

amino) purine (2-iP), 6-(= *Y*, = *Y* dimethyally-amino) purine riboside (2-iP-riboside), 6-furfury-laminopurine (kinetin), 6-(4-hydroxy-3-methylbut-Z-enyl-amino) purine (zeatin), and N-phenyl-N′-1,2,3-thiadiazol-5-ylurea (thidiazuron). All studies conclude that zeatin is the preferred cytokinin for induction of shoot proliferation. Likewise, 2-iP is essential to sustain growth in micropropagation of azalea, and is the cytokinin more in use commercially because the cost of zeatin can be prohibitive. The recommended level of 2-iP can vary from 2–10 mg/l depending on the azalea species being grown and the desired growth habit in culture. Apparently 2-iP antagonizes apical dominance of azalea causing axillary buds to break and resulting in proliferation of shoot tips (personal experience). Kinetin and BA were least effective in stimulating shoot tip proliferation, which implies that azaleas do not require or respond to these cytokinins. A new phenylurea, thidiazuron, was studied by Briggs et al. (1987) as a possible substitute for 2-iP in micropropagation of azalea. Thidiazuron was very effective in promoting shoot tip multiplication at surprisingly low concentrations (0.01–0.1 mg/l). The authors recommend caution in the commercial use of this compound because it has not been proved whether or not multiplied shoots from thidiazuron will be true to type after growing on. Another phenylurea recently tested by Fellman and Read (1984) 4-pyridyl urea demonstrated a cytokinin-like response on azaleas at concentrations 1/10 to 1/100 the recommended 2-iP normally used in micropropagation of azalea (Fig. 3).

Auxins can be used in conjunction with cytokinins for multiplication of azaleas in culture. Economou and Read (1984) found that an addition of indoleacetic acid

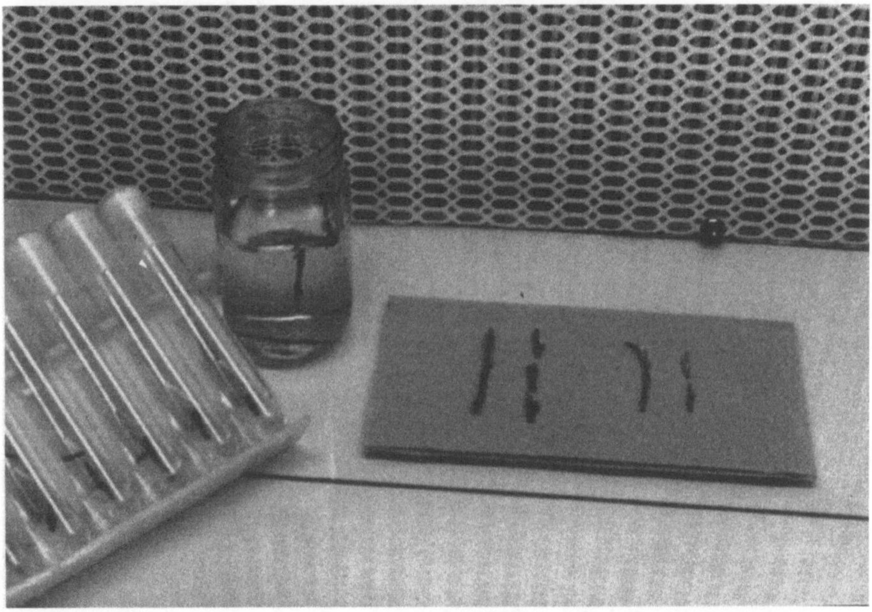

Fig. 3. Initiation of azalea in culture. (Photo S. McCulloch)

Table 2. Applications for plant culture media (after George et al., 1987)

Scientific name	Stage	Explant	Results	Medium	Plant growth regulators (mg/l)	Reference
Rh. Exbury hybrids	1	Shoot tips 10–15 mm	Establishment	Zimmerman and Broome (1980)	(15)2-iP + (4)IAA + (80)Adsh	Fordham et al. (1982)
	2	Shoots (stg. 1)	Adventitious and axillary shoot proliferation	Fordham et al. (1982)	(16.2)Z + (80)Adsh	Fordham et al. (1982)
	3	Shoots (stg. 2)	90% rooting	Fordham et al. (1982)	None	Fordham et al. (1982)
Rh. hybrids	1	Shoot tips	Optimal shoot formation	Anderson (1978; 1980)	(15–17)2-iP + (80)Adsh	Norton and Norton (1985)
Rh. simsii var.	1	Shoot tips 0.5 mm	3–5 axillary shoots	Preil and Engelhardt (1977)	(5)BAP	Preil and Engelhardt (1977)
Rh. simsii var.	1	Shoot tips 0.5 mm	3–5 axillary shoots	0.1 × Murashige and Skoog (1962)	(5)BAP	Preil and Engelhardt (1977)
	2	Shoots (stg. 1)	Shoot growth	Preil and Engelhardt (1977)	(5)BAP	Preil and Engelhardt (1977)
	3	Shoots (stg. 2)	Rooted plantlets	Preil and Engelhardt (1977)	(2)IAA	Preil and Engelhardt (1977)
Rh. spp.	1	Shoot tips	Axillary shoot proliferation	Economou and Read (1984)	(5–20)2-iP	Economou and Read (1984)
	2	Explants (stg. 1)	Shoots on reculture	Economou and Read (1984)	(1)IAA + (10)2-iP	Economou and Read (1984)
	3	Explants (stg. 1 & 2)	Shoots rooted extra vitrum	Extra vitrum	None	Economou and Read (1984)

Rh. spp.	1	Shoot tips 1.5–2 cm	Axillary shoot proliferation	Economou and Read (1984)	(10)2-iP	Economou and Read (1986a)
	2	Explants (stg. 1 minus shoots)	Axillary shoot proliferation	Economou and Read (1984)	(10)2-iP	Economou and Read (1986a)
	3	Shoots (stg. 2)	Microcuttings rooted extra vitrum	Extra vitrum	None	Economou and Read (1986a)
Rh. spp.	1	Stem tips	Axillary and adventitious shoots	Lloyd and McCown (1981) WPM	(1.6)2-iP	McCown and Lloyd (1983)
	2	Shoots (stg. 1)	Shoot proliferation	Lloyd and McCown (1981) WPM	(0.8–3.2)2-iP	McCown and Lloyd (1983)
	3	Shoots (stg. 2)	Shoots rooted plantlets	Extra vitrum	No pretreatment	McCown and Lloyd (1983)
Rh. spp.	1	Stem cuttings 5 cm	Axillary shoot growth	Kyte and Briggs (1979)	(1)IAA + (5)2-iP + (80)Adsh	Kyte and Briggs (1979)
	2	Shoots (stg. 1)	Shoots rooted	Kyte and Briggs (1979)	None	Kyte and Briggs (1979)

Fig. 4. Stages of initiation through establishment of azaleas in culture. (Photo S. McCulloch) *Left to right* Meristem and nodal explant growing on WPM salts with 2 mg/l zeatin; initial shoot growth from nodal explant 2–6 weeks from initiation growing on WPM salts with 0.5 mg/l zeatin; rejuvenated shoots actively multiplying 2–3 months from initiation growing on WPM salts with 2 mg/l 2-iP

(IAA) at 0.5–1 mg/l in the medium increased the number of shoots. Perhaps the auxin enhances the growth of short shoots proliferated by the cytokinin, therefore producing more harvestable shoots.

Rooting azaleas in vitro is attained by transferring multiplied shoots onto a medium containing half the salt concentration with the addition of 2–7 mg/l indole-3-butyric acid (IBA) and 0.6% activated charcoal. Many commercial laboratories do not root azaleas in culture but send micropropagated shoots directly to the greenhouse, where the shoots quickly root and acclimate to in vivo conditions, thus saving the extra labor and cost of rooting in culture (Fig. 4).

Many studies have been conducted (Table 2) demonstrating the range of viable systems of producing azaleas through micropropagation.

2.3 Growing Conditions of Cultures

After initiation the explant can be grown under a sixteen hour photoperiod of cool white fluorescent light (50–70 $\mu mol/s/m^2$) at 25–30 °C. Subculture time should be 14 to 21 days during establishment while the explant is rejuvenated and actively proliferating. Once in a multiplying phase the media can often be adjusted for optimum growth characteristics. Often the cytokinin level can be dropped to produce quality harvestable shoots. Azaleas should be subcultured at a 6–8-week

Fig. 5. Actively multiplying culture of azalea with a subculture period of 8 weeks. Medium is WPM salts with 2 mg/l 2-iP. (Photo S. McCulloch)

period (Fig. 5). Cultures can often be stored in cool/dark conditions 3 ± 1 °C for up to 12 months and then recultured to grow on in vitro or sent to the greenhouse to be established. Cold storage can be a useful tool for any commercial tissue culture producer of azaleas, offering flexibility to production goals and seasonal conditions.

2.4 Establishment ex Vitro

Azaleas can be established in soil from either prerooted or unrooted in vitro shoots. In either case shoots must be established in a high humidity environment. Since these microcuttings lack a highly developed cuticle, a period of high humidification or misting is necessary to prevent stress and desiccation (personal experience). Shoots can be established and rooted in fog, under intermittent mist, or in closed polyethylene tents.

Beside humidification, several other factors are critical to successful establishment of microcuttings. Factors such as: soil porosity, fertility, air and soil temperature, light intensity, photoperiod, pest and disease management, water quality, and planting techniques need to be monitored and corrected to achieve success. Shoots are stuck into a well-drained sterile soil mix. Normally a peat moss-based mixture is used with Ericaceous plants. Sixteen to 25 shoots are stuck into a

Fig. 6. A commercial laboratory's room for growing azalea and other wood plant cultures. (Photo S. McCulloch)

10-cm pot filled to a depth of 10 cm. Bottom heat of 20 °C is beneficial to rooting. Aerial greenhouse temperatures should not exceed 35 °C. Light levels should be approximately 50–70 μmol/s/m^2. Supplemental light may be necessary to compensate for low light levels in winter months. A 16-h photoperiod should be maintained to keep the plants actively growing (Fig. 6).

Microcuttings appear to be juvenile or seedling-like in morphology and rootability. Rooting is generally very fast, approximately 14 to 28 days. Evergreen azaleas generally root faster and in higher percentage than deciduous azalea. Auxins are not needed to achieve acceptable levels of rooted plants. Once microcuttings are rooted, they grow quickly and begin to form larger, more mature leaves.

2.5 Flower Bud Culture of Azalea

Flower bud culture of azalea has been used when available explant material proved too difficult to disinfest or establish in culture, (Dai et al. 1987). The protocol involves a typical sterilization procedure followed by aseptic dissection of the flower bud tissue. Bud scales are peeled back and the floret is excised including the pedicel. Shoot induction occurs on any previously mentioned standard salt concentration with approximately 4 mg/l IAA and 15 mg/l 2-iP. As shoots differentiate

Fig. 7. Transplanted tissue culture azaleas. (Photo S. McCulloch)

from the pedicel and ovary tissue they can be transferred to the defined shoot multiplication medium and assimilated to a normal production system. Flower bud culture has the advantage over shoot tip culture because florets are available from late autumn through early spring providing a long season for culture initiation. Also, florets are abundant on a mature stock plant, providing potentially hundreds of initial explants (Fig. 7).

2.6 Doubling Chromosome Numbers

Colchicine has been used to double chromosome numbers of many plants but has not been commonly applied to azalea improvement. Colchicine is most effective when applied to seeds, seedlings, or stem tissue. The advantage of higher ploidy levels with azaleas is larger flowers of thicker texture, resulting in a flower with greater lasting qualities and overcoming sterility in wide crosses. There is need for such improvements, and tissue culture could be used to increase the ploidy level with colchicine if a satisfactory method could be established. A protocol has been developed by Goldy and Lyrene (1984) for doubling the chromosome number of *Vaccinium* and could probably be adapted for studies with azalea.

2.7 Commercial Aspects of Micropropagation of azaleas

Azaleas can be produced and sold for a profit if several factors are managed correctly. Briggs Nursery currently grows and sells 400000 micropropagated azaleas each year. The average sales price is approximately $0.75 (US). The customers that purchase these plants do so not only because these plants are micropropagated, but because it is an acceptable quality product offered at an acceptable price.

A tissue culture lab must achieve high productivity levels if it is to be profitable. With azaleas we are able to achieve a 6× multiplication rates. Beyond the laboratory the plants root quickly with very high (90–95%) survival rates in the greenhouse. Contamination within the laboratory is monitored weekly and adjustments are made if needed. Every day space in both the laboratory and greenhouse are maximized fully.

Labor cannot be ignored when discussing commercial production of micropropagated plants. This is a very labor-intensive process, with perhaps 60–85% of the total cost going to labor. By concentrating not only on the production process, but also on the people, a laboratory will be more than likely to succeed in producing a profit.

3 Summary and Conclusions

Tissue culture of azaleas has proven to be a useful and valuable tool for propagation of these colorful ornamental shrubs. Using an appropriate low salt medium and proper cytokinin, shoots can be established and multiplied in quantity. These microshoots can then be rooted in vitro or in the greenhouse under regulated conditions. The cuttings grow quickly, very much like seedling azaleas. After growing nearly one million azaleas using tissue culture, the authors have noticed very few problems. What aberrations have occurred have been mostly foliage variegation. In the future this commercially important method for propagating azaleas and other members of the Ericaceae will continue to be refined and improved.

4 Protocol

The best explant material is shoot tip cuttings from actively growing stock plants grown in a greenhouse. These cuttings can then be easily surfaced sterilized and cut into nodal segments and placed onto a medium of WPM salts with 0.5–2 mg/l zeatin or 2-iP. Cultures should be placed under cool white fluorescent light (50–70 $\mu mol/s/m^2$) with a 16-h photoperiod at 25–30 °C. Bud expansion and shoot growth should occur within 7–14 days.

References

Anderson WC (1975) Propagation of rhododendrons by tissue culture. 1. Development of a culture medium for multiplication of shoots. Comb Proc Int Plant Prop Soc 25: 129–135

Anderson WC (1978) Tissue culture propagation of rhododendrons. In Vitro 14: 334 (Abst 03)

Anderson WC (1984) A revised tissue culture medium for shoot multiplication of *Rhododendron*. J Am Soc Hortic Sci 109 (3): 343–347

Bailey, L. H. (1976) Hortus Third. Macmillan Publishing Co,. Inc. New York. Collier Macmillan London 949 p

Blazich FA, Acedo JR (1988) Micropropagation of Flame azalea. J Environ Hortic 6(2): 45–47

Briggs BA, McCulloch SM, Edick LA (1987) Micropropagation of azaleas using thidiazuron. Acta Hortic 226(1): 205–208

Dai C, Lamberth VN, Taven R, Mertz D (1987) Micropropagation of *Rhododendron prinophyllum* by ovary culture. Hortscience 22(3): 491–493

Economou AS, Read PE (1984) In vitro shoot proliferation of Minnesota deciduous azaleas. Hortscience 19(1): 60–61

Economou AS, Read PE (1986a) Influence of light duration and irradiance on micropropagation of a hardy deciduous azalea. J Am Soc Hortic Sci 111(1): 146–149

Economou AS, Read PE (1986b) Microcutting production from sequential reculturing of hardy deciduous azalea shoot tips. Hortscience 21(1): 137–139

Fellman CD, Read PE (1984) Effects of 4-pyridyl urea on in vitro growth of hybrid deciduous azaleas. Hortscience 19(3): 408 (Abstr)

Fordham I, Stimart DP, Zimmerman RH (1982) Axillary and adventitious shoot proliferation of Exbury azaleas in vitro. Hortscience 17(5): 738–739

George EF, Puttock DJM, George HJ (1987) Plant Culture Media, vol 1: 169–170, 300–355

Goldy RG, Lyrene PM (1984) In vitro colchicine treatment of 4 × blueberries *Vaccinium* sp. J Am Soc Hortic Sci 109(3): 336–338

Harbage JF, Stimart DP (1987) Adventitious shoot regeneration from in vitro subcultured callus of Rhododendron Exbury hybrids. Hortscience 22(6): 1324–1325

Kyte L, Briggs B (1979) A simplified entry into tissue culture production of rhododendrons. Comb Proc Int Plant Prop Soc 29: 90–95

Lloyd G, McCown B (1981) Commercially feasible micropropagation of mountain laurel, *Kalmia latifolia*, by use of shoot-tip culture. Int Plant Prop Soc 30: 421–427

McCown BH, Lloyd GB (1983) A survey of the response of *Rhododendron* to in vitro culture. Plant Cell Tissue Organ Cult 2: 27–85

Norton ME, Norton CR (1985) In vitro propagation of Ericaceae: a comparison of the activity of the cytokinins N^6-benzyladenine and N^6-isopentyladenine in shoot proliferation. Sci Hortic 27 (1985): 335–340

Preil W, Engelhardt M (1977) Meristem culture of azaleas (*Rhododendron simsii*). Acta Hortic 78: 203–208

Zimmerman RH, Broome OC (1980) Blueberry micropropagation. In: Proc Conf Nursery Production of Fruit Plants through Tissue Culture – Applications and Feasibility. USDA–SEA Agric. Res Results ARR-NE-11, pp 44–47

XXI Micropropagation of Rose (*Rosa* L.)

W.A.H. HORN[1]

1 Introduction

1.1 Botany, Origin, Chromosome Number

The genus *Rosa* L., a member of the family Rosaceae, subfamily Rosoideae, comprises more than 100 species, mostly in the temperate parts of the northern hemisphere, having persistent or deciduous leaves and attractive flowers. The modern garden roses have been derived from systematic crossing which yielded great improvements after about 1800. As progenitors of modern garden roses, six species are of importance. They fall into two groups, contrasting in their chromosome number and their geography. Diploid species (2n = 14) from East Asia are *R. odorata* (Andr.) Sweet, *R. multiflora* Thunb., *R. chinensis* Jacq., and *R. moschata* J. Herm. (Mill.), of which the former were introduced into breeding work only after 1800. *R. gallica* L. and *R. foetida* J. Herm. are European and Near East species with a tetraploid chromosome number (2n = 4x = 28). Modern roses, with their great variety of color, habit, and hardiness, began to appear during the 19th century from crosses between these species and their hybrid derivatives (Wylie 1954, 1955).

The rose classes of floriculture are the so-called hybrid tea, floribunda, and hybrid polyantha roses, which are mostly tetraploid. Other groups of cultivated roses are grown as shrubs and ground cover in garden and landscape: climbing roses were derived from *R. multiflora*, *R. setigera* Michx., *R. wichuraiana* Crep., and several other species. They also originated as bud mutations of florists' roses. Many old roses, generally shrub roses and species hybrids, come from *R. rugosa* Thunb., *R. pimpinellifolia* L., *R. rubiginosa* L., and others. Ground cover roses also include species and interspecific hybrids. The miniature roses, cultivated as pot plants and for the rockery, originated from crosses between *R. chinensis* with hybrid tea roses and others.

Rose rootstocks are generally selections of the pentaploid *R. canina*, *R. caesia* Sm. (R. laxa Hort.), *R. rubiginosa*, and the diploid *R.multiflora*, *R.* × *noisettiana* Thory, *R. odorata*, and *R. chinensis*. An important industry connected with roses is the production of essential oils from *R.* × *damascena* Mill., a tetraploid hybrid species derived from crosses between *R. moschata* and *R. gallica*.

[1] Lehrstuhl für Zierpflanzenbau, Technische Universität München-Weihenstephan, W-8050 Freising 12, FRG

Biotechnology in Agriculture and Forestry, Vol. 20
High-Tech and Micropropagation IV (ed. by Y.P.S. Bajaj)

1.2 Distribution and Importance

Roses are grown all over the globe. They are the most important commercially produced cut flowers in the world. More than 500 million blooms are sold annually in the USA, with a wholesale value of $ 196 million. More than 2000 million blooms are sold in West Europe alone over the largest auction markets (AIPH 1990) and 225 million in Japan. About 3500 ha roses have been planted for commercial protected cut flower production in the European Community countries (EC). Rose bushes and rootstocks are produced on ca. 2500 ha. In West Germany ca. 25 million rose plants are grown yearly for sale, 80% of them hybrid tea, polyantha, and floribunda roses. The number of climbers and shrubs amounted to about 2 million. Around 50 million rootstocks are grown every year in West Germany alone (AIPH 1990).

In international trade, the USA imported 427 million cut rose flowers from Columbia in 1989, and Israel exported ca. 34 million to Europe in 1988/89. Even developing countries exported 76 million. Approximately 1250 million cut flowers were imported into EC countries in 1989, while EC domestic trade amounted to 991 million. Import into the EC and other European countries was valued at about 400 million Swiss francs, and the export from these European countries totalled more than 445 millions Swiss francs (AIPH 1990). A most valuable product of rose is "attar of roses", a highly fragrant essential oil, manufactured from several thousand ha of roses grown for oil in Bulgaria, the USSR, Turkey, and Morocco.

1.3 Conventional Propagation

Most recent rose cultivars are protected by breeders' rights. Breeders have to be paid royalties for any plants produced commercially. Roses can be propagated by seed, cuttings, budding, and grafting. Seed propagation is used in breeding new cultivars, propagation of species, and for production of rootstocks from *R. caesia*, (*R. laxa* Hort.), *R. canina*, and *R. multiflora*.

It is generally necessary to stratify seed to obtain satisfactory germination. Certain strains have been selected which perform better than unselected material. Other rootstocks such as *R. odorata* (Indica Major), *R.* × *noisettiana* Manettii, and *R. multiflora* can be propagated by hardwood cuttings.

Species are generally grown from seed. Budding on a rootstock is by far the most important propagation method used for the production of roses. Grafted plants are used on a small scale for commercial cut flower production. To some extent, vigorously growing polyantha roses and climbers are propagated by hardwood cuttings. Miniature roses are generally propagated by softwood cuttings. After planting in spring, the rootstocks are usually ready for budding in summer or autumn. Plants budded in summer are ready for sale in the autumn of the following year. Planted in greenhouses, they start producing heavy flower crops for sale in the following season. In ground cover roses, which are produced by budding in large amounts, efforts are being made to replace budding by propagation through cuttings. There are, however, open questions with regard to the performance of budded cultivars as compared to the same cultivars on their own roots.

1.4 Need for in Vitro Culture of Roses

Propagation by vegetative means is thus the predominating technique in roses. Taking of cuttings and budding outdoors are confined to the warmer periods of the year. In mild climates cuttings can be taken throughout the summer and autumn, or from plants grown in glasshouses throughout the year. Hardwood cuttings are made from the previous season's shoots, softwood cuttings from the current season's growth. Such dependence upon climatical conditions could be overcome by micropropagation.

Cuttings should be selected from flowering shoots, so that the propagator can be sure that his stock plant is true to type. Certain rootstocks are produced on specially maintained stock beds. All mother stock used for cutting or bud production should be taken from plants which have been freed of viruses and similar diseases and verticillium by heat treatment. To enable rapid multiplication and the maintenance of disease-free plants, the establishment of pathogen-free plants is of utmost importance for rose propagation. Furthermore, there is a need to replace expensive techniques such as budding by methods which require less labor. Another reason for replacing budded roses is to avoid frost damage to the cultivar, otherwise the harder rootstock will overgrow the original cultivar. Micropropagation is also useful for rapid propagation of new cultivars. Induction of adventitious shoots and regeneration from callus cultures are of importance for somaclonal variation and therefore for breeding. Also, explants can be treated with mutagens, after or prior to explanting, in order to obtain new genotypes by mutation.

2 Brief Review of Tissue Culture Work

2.1 General Remarks

There are more than 200 references that deal in one way or another with tissue culture in roses. One of the first papers dates back to 1945, where Nobécourt and Kofler succeeded in obtaining callus and roots on explanted buds. Research reports on callus and suspension cultures were given by Nickell and Tulecke (1959) and Weinstein et al. (1962), followed by about 100 further papers, the most recent being by Badzian et al. (1991), Woltering (1990) and Short and Roberts (1990). There are reports on embryo culture starting with Lammerts (1946) to Graifenberg (1973), and on somatic embryogenesis by De Wit et al. (1990). This chapter deals with papers on micropropagation only. The first papers dealing with micropropagation of roses were those of Jacobs et al. (1969, 1970a,b) and Elliott (1970) (Table 1).

Starting from shoot tips of Tropicana (syn. Superstar), these authors succeeded in obtaining shoot development. Elliott (1970), Graifenberg et al. (1975), and Walter et al. (1979) obained complete and viable plants on nutrient media containing cytokinin and auxin in relatively small amounts. Skirvin and Chu (1979a,b) and Hasegawa (1979) made changes in the media for initiation, multiplication, and rooting, which resulted in satisfactory multiplication and rooting rates. Nearly all later investigators followed this practice. A sterile stage III (rooting) was

Table 1. List of authors of some papers on rose micropropagation

Author	
Alderson et al. (1988)	Hyndman et al. (1982a,b)
Aldrufeu et al. (1982)	Jacobs et al. (1969, 1970a,b)
Avramis et al. (1982a,b)	Khosh-Khui and Sink (1982a,b)
Bini et al. (1983)	Lloyd et al. (1988)
Bjarnason et al. (1985)	McCown (1980)
Bressan et al. (1980, 1982)	Martin et al. (1981)
Collet (1985)	Mederos and Enriquez (1987)
Curir et al. (1986)	Pittet and Moncousin (1982)
Davies (1980)	Reist (1985a,b, 1989)
Delbard (1982)	Rosten and McCown (1981)
De Proft et al. (1987)	Sauer et al. (1985)
Dubois et al. (1988)	Skirvin and Chu (1979a,b)
Elliott (1970)	Tweddle et al. (1984)
Graifenberg et al. (1975)	Valles and Boxus (1987a,b)
Hasegawa (1979, 1980)	Walter et al. (1979)
Horn et al. (1988)	

partly omitted, and rooting was obtained already during multiplication (Valles and Boxus 1987b) or in stage IV (weaning in substrate or other growing media) (Pittet and Moncousin 1982).

Roses for essential oil, root stock, garden, and greenhouse have been micropropagated in this way, sometimes using a large number of cultivars. Shoot tips, nodes, and buds were used as explant (Table 2). Depending upon type of explant, one to three primary shoots develop sometimes prior to transfer to stage II nutrient medium (Hasegawa 1979, 1980; Davies 1980; Khosh-Khui and Sink 1982a, b). In all cases explants were used which included primary shoot meristems. Not until the reports of Tweddle et al. (1984), Valles and Boxus (1987a), and Lloyd et al. (1988) were shoots of adventitious origin utilized in a few genotypes.

2.2 Nutrient Media

The basic medium for initiation plus multiplication was MS with a few exceptions. Avramis et al. (1982b), Curir et al. (1986), and Valles and Boxus (1987b) found a higher multiplication rate when the ammonium content of the medium was considerably reduced. Davies (1980) investigated the effect of different nutrient salt concentrations on the multiplication of seven cultivars. He also reduced $(NH_4)_2SO_4$, but found that unchanged MS salts were the best. Bressan et al. (1982) mentioned a negative influence of low salt concentrations on vigor and viability of shoots. The rooting medium consisted of MS with major elements reduced to one-quarter to one-third strength (Badzian et al. 1991). Hyndman et al. (1982a) demonstrated that a decrease in KNO_3 and NH_4NO_3 concentration was the decisive factor for improving the rooting percentage. The number of roots increased when the NO_3/NH_4 ratio was shifted from 0.1 to 3.0 (Hyndman et al. 1982b). The medium was supplemented with organic compounds, as recommended by Murashige and Skoog (1962) or with modifications, especially with respect to the vitamin concentration. Edamin was mostly omitted. The agar concentration ranged between 0.35 and 2.1% but mostly between 0.6 and 0.8%.

For the sucrose concentration in the culture medium, Langford and Wainwright (1987, 1988) found decreased levels of CO_2 uptake and decreasing vitrification with increasing concentration of sucrose. The main hormones used were NAA as auxin and BA as cytokinin source (Table 2). Comparative investigations on different cytokinins (e.g., Hasegawa 1979) were made and all showed that BA (2–3 mg/l) is the most suitable cytokinin for multiplication. Combining BA with low rates of auxins improved its effectiveness (Bressan et al. 1982; Davies 1980; Khosh-Khui and Sink 1982a,b; Skirvin and Chu 1979a). Only Walter et al. (1979) mentioned good results when using Kinetin (0.21–8.9 mg/l). Observations on 2-iP were inconsistent. While Hasegawa (1979) considered it unsuitable, Bini et al. (1983) using Cheng (1978) medium, found proliferation in axillary shoots.

Comparisons of auxins used in multiplication media showed a promoting effect of NAA in hormone combinations of 0.1 NAA and 2.0 BA mg/l (Skirvin and Chu 1979a,b,c; Khosh-Khui and Sink 1982a,b). Higher doses of NAA decreased the number of regenerating shoots (Hasegawa 1979; Davies 1980). There are indications that IAA acted in a similar way, and that it led to greater shoot size. Good results using IAA were reported from Bressan et al. (1982), Martin et al. (1981), and Leffring et al. (1984). Inconsistent results have also been reported with regard to IBA, which is recommended for initiation and multiplication by Pittet and Moncousin (1982) and Valles and Boxus (1987b). In many cases, however, media without auxins gave the best results in initiation and multiplication.

It should be mentioned that the addition of GA_3 to the multiplication medium can be of advantage for certain cultivars. Davies (1980) and Valles and Boxus (1987b) add 0.1 to 0.3 mg/l, and also Martin et al. (1981) and Delbard (1982) report positive effects. In contrast, Pittet and Moncousin (1982) described GA_3 as having no effect.

For root induction media NAA rates of 0.03 to 2 mg/l have promoted root formation (e.g., Hasegawa 1980; Avramis et al. 1982b; Khosh-Khui and Sink 1982a,b). A supplement by IBA is described by Leffring et al. (1984), Valles and Boxus (1987b), Curir et al. (1986), in WP medium (Alderson et al. 1988), and in connection with PM medium (Fiorino and Leva (1981) by Bini et al. (1983). Shoots dipped in an IBA solution and then transferred to hormone-free rooting medium gave best results. Similar observations were made by Collet (1985). Using IAA and his own rooting medium, he obtained high rooting rates within 2 weeks in 11 cultivars and two rootstocks. IAA in the rooting medium showed a positive effect, as reported by Hasegawa (1979), Khosh-Khui and Sink (1982b), Mederos and Enriquez (1987), Curir et al. (1986), and Sauer et al. (1985). Apparently concentrations between 0.8 and 2 mg/l are suitable, but cultivar effects are obvious.

Success of rooting and weaning is to a certain extent dependent on the cytokinin content of the multiplication medium (Bressan et al. 1982; Podwyszynska and Hempel 1988). Generally, cytokinins in the the rooting medium are inhibitory to root formation. Supplementation of basal rooting medium with organic substances usually followed the recommendations of Murashige and Skoog (1962). Pittet and Moncousin (1982) used the medium of Linsmaier and Skoog (1965), which contains only inositol and thiamin HCl and obtained good proliferation. Omission of vitamins from the MS medium gave inferior results (Skirvin and Chu

Table 2. Types of explants and hormones used, and success rate in rose micropropagation

Authors	Explant[a]			Hormone		Max. % success	
	Type	Size (mm)	Source	Auxin	Cytokinin	Rooting sterile	Weaning
Avramis et al. (1982a,b)	ST AB			NAA	BA	100	100
	ST AB NE		G V	NAA	BA	90	
Bressan et al. (1982)	ST	4–6	G	IAA	BA		80
Bini et al. (1983)	AB		G	IBA	Kin, BA, 2-iP	85	Low
Davies (1980)	AB NE	2–10	G O	NAA, IBA	BA	100	100
Delbard (1982)	ST AB		G			96	
Hasegawa (1979)	ST AB	3–7	G	NAA, IAA	Kin, BA, 2-iP	55	80
(1980)	ST	1–15		NAA, IAA, IBA	Kin, BA, 2-iP	96	90
Horn et al. (1988)	ST NE	3–30	G O V	NAA, IAA, IBA	BA	93	100
Kosh-Khui and Sink (1982a,b)	NE ST AB	1–50	G	NAA	BA	82	
	ST	5–10	G	NAA, IAA, IBA	BA		94
Martin et al. (1981)	ST AB		G O	IAA	BA	90	95
Pittet and Moncousin (1982)	NE	10		NAA, IAA	BA		95
Sauer et al. (1985)	AB		G	NAA	BA		90
Skirvin and Chu (1979b)	ST	20	G	NAA	BA	80	50
Walter et al. (1979)	AB	1–3	G	IAA	Kin		100
Leffring et al. (1984)	AP			IAA, IBA	BA		
Collet (1985)				IBA, NAA, IAA	BA	100	
Curir et al. (1986)	AB			IAA	BA		
Valles and Boxus (1987a,b)	AB			IBA	BA		

[a] ST = shoot tip, AB = axillary bud, NE = node,
G = greenhouse, O = outdoors, V = tissue culture.

1979a) in multiplication. Bini et al. (1983) found a positive correlation between BA and ascorbic acid. Rooting was promoted by thiamin HCl (0.1–0.3 mg/l) (Khosh-Khui and Sink 1982b), phenylalanin, and tyrosin (Avramis et al. 1982b) and nicotinamid and pyridoxine (Khosh-Khui and Sink 1982a; Bini et al. 1983). The sucrose concentration can affect root formation. Davies (1980) mentioned optimum results in media containing 4–5% sucrose, Hyndman et al. (1982b) for 7% sucrose. Nevertheless, most authors have used the standard concentration of 3%.

Pittet and Moncousin (1982) treated shoots from the multiplication stage by dipping them in NAA solution, and obtained high rooting rates in vermiculite under nonsterile conditions. Aldrufeu et al. (1982) used autoclaved inert substrates such as sand, vermiculite, and cellulose containing half-strength MS salts, inositol, and vitamins, as well as IAA. They obtained 100% rooting and a good root system in all substrates.

2.3 Light and Temperature

Usually 16-h light from fluorescent lamps of the type Gro-lux or cool white were used. In trials by Davies (1980), a 16-h period gave better growth and multiplication than an 8-h period, and the results of Bressan et al. (1982) showed improvements following an increase in light period between 4 and 16 h. Light is essential for satisfactory shoot and root growth. With regard to light intensity 1 to 3 klx (about 16-46 $\mu mol/m^2 s$) are sufficient. Bressan et al. (1982) described 17 $\mu mol/m^2 s$ as optimal when cool white lamps are used. Khosh-Khui and Sink (1982b) showed that 3000 lx cool white (46$\mu mol/m^2 s$) inhibited formation of roots, which is in contrast to Bressan et al. (1982), where 65 $\mu mol/m^2 s$ proved to be best.

Most investigators chose a temperature of around 25 °C. Comparative trials were carried out by Khosh-Khui and Sink (1982b), Bressan et al. (1982), Davies (1980), and Alderson et al. (1988). Between 11 and 21 °C, nearly identical propagation rates were obtained. The duration of subcultures was, however, prolonged, and 21 °C appears to be optimal. Similarly, rooting percentage differed little between 15 and 31 °C. Again, best results were obtained at 21 °C, and plants cultured at this temperature were also the most successful in the weaning stage. Alderson et al. (1988), however, observed an effect of the temperature during multiplication stage on rooting as well as interactions between cultivars, multiplication temperature, and rooting temperature.

2.4 General Procedures

Explants are surface disinfected with a solution containing Cl at concentrations between 0.1 and 2% (mainly around 0.5%) plus 1% Tween 20 for 10 to 20 mins. This should be followed by three 5-min rinses in sterile distilled water. Removal of outer scales prior to disinfection resulted in lower contamination rates when using explants from outdoor material (Bjarnason et al. 1985; Horn and Abd El-Maksound, unpubl.). The period on shoot proliferation medium lasts 4–5 weeks, subcultures about 4 weeks. Plants develop roots within 2 to 3 weeks after transfer to rooting medium. Multiplication rate should be three to five per passage.

Percentage success varies considerably, maximum figures are given in Table 2; 20 to 25% (Skirvin and Chu 1979) to 80% (Davies 1980) of explants start regenerating shoots. Increasing the duration of cultures usually does not improve the rate of success. Weaning of plantlets rooted in vitro takes place under special environmental conditions, such as high humidity. Generally peat amended with perlite, vermiculite, or sand was used. Pure vermiculite or perlite have also been used. Illuminance can be increased to 10 klx (about 150 $\mu mol/m^2 s$). The weaning stage lasts between 1 and 4 weeks. The success rate should be around 90%.

Clear interactions exist between certain steps and genotypes, as found in many plants, including roses. About 13 species and some 60 cultivars have been cultured in vitro. They represent all the different rose types in cultivation. Different cultivars have different requirements, as has been noted in every report where more than one cultivar has been tested. There is, however, a basic technique what can be generally applied.

3 Experiments in Micropropagation

Results and experiences of our own work are reported here.

3.1 Materials and Methods

In the different experiments, up to 14 cultivars were used (Table 3). Leaves and thorns were removed from shoots before single nodes were cut 1–1.5 cm below or above the bud. Surface sterilization involved soaking the explants in a NaOCl solution containing 1% active Cl and 0.1% Tween 20 for 15 mins. Nodes were

Table 3. List of rose cultivars used

Code	Cultivar	Group[a]	Source[b]
AT	Athena	TH	G
CH	Champagner	FL	G
FL	Flamingo	TH	G
GT	Golden Times	FL	G
KA	Kardinal	TH	G
KG	Gabriella	FL	G
LO	Lorena	FL	G
ME	Mercedes	FL	G
PA	Pasadena	TH	G
AL	Alexander	TH	O
LM	Lilli Marleen	FL	O
ND	New Dawn	Climber	O
RG	Red Gold	FL	O
WQ	White Queen Elisabeth	FL	O

[a] FL = floribunda, TH = hybrid tea.
[b] G = greenhouse, O = outdoors.

rinsed six times in deionized water after sterilization. The basic nutrient medium containing 40 mg/l sucrose consisted of major, minor, and micro elements as given by Murashige and Skoog (1962) and modified by Davies (1980). Major elements were reduced to half-strength for rooting. The growth substances added for initiation and multiplication media were: 0.1 GA_3, 2 BA and 0.004 NAA (mg/l). Cultures were kept in a growth room illuminated with Gro-lux lamps for 16 h (400 lx) at 22/24 °C. A factorial randomized block design was used for most experiments. Evaluation was done generally by analysis of variance, where necessary after transformation (Steel and Torrie 1960). Means were tested by Duncan's multiple range test. In some cases the χ-square test was employed.

3.2 Initiation Stage

Single node explants from nine greenhouse-grown cultivars were used to investigate the effects of cultivar and bud size on success of culture initiation. Nodes were subdivided into four classes according to bud size. (Class I: bud < 5 mm, class II: bud 0.5–2.0 mm, class III: bud 2–3 mm, and class IV: bud > 3.0 mm). The mean contamination percentage amounted to 13% (n = 860), with class I buds 19% and class IV buds 8.4%. It would appear that class I nodes are more prone to contamination ($\chi^2 = 7.3$, $p < 0.01$) and need more stringent sterilization. Differences in percent of contaminated explants were also observed between cultivars (Table 4), e.g., Kardinal showed a 20% contamination rate whereas Athena had only 6% contaminated cultures. Contamination rates were very high when explants were taken from outdoor plants. Even disinfection with ethanol followed by $HgCl_2$ resulted in contamination rates between 50 and 90% in such cases. Removal of bud scales could not improve this situation generally, though in some cases (Alexander, Red Gold), buds without scales gave less contaminated cultures ($\chi^2 = 8.2$ or 7.1, $p < 0.01$) (Horn and Abd El-Maksoud, unpubl.). It proved advantageous to use upper buds from stems of trickle-irrigated greenhouse stock wherever possible. If, however, in vitro stock plants are available, contamination rate is usually only 0 to 10%.

Table 4. Shoot multiplication rate per explant

cv.	% Contam.	$\bar{x}$ Shoot no. after 3 wks	CV	% Nodes with > 1	> 2 shoots
AT	6.1	2.4	12.5	77	52
CH	14.4	2.8	10.7	93	84
FL	10.4	2.9	10.3	94	82
GT	14.7	2.2	13.6	72	41
KA	20.5	2.7	11.1	98	56
KG	9.7	3.1	12.9	94	83
LO	16.8	1.8	16.6	58	23
ME	11.3	2.4	12.5	67	52
PA	12.1	2.7	18.5	84	63
$\bar{x}$	12.9	2.5		82	61
CV		8.0		n = 748	

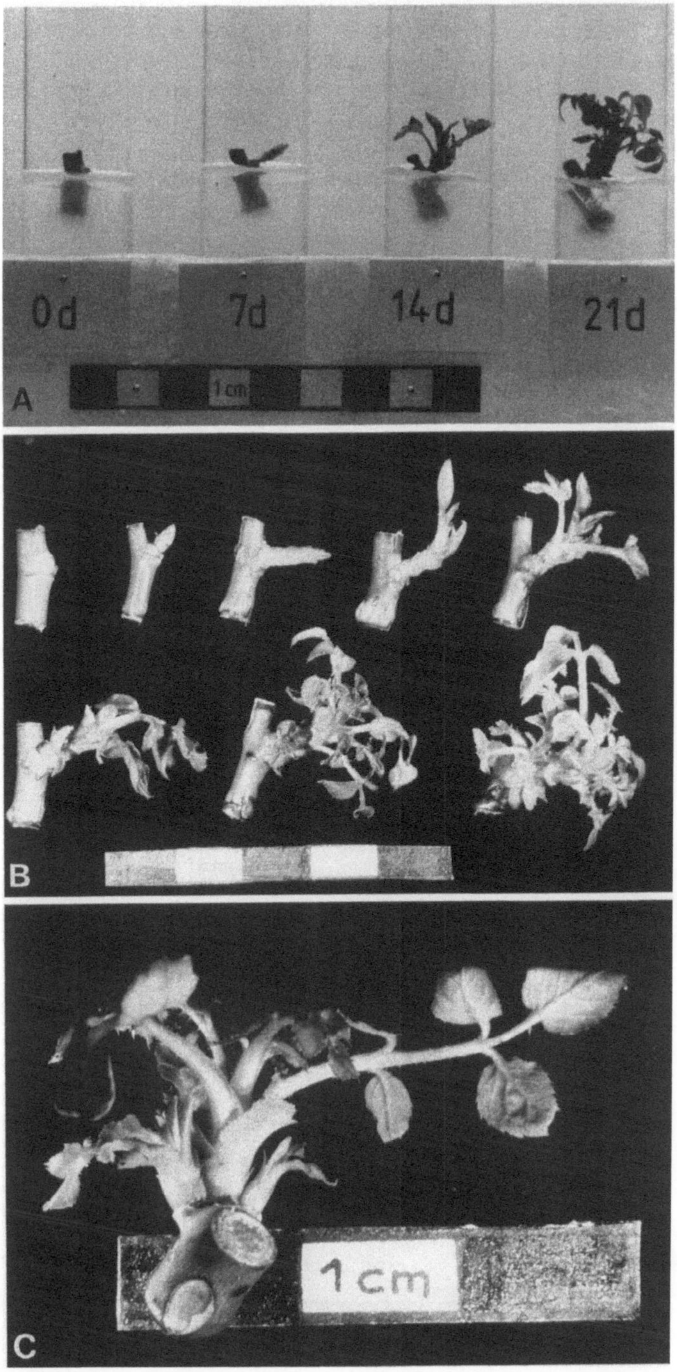

Fig. 1. A Stage 1 culture of nodes of New Dawn. **B.** After 0, 3, 7, 10, 14, 17, 21, 24 days. **C.** Single node with three shoots after 21 days. Fresh medium see Sect. 3.2

Single node explants formed between one and five shoots at an average of 2.5 shoots (coefficient of variation, CV = 8) within 3 weeks (Fig. 1). The bud size had no effect on the shoot number after 3 weeks, though a tendency was recognizable that larger (top) buds regenerated somewhat more shoots than the smaller, lower buds. There are differences between cultivars; Gabriella giving 3.1, and in contrast Lorena only 1.8 shoots. Cultivars were also different regarding the frequency of nodes forming more than one shoot. In Lorena only 23% of the nodes developed more than two shoots in comparison to more than 80% by Champagner, Flamingo, and Gabriella (Table 4). Shoot numbers were still increasing after maintaining stage I cultures for 5 weeks. In Alexander and Lilli Marleen, for instance, 2.27 and 1.77 respectively could be counted after 3 weeks, and 3.13 and 2.20 after 5 (Fig. 2). Survival rate after 5 weeks amounted to 83 and 89% respectively. The number of shoots forming cultures after 5 weeks, however, was the same as after 2 weeks.

It may be risky to maintain start cultures longer than 3 weeks, as their general condition deteriorates and the number of yellow leaflets per explant and vitrification increase (Fig. 2, Table 5). It is quite evident that the degree of vitrification depends on the cultivar. Alexander, and to a lesser degree Champagner, are prone to vitrification while Lilli Marleen very rarely showed any symptoms. Duration of stage I may differ between cultivars, which is also true for the following stages.

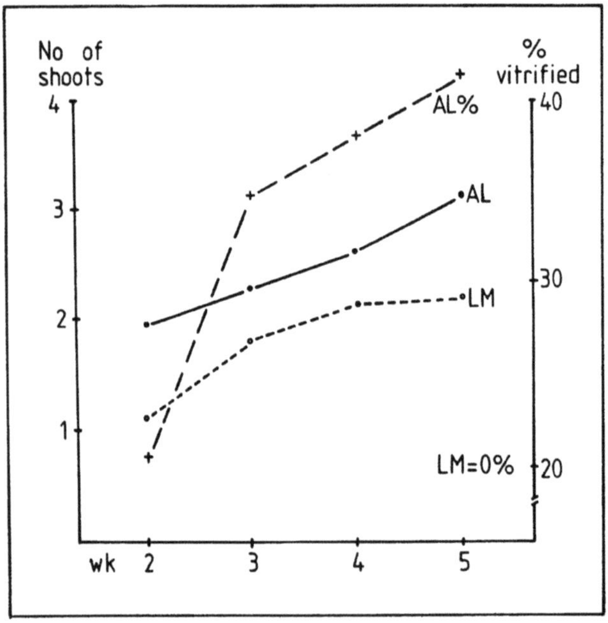

Fig. 2. Effect of cultivar and culture duration on vitrification. · = number of shoots: + = vitrification (%). Fresh medium see Sect. 3.2

Table 5. Observations of stage I cultures. (Horn and Abd El-Maksoud, unpubl.)

	cv.	After 2 wks	3 wks	4 wks	5 wks	6 wks
No. of shoots	CH	–	–	1.43	1.53	1.73
	KA	–	–	0.98	1.03	1.17
	LO	–	–	1.17	1.23	1.27
Condition of explant (1–9)	AL	8.40	8.23	7.73	7.33	–
	LM	8.37	8.35	8.24	7.90	–
No. of yellow leaves	AL	0.33	0.53	1.0	1.40	–
	LM	0.20	0.38	0.41	0.44	–
	CH	1.02	2.57	4.73	5.83	7.53
	KA	0.43	1.43	2.07	2.43	3.33
	LO	0.42	1.63	2.23	2.57	2.87

3.3 Multiplication Stage. Effect of Genotype

Multiplication stage was usually started after 3 to 5 weeks when shoots were removed from the initial explant, trimmed to about 1 cm in length and subcultured on fresh medium of the same composition as in stage I. After two subcultures with a total duration of 16 weeks, each shoot had formed 3.9 side shoots (coefficient of variation, CV 16.1), so that each source explant gave (2.5 × 3.9) 9.75 shoots within about 20 weeks. Contamination rate in subcultures averaged 2.6%. Subculturing for a period of altogether 39 weeks showed no decrease in multiplication rate per subculture, but an increase in contamination rate to 5.4%.

There were clear effects of genotypes. After five subcultures Kardinal showed the highest, Champagner and Lorena an intermediate propagation rate (n = 300 explants per subculture, Table 6, Fig. 3). In other trials also Lilli Marleen, Alexander, and White Queen Elizabeth proved to be successful in propagation. Taking into consideration other points, such as general condition of culture, speed

Table 6. Effect of cultivars on development in multiplication stage

cv. \ Trait	Development (1–9)	No. of yellow leaves	FW shoots (g)	No. of shoots[a]
AT	3.8	1.2	0.17	1.5
CH	5.6 a	1.7 b	0.30 a	2.6 a
FL	3.6 d	1.4 bc	0.22 bc	1.7 c
GT	4.0 cd	2.3 a	0.22 bc	1.6 c
KA	5.1 b	0.9 d	0.32 a	2.5 a
KG	4.0 cd	1.3 cd	0.22 bc	1.8 c
LO	4.8 b	1.0 d	0.24 b	2.2 b
ME	4.0 cd	1.1 cd	0.18 c	1.8 c
PA	4.2 c	1.7 b	0.19 bc	1.7 c
$\bar{x}$	4.4	1.4	0.24	2.0
CV	5.2	35.2	25.6	16.1

[a] Per subculture and explant.

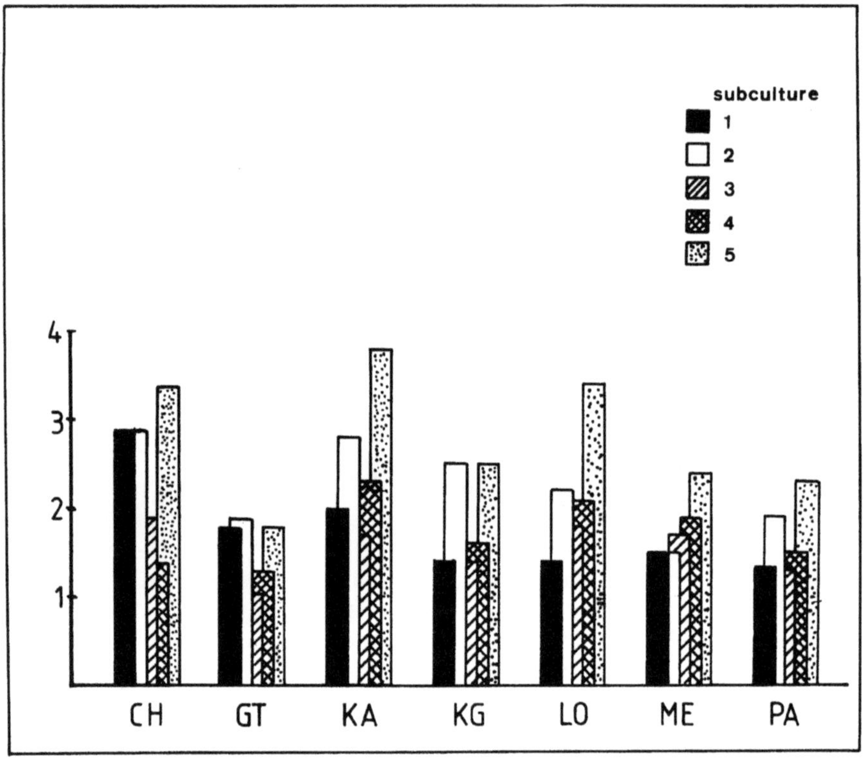

Fig. 3. Effect of cultivar and subculture on multiplication. Fresh medium see Sect. 3.3

of development, and fresh weight of shoots regenerated, ranking of cultivars resulted again with Kardinal and Lilli Marleen as most easily in propagation, followed by the other cultivars. Athena, Mercedes, Pasadena, and Golden Times were the most difficult to propagate.

Subcultures aiming at multiplication can make use of nodal explants. Shoots from cultures are then cut again into single nodes or into a basal and a top part. One trial with single nodes taken from shoots grown in vitro yielded 88 node cultures from stage I and 303 explants for the following subculture within 4 weeks. Nodal explants were superior to shoot tips when cutting shoots into a 1-cm tip and a base part with two to four nodes (Table 7). Growth, number, length, and fresh weight of shoots originating from subcultured shoots with long internodes were better than from compact shoots with short internodes (Horn et al. 1988). The effect of explant on multiplication rate was investigated by taking the position of the node at the shoot into consideration. Significant differences were found (Table 8) on scoring the general condition of culture, growth, shoot fresh weight, and number of shoots per explant. Counting from the apex to the base node, the fifth and sixth proved superior to the tip (apex plus one fully developed leaf) and nodes one to three regarding growth and shoot weight. Nodes performed better than shoot tips.

Table 7. Performance of tip versus nodal explants in multiplication stage

Trial no. (cv.)	Trait / Explant	No. of yellow leaves	Shoot Growth (1–9)	Shoot FW (g)	Shoot No.
B	Tip	1.4 a	3.8 a	0.14 b	1.5 b
(AT, CH, FL, GT, KA, KG,	Base	0.6 b	4.1 a	0.16 a	1.7 a
LO, ME, PA)	x̄	1.0	3.9	0.15	1.6
n = 540	CV	43.2	7.2	25.1	14.2
C	Tip	1.2 a	4.3 a	0.17 b	1.5 b
(CH, FL, GT, KA, KG, LO,	Base	1.2 a	4.5 a	0.23 a	2.1 a
ME, PA)	x̄	1.2	4.4	0.20	1.8
n = 480	CV	52.3	9.2	12.9	39.9
D	Tip	2.6 a	4.6 b	0.28 b	2.4 b
(AT, CH, FL, GT, KA, KG,	Base	2.5 a	5.1 a	0.38 a	2.9 a
LO, ME, PA)	x̄	2.6	4.9	0.33	2.6
n = 540	CV	35.0	8.0	11.4	30.6
E	Tip	1.3 a	3.9 b	0.22 a	1.9 b
(AL, LM, ND, RG, WQ)	Base	1.0 a	4.7 a	0.25 a	2.5 a
	x̄	1.1	4.3	0.23	2.2
n = 300	CV	76.5	11.9	10.2	39.8
x̄	Tip	1.6	4.2	0.20	1.8
n = 1860	Base	1.3	4.6	0.26	2.3

Table 8. Effect of node position (bud size) on multiplication (6 weeks)

Trait / Node position	Condition (1–9)	Development (1–9)	Shoot FW (g)	Shoot No.
Shoot tip	6.4 a	3.2 e	0.055 d	2.10 a
1	6.7 a	3.8 d	0.094 c	2.20 a
2	6.4 a	4.6 cd	0.100 bc	2.33 a
3	6.4 a	4.7 bd	0.140 ac	2.53 a
4	6.6 a	5.1 ac	0.154 ab	2.33 a
5	6.9 a	5.9 ab	0.158 a	2.57 a
6	6.0 a	6.1 a	0.140 ac	2.50 a
7	4.1 b	4.3 cd	0.132 ac	2.03 a
8	3.8 b	4.3 cd	0.111 bc	2.02 a
x̄	5.9	4.7	0.12	2.3
CV	11.2	22.8	18.6	9.4

n = 135.

3.4 Effect of Sucrose Level

Multiplication rate is also dependent on the level of sucrose in the nutrient medium. The effect of sucrose concentration from 10 to 70 g/l is summarized in Table 9. Generally, it can be concluded that concentrations below 2% and above 4% are inferior; 3 and 4% perform best. Senescence (yellow leaves) is retarded at the 2 to 4% levels. Comparing 3 and 4% sucrose in cultures of five other varieties

Table 9. Effect of sucrose level on growth and multiplication rate (AT, CH, FL, GT, KA, KG, LO, ME, PA), mean after 6–9 weeks culture

Sucrose level (g/l)	Condition	No. of yellow leaves	Growth (1/3/5/7/9)	Shoot		
				FW	No.	Weight
10	7.3	1.7	4.6	0.16	1.1	1.4 b
20	8.1 a	0.8 c	4.6 bc	0.15 c	2.0 ab	1.4 b
30	8.0 a	0.9 c	4.7 ab	0.16 bc	1.9 ab	1.4 b
40	7.9 ab	1.0 c	4.9 a	0.21 a	2.1 a	1.6 a
50	7.6 b	1.5 b	3.5 cd	0.17 b	1.7 bc	1.5 ab
60	7.4 c	1.7 ab	3.6 c	0.16 bc	1.6 c	1.4 b
70	7.5 c	1.9 a	3.4 d	0.15 c	1.6 c	1.4 b
$\bar{x}$	7.7	1.3	4.4	0.16	1.8	1.4
CV	4.0	58.6	7.5	28.6	38.3	17.1

n = 270.

Table 10. Interaction between cultivar and sucrose level

cv.	Sucrose (g/l)				
	20	30	40	50	60
Growth (1–9)					
AT	4.6 hk	5.0 eg	4.5 hl	3.5 qr	3.5 qr
CH	5.1 df	5.3 cd	5.6 ab	4.5 hl	4.1 lp
KA	4.9 fh	5.2 ce	5.9 a	5.2 ce	4.6 hk
LO	5.9 a	5.1 df	5.0 eg	4.9 fh	4.8 fi
ME	3.9 nr	3.9 nr	4.1 lp	4.1 lp	4.2 ko
Shoot FW (g)					
AT	0.14 im	0.18 fj	0.15 hm	0.13 jm	0.12 lm
CH	0.14 im	0.17 gl	0.25 ce	0.14 im	0.13 jm
KA	0.20 eh	0.21 dg	0.33 a	0.25 ce	0.20 eh
LO	0.14 im	0.17 gl	0.16 gm	0.15 hm	0.14 im
ME	0.11 m	0.12 lm	0.12 lm	0.13 jm	0.14 im
Shoot no.					
AT	1.5 fj	2.3 bc	1.4 gj	1.3 hj	1.2 ij
CH	1.8 dh	2.0 cf	2.6 ab	1.9 cg	1.9 cg
KA	1.9 eg	2.3 bc	2.9 a	2.1 be	1.6 fj
LO	2.3 bc	2.3 bc	2.1 be	1.8 dh	1.7 ei
ME	1.7 ei	1.8 dh	1.9 cg	1.9 cg	1.9 cg

n = 375.

(AL, MD, ND, RG, WQ[2]) growth and shoot number were significantly superior at the 3% level (CV = 11.9 and 28.6 respectively, n = 300). Interactions between cultivar and sucrose level could be demonstrated in several cases (Table 10). CH and KA have an especially narrow optimum range at 4%, AT at 3%; in contrast ME and LO have a wide range, within which LO performs best at 2 to 3%. In similar trials LM regenerated significantly more shoots at 3% than at 4%, and RG developed better at 3%. There was an apparent superiority at the 3% level with varieties AL, LM, ND, RG, and WQ. When propagating roses in vitro, the sucrose level should therefore be taken into account, especially in successive subcultures, as found by Langford and Wainwright (1987).

3.5 Effect of Temperature

One of the most interesting observations resulted from the investigation of temperature effects on rose micropropagation. Temperature proved a deciding factor in multiplication when using nodal explants. When comparing cultures kept initially at 24 °C for 2 weeks, after a period of 10 weeks at 12, 15, 18, and 24 °C, the condition and the number of shoots decreased with increasing temperature while percentage of dead explants and fresh weight of shoots increased with temperature. Although the best overall results were obtained at 18 °C (Table 11), there were significant interactions between temperatures and cultivars. PA, FL, RG, and LO do not react significantly to temperature, whereas AL, KG, ME, and LM react clearly. A study of the interactions shows that AL and PA have their optimum at 12 °C (thermonegative), LO, ND, and CH at 24 °C (thermopositive). An intermediate group with an optimum at 18 °C contains LM, ME, KG, KA, WQ, and FL (Fig. 4). In another experiment using cultivars Frisco, Europa, Cadillac, Kardinal, and Kiss, it could be shown that on average multiplication rate was best at 21 and 24 °C, whereas rooting was optimal at 18 or 15 °C. The cv. Europa, however, reacted "thermonegatively" for multiplication rate like AL and PA. It may be

Table 11. Effect of temperature on multiplication of nodal explants (12 cvs.)

Trait / Temperature (°C)	Condition (1–9)	%Dead cultures	Shoot		
			No.	Length (cm)	FW (g)
12	8.0 a	0	3.0 ab	1.60 b	0.23 b
15	7.3 b	2.5	2.9 ab	1.69 ab	0.24 b
18	7.4 b	2.5	3.1 a	1.97 a	0.31 ab
24	6.3 c	6.7	2.6 b	1.94 a	0.34 a
$\bar{x}$	7.3	2.9	2.9	1.80	0.28
CV	6.8		23.4	29.1	50.4

n = 120/temperature.

[2] For abbreviation see Table 3.

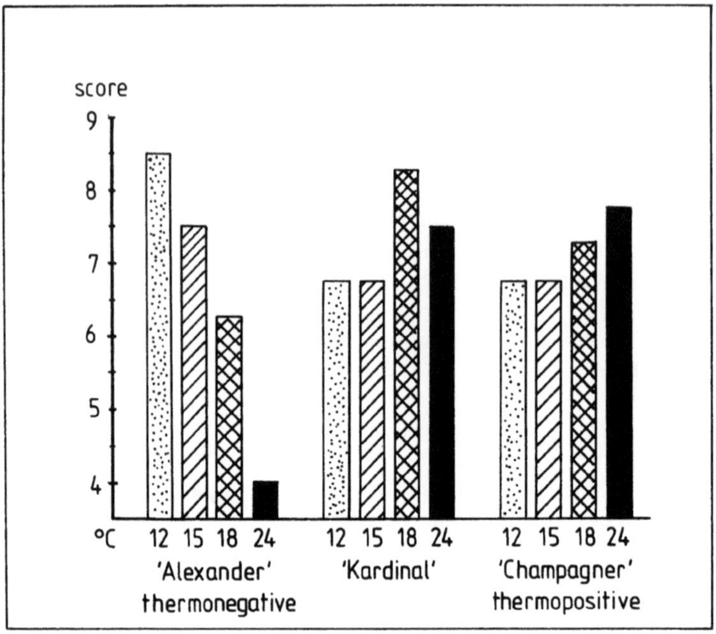

Fig. 4. Reaction of cultivars to temperature in multiplication stage (9 = best score). Fresh medium see Sect. 3.5

mentioned that a maintenance temperature of 12 °C is tolerated by all cultivars tested.

3.6 Ethylene

It was found in our trials that sealing resulted in a clear increase in ethylene level in the culture tube. Tubes with lids sealed with nescofilm (ethylene 6.5 pmol/ml) performed less well than those with unsealed lids (0.8 pmol/ml). It could be demonstrated that the amount of ethylene increased greatly (92.5 pmol/ml) if using gas-proof stoppers. Senescence (yellow leaves, general condition) increased accordingly (Table 12). Shoot growth was also reduced. Surprisingly, shoot number and fresh weight improved with increasing ethylene level, a fact which may be explained by interactions between auxin and ethylene metabolism. Ethylene effects in flasks with in vitro cultures have been reported previously by Thomas and Murashige (1979), Wulster and Sacalis (1980), Gavinlertvatana et al. (1982), and DeProft et al. (1987). No correlation between vitrification and ethylene content could be found in our experiments.

3.7 Stage III/IV. Rooting/Weaning

The effect of auxins on rooting percentage in vitro and number of roots per plantlet is shown in Table 13. Statistically, there are hardly any significant differences in

Table 12. Effect of sealing the culture tube on growth of explants (Horn et al. 1988)

Culture vessel	n	Yellow leaves		Shoot		Condition score 1–9
		No.	%	No.	FW	
Plastic container[a]	162	1.8 b	20.2 b			6.8 a
Glass jar, sealed	162	3.0 a	34.8 a			4.6 b
CV		36.3	36.6			8.9
Glass tube, unsealed	135	3.0 b	—	1.8 b	0.26 b	6.7 a
Glass tube, sealed	135	6.6 a	—	2.1 a	0.34 a	4.7 b
C		29.4		24.6	28.2	7.4
Glass tube[b], unsealed	10	1.7 a	7.3 a	3.7 a	—	—
Glass tube, sealed	10	8.5 a	33.2 a	4.2 a	—	—
CV		27.9	42.0	8.3		
Glass tube[b], unsealed	20	0.8 a	—	4.2 a	0.53 b	6.0 a
Glass tube, sealed	20	2.1 a	—	4.7 a	0.74 b	5.0 a
Gas-proof stopper	20	3.9 a	—	5.8 a	1.17 a	5.0 b
CV		39.6	—	15.1	11.5	0.6

[a] With fitting lid but not gas-proof.
[b] Ethylene concentration (pmol/ml); 0.8 (unsealed), 6.5, and 92.5 (stopper).

Table 13. Effect of auxins on rooting[a]

NAA	Hormone mg/l IAA	IBA	% Shoots with roots	No. of roots per plantlet	Plantlets established successfully (%)
–	–	–	80.0	5.9 bc	100.0
0.1	–	–	83.3 a	6.1 b	88.0
0.1	0.025	–	83.3 a	5.8 bc	88.0
0.1	0.05	–	80.0 a	6.3 ab	73.9
0.1	0.1	–	66.7 ab	6.5 ab	61.9
0.1	0.1	+ 2gAK/l[b]	83.3 a	4.8 bc	80.0
0.1	–	0.025	76.7 ab	6.6 ac	81.8
0.1	–	0.05	73.3 ab	8.4 a	77.3
0.1	–	0.1	53.3 ab	3.6 c	64.7
$\bar{x}$			75.6	6.0	80.4
CV			23.4	31.3	
Alexander			63.0 b	6.6 a	75.3
White Queen Elizabeth			88.1 a	5.4 b	84.0

[a] n = 270 and 204 resp.
[b] AK = charcoal.

rooting percentage. There is, however, a clear tendency pointing to superior results when no auxins or only 0.1 mg/l NAA are applied. Comparing addition of IAA or IBA, IBA and 0.1 mg/l IAA inhibit rooting to a certain extent. Addition of auxins decreased the length of roots. There are significant differences between varieties regarding the reaction to auxins, and interactions (Fig. 6). These results could be confirmed in later experiments using other cultivars.

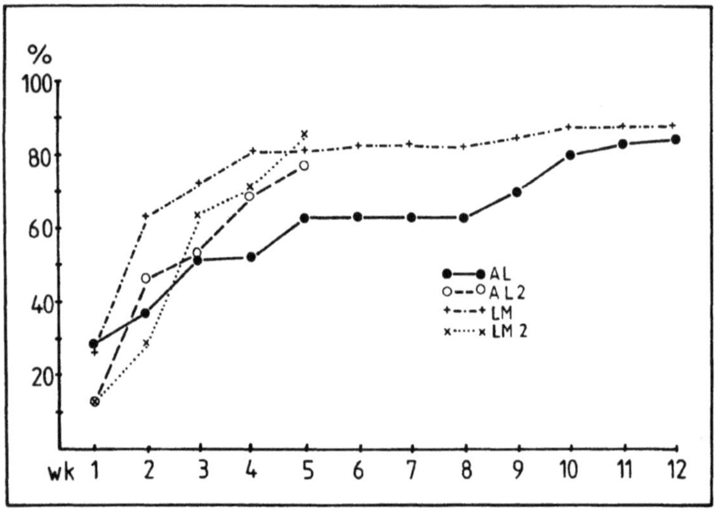

Fig. 5. Course of rooting in vitro (two experiments). Fresh medium see Sect. 3.7

In Fig. 5 the course of rooting is given for Alexander and Lilli Marleen. One week after transfer to rooting medium, about 27% of the shoots were rooted. That percentage was doubled within 4 weeks after transfer of Alexander and within 2 weeks after transfer of Lilli Marleen. Eighty percent rooting was reached after 10 or 4 weeks respectively. It may be worthwhile, therefore, to treat any cultivar according to its special requirements for rooting. Rooting percentage in the cultivars Champagner, Kardinall, and Lorena after 6 weeks came to 59, 88, and 52% respectively. Mean number of roots per plantlet ranged between 1.4 and 10.6.

Rooting under nonsterile conditions in oasis (polyurethane foam cubes) together with an MS nutrient solution with salts reduced to one-third strength, and keeping shoot bases in this substrate for 1 to 2 weeks resulted in rooting of up to 100%, depending on the cultivar. The use of oasis soaked with a sugar-free liquid medium containing NAA (0.03–0.1 mg/l) produced inferior results. The best results were obtained with plantlets rooted on sterile hormone-free medium or on media with 0.1 mg/l NAA only (Fig. 6). The plantlets with roots of 1–2 cm in length were transferred from the tube to small pots containing peat moss without fertilizer. This can take place after about 12 weeks from starting the culture. Before transfer, all agar has to be removed from the plants, which were dipped in a fungicide solution. Pots and peat were sterilized. Weaning of plantlets was carried out under humid conditions and a temperature of 25 °C in the greenhouse, which was shaded if necessary. Between 10 and 100% (mean 80%) of rooted plantlets could be established in nonsterile conditions within 4 weeks. They were then transplanted to 10-cm pots containing peat moss with 1.5 kg of a complete fertilizer per m^3. Weaning is the most critical stage, and attention has to be focused on the requirements of the different cultivars.

Micropropagated plants, growing on their own roots, have performed well in the open over several years, including two severe winter periods. Comparative

Fig. 6. Effect of auxins on rooting in vitro (cv. Kardinal) (Sect. 3.7)

trials have been carried out between micropropagated plants and plants budded on rootstocks. Data have been published by other authors regarding performance in greenhouse production of both plant types (Tweddle et al. 1984; Reist 1985a,b,1989; Brun and Tramier 1988; Dubois et al. 1988). Annual yield of flowers and average stalk length appears to be somewhat superior in grafted plants, but appropriate disbudding techniques adapted to the development of micropropagated plants may correct these results. Other personal communications mention superior vigor, better branching, and earlier flowering of plants originating by micropropagation.

Comparison of the performance of rose plants on their own roots as compared to budded plants shows further that cultivars may differ in their suitability to be grown in one or the other way (De Dood 1986a,b; Meiss 1986; Wagenknecht and Radke 1987), and that roses perform successfully on their own roots. Micropropagated roses are supplied by commercial establishments fully weaned and hardened off throughout the year. Plants at least 8–10 cm in height with a well-developed root system can be handled similarly to a rooted cutting. To encourage new basal shoots it is recommended to plant deeply. A saleable container-grown plant can be produced in 3 to 4 months.

4 Protocol

1. Grow your stock plants under hygienic conditions or use in vitro stock.
2. Initiation of culture. Use shoot tips or nodal explants with buds larger than 1 mm. Surface sterilization in a NaOCl solution containing 1% active Cl plus 0.1% Tween for 15 min. Rinse several times in sterile water. Place explant in a culture tube containing solid medium.

3. Remove shoots about 1 cm in size from the initial explant for subculture and multiplication. Plant shoots in new culture tubes using the same medium as before. Keep at 24 °C for 2 weeks, then change temperature to varietal requirement.
4. Transfer shoots at least 1 cm in size from multiplication stage to culture vessels containing hormone-free medium for rooting.
5. Rooted plantlets are transferred to peat (pH 6) without salts and weaned for at least 1 week under humid conditions in a greenhouse. Remove all agar from the roots prior to transplanting. Soak plantlets and soil in a suitable fungicide solution. Plants are ready to transplant again after a period of about 4 weeks.

Alternatively, shoots from stage 4 are dipped in NAA solution (10 mg/l) for 1 h and transferred to autoclaved vermiculite (plus sand or peat) for rooting and weaning.

Medium: Murashige and Skoog (1962) containing

0.1	GA_3 mg/l	sucrose	3 to 4%
2	BA mg/l	agar	6 g/l
0.004	NAA mg/l	pH	5.5

Salts were reduced to half-strength in rooting medium.

Acknowledgments. The contribution of Dr. G. Schlegel to planning and carrying out our experiments is gratefully acknowledged. Thanks are also due to Dr. MacCarthaigh for his critical reading of the manuscript.

References

AIPH (Assoc Int Product Hortic) (1990) Jahrb Int Gartenbaustat 38, Hannover
Alderson PG, McKinless J, Rice RD (1988) Rooting of cultured rose shoots. Acta Hortic 226: 175–182
Aldrufeu A, Pages M, Messeguer J, Mele E (1982) In vitro rhizogenesis of *Rosa* sp. in different substrates. Acta Hortic 150: 315–323
Avramis T, Hugard J, Jonard R (1982a) Exaltation des potentialités rhizogenes de pousses feuillées de rosier multipliées in vitro. C R Acad Sci Paris 294: 679–682
Avramis T, Hugard J, Jonard R (1982b) La multiplication in vitro du rosier porte-greffe *Rosa indica* major, C R Acad Sci Paris 294: 63–68
Badzian T, Hennen GR, Fotyma-Kern J (1991) In vitro rooting of clonal propagated miniature rose cultivars. Acta Hortic 289: 329–330
Bini G, Leva ARC, Nicese FP (1983) Richere sulla micro-propagazione della rosa. Riv Ortoflorofrutt Ital 67: 1–13
Bjarnason EN, Hanger DC, Moran JR, Cooper JA (1985) Production of *Prunus* necrotic ringspot virus-free roses by heat treatment and tissue culture. N Z J Agric Res 28: 151–156
Bressan PH, Kim YJ, Hasegawa PM (1980) Plant regeneration from cultured rose shoot tips. HortScience 15: 432–433
Bressan PH, Kim YJ, Hyndman SE, Hasegawa PM, Bressan RA (1982) Factors affecting in vitro propagation of rose. J Am Soc Hortic Sci 107: 979–990
Brun R, Tramier PH (1988) Culture du rosier sur laine de roche. P H M Rev Hortic 289: 43–51
Campos PS, Pais MSS (1990) Mass propagation of the dwarf rose cultivar 'Rosamini'. Scientia Hortic 43: 321–330
Cheng TY (1978) Propagation of woody plants by tissue culture techniques. Am Nurseryman (May)
Collet GF (1985) Enracinement amélioré lors de la production in vitro de rosiers. Rev Suisse Vitic Arboric Hortic 17: 259–263
Curir P, Damiano C, Cosmi T (1986) In vitro propagation of some rose cultivars. Acta Hortic 189: 221–224

Davies DR (1980) Rapid propagation of roses in vitro. Sci Hortic 13: 385–389
De Dood J (1986a) Tien cultivars op eigen wortel en onderstam. Vakbl Bloem 41 (2): 40–41
De Dood (1986b) CO_2-vergelijking met drie cultivars bij zetlingen, stekken en weef selkweek-planten. Bloem Onderz Ned Aalsmeer, 159
Delbard H (1982) Micropropagation of roses at Delbard nurseries. Plant Prop 28: 7–8
De Proft MP, Van den Broek G, De Greef JA (1987) Involvement of ethylene on senescence and vitrification of in vitro cultured miniroses. Acta Hortic 212: 217–222
DeWit JC, Esendam HF, Konkanen JJ, Tuominen U (1990) Somatic embryogenesis and regeneration of flowering plants in rose. Pl. Cell Rep 9: 456–458
Douglas GC, Rutledge CB, Casey AD, Richardson DHS (1989) Micropropagation of floribunda, ground cover and miniatur roses. Plant Cell Tissue Organ Cult 19: 55–64
Dubois LAM, Roggemans J, Soyeurt G, DeVries DP (1988) Comparison of the growth and development of dwarf rose cultivars propagated in vitro and in vivo by softwood cuttings. Sci Hortic 35: 293–299
Elliott RF (1970) Axenic culture of meristem tips of *Rosa multiflora*. Planta 97: 183–186
Fiorino P, Leva AR (1981) La multiplicatione del melo con la tecnica della micropropagazione. I. Indagine sulla ottimizzazione dei substrati nutritivi durante la fase di moltiplicatione. Atti Congr I fitoregolatori in agricoltura Firenze: 147–156
Gavinlertvatana P, Read PE, Wilkins HF, Heins R (1982) Ethylene levels in flask atmospheres of *Dahlia pinnata* Cav. leaf segments and callus cultured in vitro. J Am Soc Hortic Sci 107: 3–6
Graifenberg A (1973) Coltura in vitro di embrioni e di parti de seme in *Rosa canina*. Riv Ortoflorofrutt Ital 5: 374–380
Graifenberg A, Giustiniani L, Papandreu A (1975) La coltura in vitro di apici vegetativi di *Rosa multiflora* Thunb.. Agric Ital 104: 119–128
Hasegawa PM (1979) In vitro propagation of rose. HortScience 14: 610–612
Hasegawa PM (1980) Factors affecting shoot and root initiation from cultured rose shoot tips. J Am Soc Hortic Sci 105: 216–220
Horn W, Schlegel G, John K (1988) Micropropagation of roses. Acta Hortic 226: 623–626
Hyndman SE, Hasegawa PM, Bressan RA (1982a) The role of sucrose and nitrogen in adventitious root formation on cultured rose shoots. Plant Cell Tissue Organ Cult 1: 229–238
Hyndman SE, Hasegawa PM, Bressan RA (1982b) Stimulation of root initiation from cultured rose shoots through the use of reduced concentration of mineral salts. HortScience 17: 82–83
Jacobs G, Allan P, Bornman CN (1969) Tissue culture studies on rose: use of shoot tip explants I. auxin: cytokinin effects. Agroplantae 1: 179–188
Jacobs G, Allan P, Bornman CN (1970a) Tissue culture studies on rose: use of shoot tip explants II. cytokinin:gibberellin effects. Agroplantae 2: 25–28
Jacobs G, Allan P, Bornman CN (1970b) Tissue culture studies on rose: use of shoot tip explants. III. auxin: gibberellin effects. Agroplantae 2:45–50
Khosh-Khui M, Sink KC (1982a) Micropropagation of New and Old World rose species. J Hortic Sci 57: 315–319
Khosh-Khui M, Sink KC (1982b) Rooting enhancement of *Rosa hybrida* for tissue culture propagation. Sci Hortic 17: 371–376
Khosh-Khui M, Sink KC (1982c) Callus induction and culture of *Rosa*. Sci Hortic 17: 361–370
Lammerts WE (1946) Use of embryo culture in rose breeding. Plant Garden 2: 111
Langford PJ, Wainwright H (1987) Effects of sucrose concentration on the photosynthetic ability of rose shoots in vitro. Ann Bot 60: 633–640
Langford PJ, Wainwright H (1988) Influence of sucrose concentration on the photosynthetic ability of in vitro grown rose shoots. Acta Hortic 227: 305–310
Leffring L, Skovbogaard M, Esendam H (1984) Weefselkweek bij roos. Bloem Onderz Ned, Alasmeer, 364–365
Linsmaier EM, Skoog F (1965) Organic growth factor requirements of tobacco tissue cultures. Physiol Plant 18: 100–127
Lloyd D, Roberts AV, Short KC (1988) The induction in vitro of adventitious shoots in *Rosa*. Euphytica 37: 31–36
Martin C, Carré M, Vernoy R (1981) La multiplication végétative in vitro des végétaux ligneux cultivés: cas des rosiers. C R Acad Sci Paris 293: 175–177
McCown BH (1980) Micropropagation of hardy rose species and hybrids. HortScience 15: 417 (Abstr no 326)

Mederos S, Enriquez MJ (1987) In vitro propagation of Golden Times roses. Factors affecting shoot tips and axillary bud growth and morphogenesis. Acta Hortic 212: 619–624
Meiss W (1986) Rosenunterlagen im Vergleich. Taspo Magaz (3): 10–12
Murashige T, Skoog F (1962) A revised medium for rapid growth and bioassays with tobacco tissue cultures. Physiol Plant 15: 473–497
Nickell LG, Tulecke W (1959) Responses of plant tissue cultures to gibberellin. Bot Gaz 120: 245–250
Nobécourt P, Kofler L (1945) Culture de tissus de tige de rosier. C R Hebd Séanc Acad Sci Paris 221: 53–54
Pittet H, Moncousin C (1982) La micropropagation in vitro du rosier. Rev Hortic Suisse 55: 67–70
Podwyszynska M, Hempel M (1988) The factors influencing acclimatization of *Rosa* hybr. plants multiplied in vitro to greenhouse conditions. Acta Hortic 226: 639–642
Reist A (1985a) Culture in vitro (civ) en pépinière de rosiers. Premiers resultats d'exploitation de plants civ pour la coupe. Rev Suisse Vitic Arboric Hortic 17: 173–176
Reist A (1985b) Culture in vitro en pépinière de rosiers: une alternative au bouturage ou au greffage des varietés? Rev Suisse Vitic Arboric Hortic 17: 361–364
Reist A (1989) Comportement au champ de rosiers propagés in vitro: bilan d'un essai comparatif de 5 ans. Rev. Suisse Vitic Aboric Hortic 21: 19–23
Rosten AR, McCown BH (1981) Micropropagation of species roses. HortScience 16: 452 (Abstr no 393)
Sauer A, Walther F, Preil W (1985) Different suitability for in vitro propagation of rose cultivars. Gartenbauissuscraft 50: 133–138
Short KC, Roberts AW (1990) *Rosa* spp. (roses): in vitro culture, micropropagation and the production of secondary products. In: Bajaj YPS (ed) Biotechnology in agriculture and forestry, vol 15. Medicinal and aromatic plants III. Springer, Berlin Heidelberg New York Tokyo 376–397
Skirvin RM, Chu MC (1979a) In vitro propagation of Forever Yours rose. HortScience 14: 608–610
Skirvin RM, Chu MC (1979b) Propagation of roses with tissue culture. III Res 21: 3
Skirvin RM, Chu MC (1979c) Root and shoot formation from Forever Yours rose in vitro. HortScience 14: 457(Abstr no 455)
Skirvin RM, Chu MC, Young HJ (1989) Rose. In: Ammirato PV, Evans DA, Sharp WR, Bajaj YPS (eds) Handbook of plant cell culture, vol 5. Ornamental species. McGraw-Hill, New York pp 716–743
Steel RGD, Torrie H (1960) Principles and procedures of statistics. McGraw Hill, New York
Thomas DS, Murashige T (1979) Volatile emissions of plant tissue cultures. V. Identification of major components. In vitro 15: 654–658
Tweddle D, Roberts AV, Short KC (1984) In vitro culture of roses. In: Novak FJ, Havel L, Dolezel J (eds) Plant tissue and cell culture application to crop improvement, Czech Acad Sci, Prague, 529–530
Valles M, Boxus P (1987a) Regeneration from *Rosa* callus. Acta Hortic 212: 691–696
Valles M, Boxus P (1987b) Micropropagation of several *Rosa hybrida* L. cultivars. Acta Hortic 212: 611–617
Wagenknecht J, Radke A (1987) Anzucht wurzelechter Rosen und Vergleich ihrer Leistungsfähigkeit mit veredelten Pflanzen im Gewächshaus. Gartenbau (DDR) 34: 249–251
Walter RJ, Kamp M, Smith RH (1979) In vitro propagation of *Rosa chinensis* Jacq. var. *minima* Red Cascade, J Rio Grande Val Hortic Soc 33: 125–127
Weinstein LH, Tulecke W, Nickel LG, Laurencot HJ (1962) Biochemical and physiological studies of tissue cultures and the plant parts from which they are derived. III. Paul's Scarlet rose. Contrib Boyce Thompson Inst 21: 371–386
Woltering EJ (1990) Beneficial effects of carbon dioxide on development of gerbera and rose plantlets grown in vitro. Sci Hortic 44: 341–345
Wulster G, Sacalis J (1980) Effects of auxins and cytokinins on ethylene evolution and growth of rose callus tissue in sealed vessels. HortScience 15: 736–737
Wylie AP (1954) The history of garden roses I. J R Hortic Soc 79: 555–571
Wylie AP (1955) The history of garden roses II., III. J R Hortic Soc 80: 8–24, 77–87

XXII Micropropagation of *Ruscus racemosus* L.

P. CURIR and T. SCHIVA[1]

1 Introduction

1.1 Botany of the Plant

Ruscus (an ancient Latin name used by Virgil), vulgar name butcher's broom, is a controversial genus including various number of species depending on different authors' taxonomy (Bailey 1958). The *Dictionary of Gardening* (Synge 1956) classification considers the genus *Ruscus* as composed of the four species *Ruscus hyrcanus*, *R. aculeatus*, *R. hypoglossum*, and *R. hypophyllum*.

Ruscus racemosus, the most interesting species for ornamental purposes, is, according to Synge, a synonym for *Danae racemosa*, which is considered as a genus of a species related to *Ruscus*.

Ruscus (family Liliaceae) consists of species that are erect shrubs, woody at the base of the stem, hardy and evergreen, all native to the Mediterranean region (Bailey 1958). The stems bear cladodes, sessile and persistent, with small dioecious flowers, generally on the upper or lower surface of the cladodes, or as racemes at the axil of the cladodes; hermaphroditic flowers have, however, also been described (*Ruscus aculeatus*; Synge 1956). The fruit is a globe-shaped berry. Since the genus is very hardy, the related plants show wide adaptability to almost any type of soil; shady places are preferable for its growth.

Ruscus racemosus L., also referred to as *Danae racemosa* Moench, is a spontaneous evergreen plant originating in the Mediterranean areas of Asia Minor but endemic in southern Europe and cultivated for ornamental purposes (Fig. 1).

Tuft height may reach 1 m; the underground part of the plant is a rhizome, bearing aerial shoots, buds, and roots. The rhizome apical bud each year develops an aerial stem bearing cladodes (and flowers when the reproductive stage has been reached); rhizome growth originates from a lateral bud. The cladodes are alternate, sessile, elliptic, dark green, and leathery; the flowers are incospicuous with a short peduncle, gathered (from four to six) into a racemic, terminal inflorescence, yellowish in color and located at the cladode's axils. The flowers are composed of a sepaloidal perigonium formed by six short lobes; there are six anthers; the ovary is inferior and the style has a tri-lobated stigma. The pollen is ellipse-shaped, with a longitudinal furrow. The fruit is a small berry, red when ripe, containing a single seed with a hard albumen and rudimentary embryo. The plant is a shrub, liking

[1] Istituto Sperimentale per la Floricoltura, 18038 Sanremo, Italy

Biotechnology in Agriculture and Forestry, Vol. 20
High-Tech and Micropropagation IV (ed. by Y.P.S. Bajaj)

Fig. 1. *Ruscus racemosus* in vivo-cultivated plant

fresh and shady places, and it tolerates neither too high (above 22–24 °C) nor too low (less than − 5 °C) temperatures.

The flowering period is around the end of spring-beginning of summer, and the fruits remain on the plants throughout the autumn and winter, increasing the decorative effect and commercial value.

Seedling development is an extremely slow process, the seed requiring some months to germinate. Moreover, the seedling can produce the first true aerial stem only during the 3rd year after seed germination, the aerial production in the previous period being limited to short cladodes not inserted on a main axis. The chromosome number of the plant is 2n = 40 (Sato 1942).

1.2 Distribution and Importance

The distribution of *Ruscus* is not limited to the areas of origin but is also widespread over Europe from Madeira to the Caucasus regions, as far as Asia Minor. In particular, *Ruscus racemosus* was introduced into Europe from Asia

Minor in 1713 (Accati 1983), and during the past decades it has increased in importance as a plant producing cut green foliage. About 20–30 years ago floriculturists of the Mediterranean area began to utilize the wild plants for commercial purposes: as a consequence, the distribution of this plant extended from these regions to other areas, where it is now cultivated (e.g., some warm parts of the United States, Portugal, Spain, etc.). The green foliage is generally used in floristic decorations, and it is sold fresh, sometimes also processed for keeping. In Italy, statistics quote more than 100 ha under cultivation in the Ligurian region, and 50–60 ha in other regions (central and southern Italy). Until now, Italy has been the most important producer of this foliage, though in Spain, Portugal, and Greece some plantations have recently been made. The demand comes chiefly from Northern and Central Europe, where the cold climate does not allow this kind of cultivation. The reasons for the success in cultivating this species for ornamental purpose are:

1. relatively long vase life (up to 1 month) of the cut foliage;
2. low labor requirement;
3. low cultivation costs;
4. ability to resist frost and general hardening;
5. good demand on the market.

Indeed, *Ruscus* plants are quite simple to grow without great problems; only one insect (the coleopter *Otiorrhynchus sulcatus* L.) may provoke severe losses, but its control is easy, provided that the right program of treatment is carried out.

Some *Ruscus* species have been studied with a view to extracting metabolites of pharmacological interest, such as sterols and sitosterols, from the plant's roots or rhizomes; in particular, the *Ruscus aculeatus* rhizome possesses diuretical properties, and it is included in the French pharmacopeian register (Lodi 1966).

2 Conventional Propagation and Breeding

Plants may be grown from seeds, but their development is then quite slow; by this method, the first crop of commercial foliage can be expected only after 5 to 6 years. Multiplication is thus usually carried out by the division of tufts from the old plants, which permits a rapid start of cropping. From a 10- or 12-year-old plant, it is possible to obtain no more than three or four new plants, so that this type of propagation is satisfactory provided that a large number of adult plants are available as starting material. Where large, old cultivations are not available, propagation by conventional methods is unsatisfactory as a method for commercial use.

To obtain mass multiplication, it is important to start from selected genotypes of top commercial value. As the first cultivations were developed by propagating the best genotypes, the genetic variability of the agronomical traits is limited.

When propagation by seed is used, the seed is collected in winter, when the berry that contains it is almost completely dried: it is then easy to remove the seed from the fruit. Once harvested, the seeds have to be washed thoroughly and then left to dry slowly under aerated conditions avoiding too rapid desiccation, with consequent loss or reduction of germination potentialities. From harvest to sowing, however, it is necessary to wait a certain number of months (7 to 9 or more) for the seed to be suitable for germination: during this post-harvest period, achievement of embryo maturity is important, and it takes place together with the activation of peroxidase enzymatic pathways that are subject of current research at our Institute. The seeds to be used are usually sown at the beginning of spring, but germination takes place only some months later, during the summer: seedling development and also growth are very slow processes, and commonly 5 to 6 years are necessary for the seedling to reach the size of a productive plant. The best substrate for sowing seeds and for seedling growth is a mixture of sand, peat moss, and beech leaves. Peat moss, in particular, appears to be more effective than other substrates, such as manure, because it releases the mineral elements the seedling root needs only very slowly; beech leaves are also rich in microelements. Three years after germination, the seedling is usually transferred to the field.

Propagation by tuft division gives a low multiplication rate (average number of plants from a single tuft is around three) but, on the other hand, it results in new plants ready to produce a crop much earlier than seedlings. The procedure for this propagation method is very simple. During dormancy (September-March), the mother plant rhizome is split into two or more pieces, each maintaining its own roots, buds, and aerial stems. The rhizome fragments thus obtained are treated with fungicide powder, and are then transferred into a substrate of the same composition as previously described for seedlings. In the absence of apical bud dominance, quiescent rhizome buds develop into new aerial sprouts and a wide production of new adventive roots begins at the same time. As a consequence of tuft division, generally a large number of pre-existing roots die, but they are replaced very soon by a new, abundant root system.

3 Micropropagation

3.1 General

Very few studies have been carried out on *Ruscus* micropropagation, and till now no other research appears to have been conducted on *Ruscus racemosus* in vitro cultivation. The few available reports concern another species, *Ruscus hypophyllum*, and we will briefly refer to the most significant results obtained in in vitro propagation of this species.

The paper concerning *Ruscus hypophyllum* in vitro micropropagation (Ziv 1983) reports the utilization of different types of explants for starting the culture: the author uses rhizome apical or lateral buds, inflorescence explants, and aerial stem segments. The nutrient substrate tested is basically a MS medium (Murashige and Skoog 1962) with modifications of the original vitamin formula, and with a different hormonal composition. The culture media utilized are liquid or agar-solidified. Liquid media are particularly useful when applied as a pre-conditioning medium to improve multiplication rates and to remove development inhibition in the explants due to the influence of apical dominance. Solid medium is described by the author (Ziv 1983) as being necessary for obtaining an optimal multiplication rate from pre-treated material (liquid medium).

Ruscus hypophyllum is described as a species showing good adaptability to in vitro culture. In particular, it shows a high ability for adventive bud regeneration, and a lower multiplication capability via axillary bud proliferation. According to Ziv (1983), for best bud regeneration response on solid medium, the explant tissues require high levels (5 mg/l) of 6-benzyl-aminopurine (BAP) together with 1 mg/l of 2,4-dichlorophenoxyacetic acid (2,4-D): these two hormones, on the other hand, have been demonstrated to act synergically to produce the most effective results in the previous pre-conditioning liquid culture at a concentration of 1 mg/l each. If pre-conditioning liquid culture is not carried out and the material is directly inoculated onto gelled media, other authors (Sumita and Sumitra 1985) report the possibility of rapidly obtaining abundant callus production, particularly from stem pieces cultivated in the presence of 1 mg/l 2,4-D and 0.1 mg/l BAP: an optimal regeneration rate for new shootlets from calli by this method is achieved by using a reduced 2,4-D concentration (0.05 mg/l) and increasing the BAP concentration to 0.5 mg/l. According to Sumita and Sumitra (1985), rooting of the regenerated explants can be obtained by transferring them onto half-strength MS medium with 2 mg/l indole-3-butyric acid (IBA); in the absence of the auxin, the same propagules show an extremely low rooting response.

Sumita and Sumitra (1985) report no difficulty in explant acclimatization to in vivo growth conditions for the in vitro regenerated and rooted explants, provided that they are transferred from the rooting medium to the same medium freshly prepared, but without hormones.

In vitro propagation of *Ruscus racemosus* has been a subject of research for many years at the Experimental Institute for Floriculture in Sanremo. Because of the low genetic variability in the germplasm cultivated, the importance of this propagation method consists in its great effectiveness as compared with traditional multiplication (in vivo) systems.

While reporting the large crop obtained by means of in vitro multiplication techniques, it is, however, necessary to remember that *Ruscus* micropropagation cannot, at least at present, accelerate the growth and development of this plant; the micropropagated plants, once transferred to the field, show the same slow growth pattern as the seedlings. Micropropagation, for this reason, can be preferable only because of the high quantity of plants that it is possible to produce with no significant seasonal fluctuations, under good phytosanitary conditions. No further advantages or improvements are evident in the micropropagated plants once they are acclimatized to in vivo growth conditions.

Micropropagated *Ruscus* plants, when ready to be transferred to in vivo growth conditions, seem to have the leaf features of an adult plant (Fig. 2): in fact, the plantlet has small rhizome-bearing cladodes (very short) and a certain number of buds, just like the adult plants. Although very young, the ex vitro plantlet does not show the typical seedling developmental pattern, characterized by the production of a single, elongated cladode during the first year.

It should be mentioned that the in vitro culture of this species induces a faster growth in the rhizome bud meristem subjected to micropropagation, in comparison with the developmental behavior of the seed apical meristem germinating

Fig. 2. *Ruscus racemosus* plants. *From left to right* ex vitro plantlet after in vivo acclimatization and plants from seeds, 1, 2, and 3 years old respectively

in vivo. This positive effect on meristem growth and development, however, is unfortunately not maintained by the micropropagated plant under in vivo conditions. In any case, this observation may indicate that it is possible to speed up the slow growth; but the conditions necessary to obtain this response are still unclear, so that in vivo forcing procedures applied to the plantlets may be effective.

3.2 Explants, Sterilization, Establishment of Culture

The possibility of using explants derived from in vitro-germinated seeds has been studied. The seed is contained in a globular berry, from which it is easily removable when the fruit is ripe. The seeds show hard integument, and it can be useful to treat them with concentrated H_2SO_4 for half an hour, or to leave them submerged in sterile water for several days, to soften the outermost coats and to increase their permeability to water. This fact, together with incomplete embryo maturation, is responsible for the very slow germination, even after the above treatments. This phenomenon takes place also when the seed is germinated on a gelled medium (Brooks and Hough 1958) and under right growth conditions (25 °C temperature, 12-h photoperiod). The time required for in vitro seed germination ranges from 4 to 9 months, which is similar to that required for germination in vivo. This variability depends on different uncontrollable factors, such as embryo immaturity or dormancy, presence of endogenous inhibitors, low moisture in the seed's inner tissues, and genotypic behavior. After chemical scarification or swelling in water, seeds were sterilized with 70% ethanol for 10 min, rinsed in double distilled water, and inoculated onto Brooks and Hough gelled medium (1958). The germination begins with the emergence of a thick root from the micropyle, followed, some weeks after, by the emission of a unique short cladode inserted on a small, tubular stem like a rhizome; the apical bud is at the insertion point of the cladode on the rhizome. This bud can be utilized, once excised from the original tissue, as primary explant to start in vitro culture.

Because of these disadvantages in using seeds as starting material, we found it more efficient to use other kinds of explants. The use of buds from adult rhizomes of in vivo-cultivated plants (Fig. 3) appears very suitable since it immediately produces explants able to be inoculated in vitro, without a long wait for in vitro germination; moreover, it is possible to take the material from the best in vivo plants, chosen for their good agronomical behavior and production ability.

Ruscus rhizomes grow one to several centimeters under the soil surface, and frequently they or their roots are wounded during the work carried out by the farmers (for example, weeding), so that the rhizome tissues often contain bacterial contaminants. Although these bacteria are not dangerous for the health of the in vivo-cultivated plants, they usually prevent establishing or starting any in vitro culture. For this reason, it is absolutely necessary to utilize, as explant donors, selected mother plants, healthy, unstressed, and with rhizome and roots undamaged. In any case, the absence of bacterial contamination should be tested in the explants on diagnostic peptone-containing media before using them for the in vitro propagation; most effective results can be obtained using a bacterial contamination detecting medium commercially known as Brain Heart Infusion, produced by Oxoid, Basingstoke, Hants, UK (Beruto et al. 1983).

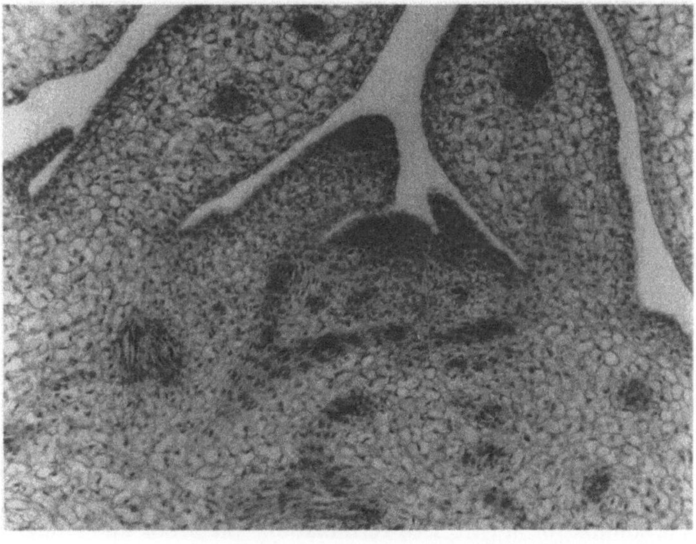

Fig. 3. Section of *Ruscus racemosus* rhizome vegetative bud (× 400)

From healthy rhizomes it is possible to obtain a large number of buds to use as primary explants. The rhizomes are cut into pieces 15 cm^3 wide approximately, and the fragments are washed under running tap water for at least 2 h. Each piece, usually bearing two or three buds, is then dipped for a few seconds into a water solution of 70% ethanol, and subsequently sterilized for 12 min in NaOCl solution, 2% active chlorine. After sterilization, the rhizome pieces are rinsed three times with double distilled sterile water to remove any chlorine residue. Buds are excised aseptically from the rhizome fragments, and transferred directly onto the gelled multiplication medium. Particular attention should be paid to the choice of buds on the rhizome surface to be inoculated: flowering buds must be discarded, since the aerial stem developing from these exhibits strong apical dominance, as a consequence of which no other vegetative meristems can develop; moreover, till now all attempts at obtaining a reversion of these buds to the vegetative phase by different hormonal treatments have been ineffective. Vegetative buds, on the contrary, can rapidly produce a rhizomatous structure with a variable number of leafy shoots or buds on the culture media; 7–8 weeks after culture initiation, from each single bud inoculated, a small rhizome with plagiotropic growth develops on the gelled medium surface: this new small rhizome can be split into pieces (each one with two or three shoots or buds) which, once transferred onto fresh multiplication medium, are able to increase their size every month and to produce new buds or shoots, which can be split in turn, thus continuing the in vitro propagation. In this way, it is possible to achieve satisfactory multiplication rates.

From rhizome tissue, on the other hand, good material for callus production results; it is possible to utilize rhizome parts with no buds, excised from in vivo-grown plants, or to use the new in vitro-formed rhizomes produced by the explanted buds. In the latter case, it is possible to omit the sterilization phase. Cladodes appear to be less sensitive to in vitro hormonal stimuli because of their cutinization: they are, in fact, able to produce callus from the area of insertion on the aerial stem, while the cells on the remaining cladode part show a limited ability to dedifferentiate into callus tissue. The medium utilized in our laboratory to induce callus formation from the explants was MS medium plus 2 mg/l 2,4-D. Research with the aim of obtaining new plantlet regeneration from cultured calli appears at present unsuccessful.

It is also necessary to remember here that in most cases it can be useful, before inoculation, to dissect the explants from in vivo- and in vitro-cultivated plants on sterile absorbent paper moistened with an antioxidant, such as ascorbic or citric acid, to limit the browning phenomena caused by endogenous phenol oxidation; the same benefits result from a preliminary washing of the explants, under sterile conditions, for 1 h in a water solution containing one of the previously mentioned substances, at a concentration ranging from 1 to 2 g/l.

3.3 Media: Multiplication and Rooting Response

Three different media were used for genotypes (clones) of *Ruscus racemosus*. The first, suitable for in vitro seed germination, was Brook and Hough medium slightly modified in microelement composition (Table 1); the second and third were MS, and an experimental medium, respectively (Table 1), the latter giving good results

Table 1. *Ruscus racemosus* multiplication media[a]. (Curir et al. 1988)

	Modified Brooks and Hough	Murashige and Skoog	Experimental
Micronutrients	BH	MS	[b]
Micronutrients	BH except $FeSO_4 \cdot 7H_2O$ and EDTA substituted with Fe Chelates of MS, and MoO_3 substituted with $Na_2MoO_4 \cdot 2H_2O$ 0.25 mg/l	MS	MS
Vitamin mixture	BH	MS	[c]
Hormonal substances (mg/l)	BAP 0.3	BAP 0.3	BAP 0.3
Sugar (g/l)	Sucrose 10	Sucrose 30	Sucrose 30
Agar (g/l)	5.0	8.0	8.0
pH	6.0	5.8	5.8

[a] Abbreviations: MS = Murashige and Skoog; BH = Brook and Hough; EDTA = ethylenediaminotetracetic acid.
[b] In mg/l; $MgSO_4 \cdot 7H_2O$: 300; $Ca(NO_3)_2 \cdot 4H_2O$: 800; KNO_3: 1200; NH_4NO_3: 300; KH_2PO_4: 250; KCl: 40; $NaH_2PO_4 \cdot H_2O$: 120.
[c] In mg/l: Ca-pantothenate: 1; nicotinic acid: 5; pyridoxine HCl: 1: thiamine HCl: 1; biotine: 0.5; myo-inositol: 100; folic acid: 0.5; L-cysteine: 2; ascorbic acid: 100.

Table 2. *Ruscus racemosus* in vitro explants: average multiplication rates on different media. Each mean is from the observations made on 100 explants

Clones	A		B		C	
Media	MS	exp	MS	exp	MS	exp
Subcultures						
1st	1.5	1.6	1.5	1.7	1.4	1.9
2nd	2.1	3.2	2.3	3.5	2.9	3.8
3rd	2.7	3.6	2.9	4.3	3.5	5.2
4th	3.4	4.2	3.8	4.9	3.9	6.7
5th	3.6	4.2	3.9	5.1	4.0	6.8
6th	3.7	4.5	4.1	4.8	4.2	6.7
7th	3.5	4.9	4.0	5.2	3.9	6.9
8th	3.3	4.4	4.0	4.4	3.2	5.6

– Differences between media effects (MS vs. exp): *
– Differences among the clones: A vs. B: n.s.
A vs. C: *
B vs. C: n.s.
* = Significant differences according to the Tukey statistical test for P = 0.05.
n.s. = No significant differences.
MS = Murashige and Skoog medium.
exp = experimental medium (Curir et al. 1988).

(Table 2). Both media contained 0.3 mg/l BAP to stimulate bud meristem proliferation, but they differed mainly in composition and concentration of macroelements and salts and, in the experimental medium, in organic elements. Standard growth conditions were 23 °C temperature, 16-h photoperiod, 3000 lx irradiation; light does not seem to exert a great influence on the explants' growth, as this species likes the shade. The duration of one subculture was about 1 month; the proliferating explants were divided monthly, and the propagules thus obtained were re-inoculated onto fresh medium, after counting the number of new buds or shootlets produced since the beginning of each subculture (Fig. 4).

The results of the trials can be summarized as follows. The experimental medium gave the highest multiplication rate (Table 2). There is a lag phase immediately after inoculation (first subculture), during which the buds show no appreciable activity, because of a period of adaptation to the new growth conditions. At the end of this lag phase, the buds begin to multiply, initially (for the first two or three subcultures) at a slow rate, then increasing progressively, stabilizing around values of four to six new buds/explant for the experimental medium, and around three or four new buds/explant for the MS medium (Table 2). The experimental medium has multiplication rates higher than MS, but it also produces shorter new shoots, that grow slowly. Statistical analysis of the data shows that there is clone-medium interaction, i.e., propagative responses on the two media differ, depending on the clones used. Thus, as in other species, in *Ruscus racemosus* there is a different genotypic response to in vitro growth conditions. Trials with Brooks and Hough medium used as multiplication medium have also been carried out, but no acceptable results have been obtained.

Fig. 4. *Ruscus racemosus* in vitro multiplication phase on MS medium

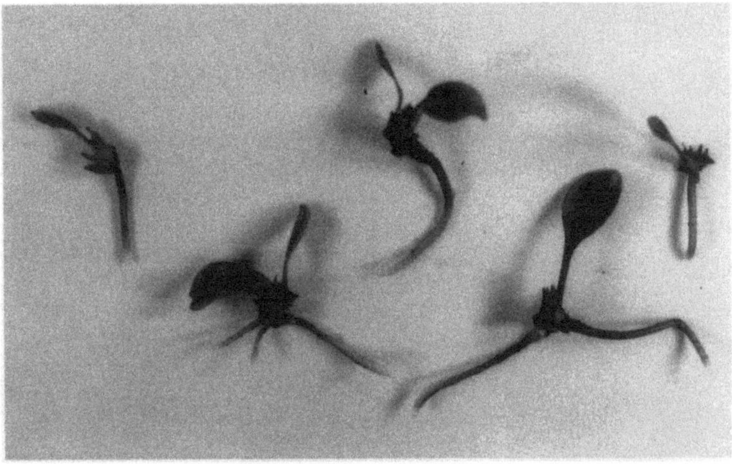

Fig. 5. *Ruscus racemosus* in vitro-rooted explants

Ruscus is a plant with no in vitro rooting problems. When the propagules, each one consisting of a small rhizome with leafy shoots, are left unsplit on the same multiplication medium for at least 45–50 days, they can root without auxin, simply by aging (Fig. 5). Development of adventitious root, however, occurs only if the explants have more than two well-developed leafy shoots, as is reported for the closely related species *Ruscus hypophyllum* (Sumita and Sumitra 1985). The rooting rates are always close to 100%. The addition of a solution containing NAA 2 mg/l with or without 30 g/l sucrose to the gelled medium surface (double layer culture) shortens the rooting time (rooting period in this case, 30 days); moreover, the roots obtained by double layer culture appear more abundant and better than those obtained otherwise.

Recently, physiological aspects of rooting have also been investigated for this plant cultivated in vitro. We found a high flavonoid level (particularly quercetin and quercetin glycosides) in the explants during the multiplication phase, that increased during the rooting phase: it is possible that one or more of these substances is related to the physiological rooting state, or even that one of these substances acts as a rooting factor/cofactor such as in the case of *Eucalyptus* cultivated in vitro (Curir et al. 1990). There is no doubt that phenolase activities were also involved in the different micropropagation phases in *Ruscus racemosus* (Curir et al. 1989): the browning of the tissues, especially immediately after the splitting of the explants, occurs rapidly. Although in other micropropagated species the browning phenomenon can represent a serious problem (Durand-Cresswell et al. 1982), in this case they are not harmful for the development of the explants. During the rooting phase of *Ruscus*, in particular, phenolase activity increased in clear relation to rooting response. The ability of phenolase to modify the endogenous phenol structure, making them able to act as auxin synergists, could also be considered (Haissig 1973).

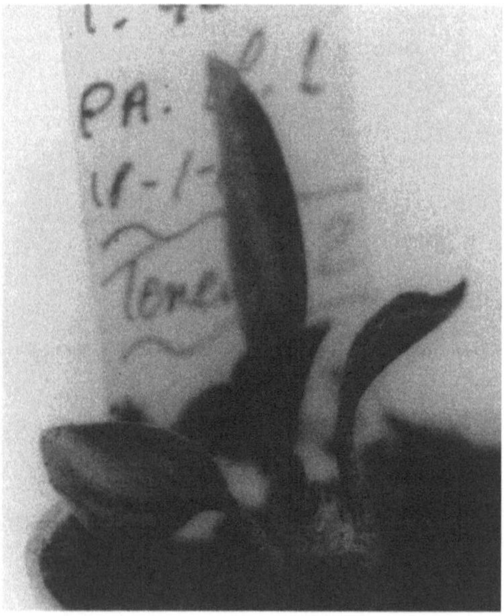

Fig. 6. *Ruscus racemosus* ex vitro plantlet acclimatized to in vivo growth conditions

3.4 Transfer to the Greenhouse, Acclimatization

The in vitro-rooted explants need a period of acclimatization before being transferred to in vivo growth conditions. This period is generally no longer than 1 month: the rooted explants are transferred into the greenhouse, potted in Jiffy pots containing perlite, beech leaves, and peat moss 1 : 1 : 1 (Fig. 6). Other substrates tested, such as different combinations of perlite, vermiculite, peat moss, and sand, were not so effective for acclimatization.

High moisture is supplied by a sprinkling mist system, to prevent dessication of tissues and leaves. Acclimatization carried out under tunnels made of plastic covers (polyethylene) gives good results, provided that the substrates containing the plantlets are kept constantly wet. During the acclimatization period, it is particularly useful to irrigate the potted plants with a 0.1% salinity nutrient solution composed of N:P205:K20 in the ratio 1:0.3:0.6.

After 1 month, the plantlets are ready to be transferred to in vivo conditions. In this phase, periodical (monthly) foliar sprays with water solutions containing BAP 1000 mg/l are useful to stimulate faster plantlet bud development.

4 Summary and Conclusions

Ruscus racemosus is a species cultivated for green foliage production that has recently shown important development. This increase of acreage depends on the agronomical characteristics of the plant and, of course, on the market demand.

Growers are faced with the problem of the slow growth of seedlings, that takes 6 to 7 years until cropping. The usual multiplication procedures are based on vegetative propagation starting from old plants (tuft division). In this way, it is possible to increase cultivation only two- or threefold, which does not meet the needs of commercial production. For this reason, in vitro multiplication appears an interesting method for mass propagation to supply the demand of growers.

5 Protocol

Some protocols for in vitro multiplication have been tried. The basic steps of this method have been set up, and the results appear satisfactory to implement industrial propagation. The multiplication rates are around four to five (average data) per month, and rooting steps take place easily with aging of the culture. Among the tested in vitro propagation protocols we suggest the following, used routinely for the multiplication of this species in our laboratories.

1. Select and harvest in vivo-cultivated rhizomes.
2. Wash rhizomes for 3 h under running tap water, to remove coarse soil particles.
3. Wash the material for a further 1–2 h in distilled water containing a few drops of detergent (e.g., Tween 20), and rinse thoroughly with distilled water.
4. Split rhizomes into fragments with vegetative buds.
5. Surface sterilize rhizome pieces with ethanol 70% for a few seconds; subsequent treatment with NaOCl solution 2% active chlorine for 12 min; rinse sterilized pieces thoroughly (three to five times) with double distilled water.
6. Rhizome buds, excised from rhizome pieces with a scalpel blade, are inoculated directly onto the multiplication medium, composed of Murashige and Skoog (1962) macro-, microelements, and vitamins, sucrose 30 g/l, Difco Bacto agar 8 g/l, pH 5.8. Growth conditions: 23 °C temperature, 16-h photoperiod, 3000 lx irradiation.
7. Renew exhausted medium each 10 days for the first 2 months to shorten the period needed for adaptation.
8. Split the small tufts (propagula) originating from the initial explant, and transfer the leafy shoots onto fresh multiplication medium. This step can be repeated every 1–2 months.
9. Roots the micropropagated explants (propagula) simply by the aging of the culture, i.e., leave the explants undivided on the same multiplication medium for 1–2 months; to hasten the rooting process, add to the gelled medium surface a thin layer (0.5 cm) of a solution containing naphthalene acetic acid (NAA) at a concentration of 2 mg/l.
10. Transfer rooted explants to the acclimatization greenhouse in a substrate composed of perlite, beech leaves, and peak moss 1 : 1 : 1 : (v/v), at temperature 20–24 °C, under intermittent mist spraying for 30 days (90% relative humidity).

Unfortunately, in vitro plantlets once transferred to in vivo growth conditions show slow development, similar to plants obtained from seeds. At present, in vitro propagation is a useful method to obtain mass propagation but it remains ineffective for reducing the time until first cropping.

Genetic variability in the plants grown for foliage production, when biometric characteristics are taken into consideration, appears quite low. Nevertheless, different genotypes subjected to in vitro culture show different behavior, suggesting a specific genotype response to this method of propagation.

References

Accati E (1983) Alcuni aspetti relativi alla coltura di *Danae racemosa* Moench (*Ruscus racemosus* Moench). In: CCIAA (ed) Atti Convegno Giornata di Studio sulle Piante da Fronda Recisa. Zerbone Press, Bordighera, pp 53–59

Bailey LH (1958) *Ruscus* genus. In: Bailey LH (ed) The standard cyclopedia of horticulture, vol 3. McMillan, New York, pp 3037–3038

Beruto M, Curir P, Lercari GB, Giusta R (1983) Propagazione in vitro di *Cordyline terminalis* L. Calypso. Ann Ist Sper Flor S. Remo 14(1): 71–77

Brooks MJ, Hough LF (1958) Vernalisation studies with peach embryos. Proc Am Soc Hortic Sci 71: 95–102

Cooney R, Sovetts C, Sattler R (1987) Phylloclades development in the Asparagaceae: an example of homoeosis. Bot J Linn Soc 94(3): 327–371

Curir P, Damiano C, Esposito P, Ruffoni B (1986) Propagazione in vitro di *Ruscus racemosus* Moench. Ann Ist Sper Flor S. Remo 17(1): 95–103

Curir P, Damiano C, Esposito P, Ruffoni B (1988) In vitro propagation of *Ruscus racemosus* Moench. Acta Hortic 226: 217–222

Curir P, Termini A, Ruffoni B, Marchesini A (1989) NAA effects on phenolase activity in *Ruscus racemosus* Moench during in vitro propagation. Acta Hortic 251: 135–139

Curir P, Van Sumere CF, Termini A, Barthe Ph, Marchesini A, Dolci M (1990) Flavonoid accumulation is correlated with adventitious roots formation in *Eucalyptus gunnii* hook micropropagated through axillary bud stimulation. Plant Physiol 92: 1148–1153

Durand-Cresswell R, Boulay M, Franclet A (1982) Vegetative propagation of *Eucalyptus*. In: Bonga JM, Durzan DJ (eds) Tissue culture in forestry. Nijhoff-Junk, Dordrecht, pp 150–181

Farina E (1987) I *Ruscus* ed i *Cyperus* come specie da fronda recisa. In: ACE INT (ed) Atti Convegno di Padova sul verde come complemento al fiore reciso. Rogenese press, Garbagnate, pp 27–32

Fisher JB (1972) Control of shoot-rhizome dimorfism in the woody monocotyledon *Cordyline* (Agavaceae). Am J Bot 59 (10): 1000–1010

Haissig BE (1973) Influences of auxins and auxins synergists on adventitious root primordium initiation and development. NZJ For Sci 4(2): 311–323

Hussey G (1977) In vitro propagation of some members of the Liliaceae, Iridaceae and Amaryllidaceae. Acta Hortic 78: 303–309

Lodi G (1966) *Ruscus aculeatus*. In: EDAGRICOLE (ed) Piante officinali italiane. Azzoguidi press, Bologna, pp 598–599

Murashige T, Skoog F (1962) A revised medium for rapid growth and bioassays with tobacco tissue culture. Physiol Plant 15: 473–497

Nicholson G, Mottet S (1896–1897) Le genre *Ruscus*. In: Doin O, Vilmorin–Audrieux (eds) Dictionnaire pratique de horticulture et de jardinage, vol 4. Librairie Agricole, Paris, pp 585–586

Nooh AE, El-Kiey T, Khattab M (1986) Studies on the keeping quality of cut green *Ruscus hypoglossum* L. and *Nephrolepis exaltata* Schott. Alexandria Univ Publ 181: 223–229

Sato D (1942) Chromosome number in *Ruscus racemosus*. Jpn J Bot 12: 57

Sumita J, Sumitra S (1985) In vitro regeneration of *Ruscus Hypophyllum* L. plants. Plant Cell Tissue Organ Cult 5: 79–87

Synge PM (1956) *Ruscus* genus. In: Chittenden FJ (ed) Dictionary of gardening, Royal Horticultural Society – 2nd edn, vol 4. Clarendon Press, Oxford, pp 1840

Ziv M (1983) The stimulatory effect of liquid induction medium on shoot proliferation of *Ruscus hypophyllum* L. . Sci Hortic 19: 387–394

XXIII Micropropagation of African Violets (*Saintpaulia* spp. and cvs.)

F. JUNGNICKEL[1] and S. ZAID[2]

1 General Account

1.1 The Plants

In the *Gesneriaceae*, the genus *Saintpaulia* H. Wendl. includes some 20 spp. or other taxa of perennial herbs (e.g., Burtt 1958; 1962; 1964; VanWilson 1979/1980; Zalesskij 1983), the stems of which are either short and thick and slowly elongating or procumbent, rooting at the nodes with distinct internodes. The more or less fleshy petiolate leaves show suborbicular to elliptic blades and are located opposite in the caulescent growth habit. The rosulate species, however, form a dense terminal crown of leaves. Leaf indumentum contributes to the differentiation between species. Short-stalked or subsessile glands are without taxonomic value in the genus. The eglandular hairs, however, may be long or short, erect or appressed, and pure as well as mixed types in both posture and distribution are found (Burtt 1956, 1958). *Saintpaulia* spp. from wild sources are remarkably constant in flower-form; at the mouth of the short corolla tube, two large yellow anthers are always clearly visible. Normally, *Saintpaulia* flowers show an irregular limb with two upper lobes and three lower ones (Fig. 1), but regular flowers with five equal corolla lobes and five fertile stamens may occur abnormally in species, and this type can be found in distinct cultivars. Saintpaulias show enantiostyly, which means that the style is turned to one side, but both directions are found in different flowers of the same plant (Mabberly 1987). The fruits vary from ovoid to linear-cylindric capsules. Peduncles occur in the leaf axils and bear one to several flowers borne cymosely with two flowers opening serially at each branching order in the inflorescence, and small linear bracts are found. In the flower anthocyanins (Khokhar et al. 1982; Hoshino 1986), both light and dark forms of *S. ionantha* H. Wendl. occur in the wild. The two upper lobes of *S. goetzeana* Engl. are said to be white and the three lower ones are marked with pinkish violet.

Most chromosome records show 2n = 30 and in some cvs. 2n = 60 (e.g. Rogers 1954; Ehrlich 1958; Fussell 1958; Ratter 1975); reports of 2n = 28, 56, 112 (Sugiura 1936; Holzer 1952; Hughes et al. 1975; Espino and Vazquez 1981) are believed to be erroneous (Wilson 1951; Ehrlich 1958; Fussell 1958), particularly since *Saintpaulia* chromosomes are small and difficult to count. Milne (1975) reported that the

[1] Friedrich-Schiller-University Jena, Institute of General Botany, von-Hase-Weg 3, O-6900 Jena, FRG
[2] Damascus University, Faculty of Sciences, Dept. of Botany, Damascus, Syria

Biotechnology in Agriculture and Forestry, Vol. 20
High-Tech and Micropropagation IV (ed. by Y.P.S. Bajaj)

Fig. 1A. The flower shape of this still unnamed breeding clone represents original *Saint-paulia* features (color of the clone: white with a pink corolla tube). **B** The cv. Romanze shows some variability in individual flower shape (color of the cv.: blue). **C** The cv. Menuett represents another type of flower shape (color of this cv.: pink **A–C** courtesy of S. Prkno, Bautzen, FRG.

hybrid *S. confusus* B. L. Burtt × *S. orbicularis* B. L. Burtt shows perfect chromosome pairing at meiosis and normal fertility, indicating an absence of barriers to gene exchange between the parental species. Sugiura (1940) and Ratter (1975) discuss that the basic ancestral number of most of the Old World *Gesneriaceae* (including *Saintpaulia*) was x = 8 or 9, and according to Ratter (1975), the numbers n = 15–18 are of tetraploid origin.

1.2 Distribution and Importance

Saintpaulias are characterized by high local endemism in the crystalline mountains of East Africa, especially Tanzania: ten species are found in the East Usambara Mountains and outlying areas, and four in the West Usambara Mountains. The total genus occupied an area no greater than about one-fifth of the British Isles (Burtt 1958). Usually, they are found in wet places in forests under diminished lighting conditions and on rocks covered with moss. *S. rupicola* B. L. Burtt, however, is found in limestone out-crops in more open country in southeast Kenya (Mabberly 1987).

On its original stands, *Saintpaulia* spp. have become rare, but under the common name African Violets, they have attained amazing horticultural popularity, and numerous cultivars are now internationally important ornamental pot plants (for instance, see VonDerBeek 1990a, b). For the history of the introduction of *Saintpaulia* in 1892 by Baron St. Paul see Burtt (1947) or Roberts (1952). The first discovered species was *S. ionantha* (Wendland 1893), and "This species is the basis of the African violet mania .... The number of named variants that have been raised from seed is a remarkable witness to the hidden potential of a species which is not unusually variable in the wild." (Burtt 1958). Some smaller species are also popular as novelties now (VanWilson 1979/1980; Mabberly 1987). Since there are "secrets of the trade" in all businesses, some hybridizers will not reveal the parentage of their hybrids (Johnson 1987), and this often results in an uncertain pedigree of cultivars. In addition, the above-mentioned lack of genetic barriers between *Saintpaulia* spp. is regarded pragmatically by local, regional, or national African violet societies. They are only concerned with standard plants, semi-miniature plants, miniature plants, and trailers with either single, double, or semi-double blossoms having numerous flower colors including two-tone, multicolor, or edged types. Flower colors, as well as green or variegated foliage, might be of homohistic or chimeric nature. Some African violet societies have their own publications. For instance, many *Saintpaulia* enthusiasts contributed to the fact that in 1987 the African Violet Society of America, Inc. could celebrate 40 years of continuous publication of the *African Violet Magazine*.

1.3 Conventional Propagation and Breeding

1.3.1 Conditions of Cultivation

Besides their aesthetic appeal, saintpaulias enjoy great popularity because their optimum day and night temperature of about 20 °C (Andersen 1983) is that of rooms where human beings live or work. In addition, about one-tenth of the full daylight was found to be the most suitable lighting, and such a condition can be found in most homes. However, saintpaulias show greatest dry matter production at high relative humidity (Gislerød et al. 1986; Amberger-Ochsenbauer 1990). Thus, the air condition may be a limiting factor, especially in the case of central heating. Window gardens are, however, recommended (Naylor 1940). At the shows of African violet societies, enthusiasts compete for amateur awards and each plant is judged on its own individual merit (Green 1987).

In commercial pot plant production (Bjerre and Amsen 1983; TerHell and Ludolph 1990), saintpaulias are grown in greenhouses which are shaded in the northern hemisphere from March to October. Moving belts (Anonymous 1971) or plant rotators (Bosse et al. 1971) may increase daylight utilization as well as greenhouse capacity. Despite a possible light acclimatization (Conover and Poole 1981), Holtkamp (1987) recommends artificial lighting (40 W/m^2 for 12 h) from November to January, when natural daylight is below 4000 lx. Plant quality was best at 40 W/m^2; however, the most economical treatment was 20 W/m^2 (Andersen and Aabrandt 1989). For combined actions of lighting and other factors see Michel (1988). Flowering came earlier and the colors were more intense under sodium than under mercury lamps (Hagen and Nilsen 1978), but saintpaulias are fairly insensitive to ozone (Cathey and Heggestad 1982), which might be produced by some types of lamps. Above the optimal range, light affected rooting and yield adversely (Carpenter 1975). Saintpaulias utilize additional CO_2, which might be supplied by the atmosphere (500–1000 ppm: Mortensen 1985; Granger and Laury 1987; Holtkamp 1987) or via the irrigation water (Molitor et al. 1986). The interaction of CO_2 with irradiance and temperature is reviewed by Heins et al. (1984). Peat or weakly acidic soil mixtures with a high peat content are the most desirable substrates (Fischer 1977; Worrall 1981; Beutel 1990). Since saintpaulias do not like salts (Nölle et al. 1975) at concentrations higher than 1.2 g/l, nutrients should be supplied carefully, depending on the depletion (Andersen 1984; Sanderson et al. 1985; Hendriks et al. 1986; Poole et al. 1986). Slow-release fertilizers are recommended (Milde 1980; Schwemmer 1990). Since saintpaulias do not make great demands on the water quality (Hörnis et al. 1984), successful hydroculture was also reported (Schwemmer 1980; Fischer and Hentig 1982). Transferring a plant from soil to hydroponic culture or vice versa, however, resulted in the replacement of the original roots by new ones with the structure typically found in the new medium (Wild 1974). Holtkamp (1987) gives an optimum pH between 5.5 and 6. For recommendations see also Lutz (1990).

Because of their fairly short period of cultivation, saintpaulias are particularly energy efficient pot flowers when propagated in March, and moderate energy requirement was found in the cases of July/September propagation (Kühn 1984). Promising results were also reported from winter cultivation (Steinbuch 1982). Suboptimal temperature, however, elongates this period considerably and a dose/lethality-nomogram was obtained in cooling tests at 1 to 8 °C for 1 to 96 h (Bodner and Larcher 1987; Larcher and Bodner 1980, 1987). In greenhouses with insufficient heating capacity this might occur, especially in arctic winters. If plants survive, leaf scorching and malformation is induced by temperatures falling to 7 °C (Axelsson 1975; Maekawa et al. 1987, 1990). Chlorophyll fluorescence was found to be a sensitive marker for chilling susceptibility (Larcher and Neuner 1989).

However, saintpaulias are fairly resistant to the stress factor ethylene (Woltering 1985). Since heating costs, yields, and profits are strongly correlated (Bazzochi et al. 1983), numerous papers exist which deal with energy saving by double glazing, triple Plexiglas, plastic films, insulation sheets, movable curtains, etc. (e.g. Onstenk 1976; Konradsen et al. 1978; Biermann 1984; Röber et al. 1984). Close spacing (Buurman 1973; Bjerre 1982) or growing miniature saintpaulias (Steven 1985) contributes to better utilization of glasshouse area and fuels.

1.3.2 Propagation

Homozygous *Saintpaulia* cultivars or species may be propagated true to type by seed. Advantages of using pellet seed are discussed by Schmidt (1981). Since, in the natural environment of the species, specific insects open the anthers, artificial pollination is necessary if seeds are expected in cultivated plants.

Except for chimeras, most cultivars can be propagated easily and true to type by adventitious shoot formation (Escher and Rehrmann 1990) at the base of leaf cuttings, which do not bear any axillary bud. The average number of plantlets produced per leaf was 3.8 with shorter petioles and 3.1 with longer ones (Batson 1972). However, in the latter case, the leaf can be used again as a new cutting. Hentig and Fischer (1987) recommend rooting at 21 °C. Number and length of roots were also always greatest in cuttings with the shortest petioles (Nowaczyk and Borys 1976). For fertilization of saintpaulias see Hendriks and Scharpf (1988). High nitrate content in a liquid medium reduced growth, but potassium at 50 mg/l mitigated this effect (Edwards and Thomas 1985). The quality of stock plants was influenced by light quality, but in cuttings, higher light intensities stimulated adventitious shoot development (Schneider-Moldrickx and Amberger 1982). Röber and Fischer (1980) obtained good results with several substrates except polyurethane granules. In the summer, individualized plantlets must be shaded considerably (Tooze 1986). Rooting of cuttings was increased by auxins (Ailincai 1974; Woszynska and Borys 1976a, b) or by encapsulated chlormequat (Read et al. 1973), but also by laser treatment (Bazzochi and Rossi 1980). In the case of periclinal chimeras, ramification of the plants might be induced with maleic hydrazide (Lemattre 1975) or by removal of the shoot tips to release lateral buds from apical dominance (Larson 1985). Growth regulator sprays are also recommended (Sanderson 1987; Sanderson and McGuire 1988).

1.3.3 Diseases and Pests

Most important *Saintpaulia* diseases are briefly reviewed by Krebs (1990). Flower infection with *Botrytis cinerea* can be controlled, for example, by Ronilan (vinclozolin) smoke treatment (Rewinkel-Jansen et al. 1985). Mildew (*Oidium* spp., Erysiphaceae) was controlled by Nimrod (bipirimate), but some fungicides damaged the flowers (Hemer 1978). Footrot in Saintpaulias caused by *Phytophthora nicotianae* var. *parasitica* and *Pythium spinosum* can be prevented by nursery hygiene and by ensuring that the potting soil is not too wet and does not have too high a salt content (Dirkse 1981, Folk 1983). For biological control of *Phytophthora*, *Pythium*, and *Phomopsis*, see Salka and Thinggaard (1988) and Thinggaard (1989). Heimann (1977) demonstrated the prophylactic effects of Previcur (prothiocarb). Busch and Smith (1978) recommend Aliette (aluminumethylphosphite) and Ridomil (metaxadine) to prevent this disease. Droplet size influences the effectiveness (Lang 1987). DeWaard and VanRooijen (1978) obtained control of rots for at least 6 weeks by a single application of Aliette or furalaxyl either as a soil drench or incorporated into the potting soil. Aliette was not fungicidal, but fungistatic (Krebs 1983). Combined drenching and pre-planting treatments with metalaxyl gave excellent control of

Phytophthora (Raabe and Hurlimann 1978; Ploetz and Engelhard 1980). Dexon (0.04%) did not damage the plant but kept inoculated plants free from crown and root rot (Plate and Kröber 1972). Except sodium hypochlorite, all of seven disinfectants and a cleaning agent killed the above fungus after a 16-h treatment (Wohanka and Doll 1985). Black spot is attributed to *Diplocarpon rosae* (Anonymous 1975a), and Holtkamp (1987) draws attention to *Rhizoctonia solani* or the combined attack of *Cylindrocarpon* spp. with nematodes, and application of Benomyl and Roveral is recommended.

The bacterium *Erwinia carotovora* var. *chrysanthemi* causes a severe blight of the crown, petioles, and leaves, and resulted in the loss of over 50% of the plants in some nurseries (Köhn 1974). Since there is no certain possibility of control (Henseler 1976), the spread of this disease through vegetative propagation can be prevented by rigorous selection of leaf cuttings from healthy, indexed mother plants (Lemattre and Narcy 1975).

Infestation of *Saintpaulia* cuttings with the root-knot nematode *Meloidogyne arenaria* caused massive callus overgrowth which developed on the root collar, and infested roots rotted rapidly (Maas and Cevat 1973). Brown spots on the lower surface of the leaves indicate another nematode, *Aphelenchides olesticus* (Holtkamp 1987). The strawberry stem nematode, *Ditylenchus dipsaci*, produces numerous malformed, almost hairless leaves on slowly growing saintpaulias. Holtkamp (1987) recommends Vydate, E 605 or Metasystox against nematodes, including *Heterodera marioni* at the roots. The mite *Tarsonemus pallidus* disturbs growth and colour of the leaves, but as contrasted with effects of *Ditylenchus*, leaf hairs are developed intensively. Umgelter (1979) draws attention to (unspecified) root aphids. Recently, Thrips (*Thrips tabaci*/*Franklinella occidentalis*) became one of the most unpleasant pests in saintpaulias. Especially the anthers are destroyed, and pollen on the petals indicates its occurrence (Tusnádi 1986; Holtkamp 1987). Böhmer and Eilenbach (1987) reported on resistant strains, but effective control was realized with synthetic pyrethroids (VanDeVrie 1985). In Thailand, a new species of mealy bugs of the genus *Rhizoecus* was found on saintpaulias by Williams (1985). Symptoms of most important pests on saintpaulias are demonstrated by Böhmer (1990).

In regard to fungicides and pesticides, saintpaulias are usually more sensitive than other ornamentals to spray treatments (e.g., Richter and Winkler 1973; Ignasse 1984; Langnes and Gjaerum 1985). Measurements against fungal infection of peat culture substrates are also recommended (Röber et al. 1985).

1.3.4 Breeding

Different susceptibility or resistance of *Saintpaulia* cvs, was reported concerning powdery mildew (*Oidium* spp.: Strider 1980), *Phythophthora nicotianae* var. *parasitica* (Strider 1978; 1979a), *Erwinia caratovora* var. *chrysanthemi* (Anonymous 1973; Köhn 1975; Roberts 1977) or the nematode *Aphelenchoides ritzemabosi* (Strider 1979b; Jones and Strider 1985), and breeders take this into consideration (Fig. 2). The number of flowers and pedicels per plant or buds per peduncle as well as plant diameter, flower size, and the time taken from potting to the plants being

Fig. 2. Evaluation of Saintpaulia breeding clones. Courtesy of Zierpflanzen, Dresden-Pillnitz, FRG.

marketable, are important breeding aims (Schnell 1972; Loeser 1973, 1983), but different groups of cvs. (see Sect. 1.2) as well as the demands of the market (DeGelder 1991) must be taken into account. For example, in 1986, 5.3% more standard-types but 37% more miniatures were sold in comparison to 1985 (Elgner 1987). Petals of modern varieties do not fall early, thus ensuring full blossoms over a long period of time. Many of the very first miniatures, however, still showed this disadvantageous feature (Fig. 3), and they were called wild types. Especially semi-miniature types may show very short petioles, which trace into soft leaf blades. Such so-called girl types differ distinctly in foliage from the boy types, and numerous other features allow full specific descriptions of thousands of cultivars, most of which are patented or protected by breeders' rights systems (Barton 1982).

Besides conventional hybridizing (Green 1987; Prkno 1988, pers. commun.), variability in plants obtained by vegetative propagation (for review see Buiatti and Roselli 1979) is also a considerable source of new varieties (Engler 1988, pers. commun.). Variability in adventitious shoots of common leaf cuttings may be increased by mutagenic irradiation or chemicals (e.g., Warfield 1973; Hentrich and Beger 1974; Pohlheim and Pohlheim 1976; Pohlheim 1980; Kelly and Lineberger 1981; Duron and Dixon 1982; Leenhouts et al. 1982; Relichova 1984).

Attempts to develop *Saintpaulia* cvs. which will grow well at 14 °C or even at lower temperatures (Peper et al. 1977a,b; Englert 1980; Sterling and Molenaar 1982; Zimmer 1985) seem to have been unsuccessful until now. For this purpose, it might be better to remember some cold-resistant Eurasian tertiary relicts (Ferns 1979, Jungnickel 1984, 1989, Halliwell et al. 1987).

Fig. 3. Axenic adult plant of the miniature cv. Hänsel, grown on phytohormone-free nitrate medium according to Jungnickel (1988a). (Figs. 3–12: Friedrich Schiller Univ., Jena)

2 Tissue Culture Studies (Table 1)

Numerous commercial growers accept only plantlets propagated through tissue culture. For instance, Müller (1988) had only positive results. The plants grow more rapidly, and by normal measurements of nursery hygiene, they need less chemicals. The cultivation period of transferred and individualized tissue culture plants for reaching sale maturity may take 4–6 weeks for the miniatures, but 6–8 weeks for the standard types (Müller 1988). However, considerable greenhouse areas, heating costs, and plant protection measurements are necessary to cultivate sufficient mother plants throughout the year for conventional propagation, and nobody can compete with the rapid changes in the total assortment which have now become possible through axenic mother plants and propagation techniques. Once somebody has cultivated saintpaulias in vitro, he can be sure that this is an ideal subject for demonstration; most morphogenetic reactions can be demonstrated more easily or even better with saintpaulias in comparison to tobacco cultures. Tissue culture-derived haploids or dihaploids can be used immediately as smaller, but marketable varieties, which might be the basis of further somatic variations. Mutagenesis, however, can also be realized more easily under axenic conditions. Somaclonal variation as well as protoplast fusion or direct gene transfer are inconceivable without axenic conditions. These methods, however, may provide for fundamental breakthroughs, especially if other gesneriads are involved which have shown features unknown until now in the African violets (e.g., yellow color or cold resistance).

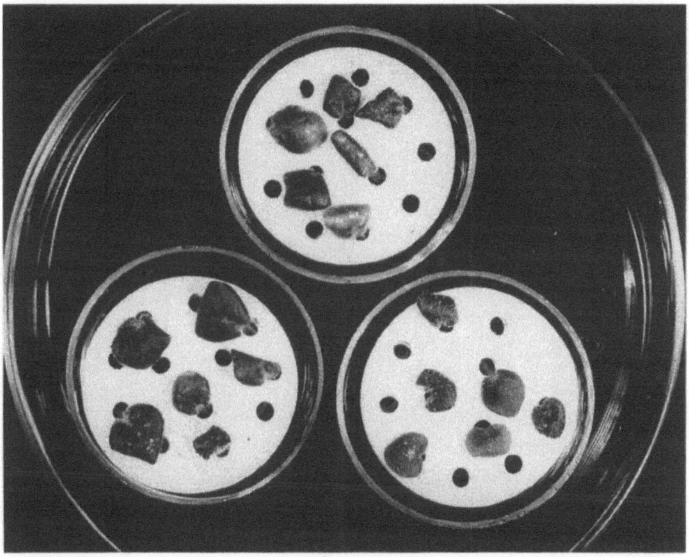

Fig. 4. In vitro mutagenic treatment of intact axenic leaves (leaf blade area about 1 cm^2). Only the lowermost parts of the petioles are dipped into the solution of a mutagenic agent (Jungnickel 1977a).

Table 1. Selected references on the use of tissue culture techniques in saintpaulias

Explant	Aim of the study	Medium	Important additives (mg/l)	Response/observation. data (mg/l)	Reference
Meristem tip	Virus eradication	MS or Berthelot and Knop		Both media were useful	Paludan (1971)
Lamina or petioles	Organogenesis	MS	Various combinations of kin, IAA, NAA	Isolated petioles without morphogenetic capacity, leaves form shoots and later on roots	Kukułczanka and Suszynska (1972)
Lamina	Organogenesis	MS	Morphactin 1, 0.1, 0.01 plus kin, IAA, NAA combinations	Morphactin disturbs morphogenesis but increases callusing	Kukułczanka and Kwiecien (1973)
Petioles	Organogenesis		*Agrobacterium tumefaciens* T37 and B6	Agrobacteria cause callusing and shooting, but no response in the controls	Rao and Morel (1973)
Lamina, buds	Organogenesis, in vitro storage, micropropagation	Specified nitrate medium	Phytohormone-less or BA max. 0.2	Germ plasm preservation and adventitious root/shoot formation without exogenous phytohormones	Jungnickel (1974)
	Micropropagation			Review	Stokes (1974)
Anthers	Haploid production	Media specified	IAA, kin, IBA	18% haploids obtained from four cvs.	Hughes et al. (1975)
Lamina, epidermis, subepidermal tissue, petioles	Commercial propagation and storage of pathogen-free cvs.	B5	Various BA and NAA	Propagation and maintenance of valuable cvs., no response without phytohormones	Flores et al. (1976)

Petioles	In vitro regeneration	LS	BA 0.4 IAA 0.2 2,4-D 0.1	Shooting via callusing, rooting on IAA-medium	Grunewaldt (1976)
Seeds, lamina	In vitro storage and propagation	specified nitrate medium	Phytohormonefree or BA max. 0.2	6 years of axenic storage, phytohormones, vitamins or NH_4 never added	Jungnickel (1976a)
Petioles	In vitro regeneration	See above	Various BA and NAA	0.03 NAA mimics action of lamina, roots < 0.01 but < 1 NAA, shoots 0.01–1 BA	Jungnickel (1976b)
Leaf sections	Micropropagation	MS, specif. rooting medium	BA 5 mg/l NAA 1 mg/l AdS 57 mg/l	Rooted plants flowered after 4 months, one leaf resulted in 500 plants after 7 months	Start and Cumming (1976)
Lamina	Micropropagation	MS	Various growth substances	Successful propagation of 13 cvs. but MS in some cvs. unsuccessful	Cooke (1977)
Lamina, petioles	Plant regeneration	LS	Various growth substances	Regenerative tissue arose from epidermis	Grunewaldt (1977)
Complete leaf cuttings from axenic plants	In vitro mutagenesis	Specified medium	N-nitroso-N-methylurea	Most common mutants: diminished apical dominance or chlorophyll deficiency	Jungnickel (1977a)
Axenic chimera clone	Chimera separation, regeneration from root tissue	See above		Two nonchimeric strains obtained and investigated	Jungnickel (1977b)
Lamina	Organogenesis	MS	Various phytohormones	Role of initial and adjacent cells outlined	Vazquez et al. (1977)
Petiole cross-sections	Commercial micropropagation	MS	BA 0.01 NAA 0.1	In 3–4 months 5000 commercial plantlets from a single petiole	Bilkey et al. (1978)

Table 1. (*Continued*)

Explant	Aim of the study	Medium	Important additives (mg/l)	Response/observation. data (mg/l)	Reference
Lamina	Micropropagation	MS modified	Various BA, kin and NAA	Organogenesis via callusing	Moncousin (1978)
Floral parts	Morphogenesis	MS	Various BA, kin and NAA	Organogenesis without or via callusing from ovary, sepal, and petal tissue initiated from epidermis	Vazquez and Short (1978)
Petioles	Micropropagation	MS	Kin 0.5 NAA 0.5	Organogenesis without callusing, rootless plantlets transferred	Harney and Knap (1979)
Lamina	Commercial micropropagation	NS	NS	Plantlets look like seedlings	Englert (1980)
Leaves	In vitro mutagenesis, spontaneous variants		N-nitroso-N-methylurea	Mutants induced, two flower color variants	Grunewaldt (1980)
Lamina pieces	Micropropagation, economic aspects	MS	BA 2 NAA 1	Calculation of the profit-earning capacity of in vitro technique	Jacob et al. (1980)
Petiole sections with or without epidermal tissue	Regeneration capacity	MS, B5	BA 0.5 NAA 0.1	B5 superior to MS, subepidermal tissue caused increased plant vigor	Bilkey and Cocking (1981)
Leaves	In vitro polyploidization	MS	BA 1 NAA 1	Tetra-. octo-, and mixoploids obtained	Espino and Vazquez (1981)
Lamina	Micropropagation, economic aspects	MS	BA 5(?) NAA 0.1/0.01	Production costs calculated, microcuttings transferred to peat	Le and Collet (1981)

Petiole segments	Protoplast isolation after callusing	MS	2,4-D 1	Teased callus tissue released many cytoplasmic protoplasts	Bilkey and Cocking (1982)
Buds	Micropropagation	MS	BA 5(?) NAA 1	40 explants/bud produced after 60 days	Cachită-Cosma and Lazăr (1982)
	Treatment of transferred plants			Optimum lighting conditions outlined	Pivot (1982)
Leaf explants	Low temperature selection			After 16 weeks below 15 °C no regeneration	Schlegel (1982)
	Micropropagation of variegated cvs.			Stability of variegation during micropropagation	Sunblade and Meyer (1982)
Anthers	Haploid production	MS	BA 0.5 NAA 1	Most plants were haploid	Weatherhead et al. (1982)
Lamina	Morphogenetic study			Embryoids and other stages demonstrated	Zhang et al. (1982)
	Micropropagation			Report	Anonymous (1983)
Anthers, leaves	Cytological and enzymatic investigation	Medium specified	Kin 0.5 IAA 5	Differences in anther-derived plants	Bhaskaran et al. (1983)
Leaf tissue	Plastids in epidermal cells investigated	MS	IAA Kin 2 IAA 2	Plastids are released from arrested state during organogenesis	Finer and Smith (1983)
	In vitro mutagenesis			Most mutants had non-chimeric character	Grunewaldt (1983)
Chimeric tissue	Histogenetic analysis			Bilkey et al. (1978) or Sunblade and Meyer (1982) confirmed, but different explanation	Norris et al. (1983)
Petiole or Lamina	Optimal media		Sucrose: 1%, 3%, 6%	Optimum pH 5.8, sucrose 6% gave best results	Park et al. (1983)

Table 1. (*Continued*)

Explant	Aim of the study	Medium	Important additives (mg/l)	Response/observation. data (mg/l)	Reference
Petiole sections	Commercial micropropagation		BA 1 NAA 1	1000 plants obtained per petiole	Schlegel (1983)
Leaf tissue	Micropropagation of chimeras	Medium specified	kin 2	Formation of single, symmetrial rosettes, and see Norris et al. (1983)	Smith and Norris (1983)
Leaves	Combining phytohormones	MS	Various BA, NAA, 2,4-D	Shooting via callusing, combinations recommended	So (1983)
Leaf sections with a vein	Organogenesis	MS	BA 0.1 IAA 1	Organogenesis near the edges and along the veins. Epidermal and subepidermal cells involved	Baktir (1984)
Leaf cuttings versus axenic materials	Commercial aspects			Plants did not differ between the three experimental populations	Cassells and Plunkett (1984)
Axenic plants	Vitrification			Vitrification results from a burst of ethylene	Kevers et al. (1984)
	Micropropagation of variegated cvs.			Norris et al. (1983) refuted/defended	Marcotrigiano et al. (1984)
Haploid tissue	In vitro mutagenesis and selection as to cold-tolerance		Ethylmethane-sulfonate 0.25 mol/l	Plantlets from anthers are superior even to a cold-tolerant cv.	Warburton et al. (1984)
Lamina, petioles	Micropropagation	MS	Various high kin, NAA, IAA	Somaclonal variation on phytohormone media	Radojevic et al. (1984)

Lamina, petioles	Micropropagation	MS	BA 0.1 NAA 0.1 ginseng	MS + ginseng root Powder improved quality and quantity	Xu (1984)
Explants from axenic donor plants	Utilization of somaclonal variation	MS	Various BA, NAA	Different responses of cvs. to suboptimal temperature, but no really tolerant cv.	Amberger et al. (1985)
	Adventitious shooting, mutagenesis			Norris et al. (1983) refuted. Thousands of solid mutants do not agree with multicellular origin	Broertjes and Van Harten (1985)
Lamina	Micropropagation	MS or 1/2-MS	Different BA, NAA	Best results with each 0.1 BA and NAA, rooting 0.01 NAA	Chang (1985)
Leaf explants or "flower clusters"	Micropropagation of chimeras			Report on advantages of "flower clusters"	Larson (1985)
Inflorescences	Micropropagation of chimeras	MS	BA 1 NAA 1	Plantlets grew rapidly from the axils of the bracts and showed high true to type frequency	Lineberger and Druckenbrod (1985a)
Inflorescences and other explants	Micropropagation of chimeras	MS	BA 1 NAA 1	No true-to-type chimeras except in inflorescence explants	Lineberger and Druckenbrod (1985b)
Leaf discs from mature or in vitro plants	Morphogenetic capacity	MS modif.	Various phytohormones	In vitro-grown explants superior to leaf discs, GA_3 or TIBA inhibits buds	Margara and Piollat (1985)
Anthers	Haploids			Haploid plantlets obtained	Radojevic et al. (1985)

Table 1. *(Continued)*

Explant	Aim of the study	Medium	Important additives (mg/l)	Response/observation. data (mg/l)	Reference
Conventional or in vitro mother plants	Comparison			65 days after potting, more saleable plants from in vitro culture explants	Röber and Horn (1985)
Leaf cuttings versus tissue culture	Micropropagation of chimeras	MS	Each 0.2 BA and NAA	Rate of single-colour sports depends on the type of explant tissue	Ando et al. (1986)
Leaf discs	Tissue culture ability of cvs.	MS	Without or with various BA and NAA	Different morphogenetic capacity of the cvs.	Cassells et al. (1986)
Different explants	Comparison of different plants and in vitro techniques		Without or with phytohormones	Specific advantages of saintpaulias outlined	Jungnickel (1986)
In vitro plants	Germplasm preservation			Long-term storage of cvs. chimeras, and mutants	Jungnickel and Gliemeroth (1986)
Excised leaves	Bud differentiation	LS	Influence of light/darkness	Bud regeneration was inhibited by red and far-red light	Lercari et al. (1986)
In vitro plants	Transfer to greenhouse			How to overcome specific properties of in vitro plants	Reuther (1986)

Axenic leaf cuttings	Comparison of chimeras in spp.				Sorting-out of chimeras in *Saintpaulia* results in L_1-derived plants	Jungnickel (1988b)
Flower buds, leaves	Establishment in some Gesneriads				Survey including saintpaulias and other Gesneriads	Jungnickel (1988c)
Leaf-, petal-, and subepidermal tissue	Propagation and separation of chimeras	MS	BA NAA	1 1	Some types of foliar variegation is non-chimeric, but pinwheel flowers are chimeric	Peary et al. (1988)
Axenic lamina pieces	Separation and rearrangement of chimeras	Jungnickel and Gliemeroth (1986)	BA	0–1	Role of cytokinins in the case of rearrangement of mericlinal- and periclinal-chimeras outlined	Zaid and Jungnickel (1988)
Axenic lamina pieces	Morphogenetic study using L_1-specific marker chimeras	see above	BA NAA	0–1 0–1	BA causes delayed rooting, but organogenetic activation of subepidermal cells	Zaid (1989)

Abbreviations: B5, Gamborg et al. 1976; LS, Linsmaier & Skoog 1965; MS Murashige and Skoog 1962; NS = Not specified.
Supplements: 2,4-D, 2-4-dichlorophenoxyacetic acid; AdS, Adenine sulfate; BA, benzyladenine; GA_3, gibberellic acid; IAA, indoleacetic acid; kin, kinetin; NAA α-naphthaleneacetic acid TIBA, triiodobencoic acid.

Commercial micropropagation of cultivars is the most common use of in vitro techniques in saintpaulias (see Sect. 3 and Table 1). The development of healthy mother plants via meristem culture (Paludan 1971) and their axenic true-to-type preservation (Jungnickel 1974, 1976a) was reported. General advantages of axenically cultivated *Saintpaulia* tissue (Jungnickel 1986) even include a nonenzymatic

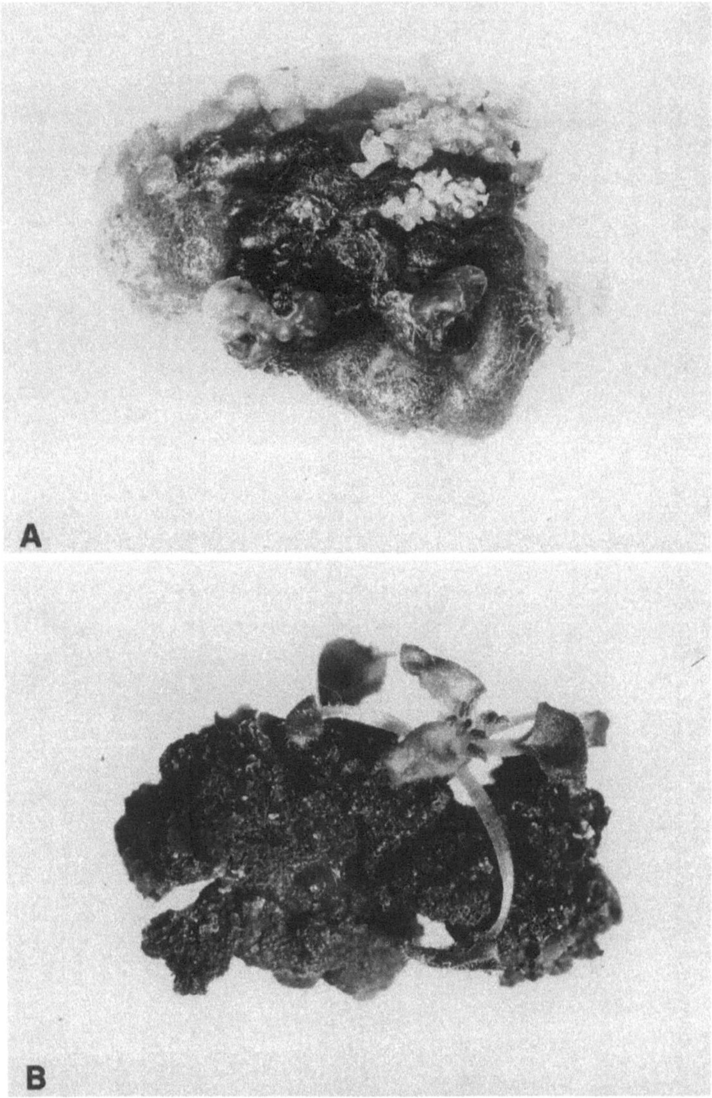

Fig. 5A, B. Shoot differentiation in a phytohormone-autonomous tumour tissue of *Saintpaulia*, 16 years after mutagenic treatment. Subsequent subcultivations on nitrate medium. **A** Early stage of organogenesis. **B** Late stage of organogenesis

method of dealing with protoplasts which were stable in the osmotic pressure of standard tissue culture media (Bilkey and Cocking 1982). Cytomorphogenetic processes were studied in epidermal cells (Finer and Smith 1983) or in flower parts (Vazquez and Short 1978), and vitrification was investigated intensively (Kevers et al. 1984). Different tissue culture ability was reported (e.g., Cassells et al. 1986; Jungnickel 1988c; Grout 1990), and embryoid formation was shown by Zhang et al. (1982). Anther culture resulted in a fairly high percentage of haploids (Hughes et al. 1975; Hughes 1977; Weatherhead et al. 1982; Bhaskaran et al. 1983; Radojevic et al. 1985), which might be the basis of miniatures (Goetz 1984) as well as of in vitro selection (Warburton et al. 1984), but the latter method was also performed using sporophytic tissue (Schlegel 1982, Amberger et al. 1985; Grunewaldt 1988).

Gene transfer via the (plasmid of) *Agrobacterium tumefaciens* strains T 37 and B 6 induced callusing and shooting in petiole fragments, whereas untreated controls did not show any organogenesis (Rao and Morel 1973). In vitro mutagenesis (Fig. 4) (Jungnickel 1977a; Grunewaldt 1980, 1983; Geier 1983) and polyploidization

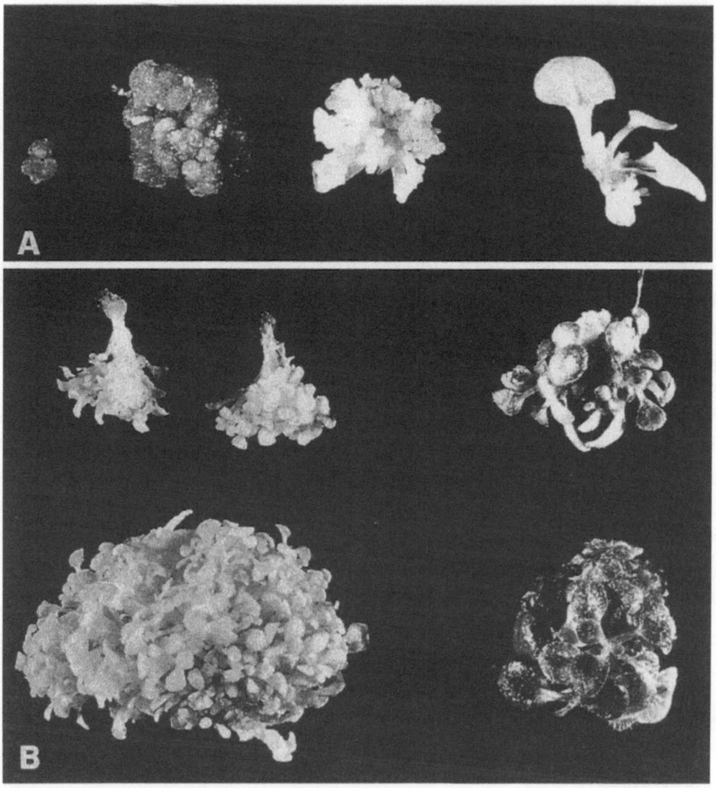

Fig. 6A, B. Development of homohistic albinotic plants from (L_3) root-derived callus. **A** Induction on heterotroph-medium (Jungnickel 1988a) with 0.3 mg/l BA + 0.1 mg/l NAA; differentiation on the same medium without phytohormones. **B** Diminished apical dominance in *Saintpaulia* mutants with (*right*) or without (*left*) chlorophyll

Fig. 7. Teratoma-like mutant of *Saintpaulia* with filamentous or tubular leaves and calloid proliferations, cultivated for 16 years on phytohormone-free medium

under axenic conditions (Espino and Vazquez 1981) contribute to new cultivars (Sehm 1974) as well as to specific strains of *Saintpaulia* with different genetically marked tissue (Jungnickel 1977b; Zaid and Jungnickel 1988; Zaid 1989). There is no basic difference between mutagenesis under in situ (see Sect. 1.3.4) or in vitro conditions; however, chlorophyll-deficient strains as well as morphogenetically aberrant ones (Figs. 5, 6) can be maintained under axenic conditions as long as desired (Jungnickel and Gliemeroth 1986) without loss of stability (Evans and Bravo 1986) or viability. Surprisingly, some of the morphogenetically aberrant mutant strains (Fig. 7, Jungnickel 1977a) are phenotypically similar to the reported effects of growth regulators (Kukułczanka and Kwiecień 1973; Anonymous 1975a; Dvorska 1979), even though these strains were not treated with growth regulators or phytohormones for several years.

3 Micropropagation

3.1 General

Saleable African violets must be true to the cultivar-specific type (Sect. 1.3.2), free of diseases and pests (Sect. 1.3.3), and cultivation conditions (Sect. 1.3.1) as well as the care for the plants should provide for production of uniform, attractive ornamentals with numerous flowers (TGL 1988). Marketability of new types or cultivars

(Goetz 1984) also depends on the question of how to obtain sufficient healthy mother plants and their true-to-type vegetative descendants. The propagation speed of axenically grown saintpaulias is fairly high (e.g. Holdgate 1977; Boccon-Gibod et al. 1978; Anonymous 1983), especially when axenic stock mother plants are used (Jungnickel 1974; 1976b; Theiler 1981). About 5000 commercial plantlets can be produced from a single petiole in 3-4 months (Bilkey et al. 1978). Jones (1983) estimated an annual world production of 2×10^6 axenically grown saintpaulias; nowadays, however, some growers exclusively use tissue culture plants for ornamental pot plant production (Sect. 1.4), and the above order of magnitude can be produced by a single unit. In the eastern FRG, the total demand of the *Saintpaulia* market was estimated to be 4×10^6 flowering pot plants per year (Hoffmann, pers commun. 1988).

3.2 Establishment of Cultures

Axenic *Saintpaulia* cultures might be established from almost all the vegetative or generative parts of the plants (Bilkey and Hildebrandt 1977). To start an in vitro storage period of 6 years, Jungnickel (1976a) used seeds of the cvs. Niobe and Grandiflora or lamina sections of a vegetatively propagated cv. Paludan (1971) and Cachită-Cosma and Lazăr (1982) preferred meristem tips or buds from conventional mother plants, but Jungnickel (1974) and Theiler (1981) stress the advantage of maintaining axenic mother plants. According to Le and Collet (1981), a single in vitro mother plant replaces 300 conventional mother plants as well as the greenhouse capacity and care they need throughout the year. Young flower buds (Jungnickel 1988c) or more adult flower parts (Vazquez and Short 1978) gave rise to adventitious shoots. Concerning periclinal chimeras, axillary buds of rosettes (Jungnickel 1977a) as well as young inflorescences (about 2 mm: unpubl. results) or sections of the peduncles which include the region of bracts (Larson 1985; Lineberger and Druckenbrod 1985a,b) are recommended explants to start enhanced axillary branching. Subepidermal tissue or explants consisting of thin layers (epidermis plus subepidermis) were also investigated (e.g., Flores et al. 1976; Bilkey and Cocking 1981). According to Schlegel and Schneider-Maessen (1981), preculture conditions of donor plants influenced regeneration in vitro.

Disinfection of the explants is usually performed by rinsing them for 30 s with 70–80% ethanol followed by a 2–15 min immersion in hypochlorite solutions which contain no more than 1% active chlorine and which contain about 0.1% of a surface active agent like Tween 20. Most authors give percentages of commercial hypochlorite solutions. Surface activity is combined with oxidation power in Chloramin T (Jungnickel 1988c). For other details see, e.g., Espino and Vazquez (1981); Schlegel and Schneider-Maessen (1981); Park et al. (1983); Cassells and Plunkett (1984); Lineberger and Druckenbrod (1985b).

3.3 Nutrient Media

As shown in Table 1, most investigators applied the high salt medium according to Murashige and Skoog (1962) or Linsmaier and Skoog (1965). Since saintpaulias are known to be salt-sensitive plants (Sect. 1.3.1), the MS medium is often diluted by one-half. Park et al. (1983) reported that 6% sucrose gave better results than 1 or

3%, but the borderline of osmotic pressure should be taken into consideration. Cooke already reported in 1977 that four out of 17 cvs. did not grow well on the MS shoot-induction medium. Bilkey and Cocking (1981) found that subepidermal tissue regenerated shoots on the B5 (see Gamborg et al. 1976) salts medium, but not on MS medium with the same hormone concentrations. Thin layer cultures (epidermis plus subepidermis) also induced shoots on the B5 medium (Flores et al. 1976). Nitrate has been shown to be a satisfactory sole source of nitrogen in saintpaulias (Paludan 1971, Jungnickel 1974, 1976a,b, 1977a), and even vitamins or exogenous phytohormones are not necessary in the case of long-term germplasm preservation (Jungnickel and Gliemeroth 1986). Such extremely simple media also provide for full development of adult individual plants under axenic conditions (see Fig. 3). In vitro flowering of saintpaulias has not yet been reviewed by Dickens and Van Staden (1988). As shown using media with or without ammonium ions (for their composition see Jungnickel 1988a), ammonium is also able to enhance shooting in saintpaulias; however, it suppresses rooting. Bartolo and Macey (1989) demonstrated cobalt requirement of *Saintpaulia* tissue cultures.

Some investigators applied rather high concentrations of phytohormones (for instance 5 mg/1 BA, NAA, or IAA: Cachita-Cosma and Lazar 1982; Radojevic et al. 1984), but this might be advantageous only in primary explants to overcome effects of disinfection stress, senescence,or latent infections (see Sect. 3.5). Exogenous phytohormones should be applied in as low concentration as possible without losing the desired morphogenetic effect (Jungnickel 1976b), and combinations of as little as 0.01–0.1 mg/l BA and NAA give optimal results (Bilkey et al. 1978; Xu 1984, Chang 1985).

3.4 Cultivation Conditions

The photosynthetic saturation point of greenhouse-grown saintpaulias is 14 000 lx, but assimilation of in vitro plants did not increase with light levels above 3000 lx (Reuther 1986). Almost complete light utilization (Pivot 1982) is mainly caused by the properties of a single palisade parenchyma cell layer (Neumann 1983). The fresh weight of regenerated shoots was influenced by light quality, but irradiation flux or light quality affected the number of shoots (Schlegel and Schneider-Maessen 1981). Bud regeneration was highest under continuous blue or white light and in darkness but red and far-red light inhibited budding (Lercari et al. 1986). Chang (1985) proposed 2000 lx for 8–10 h/day at 25–28 °C. As far as temperature is concerned, we observed that day/night variations between 26 and 20 °C were advantageous. This might be due in part to an exchange of ethylene contaminated air (see Kevers et al. 1984) with fresh air. Shake culture techniques (Takayama and Misawa 1982) could have similar effects.

3.5 Regeneration and Differentiation

3.5.1 Homohistic Plants

Shoot multiplication via callusing, enhanced axillary branching, or adventitious shoot formation have been used widely (Table 1). Callus formation was induced at petiole segments, lamina pieces (Kukułczanka and Suszynska 1972; Flores et al. 1976; Grunewaldt 1976; Schlegel 1983; So 1983; Ioannou 1987), or floral parts (Vazquez and Short 1978; Dallon 1987a,b) as an effect of rather high concentrations of phytohormones. Later, for instance, many shoot primordia developed around the callus ring formed by petiole sections (Schlegel 1983). Callusing was increased by morphactin IT 3456 but it retarded the regeneration of adventitious buds or inhibited root formation and removed polarity of regeneration in isolated leaves (Kukułczanka and Kwiecien 1973). After the influence of rather high phytohormone concentrations, Radojevic et al. (1984), found albino shoots as well as several other off-type plants. Zhang et al. (1986) discuss somatic embryogenesis.

Controlled axillary branching on phytohormone-free media (Jungnickel 1974) resulted in true to type germplasm preservation for a period of 6 (Jungnickel 1976a) or 16 (Jungnickel and Gliemeroth 1986) years. In contrast to strawberries (Jungnickel 1988a), saintpaulias form adventitious buds on phytohormone-free media besides axillary branching. However, axillary shoots grow more rapidly in comparison with adventitious buds. According to Margara and Piollat (1985), GA_3 as well as TIBA overcame the ability of isolated leaves to form adventitious buds. The stem of rosettes is also elongated by GA_3, and this makes stem cross sections more convenient (Jungnickel 1988a). Axillary branching is enhanced by low cytokinin concentrations, but in *Saintpaulia*, it is almost impossible to apply this phytohormone without increasing adventitious shoot formation.

Petiolate leaves from axenic *Saintpaulia* cultures form adventitious shoots (Fig. 8) on phytohormone-free media the same way leaf cuttings do in greenhouses

Fig. 8. Adventitious shoots formed at the base of an isolated axenic leaf (*right*) on phytohormone-free medium. (Jungnickel 1976a)

Table 2. Effect of α-naphthalene acetic acid (NAA) on morphogenesis in isolated axenic leaflets of *Saintpaulia ionantha*, cv. Niobe. Data: Period of time (days) until first root or shoot formation. Total period of cultivation: 50 days. Nutrient medium: basal-JG-B. (Jungnickel 1988a)

NAA (mol/l)	0	10^{-6}	3×10^{-6}	10^{-5}
Rooting	20	17	17	22
Shooting	22	37	No shoots	No shoots

Table 3. Effect of benzyladenine (BA) on morphogenesis in isolated axenic leaflets of *Saintpaulia ionantha*, cv. Niobe. Data: Period of time (days) until first root or shoot formation. Total period of cultivation: 50 days. Nutrient medium: basal-JG-B. (Jungnickel 1988a)

BA (mol/1)	0	3×10^{-7}	10^{-6}	3×10^{-6}
Rooting	20	37	No roots	No roots
Shooting	22	17	17	22

Table 4. Effect of genetically different strains of *Saintpaulia ionantha* on morphogenesis in isolated axenic leaflets. Data: period of time (days) until 50% of the explants formed roots, shoots, or callus on the phytohormone-free medium Basal-JG-B according to Jungnickel (1988a). Total period of cultivation: 43 days

Strain	Rooting	Shooting	Callusing
Wild type	17	27	no
A	19	31	no
LB	(only 20%/43 d)	35	22
Lanc	(only 32%/43 d)	32	22
zu Al 7	(only 24%/43 d)	39	28

(Tables 2, 3). First, adventitious roots are formed near the vascular bundles (tissue layer III = L_3), and later, adventitious shoots occur near the lowermost basal epidermis cells of the petiole (Jungnickel 1976a). Removal of the lamina stopped any regenerative properties of the petioles, but as compared with petiolate leaves, 0.03 mg/l NAA mimics organogenetic influence of the leaf blade in isolated petioles (Fig. 9) Kukułczanka and Suszynska (1972), however, reported that their in vitro cultures formed shoots before root formation. The latter might be due to the limited morphogenetic capacity of the MS medium (Bilkey and Cocking 1981) or could be due to different genotypes (Table 4). Margara and Piollat (1985) showed that freshly isolated leaf discs produced buds only on media with auxin and cytokinin, but leaves taken from axenically grown cultures did not need exogenous phytohormones. Because of their superior ratio between biomass and surface, disinfection stress is less significant in freshly isolated petioles than in lamina explants. Extremely low phytohormone concentrations induce adventitious roots

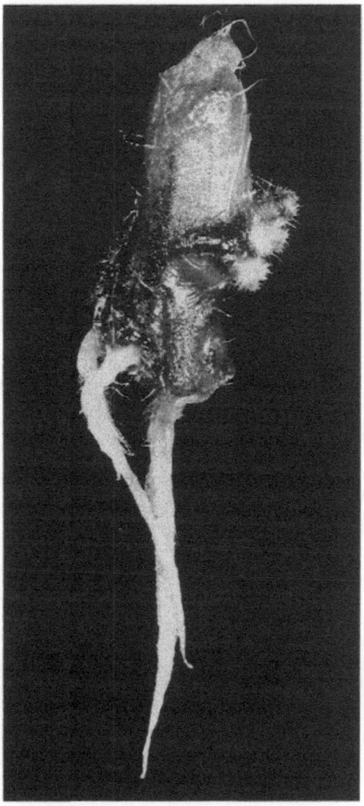

Fig. 9. Root and shoot regeneration on an isolated petiole grown on a nitrate medium supplemented with 0.03 mg/l NAA (Jungnickel 1976b)

and shoots in axenically grown petioles (Jungnickel 1976b), but phytohormone dependency of freshly isolated petioles as to organogenesis is also fairly low (Bilkey et al. 1978; Harney and Knap 1979; Xu 1984), and in this way, an intermediate callus phase can be prevented. In principle, the same is possible in sections of young leaves (e.g., Xu 1984; Chang 1985). There are histological indications that shoots are initiated from (single ? : Broertjes and VanHarten 1985) epidermis cells in each case (Grunewaldt 1977; Vazquez and Short 1978), but Baktir (1984) describes groups of epidermal and subepidermal cells. The latter result is also supported by using thin layers (Flores et al. 1976). The above question was investigated in more detail using mutagenesis and periclinal chimeras.

3.5.2 Chimeras

Broertjes and Keen (1980) presented stochastic model by which the expected relative percentage of chimera formation after the irradiation of an apex can be calculated. They suggest that the apex of an adventitious shoot is formed by only one cell of the epidermis or of the callus. In vitro mutagenesis on isolated leaves

(Jungnickel 1977a) supported the point of view (Broertjes and VanHarten 1985) that preferably solid mutants are found. Since some mutagenic agents like N-methyl-N-nitrosourea also cause extranuclear mutations (Pohlheim 1980, 1981), the resulting mericlinal (sectorial) chimeras and sorting-out processes (Hentrich and Beger 1974; Pohlheim and Pohlheim 1976) can also be understood on the basis of their single-cell origin. However, specific cultivars with variegated leaves or petals were selected (Eyerdom 1981; Hamilton 1981; Sunblade and Meyer 1982; Peary et al. 1988), which gave rise to serious misinterpretations concerning their chimera nature (Norris et al. 1983; Smith and Norris 1983; /and contrary to them/ Marcotrigiano et al. 1984; Broertjes and VanHarten 1985; Jungnickel 1988b; Peary et al. 1988; Zaid and Jungnickel 1988; Zaid 1989). In such cultivars, the epigenetic block to chloroplast development known in epidermal cells (Finer and Smith 1983) seems to act randomly in the subepidermal palisade parenchyma cells as well. Using periclinal chimeras of *Saintpaulia* with L_1 tissue, which is certainly genetically different from L_2 and L_3 tissue, Jungnickel (1977a,b, 1988b) and Pohlheim (1981) demonstrated that all of the adventitious shoots arose from epidermal tissue (see also Fig. 10), and the resulting shoots consisted exclusively of homohistic tissue.

However, all the results of Ando et al. (1986) cannot be explained on the basis of single-cell origin of adventitious shoots. They investigated edged (white corolla with pigmented outer-edge) or Geneva-edged (pigmented corolla with white outer-edge) cultivars under in situ as well as under in vitro conditions. Homohistic adventitious shoots also arose, characterized by a single color, typical to the color

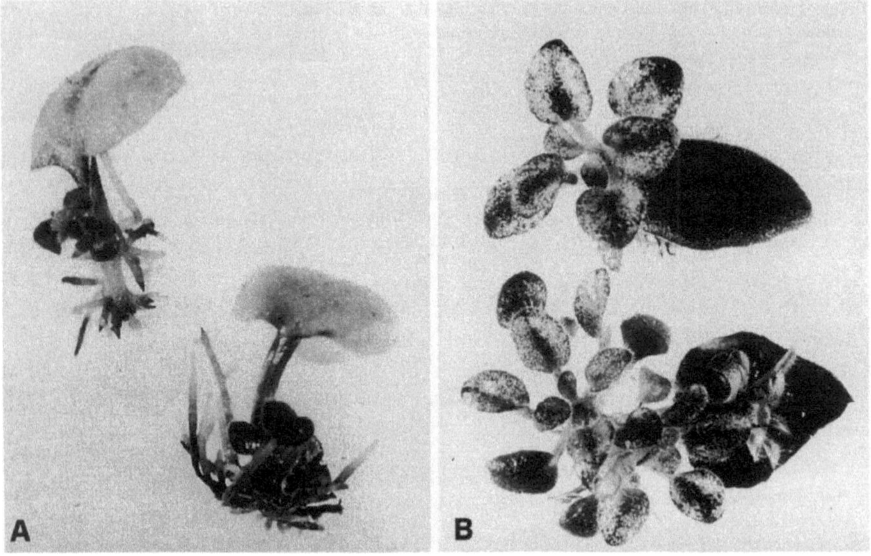

Fig. 10. A Regeneration of homohistic green shoots from a phenotypically white leaf with genotypic green L_1 tissue. Phytohormone–free nitrate medium. **B** Regeneration of homohistic variegated plants from phenotypic green leaves with genotypic variegated L_1 tissue. Phytohormone-free nitrate medium

edge of the parent cultivar. The frequency of such single-color sports, however, depended on the explant type: leaf blades, petioles, sepals, pedicles, or axillary buds on peduncles produced mainly edged types, whereas in leaf vines or in the edges of the corolla, a fairly high percentage of single-color plants was found. Ando et al. (1986) agree with the conclusion of Naylor and Johnson (1937) that "also the shoot has its origin in a single epidermis cell, adjacent epidermis cells, and parenchyma cells within the petiole contributed to its final formation". Occasional multicellular origin of chimeras has also been demonstrated in situ by Winkler (1935) and Bergann and Bergann (1984).

Since leaves or leaf pieces of saintpaulias are able to regenerate without exogenous phytohormones (Jungnickel 1976a), Zaid and Jungnickel (1988), and Zaid (1989) investigated the role of added phytohormones in adventitious shoot formation of true periclinal chimeras. Whereas in the phytohormone-free controls exclusively L_1-derived homohistic shoots were formed, benzyladenine also induced morphogenetic activity in subepidermal tissue (Fig. 11) and this resulted in a low percentage of homohistic L_2-derived shoots as well as in L_1/L_2-derived chimeras, a few of the latter even had a reverse constitution as compared with the original periclinal chimeras. In addition, Pohlheim (1983) stresses the role of periclinal cell divisions in chimeras which have layers with different intensities of growth. A similar situation might occur in specific cultivars shown to have more vigorous growth when the plants develop from subepidermal tissue (Bilkey and Cocking 1981). Under this specific genetic precondition in chimeras, it could be possible that a subepidermal cell (which is kept entrapped by the epidermis in its further development) also initiates a chimera apex formation in situ, particularly since the first appearing L_3-derived root meristems produce endogenous cytokinins which are able to form a concentration gradient towards L_1 via the L_2 layer. For information concerning general replacement of the L_1, L_2, or L_3 tissue by a neighboring layer, see Bergann (1967).

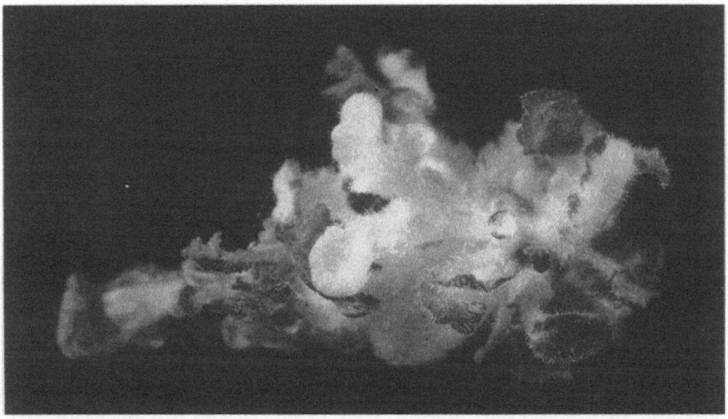

Fig. 11. Shoot regeneration in a phenotypic white leaf (histogenetic constitution L_1 green; L_2 and L_3 albinotic) under the influence of 0.3 mg/1 BA. (Zaid 1989)

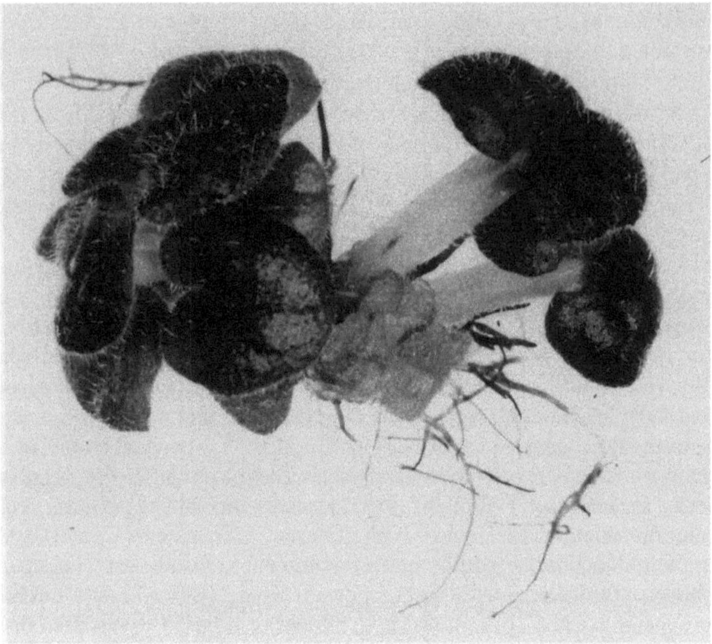

Fig. 12. Controlled axillary branching in a shoot cross-section with axillary buds on phytohormone free nitrate medium. Histogenetic constitution of the chimera: L_1 virescent; L_2 and L_3 green.

True periclinal chimeras can be maintained under axenic conditions (Fig. 12) as long as desired (Jungnickel and Gliemeroth 1986), and they are available for future investigators. During axillary branching, L_3 tissue is easily replaced by L_2 tissue (see also Broertjes and VanHarten 1985). The advantages of the axils of the bracts concerning micropropagation of chimeras are outlined by Larson (1985), and also Lineberger and Druckenbrod (1985a,b).

3.6 Transfer to the Greenhouse

General aspects of acclimatization of aseptically cultured plantlets have been outlined by Short et al. (1984) and Schulze (1988). In vitro-raised plantlets do not have well-developed stomata or cuticle, and their roots have no hairs. Two weeks after transfer, however, they increased their assimilation rate when the CO_2 content of the air was increased (Reuther 1986). When transferred saintpaulias were kept at 20–26 °C under high relative humidity, 100% success was obtained (Jacob et al. 1980). The number of transferred plantlets was greater after cultivation on a ginseng-supplemented medium (Xu 1984). It was not necessary to induce roots as the rootless shoots quickly became established in a commercial soil mixture (Harney and Knap 1979). Potting compost was disinfected by fairly low doses of gamma irradiation (Rattink 1982). Plants flowered 4–6 months (Chang 1985) or 65 days (Röber and Horn 1985), or even a shorter period of time (Müller 1988) after transplantation in the soil. We preferred to remove vitamins, phytohormones, and ammonium ions from the nutrient medium in the last in vitro stage and reduced the sugar content to a level which could be consumed completely by the plantlets. Thus, the covers of glasses can be removed some days before transfer without risk of serious contamination.

3.7 Some Commercial Aspects

Although belonging to the most popular ornamentals, saintpaulias do not occupy a specific position in the statistics of the world market of ornamentals. Grout (1990) reports on world trade being approximately $ 30 million/year. However, in the Federal Republic of Germany, their production amounted to 93 176 000 plants in 1988, and 34 600 000 plants were sold in the same year in The Netherlands. According to internal statistics of the Vereinigung Deutscher Blumenmärkte, miniatures amounted to 25–30% of the total production (Ruhnke 1990, pers. commun.). As a consequence, in Mid-Europe the average individual demand was more than one saintpaulia plant per year.

For the relations necessary between transfer and sale, see Sections 1.3.1–1.3.3. Leaf shape, number of flowering plants, number of peduncles, and number of flowers and flower buds per plant did not vary between conventionally or in vitro-grown plants (Cassells and Plunkett 1984) But Nienhaus (1991, pers. commun.) has adverse experiences. However, the above-mentioned expenses to maintain sufficient healthy mother plants of numerous varieties decrease considerably when in vitro stock mother plants are used. The cost of each stage was estimated by Le and Collet (1981) to simplify the method and to reduce the production costs. Rooting and weaning onto a horticultural substrate were successfully combined by transferring microcuttings from the test tubes directly into plastic boxes containing sterilized white peat or vermiculite.

4 Summary and Conclusions

As contrasted with the small original distribution area of *Saintpaulia* spp., numerous cultivars and followers have been distributed by men all over the world. Predominant reasons for the popularity of the pot flower plants are the considerable diversification of numerous cultivars as well as their ability to grow and to flower readily under average room conditions. The commercial value of the ornamentals increases when micropropagated healthy plantlets are grown to full bloom in a fairly short period of time. Since axenic mother plants do not cause the same expenses as conventional ones, axenic germplasm preservation of saintpaulias has also become an important horticultural routine technique.

Saintpaulias are also preferred subjects in fundamental plant research, particularly since they are able to form adventitious shoots under in situ as well as under in vitro conditions, and their uni- or multicellular origin has been studied in detail, especially in numerous chimeras. Protoplasts or haploids can be obtained easily in saintpaulias, and some haploids might already represent commercially useful varieties. Rapid substitution of the complete assortment by new ones indicates the speed of evolution in *Saintpaulia* cvs. under the influence of unconventional breeding and propagation techniques. It is expected that modern methods of biotechnology will continue to increase greatly the variability in saintpaulias, and that the plants will continue to give pleasure to many people during their daily life and work.

5 Protocol

If no axenic mother plants are available, immerse young inflorescences (2–3 mm), or in the case of homohistic plants, also flower buds, petioles, or pieces of young lamina for 30 s in 70% ethanol and transfer them for 3–6 min into a 2–3% solution of chloramin T. Wash three or four times with sterile nutrient solution and transfer the explants onto an NH_4^+-medium according to Jungnickel (1988a) or onto a half-strength LS medium. Initially, use 0.3 mg/l BA + 0.1 mg/1 NAA, but later diminish BA to 0.05–0.1 mg/l, and NAA to 0.01 – 0.03 mg/l. Cultivate the explants/plantlets between 26 and 20 °C at 2000 lx white light (14–16 h lighting, 8–10 h darkness). Before micropropagation, index all the explants for latent contamination by using common media for bacteria, yeasts and other fungi.

Acknowledgments. We wish to express our gratitude to the hybridizer S. Prkno, Bautzen, FRG, and to the breeders/growers of the Zierpflanzen Dresden, FRG for valuable discussions and the permission to use photographs from their work. The photographs taken by Mr. G. Schörlitz as well as the technical assistance of Miss C. Held, Mrs. H. Wehrberger, and Mrs. B. Bratfisch are also gratefully acknowledged. We also express our thanks to Mr. M. Bauer for linguistic advice.

References

Ailincăi N (1974) A contribution on the growing technique for *Saintpaulia ionantha* H. Wendl. Cercetări Agronomice in Moldova 3: 105–109 (in Romanian)
Amberger S, Schlegel G, Horn W (1985) Selection for chilling tolerance in *Saintpaulia ionantha.* In: Eucarpia meeting on pot plant breeding, Svendborg, Denmark, Europ Assoc Res Plant Breed, Sect Orn, pp 27–42
Amberger-Ochsenbauer S (1990) Luftfeuchtigkeit von Bedeutung bei *Saintpaulia.* Gärtnerbörse Gartenwelt 90: 1417–1419
Andersen H (1983) A temperature and fertilizer regime for *Saintpaulia.* Gartner Tidende 99: 786–787 (in Danish)
Andersen H (1984) The effect of temperature, frequency of watering, composition of fertilizer and peat type on *Saintpaulia.* Tidsskr Planteavl 88: 183–191 (in Danish)
Andersen H, Aabrandt Z (1989) Artificial light for *Saintpaulia.* Gartner Tidende 105: 608–609 (in Danish)
Ando T, Akiyama Y, Yokoi M (1986) Flower colour sports in *Saintpaulia* cultivars. Sci Hortic 29: 191–197
Anonymous (1971) Topfpflanzen–Fließband vergrößert Gewächshaus. Gartenwelt 71: 409
Anonymous (1973) Jahresbericht 1973. Bundesanstalt für Land- und Forstwirtschaft in Berlin und Braunschweig, pp. 1–157

Anonymous (1975a) Jahresbericht 1975. Forschungsanstalt für Weinbau, Gartenbau, Getränketechnologie und Landespflege, Geisenheim am Rhein, pp 1–266

Anonymous (1975b) Annual research report 1975. Institute of Food and Agricultural Sciences, Florida Univ, Gainesville, USA, pp 1–337

Anonymous (1983) Annual report for 1982. Agricultural Research Institute Nicosia. Ministry of Agriculture and Natural Resources, Cyprus, pp 1–87

Axelsson U (1975) Leaf damage with *Saintpaulia.* Vaxtskyddsnotiser 39: 39–40 (in Swedish)

Baktir J (1984) Organogenesis in in vitro cultures of African violet (*Saintpaulia ionantha*). Doğa Bilim Dergisi, D_2 (Tarim ve Ormancilik) 8: 9–12

Bartolo WCF, Macey MJK (1989) Cobalt requirement in tissue cultures of three species: *Brassica oleracea* L., *Passiflora mollissima* Bailey, and *Saintpaulia ionantha* Wendl. J Hortic Sci 64: 643–647

Barton H (1982) The international breeder's rights system and crop plant innovation. Science 216: 1071–1075

Batson F (1972) Observations on the production of plantlets of African violet (*Saintpaulia ionantha*) cvs. Lisa, Meta and Fischer's Ballett TM. Plant Prop 18 (3): 9–11

Bazzochi R, Rossi F (1980) Preliminary results on the use of lasers on plant organs. Fiori, Giugno, Univ Bologna, Italy, pp 15–16 (in Italian)

Bazzochi R, Gambini F, Giorgioni ME (1983) Analyses of heating costs in various types of greenhouse. Colture Protette 12(6): 25–34 (in Italian)

Bergann F (1967) The relative instability of chimera clones – The basis of further breeding. In: Induced mutations and their utilization. Abh Dtsch Akad Wiss, Berlin. Kl Med, pp 287–301

Bergann F, Bergann L (1984) Gelungene experimentelle Synthese zweier neuer Pfropfchimären – die Rotdornmispeln von Potsdam: + *Crataegomespilus potsdamiensis* cv. Diekto, *cv.* Monekto. Biol Zbl 103: 283–293

Beutel A (1990) Kompost als Zuschlagstoff bei Saintpaulia. Gärtnerbörse Gartenwelt 90: 1441–1442

Bhaskaran S, Smith RH, Finer JJ (1983) Ribulose bisphosphate carboxylase activity in anther-derived plants of *Saintpaulia ionantha* Wendl. Shag. Plant Physiol 73: 639–642

Biermann W (1984) Saintpaulien: Isolierglas lohnt bei den meisten Sorten. Gärtnerbörse Gartenwelt 84: 646–649

Bilkey PC, Cocking EC (1981) Increased plant vigor by in vitro propagation of *Saintpaulia ionantha* Wendl. from sub-epidermal tissue. HortScience 16: 643–644

Bilkey PC, Cocking EC (1982) A non-enzymatic method for the isolation of protoplasts from callus of *Saintpaulia ionantha* (African violet). Z Pflanzenphysiol 105: 285–288

Bilkey PC, Hildebrandt AC (1977) African violets and tissue culturing. Afr Viol. Mag 30(1): 22–23

Bilkey PC, McCown BH, Hildebrand AC (1978) Micropropagation of African violet from petiole cross-sections. HortScience 13: 37–38

Bjerre H (1982) The influence of plant density on plant quality and time of production for six species of pot plants. Tidsskr Planteavl 86: 389–398

Bjerre H, Amsen MG (1983) Pot plant production, environmental conditions and energy consumption in insulated greenhouses. Tidsskr Planteavl 87: 581–595

Boccon-Gibod J, Daux P, Pottier JC (1978) Clonal propagation of *Saintpaulia ionantha* Wendl. in vitro: determination of an index propagation's speed. In: Thorpe A (ed) Abstr 4th Int Congr Plant Tissue and Cell Culture, Calgary, Canada, p 151

Böhmer B (1990) Schädlinge an *Saintpaulia ionantha.* Gärtnerbörse Gartenwelt 90: 1434–1436

Böhmer B, Eilenbach B (1987) *Frankliniella* bildet schon resistente Stämme aus. Gärtnerbörse und Gartenwelt 87: 360–362

Bodner M, Larcher W (1987) Chilling susceptibility of different organs and tissues of *Saintpaulia ionantha* and *Coffea arabica.* Angew Bot 61: 225–242

Bosse G, Bouillon G, Reinken G (1971) Untersuchungen über das Wachstum von Zierpflanzen im Pflanzenrotor. Gartenbauwissenschaft 36: 321–338

Broertjes C, Keen A (1980) Adventitious shoots: do they develop from one cell ? Euphytica 29: 73–87

Broertjes C, VanHarten AM (1985) Single cell origin of adventitious buds. Euphytica 34: 93–95

Buiatti M, Roselli G (1979) Variability and selection in plants obtained by vegetative propagation. III. Rivista Ortoflorofrutticoltura 63: 447–467

Burtt BL (1947) Species of *Saintpaulia.* Gard Chron Ser CXXII 3: 22–23

Burtt BL (1956) A confusion in African violets. Baileya 4: 163–164

Burtt BL (1958) Studies in the *Gesneriaceae* of the old world. XV: The genus *Saintpaulia.* Notes R Bot Gard Edinb 22: 547–568

Burtt BL (1962) Studies in the Gesneriaceae of the old world. XXIV: Tentative keys to the tribes and genera. Notes R Bot Gard Edinb 24: 205–220

Burtt BL (1964) Studies in the *Gesneriaceae* of the old world. XXV: Additional notes on *Saintpaulia*. Notes R Bot Gard Edinb 25: 191–195

Busch LV, Smith EA (1978) Control of root and crown rot of African violet and of gloxinia caused by *Phytophthora nicotianae* var. *nicotianae*. Can Plant Dis Surv 58: 73–74

Buurman H (1973) Calculation of the cost price of saintpaulias. Vakbl Bloem 28 (25): 8–9 (in Dutch)

Cachită-Cosma D, Lazăr M (1982) In vitro multiplication of *Saintpaulia*. Productia Vegetală, Horticultură 31(11): 29–32 (in Romanian)

Carpenter WJ (1975) Supplementing daylight with HID lamps. Ohio Flor Assoc Bull 551: 8–9

Cassells AC, Plunkett A (1984) Production and growth analysis of plants from leaf cuttings, and from tissue cultures of disks from mature leaves and young axenic leaves of African violet (*Saintpaulia ionantha* Wendl.). Sci Hortic 23: 361–369

Cassells AC, Plunkett A, Kelleher D (1986) Screening of *Saintpaulia ionantha* Wendl. cultivars for caulogenetic potential based on the in vitro responses of young axenic leaves on auxin and cytokinin factorial media. Sci Hortic 30: 151–157

Cathey HM, Heggestad HE (1982) Ozone sensitivity of herbaceous plants: modification by ethylenediurea. J Am Soc Hortic Sci 107: 1035–1042

Chang JS (1985) In vitro rapid propagation of *Saintpaulia ionantha* Wendl. Plant Physiology Communications (Zhiwu Shenglixue Tongxun) 5: 34–36 (in Chinese)

Conover CA, Poole RT (1981) Light acclimatization of African violet. HortScience 16: 92–93

Cooke RC (1977) Tissue culture propagation of African violets. HortScience 12: 549

Dickens CWS, van Staden J (1988) The induction and evocation of flowering in vitro. S Afr J Bot 54: 325–344

Dallon J (1987a) Production of multiple propagules of African violets from floral and vegetative tissues. HortScience 22: 384

Dallon J (1987b) Expressions of polarity in in vitro propagation of African violets (*Saintpaulia ionantha* Wendl.) HortScience 22: 1149

DeGelder A (1991) Comparing *Saintpaulia* assortments. Vakbl Bloem 46(33): 28–32 (In Dutch)

Dirkse FB (1981) Footrot in *Saintpaulia* need not be a problem. Vakbl Bloem 36(7): 36–37 (in Dutch)

Duron M, Dixon B (1982) Influence de la dose initiale sur l'étendue et la durée de l'effet protecteur induit chez des feuilles de *Saintpaulia ionantha* (Wendl.) irradiées par des rayons gamma. Compt Rend Seanc Acad Sci, III. Sciences de la Vie 295: 185–189

Dvorská L (1979) The effect of growth regulators and temperature gradient on flowering of *Saintpaulia ionantha* Wendl. Acta Horticult. 91: 507–510

Edwards RA, Thomas MB (1985) The effect of N and K on the rooting of leaf cuttings grown in liquid media. Annu J R N Z Inst Hortic 13: 22–28

Ehrlich HG (1958) Cytological studies in *Saintpaulia* Wendl. (Gesneriaceae). Am J Bot 45: 177–182

Elgner N (1987) Mit einem kleinen Produkt ganz groß im Geschäft: Wie bei F. J. Müller in Harreshausen die Storytella–Minis entwickelt worden sind. Gärtnerbörse Gartenwelt 87: 1648–1649

Englert U (1980) *Saintpaulia.* Neuheiten aus Vaihingen. Gärtnerbörse Gartenwelt 80: 131–132

Escher F, Rehrmann P (1990) Stecklingserträge bei *Saintpaulia ionantha.* Gärtnerbörse Gartenwelt 90: 1423–1425

Espino FJ, Vazquez AM (1981) Chromosome numbers of *Saintpaulia ionantha* plantlets regenerated from leaves cultured in vitro with caffeine and colchicine. Euphytica 30: 847–853

Evans DA, Bravo JE (1986) Phenotypic and genotypic stability of tissue cultured plants. In: Zimmerman RH, Griesbach RH, Hammerschlag FA, Lawson RH (eds): Tissue culture as a plant production system for horticultural crops. Nijhoff Publ, Dordrecht, Netherlands, pp 73–94

Eyerdom H (1981) Flower colour sports and variations in *Saintpaulia* hybrids. Afr Viol Mag 34(5): 33–37

Ferns FEB (1979) Gesneriaceae of Alpine affinity. Bull Alpine Gard Soc 47: 123–151

Finer JJ, Smith RH (1983) Structure and development of plastids in epidermal cells of African violet (*Saintpaulia ionantha* Wendl.) in culture. Ann Bot 51: 691–695

Fischer M, Hentig WU von (1982) Blütenpflanzen in Hydrokultur. Zur Wasserversorgung während der Anzucht. Gärtnerbörse Gartenwelt 82: 471–474

Fischer P (1977) Nährstoff- und Salzgehalt in Substraten von *Saintpaulia ionantha* im Verlaufe der Kultur. Gartenwelt 77: 214–217

Flores HE, Fierro CA, Koo FKS (1976) Tissue culture of African violet *Saintpaulia ionantha* Wendl. In: University of Puerto Rico (ed) 24th Annu Congr Am Soc Hortic Sci, Tropical Region, Mayaguez, Puerto Rico, pp 449–454

Folk G (1983) The *Phytophthora* disease of African violets. Noevenyvedelem (Budapest) 19: 486–490 (in Hungarian)

Fussell CP (1958) Chromosome numbers in the *Gesneriaceae*. Baileya 6: 117–126

Gamborg OL, Murashige T, Thorpe TA, Vasil IK (1976) Plant tissue culture media. In Vitro 12: 473–478

Geier T (1983) Induction and selection of mutants in tissue cultures of *Gesneriaceae*. In: Preil W (ed) In vitro culture. Acta Hortic 131: 329–337

Gislerød HR, Mortensen LM, Selmer-Olsen AR (1986) The effect of air humidity on growth and nutrient content of some greenhouse plants. Acta Hortic 178: 181–184

Goetz W (1984) Usambaraveilchen mit Wildcharakter, eine Marktlücke für den Betrieb Wittenbreder. Dtsch Gartenbau 38: 27

Granger J, Laury JC (1987) La fertilisation carbonée des plantes en pot. Hortic Française 192: 7–11

Green BL (1987) Beginner's column. Afr Viol Mag 40(1): 5–6

Grout BWW (1990) African violet. In: Ammirato PV, Evans DA, Sharp WR, Bajaj YPS (eds) Handbook of plant cell culture, vol. 5, Ornamental species McGraw-Hill, New York pp 181–205

Grunewaldt J (1976) Die in vitro regeneration aus Blattstielquerschnitten von *Saintpaulia ionantha* H. Wendl. Gartenbauwissenschaft 41: 145–148

Grunewaldt J (1977) Adventivknospenbildung und Pflanzenregeneration bei Gesneriaceae in vitro. Gartenbauwissenschaft 42: 171–175

Grunewaldt J (1980) Spontane und induzierte Farbschecken von *Saintpaulia ionantha* H. Wendl. Gartenbauwissenschaft 45: 124–128

Grunewaldt J (1983) In vitro mutagenesis of *Saintpaulia* and *Pelargonium* cultivars. Acta Hortic 131: 339–343

Grunewaldt J (1988) In vitro selection and propagation of temperature-tolerant *Saintpaulia*. Acta Hortic. 226, I: 271–275

Hagen P, Nilsen J (1978) Comparison of incandescent lamps, high pressure mercury halogen and high pressure sodium vapour lamps as supplementary light sources for flower crops. Gartneryrket 68: 96–100 (in Norwegian)

Halliwell B, Mathew B, Starling B, Hall T, Hutchins G (1987) Plant awards, 1986–1987, Preliminary commendations: × *Brigandra calliantha* Schwarz, Gesneriaceae. Bull Alpine Gard Soc 55: 328–365, esp. pp356, 361

Hamilton R (1981) African violet flower and leaf types. Afr Viol Mag 34(3): 11–14

Harney PM, Knap A (1979) A technique for the in vitro propagation of African violets using petioles. Can J Plant Sci 59: 263–266

Heimann M (1977) Erfahrungen mit Previkur im Erikenbau. Gärtnerbörse und Gartenwelt 77: 990–992

Heins RD, Karlsson MG, Erwin JE, Hausbeck MK, Miller SH (1984) Interaction of CO_2 and environmental factors on crop responses. Acta Hortic 162: 21–28

Hemer M (1978) Echter Mehltau. Bekämpfung an blühenden Zierpflanzen. Gärtnerbörse Gartenwelt 78: 737

Hendriks L, Scharpf HC (1988) Düngung von Saintpaulien. Gärtnerbörse und Gartenwelt 88: 510–515

Hendriks L, Scharpf HC, Grantzau E (1986) Beim Düngen nicht nur an Wachstum denken. Gärtnerbörse Gartenwelt 86: 55–59

Henseler K (1976) Keine sichere Bekämpfungsmöglichkeit. Bakterien bedrohen Saintpaulien-Bestände. Deutsche Gärtnerbörse 76: 45–46

Hentig WU von, Fischer M (1987) Bewurzelungstemperaturen von Zierpflanzenstecklingen für Hydrokultur. Dtsch Gartenbau 41: 522–525

Hentrich W, Beger B (1974) Untersuchungen über die mutagene Effizienz von N-Nitroso-N-Methylharnstoff bei *Saintpaulia ionantha* H. Wendl. Arch Züchtungsforsch 4: 29–43

Hörnis A, Fritzsche G, Röber R (1984) Totalentsalzung: Abzuraten. Saintpaulien stellen ans Gießwasser keine sehr hohen Anforderungen. Gärtnerbörse Gartenwelt 84: 788–791

Holdgate DP (1977) Propagation of ornamentals by tissue culture. In: Reinert J, Bajaj YPS (eds) Applied and fundamental aspects of plant cell, tissue and organ culture, Springer Berlin pp. 18–43

Holtkamp H (1987) In: Holtkamp (ed) Saintpaulien Kulturtips. Rees-Haffen, FRG, pp 1–14

Holzer K (1952) Untersuchungen zur karyologischen Anatomie der Wurzel. Österr Bot Z 99: 118–155
Hoshino T (1986) Circular dichroic measurements of anthocyanins in intact flower petals. Phytochemistry 25: 829–832
Hughes KW (1977) Isolation of protoplasts from haploid plants of African violet. ASB Bull. 24: 60
Hughes KW, Bell SL, Caponetti JD (1975) Anther-derived haploids of the African violet. Can J Bot 53: 1442–1444
Ignasse JF (1984) L'utilisation des produits phytosanitaires en culture de plantes en pot. Hortic Française 165: 9–12
Ioannou M (1987) Micropropagation of African violet from petiole and leaf blade tissue. Tech Bull Agric Res Inst Nicosia, Cyprus, No 92, 4pp
Jacob M, Denee G, Coumans M (1980) La production du *Saintpaulia ionantha* in vitro: aspects économiques. Meded Fac Landbouwwet Rijksuniv Gent 45: 335–343
Johnson B (1987) Ask your hybridizer ! Afr Viol Mag 40(1) 14–15
Jones JB (1983) Tissue culture takes off. Flor Review, 24th March: 28–34
Jones RK, Strider DL (1985) African violet diseases. N C Flower Growers' Bull 29(5): 5–9
Jungnickel F (1974) Verfahren zur Erhaltung und Vermehrung von Kulturpflanzen–Zuchtstämmen. Patent DD 114207, Int Cl A O1 G 1/06
Jungnickel F (1976a) Sterilkultivierung von *Saintpaulia ionantha* H. Wendl. auf phytohormonfreiem Medium. Arch Gartenbau 24: 387–392
Jungnickel F (1976b) Regenerative properties of isolated petioles of *Saintpaulia ionantha* influenced by 6-benzylaminopurine or α-naphthalene acetic acid. Biochem Physiol Pflanz 170: 457–459
Jungnickel F (1977a) Induktion und Vermehrung von Mutanten bei *Saintpaulia ionantha* H. Wendl in Sterilkultur. Biol Zbl 96: 335–343
Jungnickel F (1977b) Some physiological properties of the virescent *Saintpaulia ionantha* strain *vis-3*. In: Abstr Int Conf Regulation of Developmental Processes in Plants. Halle, GDR, p 75
Jungnickel F (1984) Sämlingsunterschiede und erste Spontan–Varianten bei einer neuen Gesneriaceen-Hybride. Arch Züchtungsforsch 14: 193–201
Jungnickel F (1986) Zur Beziehung zwischen Modell-und Praxisobjekten bei der axenischen Mikrovermehrung Von Blütenpflanzen. Wiss Z Friedrich–Schiller–Univ, Jena, Math Nat R 35: 599–611
Jungnickel F (1988a) Strawberries (*Fragaria* spp and hybrids). In: Bajaj YPS (ed) Biotechnology in Agriculture and Forestry, vol 6: Crops II, Springer. Berlin Heidelberg, New York pp 38–103
Jungnickel F (1988b) Zur Einordnung von somaklonaler Variation und axenischer Mikrovermehrung in den Züchtungs- und Produktionsprozeß. In: Potsdamer Forschungen d Pädagog Hochschule Potsdam, Naturwiss Reihe 57: 109–111
Jungnickel F (1988c) Einige Probleme bei der Etablierung und Deponierung von Gefäßpflanzen. Wiss Z Friedrich–Schiller–Univ, Jena, Math Nat R 37: 85–90
Jungnickel F (1989) × *Brigandra calliantha*. Past, present and future prospects. Bull Alpine Gard Soc 57: 177–178, 181–182
Jungnickel F, Gliemeroth K (1986) In-vitro-Depots vegetativ vermehrter Blütenpflanzen. Wiss Z Friedrich–Schiller–Univ, Jena, Math Nat R 35: 631–638
Kelly JW, Lineberger RD (1981) Thermal neutron induced changes in *Saintpaulia*. Environ Exp Bot 21: 95–102
Kevers C, Coumans M, Coumans–Gillès MF, Gaspar T (1984) Physiological and biochemical events leading to vitrification of plants cultured in vitro. Physiol Plant 61: 69–74
Khokhar JA, Humphreys JM, Short KC, Grout BWW (1982) Anthocyanins in African violet. HortScience 17: 810–811
Köhn S (1974) Eine Nassfäule an Saintpaulien, hervorgerufen durch *Erwinia carotovora* var *chrysanthemi* Dye. Nachrichtenbl Dtsch Pflanzenschutzd 26: 161–162
Köhn S (1975) Die neue Bakteriose an Saintpaulien. Dtsch Gartenbau 29: 869–871
Konradsen P, Andersen H, Christensen O (1978) Energy-saving at Søhus 1976–77. (1–3). Gartner Tidende 7–10: 102–104, 132–134, 152–154 (in Danish)
Krebs E-K (1990) Krankheiten an Saintpaulien. Gärtnerbörse Gartenwelt 90: 1437–1440
Krebs G (1983) Saintpaulien. Bekämpfung der *Phytophthora*-Stammgrundfäule. Gärtnerbörse Gartenwelt 83: 575–577
Kühn (1984) Energiegünstige Topfblumen für die Frühjahrsversorgung. Gartenbau 31: 248
Kukułczanka K, Kwiecień J (1973) Modifications in regeneration of *Saintpaulia ionantha* Wendl. leaves, due to morphactin added to in vitro cultures. Acta Soc Bot Pol 42: 521–531

Kukułczanka K, Suszynska G (1972) Regenerative properties of *Saintpaulia ionantha* Wendl. leaves cultured in vitro. Acta Soc Bot Pol 41: 503–509

Lang PM (1987) Pflanzenschutzmittelverträglichkeit von Fungiziden und Insectiziden bei *Saintpaulia ionantha*. Zierpflanzenbau 27: 437–439

Langnes R, Gjaerum HB (1985) Effect of some fungicides on *Begonia*, *Saintpaulia* and *Verbena*. Gartneryrket 75: 266 (in Norwegian)

Larcher W, Bodner M (1980) Dosisletalität – Nomogramm zur Charakterisierung der Erkältungsempfindlichkeit tropischer Pflanzen. Angew Bot 54: 273–278

Larcher W, Bodner M (1987) Criteria for chilling stress in *Saintpaulia ionantha*. Angew Bot 61: 309–323

Larcher W, Neuner G (1989) Cold-induced sudden reversible lowering of in vivo chlorophyll fluorescence after saturating light pulses. A sensitive marker for chilling susceptibility. Plant Physiol 89: 740–742

Larson RA (1985) African violet chimeras: a practical use of micropropagation. N C Flower Growers' Bull 29(6): 9–10

Le CL, Collet GF (1981) Advantages et contraintes de la multiplication in vitro de *Saintpaulia ionantha* Wendl. Rev Suisse Vitic Arboric Hortic 13: 265–270

Leenhouts HP, Broertjes C, Sijsma HJ, Chadwick KH (1982) Radiation stimulated repair in *Saintpaulia*: its cellular basis and effect on mutation frequency. Environ Exp Bot 22: 301–306

Lemattre P (1975) Action des régulateurs de croissance, incidence sur la multiplication du *Saintpaulia*. Hortic Française 60: 5–9

Lemattre M, Narcy JP (1975) La culture du *Saintpaulia* face au problème nouveau – de la bactériose (*Erwinia chrysanthemi*) Hortic Française 62, 3–7

Lercari B, Tognoni F, Anselmo G, Chapel D (1986) Photocontrol of in vitro bud differentiation in *Saintpaulia ionantha* leaves and *Lycopersicon esculentum* cotyledons. Physiol Plant 67: 340–344

Lineberger RD, Druckenbrod M (1985a) Micropropagation of chimeral African violets. Res. Circular 284, Ohio Agric Res Dev Centre, Ohio State Univ, Wooster, USA, pp 25–28

Lineberger RD, Druckenbrod M (1985b) Chimeral nature of the pin-wheel flowering African violets (*Saintpaulia, Gesneriaceae*). Am J Bot 72: 1204–1212

Linsmaier EM, Skoog F (1965) Organic growth factor requirements of tobacco tissue cultures. Physiol Plant 18: 100–127

Loeser H (1973) Saintpaulien-Sortenvergleich 1972. Dtsch Gärtnerbörse 73: 136–139

Loeser H (1983) Saintpaulien–Sortenvergleiche 1979 bis 1982 in Heidelberg. Zierpflanzenbau 23: 100–109

Lutz W (1990) Top–Technik bestimmt die Saintpaulienproduktion in Rees-Haffen. Holtkamp setzt neue Maßstäbe in Verkauf und Produktion. Zierpflanzenbau Gartenbautechnik 30(1): 39–41, (41–43)

Maas PWT, Cevat HN (1973) Root–Knot nematodes on *Saintpaulia*. Vakbl Bloem 28(14) 15 (in Dutch)

Mabberly DJ (1987) The plant book: a portable dictionary of the higher plants. Cambridge Univ Press

Maekawa R, Torisu Y, Inagaki N, Terabun M (1987) Leaf injury caused by drop in leaf temperature of *Saintpaulia ionantha* (in Japanese). J Jpn Soc Hortic Sci 55: 484–489

Maekawa S, Torisu Y, Inagaki N, Terabun M (1990) Relation between development of leaf spot and electrolyte leakage of *Saintpaulia ionantha* leaves. J Jpn Soc Hortic Sci 58: 484–489

Marcotrigiano M, Stewart RN, Reply by Norris RE, Smith RH, Vaughn KC (1984) All variegated plants are not chimeras. Science 223: 505

Margara J, Piollat MT (1985) Evolution de l'apitude à l'organogenèse in vitro à partir des feuilles de *Saintpaulia ionantha* au cours des cultures successives. C R Séancls Acad Sci Ser III Sci Vie 301: 265–268

Michel K (1988) Licht und Wachstum. Gärtnermeister 91: 476–477

Milde H (1980) *Saintpaulia*. Vergleich von Langzeitdüngung. Gärtnerbörse Gartenwelt 80: 486–488

Milne C (1975) Chromosome numbers in the Gesneriaceae: V. Notes R Bot Gard Edinb 33: 523–525

Molitor HD, Hentig WU von, Fischer M (1986) Carborain: Wirkung bei Hydrokulturen nachweisbar. Gärtnerbörse Gartenwelt 86: 867, 870–872

Moncousin C (1978) Contribution à la mise au point d'un milieu de culture pour la multiplication in vitro de Gesnériacées. Rev Hortic Suisse 51: 295–301

Mortensen LM (1985) Nitrogen oxides produced during CO_2 enrichment. I. Effects on different greenhouse plants. New Phytol 101: 103–108

Müller FJ (1988) Storytella – Schöne Geschichten in Saintpaulien. Circular January 1988, Babenhausen 5. FRG, pp 1–5

Murashige T, Skoog F (1962) A revised medium for rapid growth and bioassays with tobacco tissue cultures. Physiol Plant 15: 473–497

Naylor EE (1940) African violets for the window garden: a note on their culture and propagation. J New York Bot Gard, 41: 166–168

Naylor EE, Johnson B (1937) A histological study of vegetative reproduction in *Saintpaulia ionantha*. Am J Bot 24: 673–678

Neumann D (1983) Anpassung höherer Pflanzen an extreme Lichtverhältnisse. Biol Rdsch 21: 137–141

Nölle HH, VanBemmel P, Härig H, Helmkampf P (1975) Ein Verfahren zur Auswertung von Düngungsversuchen bei Zierpflanzen. Landwirtsch Forsch 28: 83–94

Norris R, Smith RH, Vaughn KC (1983) Plant chimeras used to establish de novo origin of shoots. Science 220: 75–76

Nowaczyk E, Borys MW (1976) Rooting of *Saintpaulia ionantha* leaf cuttings. I. The effect of petiole length. Roczniki Akad Rolniczej Poznaniu, Ogrodnictwo, 85: 137–142

Onstenk R (1976) It is advantageous to grow Saintpaulias in a glasshouse with double glass. Vakbl Bloem 31(48): 88–89 (in Dutch)

Paludan N (1971) The establishment of virus-free meristem tip cultures of horticultural plants. Tidsskr Planteavl 75: 387–410 (in Danish)

Park CH, So IS, Kwack BH (1983) Micropropagation for in vitro culture of *Saintpaulia* at controlled pH and various sucrose levels. J Kor Soc Hortic Sci 24: 158–161

Peary JS, Lineberger RD, Malinich TJ, Wertz K (1988) Stability of leaf variegation in *Saintpaulia ionantha* during in vitro propagation and during chimeral separation of a pinwheel flowering form. Am J Bot 75: 603–608

Peper H, Döpcke H, Brandis A von (1977a) Saintpaulien–Sortiment. Der Einfluß unterschiedlicher Temperaturen Gärtnerbörse Gartenwelt 77: 1185–1187

Peper H, Döpcke H, Brandis A von (1977b) Saintpaulien–Sortiment. Prüfung bei unterschiedlichen Temperaturen. Gärtnerbörse Gartenwelt 77: 1204–1206

Pivot D (1982) Mise en culture de *Saintpaulia* issus de multiplication in vitro: utilization de l'eclairage et économie d'énergie. Rev Suisse Vitic Arboric Hortic 14: 307–312

Plate HP, Kröber H (1972) Eine neue *Phytophthora*-Krankheit an Saintpaulien. Gartenwelt 72: 513–515

Ploetz RC, Engelhard AW (1980) Crown and root rot of *Gloxinia* and other gesneriads caused by *Phytophthora parasitica*. Plant Dis 64: 487–490

Pohlheim F (1980) Periklinalchimärische Anthozyanmuster bei *Saintpaulia ionantha* H. Wendl nach NMH–Behandlung. Arch Züchtungsforsch 10: 261–269

Pohlheim F (1981) Genetischer Nachweis einer NMH–induzierten Plastommutation bei *Saintpaulia ionantha* H. Wendl. Biol Rdsch 19: 47–50

Pohlheim F (1983) Vergleichende Untersuchungen zur Änderung der Richtung von Zellteilungen in Blattepidermen. Biol Zbl 102: 323–336

Pohlheim F, Pohlheim E (1976) Herstellung von Plastommutanten bei *Saintpaulia ionantha* H. Wendl. Biochem Physiol Pflanz 169: 377–383

Poole RT, Conover CA, Ozeri Y (1986) Responses of African violets to fertilizer source and rate. HortScience 21: 454–455

Raabe RD, Hurlimann JH (1978) Control of *Phytophthora* rot of African violet. Progress report. Flower and Nursery Rep, Winter, Cooperative Extension, Berkeley, Calif: 5–6

Rodojevic L, Vapa L, Borojevic K (1984) Vegetative in vitro propagation of *Saintpaulia ionantha* and *Pelargonium zonale* hybrids. Novi Sad 32: 9–10 (in Slovenian)

Radojevic L, Vapa L, Borojevic K (1985) Plant regeneration in anther cultures of *Saintpaulia ionantha* Wendl. Novi Sad 33: 485–491 (in Slovenian)

Rao AN, Morel G (1973) Effects of *Agrobacterium* sp. on petiole cultures of *Saintpaulia*. Phytomorphology 23: 1–9

Ratter JA (1975) A survey of chromosome numbers in the *Gesneriaceae* of the old world. Notes R Bot Gard Edinb 33: 527–535

Rattink H (1982) Disinfection of potting compost by gamma rays, as an alternative. Vakbl Bloem 37(36): 44–45 (in Dutch)

Read PE, Herman VL, Gavinlertvatana P (1973) Controlled release growth regulators for plant growth control and rooting of cuttings. Comb Proc Int Plant Prop Soc 23: 289–296

Relichova J (1984) Induction of somatic mutations in some vegetatively propagated plants. Genetika a slechteni (Praha) 20: 85–94 (in Czech)

Reuther G (1986) Eigenschaften von Gewebekulturpflanzen und deren Auswirkungen auf die Weiterkultur. Zierpflanzenbau 26: 172–174
Rewinkel-Jansen MJH, VanDerBroek GJ, Drijver C (1985) Flowering pot plants should not be offered for sale when too immature. Vakbl Bloem 40(44) 38–39 (in Dutch)
Richter S, Winkler G (1973) Pflanzenschutzmittelverträglichkeit bei Zierpflanzen. Gartenbau 20: 117–118
Röber R, Fischer M (1980) Vermehrungssubstrate. Zur Anzucht von Pflanzen für die Hydrokultur. Gärtnerbörse Gartenwelt 80: 132, 134, 136
Röber R, Leinefelder J, Achatz A, Hege H (1984) Doppelsteg-Eindeckung hat Vorzüge. I., II. Dtsch Gartenbau 38: 1565–1567, 1666–1668
Röber R, Lohweg, E, Haas HP (1985) Warum verpilzen Torfkultursubstrate? Dtsch Gartenbau 39: 1098–1100
Röber W, Horn W (1985) Vergleichsanbau in vitro und Blattstecklinge. Saintpaulien–Jungpflanzen aus Gewebekultur sind Blattstecklingen gleichwertig. Gärtnerbörse Gartenwelt 85: 199–200
Roberts BJ (1977) Susceptibility of certain *Saintpaulia* species and cultivars to bacterial blight. Plant Dis Rep 61: 1048–1050
Roberts EP (1952) A key to the genus *Saintpaulia*. Afr Viol Mag 6: 24–27
Rogers OM (1954) Some chromosome counts in the Gesneriaceae. Baileya 2: 14–18
Salka J, Thinggaard K (1988) *Phytophthora* rot in *Saintpaulia*. Biological control. Gartner Tidende 104: 500–501 (Danish)
Sanderson KC (1987) Increasing crown and leaf development of African violet cuttings during propagation with growth regulator sprays. HortScience 22: 1162
Sanderson KC, McGuire J (1988) Growth regulator sprays during propagation increase African violet crowns and leaves. HortScience 23: 1085
Sanderson KC, Martin WCJr, Shu LJ, Evans CE, Patterson RM (1985) Fertilizer and irrigation effects medium leachate and African violet growth. HortScience 20: 1062–1065
Schlegel G (1982) Influence of low temperature on regeneration of *Saintpaulia ionantha* H. Wendl. In: Abstr 21st Intern Horticult Congr Hamburg, FRG, Int Soc Hortic Sci, vol II, Abstr No 1719
Schlegel G (1983) Gewebekulturen bei *Saintpaulia* und *Kalanchoe* auf der IGA 83. Dtsch Gartenbau 37: 1852–1855
Schlegel G, Schneider–Maessen R (1981) Einfluß von Licht auf die Regeneration von *Saintpaulia ionantha* H. Wendl. in vitro. Gartenbauwissenschaft 46: 106–114
Schmidt P, (1981) Granulate im Zierpflanzenbau. Drei bis zehn Samen je Granulatkorn – Direktaussaat in Verkaufseinheiten möglich. Dtsch Gartenbau 36: 2136
Schneider-Moldrickx R, Amberger S (1982) Propagation of *Saintpaulia* under artificial lighting. Acta Hortic 128: 151–159
Schnell HE (1972) Sortenvergleich bei Saintpaulien. Gartenwelt 72: 478, 480
Schulze D (1988) *Saintpaulia ionantha* H. Wendl. – In vitro propagation and acclimatization in a commercial laboratory. Acta Hortic 226, II: 619–622
Schwemmer E (1980) *Saintpaulia ionantha* in Hydrokultur. Dtsch Gartenbau 34: 1046–1047
Schwemmer E (1990) Depotdünger bei *Saintpaulia* in Hydrokultur. Gärtnerbörse Gartenwelt 90: 1429–1431
Sehm E (1974) Neue Topflanzensorten – ein Beitrag der Züchtung zur besseren Versorgung der Bevölkerung. Gartenbau 21: 152–154
Short KC, Wardle K, Grout BWW, Simkins J (1984) In vitro physiology and acclimatization of aseptically cultured plantlets. In: Novak FJ, Havel L, Dolezel J (eds) Plant Tissue and Cell Culture – Application to Crop Improvement, Czechoslovak Acad. Sci, Prague, pp 475–486
Smith RH, Norrris RE (1983) In vitro propagation of African violet chimeras. HortScience 18: 436–437
So IS (1983) Studies on in vitro culture of *Saintpaulia ionantha* Wendl. I. Effect of different combinations of growth regulators on organogenesis. J Kor Soc Hortic Sci 24: 86–91
Start ND, Cumming BG (1976) In vitro propagation of *Saintpaulia ionantha* Wendl. HortScience 11: 204–206
Steinbuch F (1982) Promising results from winter cultivation of *Saintpaulia*. Vakbl Bloem 37(21): 48–49 (in Dutch)
Sterling EP, Molenaar WH (1982) First results of research on the suitability of pot plants for transport. Vakbl Bloem 37(42) 78–81 (in Dutch)
Steven HH (1985) Mini–Ballett–Saintpaulien suchen sich neuen Markt: Echte Neuzüchtungen für den 5,5- oder gar den 4-cm-Topf von Fischer. Gärtnerbörse und Gartenwelt 85: 187–188

Stokes MJ (1974) Plant propagation by means of aseptic techniques. Comb Proc Int Plant Prop Soc 24: 196–206
Strider DL (1978) Reaction of African violet cultivars to *Phytophthora nicotianae* var. *parasitica*. Plant Dis Rep 62: 112–114
Strider DL (1979a) Control of *Aphelenchoides ritzemabosi* in African violet. Plant Dis Rep 63: 378–382
Strider DL (1979b) Evaluation of Melodie and Optimara African violets for resistance to *Phytophthora* rot. Plant Dis Rep 63: 382–384
Strider DL (1980) Resistance of African violet to powdery mildew and efficacy of fungicides for control of the disease. Plant Dis Rep 64: 188–190
Sugiura T (1936) Studies on the chromosome numbers in higher plants with special reference to cytokinesis. Cytologia 7: 544–595
Sugiura T (1940) Studies in the chromosome numbers in higher plants. III., IV., V. Cytologia 10: 205–212, 324–333, 336–370
Sunblade TM, Meyer MM (1982) In vitro culture of variegated *Saintpaulia* and *Episcea* plants. HortScience 17: 104
Takayama S, Misawa M (1982) Mass propagation of ornamental plants through tissue culture. In: Fujiwara A (ed) Plant tissue culture 1982, Maruzen Tokyo, Japan Assoc Plant Tissue Culture pp 681–682
TerHell B, Ludolph D (1990) Die Zukunft liegt in der Qualität. Kundenseminar bei Saintpaulien Fischer in Hannover. Zierpflanzenbau Gartenbautechnik 30(1): 44–45
TGL (1988) Zierpflanzen, *Saintpaulia ionantha* (Usambaraveilchen) Fachbereichsstandard DDR, TGL 26023, 28. 1. 1988, Quedlinburg, GDR
Theiler R (1981) Stand der Gewebekulturarbeiten an der Eidg. Forschungsanstalt Wädenswil. Obstbau 6: 97–103
Thinggaard K (1989) Biological control of root-pathogenic *Phytophthora*, *Pythium* and *Phomopsis* in greenhouse crops. Växtskyddsnotiser 53: 1–2, 25–29 (in Danish)
Tooze SA (1986) Shadowing of young saintpaulias. Diminished light has little influence on rapid growth. Vakbl Bloem 41(17): 56–57, 59 (in Dutch)
Tusnádi CK (1986) Tobacco thrips on African violets. Kertészet és Szölészet 35(22): 9 (in Hungarian)
Umgelter H (1979) Versuche der Landesanstalt für Pflanzenschutz Stuttgart. Dsch Gartenbau 33: 456–458
VanDeVrie M (1985) Recognition and control of thrips in ornamental crops. Vakbl Bloem 40(47) 56–57 (in Dutch)
VanWilson HP (1979/1980) *Saintpaulia* species: a new interest. Am Hortic (Dec/Jan): 35–39
Vazquez AM, Short KC (1978) Morphogenesis in cultured floral parts of African violet. J Exp Bot 29: 1265–1271
Vazquez AM, Davey MR, Short KC (1977) Organogenesis in cultures of *Saintpaulia ionantha*. Acta Hortic 78: 249–258
VonDerBeek A (1990a) Saintpaulien Holtkamp – wo Jungpflanzen produziert werden. Gärtnerbörse Gartenwelt 90: 1413–1416
VonDerBeek A (1990b) Neue Sorten bei Fischer's Ballett–Saintpaulien. Gärtnerbörse Gartenwelt 90: 1433
DeWaard MA, VanRooijen CI (1978) Chemical control of root and crown rot in *Saintpaulia*. Meded Fac Landbouwwet, Rijksuniv Gent 43: 951–959
Warburton J, Grout BWW, Short KC (1984) In vitro selection for cold-tolerant cell lines of *Saintpaulia ionantha* (Wendl). In: Novak FJ, Havel L, Dolezel J (eds) Plant Tissue and Cell Culture – Application to Crop Improvement. Czech Acad Sci, Prague, pp 355–356
Warfield D (1973) Induction of mutations in African violet (*Saintpaulia ionantha* Wendl.) by ethyl methanesulfonate. HortScience 8: 29
Weatherhead MA, Grout BWW, Short KC (1982) Increased haploid production in *Saintpaulia ionantha* by anther culture. Sci Hortic 17: 137–144
Wendland H (1893) *Saintpaulia ionantha*. Gartenflora 42: 321–323
Wild W (1974) Morphologische Unterschiede zwischen Wurzeln aus Erdkultur und Wurzeln aus Hydrokultur. Schweiz Gärtnerztg 77: 449–452
Williams DJ (1985) Mealy bugs of the genus *Rhizoecus* (*Hemiptera, pseudococcidae*) on African violets (*Saintpaulia* spp.) with description of a new species. B. Ent. Res. 75: 621–624
Wilson GR (1951) A note on the cytology of saintpaulias. Afr Viol Mag 5(2) 18–19
Winkler H (1935) Chimären und Burdonen. Der Biologe, München 4(9): 279–290

Wohanka W, Doll W (1985) Was leisten Desinfektionsmittel in der Praxis? Dtsch Gartenbau 39: 1482–1483

Woltering EJ (1985) Blüten- und Blattfall durch Streßfaktor Äthylen. Gärtnerbörse Gartenwelt 85: 1303–1305

Worrall RJ (1981) Comparison of composted hardwood and peat-based media for the production of seedlings, foliage and flowering plants. Sci Hortic 15: 311–319

Woszczynska K, Borys MW (1976a) Rooting of *Saintpaulia ionantha* leaf cuttings. II. The effect of concentration, duration of treatment with growth regulators and time of striking. Roczniki Akad Rolniczej Poznaniu, Ogrodnictwo 85: 171–179

Woszczynska K, Borys MW (1976b) Rooting of *Saintpaulia ionantha* leaf cuttings. III. The effect of sequenced application of growth regulators. Roczniki Akad Rolniczej Poznaniu, Ogrodnictwo 85: 181–188

Xu LQ (1984) The promotive effects of ginseng on rapid clonal in vitro propagation of African violet (*Saintpaulia ionantha* Wendl.). Acta Bot Sin 26: 489–498

Zaid S (1989) Morphogenetische Untersuchungen an axenisch kultivierten Gesneriaceen. Thesis, Friedrich-Schiller-Univ, Jena, pp 1–108

Zaid S, Jungnickel F (1988) Entstehen Adventivsprosse bei Gesneriaceen aus einer oder aus mehreren Zellen? In: Potsdamer Forschungen d Pädagog Hochschule Potsdam, Naturwiss Reihe 57: 211–212

Zalesskij DM (1983) (ed) Saintpaulias, Leningrad Univ (in Russian)

Zhang PF, Ni DX, Wang KJ, Bao ZH (1982) Morphological studies on embryoid formation in leaf culture of *Saintpaulia ionantha.* Acta Bot Sin 24: 282–284

Zhang PF, Ni DX, Wang Q, Wang KJ (1986) Hormonal control of somatic embryogenesis and callus growth of *Saintpaulia ionantha* in vitro. Acta Bot Sin 28: 446–449

Zimmer K (1985) Temperatur-Untersuchungen an Saintpaulien. Dtsch Gartenbau 39: 902–904

XXIV Micropropagation of Cineraria (*Senecio* × *hybridus* Hyl.)

U.E. Gertsson[1]

1 Introduction

Senecio × *hybridus*, commonly known as cineraria, (family Asteraceae) originated in the Canary Islands. It can grow and flower with low light intensities and at low temperatures. It is therefore suitable as a flowering pot plant for winter production and is commonly grown in northern latitudes. Usually *Senecio* × *hybridus* is raised from seed. As seedling plants are heterogeneous, for example, in flower color and developmental rates, it would be valuable to have a method by which selected clones can be propagated. It is possible to produce multiple shoots from seeds sown in vitro (Cockrel et al. 1986). For most purposes it is an advantage to make the selection in a flowering population. If the selected plants are propagated vegetatively, a progeny clone with uniform development and appearance should be obtained from each plant. One prerequisite for this is to have a large-scale propagation method that starts from mature plants.

The growth and development of micropropagated plants can be regulated by treatments during the in vitro phase. The flowering of the *Senecio* cultivar described below, Gmünder Zwerg, is promoted by maintaining a low temperature for a few weeks (Hildrum 1969; Larsen 1985). Therefore, it is interesting to study how low temperature treatment during the micropropagation can produce an early flowering crop.

2 In Vitro Studies

Papers on micropropagation have been published by Cockrel et al. (1986) and Gertsson (1986, 1988a,b,c). An important group of secondary compounds found in the genus *Senecio* is pyrrolizidine alkaloides. In vitro-cultured roots of some species are capable of synthesizing alkaloids, specially senecionine N-oxide (Toppel et al. 1987). These roots accumulate quantities which are comparable to those found in soil-grown plants. Therefore in vitro grown roots are an excellent system to study alkaloid biochemistry.

[1] The Swedish University of Agricultural Sciences, Department of Horticulture, Box 55, 230 53 Alnarp, Sweden

Biotechnology in Agriculture and Forestry, Vol. 20
High-Tech and Micropropagation IV (ed. by Y.P.S. Bajaj)

When using generative stock plants for propagation purposes petiole pieces are suitable as explants. Some investigations, into the physiology of these explants were published earlier (Gertsson 1986, 1988c) and are summarized below.

2.1 Morphogenesis

The different kinds and combinations of auxin and cytokinin used influence the way in which the explants develop (Table. 1). It is obvious that IAA promotes shoot formation more than NAA does, and that more shoots are developed by BAP than by kinetin, zeatin, or 2-iP. Root formation is better stimulated by NAA than by IAA. When zeatin or 2-iP is used, more explants form roots than with kinetin or BAP. In general, root and callus formation show similarities in their relationship with the growth regulators. The concentration of growth regulators is decisive for explant development. For example, when 5 mg/l of both BAP and IAA is used, only shoots are formed; however, if the BAP concentration is lowered to 0 or 0.1 mg/l, roots and callus form instead.

2.2 Polarity

Most explants grown with their morphological base pointing downwards expand at the lower end, whereas explants grown base upwards show a tendency to increase in size at both ends. With inverted (base upwards) explants it is found that an increase in auxin concentration or a decrease in cytokinin concentration in the medium will increase the proportion of explants with growth oriented only at the lower end. Callus growth is more prolific on base-up than on base-down explants. The number of organs formed is also influenced by the explants' orientation. More explants form shoots when grown with the morphological base pointing downwards or sideways than when oriented upwards. The number of shoots formed is influenced in the same way. Largest root numbers are obtained on inverted explants.

Table 1. Effect of kind of cytokinin (4.6×10^{-6} M) and auxin (28.5×10^{-6} M) on explant development. (Gertsson 1986)

	Percentage of explants with					
	Shoots		Roots		Callus	
	IAA	NAA	IAA	NAA	IAA	NAA
Kinetin	15	2	13	5	0	7
BAP	36	12	3	12	0	3
Zeatin	3	3	26	86	0	53
2-iP	5	3	59	95	18	62

2.3 Nitrogen Uptake

Nitrogen uptake by the petiole explants is influenced by the content of auxin, nitrogen, and potassium in the medium. An increased concentration of nitrogen in the medium gives an increased nitrogen content in the explants–except at high concentrations, where no further increase in explant content results. A reduced potassium concentration may increase the nitrogen content of the explants. Conversely, an increase in auxin concentration may reduce the nitrogen content. Moreover, the nitrogen content will be greater when IAA rather than NAA is used. On the other hand, neither the ratio between ammonium and nitrate, nor the cytokinin content of the medium affects the nitrogen uptake.

3 Micropropagation

3.1 Shoot Formation from Petiole Explants

The propagation method described below was developed in our laboratory for the *Senecio* × *hybridus* cultivar Gmünder Zwerg (Gertsson 1986, 1988a, c). Petioles from both vegetative and generative stock plants can be used. Whole petioles are surface sterilized for 30 s in 70% ethanol and then in 6% Ca-hypochlorite for 12 min. After three washes in sterile distilled water they are cut into 4-mm-long pieces and distributed into test tubes with solid medium (Fig. 1).

Fig. 1. Petiole explants immediately after excision (*left*) and with shoots (*center*) or callus and roots (*right*) 5 weeks later

The medium always contains the same micronutrients (Welander 1979) and organic constituents (Bourgine and Nitsch 1967), plus 30 g/l sucrose and 6 g/l Difco Bacto agar. The best proliferation of adventitious shoots is obtained when the macronutrients are according to Murashige and Skoog (1962) with reduced concentrations of nitrogen (30 mg/l) and potassium (15 mg/l) (Table 2). This

Table 2. Media used for shoot production

Components	Shoot proliferation from petiole explants (mg/l)	Multiple shoot formation during subculture (mg/l)
Macronutrients		
NH_4NO_3	825	1650
KNO_3	950	1900
$CaCl_2.2H_2O$	440	440
KCl	328	–
$MgSO_4.7H_2O$	370	370
KH_2PO_4	170	170
Micronutrients		
Fe-Na-EDTA	40.3	40.3
$MnSO_4.H_2O$	19.0	19.0
$ZnSO_4.7H_2O$	12.4	12.4
H_3BO_3	10.0	10.0
KJ	0.8	0.8
$NaMoO_4.H_2O$	0.25	0.25
Organic constituents		
Inositol	100	100
Glycine	2.0	2.0
Nicotinic acid	5.0	5.0
Pyridoxine	0.50	0.50
Thiamine	0.50	0.50
Folic acid	0.50	0.50
Biotin	0.05	0.05
BAP	1	0.1–1
IAA	5	0.1
TIBA	0.5	–
Sucrose	30000	30000
Agar	6000	6000
pH	6.0	6.0

Table 3. Effects of various concentrations of BAP and IAA on percentage of petiole explants forming shoots. (Gertsson 1988a)

BAP	IAA, mg/l		
mg/l	2	4	8
	%	%	%
0.5	12	19	19
1	14	17	21
2	9	19	4
4	9	6	13

Table 4. Influence of TIBA on shoot formation from petiole explants. The media contained 1 mg/l BAP and 5 mg/l IAA. (Gertsson 1988c)

	TIBA, mg/l		
	0	0.01	0.5
Explants with shoots, %	22	26	56

Table 5. Effect of dark treatment on shoot formation from petiole explants. The dark treatments were carried out at the beginning of a 6-week growth period immediately after explant exision. The media contained 1 mg/l BAP and 5 mg/l IAA. (Gertsson 1988c)

	Duration of dark treatment, days				
	0	1	7	14	21
Explants with shoots, %	40	45	65	74	62

decrease in nitrogen and potassium content increases the number of shoot-forming explants threefold (Gertsson 1988c). The standard addition of growth regulators is 1 mg/l of BAP and 5 mg/l of IAA (Table 3). Shoot production is further stimulated by adding 0.5 mg/l of TIBA to the medium (Table 4). The pH is usually adjusted to 6.

The test tubes are placed in 18–21 °C under cool white fluorescent lamps for a 20 h photoperiod, the light intensity being about 5 klx. The shoot proliferation is markedly increased when the explants are grown in darkness during the first 2 weeks after explant excision (Table 5). After a total culture period of 6 weeks the shoots formed on the petiole pieces are of a suitable size to be transferred to fresh medium.

3.2 Multiple Shoot Formation

In the second propagation step the shoots formed on the petiole explants are excised and subcultured on fresh medium in order to produce more shoots. Every subculture lasts 5 weeks (Fig. 2). The same growth chamber climate is used as for the petioles. Except for the BAP, IAA, and TIBA concentrations and the macronutrients even the medium used is the same (Table 2). TIBA has not yet been tested for this purpose and is therefore not used.

Macronutrients according to Murashige and Skoog (1962) are suitable. The shoot number does not increase if the nitrogen concentration is reduced, as is the case for petiole explants. A high IAA concentration is necessary to obtain shoot formation from petiole explants. Later, when the shoots produced are subcultured, the IAA content must be reduced to about 0.1 mg/l in order to avoid inhibition of shoot formation.

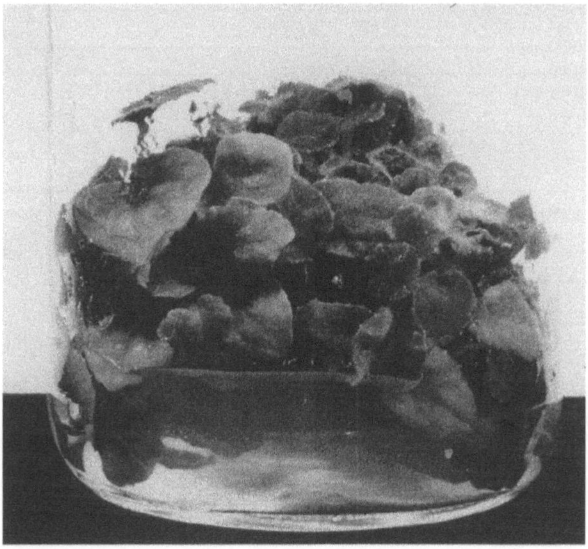

Fig. 2. Multiple shoot formation in separated shoots 5 weeks after transfer to fresh medium

As cytokinin, BAP is usually used, but kinetin is also suitable. The optimal BAP concentration is influenced by three factors, namely the clone, the IAA concentration, and the BAP concentration used in the preceding subcultures. A high BAP concentration (e.g., 1 mg/l) is required in the first subculture in order to achieve prolific shoot production in this subsequent subcultures. Even when the BAP concentrations are kept low (e.g., 0.1 mg/l) in the subsequent subcultures, shoot production will be high. Later in the propagation scheme the BAP content has to be increased to maintain a high level of shoot production. It is difficult to restore this ability once it has been lost. Since shoots produced on medium with a low BAP content are larger than those grown on a BAP-rich medium, it is advisable to alter the BAP concentration according to the condition of the shoots and the objective of the current subculture. For example, the medium used in the last subculture before rooting must not contain any BAP–or only a trace. The effects on shoot formation, of varying BAP concentrations in subsequent subcultures, are shown in Fig. 3.

3.3 In Vitro Storage

It is possible to store shoots in vitro for at least 6 months. A suitable climate is 10 °C and illumination with cool white fluorescent lamps for a 20-h photoperiod, the light intensity being about 400 lx. Shoots stored at lower light intensity do not survive. During the storage period the shoots are not transferred to fresh medium. Usually nearly all shoots survive and, although rather straggly, they are green and healthy and can be rooted in soil. The shoot production rate is low during storage, but when the stored shoots are subcultured in normal conditions they produce

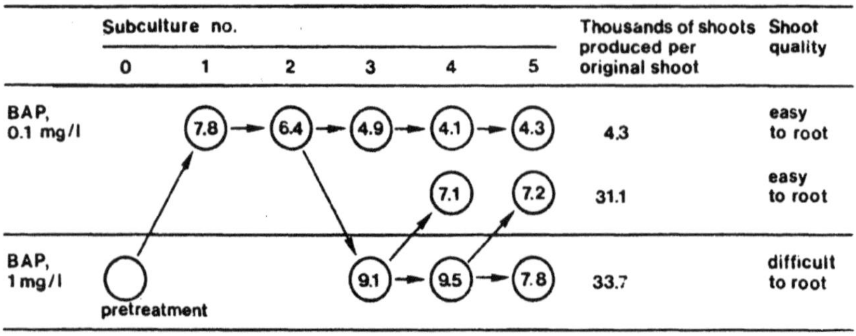

Fig. 3. Effects of repeated subculture at different BAP concentrations on the number of shoots produced per shoot, each subculture lasting 5 weeks. Mean values are given, and the LSD is 1.7 ($P = 0.05$). (Gertsson 1988a)

about the same number of shoots as in a corresponding subculture of nonstored shoots.

3.4 Plant Establishment in Soil

Shoots produced during the subcultures can be separated from one another and rooted in a well drained compost or soil without any preparatory treatment. Some 80–100% of the large shoots grown on media with no or a low BAP content (0.1 mg/l) generally form roots and survive, compared with about 50% of the small shoots raised at a high BAP concentration (1 or 5 mg/l).

4 Growth and Development of Micropropagated Plants

The development of micropropagated *Senecio* plants is influenced by three main factors, namely the clone, the BAP concentration, and the temperature applied in vitro. The relations between these factors and plant characteristics are schematically illustrated in Fig. 4. The account below is based on results presented earlier (Gertsson 1988b).

4.1 Cytokinin Concentration

Plants originating from shoots grown at the BAP concentrations of 0. 0.1, 1, and 5 mg/l in the last subculture have been compared (Fig. 5). The number of flowering shoots per plant increases with increasing BAP concentration in the interval 0 to 1 mg/l. At 1 mg/l BAP, this increase in shoot number results in plants with delayed flower development and large numbers of flowers and leaves. However, even when plants with only one shoot are compared, developmental time and leaf number increase with increasing BAP concentration from 0 to 1 mg/l. Plants grown from

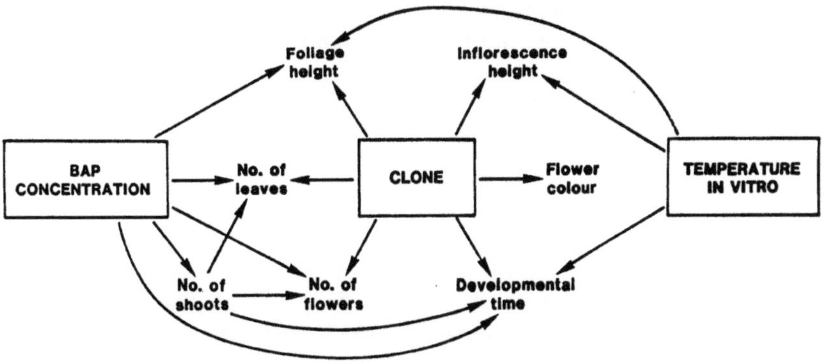

Fig. 4. Schematic diagram showing how choice of clone, BAP concentration, and temperature in vitro influences the plant's development. (Gertsson 1988b)

Fig. 5. Appearance of plants from the same clone, propagated on media with 0, 0.1, or 5 mg/l BAP in the last subculture in vitro

5 mg/l BAP also develop slowly and their growth is inhibited, resulting in few flowers per shoot, low foliage height, and few leaves. Neither the inflorescence height nor the flower's color is dependent on the BAP concentration used.

4.2 Temperature in Vitro

The development of two groups of plants, treated differently during the last 15 weeks in vitro, has been compared (Gertsson 1988b). The medium contained 0.1 mg/l of BAP and IAA. The first group of plants were grown at 21 °C in 4–5 klx, and the shoots were subcultured every 5th week. In the second group, the shoots were stored at low temperature, (10 °C) in 400 lx, and were not transferred to fresh medium. It was observed that plants raised at 10 °C in vitro develop more rapidly than plants raised at 21 °C. Furthermore, the low temperature results in plants with decreased foliage height and increased inflorescence height. On plants grown at

10 °C, generally all the shoots flower, compared with about 80% generative shoots on the plants raised at 21 °C. Since the total number of shoots is greatest for plants grown at 21 °C, the number of flowering shoots is evidently not influenced by temperature. Neither the number of leaves nor of flowers is affected by decreased temperature.

4.3 Comparison Between Micropropagated Plants and Seedlings

Usually seedling plants need the same development time until flowering as do the micropropagated clones with the slowest flower development. Selected, quick developing clones can flower about 5 weeks earlier. Generally speaking, micropropagated plants have lower foliage and higher inflorescences than do seedlings. These differences are most pronounced after cold treatment in vitro. Many micropropagated plants have more than one shoot, whereas seedlings always have one shoot each. Seedlings have more leaves at anthesis than micropropagated plants with one shoot, but the same number of flowers.

4.4 Clone

The color of flowers of the progeny clone is always the same as that of the stock plants, and it is not affected by any of the treatments in vitro. All the other plant characteristics can be modified by the micropropagation. Since the clones react differently to the in vitro treatments, micropropagation results in clones between which there are greater differences in plant development and appearance than between their seedling stock plants. The number of shoots is the only parameter that does not vary between different clones. Therefore, it is important to select clones that are both easy to propagate and have a pleasing appearance, many flowers, and rapid flower development after micropropagation. The probability of obtaining a clone with short development time is greatest when an early-flowering stock plant is chosen. Except for the flower's color, it is possible to predict the other plant characteristics of a micropropagated clone by studying the seedling stock plants.

4.5 Large-Scale Production

The in vitro-produced shoots can be easily rooted in soil either in a growth chamber or in a greenhouse. There are no problems with the transfer to the greenhouse climate and the micropropagated plants can be grown in the same way as seedling plants.

5 Summary and Conclusions

It is possible to propagate *Senecio* × *hybridus* Gmünder Zwerg on a large scale in vitro. By selecting suitable clones, uniform populations of plants with desired bloom colors, good appearance, and quick development can be obtained. It is possible to grow populations which develop flowers more than 1 month earlier than do seedling populations. To achieve this, it is necessary to have a low concentration of cytokinin–or to omit it altogether–in the last subculture before rooting. Flower development can be further accelerated if a low temperature treatment is started in vitro. This is probably due to an early vernalization. Micropropagated plants have compact foliage, and when they produce more than one shoot, they also have more flowers than do seedlings.

When the propagation system was developed, the experiments were concentrated on ascertaining the various factors that influence it. More research is needed to establish the correlations between these factors and to describe their influence in greater detail; there are numerous problems, both fundamental and practical, whose solution would allow further advances. For example, anatomical studies may give some ideas about what happens during the 2 weeks of dark treatment following explant excision. It is obvious that the inherent polarity of explants is modified by the growth regulators in the medium. However, our knowledge about this is not yet complete. It would also be interesting to study the fluctuations in endogeneous cytokinin content in relation to various concentrations in the medium during the multiple shoot formation. Furthermore, it may be possible to increase shoot proliferation at low temperatures. Information on the reaction of other cultivars would also be valuable.

6 Protocol

6.1 Protocol for Selection of Stock Plants

A) Selection of stock plants with desirable flower colors and early flower development. The flower color of the progeny clone is always the same as that of the stock plant. The probability of obtaining clones with short development times increases when the stock plants have flowered early.

B) Selection of the clones which could be successfully propagated. These clones are then propagated and it is important to omit or have low concentrations of BAP in the last subculture

C) Selection of micropropagated clones with pleasing appearance, many flowers, and short development time.

D) Micropropagation of ultimately selected clones.

6.2 Protocol for Micropropagation

A) Adventitious shoots are formed at 4–mm–long pieces of petioles, treated as described in section 3.1 above. The medium is according to Table 2.

B) Repeated subcultures of excised shoots according to section 3.2 above. The medium is presented in Table 2. A high BAP concentration is necessary in the first subculture to initiate prolific shoot production. However, in subsequent subcultures, the BAP concentration can be varied between 0.1 and 1 mg/l, depending on the condition and pretreatment of the shoots and the purpose of the subculture.

C) Separated shoots are rooted in soil as cuttings. It is easier to root the shoots when the BAP concentration in the last subculture did not exceed 0.1 mg/l

References

Bourgine JP, Nitsch JP (1967) Obtention de *Nicotiana* haploides a partir d' étamines cultivées in vitro. Ann physiol Veg 9: 377–382

Cockrel AD, McDaniel GL, Graham ET (1986) In vitro propagation of florists' cineraria. HortScience 21: 139–140

Gertsson UE (1986) Influence of explant orientation and media composition on organ formation growth and nitrogen content in petiole explants of *Senecio × hybridus* Hyl. J Hortic Sci 61: 121–128

Gertsson UE (1988a) Large-scale in vitro propagation of *Senecio × hybridus* Hyl. J Hortic Sci 63: 131–136

Gertsson UE (1988b) Development of micropropagated plants from different clones of *Senecio × hybridus* in relation to BAP concentration and temperature in vitro. J. Hortic Sci 63: 489–496

Gertsson UE (1988c) Influence of macronutrient composition, TIBA and dark treatment on shoot formation and nitrogen content in petiole explants of *Senecio × hybridus*. J Hortic Sci 63: 497–502

Hildrum H (1969) Factors affecting flowering in *Senecio cruentus* D. C. Acta Hortic 14: 117–123

Larsen R (1985) Temperature and flower induction of *Senecio × hybridus* Hyl. Swed J Agric Res 15: 87–92

Murashige T, Skoog F (1962) A revised medium for rapid growth and bio-assays with tobacco tissue cultures. Physiol Plant 15: 473–497

Toppel G, Witte L, Riebesehl B, Borstel K, Hartmann T (1987) Alkaloid patterns and biosynthetic capacity of root cultures from some pyrrolizidine alkaloid producing *Senecio* species. Plant Cell Rep 6: 466–469

Welander T (1979) Influence of medium composition on organ formation in explants of *Begonia × hiemalis* in vitro. Swed J Agric Res 9: 163–168

XXV Micropropagation of Lilac (*Syringa vulgaris* L.)

R.L.M. PIERIK, H.H.M. STEEGMANS, and P.A. SPRENKELS[1]

1 Introduction

The genus *Syringa*, containing about 30 species, belongs to the family Oleaceae. It originated from Central and S.E. Europe, the Himalayas, N.E. Asia, and Japan. *Syringa vulgaris* L., the most important species of lilac used for ornamental purposes, is native to Central and S.E. Europe. In 1563 it was imported from Constantinople to the Netherlands (Bailey 1917).

Lilac is a shrub normally planted for ornamental purposes, but a very important commercial application in The Netherlands and some other European countries, is the cultivation as "forced shrub" for the production of flowering branches during winter. In The Netherlands the cultivation of shrubs for forcing has been concentrated for over 100 years in the region around Aalsmeer near Amsterdam. About 75 lilac growers in the Netherlands possess one million lilac bushes for forcing. Due to the loss of about 10% of the forcing stock per year, new stock is constantly required. In other European countries, lilac is less frequently cultivated for forcing in greenhouses. Although forced shrubs represent a relatively small section of the Dutch flower industry, they are an interesting and important product group, especially for export. Forced lilacs form an essential supplement in the supply of cut flowers, particularly in winter. In 1986 about 11 million lilac branches were produced in The Netherlands from an area of 87 ha. There is a turnover of more than 7 million $US, approximately three-quarters of which is received from exports, mainly to Europe and the USA (van den Berg 1986; Anonymous 1987b).

Lilacs are difficult to force into flower. A great deal of work is being carried out to improve the product and lengthen the supply period, but results can only be expected in the long term. At present, the supply period runs from October to May. More than 90% of the lilac forced in The Netherlands consists of white cultivars, which are propagated by budding or grafting on the wild lilac rootstock (*Syringa vulgaris* L.). The most important white cultivar for forcing is Madame Florent Stepman which accounts for approximately 94% of all supplies. This cultivar is especially important since no deterioration occurs during transport (export). The white cultivar Marie Legraye represents approximately 1% of the total lilacs available for forcing. Colored cultivars, mainly blue or pale lilac represent the remaining 5%. Within this last colored group the most important cultivars are

[1] Department of Horticulture, Agricultural University, Haagsteeg 3, 6708 PM Wageningen, The Netherlands

Biotechnology in Agriculture and Forestry, Vol. 20
High-Tech and Micropropagation IV (ed. by Y.P.S. Bajaj)

Andenken an Ludwig Späth, Hugo Koster and Herman Eilers (Anonymous 1987a; van den Berg 1988).

There is hardly any literature on production of lilac shrubs for gardens in The Netherlands and other countries. However, there is no doubt that lilacs are very popular in gardens, particularly in temperate climates. Garden cultivars are generally not grafted on *Syringa vulgaris*, but on other species: *S. josikaea*, *S. reflexa*, and *S. tomentella*.

For garden use, as well as for forcing, budded or grafted plants are often used (Lima and Ticknor 1986). Budded or grafted plants have the advantage that they can be forced earlier than plants from cuttings (Bailey 1917). Other advantages are that flowering is more abundant and there is less wastage during 2–3 years of forcing (van den Berg, pers. commun.). For these reasons attention is paid to the selection of good rootstocks and their cloning in vitro, as well as to cloning of cultivars on their own root systems (see Sect. 3). It is thought that budded or grafted plants perform better during forcing than plants on their own root systems, although this has not yet been proved. *Syringa vulgaris* is the most suitable rootstock for budding or grafting; other *Syringa* species are less suitable since they do not produce such a good root ball, sometimes produce crooked branches and often show less flower bud development (van den Berg, pers. commun.). For garden lilacs these considerations are probably not so important.

2 Literature Review (Table 1)

Minocha (1980) and Uhring (1980) were the first to attempt in vitro vegetative propagation of lilac using callus culture. They and later Hildebrandt and Harney (1983) had little success because the callus did not differentiate into shoots. Since this early research, two systems of micropropagation have been described in the literature: the axillary bud method (Hildebrandt and Harney 1983) and the single-node method (Einset and Alexander III 1984; Pierik and Steegmans 1985; Pierik et al. 1986, 1988; Welander 1987). These two methods are reviewed below.

The axillary bud method was first applied by Hildebrandt and Harney (1983) with *Syringa vulgaris* Vesper. After isolation of shoot tips on a revised MS (Murashige and Skoog 1962) medium with low auxin and cytokinin concentrations, axillary bud formation could be induced on a multiplication medium with 7.5 mg/l BA and 0.1 mg/l NAA. Two-year-old material was used, which probably accounted for the lack of difficulties (rejuvenation of the plant material, rooting problems with adult material) in cloning lilac in vitro. Their plant material was cultured at 27–28 °C, a temperature detrimental to a so-called temperate shrub such as lilac, according to Pierik et al. (1988). Hildebrandt and Harney (1983) concluded that rooting of lilac shoots in vivo (in moistened vermiculite) was more successful (81% rooting) than in vitro (58% rooting); in vivo-rooted plants could be transplanted to soil 4–6 weeks earlier than in vitro-rooted plants.

More literature is available on micropropagation by single-node culture. Einset and Alexander III (1984) explored the single-node method with the hybrid *Syringa* × *hyacintiflora* Exel. They found that cytokinin (2-iP, kinetin, BA, and thidiazuron)

Table 1. Literature on micropropagation of lilac (*Syringa* species or hybrids)

Plant species (hybrid)	Shoot multiplication system applied	Growth regulator(s)	Reference
S. hyacintiflora Exel	Single-node culture	2-iP	Einset and Alexander III (1984)
S. vulgaris Vesper	Axillary buds	BA, NAA	Hildebrandt and and Harney (1983)
S. vulgaris	Callus culture (not successful)	BA	Minocha (1980)
S. vulgaris Madame Lemoine	Single-node culture	BA	Patience and Alderson (1987)
S. vulgaris	Single-node culture	2-iP	Pierik and Steegmans (1985)
S. vulgaris	Single-node culture	2-iP, zeatin	Pierik et al. (1986)
S. vulgaris	Single-node culture	2-iP, zeatin (riboside)	Pierik et al. (1988)
S. vulgaris	Callus (not successful)	2-iP, BA, IBA, 2,4-D	Urhing (1980)
S. chinensis Saugeana	Single-node culture	BA, 2-iP, NAA	Welander (1987)

can induce stem elongation, whereas auxin inhibits growth. They also kept their cultures at 27 °C, the same as Hildebrandt and Harney (1983). In most experiments, Einset and Alexander III (1984) used the rather high concentration of 6 mg/l 2-iP to induce stem elongation in *S. vulgaris*, but concluded that other cytokinins can also be used with about the same effectiveness. This is in contradiction to Pierik et al. (1988). Einset and Alexander III (1984) concluded that rooting in vivo (in vermiculite) is better than in vitro.

Pierik and Steegmans (1985) and Pierik et al. (1986, 1988) were the first to observe that there are a number of limiting factors for effective stem elongation in *S. vulgaris*: a relatively low irradiance is required, as well as, in principle, a relatively low temperature. The effectiveness of various cytokinins is very variable. Later, Pierik et al. (1986, 1988) concluded that the cytokinins 2-iP and zeatin (riboside) are extremely efficient for induction of stem elongation. They also demonstrated that rejuvenation, obtained by repeated subculturing of shoot tips, is as essential factor for stem elongation and optimal rooting. The important role of mineral nutrition was recently described (Pierik et al. 1988). Optimal stem elongation was obtained on media with an MS strength of 1.5; at this strength a physiological disease, called leaf curling, could be completely overcome in all rootstocks and cultivars. They concluded that there is no auxin requirement for rooting in juvenile lilacs. Patience and Alderson (1987) also came to the same conclusion.

Welander (1987), working with nodal explants of 3-year-old *Syringa* × *chinensis* Saugeana concluded that axillary branching and shoot elongation were optimal when an auxin in a low concentration (NAA 0.1 mg/l) was combined with BA or 2-iP at 3 mg/l, whereas kinetin was not very effective. The promotive effect of NAA on multiplication (Welander 1987) was not found by Einset and Alexander III (1984) or by Pierik et al. (present results).

3 New Experimental Results

3.1 Material and Methods

Actively growing shoot tips were selected in the spring from greenhouse-grown rootstocks and cultivars of various ages. Two rootstocks (*Syringa vulgaris* L. of seed origin) were chosen for experimental use, they will be referred to as A2 and A3 (they were selected by a lilac forcer in Aalsmeer from 50-60-year-old shrubs). In a number of experiments (see Sect. 3.3) the following cultivars were ålso used: Madame Florent Stepman (50 years old), Marie Legraye (30 years old), Hugo Koster (15 years old), Herman Eilers (45 years old) and Maréchal Foch (50 years old); all these cultivars were budded or grafted on *S. vulgaris* L. Repeated subculture to obtain rejuvenation of this plant material is described in Sections 3.2 and 3.3.

In most experiments shoot tips and single nodes (Fig. 1) were grown on the following basic culture medium (Pierik et al. 1988): MS (Murashige and Skoog 1962) macro-salts, MS micro-salts (without Fe) at full strength, NaFeEDTA 37.5–40 mg/l, sucrose, Daishin agar 0.8%, pH 6 before autoclaving, and distilled

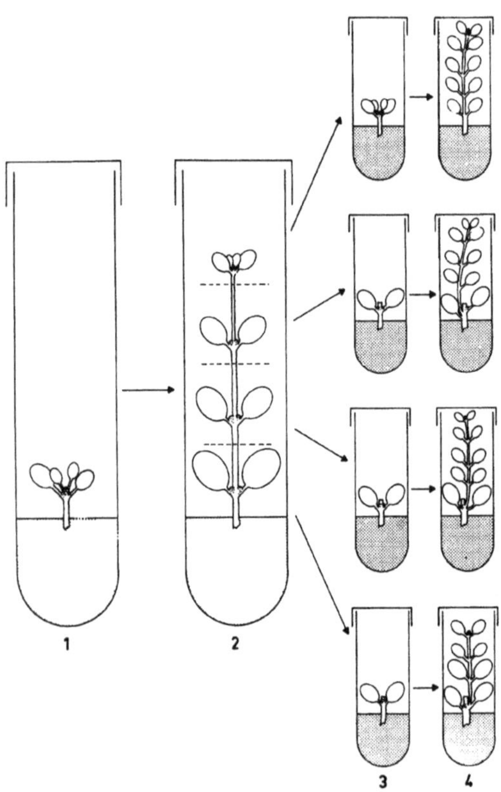

Fig. 1. Schematic representation of the single-node method. (Pierik 1987)

Table 2. Specification of the optimal basic culture media for single-node culture and stem elongation for the various rootstocks and cultivars. (Pierik et al., unpubl.)

Rootstock (rs) or cultivar (cv.)	Murashige and Skoog (1962) macro-salts	Sucrose conc. (%)	2-iP conc. (mg/l)	Zeatin-riboside conc. (mg/l)
rs A2 and A3	1.50	3.5	2.0	-
cv. Madame Florent Stepman	1.50	3.5	–	0.5
cv. Marie Legraye	1.50	3.0	1.0	–
cv. Hugo Koster	1.25	3.0	1.5	–
cv. Herman Eilers	1.50	3.0	–	0.5
cv. Maréchal Foch	1.50	3.0	–	0.5

water. Optimal basic media for stem elongation and single-node culture for rootstocks and cultivars are further specified in Table 2. Pyrex test tubes (diameter 20 mm) were used each containing 12 ml of medium. All media were autoclaved at 112 °C for 20 min. To avoid dehydration during the 6–8 week culture period and to reduce the number of infections, the test tubes were closed with cotton plugs and aluminum foil, with cotton plugs and Vitafilm (Good Year), or with Cap-O-Test caps and Vitafilm. Shoot-tip explants from greenhouse-grown material were initially surface sterilized as follows: dipped for a few seconds in alcohol 70% (v/v), 20 min in 1% NaOCl (with a few drops of Tween 20) and then rinsed three times (for 3, 5, and 15 min respectively) in sterile tap water.

Cultures were incubated in a growth chamber at 21 °C under a schedule of 16-h photoperiod provided by fluorescent tubes (Philips TL 54/38W, 4–5 Wm^{-2} and 8 h darkness (Fig. 2). In some experiments cultures were grown at various irradiances and temperatures.

All experiments had, in principle, one variable factor (nutritional, hormonal, physical, etc.). Each treatment consisted of 24 test tubes, each containing one shoot tip or single node. Experiments on stem elongation (Fig. 1) were in most cases carried out with single-node explants, whereas in the case of rooting experiments with shoot cuttings, all starting material was derived from repeatedly subcultured and rejuvenated plants.

Rooting experiments were not carried out in vitro, but in rockwool plugs (Rockwool/Grodan, P.O. Box 1160, 6040 KD Roermond, The Netherlands). These plugs were saturated with tap water and placed in glass containers with a thin layer of tap water, under nonsterile conditions.

All experiments reported were repeated at least once during the years 1985–1988. Data represented in the figures are the means of at least two experiments.

After 6 weeks of in vitro culture, stem elongation of the rootstocks was evaluated by measuring stem length (SL) and counting the number of leaf pairs (LP). The cultivars were evaluated later (to be specified in Sect. 3.3) because they grew much slower than the rootstocks. The parameter MR (Multiplication Rate) represents the number of single nodes produced, that are suitable for subculturing. Rooting was evaluated by determining the percentages of rooted cuttings after 5–7 weeks in rockwool.

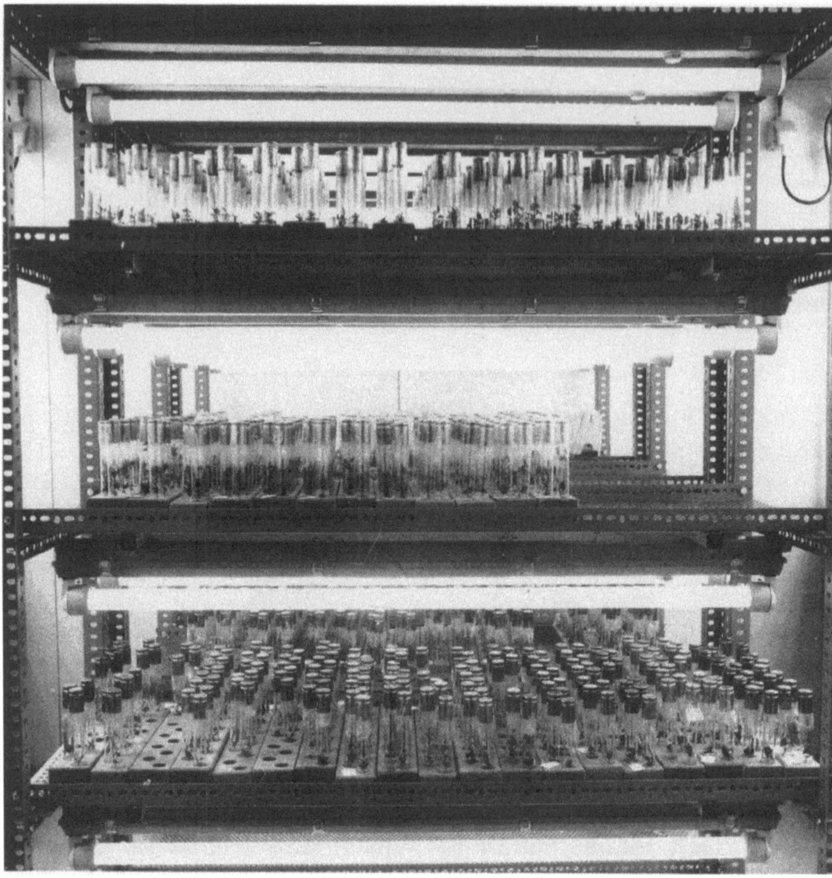

Fig. 2. Culture room in which the lilacs were grown; the fluorescent light fittings are mounted above the shelves. (Pierik et al., unpubl.)

3.2 Results with Newly Selected Rootstocks

In 1985, with the help of a commercial grower/forcer in the lilac growing area of Aalsmeer, The Netherlands, two rootstocks (A2 and A3) were selected with excellent qualities: they produced no suckers, had strong stems, no side-shoot formation, were disease-resistant and healthy, had a compact root system and had given excellent results for at least 50 years of forcing. The experiments with A2 and A3, described in this section, were based on experiments described by Pierik et al. (1988) with a rootstock K8. The reasons for using the single-node method instead of the axillary bud method were described previously (Pierik et al. 1988).

After isolation of shoot tips from adult plants, it soon became clear that rejuvenation would be necessary to obtain rapid growth and ultimately easy adventitious root regeneration. Rejuvenation was obtained by repeated subculturing the shoot tips in vitro; we found that the relatively old rootstocks A2 and A3, and the cultivars mentioned in Section 3.3, were much more difficult to rejuvenate than the 2-year-old rootstock called K8 (Pierik et al. 1988); it took many more subcultures to rejuvenate the older starting material. The rootstocks A2 and A3 reacted similarly; the MR rate was generally slightly higher in A3 than in A2.

The results with rootstock A2 and A3 are always shown together in one figure (A2 left, A3 right); however, this does not mean that the graphs of these two rootstocks are fully comparable, because each graph is the average of at least two experiments, and the experiments with A2 and A3 were not carried out simultaneously.

The influence of the sucrose concentration is shown in Fig. 3. In the range 2.5–4% the multiplication parameters are hardly influenced by the sugar concentration. In rootstock A3 where a sucrose concentration of 4.5% was also tested SL, LP, and MR decrease particularly by increasing the sucrose concentration from 4 to 4.5%. It can be concluded that a sucrose concentration of 3.5% is optimal. The sugars glucose or fructose (not shown graphically) were less effective than sucrose.

A MS macrosalt strength of 1.5 gave the best results as is shown in Fig. 4. The most notable effect of increasing the MS strength from 0.5 to 1.5 was the influence on leaf curling (LC), a physiological disease occurring in most lilacs in vitro. The higher the MS concentration the lower the leaf curling percentage.

Since the color of the lilac shoots at 25 mg/l NaFeEDTA was rather pale and yellow, the effect of the compound was tested (Fig. 5). All parameters reached an optimum value at 37.5–40 mg/l NaFeEDTA, whereas SL, LP, and MR strongly decreased on increasing the concentration from 37.5 to 75 mg/l.

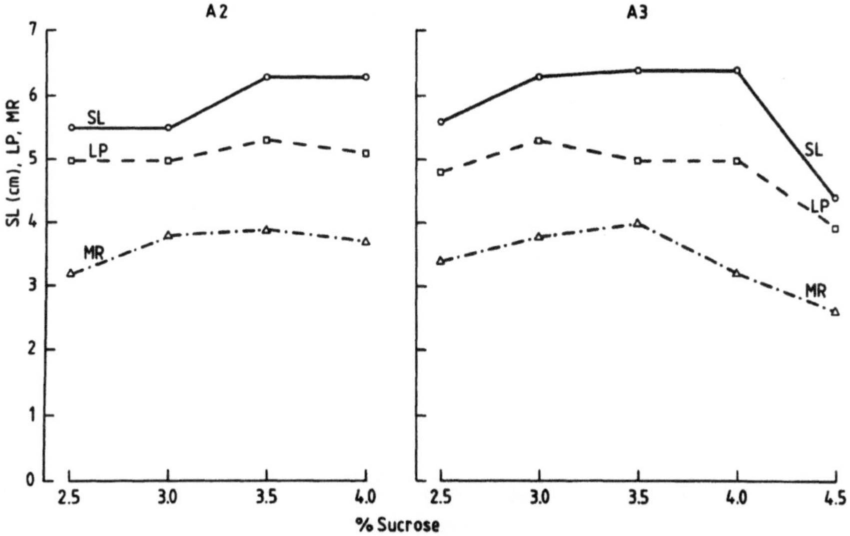

Fig. 3. The influence of sucrose concentration on stem elongation of the lilac rootstocks A2 and A3 (Pierik et al., unpubl.). *SL* shoot length; *LP* number of leaf pairs; *MR* multiplication rate. (cf. Sect. 3.1)

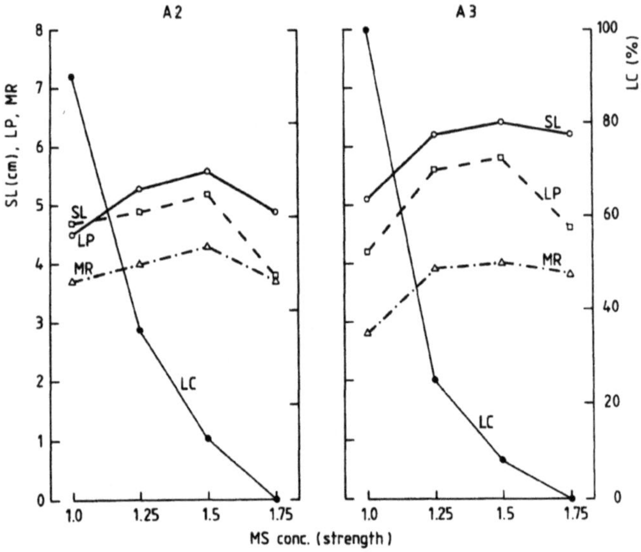

Fig. 4. The influence of the MS macrosalt concentration (strength) on stem elongation of the lilac rootstocks A2 and A3. Abbreviations see Fig. 3. (Pierik et al., unpubl.). *LC* leaf curling

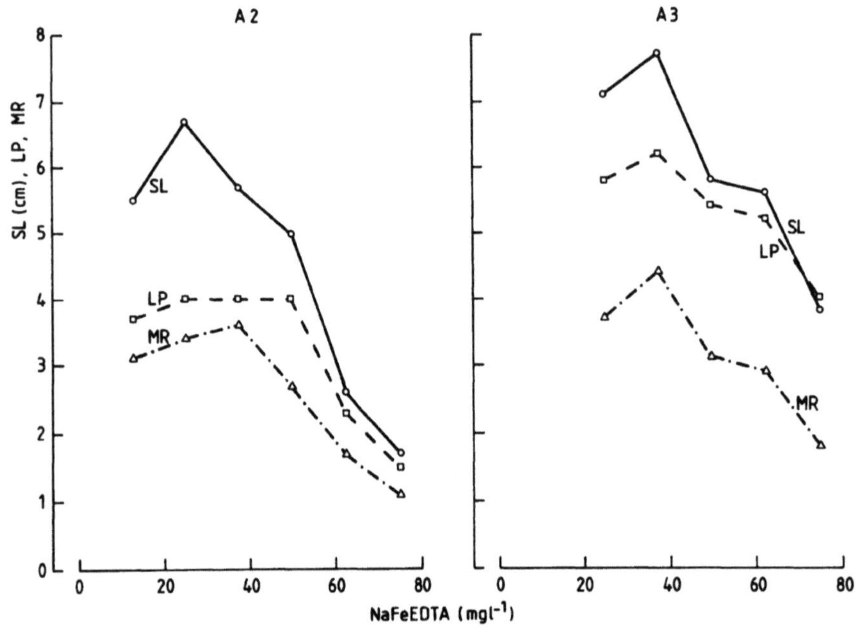

Fig. 5. The influence of the NaFeEDTA concentration on stem elongation of the lilac rootstocks A2 and A3. Abbreviations see Fig. 3. (Pierik et al., unpubl.)

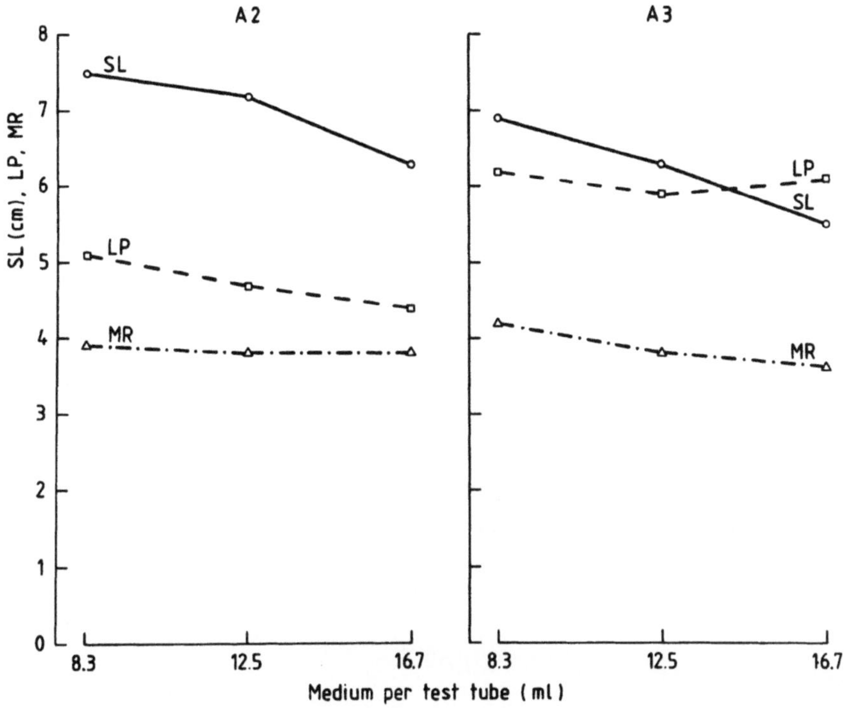

Fig. 6. The influence of medium volume per test tube on stem elongation of the lilac rootstocks A2 and A3. Abbreviations see Fig. 3. (Pierik et al., unpubl.)

The influence of the amount of medium per test tube (Fig. 6) is very small, so that it is not necessary to use more than 8–10 ml. This is particularly important when using very expensive cytokinins such as zeatin or zeatin riboside.

A comparison of two agar brands and various agar concentrations (Fig. 7) shows that shoot length is clearly promoted by using Daishin agar rather than Difco Bacto agar. The number of leaf pairs and the multiplication rate are only slightly influenced by the Daishin agar when compared to Difco Bacto agar. The MR is highest at 0.8% Daishin agar.

Irradiance has a strong influence on SL, LP, and MR (Fig. 8), and is optimal at 4 $W m^{-2}$; at higher irradiances all parameters decrease with increasing irradiance. At 3 $W m^{-2}$ the plants grow too quickly and are too spindly, and their thin stems are less suitable for subculture. Preliminary experiments with A2 and A3 have shown that the fluorescent tube TL57/38W (Philips) results in a slower growth with yellowing of the leaves; TL54/38W obtained from the same company results in increased growth with no yellowing of the leaves.

The influence of temperature and irradiance (with fluorescent tubes TL57/38W) was only examined with the rootstock A2. Figure 9 shows that at 25 °C SL, LP, and MR are increased at all irradiances in comparison to 21 °C, except SL at the highest irradiance. However, all shoots at 25 °C were yellow and unhealthy. When the temperature was further increased from 25 to 27 °C, very strong yellowing oc-

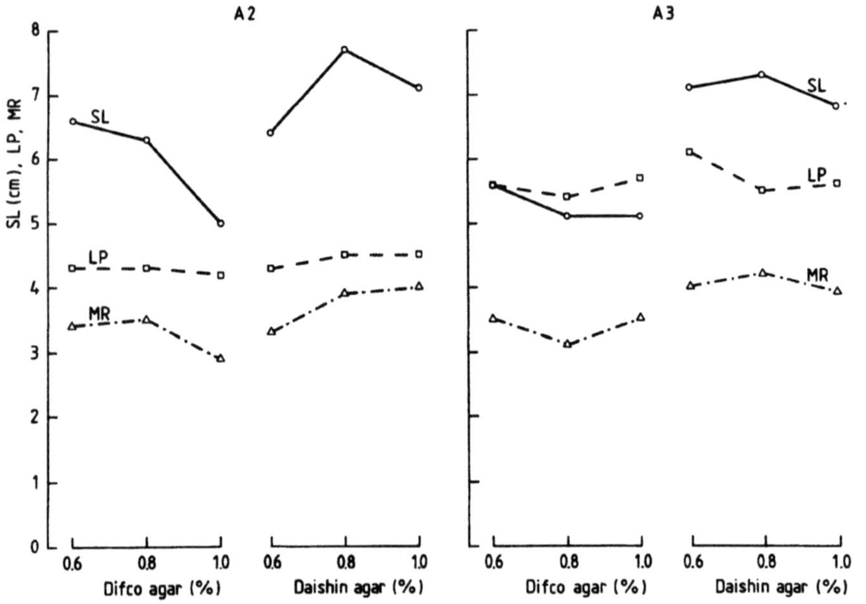

Fig. 7. The influence of the brand of agar and agar concentration on stem elongation of the lilac rootstocks A2 and A3. Abbreviations see Fig. 3. (Pierik et al., unpubl.)

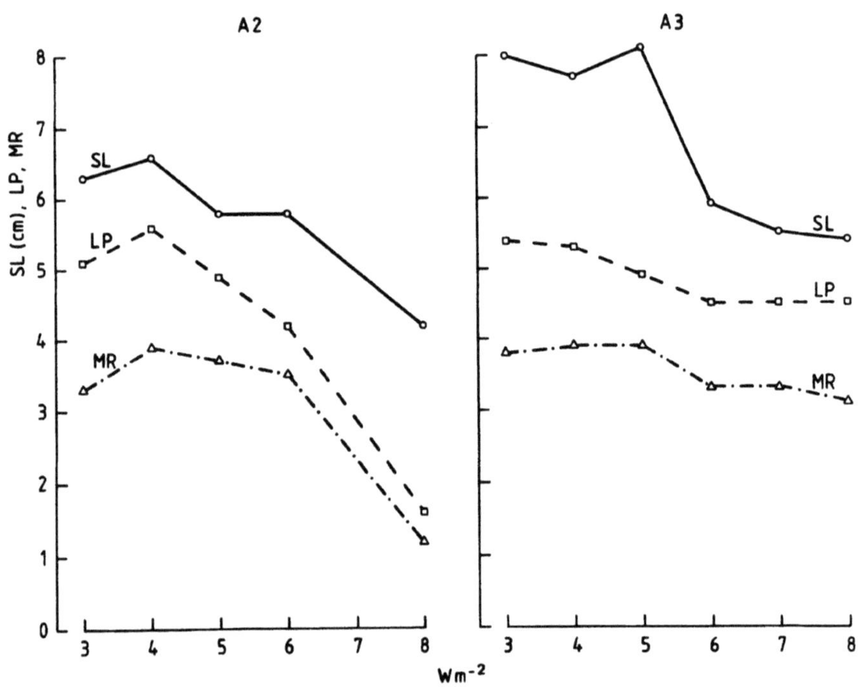

Fig. 8. The influence of the irradiance on stem elongation of the lilac rootstocks A2 and A3. Abbreviations see Fig. 3. (Pierik et al., unpubl.)

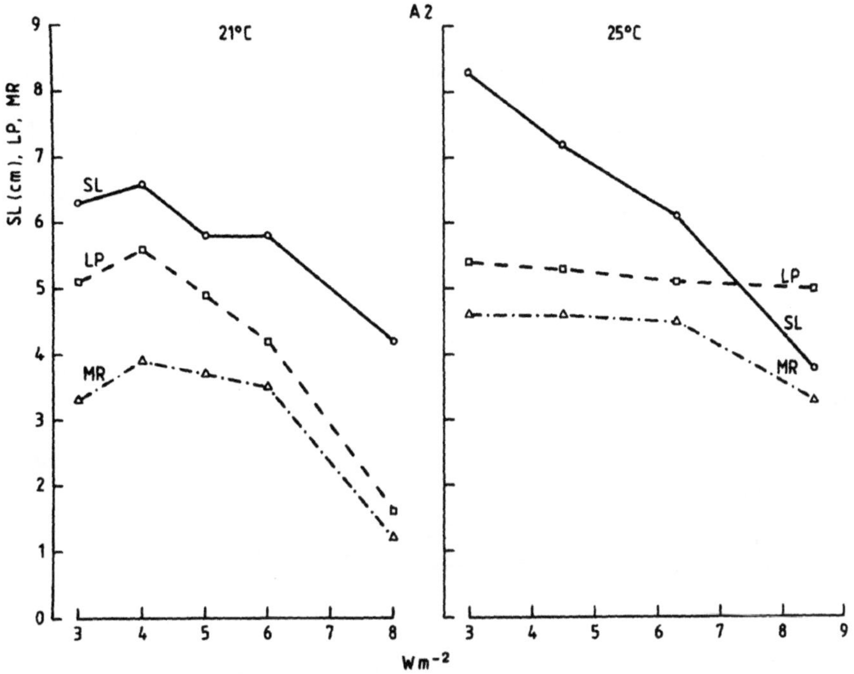

Fig. 9. The influence of temperature and irradiance on stem elongation of the lilac rootstock A2. Abbreviations see Fig. 3. (Pierik et al., unpubl.)

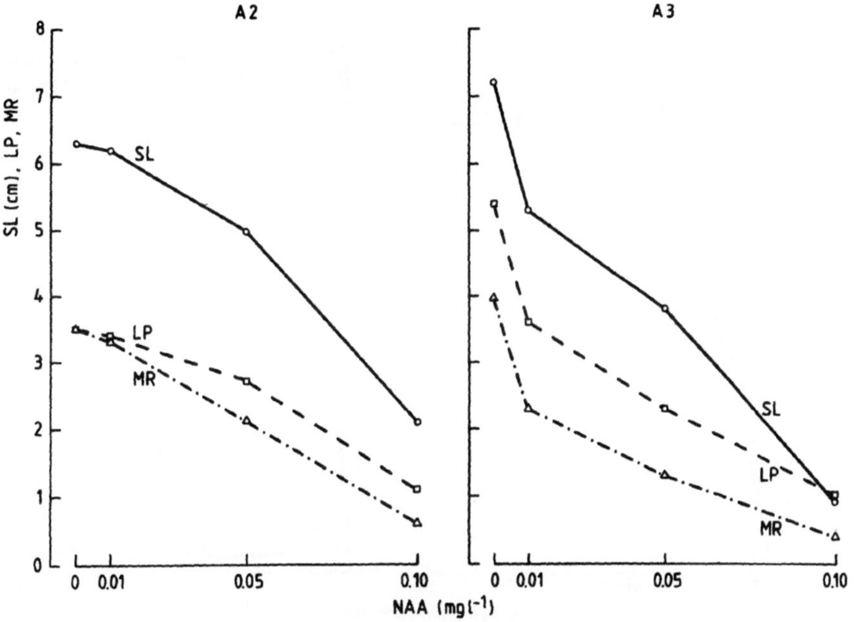

Fig. 10. The influence of the NAA concentration on stem elongation of the lilac rootstocks A2 and A3. Abbreviations see Fig. 3. (Pierik et al., unpubl.)

curred. Unfortunately, a temperature of 23 °C, probably the optimum for multiplication, was not tested.

Since Hildebrandt and Harney (1983) and Welander (1987) found that NAA promoted stem elongation in lilac, we also tested the effect of this strong auxin. Figure 10 clearly shows that SL, LP, and MR are decreased with increasing concentrations of NAA.

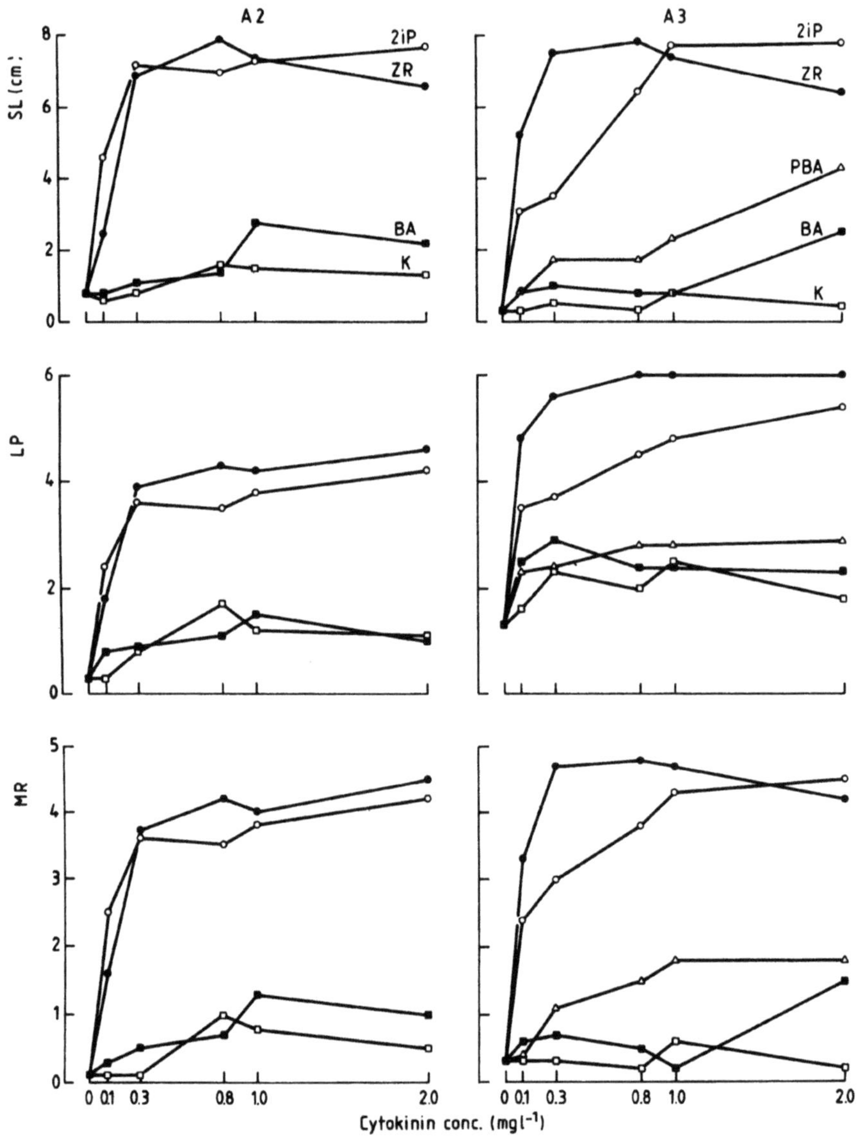

Fig. 11. The influence of several cytokinins in various concentrations on stem elongation of the lilac rootstocks A2 and A3. See also Fig. 12. (Pierik et al., unpubl.)

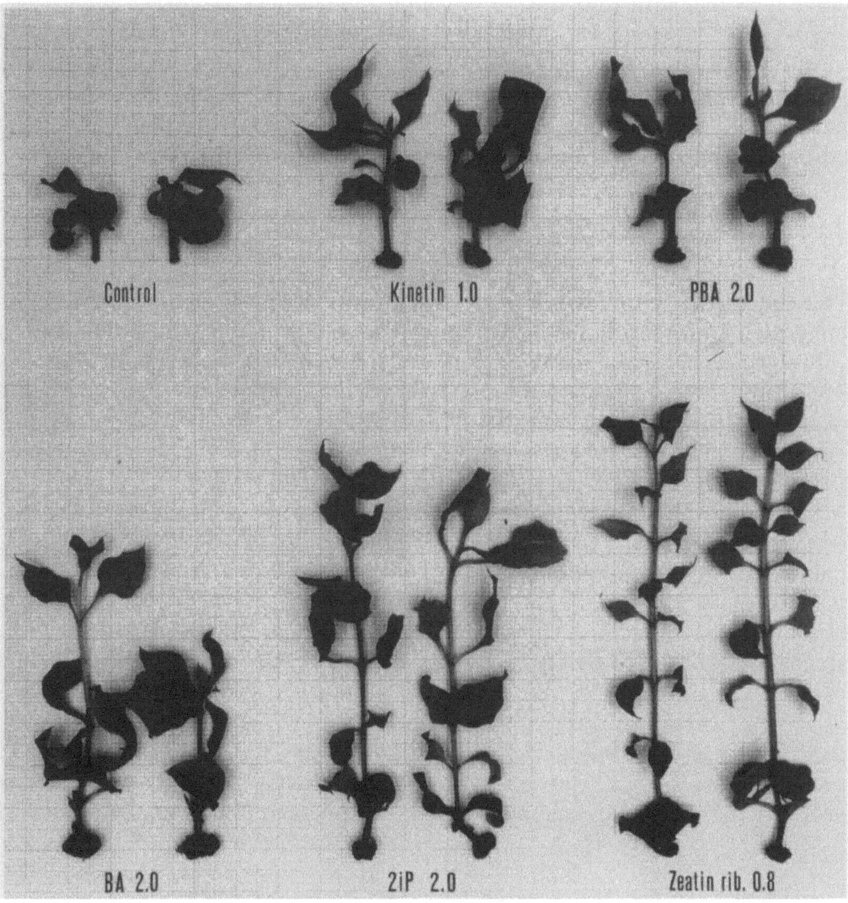

Fig. 12. Influence of five cytokinins (each in optimal concentration in mg/l for multiplication) on stem elongation of rootstock A3. (Pierik et al., unpubl.). See also Fig. 11

The influence of several cytokinins (kinetin, BA, PBA, 2-iP, and zeatin riboside) was tested in the concentration range 0–0.1- ⋯ 2 mg/l (Fig. 11). The effects of the cytokinins for the rootstocks A2 and A3 on SL, LP, and MR are similar. Kinetin is a very inefficient cytokinin for lilac stem elongation; BA and PBA have a very weak stem-elongation action, whereas 2-iP and zeatin are very efficient cytokinins. The influence of PBA was only tested for rootstock A3 (Fig. 12).

3.3 Results with Difficult-to-Propagate Cultivars

Research was initiated on cloning a few difficult-to-propagate cultivars in vitro. Reasons for propagating cultivars on their own root systems are: reduction of high labor and other costs (rootstock production and grafting), avoidance of heterogeneity which is unavoidable (rootstocks are heterogeneous), the avoidance of frequently appearing (certainly after a long time, after many forcings) grafting problems (poor vascular connection between stock and scion), the avoidance of virus transfer via the rootstock, provided that virus-free starting material is used in vitro. This last reason is especially important for export to foreign countries.

3.3.1 Cultivar Madame Florent Stepman

This cultivar has been used in many experiments since it is the most important of those used for forcing in the Netherlands. The results on stem elongation with this cultivar can be summarized as follows. Rejuvenation took place after eight subcultures. Stem elongation was much slower than with the rootstocks A2 and A3 (these rootstocks had a shoot MR of approximately 4 after 6 weeks) even under optimal conditions. Sucrose appeared to be optimal at 3.5% (Table 2), whereas glucose or fructose were less effective than sucrose. Shoot length was highest at a MS strength of 1.5 in comparison to lower concentrations, but leaf color at MS 1.5 was slightly yellow. At low MS concentration. (0.5 strength) the leaves were pointing diagonally upwards while at 1.5 MS strength they were perpendicular to the main stem. This has the effect that by single-node culture the explants are more likely to come out of the medium (Fig. 13). Daishin agar resulted in a stronger stem elongation than Difco Bacto agar. Irradiances between 3 and 6 W/m^2 resulted in optimal stem elongation, but at higher irradiances stem elongation declined. Madame Stepman preferred a temperature of 21 °C in comparison to 25 °C; a temperature of 23 °C was not tested. Yellowing of the leaves could be prevented by increasing the NaFeEDTA concentration, but stem elongation was decreased proportionally by increasing the NaFeEDTA concentration from 25–75 mg/l. The

Fig. 13. Influence of the MS macrosalt strength on stem elongation of the lilac cv. Madame Florent Stepman. At a strength of 0.5 the leaves pointed diagonally upward; at other concentrations they were perpendicular to the main stem. (Pierik et al., unpubl.)

most pronounced stem elongation and MR were obtained with zeatin riboside or IPA in the medium, whereas zeatin and 2-iP were far less effective; the other cytokinins (kinetin, BA, and PBA) induced little stem elongation.

3.3.2 Cultivar Marie Legraye

Rejuvenation of this cultivar was obtained after six subcultures. It is an easy-to-clone cultivar with a relatively high MR, certainly in comparison to the cultivar Madame Florent Stepman, and the cultivars used later. Stem elongation was again poor on kinetin- and BA-containing media, whereas 2-iP and zeatin (riboside) were effective in inducing stem elongation. Multiplication rate appears to be promoted by increasing the temperature from 21–25 °C.

3.3.3 Cultivar Hugo Koster

To obtain rejuvenation it was necessary to subculture shoot tips five times. In comparison to the rootstocks A2 and A3 (Sect. 3.2) Hugo Koster elongated slowly and it took 8–9 weeks to reach a stem length of 5–6 cm; the number of single nodes per stem that were suitable for subculturing was rather low (three to four) because the high number of leaf pairs formed grew too close together. Optimal media conditions for stem elongation are described in Table 2. The cytokinins kinetin, BA, and PBA were inefficient in inducing stem elongation, whereas 2-iP (1.5 mg/l), and zeatin riboside (1.0 mg/l) were equally effective. The most effective cytokinin was IPA (at 1–2 mg/l).

3.3.4 Cultivar Herman Eilers

This cultivar took ten subcultures before rejuvenation occurred; rejuvenation was expressed by the capacity for forming adventitious roots. Stem elongation was hardly induced by 2-iP or IPA(2–4 mg/l); zeatin riboside at 0.5 mg/l was only able to induce a little stem elongation, but very inefficiently and slowly.

3.3.5 Cultivar Maréchal Foch

After eight subcultures this cultivar was rejuvenated. It shows weak growth with hardly any stem elongation. Both 2-iP and zeatin riboside were unable to induce stem elongation. Shoot tips reacted slightly more favorably than single nodes.

3.4 Rooting Experiments and Flowering

Shoots originally produced in vitro were rooted in vitro on media without cytokinin and with auxin (Pierik and Steegmans 1985); auxin requirement of the rejuvenated shoots appeared to be very low (Fig. 14). In vitro the rooting percent-

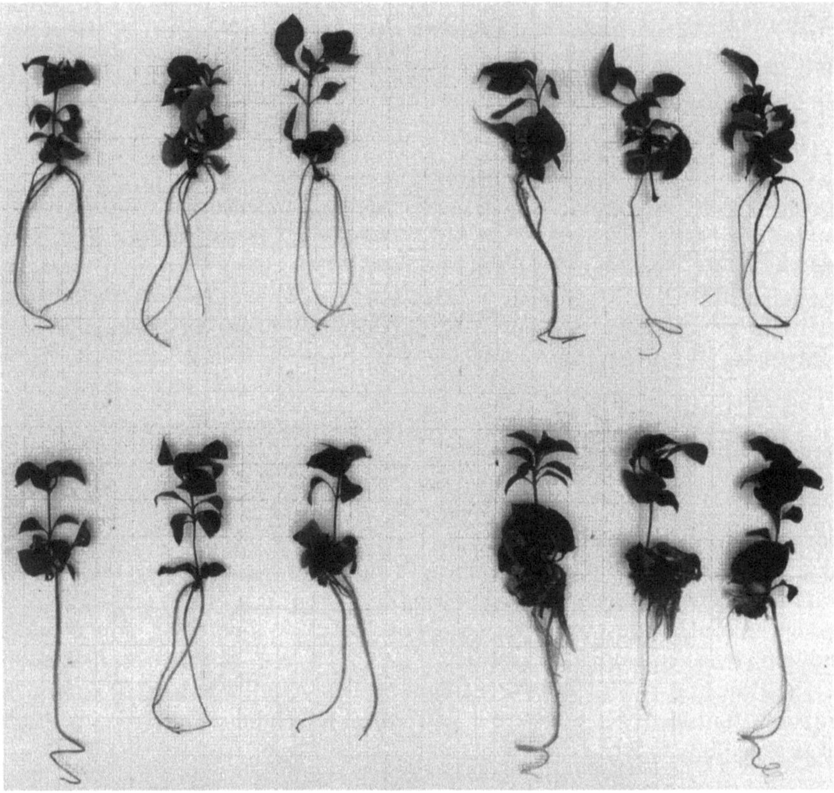

Fig. 14. Adventitious root formation in vitro in shoot cuttings of the lilac cv. Marie Legray, as influenced by the IBA conc. *Top left* no auxin; *top right* IBA 0.01 mg/l; *bottom left* IBA 0.1 mg/l; *bottom right* IBA 1.0 mg/l. (Pierik et al., unpubl.)

age never reached 100%, and root initiation and development were poor and slow. For these reasons rockwool plugs were utilized (Pierik et al. 1988). At present these plugs are the best and cheapest way of rooting lilac shoots. A small unit (a glass jar, a so-called Weckpot) is generally used for rooting on the laboratory scale; each glass jar can contain a block of 9–12 microplugs (Fig. 15) saturated with tap water, with a thin layer of tap water at the bottom of the jar; no mineral nutrition is required, otherwise the plants become too elongated (see Fig. 16). During rooting in rockwool, transpiration should be restricted by placing the plugs in glass containers and by keeping the mini-cuttings at low irradiance. Acclimatization of rockwool-rooted plants was very successful (Fig. 17); in the greenhouse the irradiance should be lowered and the relative humidity increased during the first 4 weeks.

The first flowering of the cultivars Madame Florent Stepman and Marie Legraye was observed after a 3-year period of continuous greenhouse culture.

Fig. 15. Rockwool unit with nine microplugs fitted in a plastic frame. Shoot cuttings of the lilac cv. Madame Stepman 6 weeks after taking the cuttings are rooted for 100%. (Pierik et al., unpubl.)

Fig. 16. Rooting of shoot tips of the lilac rootstock A3 as influenced by the mineral nutrition (numbers indicate the weeks after the beginning of the rooting experiment that mineral nutrition (MS macrosalts 0.5 strength) was applied. Photograph taken 6 weeks after isolation of the shoot cuttings. (Pierik et al., unpubl.)

Fig. 17. Acclimatized plants of the rootstock K8 (Pierik et al. 1988) in the greenhouse. For explanation see text

4 General Discussion and Conclusions

In vitro propagation overcomes the many shortcomings of the conventional technology of lilac propagation by cuttings and grafting. Once vigorously growing and rejuvenated shoots are obtained in vitro, micropropagation can continue throughout the year (Einset and Alexander III 1984). Easy-to-clone *Syringa vulgaris* rootstocks, such as A2 and A3 (see Sect. 3.2) can be propagated in vitro very rapidly when needed for forcing: with a MR of about 4.0 per 6 weeks at least 260 000 plants can be obtained in 1 year from one rejuvenated shoot tip or single node. Difficult-to-clone cultivars with a much lower MR (e.g., 4 per 8 weeks; see Sect. 3.3) can yield 16 000 plants in 1 year. The MR remains strongly dependent on the genotype. It is not yet known what MRs can be obtained when the axillary bud method is applied.

The reasons for using the single-node method were based on the fact that the axillary branching method strongly induced leaf curling; the single-node method is also generally used in commercial tissue culture laboratories in the USA (Constantin, pers. commun.). Results presented in Section 3.2 clearly demonstrate that the success of the single-node method is strongly dependent on a number of factors: an effective cytokinin (preferentially zeatin riboside, zeatin, IPA, or 2-iP), a low irradiance (3–6 W/m^2), a relatively low temperature (21 °C), a high MS-macrosalt concentration (1.5 strength), a relatively high sucrose concentration (3.5%), and the genotype. Inefficient cytokinins such as kinetin, BA, and PBA should not be used with the single-node method.

It can be concluded from the work of Hildebrandt and Harney (1983) and Welander (1987) that auxin is essential for multiplication. Our work with the rootstocks A2 and A3 has demonstrated that there is no auxin requirement for stem elongation of these rootstocks. This discrepancy remains unexplained. Possibly gibberellins are also involved. Perhaps only a few cultivars have an auxin requirement, but this has not yet been thoroughly examined.

Rejuvenated shoots were easy to root on artificial culture substrates such as rockwool and vermiculite without auxin. Acclimatization of shoots rooted in rockwool to greenhouse conditions was successful, so in vitro rooting of lilacs will no longer be necessary.

It is not yet known how far lilac grown in vitro on its own root system is suitable for forcing. It has proved difficult to induce flowering in the first group of bushes of the cv. Madame Florent Stepman (a lilac well known for forcing) on their own root systems (Eveleens, pers. commun.). Two lilac cultivars (except the cv. Madame Florent Stepman; Verhoeven, pers. commun.) from tissue culture were significantly taller and broader than plants from cuttings (Lima and Ticknor 1986), whereas micropropagated plants grew faster and flowered at an earlier age than lilac plants from rooted cuttings. On the other hand, it must be remembered that lilac on its own root system is possibly much more susceptible to soil pathogens than those on a rootstock (Eveleens, pers. commun.). This is especially true for the cultivar Madame Florent Stepman.

Acknowledgments. We are grateful to Mrs. C.I. Kendrick for her help in correcting the English. Numerous M.Sc. students have also made a great contribution to the experimental part of this article (Sect. 3) through their research during the last years of their studies; without the support of these students writing this article would have been impossible. We are especially grateful to the following students: H. Al Habbal, A.R. Bril, M. de Bruijn, A.A. Elias, L. Hof, M.E.C.M. Hop, H.F. van Huizen, H. Molenberg, F.M. Praasterink, O.T.J. Stiekema, A.J. van der Velde and F.A. Verheij. We are grateful to Mr. P.A.W. Verhoeven and Mr. P. Eveleens for their suggestions and criticism during the preparation of the manuscript.

We are also grateful to Mr. J. Maarsse and Mr. H.J. Keesen, Aalsmeer, for the supply of plant material.

References

Anonymous (1987a) Westland product info. Floraprint, Lisse, The Netherlands, 32/5–32/14

Anonymous (1987b) Tuinbouwcijfers 1987 (statistics on horticultural production). Landbouw-Economisch Inst, The Hague, 60

Bailey LH (1917) The standard cyclopedia of horticulture. Macmillan, New York, 3297–3302

Berg van den AJ (1986) Ontwikkeling in trekheesterteelten nog lang niet ten einde. Vakbl Bloem 41(41): 16–23

Berg van den AJ (1988) Teelt van seringen (lilac culture). Bloemeninformatie, Consulentschap in Algemene Dienst, Aalsmeer, The Netherlands 33: 1–94

Einset JW, Alexander III JH (1984) Multiplication of *Syringa vulgaris* species and cultivars in tissue culture. Comb Proc Plant Prop Soc 34: 628–636

Hildebrandt V, Harney PM (1983) In vitro propagation of *Syringa vulgaris* Vesper. HortScience 18: 432–434

Lima H, Ticknor R (1986) The effect of propagation method on forced potted plants and on pathogen resistance in *Syringa vulgaris*. Comb Proc Plant Prop Soc 36: 126–131

Minocha SC (1980) Lilacs in test tubes: potential for cloning of lilacs by cell and tissue culture. Lilacs 8: 12–19

Murashige T, Skoog F (1962) A revised medium for rapid growth and bioassays with tobacco tissue cultures. Physiol Plant 15: 473–497

Patience PA, Alderson PG (1987) Development of a system to study peroxidases during the rooting of lilac shoots produced in vitro. Acta Hortic 212: 267–272

Pierik RLM (1987) In vitro culture of higher plants. Martinus Nijhoff, Dordrecht, 1–344

Pierik RLM, Steegmans HHM (1985) Vegetatieve vermeerdering van sering in kweekbuis. Vakbl Bloem 40(23): 44–45

Pierik RLM, Steegmans HHM, Molenberg H (1986) Vegetative propagation of *Syringa vulgaris* L. in vitro. Abstr Int Congr Plant Tissue Cell Culture, Univ Minnesota, p 434

Pierik RLM, Steegmans HHM, Elias AA, Stiekema OTJ, Velde van der AJ (1988) Vegetative propagation of *Syringa vulgaris* L. in vitro. Acta Hortic 226: 25–40

Uhring J (1980) Organ and tissue culture of lilac. HortScience 15: 418 (Abstr)

Welander NT (1987) Propagation of *Syringa chinensis* cv. Saugeana by in vitro culture of nodal explants. J Hortic Sci 62: 89–96

XXVI Micropropagation of Virus-Free Ornamentals in the USSR

A.E. Rybalko[1]

1 Introduction

Commercial floriculture is of great importance in the USSR. Methods for obtaining initial virus-free commercial plant specimens are of great value for improving the quality of plant material. Biotechnology can bring about the elimination of pathogens of fungal, bacterial, and viral diseases. The experience of other countries, widely used in the USSR, and the increasing production of materials for immunoassays within the USSR itself should made it possible to produce certified planting stock, from which the main viral pathogens have been eradicated. More than 20 commercial tissue culture laboratories (Zhola 1986) have been set up in the USSR with the aim of introducing biotechnology into commercial floriculture.

The main commercial ornamental cultures are vegetatively propagated plants, e.g., carnation, chrysanthemum, rose, gerbera, orchid, freesia, gladiolus, and tulip. Accumulating from year to year, virus infections result in the degeneration of the species, decreasing yield, and deterioration in reproductive ability and ornamental quality. Many viruses transmitted by the planting material are very harmful, and often latent (symptomless). A great proportion of the above-mentioned species are almost completely infected with viruses. We are mainly dealing with complex virus infections, caused by the agents of different taxonomic groups, for which reason great attention is paid to research on the virus populations on the plants and to evolving techniques to extract viruses from chronically infected specimens. However, the extraction of viruses does not make the plants resistant to them, so that methods of preventing secondary mass infection are of prime importance. The eradication of viruses from ornamental cultures represents a complex measures to promote chronically infected specimens (i.e., restoration of species), to introduce methods of intensive propagation, and to defend against repeated mass infections with viruses and other pathogens. Tissue culture techniques should also be utilized to select new species, with new qualities, including resistance to pathogens.

This chapter describes methods and commercial applications to free ornamental crops from virus diseases by means of micropropagation.

[1] The Enterprise of the Meristematic Cultures, Ministry of Housing and Municipal Economy RSFSR Sochi, A-354, 354354, USSR

Biotechnology in Agriculture and Forestry, Vol. 20
High-Tech and Micropropagation IV (ed. by Y.P.S. Bajaj)

2 Viruses of Ornamentals in the USSR

2.1 Structure of Virus Populations on Ornamentals

The structure of virus populations on carnation, chrysanthemum, tulip, freesia, narcissus, hyacinth, orchid, and other flowers is under study in the USSR. Particular attention is being paid to the complex study of ornamental viruses. Six carnation viruses and three tulip viruses were registered in Moldavia (Voloshchuk 1985). Their damage potential, distribution, and symptoms have been described. The presence of viruses was confirmed with the help of electron microscopy. Lengthy and large-scale investigations of pathogens of viral diseases are being carried out at the Enterprise of the Meristematic Cultures, with the aim of studying virus infection on different cultures, their relationships, the extraction of pathogens, the production of preparations for immunoassay, and subsequently, the development and introduction of an effective preventive system.

A number of studies of virus diseases in ornamental crops have been published in the USSR: a monograph by Makutenajte-Navalinskene (1981), an atlas of plant diseases by Zirka (1984), and a review by Rybalko (1984b, 1987).

Numerous research programs have been carried out on the detection of carnation mottle virus (CMV). In Latvia, CMV was extracted from infected carnations and purified by differential centrifugation. The serum had a titer of 1:256–512 on double-diffusion precipitation in agar gel (Kilevitz 1972). At the same time, the virus was extracted in Lithuania (Makutenajte 1974).

In the Ukraine the virus was also detected by electron microscopy and described. The extraction of pure preparations of this virus was made by sedimentation of PEG 6000 and gel filtration on Sepharose 4B. The titer of the serum was 1:128. Viruses on carnation were studied in detail in Moldavia (Voloshchuk 1984). The harmful effect of CMV was not only observed to affect the rooting of grafts of various carnation cultivars and the quality and quantity of flower production, but the increase in damage caused by viruses in a mixed infection was also shown. At the same time, the distribution and injurious properties of other carnation viruses were studied: carnation ringspot, carnation latent virus, carnation vein mottle virus, carnation etched ringspot virus, carnation necrotic fleck virus, and the viruses from tulip: tulip-breaking virus, tobacco mosaic virus, cucumber mosaic virus, and tobacco rattle virus. From roses the following viruses were extracted: tobacco mosaic, prunus necrotic ringspot, rose mosaic, cucumber mosaic, and arabis mosaic (Voloshchuk and Teploukhova 1986).

In Latvia, tomato aspermy virus has been studied over a long period on chrysanthemum, which is propagated in rotation with tomatoes, which are used as a source of the virus (Ignash 1978). Three strains of this virus were extracted and studied in Latvia, methods of obtaining reagents for the ELISA test were perfected, and the materials for immunodiagnostic sets were delivered to Biolar (Dzirkale 1987). At present, chrysanthemum virus B is being studied (Rybalko and Kishko 1984).

2.2 Methods of Identification of Viruses of Ornamentals

2.2.1 Electron Microscopy

This method has been applied for a long time for the detection of viruses, including plant viruses (Protsenko 1965). We use electorn microscopy for the detection of viruses in plant sap in the process of plant promotion (Fig. 1). It was tested on a great number of objects, and enabled the selection of virus-free strains in tube culture. Long-term commercial research of virus-free specimens, initially propagated by micrografting in aseptic culture, showed complete regeneration. This work was carried out on a large number of potato, carnation, and chrysanthemum cultivars in 1975–1985, when the ELISA test was not yet commercially applied in the USSR. This method, if handled properly, detects all viruses present on the plant, and additionally when modified by immunosorbent electron microscopy, it proves to be more sensitive than the ELISA test. It was used for testing meristems of carnation and orchid (Fig. 2).

2.2.2 Precipitation in Agar Gel

While the ELISA test was not in use, agar gel double-diffusion serology was used for the detection of carnation mottle virus in the USSR. To increase the rate of reaction, improve the arrangement of immunodiagnostic plates, conserve reagents, and improve the methods of large-scale assay, we use the method of radial diffusion (Fig. 3; Kharuta et al. 1984).

The serum in working dilution is mixed with 0.5 of the final concentration of gel with 0.05 M Tris-HCl buffer, pH 7.2; in the second step, agar is dissolved to 1% and agarose to 0.8% concentration at 80 °C. Reagents are warmed to 50 °C, then mixed

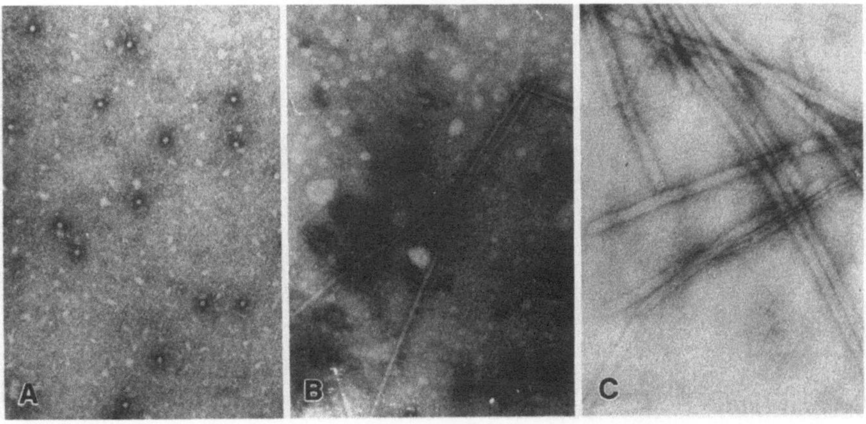

Fig. 1A–C. Electron micrograph of viruses in infectious plant sap: **A** carnation; **B** chrysanthemum; **C** cymbidium

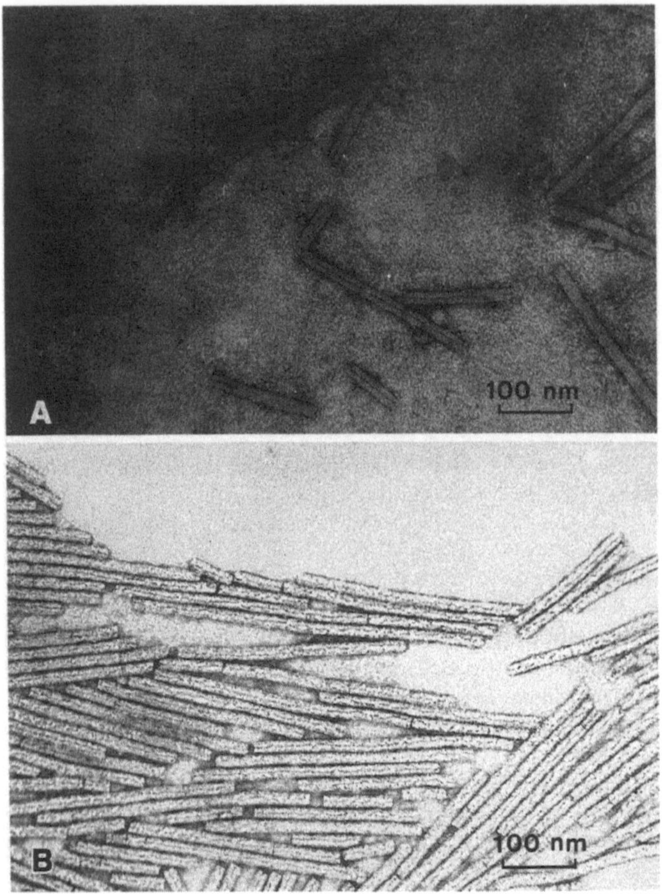

Fig. 2A,B. Virus, taken from orchid. **A** Electron microscopy; **B** immunoelectron microscopy (purified preparation)

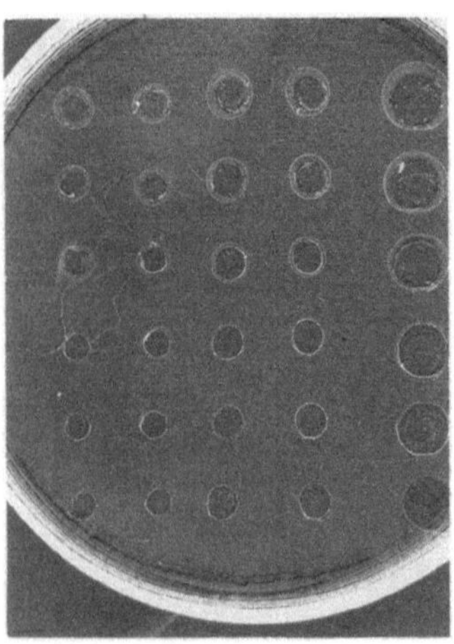

Fig. 3. Radial immunodiffusion of CMV in agar gel

and maintained for 10 min in a thermostatic water bath at 50 °C and poured into a Petri dish. As a rule, 10 ml of gel is required for a 100-ml Petri dish. When the gel hardens, the sap of the tested plants is added, then the dishes are thermostated at 37 °C.

2.2.3 Enzyme-Linked Immunosorbent Assay (ELISA)

This method has been used for enzyme-linked immunosorbent assay of ornamental cultures since the identification of carnation mottle virus (Odinets et al. 1983). In the study cited, different methods of CMV identification were applied: double (DID) and single radial (RID) immunodiffusion in agar gel, virobacterial agglutination (ABV test), and enzyme-linked immunosorbent assay (ELISA). Antiserum from pure virus preparations with a titer in double diffusion of 1:512 at an antigen concentration 50 mg/ml was used. Rabbits were immunized three times with 1 mg of the virus preparation with Freund's adjuvant at 1-week intervals. The sensitivity of these methods for CMV identification on the virus-infected plants in this study was: DID 29%, RID 40%, ABV 50%, and ELISA 58%.

This technique was also modified for the detection of yellow bean mosaic and freesia mosaic viruses taken from freesia in tissue culture (Odinets et al. 1985b), for the detection of tulip-breaking virus (Odinets and Kopylov 1986), and cucumber mosaic virus on gladioli (Odinets et al. 1985a), and chrysanthemum virus B (Odinets, 1987), narcissus mosaic virus and tobacco rattle on narcissus (Odinetz, 1986). The ELISA test was applied for the identification of different viruses with horseradish peroxidase (Biolar USSR).

At present, the ELISA test is a method commonly used in Latvia to assay carnation plants in test tubes (Zhola et al. 1985). At the Enterprise of the Meristematic Cultures, ELISA is used to detect carnation mottle, chrysanthemum virus B, arabis mosaic, *Cymbidium* mosaic, and bean yellow mosaic virus.

2.3 Mass Virus Identification in Induced Phytocenosis in Ornamentals

A variety of reagents for immunoassays on respective plant viruses must be available to organize a system of seed growing for ornamental crops on a virus-free basis throughout the country. To date, these preparations have not been mass-produced in the USSR. To supply the concerns, which produce virus-free planting material, the Enterprise of the Meristematic Cultures began mass production of extremely active sera to ornamental crop viruses (Kharuta et al. 1984). Rabbits are used to produce the sera, which are immunized with purified preparations of the respective viruses according to their properties, characterized by electron microscopy, sedimentation analysis, and UV spectral analysis (Kishko et al. 1984).

First, carnation mottle virus was procured and purified in Latvia (Kilevits 1972), the procedure for purification from *Chenopodium quinoa* was by differential centrifugation of the partly purified preparations; the antiserum had a titer of 1:256–512. Later, the virus was extracted and purified from carnation by gel filtration on Sepharose 4B and sedimentation by PEG-6000. The titer of this preparation was 1:128 (Rybalko et al. 1978b). In Moldavia, carnation mottle virus was purified by ultracentrifugation in a 20% sucrose gradient for 90 min at 39 000 g. Antiserum with a titer 1:2048 was obtained by mixed intravenous and intramuscular injections in rabbits with Freund's adjuvant (Voloshchuk and Kosakovskaya 1984). A serum highly active to CMV was obtained at Moscow State University. At an antigen concentration of 50 g/ml it gave a single line of

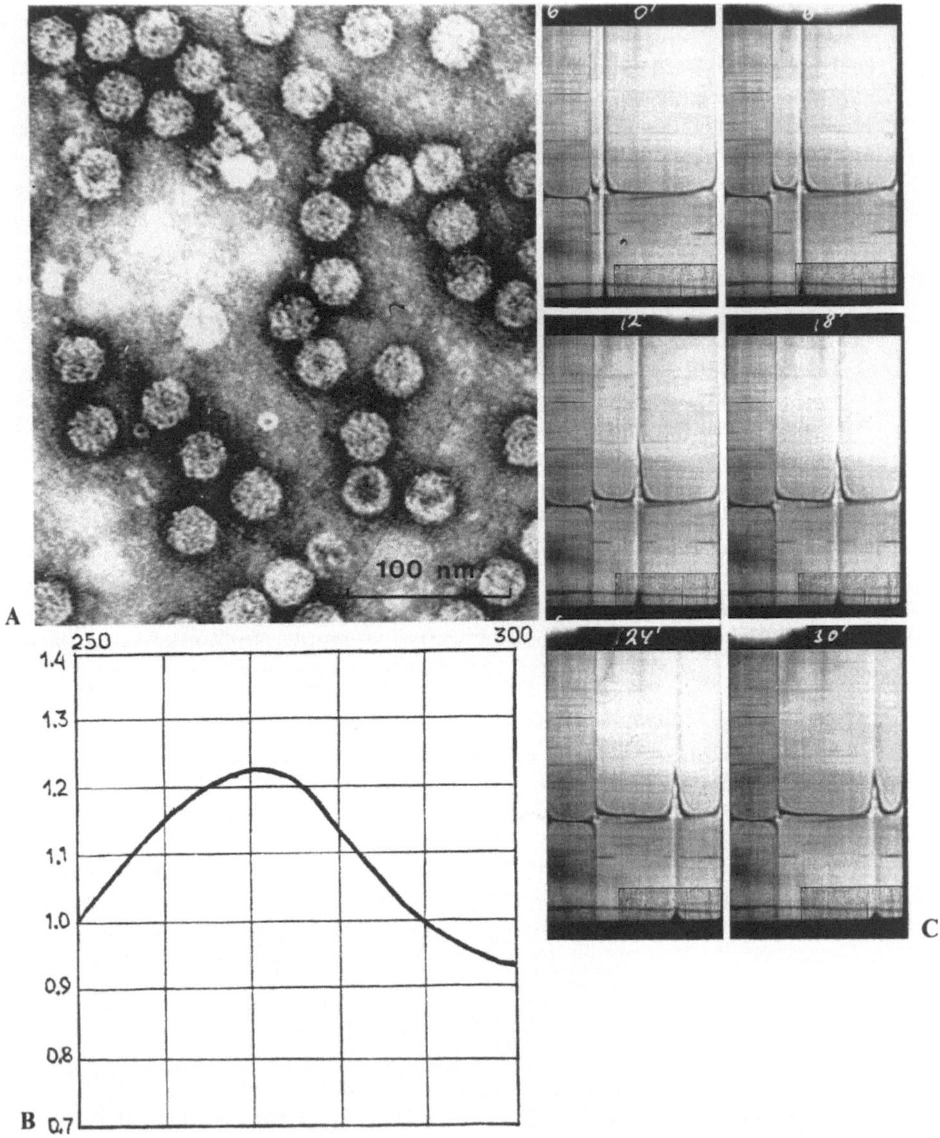

Fig. 4A–C. Purity criteria for CMV preparation to obtain diagnostic serum. **A** Electrograph; **B** spectrophotometry; **C** sedimentograms

precipitation on double-diffusion testing at a dilution of 1:512 (Odinets et al. 1983). Sera for other viruses were obtained here for carnation, chrysanthemum, tulip, freesia, and narcissus plants (Odinets 1987).

At the Enterprise of the Meristematic Cultures commercial lots of diagnostic serum for CMV were obtained with a stable titer of 1:512–1024 to 2048. Here, the optimal scheme of rabbit immunization was elaborated, which is used as the basis for the production of serum for different viruses. Every batch of serum is supplied with a "certificate of purity" (Fig. 4). According to this scheme, highly active sera were obtained for chrysanthemum virus B, *Cymbidium* mosaic virus, and also for viruses taken from gerbera, anthurium, and poinsettia. It is intended to enlarge the assortment of sera produced. At present, research is being carried out on the production of diagnostic material for virus polyphages: cucumber mosaic, alfalfa mosaic, arabis mosaic, bean yellow mosaic, tomato aspermy, and tobacco rattle virus. Standards of production are being developed that should make it possible to produce only sera of high quality.

3 Production of Virus-Free Plants by Tissue Culture

Research on the use of tissue culture in ornamental crops was initiated in the USSR in the mid-1970s. The first study was made in Latvia on carnation (Peterson et al. 1974). Shoot tips (0.18–0.4 mm) were placed on Van Hoof's medium (1971) for regeneration. The plants from which explants were excised were incubated for 1 month at 37 °C. Shoots regenerated within 20 days on Van Hoof's medium with 2 mg/l NAA and without gibberellin. Rooted plants were transferred to soil. In this experiment on virus identification on *Chenopodium amaranticolor* and *Chenopodium quinoa*, 8.4% of the plants taken from meristems were free of carnation latent virus and 5.4% were free of CMV.

The shoot tips method, combined with thermotherapy, was also used at other laboratories, and is still widely used to date. The most frequently used nutrient medium is a Murashige and Skoog's formula (1962). Thermotherapy is carried out in a special chamber for 6–7 weeks at 35 °C or 3 weeks at 37.5 °C (Mitrophanova and Burdejnaya 1977). In the Ukraine, research on the elimination of viruses from species and genotypes by regeneration of meristems (100 μm) has been carried out. This technique was called the two-phase method. Here, Van Hoof's medium was also used. A high percentage of meristem survival (up to 86.6%) was achieved by improving the explanation technique, optimizing the light regime in aseptic culture production, and obtaining an optimal nutrient medium composition (Rybalko and Kharuta 1978). According to this technique, chrysanthemum plants, free of chrysanthemum virus B, were obtained from meristems (Rybalko 1977). Later, this technique for the promotion of chrysanthemums was used on a large scale in combination with shoot tip culture and thermotherapy in Estonia (Rosenberg 1984), in the Nikitsky Botanic Garden (Mitrofanova et al. 1985), and at Meristem Cultures in Latvia (Dubov and Tretjak 1986).

Attempts were made to obtain lily, free of mosaic disease (Korneeva 1977). Methods to free freesia of virus were assayed and introduced into production in

Latvia by Vitslane (1986). Promotion of plants is made by selecting freesia mosaic virus (FMV)- and bean yellow mosaic virus (BYMV)-free clones with subsequent excision of shoot tips, then cloning them in vitro.

This two-phase method was taken as a basis for developing the commercial technique of pathogen-free propagation of ornamentals, and was introduced into commercial production in 1980 (Rybalko 1984a,b, 1985, 1986). Using this technique, virus-free planting stock of different species of carnation, chrysanthemum, orchid, and gerbera is being produced by the Enterprise of the Meristematic Cultures.

There are thus two quite different approaches to applying the tissue culture technique to rid ornamentals of virus diseases: cultivation of shoot tips after thermotherapy for a certain period of time, and regeneration of plants from true meristems (100 μm).

4 Clonal Micropropagation of Ornamental Cultures

The research at the Timiryasev Institute of Plant Physiology, Academy of Sciences, USSR (Kataeva and Butenko 1983) is the basis of the development of the aseptic microclone technique and its introduction into mass production. The author divides clonal micropropagation of plants into two types: activation of plant meristems (shoot apex, axillary, and dormant shoot buds) and induction of buds or embryoids de novo. The latter consists of three methods: (1) induction of meristematic zones directly on the specific tissues of the explant (reproductive organs, epidermis, subepidermal tissues, leaf mesophyll, etc.); (2) induction from the initial callus, formed by the explant tissues; (3) induction from the transplanted callus tissue or cells of suspension culture.

It is very important to understand the mechanism of apical dominance in order to improve the micropropagation procedure. It was shown that kinetin, introduced into the nutrient medium, eliminates apical dominance and induces the proliferation of numerous shoots (Butenko 1960). This was evident in the propagation of carnation by the shoot grafting method, when up to 30 shoots were obtained by increasing the content of kinetin in the medium (Rybalko and Kharuta 1978).

At present, the division of shoots according to the number of internodes is very important in practical floriculture. We have used this method to propagate carnation and chrysanthemum since 1976 (Rybalko 1977). If the propagation coefficient is 1–5, it is possible to obtain about 3000 identical plants within a period of 5 months. The attempt was made to use cytokinins to induce proliferation of lateral buds, but it resulted in succulent shoots which were unsuitable for transplantation to soil (Smirnova 1983).

Cymbidium is propagated by protocorms. This method is used for the commercial propagation of pathogen-free *Cymbidium* plants (Kharuta 1983). In the Central Republican Botanic Garden of the Academy of Sciences of the Ukraine, this method is used to propagate *Cymbidium, Cattleya, Phalaenopsis,* and *Tunia* (Cherevchenko et al. 1982). The technique of micropropagation of gerbera has been developed by Butenko et al. (1982).

Clonal micropropagation has the following advantages:

1. Conservation of place and energy for mass production,
2. Propagation of plants under controlled conditions,
3. Safeguard against all kinds of reinfection,
4. Elimination of infectious agents,
5. Increase in selection rate,
6. Increase in propagation coefficient in some plants.

5 Introduction of Promotion Technique in Commercial Floriculture

As mentioned already, the methods of freeing different plants from viruses are based mainly on thermotherapy and regeneration of shoot tips of different size (300 μm and larger). The technique of explantation and preliminary treatment of initial material (sterilization) differs slightly in most cases. A fragment of a razor blade is used to dissect the shoot tips. Murashige and Skoog's formula (MS) is used as a nutrient medium. Our aim was to study the conditions of mass plant regeneration from meristems (100 μm), to investigate the possibilities of mass production of virus-free plants and to develop an efficient technique for the production of disease-free planting material.

The possibility of mass production of virus-free plants was supported by the elimination of X, M, and S viruses from potato (Rybalko 1978). In this technique the following modifications were made: improved and regulated explantation conditions for shoot tips; 8–10 klx illumination of the growth zone of the cultures; nutrient media that ensure a high rate of survival for meristems, and plant regeneration without an intermediate stage of callus formation. Different kinds of instruments were tested (Table 1), from which a dissecting needle for the destruction of the capsule of the crystalline lens was the most convenient. The use of this instrument halved the time of excision per explant.

In January 1976, 30 meristems of 5 carnation species (William, White, Lady, Lena, Flamingo) were placed on various nutrient media. On the 3rd day after planting, explants increased in size, and after being placed under fluorescent light

Table 1. Influence of the kind of instrument used on the excision rate of differently sized shoot tips[a]

Instrument	Time of transplantation (min)		
	Explant size (150 μm)	Explant size (250 μm)	Explant size (375 μm)
1. Microknife from razor blade	108	100	88
2. Dissecting needle	100	96	92
3. Dissecting needle to destroy the capsule of the crystalline lens	64	64	56

[a] 33 Shoot tips were transplanted per experiment.

Table 2. Influence of medium composition on the regeneration of shoots from meristem of different carnation species[a]

Species	Murashige-Skoog's medium		Van Hoof's medium	
	Regenerated shoots	(%)	Regenerated shoots	(%)
William	22	73.3	28	93.3
White	19	63.3	26	86.7
Lady	23	76.7	22	73.6
Lena	18	60.0	28	93.2
Flamingo	22	73.3	26	86.7
Average per species	104	69.2	130	86.6

[a] 30 Meristems were planted per experiment.

began to grow rapidly on MS and Van Hoof's media, forming green leaves and shoots. The formation of shoots suitable for further transplantation on the secondary medium took place on Van Hoof's medium within 2 weeks, i.e., 1 week earlier, than on MS.

The highest percentage of survival was also on Van Hoof's medium (Table 2). Average shoot survival of all five species on Van Hoof's medium was 86.7% in comparison to 69.3% on MS. The regeneration of explants occurred on Morel's medium (1964), as only callus was formed.

Van Hoof's medium was also used for mass propagation of disease-free species. In 1976, 1340 explants of different sizes (75–250 μm) were planted for 3 months, and 84% of them regenerated. There was no difference in the survival rate of the explants. The most intensive growth rate was in the period from April until July. Various auxins were added to liquid and solid Van Hoof's medium. The medium, supplemented with auxins, promoted callus formation and lowered survival of explants; increasing the auxin dose inhibited the growth of explants. Addition of indoleacetic acid (β-IAA) and naphtaleneacetic acid (α-NAA) at 1 mg/l showed callus and root formation. Explants 75 μm in size survived better in liquid medium. There were difficulties in planting these explants on the surface of the agar medium, even if only slightly immersed, as the explants died. Shoots rooted initially on the medium with auxins did not survive when transplanted on new nutrient medium. Thus, Van Hoof's medium without auxins was the best for regeneration of shoots suitable for transplanting onto the secondary medium.

Electron microscopy showed that plants regenerated from explants of ca. 100 μm were 60% free of CMV. This method is the basis of the technique for mass production of planting stock of ornamental cultures, mainly used for carnation. For the production of pathogen-free planting stock, the protocol is as follows (Fig. 5).

5.1 Regeneration of Meristems on Van Hoof's Medium (1971)

This is done in two steps, regeneration of initial shoots and root induction.

Phase 1. The initial material, consisting of defoliated pieces of shoots of a carnation cultivar with exposed buds, is surface-sterilized with 0.1% solution of

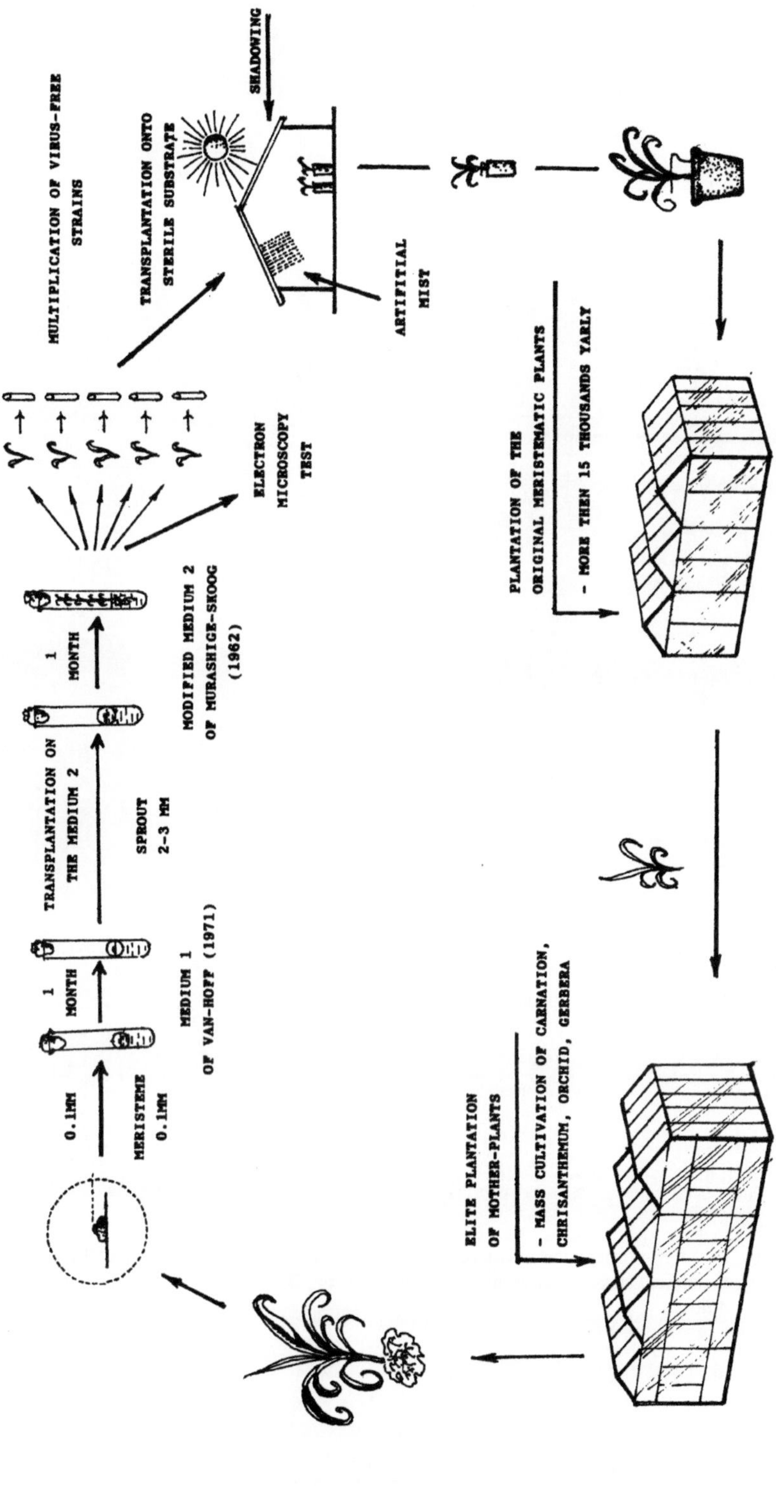

Fig. 5. Scheme of two-phase method of plant regeneration from meristems (0.1 mm)

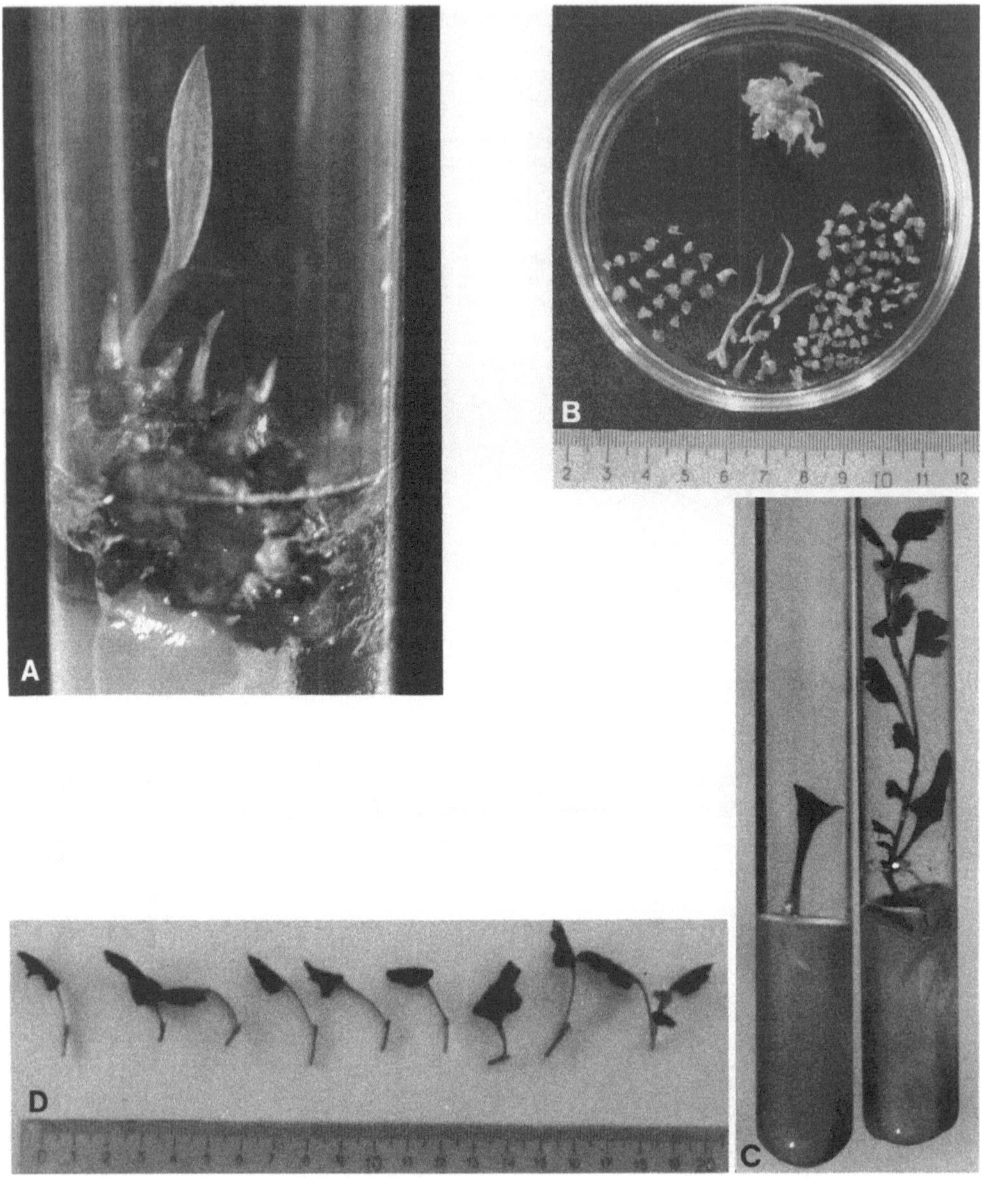

Fig. 6A–D. Micropropagation of ornamental cultures. **A** *Cymbidium* protocorms; **B** division of protocorms; **C** different stages of chrysanthemum propagation; **D** division of chrysanthemum according to number of nodes ethanol mercury chloride and N-etylpiridinium chloride (1:2) for 5 min and rinsed five times with sterile distilled water; the work is carried out in a sterile chamber. Sterile pieces are pressed together between clamps, and the primordia covering the apical dome are removed with the help of a stereomicroscope and a sterile needle (Fig. 6). Needles are changed frequently, using ethanol-flame sterilization. The apical meristem is dissected without leaf primordia and with the tip of the needle is transferred onto the surface of the initial Van Hoof's medium (1971) in the test tube. Medium composition is as follows (mg/l): $NaNO_3$ 1800; $(NH_4)_2SO_4$ 790; $Ca(NO_3)_2$ 200; KNO_3 80; NH_4NO_3 200; $MgSO_4 \cdot 7H_2O$ 360; Na_2SO_4 200; KCl 910; NaH_2PO_4 316; Na_2EDTA 37.3; $FeSO_4 \cdot 7H_2O$ 27.8; $ZnSO_4 \cdot 7H_2O$ 1; H_3BO_3 1; $MnSO_4 \cdot 4H_2O$ 0.1; $CuSO_4 \cdot 5H_2O$ 0.03; $AlCl_3$ 0.03; $NiCl_2 \cdot 6H_2O$ 0.03; pyridoxine 1; nicotinic acid 0.1; mesoinosit 100; gibberellin 0.1; kinetin 0.5; adenine 40; sucrose 20 000; agar 7000. Test tubes are placed in a phytostatic room under fluorescent illumination (8–10 klx). In 3–4 weeks cultures regenerated shoots 2–3 mm in size.

Phase 2. For root induction, shoots regenerated in phase 1 are transferred to the secondary MS medium with some changes (mg/l): KNO_3 1900; NH_4NO_3 1650; $MgSO_4 \cdot 7H_2O$ 370; KH_2PO_4 170; $CaCl_2 \cdot 2H_2O$ 440; Na_2EDTA 37.3; $FeSO_4 \cdot 7H_2O$ 27.8; $ZnSO_4 \cdot 4H_2O$ 8.6; H_3BO_3 6.2; $MnSO_4 \cdot 4H_2O$ 22.3; $CuSO_4 \cdot 5H_2O$ 0.025; KJ 0.83; $Na_2MoO_4 \cdot 2H_2O$ 0.25; $CoCl_2 \cdot 6H_2O$ 0.025; thiamine 0.2; kinetin 0.04; gibberellic acid 0.5; sucrose 20000; agar 7000. The cultures are grown under fluorescent light.

5.2 Selection of Virus-free Strains

Regenerated plants with five to six pairs of leaves are tested for the presence of viruses. For this purpose we use electron and immunoelectron microscopy. One or two leaves are removed under aseptic conditions and then homogenized on a slide plate in a drop of 2% phosphotungstate, pH 7.2. The homogenate is covered with a Formvar-coated specimen grid and sustained for 30–40 s. The superfluous moisture is removed by filter paper, and the sample is examined through the electron microscope.

5.3 Micropropagation

Virus-free strains are propagated by the method of nodal grafting in test-tube culture (carnation, chrysanthemum) as shown in Fig. 7. Generally, the plants develop within 1 month. The growth coefficient is estimated to be $5 \times (n - 1)$, where n denotes the number of months. A laboratory assistant can plant 400 micrografts (preparatory work included) per day.

5.4 Transplanting into Soil

Virus-free mericlones are transplanted into polyethylene pots, 7 cm in diameter and 10 cm in height, filled with sterile substrate (mainly peat).

Fig. 7A–D. Propagation of virus-free carnation plants. **A** Different stages of plant propagation; **B** transplantation of plants into soil; **C** acclimatized plant; **D** mother plant

For acclimatization the planted mericlones are sustained in the greenhouse under a constantly watered cheesecloth cover. The greenhouse is supplied with ventilation and special shading and protected from insects. The mericlones remain in the pots for 4–5 weeks, and here they are tested for the presence of viruses and *Fusarium*. Virus- and *Fusarium*-free plants are transplanted into 3–5 l containers for individual growing. All plants are propagated by the clonal method (progeny of one meristem). During the whole period of propagation, the plants are tested for the presence of viruses and *Fusarium*.

Long-term virological assays confirmed the efficiency of electron microscope selection of virus-free strains in sterile culture. The ELISA test of the average sample of grafts of the second order (Fig. 8) showed that all of them were virus-free. Such a technique makes it possible to obtain a great number of meristematic plants: thus, in 1981–1988, 193 000 mericlones were produced (Fig. 10), and the cost of production was not high (Table 4).

5.5 Propagation of Plants in the Nursery

In the nursery, rooted shoots are planted on the plant beds, approximately 20 per m^2. Every plant is supported by a metal wire. A supporting net, usually used

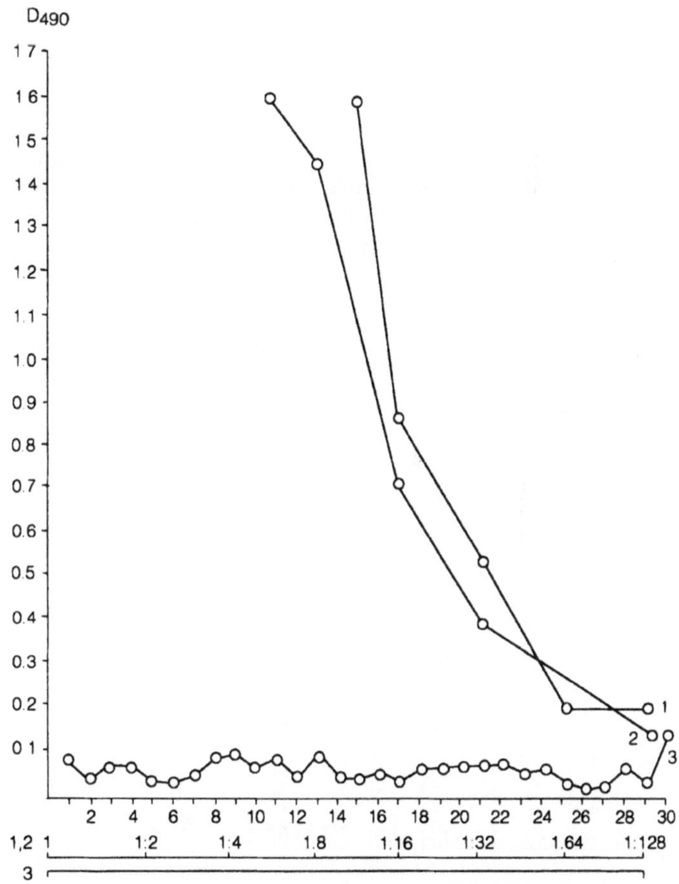

Fig. 8. Analysis of an average sample of carnation grafts of cv. Scania by ELISA in the nursery of meristem mother plants. *1*, *2* dilution of CMV; *3* plant samples

for carnations, helps in transferring plants in the process of growing (quality control, rearrangement). The characteristics of every plant used were tested by leaving the shoot to flower, but this prevented obtaining juvenile grafts. Since 1984, the best quality plants are selected in the meristem nursery, e.g., according to properties of a certain cultivar, good ornamental qualities, or whether they are pathogen-free. These should be the donors for the mass selection of meristems with subsequent clonal propagation. This material is the foundation for the mother stock of a given cultivar. Such a procedure is planned annually, which makes it possible to rejuvenate the mother stock and minimizes the danger of reinfection by pathogens of a viral, mycoplasmic, bacterial, and fungous nature. As mycoplasmic control confirms, elite grafts of carnation are practically free from such dangerous diseases as *Fusarium*, which causes great damage to carnation culture throughout the world.

5.6 Preventing Infection and Pests

Meristem nurseries are protected from pests and diseases by maintaining a preventive system: strict isolation of plantations, individual hygiene in the handling of the mother plants (working clothes, tool and hand sterilization in sample selection), sterilization of the substrate, individual growing of each plant, elimination of insects which transmit viruses, preventive treatment with fungicides. Keeping to this system guarantees obtaining practically pathogen-free elite plants. Thus, according to long-term data of mass immunoassay, only 5% of the plants were infected with carnation mottle virus (Fig. 9) in carnation plantations (in

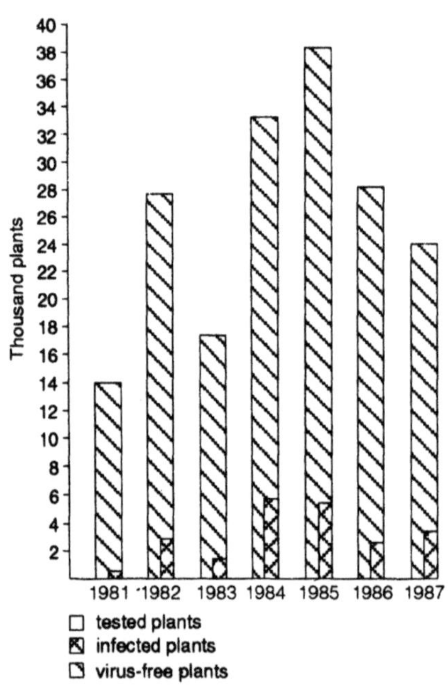

Fig. 9. Long-term serologic data of regenerated carnation mother plants

1981–1987 more than 180000 plants were tested). Carnation ringspot, carnation vein mottle, carnation latent and carnation etched ringspot viruses were rarely identified (sera were received from F.A. Hakkaart and D.Z. Maat from the Netherlands, and from K. Ortel from the former GDR).

6 Production of Planting Stock

The production of planting material began in the Ministry of Housing and Municipal Economy RSFSR in 1981. In 1981–1988, 193000 promoted mericlones of carnation, orchid, chrysanthemum, and gerbera were produced. The growth of mericlone production per year is shown in Fig. 10.

The procedure for mericlone production is as follows: 20–30 meristems are chosen for the promotion of a certain species. After regeneration, virus-free strains (progeny of one meristem) are selected by the two-phase method and by micrografting in vitro. When sufficient plants are available, grafts are planted by strains from the nursery of the original meristem plants. In 1981 the productivity of various strains of cv. Shocking was tested. If the plantlets were planted out of the test tubes in March, then the mass selection of grafts was started at the end of July, 25–52 grafts being taken form one plant. As shown in Fig. 11, the difference between the strains was inconsiderable.

Mericlones are mainly used for the internal needs of the Enterprise. The system has been designed for plant acclimatization upon transfer to soil. The use of rooted micrografts, formed in axenic culture, was more effective than planting grafts from test tubes (Table 3).

To date, mericlones have been used for the internal needs of the Enterprise and had the status of a semi-complete production. This was considered unsuitable, as the production of commercial planting material was small (500000 Rbl.), and could be carried out with fewer mericlones. Mericlones were therefore given the status of commercial material, and standards and price lists were prepared.

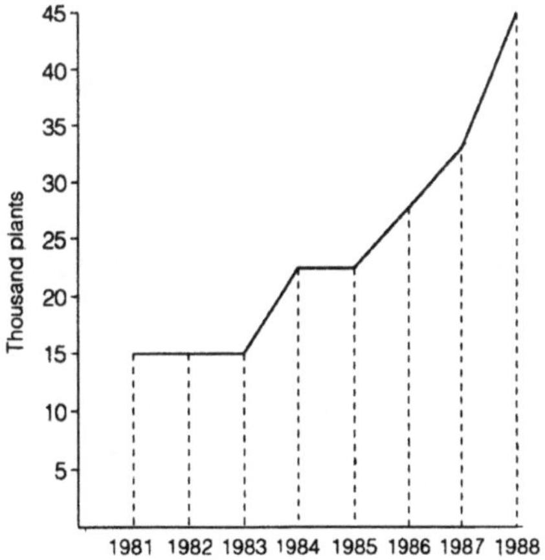

Fig. 10. Growth of production of virus-free mericlones at the Enterprise of the Meristematic Cultures

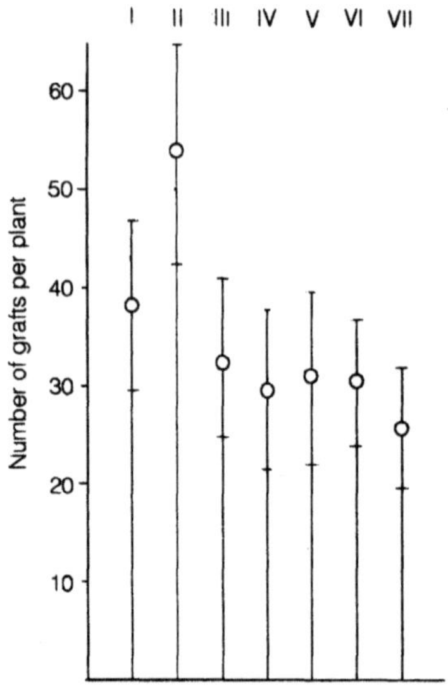

Fig. 11. Productivity of seven virus-free clones (progeny of one meristem) of carnation cv. Shocking in July–December 1981

Table 3. Survival of mericlones transferred to soil

	Planted	Survived	Survival (%)
1. Carnations from test tubes	2254	1775	78.8
2. Carnation minigrafts	2421	2222	91.8
3. Chrysanthemums from test tubes	612	571	93.3
4. Chrysanthemum minigrafts	2453	2442	99.6

Table 4. Structure of costs at different stages of mericlone propagation

	Cost of the unit of production (Kop)					
	Laboratory stage		Greenhouse stage		Total, according to item	
	Rbl, Kop	%	Rbl, Kop	%	Rbl, Kop	%
1. Labor	0–17.6	36.9	0–11.3	39.4	0–28.9	37.8
2. Cost of laboratory equipment	0–13.8	28.9	0–03.3	11.5	0–17.1	22.4
3. Other expenditures	0–16.3	34.2	0–14.1	49.1	0–30.4	39.8
Total production cost	0–47.7	—	0–28.7	—	0–76.4	—
Percentage of total production cost	62.4%		37.6%		100.0%	

At present, the Enterprise is contracted to introduce scientific methods into other enterprises in the USSR. According to such a contract, a client receives scientific data, the initial material (virus-free mericlones), and material for immunoassay. Due to uniform production methods, the cost of a mericlone is not too high, and can be compared to that of carnation grafts (Table 4).

Analysis of the cost of mericlone production shows that the greater part falls in the laboratory stage, ca. 2/3 (61.4%), and 1/3 (37.6%) in the stage of acclimatization in the greenhouse (Rybalko 1991). At both stages, the major part of the costs is for labor and laboratory equipment. In the future, automation of the techniques is planned, to reduce costs at the laboratory and greenhouse stages, preparation of laboratory dishes, preparations of media, application of new effective methods of micropropagation and, at the greenhouse stage, application of inert substrates and automation of climate factors, regulation.

7 Conclusions

Research into developing and introducing methods of freeing ornamentals from virus diseases is being carried out in the USSR. This work is accompanied by virologic investigations, which include the identification of viruses on different species and the introduction of new assay methods, the ELISA test in particular.

The necessity for this is based on the fact that the important ornamental cultures, representing the main part of flower production (carnation, chrysanthemum, orchid, tulip, and other plants), are almost completely infected with viruses, and are subject to a complex infection.

As in other countries, the technique is based on a combination of thermotherapy and shoot tip culture. Numerous studies are devoted to the main ornamental culture, carnation. Recently, studies on chrysanthemum, freesia, and orchid have also been initiated.

The more effective the method of promotion is, the greater the number of virus-free plants we can obtain, and the cheaper production becomes. The two-phase method of plant regeneration from meristem, developed and tested by us over 10 years, ensures the production of virus-free plants without the additional stage of thermotherapy. These methods have been introduced into production, and provide mass production of mericlones at comparatively low costs. In the future the production of mericlones will be extended and the costs reduced. This will provide the possibility of using mericlones in commercial production. "Possibilities of cultured cells are infinite, but the fate of cell technologies will be decided by economic criteria" (Butenko 1986).

References

Butenko RG (1960) Application of method of isolated shoot tip culture for studies on plant growth and organogenesis. Fisiol Rast 7, 6: 715–723

Butenko RG (1983) In vitro technology in agriculture. Selsklkhoz Biol 5: 3–7
Butenko RG (1986) Cell technology in production of economically important substances of plant nature. In: Plant cell culture and biotechnology. Inst Plant Physiol Acad Sci USSR. Nauka, Moscow, pp 3–20
Butenko RG, Kataeva NV, Ksenofontova DV (1982) Methodic instruction on clonal micropropagation of gerbera and freesia. Inst of Mountain Gardening and Floriculture, Min of Agric USSR, Sochi, p 34
Cherevchenco TM, Kushnir GP, Lavrentjeva AN, Budal VT, Kovalskaya KA (1982) Methodic instruction on mass propagation of orchids. Acad Sci Ukr SSR, Central Bot Gard, Kiev, p 54
Dubov DL, Tretjak SA (1986) Promotion of chrysanthemum and technology of propagation of mother stock, obtained in tissue culture and by rooting of cuttings. Abstr Conf Progressive innovation, Introduction of Scientific Achievements and the Ways of Improving of Labour Organization and Management in Floriculture. Min of Housing and Municipal Economy Latv: SSR, Riga, pp 39–41
Dzirkale LT (1987) Comparative biologic, physical and chemical characteristics. Thesis D, Biol Acad Sci Ukr SSR, Kiev, p 21
Ignash YaR (1978) Some results of studies on tomato aspermy strains. In: Virus diseases of agricultural crops and their control. Abstr All-Union Conf, Sept 1978, Leningrad, VASKhNIL, Moscow, p 154
Kataeva NV, Butenko RG (1983) Clonal micropropagation of plants. Nauka, Moscow, p 96
Kharuta LG (1983) Studies on mass propagation of *Cymbidium* in aseptic culture on virus-free basis for seed growing. In: Plant cell culture and biotechnology. Abstr Reports to the 4th All-Union Conf, 3–6 Oct, Kishinev, Shtiinca Publ, Kishinev, pp 130–131
Kharuta LG, Rybalko AE, Ryndina NA (1984) Mass diagnostics of virus diseases in meristematic mother stocks of ornamentals. Nauch Tr NII Gornoho Sadovodstva i Tsvetovodstva Sochi 31: pp 34–41
Kilevits M (1972) Identification, purification and serology of carnation mottling. In: Pests and diseases of agricultural crops. Tr Latv Skh Akad 69: 65–72
Korneeva IT (1977) Production of virus-free planting material of *Lilium* in meristem culture. Abstr Reports to the Conf Propagation of ornamental cultures in meristem culture, 1977, Riga, p 15
Makutenajte MK (1974) Virus disease of horticultural (ornamental) plants in Lithunian SSR and their economical importance. In: Viruses and virus diseases of plants. Naukova Dumka, Kiev, pp 292–296
Makutenajte-Navalinskene NK (1981) Virus and mycoplasmic diseases of ornamental plants. Inst Bot Acad Sci Lithuan SSR, Vilnus, Mokslas, p 72
Mitrophanova OV, Zlenco IL, Soboleva LB, Feophylova GV (1985) Production of virus-free planting material of chrysanthemum. Tsvetovodstvo 4: p 11
Morel G (1964) Regeneration des varietes viroses par la culture des meristems apicaus. Rev Hortic 136(2261): 733–740
Murashige T, Skoog F (1962) A revised medium for rapid growth and bioassays with tobacco tissue culture. Physiol Plant 15: 473–497
Odinets AG (1986) Diagnostics of narcissus mosaic and tobacco rattle viruses in nacrissus by the ELISA test. In: Biocenotical relations of organisms in plantings of artificial phytocenosis. Abstr of Regional Conf, 25–26 June 1986, Kishinev, Min Housing and Municipal Economy Mol SSR, Kishinev pp 32–33
Odinets AG (1987) Elaboration and application of the ELISA test for diagnostics of virus diseases of ornamental culture. Thesis D Biol 1987, Moscow p 22
Odinets AG, Atabekova TI, Atabekov IG (1985a) Diagnostics of cucumber mosaic virus by the ELISA test. Dokl VASKhNIL 7: 19–21
Odinets AG, Bojkov SV, Vitslane LF, Atabekov IG (1985b) Diagnostics of yellow bean mosaic and freesia mosaic viruses by ELISA test. Selsk Khoz Biol 11: 39–42
Odinets AG, Kopylov SS (1986) Diagnostics of tulip breaking virus by the ELISA test. Selsk Khoz Biol 9: 56–60
Odinets AG, Kulinich AV, Chirkov SN, Kozitsky YuN, Atabekov IG (1983) Comparative testing of immunological methods for mass diagnostics of carnation mottle disease. Selsk Khoz Biol 5: 37–41
Peterson L, Zhola I, Done D (1974) Propagation of carnation in meristem culture for production of virus-free initial material. In: Plant and insect viruses. Tr Latv SKhA Elgava 82: 53–57
Protsenko AE (1965) Detection of plant viruses by electron microscopy. In: Morphology and classification of phytopathogenic viruses. Nauka, 1966, Moscow, pp 7–19
Rosenberg V (1984) In: Estonian Institute of Agriculture and Melioration. Tsvetovodstvo 2: 11
Rybalko AE (1977) Production of initial virus-free planting material remontant carnation and chrysanthemum in the Ukrainian SSR. In: Propagation of ornamental in meristem culture. Abstr All-Union Conf MZhKKH Latv. SSR, 1977, Riga, pp 26–27

Rybalko AE (1978) Studies of mass regeneration of virus-free plants in meristem culture. In: Virus diseases of agricultural crops and their control. Abstr All-Union Conf, 3–6 Oct, Leningrad. VASKhNIL, Moscow, pp 10–11
Rybalko AE (1984a) On promotion of plants from virus diseases in tissue culture (ornamental cultures). Selsk Khoz Biol 4: 80–81
Rybalko AE (1984b) Viruses of ornamentals and their control. Review. TsBNTI MZhKKh RSFSR, Moscow, p 42
Rybalko AR (1984c) Production of virus-free planting material for commercial floriculture in the state farm Uzhnyje Kultury. In: Commercial propagation of ornamental culture in the south USSR. Nauch Tr NII Gornoho Sadovodstva i Tsvetovodstva. MSKh SSSR, Sochi, 31, pp 32–34
Rybalko AE (1984d) Experimental grounds and principles of introduction into production of technology of promotion of plants from viruses. In: Theory and practice of usage of immunity of agricultural crops to virus diseases. Proc 8th All-Union Conf, May 1984, Vilnus Inst Bot AN Lithuan SSR, VASKhNIL, Vilnus, pp 158–160
Rybalko AE (1985) Preventive measure against virus diseases of ornamental culture. Zashch Rast 2: 32–33
Rybalko AE (1986) System of promotion of introduced ornamental cultures from phytopathogenic viruses. In: Biocenotical relations of organisms in artificial phytocenosis. Proc Regional Conf 25026, June 1986, Kishinev, Min Housing and Municipal Economy Mol SSR, Kishinev, pp 32–33
Rybalko AE (1987) Promotion and clonal micropropagation of ornamental culture. Review MZhKKh RSFSR. Ser Greening of Settlement 3(51), Moscow, p 47
Rybalko AE (1991) The economical description of production of virus-free mericlones of flowering cultures. In: Biology of cultivated cell and biotechnology of plants. Nauka, Moscow, pp 227–232
Rybalko AE, Kharuta LG (1978) Application of meristem culture for promotion of remontant carnation from virus diseases. Fiziol Biokhim Rast Kiev Inst Fisiol Rast AN UKrSSR: 19, 5, 535–540
Rybalko AE, Kharuta LG, Kozar FE, Sivers NA, Kopejchenko II (1978a) Methodic instruction on protection from virus diseases in mother stocks of remontant carnation. Min Housing and Municipal Economy Ukr SSR, p 13
Rybalko AE, Kishko YaG (1984) Carlaviruses of plants: distribution and their control. Proc 6th Conf Ukrain Microbiol Soc June, 1984, Donetsk, Naukova Dumka, pp 195
Rybalko AE, Sivers NA, Kharuta LG (1978) Studies on biological properties of spherical virus, identification in commercial plantations of remontant carnation. In: Proc All-Union Conf Virus diseases of agricultural crops and their control, Sehn, Leningrad, 1979, VASKhNIL, Erevan, Moscow, p 132
Smirnova NS (1983) Micrografting of remontant carnation. In: Plant cell culture and biotechnology. Proc 4th All-Union Conf Oct 3–6, Kishinev, Stiinsa, pp 121–124
Van Hoof P (1971) Methods de culture in vitro de meristems pour l'obtention d'oeillets non-viroses. Parasitica 70: 27–35
Vitslane LF (1986) Promotion of planting material of freesia and propagation of gerbera in tissue culture. Abstr Conf Progressive innovations. Introduction of Scientific Achievements and the Ways of Improving of Labour Organisation and Management in Floriculture. Min Housing Municipal Economy Latv. SSR, Riga, p 68
Voloshchuk LF (1984) Virus diseases of carnation in Moldavia. In: Virus diseases of cultivated plants in Moldavia. Kishinev, Stiinsa, p 65
Voloshchuk LF (1986) Virus diseases of carnation and tulip in MSSR. Grounds on their control. Thesis D Biol, All-Union Inst Plan Protection, Leningrad-Pushkin, p 38
Voloshchuk LF, Kosakovskaja OI (1984) Serodiagnostics of carnation mottle virus. In: Virus diseases of cultivated plants in Moldavia. Shtiintsa, Kishinev, p 81
Voloshchuk LF, Teploukova TN (1986) Main virus diseases of ornamental crops. Methodic instructions of All-Union Inst Plant Protection, Leningrad-Pushkin, p 34
Zhola IYa (1986) Advantages and perspectives of application of tissue culture in development of floriculture. Abstr Conf Progress innovations. Introduction of Scientific Achievements and the Ways of Improving of Labour Organisation and Management in Floriculture. MZhKKh Latv. SSR. Riga, pp 73–74
Zhola IYa, Ushkuare A, Odinets AG (1985) Application of ELISA test for early diagnostics of mottle virus in meristem carnation in the test tubes. In: Virus diseases of plants and insects of Latv. SSR, Trudy LSKhA, Elgava, Issue 222: 39–46
Zirka TI (1984) Atlas of virus and microplasmic diseases of ornamental cultures. Acad Sci Ukr SSR, Centr Resp Bot Garden, Naukova Dumka, Kiev

XXVII Micropropagation of Some Parasitic Plants

S.N.C. Okonkwo[1]

1 General Account

1.1 Botany/Morphology/Chromosome Number

Some families of flowering plants include genera which are parasitic on other flowering plants (Kuijt 1969). The major attribute of parasitic flowering plants is that, at least during a part or whole of their life cycle, they depend on a host plant for some or all of their nutrients. Parasitic flowering plants are generally classified into (1) root parasites, and (2) stem parasites, depending on which organ of the host plant that is attacked. For the purpose of this account, major emphasis will be laid on root parasites, while only brief reviews will be given of work done on stem parasites.

In parasitic plants, the connection between the host and parasite vascular conductive tissues is ensured through the formation by the parasites of a specialized morphological and physiological bridge, the haustorium, which establishes the main channel for flow of nutrients from host to parasite.

Root parasites are found in diverse dicotyledonous families including Balanophoraceae, Hydnoraceae, Krameriaceae, Lennoaceae, Santalaceae, Scrophulariaceae, and Orobanchaceae (Kuijt 1969). Most of the root parasites are weeds as well as parasites in that they produce an aerial, chlorophyllous shoot system during part of their life cycle, even though they are dependent on the host for some of their nutrient needs. The parasites in this category are referred to as semi- or hemiparasites. Others are total parasites, deriving all their food supply from the host, and are thus termed holo-parasites. Root parasites are further classified into facultative and obligate types. Facultative parasites, exemplified by many members of the Scrophulariaceae, do not normally depend on a host in order to complete their life cycles. However, during their growth, any host plant in their vicinity is attacked and their performance is thus enhanced (Chuang and Heckard 1971; Nwoke and Okonkwo 1974). Besides, seeds of facultative parasites do not normally require a host stimulus for germination. They germinate either unaided in darkness, or they are stimulated to germinate by light, e.g., *Buchnera hispida* (Okonkwo and Nwoke 1974; Nwoke and Okonkwo 1980) *Sopubia delphinifolia* (Shivanna and Rangaswamy 1976). On the other hand, obligate root parasites such as most members of the Orobanchaceae and some Scrophulariaceae are unable to complete their life cycles in the absence of a host. Their seeds usually require a stimulus from

[1]Department of Botany, University of Nigeria, Nsukka, Nigeria

Biotechnology in Agriculture and Forestry, Vol. 20
High-Tech and Micropropagation IV (ed. by Y.P.S. Bajaj)

the host to germinate. Besides, a haustorial attachment to the host root must be established after seed generation before the shoot can develop, otherwise the young parasite seedling perishes. Stimulation of shoot growth and normal seedling development of the parasite appear to depend on the nutrients passing from the host to the parasite, and lead subsequently to flowering and seed set.

Stem parasites are equally to be found in a number of unrelated plant families including *Cassytha* spp. (Lauraceae), *Cuscuta* spp. (Convolvulaceae), *Tapinanthus bangwensis*, *Phoradendron* spp. (Loranthaceae), and *Viscum* spp. (Viscaceae). As in the case of root parasites, stem parasites include semi- or hemi-parasites which have chlorophyllous shoots, e.g., *Tapinanthus* spp. (Onofeghara 1971) and *Phoradendron* spp. (Hull and Leonard 1964), as well as holo-parasites, e.g., *Arceuthobium* spp. (Kuijt 1969).

1.2 Distribution and Importance

Root parasites are widely distributed in the tropical and subtropical regions of the world, including Africa, Asia, India, and North America. Considerable interest has been focussed on root parasitic weeds in recent years in view of some of their members having been shown to be serious pests of grain cereal and legume crops, especially in the savanna grasslands and Sahelian zones of Africa. In West Africa, for example, several species of *Striga* (witchweed) are parasitic on these crops. The giant witchweed (*Striga hermonthica*) (Fig. 1) is a virulent parasite of *Sorghum*, *Pennisetum*, *Zea mays*, and *Oryza*. Also, the cowpea witchweed (*Striga gesnerioides*) is a parasite of *Vigna unguiculata* and occasionally of *Arachis hypogaea* (Okonkwo and Nwoke 1978). The related root parasitic weed, also in the family Scrophulariaceae, *Alectra vogelii*, is a serious pest of cowpeas and groundnuts (Nwoke and Okonkwo 1978). On the other hand, *Striga asiatica* is a serious parasite of maize and sorghum in South Africa (Saunders 1933), in warm southern United States of America (Musselman 1980), and in India (Kumar 1940), but is not very significant is West Africa. Previous researches have recorded over 50% reduction in yield of sorghum due to witchweed attack (Butler 1956; Ogborn 1970), but more recent estimates put the losses up to 91% in sorghum (Ogborn 1979) and 95% in maize (Parkinson 1985). Both *Striga gesnerioides* and *Alectra vogelii* are reported to cause yield losses in cowpeas and groundnuts (Lagoke et al. 1986). Some root parasites, e.g., members of the Santalaceae, are cherished for their yield of commercial sandalwood (Kuijt 1969; Bhatnagar 1987).

Stem parasites, on the other hand, are widely distributed in various parts of the world, and occur in both tropical and temperate regions. Their members, especially genera in the family Loranthaceae, attack economically important tree species, including timber trees and woody cash crops, causing serious losses in yield (Kuijt 1969; Onofeghara 1971).

1.3 Need for in Vitro Culture of Parasites

In vitro culture of embryos facilitates experiments to determine the factors that regulate the growth and development of the organs of the seedling plant. It also

Fig. 1. Whole plant of flowering giant witchweed (*Striga hermonthica*), growing in a sorghum field. (× 0.3)

aids the study of the biochemistry and metabolism of germination, which are difficult to assess in embryos enclosed in seeds. Seeds of root parasites are generally microscopic and consequently are not amenable to embryo excision. However, the tissues surrounding the embryo proper are so thin that culture of the entire seed is the norm (Fig. 2). Besides, culture of isolated parasite seedlings in the absence of the host provides a unique opportunity for assaying various growth regulatory functions in the parasites, and the effects of specific metabolites and substrates on the development of the parasitic plants.

Effective control of parasites in crops deserves attention wherever they are a major problem impeding food production. Several methods of control have been attempted. These include, for root parasites, hand-weeding, "trap" and "catch" cropping, use of nitrogen fertilizers, herbicides, resistant crop cultivars and germination stimulants to stimulate suicidal germination; and for stem parasites herbicides, and felling and burning of infected trees (Kuijt 1969; Musselman 1980;

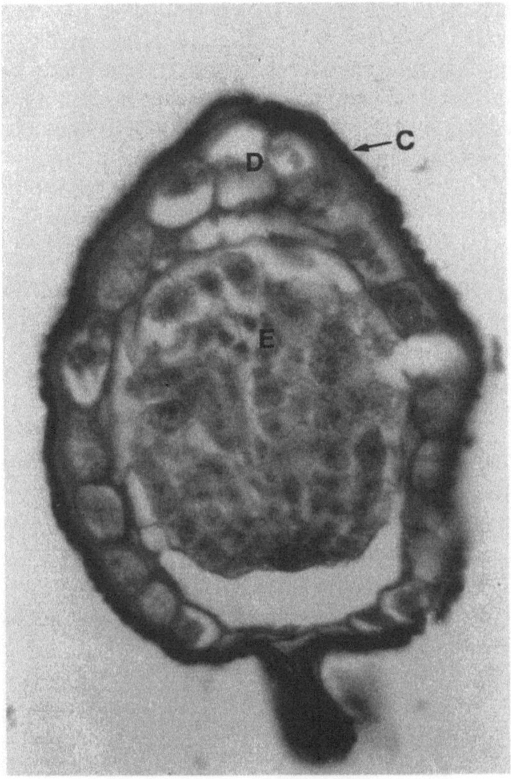

Fig. 2. L.S. seed of *Alectra vogelii* showing thin endosperm surrounding the minute embryo (× 364). *C* Seed coat; *D* endosperm; *E* embryo. (After Nwoke and Okonkwo 1978)

Ayensu et al. 1984). Of all these, the use of resistant crop cultivars seems the most promising method, especially for the smallholder farmer. However, an integrated control package comprising all the various methods would be the most effective approach to control.

In vitro culture of parasites (Okonkwo 1966b, c) has opened a novel avenue for examining in more detail the effects of nitrogen and herbicides on their physiology and biochemistry. One of the promising methods for control of the parasitic weed, witchweed, is through the use of nitrogenous fertilizers. Last (1960) and Ogborn (1984) have reported that application of nitrogen fertilizers reduced the severity of witchweed attack, and increased the yields of the affected host grain crops. However, these must be applied at high dosages of 100–200 kg/ha. The mode of action of the nitrogen fertilizer has not been clearly understood. It is possible that certain forms and levels of nitrogen may be beneficial to the host but harmful to the attached parasite plants in an unknown way. It could also be that the nitrogen fertilizers do not affect the parasites directly, but instead boost crop growth to a level that it can tolerate the parasite. It has been reported (Stewart et al. 1984; McNally and Stewart 1987) that *Striga* obtains most of its nitrogen requirements in organic form from the host. The authors further showed that *Striga* plants have minimal capacity for assimilating nitrate nitrogen. Pesch and Pieterse (1982) reported that *Striga* seed-germination and radicle elongation were inhibited by urea.

The nutrient agar culture technique seems ideal for offering various sources and doses of nitrogen to isolated cultured seedlings of parasites in order to assess their effects on them. The technique is also adequate for exposing parasitic weed and stem parasite embryos and seedlings to various herbicides as a means of screening the herbicides for their effects on the parasites.

2 Review of Tissue Culture Work on *Striga* and Related Root Parasites (Table 1)

2.1 Culture of Hemi-Parasites

Investigations of the structure of the mature seeds of root parasites have revealed that the enclosed embryos are at different stages of development depending on the genus and species. In some chlorophyllous hemi-parasites like *Striga hermonthica* and *S. gesnerioides*, the embryos are differentiated to the cotyledonary stage with bipolar structure showing plumular and radicular poles in the mature seed (Okonkwo and Nwoke 1978; Fig. 3; Nwoke and Okonkwo 1974). In others like

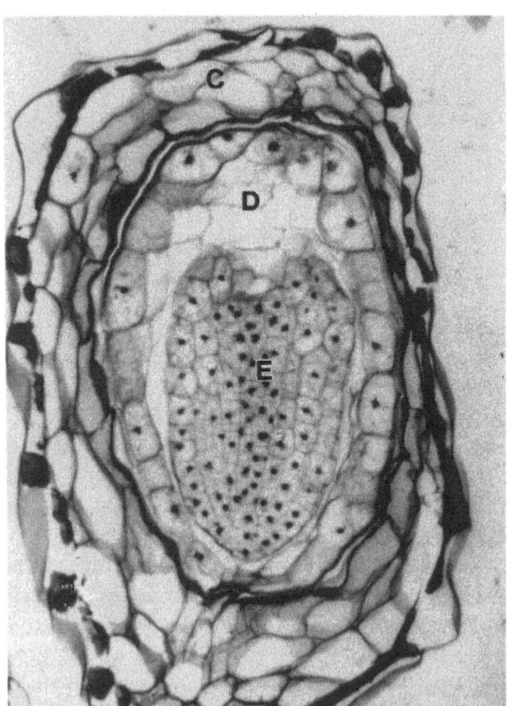

Fig. 3. L.S. seed of cowpea witchweed (*Striga gesnerioides*), showing embryo with plumular and radicular poles, thin endosperm, and many-celled seed coat. (× 300). *C* Seed coat; *D* endosperm; *E* embryo. (After Okonkwo and Raghavan 1982)

Table 1. Review of in vitro culture studies on parasitic plants

Plant species	Family	Medium[a]	Results/remarks	Reference
Root parasites				
Aeginetia indica	Orobanchaceae	Mineral salts, + sucrose (2000) + water melon juice (15%)	Normal plants chlorophyllous shoots in light; flowered in 4–5 months	Tashima et al. (1974); French and Sherman (1976)
Alectra vogelii	Scrophulariaceae	WM + CM (20%) or MS	Normal plants. No flowering	Okonkwo (1975)
A. sessiflora var. *senegalensis*	–	MS	Normal plants. No flowering	Okonkwo (this work)
Cistanche tubulosa	Orobanchaceae	WM + CM (15%) + CH (400). THS	Callus after germination. Shoot bud on transfer to THS. No plantlets	Rangan (1965); Rangan and Rangaswamy (1968)
Exocarpus cupressiformis	Santalaceae	WM + IAA(1) + Kin(1) + CH (400)	Callus, root, and shoot primordia	Bhojwani (1969)
Melampyrum lineare	Scrophulariaceae	Mineral salts with or without sugar	Stunted plants, flowered and set seeds	Cantlon et al. (1963)
Orobanche aegyptiaca	Orobanchaceae	WM or NM or MS or THS + IAA(0.5) or kin (5) or GA_3(5)	Germination with radicle forming callus or bipolar germination with formation of normal seedlings. No flowering	Rangaswamy (1963); Kumar and Rangaswamy (1977)
Santalum album	Santalaceae	WM + IAA (1)	Normal plants, no flowering	Rangaswamy and Rao (1963); Sahai (1983)
Sopubia delphinifolia	Scrophulariaceae	WM or mineral salts + high light intensity	Normal plant, no flowering	Shivanna and Rangaswamy (1976); Sahai (1983)
Striga angustifolia	–	WM	Shoot development, no flowering	Rangaswamy and Rangan (1969)
Striga asiatica (red-flowered American variant)	Scrophulariaceae	LS + kin (5) + glut (500)	Germination with normal plants. No flowering	Yoshikawa et al. (1978)

Table 1. (*Continued*)

Plant species	Family	Medium[a]	Results/remarks	Reference
Striga asiatica (red-flowered American Variant)	–	Mineral salts and sugar (details not specified)	Germination with normal plants. flowering after 3 months	Riopel and Baird (1987)
Striga asiatica (white-flowered Indian variant)	–	MS	Normal plants. Flowering after 105 days	Chidley and Drennan (1987)
Striga asiatica (red-flowered American variant)	–	MS	Normal plants. Flowering after 2 months	Okonkwo (this work)
Striga euphrasioides	–	WM + IAA (1) + kin (10)	Callus. No shoot development	Rangaswamy and Rangan (1966)
Striga gesnerioides	–	MS + CH (200) + Inos (100)	Normal plants; Flowering in 2–2 1/2 months	Okonkwo (1982)
Striga hermonthica	–	MS	Normal plants. Flowering after 4 months	Okonkwo (1966b,c)
Stem parasites				
Amyema sp.	Loranthaceae	WM + IAA (0.5) + CH (500)	Embryoids from papillate outgrowths. Haustorial disc developed	Bajaj (1963, 1967); Johri and Bajaj (1964)
Amylotheca sp.	–	WM + IAA (0.5) + CH (500)	No embryoids from papillate out-growths. Embryo with a pair of leaves, holdfast, 2–5 haustorial discs in 25 weeks	Johri and Bajaj (1964)
Arceuthobium pusillum	Loranthaceae	HM + 2,4-D (0.1–1) + CM (25%) + CH (500)	Germination with radicle. Holdfast and callus formation, and embryoids	Bonga (1971); Harvey (1967)

Cassytha filiformis	Lauraceae	WM + IAA (1)	Normal plants. No haustoria	Rangaswamy and Rangan (1963); Rangan and Rangaswamy (1969)
Cuscuta reflexa	Convolvulaceae	WM	Callus and adventitious embryos. Haustoria-like organs formed on cultured stem explants	Maheshwari and Baldev (1962); Baldev (1959)
Dendrophthoe falcata	Loranthaceae	WM + IAA (1) + kin (1) + CH (400)	Callus from radicular end of embryo, shoot bud formation, accessory buds. Cultured endosperm formed triploid shoot buds, and haustoria	Johri and Bajaj (1963); Bajaj (1968); Nag and Johri (1976); Johri and Bajaj (1965); Bhojwani (1968) Johri and Nag (1968)
Nuytsia sp.	–	WM + IBA (5) + kin (5) + CH (2000)	Callus and embryoids	Johri and Nag (1974)
Phoradendron sp.	–	WM + IAA (0.5) + CH (500) + kin (1)	Tips of cotyledons proliferated. Massive holdfast formed. No haustorial discs	Bajaj (1970)
Scurrula pulverulenta	–	WM + BA (10^{-5} M)	Well-developed haustoria and plumular leaves	Bajaj 1963; Johri and Bhojwani (1970)
Tapinanthus bangwensis	Loranthaceae	WM + kin (1) + IAA (1) + GA_3 (0.5)	Leafy shoots produced	Onofeghara (1972)
Taxillus sp.	–	WM + IAA (0.5) + CH (500)	Embryoids from callus derived from embryo	Johri and Nag (1974)

[a]Only representative successful medium for each genus is given. Other media and variations can be obtained from the original papers.
Abbreviations: Basal media: HM: Harvey (1967); MS: Murashige and Skoog (1962);
THS: Tepfer, Greyson, Craig, and Hindman (1963);
WM: White (1943); NM: Nitsch (1951); LS: Linsmaier and Skoog (1965).
Supplements: CH: casein hydrolysate; CM: coconut milk; kin: kinetin; IAA: indoleacetic acid.
(mg/l) GA_3: gibberellic acid; BA: benzyladenine; Inos: inositol; 2,4-D: 2,4-dichlorophenoxyacetic acid; Glut: L-glutamine.

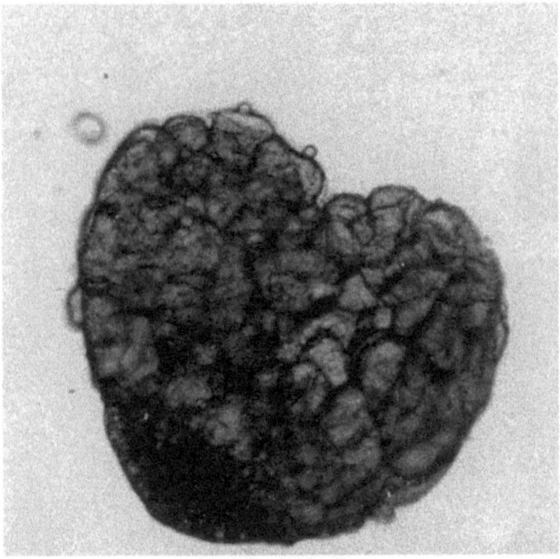

Fig. 4. Isolated heart-shaped embryo from mature seed of *Alectra vogelii* (× 455). (After Nwoke and Okonkwo 1978)

Alectra vogelii, the embryo is at the heart-shaped stage (Nwoke and Okonkwo 1978; Fig. 4). However, in nonchlorophyllous, holo-parasites, e.g., *Orobanche* sp., the seed, at the time of shedding, harbors an undifferentiated embryo that is virtually a globular mass of cells (Rangan 1965; Rangaswamy 1967). One aspect of the morphogenesis of root parasites which has been successfully studied through aseptic culture techniques is seed germination. Although seeds of obligate root parasites do not germinate under normal circumstances in the absence of a host plant, aseptic culture of the seeds has demonstrated that the germination stimulant can be replaced by a variety of chemical supplements to the culture medium. In author's work (1966b, c) on *Striga hermonthica*, the germination was achieved by incorporation of the host root exudate in the medium. Other workers have stimulated germination of seeds in other genera and species using other supplements including kinetin and related substituted purines (Worsham et al. 1959; Williams 1961); gibberellic acid (Williams 1961; Nash and Wilhelm 1960); strigol (Cook et al. 1966, 1972); coumarin derivatives (Worsham et al. 1962); casein hydrolysate, coconut milk, or both (Rangaswamy 1967; Rangaswamy and Rao 1963; Rangan 1965; Rangan and Rangaswamy 1968); ethylene and ethrel (= 2-chloroethylphosphonic acid) (Egley and Dale 1970); and bromine and chlorine (Okonkwo and Nwoke 1975). In most cases studied, germination of seeds of obligate root parasites ends with the emergence and elongation of the radicle (Fig. 5).

The tissue culture technique has provided a novel avenue for morphogenetic studies on root parasitic weeds. However, initial attempts to culture seeds and seedlings of parasitic species were unsuccessful. Worsham et al. (1959) and Williams (1960) reported that when *S. asiatica* seeds were germinated and incubated in the natural host root exudate under nonsterile conditions, no shoot development occurred. On the other hand, when the seeds were stimulated to germination with kinetin and gibberellic acid and the resulting seedlings were left in the stimulant chemcial solutions, the shoot tips became swollen and elongated beyond the

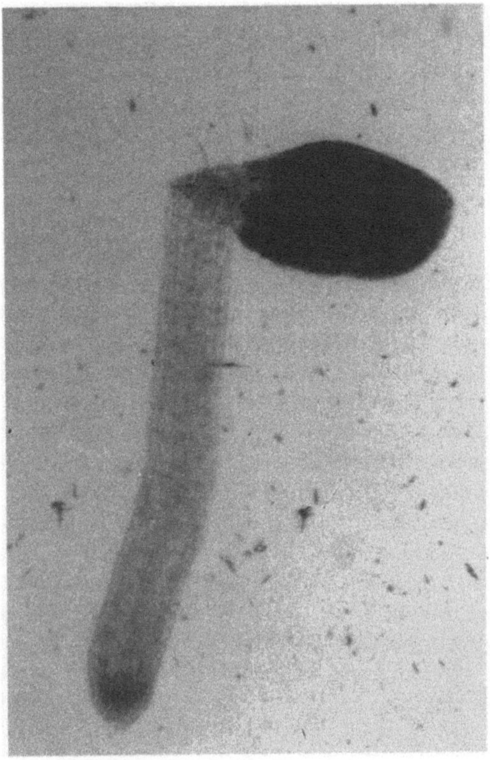

Fig. 5. Germinated seed of cowpea witchweed (*Striga gesnerioides*), showing elongated radicle with no shoot emergence. (×92)

cotyledons, but showed no further growth. Williams (1960) therefore suggested that *Striga* seeds might lack growth factor(s) similar to kinetin or gibberellic acid, and that they probably obtain their supply from the host in vivo. Subsequent attempts to prolong the growth of *S. asiatica* seedlings in culture by supplying them with mineral salts, vitamins, and organic compound additives like thiamine, nicotinic acid, pyridoxine, adenine, and kinetin, were unsuccessful. Williams (1961) did not report his methods of experimentation, and it was uncertain whether he employed aseptic methods, since he stated that "parasite seeds could not be surface sterilized without damage to the embryo". Subsequent studies (Okonkwo 1966b; Okonkwo and Raghavan 1982) demonstrated that *Striga* seeds could be surface sterilized with 1% solution of sodium hypochlorite for 15 to 20 min, and stimulated to germinate with host root exudate. Employing the aseptic culture technique, Okonkwo (1966b, c) investigated the growth of *S. hermonthica* seedlings in nutrient media of varying compositions. In the initial studies, he incorporated filter sterilized host root exudate in the agar nutrient media in order to stimulate germination of the seeds. By experimenting with low (Knop 1865) and high (Murashige and Skoog 1962) mineral salts media, supplemented with various additives such as sugar, vitamins, casein hydrolysate, meso-inositol, and hormones, he was able to demonstrate, for the first time, that *S. hermonthica* germinated seeds could develop shoots in culture in the absence of a host. In subsequent cultures, he was able to stimulate germination of preconditioned *Striga* seeds with sterilized host root exudate and transfer the germinated seedlings individually on to agar nutrient media, thus showing unequivocally that shoot morphogenesis was not triggered by some substance contained in the host root exudate. Shoot formation and development of normal *Striga* seedlings and plantlets was achieved on a medium containing only mineral salts and sugar. Supplementation of the mineral salts-sugar medium with vitamins, or with vitamins, casein hydrolysate (CH), and meso-inositol did not improve growth, suggesting that the seedlings had competence to synthesize their vitamins and other growth-supporting organic compounds. Addition of hormones to the mineral salts-sugar media triggered callus formation from the seedlings, again suggesting that the seedlings were capable of synthesizing their own hormones. The results thus provided an insight as to what *S. hermonthica* obtains from the host. Certainly, the minimum requirements to elicit shoot growth are a combination of mineral salts and a suitable sugar. Furthermore, it was demonstrated (Okonkwo 1966b) that the high mineral salts medium of Murashige and Skoog (MS) (1962) supported better growth than the low mineral salts medium of Knop (1865). This enhanced growth on MS medium could be due to its high mineral salt content, its content of ammonium nitrogen, or other variations. It would also appear that *S. hermonthica* synthesizes its own hormones, contrary to the suggestion by Williams (1961). Thus, this parasitic weed would seem to depend on the host for water, mineral salts, and sugar, but not for more elaborate substances like growth regulators and vitamins.

When giant witchweed (*S. hermonthica*) plants were cultured on agar nutrient media, they produced adventitious roots with root hairs, contrary to reports on *S. asiatica* by Stephens (1912a, b) and Saunders (1933). It may be inferred, therefore, that in nature, *S. hermonthica* roots are capable of some independent absorption of water and mineral salts from the soil (Ismail and Obeid 1976). Since the parasite shoot does not develop in vivo from the germinated seedling, unless the penetration of the host is accomplished and connection is made with the host root vascular tissues (Okonkwo 1966a), it seems likely that in the absence of the host, the meager organic food reserve in the seed is used up in the formation of the radicle. Even if the radicle absorbs water and mineral salts from the soil medium, no sugar is available for absorption so as to trigger shoot growth, in combination with the absorbed mineral salts. On the other hand, based on these experiments, Okonkwo (1966a, b) suggested that as soon as the parasite radicle penetrates the host-root conductive tissues, a supply of mineral salts and sugar to the parasite stimulates shoot growth. The callusing of seedlings on media containing hormones, apart from indicating that the seedlings can synthesize their own hormones, further suggests that absorption of additional supplies from the medium leads to supra-optimal levels. Thus, the reported failure of shoot growth from germinated seeds of other chlorophyllous root hemi-parasites on nutrient culture media (Rangaswamy and Rangan 1966) may well be due to the presence of growth regulators in such media. In other studies (Okonkwo 1966c), it was shown that *S. hermonthica* seedlings cultured on nutrient media were able to metabolize a number of carbohydrates including mono-, di-, and tri-saccharides, as well as some pentoses. This indicates the versatility of *Striga* in enzyme machinery for handling a variety of energy sources for its metabolism.

It is noteworthy that in Okonkwo's (1966b, c) cultures, no haustoria were formed on the roots of seedlings cultured in vitro. Under natural conditions, the roots would have formed haustoria in contact with the host roots. This provides indirect evidence that the stimulus for the initiation of haustoria comes from the host. More direct evidence that the haustorium-inducing substance comes from the host has emanated from the work of Riopel (1979) and Riopel and Musselman (1979) on the facultative hemi-parasite, *Agalinis purpurea*. These inducing substances have been identified as phenolic chemical compounds, namely, xenognosin and vanillin, which are active on *Striga asiatica* and *Agalinis purpurea* (Lynn et al. 1981; Riopel and Baird (1987) and on *Striga hermonthica* (MacQueen 1984).

In other studies, Yoshikawa et al. (1978) reported shoot formation and normal seedling and plantlet development in *S. asiatica* on LS (1965) nutrient media consisting of mineral salts, agar, sugar, vitamins, amino acids and hormones (specifically cytokinins and auxins). No shoot development occurred on media lacking hormones. Thus, they inferred that stimulation of shoot development in vivo would be dependent on a supply of mineral salts, carbohydrates (sugar), and hormones from the host. Yoshikawa et al. (1978) referred to the possibility of Okonkwo's (1966b) media containing unspecified organic growth substances carried in the host root exudate or extract which he added to the culture media. However, as stated above, Okonkwo (1966b) added host root exudate to culture media only in preliminary studies. In subsequent experiments with *S. hermonthica* and other obligate root parasitic weeds, e.g., *Alectra vogelii* (Okonkwo 1975), and *S. gesnerioides* (Okonkwo 1982), he never added the host root exudate to the culture media used for growing germinated parasite seedlings.

In culture studies on *S. hermonthica*, it was shown (Okonkwo 1966b) that the plants became chlorophyllous in light, and flowered after 4 to 5 months in culture. This is evidence that flowering in this species of *Striga* can take place independently of the host stimulus.

Although *Striga asiatica* is not common in Nigeria, the red-flowered variety is occasionally found parasitic on maize and on wild grasses in derived savanna. The yellow-flowered variants also occur on maize and on hungry rice in Plateau and Gongola States of Nigeria (Parkinson 1985). In current studies, we have cultured seedlings of *S. asiatica* from seeds obtained from the 1984 collection from Whiteville Methods Development Centre, Whiteville, North Carolina USA. Germination occurred readily in 1 mg/l solution of GR24 (a strigol analog) and seedlings were ready for transfer to nutrient media after 2 to 3 days of exposure to the stimulant. The seedlings were cultured on MS medium. A typical result (Table 2) shows clearly that mineral salts alone could not support shoot differentiation from the seedlings. As in the case of *Striga hermonthica*, a minimum of mineral salts and sugar was required before shoots could develop. Supplementation of the mineral salts-sugar medium with vitamins, or with vitamins, CH, and inositol, did not improve seedling performance. Thus, like *S. hermonthica*, the best growth is on mineral salts and sugar. Addition of the cytokinin, benzylaminopurine (BAP) to the MS mineral salts-sugar-vitamins medium tended to depress shoot length and root formation, but increased shoot number per seedling by stimulating adventitious shoots from the seedling base, and therefore shoot fresh weight (Table 2). These results, as well as those of Riopel and Baird (1987) and Riopel (pers. commun.), are contrary to those of Yoshikawa et al. (1978). In our studies, *S. asiatica* seedlings

Table 2. Growth of *Striga asiatica* seedlings on MS medium. Mean of 20–25 seedlings ± S.E. after growth in darkness for 30 days at 25 ± 2 °C

Culture medium[a]	Shoot length (mm)	No. of pairs of leaves	Shoot no. per seedling	Primary root length (mm)	No. of laterals from primary root	Fresh wt. of shoot (mg)	Fresh wt. of roots (mg)	Total plant fresh wt. (mg)
MS + sucrose	39.5 ± 1.2	18	1	16.2 ± 0.8	11	11.6 ± 0.9	63.0 ± 4.6	74.6 ± 5.1
MS + sucrose + vitamins	41.3 ± 0.2	19	1	20.6 ± 0.9	17	11.6 ± 1.1	70.9 ± 8.7	82.5 ± 9.2
MS + sucrose + vitamins + BAP	23.0 ± 0.8	14	3	0.0 ± 0.0	0.0 ± 0.0	27.0 ± 3.0	53.6 ± 4.0	80.6 ± 5.6
MS + sucrose + vitamins + CH + INOST	29.8 ± 1.5	17	1	14.4 ± 0.9	12	9.5 ± 0.9	85.2 ± 10.2	94.7 ± 10.4

[a]Sucrose tested at 20 g/l; CH (= casein hydrolysate) tested at 200 mg/l; INOST (= meso-inositol) tested at 100 mg/l; BAP (= 6-benzylaminopurine) tested at 2 mg/l.

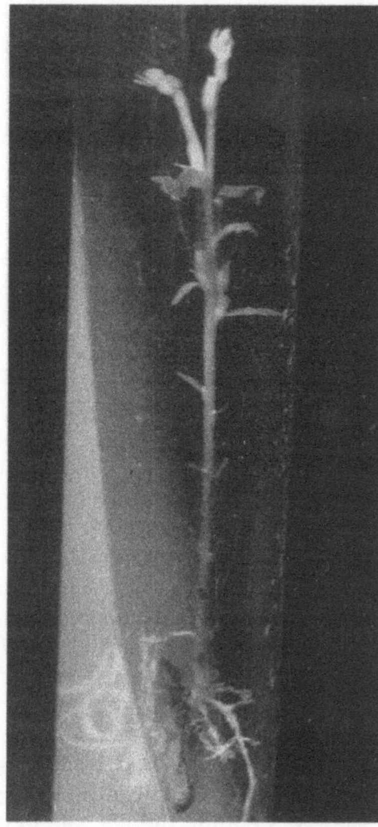

Fig. 6. Two-month-old seedling of *Striga asiatica* cultured from germinated seed on nutrient agar medium showing flowering. (× 1.5)

became chlorophyllous on transfer to light, and flowered after 2 months in culture (Fig. 6). Riopel and Baird (1987) also achieved flowering in culture, but after 3 months' duration. Besides, Chidley and Drennan (1987) have reported successful cultivation of the white-flowered Indian variant of *S. asiatica*, and achieved flowering in a simple medium of mineral salts and sugar.

Okonkwo (1982) also investigated the developmental pattern of the cowpea witchweed (*Striga gesnerioides*). He planted the seedlings derived from aseptic stimulation of the seeds by cowpea root exudate, on nutrient media of varying compositions. In both MS and WM media the seedlings developed shoots and roots on a mineral salts-sugar combination. In both media, growth was better when the basal mineral salts-sugar medium was supplemented with vitamins, or with vitamins, CH, and insitol. However, the seedlings performed better on WM than on MS medium. Normal seedlings with hairy shoots and vestigial leaves closely adpressed to the stem were formed, and they became chlorophyllous on exposure to daylight. They flowered 8 to 10 weeks after inoculation on nutrient media whether kept in light or in darkness.

The other species of *Striga* whose micropropagation has been studied is *S. euphrasioides* (= *S. angustifolia*). In this species, seed germination was shown to be stimulated by light (Rangaswamy and Rangan 1966), or by 10 mg/l kinetin after preconditioning. When the seeds were cultured in WM medium containing mineral salts, sugar, vitamins, and kinetin, the released embryos enlarged considerably and the cotyledons and hypocotyl proliferated forming a friable callus. Thus, no shoot morphogenesis occurred. This could be the result of the cytokinin in the medium. In later studies (Rangaswamy and Rangan 1969), *S. angustifolia* seeds germinated in WM media, on stimulation by light, and produced chlorophyllous plants, but flowering was not achieved.

Of the other genera of root parasites related to *Striga* and in the family Scrophulariaceae, two have been studied in sterile culture, namely, *Alectra* and *Sopubia*. *Alectra vogelii*, a parasite of cowpea and groundnut (Okonkwo 1975), was successfully cultured on WM and MS media. Best growth of seedlings was observed on WM mineral salts supplemented with sugar and coconut milk (20% v/v), or MS mineral salts supplemented with sugar. Minimal requirement for shoot development was mineral salts and sugar. Although chlorophyllous shoots produced in illuminated cultures grew normally, flowering failed to occur.

Another species of *Alectra*, *A. sessiflora* var. *senegalensis*, a parasite of wild grasses, has been successfully reared on nutrient media. As shown in Table 3, germinated seedlings were grown on MS and WM media. Again, a medium of mineral salts and sugar was required to stimulate shoot development and further growth of the plants. Further supplementation of the medium with vitamins, or with vitamins, CH, and inositol did not significantly improve shoot growth. However, root production and growth were greatly enhanced in this enriched medium (either WM or MS) in the form of profuse development of laterals and their callusing. Seedlings transferred to fresh media after 35 days continued to grow, and developed green shoots in light. Maintenance of the cultures in light for over 2 or 3 months did not lead to flowering of the plants. Instead, many plantlets regenerated from the callused roots and basal parts of the stem.

In *Sopubia delphinifolia*, Shivanna and Rangaswamy (1976) observed continued growth with normal shoot morphogenesis when seedlings were cultured on media containing mineral salts and a suitable carbohydrate energy source (usually sucrose). The result indicated their dependence on the host, like *Striga hermonthica*, for mineral salts and carbohydrates only. Sahai (1983) additionally has shown that high light intensity can support *Sopubia delphinifolia* seedling growth and replace the requirement for a carbohydrate. Although all cultured plants became chlorophyllous in light, no flowering occurred.

A few of the facultative hemi-parasites which can complete their life cycles in the absence of a host have been cultured on nutrient media. Cantlon et al. (1963) cultured *Melampyrum lineare* seedlings on agar-mineral salts media, with or without sucrose. The seedlings remained stunted, but flowered and set viable seeds, and, as stated above, Shivanna and Rangaswamy (1976) obtained shoot development in culture without flowering in *Sopubia delphinifolia*. Similar results were obtained by Sahai (1983) with *Sopubia delphinifolia* and *Santalum album*. Similarly, regeneratioin of shoot and root primordia from callus derived from the radicular pole of the embryo has been reported for seeds of *Exocarpus cupressiformis* (Santalaceae) cultured on a medium containing CH and IAA (Bhojwani 1969).

Table 3. Growth of *Alectra sessiflora* var. *senegalensis* seedlings on WM and MS media. Mean of 20–30 plants ± S.E. after growth in darkness, at 25 ± 2 °C for 35 days. Mineral salts alone supported no seedling development

Culture medium[a]	Shoot length (mm)	No. of pairs of leaves	Primary root length (mm)	No. of laterals from primary root	No. of adventitious root from shoot	Fresh wt. of shoot (mg)	Fresh wt. of roots (mg)	Total plant fresh wt. (mg)
WM + sucrose	0.9 ± 0.1	2.4 ± 0.2	4.5 ± 0.6	1.5 ± 0.2	0.4 ± 0.1	–	–	0.5 ± 0.1
WM + sucrose + vitamins	1.9 ± 0.1	3.2 ± 0.1	6.2 ± 1.0	2.3 ± 0.3	0.4 ± 0.1	1.5 ± 0.3	2.2 ± 0.4	3.7 ± 0.6
WM + sucrose + vitamins + CH + INOST	2.2 ± 0.2	3.1 ± 0.2	26.1 ± 2.6	13.2 ± 2.6	0.5 ± 0.3	2.3 ± 0.3	61.4 ± 4.9	63.7 ± 4.9
MS + sucrose	6.1 ± 0.6	5.1 ± 0.3	–	–	3.8 ± 0.5	3.2 ± 0.6	60.3 ± 6.1	63.5 ± 6.0
MS + sucrose + vitamins	3.2 ± 0.3	4.0 ± 0.2	–	–	2.7 ± 0.4	2.0 ± 0.4	49.4 ± 4.5	51.4 ± 4.5
MS + sucrose + vitamins + CH + INOST	6.1 ± 0.5	5.7 ± 0.2	–	–	3.7 ± 0.4	7.6 ± 1.0	109.7 ± 12.2	117.3 ± 12.9

[a]Sucrose tested at 20 g/l; CH (= casein hydrolysate) tested at 200 mg/l; INOST (= meso-inositol) tested at 100 mg/l.

2.2 Culture of Holo-Parasites

Post-germination shoot development from embryos of three holo-parasites, *Orobanche aegyptiaca* (Rangaswamy 1963), *Cistanche tubulosa* (Rangan and Rangaswamy 1968), and *Aeginetia indica* (Tashima et al. 1974; French and Sherman 1976), cultured on aseptic nutrient media, has been achieved. *Cistanche tubulosa* seeds cultured in a medium containing CM formed callus which originated from the radicular pole of the globular embryo. Shoot buds differentiated from the callus on transfer to a medium supplemented with IAA, GA, and kinetin, and maintained in the dark (Rangan 1965; Rangan and Rangaswamy 1968).

In *Orobanche aegyptiaca*, Kumar and Rangaswamy (1977) demonstrated that morphogenesis of the embryo in vitro depends on the composition of the medium. They found that hormones and complex additions to the medium were essential to elicit embryo morphogenesis. When seeds were cultured in a medium containing coconut milk or yeast extract, a typical monopolar germination occurred, followed by seedling formation through the activity of only the radicular end of the embryo. In this mode of development, following germination, the radicular pole of the embryo formed a mass of callus from which both root and shoot buds were initiated, while the plumular pole remained quiescent. In contrast, in a medium containing IAA, GA, kinetin, or strigol, seeds germinated by the normal bipolar system, with the root being formed from the radicular end, and a shoot from the plumular end of the embryo. Thus, a root-shoot axis was established leading to the production of normal seedlings. It is clear therefore that although in vivo the plumular pole of *O. aegyptiaca* embryo remains morphogenetically dormant, it can be stimulated to embark on morphogenesis by hormones. Thus, the requirements of complex media for development of normal shoot and seedlings of the holo-parasites must be related to the undifferentiated status of their embryos. The precise role of growth regulators and organic compound additives to the medium in stimulating shoot development and controlling morphogenetic processes are aspects for further investigations (Sahai and Shivanna 1982).

Similarly, shoot development and growth in *Aeginetia indica* were possible only on a complex medium (Tashima et al. 1974; French and Sherman 1976). Although *Aeginetia* plants flowered in culture (Tashima et al. 1974), there was no similar report for *C. tubulosa* and *O. aegyptiaca.*

3 Culture of Stem Parasites

Of the stem parasites, the bulk of the aseptic culture studies have been done on members of the Loranthaceae (mistletoes). The mistletoes that have been studied include various species of *Dendrophthoe, Taxillus, Scurrula, Amyema, Amylotheca* and *Nuytsia*, (see Bhatnagar 1987), *Tapinanthus* (Onofeghara 1971), and *Phoradendron* (Bajaj 1970). Mistletoe seed consists of an embryo surrounded by a massive endosperm, but lacks a seed coat. The mature embryo is peg-shaped and the blunt knob-like pole is known as the radicular end.

Seeds of mistletoes do not require a host stimulus to germinate, as they can germinate in contact with inanimate objects. However, during germination the radicular end of the embryo proliferates when in contact with a host and develops a hold fast. From the holdfast arises the haustorium which penetrates the host tissue. One interesting point that can be studied in vitro is whether the haustorium is formed in response to a stimulus from the host, as in the case of root parasitic weeds, or not.

When mature embryos of *Dendrophthoe falcata* were cultured in vitro, the radicular end of the embryo bent at right angles and a small protuberance grew out of the tip to develop into a haustorium (Bajaj 1963; Nag and Johri 1976). When the mature embryos were cultured in WM supplemented with CH and IAA, they developed several accessory shoot buds (Johri and Bajaj 1963). On WM supplemented with IAA and kinetin, the radicular end proliferated and buds differentiated. Supplementation of the medium with CM accelerated germination and the embryos developed massive holdfasts.

Globular embryos of *D. falcata* when grown on WM plus 500 mg/l CH, developed into mature embryos with free cotyledons (Johri and Bajaj 1965), contrary to what one obtains in nature where the cotyledons are closely appressed, resembling a monocotyledonous condition. In some cultures the globular embryos proliferated to form a callus which later differentiated into embryoids.

Bajaj (1963) studied morphogenetic response of various segments of the embryos of *Dendrophthoe*. He demonstrated that excised cotyledons in culture yielded only callus, the radicular portion formed holdfast and haustorium, and excised plumular region gave rise to shoots.

Scurrula pulverulenta mature embryos with endosperm were cultured in WM where they germinated and had well-developed haustoria and plumular leaves (Johri and Bhojwani 1970). This observation is contrary to earlier reports on some other mistletoes where the presence of endosperm was reported to inhibit or delay the development of plumular leaves (Bajaj 1967, 1968). The embryo callus underwent shoot bud differentiation when cultured on media supplemented with zeatin. Also, when added individually, kinetin and IAA brought about differentiation, but together they inhibited differentiation.

In *Amyema*, the papillate outgrowths present at the radicular end were stimulated to grow, and in a number of cultures these outgrowths developed into embryoids which later germinated (Bajaj 1963, 1967). In *Amylotheca*, papillate outgrowths occurred similar to those on *Amyema*. However, they did not develop into embryoids (Johri and Bajaj 1964). In more recent studies (Hall et al. 1987), the effects of hormones on development of *Amyema* seedlings were assessed. Zeatin and other cytokinins promoted leaf expansion but did not promote callus formation. On the other hand, gibberellin enhanced leaf elongation. Auxin (IAA) and auxin-amino acid conjugates promoted callus formation.

In *Nuytsia*, the cultured mature embryo produced callus which later differentiated embryoids, root and shoots (Nag and Johri 1976).

Similar results have been obtained with the tropical mistletoe, *Tapinanthus bangwensis* by Onofeghara (1971, 1972). He cultured mature embryos of *Tapinanthus* on WM medium, and found that the hormones kinetin, IAA and GA_3 enhanced growth of the embryos over the basal medium. Kinetin and GA_3 particularly promoted production of accessory leaves.

A noteworthy feature observed in *Nuytsia*, *Dendrophthoe*, and *Taxillus*, is that the callus and even the in vitro-formed organs did not show any change in the normal diploid chromosomal complement even after several passages (Johri and Nag 1974).

Endosperm tissues of mistletoes have also been studied in vitro. In *Scurrula*, *Dendrophthoe*, and *Taxillus*, organogenesis from the cultured endosperm was observed. Mature endosperm of *Scurrula* developed buds in the presence of cytokinin. The shoot buds were green, and histological studies revealed that the buds originated superficially from uncallused endosperm. On subculturing, the buds proliferated, forming a compact, slow-growing callus which later organized shoot buds and/or haustoria (Bhojwani 1968). In *Dendrophthoe* also the cultured endosperm tissue differentiated shoot buds and haustoria. These organs differentiated either directly or after callusing of endosperm (Johri and Nag 1968). Both cytokinin and an auxin were required for shoot bud differentiation, unlike *Scurrula* (Bhojwani 1968) and *Taxillus* (Johri and Nag 1968), where only a cytokinin was needed. The shoot buds differentiated from endosperm tissue of various mistletoes were invariably triploid.

The seeds/embryos of dwarf mistletoe (*Arceuthobium pusillum*) cultured in vitro (Bonga 1971), germinated and formed an elongated radicle, followed by formation of a holdfast from the radicular apex. Subsequently, the holdfast proliferated to form a callus part of which developed embryo-like structures.

Thus, the in vitro findings support the in vivo observations that mistletoe seeds can germinate in the absence of a host. Also, the differentiation of haustoria is not solely dependent on contact with a living host (Kuijt 1969).

The other stem parasites which have been cultured in vitro include *Cuscuta* sp. (dodder) and *Cassytha* sp. *Cassytha filiformis* seeds when cultured in vitro germinated but the radicle soon ceased growth and dried up, thus, a primary root was not formed. Instead, adventitious roots developed from the hypocotyl (Rangan and Rangaswamy 1969). Similarly, the mature embryo of *Cuscuta reflexa* failed to develop a primary root in vitro, neither did it develop any adventitious roots (Maheshwari and Baldev 1962). However, the radicular end of the mature embryo formed a callus which eventually differented adventitious embryos. In *Cassytha*, haustoria were never developed in axenic cultures, but Baldev (1959) reported the formation of haustoria-like outgrowths from the stems of cultured *Cuscuta reflexa*. Zimmerman (1962) demonstrated that the in vitro cultured seedlings of *Cuscuta pentagona* required far-red light to develop haustoria.

4 Summary and Conclusions

Considerable insight has been gained into the factors that control the growth and development of embryos of root- and stem parasites through seed and embryo culture techniques. However, there is need for more detailed analytical studies aimed at relating morphogenesis of embryos with specific nutritional and/or hormonal factors in the medium. In any case, it must be conceded that the preliminary knowledge of the mode of growth of embryos in culture and their responses to different supplements to the medium will form a basis for further investigations on the chemical control measures of the parasitic plants (Okonkwo 1987a, b).

Studies to date show clearly that embryos and seeds of facultative root hemi-parasites and stem parasites germinate readily in vitro, without the aid of any chemicals coming from external sources. On the other hand, for the obligate root hemi-parasites, where there is a requirement for a stimulant to bring about embryo and seed germination, development of shoot systems and further growth of the germinated seedlings seem to depend on only mineral salts and sugar, but not on hormones or other nutrients. Studies to date have therefore demonstrated the similarity in this regard among *Striga hermonthica*, *S. gesnerioides*, *S. asiatica*, *Alectra vogelli*, and *A. sessiflora* var. *senegalensis*. Results varying from the above may be open to reinvestigation and/or reinterpretation. It is of interest that this grouping spans parasites that attack grain cereals (monocotyledons) and those that attack broadleaved plants (mainly dicotyledons). On the other hand, embryos of holo-parasites require to be stimulated to germinate by a number of chemicals, and subsequent development and shoot morphogenesis would be supported only on complex media comprising mineral salts, sugar, amino acids, and hormones.

The in vitro culture of root and stem parasites has opened a new avenue for examining in more details the effects of nitrogen on their physiology and biochemistry. The technique is also adequate for screening herbicides for their effects on parasites.

Expansion and intensification of research in these and other areas bordering on the biology and control of parasitic plants seem now more urgent than ever before, as there are worrying reports that these parasitic plants, especially the root parasitic weeds, are now invading new crop host plants and expanding into wetter areas of Nigeria and the rest of the West African tropics (Parkinson 1985; Lagoke et al. 1986; Okonkwo 1987a,b).

References

Ayensu ES, Doggett H, Keynes RD, Marton-Lefevre J, Musselman LJ, Parker C, Pickering A (eds) (1984) *Striga* – biology and control. ICSU Press, Paris

Bajaj YPS (1963) In vitro studies on the embryos, ovules and ovaries of some angiosperms. PhD Thesis, Univ Delhi

Bajaj YPS (1967) In vitro studies on the embryos of two mistletoes, *Amyema pendula* and *Amyema miguelli.* NZ J Bot 5: 49–56

Bajaj YPS (1968) Some factors affecting growth of embryos of *Dendrophthoe falcata* in cultures. Can J Bot 46: 429–433

Bajaj YPS (1970) Growth responses of excised embryos of some mistletoes. Z Pflanzenphysiol 63: 408–415

Baldev B (1959) In vitro responses of growth and development in *Cuscuta reflexa* Roxb. Phytomorphology 9: 316–319

Bhatnagar SP (1987) In vitro morphogenic responses of mistletoes. In: Weber HC, Forstreuter W (eds). Parasitic flowering plants. Marburg, p 105–108

Bhojwani SS (1968) Morphogenetic studies on cultured mature endosperm and embryo of some angiosperms. PhD Thesis, Univ Delhi

Bhojwani SS (1969) Morphogenesis of embryo in a parasitic angiosperm – *Exocarpus cupressiformis.* Curr Sci 38: 6

Bonga JM (1971) Formation of holdfasts, callus, embryoids and haustorial cells in the in vitro cultures of dwarf mistletoe *Arceuthobium pusillum.* Phytomorphology 21: 140–153

Butler EF (1956) Nigeria Dept Agric (Central) Annu Rep 1953/1954, part II: 32
Cantlon EJ, Curtis JC, Malcolm WM. Studies on *Melampyrum lineare.* Ecology 44: 466–474
Chidley VL, Drennan DSH (1987) In vitro culture of *Striga asiatica* (L.) Kuntze (Scrophulariaceae). In: Weber HC, Forstreuter W (eds) Parasitic flowering plants. Marburg, pp 829–831
Chuang TI, Heckard LR (1971) Observations on root parasitism in *Cordylanthus* (Scrophulariaceae). Am J Bot 58: 218–228
Cook CE, Whichard LP, Turner B, Wall ME, Egley GH (1966) Germination of witchweed (*Striga lutea* Lour.) Isolation and properties of a potent stimulant. Science 154: 1189–1190
Cook CE, Whichard LP, Wall ME, Egley GH, Coggon P, Luhan PA, McPhail AT (1972) Germination stimulants. II. The structure of strigol – a potent seed germination stimulant for witchweed *Striga asiatica* Lour.). J Am Chem Soc 94: 6198–6199
Egley GH, Dale JE (1970) Ethylene, 2-chloroethylphosphonic acid and witchweed germination. Weed Sci 18: 586–589
French RC, Sherman LJ (1976) Factors affecting dormancy, germination, and seedling development of *Aeginetia indica* L. (Orobanchaceae). Am J Bot 65: 558–570
Hall P, Letham DS, Barlow BA (1987) The influence of hormones on development of *Amyema* seedlings cultured in vitro. In: Weber HC Forstreuter W (eds) Parastic flowering plants. Marburg, pp 285–291
Harvey AE (1967) Tissue culture of *Pinus monticola* on a chemically defined medium. Can J Bot 45: 1783–1787
Hull RJ, Leonard OA (1964) Physiological aspects of parasitism in mistletoes. I. The carbohydrate nutrition of mistletoe. Plant Physiol 39: 996–1007
Ismail AMA, Obeid M (1976) A study of assimilation and translocation in *Cuscuta hyalina* Heyen ex Roth., *Orobanche ramosa* L., and *Striga hermonthica* Benth. Weed Res. 16: 87–92
Johri BM, Bajaj YPS (1963) In vitro response of the embryo of *Dendrophthoe falcata* (L.f.) Ettings. In: Maheshwari P, Rangaswamy NS (eds) Plant tissue and organ culture – a symposium. Int Soc Plant Morphologists, Delhi pp 292–301
Johri BM, Bajaj YPS (1964) Growth of embryos of *Amyema*, *Amylotheca* and *Scurrula* on synthetic media. Nature (London) 204: 1220–1221
Johri BM, Bajaj YPS (1965) Growth responses of globular proembryos of *Dendrophthoe falcata* (L.f.) Ettings. in culture. Phytomorphology 15: 292–300
Johri BM, Bhojwani SS (1970) Embryo morphogenesis in the stem parasite, *Scurrula pulverulenta.* Ann Bot 34: 685–690
Johri BM, Nag KK (1968) Experimental induction of triploid shoots in vitro from endosperm of *Dendrophthoe falcata* (L.f.) Ettings. Curr Sci 37: 606–607
Johri BM, Nag KK (1974) Cytology and morphogenesis of embryos and endosperm tissue of *Dendrophthoe* and *Texillus.* Cytologia 39: 801–813
Knop W. (1865) Quantitative Untersuchungen über die Ernährungsprozesse der Pflanzen. Landwirtsch Vers Stn 7: 93–107
Kuijt J (1969) The biology of parasitic flowering plants. Univ Calif Press, Berkeley and Los Angeles
Kumar LSS (1940) Flowering plants which attack economic crops. I. *Striga.* Indian Farming 1: 593–595
Kumar U, Rangaswamy NS (1977) Regulation of seed germination and polarity in seedling development in *Orobanche aegyptiaca* by growth substances. Biol Plant 19: 353–359
Lagoke ST, Manzo SK, Emechebe AM, Ajayi O (1986) Status of parasitic weed-research in the Nigerian savanna. Paper presented, 15th Annu Conf Weed Sci Soc of Nigeria, Univ Nigeria, Nsukka Nigeria
Last FT (1960) Effect of cultural treatments of the incidence of *Striga hermonthica* (Del) Benth. and yields of sorghum in the Sudan. Ann Appl Biol 48: 207–229
Linsmaier EM, Skoog F (1965) Organic growth factor requirements of tobacco tissue cultures. Physiol Plant 18: 100–127
Lynn DG, Steffens JC, Kamut VS, Graden DW, Shabanowitz J, Riopel JL (1981) Isolation and characterization of the first host recognition substance for parasitic angiosperms. J Am Chem Soc 103: 1868–1870
MacQueen M (1984) Haustorial initiating activity of several simple phenolic compounds. In: Parker C, Musselman LJ, Polhill RM, Wilson AK (eds) Proc 3rd Int Symp Parasitic Plants. International Center for Agricultural Research in Dry Areas. Aleppo, Syria, pp 118–122
Maheshwari P, Baldev B (1962) In vitro induction of adventitious buds from embryos of *Cuscuta reflexa* Roxb. In: Plant embryology – a symposium, CSIR, New Delhi, p 129–138
McNally SF, Stewart GR (1987) Inorganic nitrogen assimilation by parasitic angiosperms. In: Weber HC, Forstreuter W (eds) Parasitic flowering plants. Marburg, pp 539–546

Murashige T, Skoog F (1962) A revised medium for rapid growth and bioassays with tobacco tissue cultures. Physiol Plant 15: 473–497

Musselman LJ (1980) The biology of *Striga*, *Orobanche* and other root-parasitic weeds. Annu Rev Phytopathol: 463–489

Nag KK, Johri BM (1976) Experimental morphogenesis of the embryo of *Dendrophthoe*, *Taxillus* and *Nuytsia*. Bot Gaz 137: 378–390

Nash SM, Wilhelm S (1960) Stimulation of broomrape seed germination. Phytopathology 50: 772–774

Nitsch JP (1951) Growth and development in vitro of excised ovaries. Am J Bot 38: 566–577

Nwoke FIO, Okonkwo SNC (1974) Facultative hemi-parasitism in *Buchnera hispida* Buch-Ham. ex D. Don. Ann Bot 38: 993–1002

Nwoke FIO, Okonkwo SNC (1978) Structure and development of the primary haustorium in *Alectra vogelii* Benth. (Scrophulariaceae). Ann Bot 42: 447–454

Nwoke FIO, Okonkwo SNC (1980) Photocontrol of seed germination in the hemi-parasite *Buchnera hispida* (Scrophulariaceae). Physiol Plant 49: 388–392

Ogborn JEA (1970) Methods of controlling *Striga hermonthica* for West African farmers. Samaru Agric Newsl 12: 90–96

Ogborn JEA (1979) The potential use of germinators for *Striga* control by African peasant farmers. In: Musselman LJ, Worsham AD, Eplee RE (eds) Suppl Proc 2nd Int Symp Parasitic Weeds. Raleigh, North Carolina, pp 29–36

Ogborn JEA (1984) *Striga*: research priorities with specific reference to agronomy. In: Ayensu ES, Doggett H, Keynes RD, Martin-Lefevre J, Musselman LJ, Parker C, Pickering A (eds) Proc Workshop on biology and control of *Striga*, Dakar, Senegal, pp 195–212

Okonkwo SNC (1966a) Studies on *Striga senegalensis* Benth. I. Mode of host-parasite union and haustorial structure. Phytomorphology 16: 453–462

Okonkwo SNC (1966b) Studies on *Striga senegalensis* Benth. III. In vitro culture of seedlings. Establishment of cultures. Am J Bot 53: 679–687

Okonkwo SNC (1966c) Studies on *Striga senegalensis*. IV. In vitro culture of seedlings. Effects of various sugars and glutamine. Am J Bot 53: 687–693

Okonkwo SNC (1975) In vitro post-germination growth and development of embryos of *Alectra* (*Scrophulariaceae*). *Physiol Plant* 34: 378–383

Okonkwo SNC (1982) Nutrient factors for shoot development and seedling growth of *Striga gesnerioides* (Willd.) Vatke. Z Pflanzenphysiol 106: 381–389

Okonkwo SNC (1987a) Developmental studies on witchweeds. In: Musselman LJ (ed) Parasitic weeds in agriculture, vol 1: *Striga* CRC Press, Boca Raton, pp 64–75

Okonkwo SNC (1987b) Research techniques – laboratory. In: Musselman (ed) Parasitic weeds in agriculture, vol 1: *Striga*. CRC Press, Boca Raton, pp 257–269

Okonkwo SNC, Nwoke FIO (1974) Seed germination in *Buchnera hispida* Buch. H – Ham. ex D. Don. Ann Bot 38: 409–417

Okonkwo SNC, Nwoke FIO (1975) Bleach-induced germination and breakage of dormancy of seeds of *Alectra vogelii*. Physiol Plant 35: 175–180

Okonkwo SNC, Nwoke FIO (1978) Initiation, development and structure of the primary haustorium in *Striga gesnerioides* (Scrophulariaceae). Ann Bot 42: 455–563

Okonkwo SNC, Raghavan V (1982) Studies on the germination of seeds of the root parasites, *Alectra vogelii* and *Striga gesnerioides*. I. Anatomical changes in the embryos. Am J Bot 69: 1636–1645

Onofeghara FA (1971) Establishment of in vitro cultures of *Tapinanthus bangwensis* (Engler and K. Krause) Danser and the Effects of sugars on growth. Ann Bot 35: 729–743

Onofeghara FA (1972) The effects of growth substances on the growth of *Tapinanthus bangwensis* (Loranthaceae) in vitro. Ann Bot 36: 563–570

Parkinson V (1985) *Striga*, serious threat to maize production in Africa and research being conducted by IITA. Paper presented at FAO *Striga* Workshop, Yaounde, Cameroon, 23–27 September, 1985

Pesch C, Pieterse AH (1982) Inhibition of germination in *Striga* by means of urea. Experientia 38: 559–560

Rangan TS (1965) Morphogenesis of the embryo of *Cistanche tubulosa* Wight in vitro. Phytomorphology 15: 180–182

Rangan TS, Rangaswamy NS (1968) Morphogenetic investigations on parasitic angiosperms. I *Cistanche tubulosa* (Orobanchaceae). Can J Bot 46: 263–266

Rangan TS, Rangaswamy NS (1969) Morphogenic investigations on parasitic angiosperms. *III*. *Cassytha filiformis* (Lauraceae). Phytomorphology 19: 292–300

Rangaswamy NS (1963) Studies on culturing seeds of *Orobanche aegyptiaca* Pers. In: Maheshwari P, Rangaswamy NS (eds) Plant tissue and organ culture – a symposium. Int Soc Plant Morphologists, Delhi, pp 345–354

Rangaswamy NS (1967) Morphogenesis of seed germination in angiosperms. Phytomorphology 17: 477–487

Rangaswamy NS, Rangan TS (1966) Studies of seed germination stimulants on the witchweed *Striga euphrasioides* (Vahl) Benth. Nature 210: 440–441

Rangaswamy NS, Rangan TS (1969) Morphogenetic investigations on parasitic angiosperms. II. *Striga angustifolia* (Don) Saldhana (Scrophulariaceae). Flora Allg Bot Ztg Abt A 160: 448–456

Rangaswamy NS, Rao PS (1963) Experimental studies on *Santalum album* L. Establishment of tissue culture of endosperm. Phytomorphology 13: 450–454

Riopel JL (1979) Experimental studies on induction of haustoria in *Agalinis purpurea*. In: Musselman LJ, Worsham AD, Eplee RE (eds) Proc 2nd Symp on Parasitic Weeds. North Carolina State Univ, Raleigh, pp 165–173

Riopel JL, Baird WV (1987) Morphogenesis of the early development of primary haustoria in *Striga asiatica* In: Musselman LJ (ed) Parasitic weeds in agriculture, vol. I: *Striga*. CRC Press, Boca Raton

Riopel JL, Musselman LJ (1979) Experimental initiation of haustoria in *Agalinis purpurea* (Scrophulariaceae) Am J Bot 66: 570–575

Sahai A (1983) Experimental studies on seed germination and seedling growth in two angiosperm root parasites *Sopubia* and *Santalum*. PhD Thesis, Univ of Delhi, India

Sahai A, Shivanna KR (1982) Seed germination and seedling morphognesis in parasitic angiosperms of the families Scrophulariaceae and Orobanchaceae. Seed Sci Technol 10: 565–583

Saunders AR (1933) Studies in phanerogamic parasitism with particular reference to *Striga lutea* Lour. Dep Agric S Afr Bull 128: 1–57

Shivanna KR, Rangaswamy NS (1976) Seed germination and seedling morphogenesis of the root parasite *Sopubia delphinifolia* G. Don. Z Pflanzenphysiol 80: 112–119

Stephens EL (1912a) The structure and development of the haustorium of *Striga lutea* Lour. Ann Bot 26: 1067–1076

Stephens EL (1912b) Note on the anatomy of *Striga lutea* Lour. Ann Bot 26: 1125–1126

Stewart GR, Nour J, MacQueen M, Shah N (1984) Aspects of the biochemistry of *Striga*. In: Ayensu ES, Doggett H, Keynes RD, Marton-Lefevre J, Musselman LJ, Parker C, Pickering A (eds) *Striga*: biology and control. ICSU Press, Paris, pp 161–178

Tashima Y, Matsumoto M, Kuroki S (1974) Flower initiation of holoparasite, *Aeginetia indica* L. under non-parasitic condition of sterile culture. Proc Jpn Acad 50: 493–496

Tepfer SS, Greyson RI, Craig WR, Hindman JL (1963) In vitro culture of floral buds of *Aquilegia*. Am J Bot 50: 1035–1045

White PR (1943) A handbook of plant tissue culture. Lancaster, Pa

Williams CN (1960) Growth movements of the radicle of *Striga*. Nature 188: 1043–1044

Williams CN (1961) Growth and morphogenesis of *Striga* seedlings. Nature 189: 378–381

Worsham AD, Moreland DE, Klingman GC (1959) Stimulation of *Striga asiatica* (witchweed) seed germination by 6-(substituted) purines. Science 130: 1654–1656

Worsham AD, Klingman GC, Moreland DE (1962) Promotion of germination of *Striga asiatica* seed by coumarin derivatives and effects on seedling development. Nature 195: 199–201

Yoshikawa FA, Worsham AD, Moreland DE, Eplee RE (1978) Biochemical requirements for seed germination and shoot development of witchweed (*Striga asiatica*). Weed Sci 26: 119–122

Zimmerman CE (1962) Autotrophic development of dodder (*Cuscuta pentagona* Englm.) in vitro. Crop Sci 2: 449–450

XXVIII In Vitro Culture of *Arabidopsis thaliana*

K.A. FELDMANN[1]

1 *Arabidopsis* as a Plant Model System

Arabidopsis thaliana has become one of the plant model systems of choice for physiologists, geneticists, and molecular biologists. Dramatic improvements in culture techniques have occurred since the last systematic effort was made to fashion *A. thaliana* into a better organism for somatic cell genetic studies (Negrutiu and Jacobs 1977, 1978a,b; Negrutiu et al. 1975, 1978a,b). For detailed reviews of the biology, genetics, tissue culture as related to the production of haploids, and molecular biology of *Arabidopsis*, see Redei (1970, 1974, 1975), Meyerowitz and Pruitt (1985), and Scholl and Feldmann (1990).

A. thaliana is a small flowering plant in the family Cruciferae (Fig. 1). The species forms a semi-rosette with six to eight leaves below the principal inflorescence. It has a short life cycle, typically between 5–10 weeks in controlled growth chambers (diurnal cycle). However, under continuous light regimes or when nutrient or water stressed, many geographic races can be forced to cycle in under 4 weeks. The species has a wide geographical distribution, being represented by at least 294 ecological (geographic) races (Redei 1970). These ecotypes differ with respect to many characteristics including number and size of leaves in the rosette, flowering time (Redei 1970), leaf shape, chlorophyll content, isozyme patterns, and restriction fragment-length polymorphisms (Meyerowitz et al. 1988).

Arabidopsis embraces numerous attributes that make it particularly amenable to conventional genetic and tissue culture studies, including: (1) the availability of numerous mutants, (2) a useful genetic map (Koornneef et al. 1983), (3) cleistogamy, (4) large seed output (10000 seeds per plant), and (5) a low chromosome number ($n = 5$). The nuclear volume of *A. thaliana* is the smallest among the angiosperms (Sparrow et al. 1972, Leutwiler et al. 1984). In fact, Meyerowitz and Pruitt (1985) have shown that the haploid genome consists of 70000 kb, containing only 10–14% repetitive DNA. This attribute has played a significant role in propelling *A. thaliana* into the molecular arena.

The need for in vitro culture of *Arabidopsis* centers around experimentation in basic plant research; identifying genes with known function via complementation. Direct gene transfer into protoplasts will probably prove the most useful way to shotgun long pieces of wild-type DNA into mutants. This means that a fast and reliable system to regenerate fertile whole plants from protoplasts is required. An

[1] Central Research and Development, Plant Sciences, E.I. du Pont de Nemours & Company, Wilmington, DE 19898, USA. (Present address: University of Arizona, Plant Sciences, Forbes 303, Tucson, AZ 85721, USA)

Biotechnology in Agriculture and Forestry, Vol. 20
High-Tech and Micropropagation IV (ed. by Y.P.S. Bajaj)

Fig. 1. Five-week-old *Arabidopsis thaliana* L. Heynh. plant (ecotype Wasselewskija)

efficient protoplast system would also be useful for assessing the activity of gene cassettes in transient assays.

For verification of putative complementary genes, *Agrobacterium*-mediated transformation will be the most efficient technique. This requires a regeneration system coupled to a transformation procedure for explants. In addition, an efficient protocol for generating fertile whole plant transformants should make it possible to do insertional mutagenesis (T-DNA tagging) with this species. This will require tens of thousands of transformants which exhibit little or no somaclonal variation.

2 *Arabidopsis* as a Model System for Somatic Cell Genetic Studies

A. thaliana has been shown to be amenable to a variety of tissue culture techniques, a few of which have recently been optimized and will be the focus of this review.

Calli have been produced on well-defined callus-inducing medium (CIM; intermediate auxin: low cytokinin ratio) from a variety of explant sources including anthers (Amos and Scholl 1978; Negrutiu et al. 1978b), seedlings (Anand 1966; Feldmann 1977), leaves, roots, and stems (Negrutiu et al. 1975, 1978b; Yokoyama and Jones 1965; Acedo 1986). Once the callus was established (30 or more days) it was subdivided to organogenic medium (shoot-inducing medium, SIM [low auxin: high cytokinin]; root-inducing medium, RIM [high auxin: low cytokinin]; (Amos 1977; Negrutiu et al. 1978a; Negrutiu and Jacobs 1987b). The regeneration frequencies were typically low and extremely unreliable (Gresshoff and Doy 1972; Feldmann 1977; Negrutiu et al. 1978b). This was especially apparent for established cultures.

Thc loss in morphogenetic potential with time in culture (Negrutiu et al. 1975, 1978b; Feldman and Marks 1986; Valvekens et al. 1988) may be due primarily to chromosomal instability (Negrutiu et al. 1975, 1978b; Feldmann 1977). Negrutiu et al. (1975) reported that chromosomal instability was more pronounced in (1) callus derived from leaves or stems in contrast to seedling-derived callus, (2) callus cultured on 2,4-D rather than NAA as an auxin source, and (3) callus cultured on cytokinin concentrations greater than 1.5 mg/l. Concomitant with this chromosomal instability is a high frequency of somaclonal variants (Negrutiu et al. 1978a, and unpubl. observ.).

Recently, the regeneration of fertile whole plants from explants has been optimized or greatly improved (Feldmann and Marks 1986; Valvekens et al. 1988). Feldmann and Marks (1986) took a novel approach to regenerating *Arabidopsis* and found that explants presented with a short preculture on CIM (5–7 days) regenerate very rapidly and efficiently when transferred to a SIM (100% of cultured explants make multiple shoots in 3–4 weeks, to be discussed in detail later). Valvekens et al. (1988) followed up this observation and linked an *Agrobacterium*-mediated transformation procedure to this preculture concept.

Other regeneration protocols have recently, been published all of which have one or more shortcomings. For example, Lloyd et al. (1986), using the ecotype, Columbia, were able to induce callus and shoots on a primary SIM (1 mg/l BAP and 0.1 mg/l NAA). This regeneration procedure was coupled to *Agrobacterium*-mediated transformation. However, the procedure took months to complete, the efficiency was low, and none of the regenerated plants contained a T-DNA marker that cosegregated with an altered phenotype (i.e., somaclonal variants, R. Horsch, Monsanto Co., pers. commun.). Patton and Meinke (1988) followed up this result and employing the same protocol tested immature cotyledons (ecotype, Columbia). They observed regenerating shoots from these explants in as little as 10 days and, after just 3 weeks in culture, 30–60% of the explants had produced shoots. They also found that by increasing the NAA to 0.4 mg/l the simultaneous formation of roots and shoots was promoted. This is, however, a difficult and time-consuming tissue to utilize as an explant source.

Protoplasts can be made from suspension cultures or from excised leaves (Negrutiu et al. 1975; Damm and Willmitzer 1988). Protoplasts from various ecotypes have successfully been induced to form walls, divide, and regenerate into whole plants (Negrutiu et al. 1975; Gresshoff 1978; Aerts 1980; Xuan and Menczel 1979). Damm and Willmitzer (1988) have been able to regenerate fertile whole plants from four different ecotypes of *A. thaliana* at a reasonable efficiency (for ecotype Columbia: plating efficiency, 0.6–3%; shooting efficiency, > 80%; rooting efficiency on shoots, 70%; fertile whole plants, > 50%). Briefly, they prepared protoplasts from leaves, embedded the protoplasts in sodium alginate for 1–2 days and transferred them to CIM. Following a 5–7-week incubation in CIM, the sodium alginate beads were removed and the calli were transferred to SIM. Regenerated shoots, with some attached callus, were moved to shoot-elongation medium (SEM, MS salts and vitamins, 2% sucrose, 0.1 mg/l NAA, 1 mg/l BAP, and 0.1 mg/l GA_3) for 2–3 additional weeks. Shoots, minus callus, were transferred to RIM until they rooted, flowered, and set seed. The authors noted the absence of apparent somaclonal variants amongst more than 200 whole plants that set viable seed. However, under optimal conditions, only 50–70% of rooted plants set viable seeds (Damm and Willmitzer 1988). The high frequency of sterile plants following prolonged exposure to exogenous hormones (≥ 5 weeks on 0.2–1 mg/l 2,4-D) may be due to chromosomal heterogeneity (Negrutiu et al. 1975, 1978b). It may be possible to improve this procedure by both omitting the SEM step and exposing the regenerated shoots to RIM for a brief period followed by removal to basal medium (BM).

Only in the past 2 years have a number of these tissue culture techniques been advanced to the point that transformation (direct gene transfer or *Agrobacterium*-mediated) is now possible with a number of tissue sources (Acedo 1986; Feldmann and Marks 1986; Lloyd et al. 1986; Damm and Willmitzer 1988; Patton and Meinke 1988; Valvekens et al. 1988).

3 Optimized Regeneration System and Micropropagation

The strategy that has significantly advanced the efficiency at which fertile whole plants can be regenerated from explants of *A. thaliana* is the short preculture on CIM preceding exposure to SIM (Feldmann and Marks 1986). This work was based on a report by Christianson and Warnick (1985). They found that leaf explants in some nonregenerating genotypes of *Convolvulus arvensis* could be induced to form shoots if exposure to a SIM was preceded by a short preculture on CIM. In *Arabidopsis* this strategy has worked extremely well. The whole process of shoot and root induction and elongation, flowering and seed set, is only a couple of weeks slower via *in vitro* culture than it is through a normal seed-to-seed cycle (Fig. 2). However, it has become clear that each genotype and explant source requires a slightly different protocol for optimal regeneration frequencies (see below). The optimized conditions for a couple of races are listed in this review.

The best donor tissue from *A. thaliana* to use for tissue culture experiments is that grown aseptically. This tissue is started by sterilizing seeds for 8 min in pure

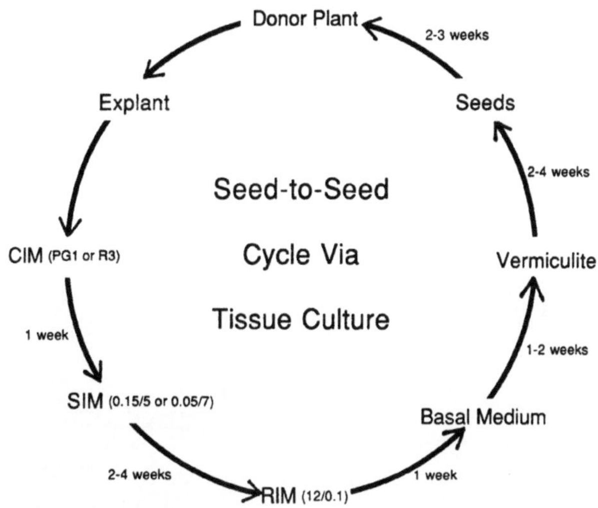

Fig. 2. Optimized protocol for the efficient regeneration of whole plants from cultured explants of *A. thaliana*. All media components are listed in Table 1.

Clorox [5.25% (w/v) sodium hypochlorite] containing 0.15% Tween 20 (v/v) as a wetting agent. After sterilization, the seeds are rinsed three times with sterile water, allowed to dry, and plated in 80 × 100 mm Pyrex jars containing 75 ml of agar-solidified (0.8%) medium (MS salts and vitamins and 1% sucrose). The pH of the medium is 6 before autoclaving. The jars are wrapped with filter tape (Carolina Biological, Burlington, NC) which allows for ethylene release (unpubl. observ.). The jars containing seeds are transferred to 4 °C (dark) for 48 h. Following vernalization the seeds are transferred to continuous florescent light (750 lx) at 22–25 °C. Alternative containers which allow for gas exchange (e.g., 500-ml Ehrlenmeyer flasks with sponge plugs) have proven suitable for growth of donor plants. Plants (WS or RLD) grown under the described conditions are optimal for culture at 16–19 days of age (data not shown).

Explants are harvested under sterile conditions taking care to minimize the possibility of desiccation. Fifty to 75 explants (leaves 3 × 4 mm; stems 4–5 mm in length) are transferred to 50 ml of CIM in 125-ml Ehrlenmeyer flasks (Table 1). The flasks containing the explants are transferred to a reciprocating or gyrotory shaker (120 rpm) at 22 °C (continuous florescent light; 750 lx). Alternatively, the explants can be placed on an agar-solidified CIM (Table 1) for preculture. Substituting agar-solidified for liquid CIM extends the time that it takes the explants to become competent and reduces the regeneration efficiency (both the number of explants that give rise to shoots and the number of shoots per explant, unpubl. results). After 5–7 days the explants in liquid CIM are blotted and transferred to 30 ml of an agar-solidified SIM (15 × 100 mm dishes; Table 1). The dishes are wrapped with filter tape. The dishes containing the explants are transferred to continuous florescent light (750 lx; 22–25 °C). After 5–6 days on SIM green nodule-like bumps are visible and by 7–8 days regenerating shoots appear from the green structures. Two weeks later the regenerated shoots (Fig. 3A) are transferred to

Fig. 3A–E. Protocol for making fertile whole plants from regenerated shoots. **A** Regenerated shoots, from precultured leaf explants, on SIM. **B** Shoots after 7 days on RIM. **C** Rooting shoots on hormone-free medium. **D** Rooted shoots being transferred to vermiculite. **E** Regenerated fertile plant (WS)

Pyrex jars (80 × 100 mm) containing 80 ml of RIM (Table 1; continuous light, 750 lx; 22–25 °C). After 5–10 days on RIM the induced shoots (Fig. 3B) are transferred to Pyrex jars (80 × 100 mm) containing 75 ml of an agar-solidified medium lacking hormones but containing 3% sucrose (REM, root elongation medium). Roots elongate rapidly on REM (Fig. 3C) and rooted shoots (Fig. 3D) can be moved to vermiculite in 1–2 weeks. After the shoots are moved to

Table 1. Media that were used for callus induction and shoot regeneration from *A. thaliana* explants

Media components	Callus induction medium[a]			Shoot induction medium			Root induction medium
	R3[b]	PG1[c]	DBM1[d]	0.05/7[e]	015/5[f]	PG3[c]	12/0.1[f]
Salts + vitamins	MS	B5	MS	MS	MS	B5	MS
			(mg/l)				
Hormones							
IAA	5.0	–	6.0	0.05	0.15	0.03	–
IBA	–	–	–	–	–	–	12.0
2,4-D	0.5	2.2	2.0	–	–	–	–
2-iP	–	–	–	7.0	5.0	–	–
Kinetin	0.3	0.05	1.5	–	–	1.0	0.1

[a]All media contained 3% sucrose and were solidified with 8 g/l agar when appropriate. All media were adjusted to pH 6 before autoclaving.
[b]Christianson (1977).
[c]Negruitiu et al. (1975).
[d]Gresshoff and Doy (1972).
[e]Christianson and Warnick (1985).
[f]Feldmann and Marks (1986).

vermiculite (presoaked with Hoagland's solution; Feldmann and Marks 1986) they are covered with Saran wrap to prevent rapid desiccation. Pots are moved to a growth chamber (12-h light cycle; 2500 lx; 20–22°C, 70–80% humidity) and the plastic wrap is gradually removed over the next week. Plants are subirrigated with Hoagland's solution as necessary. Seed can be harvested 2 to 8 weeks after transfer to vermiculite (Fig. 3E).

The short preculture (3–7 days) on CIM allows very little callus formation but it is necessary for the cells of the explant to acquire competence so that they can respond to an inductive signal, SIM in this case. If leaf explants from WS are not precultured on CIM before transfer to SIM death results in 100% of the explants (Table 2, Fig. 4C). Leaf explants of RLD plated directly on SIM (0.15/5 or PG3) gave rise to callus at the petiole end of the leaf at a low frequency (Table 2). None of these calli had given rise to shoots after 30 days on SIM (Table 4). Valvekens et al. (1988) showed that another ecotype, C24, was capable of differentiating shoots when root explants were placed directly on SIM (0.15/5 or 0.05/7, Table 1). These shoots arose from the proximal end of the root explant. Also, Lloyd et al. (1986) were able to induce leaf explants of the ecotype Columbia to differentiate callus and eventually shoots on a SIM as the primary medium. These limited data indicate that there is an interaction, between source of explant, genotype, and media composition, for the preculture requirement.

A 5-day preculture before transfer to 0.15/5 or PG3 provides enough time for 100% of the explants (WS ecotype Wassilewskija) to survive (Table 2). All CIM tested could stimulate callus formation from 100% of the leaf explants (Fig. 4B, Table 2, and data not shown). In fact, leaf explants precultured on DBM1 gave rise primarily to callus even after transfer to SIM.

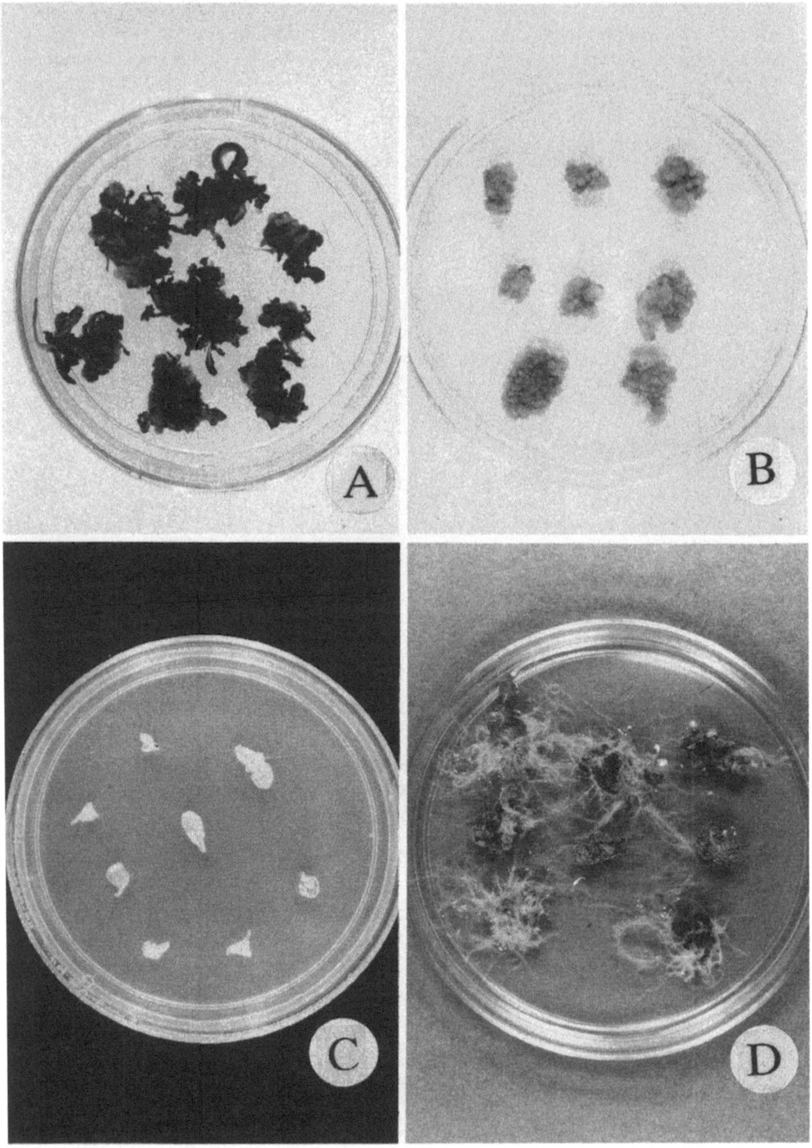

Fig. 4A–D. Morphogenetic response of leaf explants (WS) to various media. **A** Regenerating shoots on SIM (0.15/5) after a 14-day preculture on CIM (R3). **B** Callus proliferation after 4 weeks on CIM. **C** Leaf explants plated directly on SIM. **D** Typical response of precultured explants on PG3 (SIM)

Two of the CIM (R3 and PG1) were superior at affecting competency in the cells of the explant (Table 3). The data in Table 3 shows that if a short preculture is required for the proposed experiments a 5–7-day exposure to PG1 before transfer to 0.15/5 will lead to nearly 100% of the explants surviving (Table 2) and 100%

Table 2. Percent of explants that gave rise to callus on SIM after 0, 3, 5, 7, or 14 days preculture on CIM (averaged over all CIM)

	Genotype									
	WS					RLD				
SIM	0	3	5	7	14	0	3	5	7	14
0.15/5	0[a]	63	100	99	100	19	70	96	100	100
0.05/7	0	55	62	98	100	0	64	92	95	100
PG3	0	96	100	100	100	19	89	100	100	100

[a]Scored 30 days after transfer to SIM.

Table 3. Percent of live explants that gave rise to shoots after 3, 5, 7, or 14 days preculture on CIM before transfer to SIM

		Genotype							
		WS Days on CIM				RLD Days on CIM			
CIM	SIM	3	5	7	14	3	5	7	14
R3	A[a]	–[b]	32[c]	71	50	18	38	38	22
	B	–	36	53	88	17	26	59	8
	C	–	58	72	12	9	52	13	0
PG1	A	50	91	100	0	67	33	94	0
	B	33	58	71	0	52	41	88	19
	C	24	66	54	0	25	21	29	0
DBM1	A	7	31	6	4	0	21	14	0
	B	10	13	8	13	6	24	11	0
	C	35	22	5	0	6	11	10	0

[a]A = 0.15/5, B = 0.05/7, C = PG3.
[b]Not scored.
[c]Scored 30 days after transfer to SIM.

giving rise to shoots (Table 3). However, if a longer preculture is required, such as in transformation studies or experiments involving protoplast isolation and regeneration, the CIM, R3, is the best suited for providing subsequent morphogenetic potential on SIM. In fact, leaf explants of WS can be precultured on R3 for 14 days before transfer to 0.05/7 and nearly 90% of the explants will still give rise to shoots (Table 3). Explants on PG3 generally gave rise to a lower frequency of shoots and a higher frequency of roots (Table 3, Fig. 4D, and data not shown). Damm and Willmitzer (1988) also observed low regeneration frequencies for PG3 in their regeneration procedure.

For both WS and RLD the highest number of shoots per explant occurred when a seven day preculture on PG1 was followed by regeneration on 0.15/5, 3.91, and 3.22, respectively (Table 4). A 2-week exposure of leaf explants of WS to R3 before transfer to 0.15/5 or 0.05/7 also provided for a high number of shoots per

Table 4. Mean number of shoots produced per viable explant after exposure to CIM for various lengths of time before transfer to SIM (30 day observations; n = 32)

			Time on CIM before transfer to SIM				
Genotype	CIM	SIM	0	3	5	7	14
WS	R3	A[a]	0.00	–[b]	0.68	2.13	3.83
		B	0.00	–	0.50	1.69	3.13
		C	0.00	–	0.87	1.88	0.91
	PG1	A	0.00	1.33	2.91	3.91	0.00
		B	0.00	0.67	1.39	1.97	0.00
		C	0.00	0.60	2.25	1.50	0.00
	DBM1	A	0.00	0.07	1.09	0.06	0.08
		B	0.00	0.10	0.47	0.11	0.33
		C	0.00	0.40	0.31	0.09	0.00
RLD	R3	A	0.00	0.27	0.44	0.94	0.75
		B	0.00	0.89	0.37	1.34	0.21
		C	0.00	0.18	0.97	0.13	0.00
	PG1	A	0.00	1.85	0.82	3.22	0.00
		B	0.00	1.76	0.91	2.56	0.47
		C	0.00	0.25	0.21	1.25	0.00
	DBM1	A	0.00	0.00	0.45	0.19	0.00
		B	0.00	0.22	0.48	0.11	0.00
		C	0.00	0.06	0.15	0.35	0.00

[a]A = 0.15/5, B = 0.05/7, C = PG3.
[b]Not scored.

explant, 3.83 and 3.13, respectively. Leaf explants from RLD were able to differentiate shoots when precultured on PG1 for 2 weeks, but only when transferred to 0.05/7 (lower auxin:cytokin ratio).

The regenerated shoots can be rooted at high efficiency by employing the same strategy that was used to induce explants to regenerate shoots. Shoots (~ 1 cm in height; Fig. 3A) are transferred to RIM (Table 1) for approximately 1 week. In these large containers, and on RIM, the shoots elongate rapidly (Fig. 3B) but roots do not become visible. If the shoots remain on RIM for 3–4 weeks a few will eventually give rise to roots (data not shown). However, prolonged exposure to RIM causes a substantial decrease in fertility in the resulting whole plants. A 7-day preculture on RIM is sufficient time to induce the cells in the basal end of the shoot to initiate root formation. When the induced shoots are transferred to REM roots become visible in just 2–5 days (Fig. 3C). One hundred percent of the shoots form roots under these conditions (Fig. 3D). Other RIM were tested but rooting efficiency was reduced in comparison to 12/0.1 (Table 1 and data not shown). Regenerated plants harden off slowly when moved from REM to vermiculite but the vast majority survive to produce seeds (Fig. 3E). Hardening off the regenerated plants at 20 °C with a 12-h photoperiod (2500 lx) yields the greatest number of surviving plants (data not shown).

The resulting plants have not been tested rigorously, but all plants that survived transfer to vermiculite were fertile. This suggests that the plants have near-normal chromosome complements. Under the best conditions regenerated plants and progenitor cells are exposed to exogenous hormones for as little as 3–4 weeks (Fig. 2). This procedure, coupled to transformation, may prove useful for generating insertional mutants in *A. thaliana.*

4 Summary and Conclusions

The best regeneration scheme for leaf explants of two geographic races of *A. thaliana* (WS and RLD, ecotype Rschew) has been outlined (Fig. 2). Other explants and geographic races require slightly modified media for optimal regeneration efficiencies. For example, stem explants from these two genotypes form callus rapidly and at high efficiency when tested with the protocol optimized for leaves. They also stand up to the abuse that the liquid culture subjects them to, but with reduced regeneration frequencies. Preliminary evidence indicates that, in contrast to An et al. (1986), stem explants do not transform as efficiently as leaf (unpublished results) or root explants (Valvekens et al. 1988).

Leaf explants of the ecotype Columbia, while not performing well on the scheme in Fig. 2, regenerate with high efficiency when precultured on a CIM containing 0.01 mg/l NAA (MS salts and vitamins (except thiamine to 10 mg/l), and 40 gm/l sucrose) before transfer to a SIM containing 0.1 mg/l NAA and 1.0 mg/l BAP (MS salts and vitamins, 20 gm/l sucrose; Debra Warnick, Sandoz Crop Protection, pers. commun.). Valvekens et al. (1988), also utilizing a short preculture on CIM before transfer to SIM, reported high regeneration frequencies for root explants (both *Agrobacterium*-infected and untreated) from two genotypes [C24, and Landsberg (erecta)] of *A. thaliana.* They also reported that root explants of the ecotype, Columbia, transformed efficiently but failed to regenerate shoots at comparable frequencies to C24 or Landsberg (Valvekens et al. 1988).

These improved regeneration schedules allow the regeneration of fertile whole plants at a high frequency with a minimum amount of somaclonal variation. Coupling this regeneration procedure to an *Agrobacterium*-mediated transformation system should be an expedient method for moving genes into plants. Transformation may require more extended periods of culture on CIM containing selective agents. Timing of application of selection and the switch from CIM to SIM need to be carefully coordinated to balance between reasonable regeneration frequencies and minimal somaclonal variation. With reduced somaclonal variation it may be possible to distinguish mutant phenotypes, in the resulting transformants, that are due to the insert and not chromosomal changes brought about by the tissue culture process.

The liquid method of handling explants has a number of advantages. First, it is a very efficient way to handle hundreds of explants. Secondly, the medium can be changed regularly, allowing careful selection of transformed clones. Third, more explants per ml of medium can be handled, making it a less expensive method of treating explants with expensive or difficult to synthesize compounds. Finally, it is an extremely rapid method of going from a differentiated cell to a dedifferentiated suspension cell (data not shown). Five days after the explants are cultured in liquid medium the differentiated cells begin to slough off. By 7 days various cell types can be found in the substrate including trichomes, guard cells, and other epidermal cells, as well as greatly elongated cells. Following an additional week or two in culture, these dividing cells dedifferentiate and the cells of the suspension become more uniform in size and shape and are a suitable source of protoplasts (data not shown).

5 Protocol

The most suitable donor tissue is that grown asceptically. The best explant may depend on the geographic race but certainly leaves and roots are superior. A short preculture (5 days) of the explant on a CIM such as PG1 or R3 followed by transfer to a SIM (0.15/5) will, in just 2–4 weeks, yield multiple plants from each explant. Shoots can be rooted by exposing them to a RIM for 5–7 days before transfer to REM and eventually to soil. Various ecotypes and explant sources may require media adjustments for optimized results.

Acknowledgments. I thank Michael L. Christianson and M. David Marks for helpful discussions and Jeffrey C. Mauvais for valuable suggestions on the preparation of this manuscript.

References

Acedo GN (1986) Regeneration of *Arabidopsis* callus *in vitro*. Plant Cell Tissue Organ Cult 6: 109–114

Aerts M (1980) Protoplast culture as cloning technique for crown gall tumour cells. Arabid Inf Serv 17: 68–69

Amos JA (1977) Anther culture and regeneration studies in four species of *Arabidopsis*. PhD Thesis. Ohio State Univ., Columbus

Amos JA, Scholl RL (1978) Induction of haploid callus from anthers of four species of *Arabidopsis*. Z Pflanzenphysiol 90: 33–43

An G, Watson BD, Chiang CC (1986) Transformation of tobacco, tomato, *Petunia*, and *Arabidopsis thaliana* using a binary Ti vector. Plant Physiol 81: 301–305

Anand RS (1966) Preliminary studies on callus culture of *Arabidopsis thaliana*. Arabid Inf Serv 3: 15

Christianson ML (1977) Zinc sensitivity in *Phaseolus*: expression in cell culture. Environ Exp Bot 19: 217–221

Christianson ML, Warnick DA (1985) Temporal requirement for phytohormone balance in the control of organogenesis *in vitro*. Dev Biol 112: 494–497

Damm B, Willmitzer L (1988) Regeneration of fertile plants from protoplasts of different *Arabidopsis thaliana* genotypes. Mol Gen Genet 213: 15–20

Feldmann KA (1977) Chromosome stability and regenerative potential of anther and seedling callus of *Arabidopsis thaliana*: Effect of age, race and induction medium. Thesis Univ Northern Iowa, Cedar Falls, IA

Feldmann KA, Marks MD (1986) Rapid and efficient regeneration of plants from explants of *Arabidopsis thaliana*. Plant Science 47: 63–69

Gresshoff PM (1978) Interspecific hybridization of *Arabidopsis thaliana* and *Trifolium repens* (white clover). Arabid Inf Serv 15: 24–27

Gresshoff PM, Doy CH (1972) Haploid *Arabidopsis thaliana* callus and plants from anther cultures. Aust J Biol Sci 25: 259–264

Koornneef M, Van Eden J, Hanhart CJ, Stam P, Braaksma FJ, Feenstra WJ (1983) Linkage map of *Arabidopsis thaliana*. J Hered 74: 265–272

Leutwiler LS, Hough-Evans BR, Meyerowitz EM (1984) The DNA of *Arabidopsis thaliana*. Mol Gen Genet 194: 15–23

Lloyd AM, Barnason AR, Rogers SG, Byrne MC, Fraley RT, Horsch RB (1986) Transformation of *Arabidopsis thaliana* with *Agrobacterium tumefaciens.* Science 234: 464–466

Meyerowitz EM, Pruitt RE (1985) *Arabidopsis thaliana* and plant molecular genetics. Science 229: 1214–1218

Meyerowitz EM, Chang C, Dejohn AW, Bowman JL (1988) Preliminary report on a restriction length polymorphism map for *Arabidopsis.* Arabid Inf Serv 25: 11–14

Negrutiu I, Jacobs M (1977) *Arabidopsis thaliana* as a model system in somatic cell genetics. II. Cell suspension culture. Plant Sci Lett 8: 7–15

Negrutiu I, Jacobs M (1978a) Restoration of the morphogenetic capacity in long-term callus cultures of *Arabidopsis thaliana* Z Pflanzenphysiol 90: 431–441

Negrutiu I, Jacobs M (1978b) Factors which enhance in vitro morphogenesis of *Arabidopsis thaliana.* Z Pflanzenphysiol 90: 423–430

Negrutiu I, Beeftink RR, Jacobs M (1975) *Arabidopsis thaliana* as a model sysem in somatic cell genetics. I. Cell and tissue culture. Plant Sci Lett 5: 293–304

Negrutiu I, Jacobs M, Cachita D (1978a) Some factors controlling in vitro morphogenesis of *Arabidopsis thaliana.* Z Pflanzenphysiol 86: 113–124

Negrutiu I, Jacobs M, deGreef W (1978b) In vitro morphogenesis of *Arabidopsis thaliana*: the origin of the explant. Z Pflanzenphysiol 90: 363–372

Patton DA, Meinke DW (1988) High-frequency plant regeneration from cultured cotyledons of *Arabidopsis thaliana.* Plant Cell Rep 7: 233–237

Redei GP (1970) *Arabidopsis thaliana* (L.) Heynh.: a review of the genetics and biology. Bibl Genet 20: 1–151

Redei GP (1974) *Arabidopsis thaliana.* In: King RC (ed) Handbook of genetics. Plenum, New York

Redei GP (1975) *Arabidopsis* as a genetic tool. Annu Rev Genet 9: 111–127

Scholl RL, Feldmann KA (1990) *Arabidopsis thaliana* (L.): in vitro production of haploids. In: Bajaj YPS (ed) Biotechnology in agriculture and forestry, vol 12. Haploids in crop improvement I Springer, Berlin Heidelberg New York, pp 307–321

Sparrow AH, Price HJ, Underbrink AG (1972) A survey of DNA content per cell and per chromosome of prokaryotic and eukaryotic organisms: some evolutionary considerations. Brookhaven Symp Biol 23: 451–494

Valvekens D, Van Montagu M, Van Lijsebettens M (1988) *Agrobacterium tumefaciens*-mediated transformation of *Arabidopsis thaliana* root explants by using kanamycin selection. Proc Natl Acad Sci USA, 85: 5536–5540

Xuan LT, Menczel L (1979) Improved protoplast culture and plant regeneration from protoplast-derived callus in *Arabidopsis thaliana.* Z Pflanzenphysiol 96: 77–80

Yokoyama K, Jones WH (1965) Tissue culture of *Arabidopsis thaliana.* Plant Physiol 40: LXXVII

XXIX Micropropagation of Boston Ferns (*Nephrolepis* spp.)

W. Amaki and H. Higuchi[1]

1 Introduction

1.1 Distribution and Morphology

The genus *Nephrolepis*, (sword fern or Boston fern), is widely distributed throughout the tropics and subtropics, extending to America, Africa, Asia, Australia, New Zealand, and the islands of the Pacific. They are epiphytic or terrestrial ferns of perhaps 20 species (Tryon and Tryon 1982). *Nephrolepis* usually has short, erect stems, and few to many fibrous roots. Fronds are usually monomorphic, partly dimorphic (the fertile and the sterile), and those in the wild species are typically unipinnate, but in horticultural cultivars are often multi-pinnate and variously lobed, or frilled. The arrangement of sori (spore case) differs according to species, although most of horticultural cultivars with multi-pinnate fronds produce sterile spores, or hardly form sorus. Some *Nephrolepis* produce many stolons (rhizomes) at the base of the petioles. The chromosome number of *Nephrolepis* seems to be consistently n = 41 or 82 (Tryon and Tryon 1982).

1.2 Importance and Propagation

Nephrolepis gained its popularity during the Victoria era (Briggs and Calvin 1987). Now, the ferns are the most popular horticulturally. All species are satisfactory for indoor gardens, and according to cultivar, are used for interior plants, groundcover, hanging baskets, and pedestal and rockery plants (Everett 1981). *Nephrolepis exaltata* is tolerant to low temperature (4 to 10 °C), and requires medium light intensity (800 to 1600 lx) for maintenance (Briggs and Calvin 1987).

Many cultivars of *Nephrolepis* are commonly propagated by asexual methods, division of crowns, offsets (rooted stolons), and by in vitro culture, since most of them do not produce spores. It is necessary to be careful in selecting propagules, because some cultivars often produce variants which are not true to type (McConnell et al. 1981). Commercially, *Nephrolepis* cultivars have been propagated through in vitro culture, for example, in the Netherlands, 11 194 900 plants were cloned by in vitro culture in 1986 (Pierik 1987).

[1] Laboratory of Floriculture, Department of Agriculture, Tokyo University of Agriculture, 1-1-1 Sakuragaoka, Setagaya-ku, Tokyo 156, Japan

Biotechnology in Agriculture and Forestry, Vol. 20
High-Tech and Micropropagation IV (ed. by Y.P.S. Bajaj)

2 Review of Tissue Culture Work

2.1 Runner Tip Culture and Multiplying Procedures

Nephrolepis exaltata Bostoniensis, Boston fern, was the first foliage plant to be commercially propagated in vitro, and now the majority of Boston ferns are tissue culture-derived (Hartman and Zettler 1986). Although tissue culture is common in

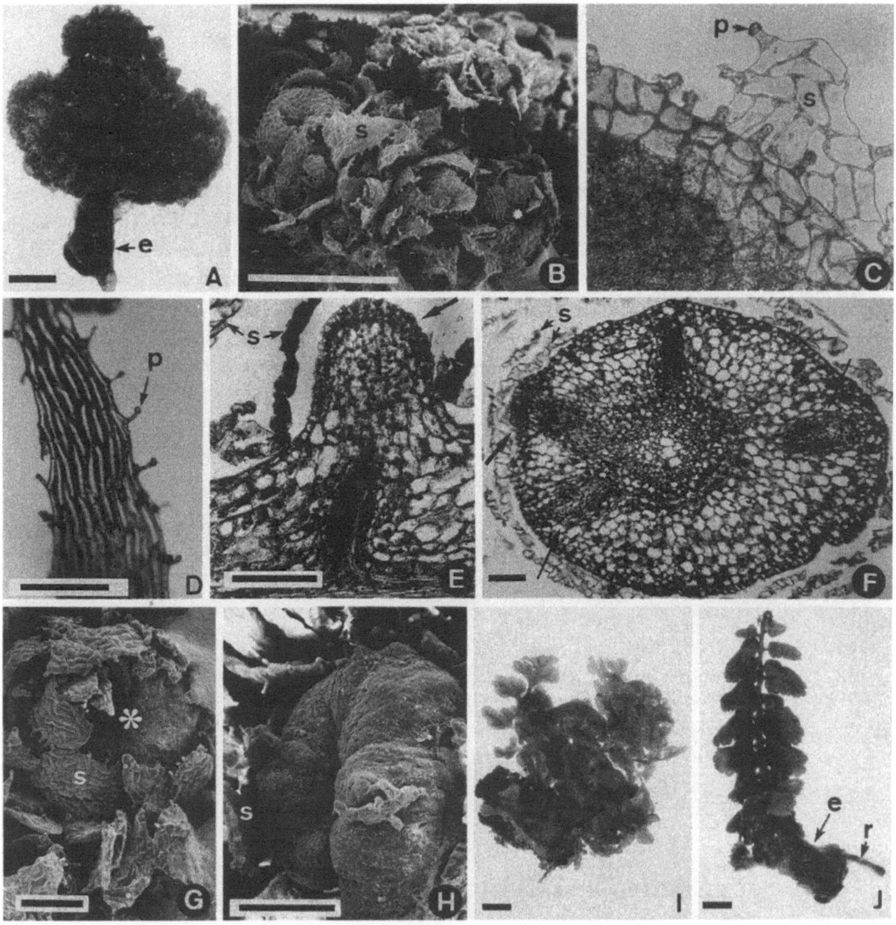

Fig. 1A–J. Structures produced from the runner tips of *Nephrolepis cordifolia*. **A** GGB formation from a runner tip explant (*e*). **B** SEM micrograph of GGBs. The surface of the GGBs was covered with scales (*s*). A meristematic tissue (*) came out at the surface of the GGBs. **C** Scales (*s*) with papillae (*p*) of GGBs. **D** Scales at the tip of the runner. **E** Longitudinal section of runner tip explant. *Arrow* shows initiation of GGBs. **F** Cross-section of a GGB. *Arrows* show meristematic tissue. **G** Developing meristematic tissue (*). **H** More developed meistematic tissue. Scales were initiated at the surface of the tissue. **I** Plantlet regeneration from GGBs. **J** Plantlet regeneration from a runner tip explant (*e*). A root (*r*) has emerged. *Bars* in **A**, **I**, **J** 2 mm; *bar* in **B** 1 mm; *bars* in **C–H** 200 μm commercial propagation of Boston ferns, the details of the procedures have scarcely been published. The results on the laboratory research of *Nephrolepis* spp. are summarized in Table 1.

Runner tips were firstly used as explant for in vitro mass propagation (Murashige 1974). Runner tips regenerate plantlets, and then the plantlets are subcultured to multiply shoots every 6 weeks (Hartmann and Kester 1983). In this multiple-shoot system, the rate of multiplication is about six times per culture. Authors have reported that another multiplying system, green globular bodies (GGB), was used as propagule to increase the rate of multiplication. This method gave a much higher rate of multiplication than that of the multiple-shoot system (Higuchi et al. 1987).

2.2 Characteristics of Green Globular Bodies (GGB)

Runner tips of *Nephrolepis cordifolia* produced GGB on Murashige and Skoog (1962) medium at quarter-strength (1/4 MS) supplemented with BA after 4 weeks culture (Fig. 1A). By observation of the external and internal structure, GGB was distinguished from callus tissues (Higuchi et al. 1987). The surface of the GGB was entirely covered with scales, which had papillae on the periphery and resembled those of the runners (Fig. 1B,C, and D). As the internal structure was observed microscopically, the initial GGB emerged from the peripheral zone of the vascular bundle of runner tip explant (Fig. 1E), and continued swelling to produce GGB. The swollen GGB had many meristems, and each meristem was connected with the central vascular bundle system (Fig. 1F). With the development of meristematic tissues in the GGB, some tissues were pushed aside and came to the surface of the GGB (Fig. 1G,H). It is possible that the GGB have a functional nature similar to that of shoot primordia (green globular aggregates of shoot primordium) in *Haplopappus gracilis* (Tanaka and Ikeda 1983) and/or protocorm-like body (PLB) in orchids (Murashige 1974). That is, the GGB's have many meristems, are rapidly multiplicated on the 1/4 MS supplemented with BA, and easily regenerated plantlets on the 1/4 MS without plant growth regulators (Fig. 1I). From these results, GGB can be used as an intermediate propagule in in vitro mass propagation. Runner tips directly produced fronds when they were cultured on the growth regulator-free medium (Fig. 1J).

3 Micropropagation by GGBs

3.1 Explant Preparation and Sterilization

Runners were cut off to a length of 1 cm from mature *Nephrolepis* plants, and sterilized for 10 min in 1% calcium hypochlorite (or 1% sodium hypochlorite) plus 0.1% Tween 20 (polyoxyethylene monolaurate). After washing in sterilized distilled water three times, they were cut into segments 3 mm long from the tip. These runner tips were used as explants.

Table 1. Summary of the work done on the in vitro propagation through runner tip culture in *Nephrolepis*

Species	Medium	Sucrose (%)	Supplements (mg/l)	pH	Culture condition	Intermediate propagule	Reference
N. exaltata Bostoniensis	MS	3	Kinetin 2 NAA 0.1 NaH_2PO_4 255	5.7	27 °C 16-h light (1000 1x)	Multiple shoot	Burr (1976)
N. exaltata Bostoniensis	1/2 MS (1/2 PO_4^{2-})	2	Kinetin 1 NAA 0.1	5.7	25 °C 16-h light (1150 μW/cm²)	Multiple shoot	Loescher and Albrecht (1979)
N exaltata *N. cordifolia* *N. acuta*	B5	2	IBA 10	5.7	25 °C continuous light (1000 lx)	Multiple shoot (green mass – GGBs?)	Padhya and Mehta (1982)
N. exaltata Bosotiensis Massii Plumosa Rooseveltii Teddy Junior Whitmannii	1/2 MS	3	IAA 0.3 –0.5	5.8	Not described	Multiple shoot	Leffring and Soede (1982)
N. acuta	MS	3	Kinetin 2 NAA 0.1 NaH_2PO_4 255	5.7	25 °C 16-h light (2500 lx)	Multiple shoot	Beck and Caponetti (1983)
N. cordifolia	1/4 MS	3	BA 0.5	5.0	23 °C continuous light (6000 lx)	GGB	Higuchi et al. (1987)

3.2 Culture Media and Culture Condition

Loescher and Albrecht (1979) reported that the shoot and root formation from runner tip of *Nephrolepis exaltata* during in vitro culture was inhibited with standard levels of MS inorganic salts, and the inhibition could be avoided with the use of diluted MS at half-strength (1/2 MS). Leffring and Soede (1982) also reported that the best result was obtained on the 1/2 MS medium when runner tips of five cultivars of *N. exaltata* were cultured on MS medium with inorganic salts diluted at 1/8, 1/4, 1/2, 3/4, or 1/1 strength. Our results on *N. cordifolia* indicated that the growth of plantlets and GGBs was more rapid on the 1/4 MS medium than on the 1/2 and 1/1 MS medium (Higuchi et al. 1987). From these results, 1/2 MS or 1/4 MS are recommended for the initial culture of runner tips in *Nephrolepis*. Borgen and Næss (1987) reported that the homogenity and the plant quality of in vitro-propagated plants was affected by the content of NaH_2PO_4 in media. Therefore, the addition of NaH_2PO_4 to the medium for plantlet regeneration might result in better plant quality.

The optimum concentration of sucrose for the growth of fronds and roots was 20 g/l (Loescher and Albrecht 1979). The effects of pH on the medium are shown in Fig. 2. Fronds and GGBs grew rapidly at relatively low pH.

The addition of NAA and BA to the medium affected the organogenesis from the runner tip explants (Table 2). When runner tips were cultured on a NAA- and BA-free medium, all regenerated plantlets. Plantlet regeneration was greatly inhibited by amounts of BA higher than 0.5 mg/l. When media contained above 0.5 mg/l of BA, runner tips produced only GGBs regardless of NAA concentration,

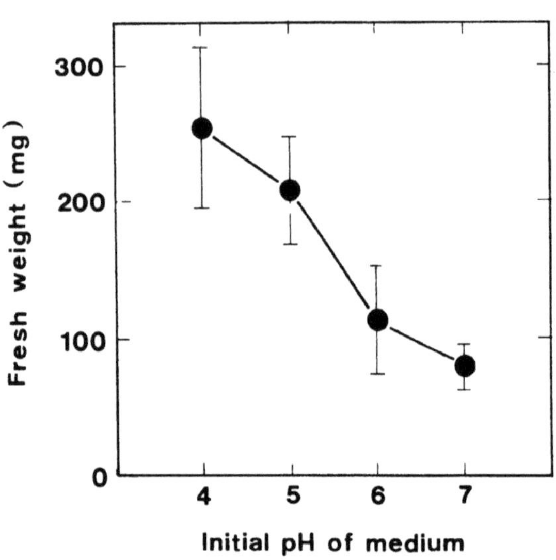

Fig. 2. Effects of initial pH of medium on the growth of GGBs during in vitro culture. GGB segments of 20 mg fresh weight were cultured on the 1/4 MS medium contained BA 0.5 mg/l for 7 weeks at 23 °C under 16-h light (6000 lx)

Table 2. The effect of BA and NAA on organogenesis in runner tips of *Nephrolepis cordifolia* in vitro. Cultures were incubated at 23 °C under continuous fluorescent light at 80 μmol/s/m^2 for 12 weeks. Each value is the mean of three replications. (Higuchi et al. 1987)

Concentration (mg/l)		Survival of explants	Differentiation of explants (%)	
BA	NAA	(%)	Plantlet	GGB
0	0	80.0a[a]	66.7a	13.3c
0	0.5	56.7a	23.3b	33.3b
0	1.0	53.3a	20.0b	33.3b
0.5	0	70.0a	0b	70.0a
0.5	0.5	73.3a	3.3b	70.0a
0.5	1.0	66.7a	0b	66.7ab
1	0	73.3a	0b	73.3a
1	0.5	63.3a	3.3b	60.0ab
1	1.0	56.7a	0b	56.7ab
F test significance[b]				
BA		n.s.	**	**
NAA		n.s.	*	n.s.
BA × NA		n.s.	*	n.s.

[a] Mean separation within columns at 1% level, Duncan's multiple range test.
[b] Significant at 1% level (**), 5% level (*), and not significant (n.s.).

Table 3. The influence of temperature during in vitro culture on organogenesis in runner tips of *Nephrolepis cordifolia.* Cultures were incubated under continuous fluorescent light at 40 μmol/s/m^2 for 12 weeks. Each value is the mean of three replications. (Higuchi et al. 1987)

Temperature (°C)	Survival of explants (%)	Differentiation of explants (%)			
		Plantlet	Frond	Root	GGB
18	49.9a[a]	0c	21.6a	12.8a	15.5a
23	56.7a	19.9a	13.7a	0b	23.1a
28	36.6b	7.3b	15.4a	0b	13.9a

[a] Mean separation within columns at 1% level, Duncan's multiple range test.

except in two cases. The GGBs proliferated continuously for at least 12 weeks. NAA without BA showed a tendency to reduce the production of plantlets and to enhance the formation of GGB. NAA effects on GGB formation were not significant, and the interaction between NAA and BA on GGB formation was also statistically insignificant.

Table 3 shows the effect of temperature during in vitro culture on the organogenesis from runner tips. The optimal temperature for plantlet regeneration was 23 °C. Runner tips cultured at 18 °C produced no plantlets, and some of them produced only roots. At 28 °C, the rate of plantlet regeneration decreased, compared with that at 23 °C.

Table 4 shows the effect of light intensity on organogenesis in runner tips. Under dark conditions, runner tips could not produce plantlets and fronds, and

Table 4. The influence of light intensity during in vitro culture on organogenesis in runner tips of *Nephrolepis cordifolia*. Cultures were incubated at 23 °C for 12 weeks. Each value is the mean of three replications. (Higuchi et al. 1987)

Light intensity (μmol/s/m²)	Survival of explants (%)	Differentiation of explants (%)			
		Plantlet	Frond	Root	GGB
0	50.0a[a]	0c	0b	19.9a	30.1a
40	56.7a	20.3b	15.7a	0b	20.7a
80	63.3a	30.8a	0b	0b	32.5a

[a] Mean separation within columns at 1% level, Duncan's multiple range test.

some of them produced only roots. The optimal intensity of continuous light for plantlet regeneration was 80 μmol/s/m². Under 40 μmol/s/m², runner tips produced plantlets, but the percentages were lower than those seen under 80 μmol/s/m².

3.3 Propagation through GGB Culture

The organogenesis from runner tips of *Nephrolepis cordifolia* could be easily controlled by a single growth regulator, BA. Runner tips produced only GGB and the GGB continued proliferation on the medium containing 0.5 mg/l BA, while runner tips easily regenerated plantlets on the growth regulator-free medium. GGB segments also regenerated plantlets on the same medium in a similar manner. For rapid multiplication, runner tips were cultured on medium containing 0.5 mg/l BA to produce GGBs, and the GGBs divided repeatedly and subcultured on the BA-containing medium to continue rapid proliferation. Figure 3 shows the effects of the cutting size of the GGBs on the rate of growth. GGBs were cut into segments of diameter about 0.5, 1, 2 or 3 mm (average fresh weight was 0.19, 0.51, 5.25, and 13.6 mg, respectively). Each segment was cultured on medium containing 0.5 mg/l BA, at 23 °C under continuous light at 80 μmol/s/m². After 4 weeks, the fresh weight of each segment increased to 7.86, 13.2, 29.6, and 54.8 mg, respectively. When the rate of proliferation was expressed as the quotient of the fresh weight after the 4-week culture divided by the initial fresh weight of the GGB segment, the highest value was obtained from the culture of the smallest segment (about 40 times). On the other hand, the rate of the largest segment was very low, only four times.

The most efficient procedure for practical proliferation of GGB might be to divide piece of GGB, 20 mg in fresh weight, into segments of 2 mg of fresh weight (about 1.5 mm in diameter). Finally, GGBs are divided into segments with a diameter of 1 mm, and each segment is cultured on the growth regulator-free medium to regenerate plantlets. This process is summarized in Fig. 4.

With the multiple-shoot method, Batlle and Aldrufeu (1987) reported that an increase in concentration of kinetin caused an increase in multiplication rate, but it also decreased the size of the regenerated plants, indicating that the smaller plantlets have problems in acclimatization. However, with the GGB method, the

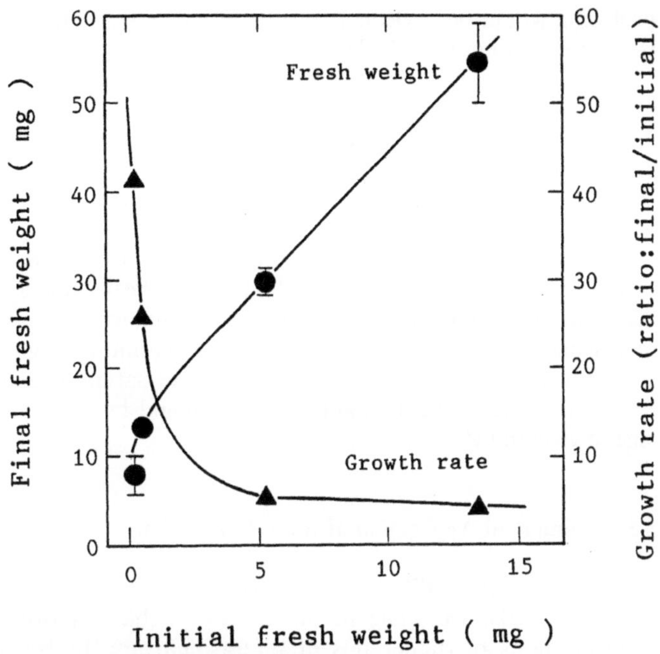

Fig. 3. Effects of cutting size (initial fresh weight) of GGBs on the proliferation measured as fresh weight and growth rate of subcultured GGBs. (Higuchi et al. 1987)

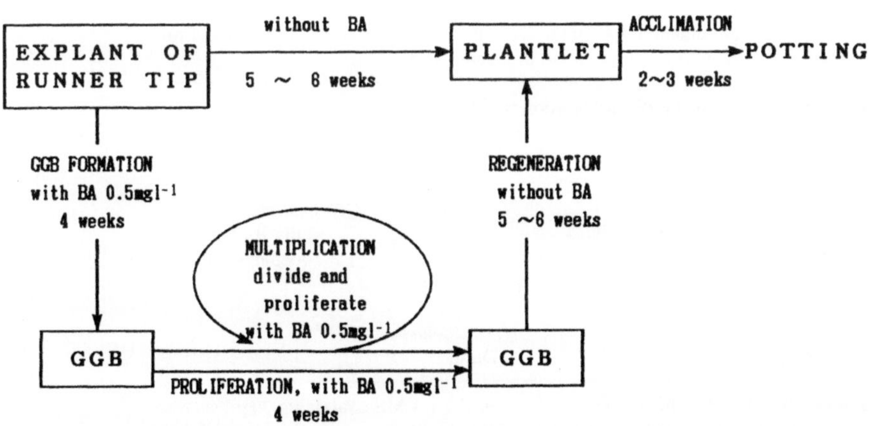

Fig. 4. A scheme for the in vitro propagation of *Nephrolepis cordifolia.* (Higuchi et al. 1987)

problems would be neglected because plantlets regenerated on the growth-regulator-free medium after GGB proliferation.

3.4 Acclimatization and Transfer to Greenhouse

When GGB segment cultured on the BA-free medium, transplantable plantlets were obtained within 6 weeks. These plantlets were transplanted into a porous media (for example, vermiculite, perlite, and peat moss) in a transparent plastic box, and protected from drying for 2 to 3 weeks under high humidity. The acclimatized plantlets were potted into a soil mix which was well drained and rich in organic matter, and transferred to a greenhouse. Night temperature in the greenhouse should be above 15 °C to avoid cessation of growth. After the establishment of the plant, the light intensity should be increased by 25 to 30 klx to ensure growth (McConnell et al. 1981).

3.5 Commercial Application of the GGB Method

It is necessary to modify the concentration of BA for the formation and proliferation of GGBs, according to the cultivar. Table 5 shows the response to BA concentrations of runner tips of several cultivars. In *N. cordifolia*, which is a common type, plantlet formation was completely inhibited and GGB production was promoted by the addition of 0.5 mg/l BA (Higuchi et al. 1987). Runner tips of some *Nephrolepis* cultivars, such as Scottii, Smithii, and Plumosa, required lower concentration of BA to inhibit plantlet formation, and they produced GGB on medium containing 0.3 mg/l BA. On the other hand, runner tips of two cultivars, Fluffy Ruffles and Todeoides, produced plantlets on medium containing 1 mg/l BA. From these results, a preliminary experiment seems to be necessary for each cultivar, prior to practical application of this GGB system.

Table 5. The requirement of BA for the GGB proliferation and the inhibition of frond formation during in vitro propagation in *Nephrolepis*

BA concentration for GGB proliferation (mg/l)			
< 0.3	0.3–0.5	0.5–1	> 1
N. exaltata	*N. exaltata*	*N. exaltata*	*N. exaltata*
Scottii	Amerschlii	Bostoniensis	Fluffy Ruffles
Smithii		Teddy Junior	Todeoides
		Piersonii	
		Victorii	
N. cordifolia		*N. cordifolia*	
Plumosa		Duffii	

Runner tips of each cultivar were cultured on the 1/4 MS containing 30 g/l sucrose, 8 g/l agar, and represented concentration of BA for 40 days at 23 °C under 16-h light (5000 lx).

From our experimental results in *N. cordifolia*, the rate of multiplication from one runner tip could be calculated as follows; a runner tip cultured on BA-containing medium produces GGBs of about 3 mm in diameter after 4 weeks. The GGBs are cut to a diameter of 1 mm, and subcultured. After 4 weeks, the GGB segments proliferate more than 20 times. If this process is repeated every 4 weeks, a single GGB of diameter about 3 mm will be increased by more than 8000 after 3 months. Each of these is cut into segments with a diameter of 1 mm, and cultured on the growth regulator-free medium to produce plantlets. Consequently, 160 000 plantlets could be obtained from one runner tip in a 6-month culture. Thus, the propagation method by GGBs is very efficient.

As far as we have observed, the regenerated plantlets from GGB segments of all the cultivars represented in Table 5 showed the typical phenotype corresponding to each cultivar.

This micropropagation method using GGBs could be applied to other genera of ferns. After slight modification, *Asplenium nidus* and *Pteris ensiformis* Victoriae can be propagated by this method (Higuchi and Amaki 1989, Amaki and Higuchi 1989). Moreover, our preliminary experiments showed the possibility of applying the GGB method to *Rumohra adiantiformis* and *Adiantum raddianum* Fritz Luthii (Amaki and Higuchi 1990).

4 Summary and Conclusions

Nephrolepis ferns are significant as ornamental foliage plants. Most of them, especially important cultivars whose fronds are multi-pinnate, are propagated vegetatively, and also by runner tip culture.

Two kinds of propagules are used at the multiplication stage [so-called stage II of Murashige (1974)], that is, multiple-shoot (Murashige 1974), and GGB (Higuchi et al. 1987). In the micropropagation method using GGBs, GGB proliferation and plantlet formation can be regulated by only one single growth-regulator, BA. Once GGBs are obtained, they rapidly proliferate on the medium containing an adequate amount of BA according to each cultivar. The proliferated GGB's are divided into small segments, and subcultured on the growth regulator-free medium to obtain plantlets. The rate of proliferation with the GGB method seems to be higher than that with multiple-shoots.

5 Protocol

Explant. Runner tip 3 mm long.

Medium. Plantlet regeneration; Murashige and Skoog (1962) medium diluted at quarter strength, added 20 g/l of sucrose and 8 g/l of agar. The pH adjusted before addition of agar is 5 (basal medium). GGB formation and proliferation; basal medium plus 0.5 mg/l of BA.

Cultural condition. Temperature; 23 °C. Lighting; using fluorescent lamps, the light intensity is 80 µmol/s/m^2.

Acknowledgements. The authors are grateful to Drs T. Nakamura and F. Miyamoto, Botanical Laboratory, Course of Liberal Arts, Tokyo University of Agriculture, for their supply of many cultivars of *Nephrolepis*.

References

Amaki W, Higuchi H (1989) In vitro propagation of *Pteris ensiformis* Victoriae. J Jpn Soc Hortic Sci 58 (Suppl 2): 514–515

Amaki W, Higuchi H (1990) A possible propagation system of *Nephrolepis, Asplenium, Pteris, Adiantum* and *Rumohra* through tissue culture. 23rd Int Horticl Congr Florence (Italy), Abstr. No. 1343

Batlle A, Aldrufeu A (1987) Some factors influencing *Nephrolepis* acclimatization. Acta Hortic 212: 367–370

Beck MJ, Caponetti JD (1983) The effects of kinetin and naphthaleneacetic acid on in vitro shoot multiplication and rooting in the fishtail fern. Am J Bot 70: 1–7

Borgen AK, Næss SK (1987) Homogenity and plant quality of in vitro propagated *Nephrolepis exaltata* Bostoniensis. Acta Hortic 212: 433–438

Briggs GB, Calvin CL (1987) Indoor plants. John Wiley & Sons, New York, pp 173–205, 458–459

Burr RW (1976) Mass propagation of ferns through tissue culture. In Vitro 12: 309–310

Everett TH (1981) The New York Botanical Garden illustrated encyclopedia of horticulture. Garland Publishing, New York, pp 2310–2314

Gamborg OL, Miller RA, Ojima K (1968) Nutrient requirements of suspension culture of soybean root cells. Exp Cell Res 50: 151–158

Hartman RD, Zettler FW (1986) Tissue culture as a plant production system for foliage plants. In: Zimmerman RH, Griesback RJ, Hamnerschlag FA, Lawson RH (eds) Tissue culture as a plant production system for horticultural crops. Martinus Nijhoof Publishers, Dordrecht, pp 293–299

Hartmann HT, Kester DE (1983) Plant propagation, 4th edn. Prentice-Hall, New Jersey, pp 566–594

Higuchi H, Amaki W (1989) Effects of 6-benzylaminopurine on the organogenesis of *Asplenium nidus* L. through in vitro propagation. Sci Hortic 37: 351–359

Higuchi H, Amaki W, Suzuki S (1987) In vitro propagation of *Nephrolepis cordifolia* Prsel. Sci Hortic 32: 105–113

Leffring L, Soede AC (1982) Influence of salt concentration on the homogenity of *Nephrolepis* plants from tissue culture. In: Fujiwara A (ed) Plant Tissue Culture 1982. Maruzen, Tokyo, pp 499–500

Loescher MH, Albrecht CN (1979) Development in vitro of *Nephrolepis exaltata* cv. Bostoniensis runner tissues. Physiol. Plant 47: 250–254

McConnell DB, Henley RW, Biamonte RL (1981) Commercial foliage plants. In: Joiner JN (ed) Foliage plant production. Prentice-Hall, New Jersey, pp 544–593

Murashige T (1974) Plant propagation through tissue culture. Annu Rev Plant Physiol 25: 135–166

Murashige T, Skoog F (1962) A revised medium for rapid growth and bioassays with tobacco tissue cultures. Physiol Plant 15: 473–497

Padhya MA, Mehta AR (1982) Propagation of fern (*Nephrolepis*) through tissue culture. Plant Cell Rep 1: 261–263

Pierik RLM (1987) In vitro culture of higher plants. Martinus Nijhoff, Dordrecht, pp 305–308

Tanaka R, Ikeda H (1983) Perennial maintenance of annual *Haplopappus gracilis* ($2n = 4$) by shoot tip cloning. Jpn J Genet 58: 65–70

Tryon RM, Tryon AF (1982) Ferns and allied plants. Springer, Berlin Heidelberg, New York, pp 656–662

Subject Index

Printing: Druckerei Zechner, Speyer
Binding: Buchbinderei Schäffer, Grünstadt